# LIPID SIGNALING AND METABOLISM

# LIPID SIGNALING AND METABOLISM

*Edited by*

James M. Ntambi

*Departments of Biochemistry and Nutritional Sciences, University of Wisconsin-Madison, Madison, WI, United States*

ELSEVIER

ACADEMIC PRESS

An imprint of Elsevier

Academic Press is an imprint of Elsevier
125 London Wall, London EC2Y 5AS, United Kingdom
525 B Street, Suite 1650, San Diego, CA 92101, United States
50 Hampshire Street, 5th Floor, Cambridge, MA 02139, United States
The Boulevard, Langford Lane, Kidlington, Oxford OX5 1GB, United Kingdom



**Library of Congress Cataloging-in-Publication Data**
A catalog record for this book is available from the Library of Congress

**British Library Cataloguing-in-Publication Data**
A catalogue record for this book is available from the British Library

ISBN: 978-0-12-819404-1

For information on all Academic Press publications
visit our website at https://www.elsevier.com/books-and-journals

*Publisher:* Andre Gerhard Wolff
*Acquisitions Editor:* Peter B. Linsley
*Editorial Project Manager:* Tracy I. Tufaga
*Production Project Manager:* Maria Bernard
*Designer:* Christian J. Bilbow

Typeset by Thomson Digital

# Contents

# Contributors


**Ahmed A. Abokor** Department of Physiology and Pharmacology, University of Toledo College of Medicine and Life Sciences, Toledo, OH, United States

**Daniela Albanesi** Instituto de Biología Molecular y Celular de Rosario (IBR) - CONICET, Facultad de Ciencias Bioquímicas y Farmacéuticas, Universidad Nacional de Rosario, Rosario, Argentina

**Ana Arabolaza** Instituto de Biología Molecular y Celular de Rosario (IBR) - CONICET, Facultad de Ciencias Bioquímicas y Farmacéuticas, Universidad Nacional de Rosario, Rosario, Argentina

**Ademola O. Ayeleso** Department of Biochemistry, Faculty of Science, Adeleke University, Ede, Osun State, Nigeria

**Christoph Benning** DOE-Plant Research Laboratory, Michigan State University, East Lansing, MI; Biochemistry and Molecular Biology, Michigan State University, East Lansing, MI, United States

**Karl-F. Bergeron** Département des Sciences Biologiques, Centre de recherche CERMO-FC, Université du Québec à Montréal, Montreal, QC, Canada

**David A. Bernlohr** Department of Biochemistry, Molecular Biology and Biophysics, The University of Minnesota, Minneapolis, MN, United States

**Kimberly K. Buhman** Purdue University, West Lafayette, IN, United States

**Chuchun L. Chang** Institute of Human Nutrition, Columbia University Irving Medical Center, New York, NY, United States

**Ting Chen** Department of Endocrinology, Union Hospital, Tongji Medical College, Huazhong University of Science and Technology, Wuhan, China

**Diego de Mendoza** Instituto de Biología Molecular y Celular de Rosario (IBR) - CONICET, Facultad de Ciencias Bioquímicas y Farmacéuticas, Universidad Nacional de Rosario, Rosario, Argentina

**Richard J. Deckelbaum** Institute of Human Nutrition, Columbia University Irving Medical Center, New York, NY; Department of Pediatrics, Vagelos College of Physicians and Surgeons, Columbia University Irving Medical Center, New York, NY, United States

**Pascal Degrace** UMR 1231 INSERM-UB-Agrosup, Team Pathophysiology of Dyslipidemia, Faculty of Sciences, Dijon, France

**Frédérik Desmarais** Département des Sciences Biologiques, Centre de recherche CERMO-FC, Université du Québec à Montréal, Montreal, QC, Canada

**Aneta M. Dobosz** Laboratory of Cell Signaling and Metabolic Disorders, Nencki Institute of Experimental Biology, Polish Academy of Sciences, Warsaw, Poland

**Agnieszka Dobrzyn** Laboratory of Cell Signaling and Metabolic Disorders, Nencki Institute of Experimental Biology, Polish Academy of Sciences, Warsaw, Poland

**Pawel Dobrzyn** Laboratory of Medical Molecular Biochemistry, Nencki Institute of Experimental Biology, Polish Academy of Sciences, Warsaw, Poland

**Anthony S. Don** Centenary Institute, The University of Sydney, Camperdown, NSW; NHMRCClinicalTrialsCentre,TheUniversityof Sydney, Camperdown, NSW, Australia

**H. Henry Dong** Division of Endocrinology and Diabetes, Department of Pediatrics, University of Pittsburgh School of Medicine, Pittsburgh, PA, United States; Rangos Research Center, Children's Hospital of Pittsburgh of UPMC, Pittsburgh, PA, United States

**Xiaoyun Feng** Division of Endocrinology and Diabetes, Department of Pediatrics, University of Pittsburgh School of Medicine, Pittsburgh, PA, United States; Department of Endocrinology and Metabolism, Shanghai General Hospital, Shanghai Jiaotong University, Shanghai, China

**Anna Filip** Laboratory of Medical Molecular Biochemistry, Nencki Institute of Experimental Biology, Polish Academy of Sciences, Warsaw, Poland

**Vincent Fong** Division of Endocrinology, Diabetes and Metabolism, University of Cincinnati, Cincinnati, OH, United States

**Sheetal Gandotra** Council of Scientific and Industrial Research-Institute of Genomics and Integrative Biology, Academy of Scientific and Innovative Research, New Delhi, India

**Rachel M. Golonka** Department of Physiology and Pharmacology, University of Toledo College of Medicine and Life Sciences, Toledo, OH, United States

**Isabel González-Mariscal** Biomedical Research Institute of Malaga-IBIMA, Endocrinology and Nutrition UGC, Regional University Hospital of Malaga, Malaga, Spain

**Hugo Gramajo** Instituto de Biología Molecular y Celular de Rosario (IBR) - CONICET, Facultad de Ciencias Bioquímicas y Farmacéuticas, Universidad Nacional de Rosario, Rosario, Argentina

**Liang Guo** The Key Laboratory of Metabolism and Molecular Medicine of the Ministry of Education, Institute of Stem Cell Research and Regenerative Medicine of Institutes of Biomedical Sciences, Department of Biochemistry and Molecular Biology of School of Basic Medical Sciences, Fudan University, Shanghai, China

**Xin Guo** Departments of Nutrition and Food Hygiene, School of Public Health, Shandong University, Jinan, China

**Abigail M. Harris** Nutritional Sciences, University of Wisconsin Madison, United States

**Ann V. Hertzel** Department of Biochemistry, Molecular Biology and Biophysics, The University of Minnesota, Minneapolis, MN, United States

**Susanne Hoffmann-Benning** Genetics and Genome Sciences, Michigan State University, East Lansing, MI; Biochemistry and Molecular Biology, Michigan State University, East Lansing, MI, United States

**Yumiko Ishii** Meakins-Christie Laboratories, Department of Medicine, McGill University, Montreal, QC, Canada

**Justyna Janikiewicz** Laboratory of Cell Signaling and Metabolic Disorders, Nencki Institute of Experimental Biology, Polish Academy of Sciences, Warsaw, Poland

**Tony Jourdan** UMR1231 INSERM-UB-Agrosup, Team Pathophysiology of Dyslipidemia, Faculty of Sciences, Dijon, France

**Babunageswararao Kanuri** Division of Endocrinology, Diabetes and Metabolism, University of Cincinnati, Cincinnati, OH, United States

**Audrey L. Keenan** Division of Renal Diseases and Hypertension, University of Colorado-Denver, Aurora, CO, United States

**Charlie Kirsh** Biochemistry, University of Wisconsin Madison, United States

**Amanda M. Koenig** Genetics and Genome Sciences, Michigan State University, East Lansing, MI, United States

**Ewelina Krogulec** Laboratory of Cell Signaling and Metabolic Disorders, Nencki Institute of Experimental Biology, Polish Academy of Sciences, Warsaw, Poland

**Sojin Lee** Division of Endocrinology and Diabetes, Department of Pediatrics, University of Pittsburgh School of Medicine, Pittsburgh, PA, United States

**Kenneth T. Lewis** Department of Molecular and Integrative Physiology, University of Michigan Medical School, Ann Arbor, MI, United States

**Ormond A. MacDougald** Department of Molecular and Integrative Physiology, University of Michigan Medical School, Ann Arbor, MI; Department of Internal Medicine, University of Michigan Medical School, Ann Arbor, MI, United States

**Denny Joseph Manual Kollareth** Institute of Human Nutrition, Columbia University Irving Medical Center, New York, NY, United States

**Oana C. Marian** Centenary Institute, The University of Sydney, Camperdown, NSW, Australia

**Douglas G. Mashek** Department of Biochemistry, Molecular Biology and Biophysics, University of Minnesota, Minneapolis, MN; Department of Medicine, Division of Diabetes, Endocrinology and Metabolism, University of Minnesota, Minneapolis, MN, United States

**Mashudu G. Matumba** Department of Biochemistry, Faculty of Natural and Agricultural Science, North West University, Mmabatho, South Africa

**Makoto Miyazaki** Division of Renal Diseases and Hypertension, University of Colorado-Denver, Aurora, CO, United States

**Antonio Moschetta** Clinica Medica Cesare Frugoni, Department of Interdisciplinary Medicine, University of Bari Aldo Moro, Bari; INBB, National Institute for Biostructures and Biosystems, Rome; IRCCS Istituto Tumori Giovanni Paolo II, Bari, Italy

**Catherine Mounier** Département des Sciences Biologiques, Centre de recherche CERMO-FC, Université du Québec à Montréal, Montreal, QC, Canada

**Emmanuel Mukwevho** Department of Biochemistry, Faculty of Natural and Agricultural Science, North West University, Mmabatho, South Africa

**Charles P. Najt** Department of Biochemistry, Molecular Biology and Biophysics, University of Minnesota, Minneapolis, MN, United States

**Hai P. Nguyen** Department of Nutritional Sciences & Toxicology, Endocrinology Program, University of California, Berkeley, CA, United States

**James M. Ntambi** Departments of Biochemistry and Nutritional Sciences, University of Wisconsin-Madison, Madison, WI, United States

**Timothy D. O'Connell** Department of Integrative Biology and Physiology, The University of Minnesota, Minneapolis, MN, United States

**Toshiaki Okuno** Department of Biochemistry, Juntendo University Graduate School of Medicine, Tokyo, Japan

**Shailendra B. Patel** Division of Endocrinology, Diabetes and Metabolism, University of Cincinnati, Cincinnati, OH, United States

**Chad M. Paton** Department of Foods & Nutrition, University of Georgia, Athens, GA; Department of Food Science & Technology, University of Georgia, Athens, GA, United States

**Elena Piccinin** Clinica Medica Cesare Frugoni, Department of Interdisciplinary Medicine, University of Bari Aldo Moro, Bari, Italy

**Shuwen Qian** The Key Laboratory of Metabolism and Molecular Medicine of the Ministry of Education, Institute of Stem Cell Research and Regenerative Medicine of Institutes of Biomedical Sciences, Department of Biochemistry and Molecular Biology of

School of Basic Medical Sciences, Fudan University, Shanghai, China

**Shen Qu** Department of Endocrinology and Metabolism, Shanghai 10th People's Hospital, Tongji University School of Medicine, Shanghai, China

**Eric Rassart** Département des Sciences Biologiques, Centre de recherche CERMO-FC, Université du Québec à Montréal, Montreal, QC, Canada

**Karen Reue** Department of Human Genetics, David Geffen School of Medicine at UCLA, Los Angeles, CA; Molecular Biology Institute, University of California, Los Angeles, CA, United States

**Carrie Riestenberg** Department of Human Genetics, David Geffen School of Medicine at UCLA, Los Angeles, CA, United States

**Yuji Shiozaki** Division of Renal Diseases and Hypertension, University of Colorado-Denver, Aurora, CO, United States

**Judith Simcox** Biochemistry, University of Wisconsin Madison, United States

**Yura Son** Department of Foods & Nutrition, University of Georgia, Athens, GA, United States

**Hei Sook Sul** Department of Nutritional Sciences & Toxicology, Endocrinology Program, University of California, Berkeley, CA, United States

**Gergo Szanda** MTA-SE Laboratory of Molecular Physiology, Department of Physiology, Semmelweis University, Budapest, Hungary

**Joseph Tam** Obesity and Metabolism Laboratory, Institute for Drug Research, School of Pharmacy, Faculty of Medicine, The Hebrew University of Jerusalem, Jerusalem, Israel

**Qiqun Tang** The Key Laboratory of Metabolism and Molecular Medicine of the Ministry of Education, Institute of Stem Cell Research and Regenerative Medicine of Institutes of Biomedical Sciences, Department of Biochemistry and Molecular Biology of School of Basic Medical Sciences, Fudan University, Shanghai, China

**Zuzanna Tracz-Gaszewska** Laboratory of Medical Molecular Biochemistry, Nencki Institute of Experimental Biology, Polish Academy of Sciences, Warsaw, Poland

**Collin Tran** Centenary Institute, The University of Sydney, Camperdown, NSW, Australia

**Laurent Vergnes** Department of Human Genetics, David Geffen School of Medicine at UCLA, Los Angeles, CA, United States

**Matam Vijay-Kumar** Department of Physiology and Pharmacology, University of Toledo College of Medicine and Life Sciences, Toledo, OH, United States

**Chaodong Wu** Department of Nutrition, Texas A&M University, College Station, TX, United States

**Jun Yamauchi** Division of Endocrinology and Diabetes, Department of Pediatrics, University of Pittsburgh School of Medicine, Pittsburgh, PA, United States

**Danielle Yi** Department of Nutritional Sciences & Toxicology, Endocrinology Program, University of California, Berkeley, CA, United States

**Takehiko Yokomizo** Department of Biochemistry, Juntendo University Graduate School of Medicine, Tokyo, Japan

**Alyssa S. Zembroski** Purdue University, West Lafayette, IN, United States

**Juan Zheng** Department of Endocrinology, Union Hospital, Tongji Medical College, Huazhong University of Science and Technology, Wuhan, China; Department of Nutrition, Texas A&M University, College Station, TX, United States

**Cuiling Zhu** Division of Endocrinology and Diabetes, Department of Pediatrics, University of Pittsburgh School of Medicine, Pittsburgh, PA, United States; Department of Endocrinology and Metabolism, Shanghai 10th People's Hospital, Tongji University School of Medicine, Shanghai, China

**Ping Zhu** Division of Endocrinology and Diabetes, Department of Pediatrics, University of Pittsburgh School of Medicine, Pittsburgh, PA, United States; Department of Endocrinology, Guangzhou Red Cross Hospital, Medical College of Jinan University, Guangzhou, China

**Hylde Zirpoli** Institute of Human Nutrition, Columbia University Irving Medical Center, New York, NY, United States

# Preface

There is a vast literature and ongoing research on lipid metabolism and signaling. The coordination of metabolic regulation involves extensive crosstalk among cells, tissues, and organs. Signaling factors that are secreted locally or into the circulation and impart systemic metabolic effects include molecules such as hepatokines and adipokines. Many of these factors regulate lipid metabolism, including de novo lipogenesis. The lipids themselves are not only constituents of cellular membranes but because of their recently appreciated functional diversity can also act as key signaling mediators, revealing new and unexpected biological, metabolic, and cellular functions. The key issue is that lipids have moved to center stage and that new technologies (especially lipidomics, metabolomics) used in lipid research produce an avalanche of new data, which is overwhelming. Anything that can help bring structure in this field and provide a concise overview of the current knowledge is greatly needed. The goal of this book is thus to provide a comprehensive overview of the field and could, despite the rapid developments, be used as a reference for years to come. It is written by world-renowned diverse expert research investigators from North and South America, Europe, Asia, Australia, and Africa studying lipid metabolism and lipid signaling mediators. The lipids are now recognized to regulate a broad range of normal biological processes as well as diseases such as obesity, diabetes, fatty liver disease, inflammation, cancer, cardiovascular, and neurodegenerative diseases. The book covers a wide range of topics including recent advances, new concepts, applications, new ideas, and next steps on the mechanisms underlying lipid metabolism and signaling in bacteria, plants, and animals. Expansion of our understanding of metabolism and the role of lipid mediators in metabolism and signaling in health and disease affords the opportunity for novel therapeutics.

Although the lipid biosynthetic pathways are conserved in bacteria, there are notably differences in the gene organization, gene regulation, and biochemical control of the enzymes that perform these reactions in Gram-positive and Gram-negative bacteria. In the first chapter of this elegant book Diego de Mendoza et al. examine this diversity to provide a timely overview of lipid synthesis and membrane homeostasis in prokaryotes. In addition to their structural functions in membranes, lipids in plants are also involved in many signaling mechanisms that influence development and stress responses. Thus, in the second chapter Hoffmann-Benning et al. briefly review key aspects of plant lipid biosynthesis and trafficking. They discuss a novel long-distance, lipid-mediated signaling mechanism for systemic stress response in plants. When we move into higher organisms, Reue et al. begins by reviewing our current understanding of the physiological mechanisms that contribute to sex differences in lipoprotein levels, obesity, microbiome composition, and development of atherosclerosis in humans and in mouse models. Then comes a chapter by Lewis and MacDougald that summarize recent advancements in our understanding of the relationship between bone marrow adipocytes (BMAs) and their neighboring bone marrow cells and how these interactions influence the local marrow niche and systemic metabolism. Sul et al. discusses the development, metabolism, and function of adipose tissues. Following, Tang et al. describe the recently, identified lipids as the inducers for biogenesis and activation of beige/brown adipocytes. Lipid receptors and signaling in adipose tissue comes next and, in this chapter, Bernlohr et al. summarizes our current awareness of fatty acid receptors expressed by adipocytes and the signaling pathways that are affected downstream of lipid binding. After this, a chapter by Simcox et al. discusses adipocyte lipolysis and lipid-derived metabolite signaling. Lipid droplets are widely recognized as the primary energy storage depot in most cell types. However, as the field of lipid droplet biology has grown, roles for these dynamic organelles have expanded beyond storage of lipids. Zembroski and Buhman describe the regulation of intracellular lipid storage and utilization and how it is essential to prevent abnormal cellular and systemic lipid levels and its associated pathological consequences. The chapter by Najt and Mashek summarizes research evaluating how lipid droplets communicate within cells and provide a context for how this communication (or miscommunication) can lead to cellular dysfunction. In a follow up chapter Sheetal Gandotra discusses the central role of lipid droplets in lipid signaling and mobilization in the immune system indicating clearly that altering the metabolic state of an immune cell particularly the myeloid cell can have important consequences for its polarized state. Mounier et al. focused on the fatty acids and their role in the inflammasome while Ishii et al. summarizes the biosynthesis of $LTB_4$ and 12-HHT, and the characterization and pathophysiological roles of BLT1 and BLT2.

The FoxO transcription factors are instrumental for integrating nutritional and hormonal signaling key functions in diverse pathways including cell metabolism, proliferation, differentiation, oxidative stress, cell survival and senescence, autophagy and aging. Dong et al. discuss the potential therapeutic benefits and possible adverse effects of pharmacological inhibition of FoxO1 activity in insulin resistant subjects with metabolic diseases. Nonalcoholic fatty liver disease (NAFLD) is a leading cause of liver-related morbidity and mortality in the United States and globally. Wu et al. present a better understanding of the interplays between nutritional and inflammatory signaling, and fat metabolism has both diagnostic and therapeutic implications for the treatment of NAFLD. Jourdan et al. discuss the accumulating evidence indicating that Endocannabinoid signaling is involved in modulating energy homeostasis, adipose tissue metabolism, glucose and insulin balance as well as hepatic lipogenesis in health and metabolic diseases. The timely chapter by Matam Vijay-Kumar et al. explores how gut microbiota modulates host lipid metabolism and how this subsequently affects pathophysiology. The chapter explores the gut microbiota on lipogenesis, beiging, metabolic endotoxemia, and circadian rhythm with a discussion on the emerging discipline of lipidomics, which is advancing the field of personalized medicine. Mukwevho et al. discuss some perspectives and insights on the metabolism of lipids, in particular, in obesity and diabetes, metabolic conditions that have increased to epidemic proportions around the world.

Muscle plays a major role in regulating the metabolism of lipids. Throughout their chapter, Son and Paton present a thorough description of physiological and pathological mechanisms controlling skeletal muscle lipid metabolic control and where possible, present a comparison and contrast between human and rodent muscle physiology. Dobrzyn et al. summarizes our recent understanding of the role of different sphingolipid derivatives in cell-signaling responses and their role in the pathogenesis of metabolic diseases, including type 2 diabetes and cardiovascular disorders. Bile acid receptors regulate lipid homeostasis, carbohydrate homeostasis, drug metabolism, energy consumption and inflammatory responses in response to bile acids and drugs. Miyazaki et al. summarize basic bile acid receptor signaling and functions, and effects of bile acid receptor modulations in the development of cardiovascular diseases. Deckelbaum et al. reviewed and compared evidence on the benefits, and in some cases adverse effects, of n-3 and n-6 PUFAs in cardiovascular disease prevention, and provide insight into potential molecular mechanisms of these benefits.

Reprogramming of lipid metabolism is now a recognized hallmark of malignancy and neurodegenerative diseases. Moschetta et al. discuss lipid metabolism and signaling in cancer and emphasize that although additional studies are needed strategies aimed at limiting the availability of lipids are necessary to block cancer growth. The chapter by Don et al. describes the current knowledge on altered lipid metabolism in Alzheimer's disease, and how this relates to genetic risk, myelin deterioration, impaired endosomal-lysosomal flux, and the neuropathological hallmarks of the disease. Finally, Patel et al. reviewed the importance of xenosterols in health and disease; discuss their significance in the pathogenesis of Sitosterolemia and the current knowledge on their impact in management of cardiovascular (CVS) and central nervous disorders (CNS).

I am extremely grateful to the many people who have contributed the various chapters to this book for their cooperation and excellent work. The contributing authors are at the forefront of many discoveries in lipid metabolism and have provided chapters with basic knowledge component coupled with unique insights into their own fields of research. The content is accessible to broad spectrum of learners with basic understanding of biochemistry and metabolism. The book also serves a gateway to further exploration of topics and provides a bridge between basic concepts and current research literature. Thus, undergraduates, graduate students, nutritionists, biologists, geneticist, professors, clinicians, endocrinologists, and teachers will find this book to be an essential resource for course and research-related studies while experienced researchers can use it as a reference guide to advance the lipid field.

I thank Springer support staff for their assistance and support during the course of this project. Peter Linsley saw the potential for this project and Tracy Tufaga who has provided considerable help as the editorial manager during the course of this project. Finally, I would like to thank my family for their support, encouragement over the years.

James M. Ntambi PhD
Madison, WI, United States

CHAPTER

# 1

# Homeostatic control of membrane lipid biosynthesis in bacteria

*Daniela Albanesi*, Ana Arabolaza*, Hugo Gramajo, Diego de Mendoza*

Instituto de Biología Molecular y Celular de Rosario (IBR) - CONICET, Facultad de Ciencias Bioquímicas y Farmacéuticas, Universidad Nacional de Rosario, Rosario, Argentina

OUTLINE



## Introduction

The plasma membrane has an essential function in cells as a barrier and a matrix of biological activities. In the case of bacteria, plasma membranes display a large diversity of amphiphilic lipids, including phospholipids and a variety of other membrane lipids, like glycolipids, sphingolipids, ornithine lipids, or hopanoids among others. Different bacterial phyla possess

*These authors contributed equally to this work.

Lipid Signaling and Metabolism. http://dx.doi.org/10.1016/B978-0-12-819404-1.00001-4

a particular content of membrane lipids possibly associated with their various lifestyles [1,2]. In all organisms the control at the level of fatty acid biosynthesis is crucial for membrane homeostasis, because the biophysical properties of membranes are determined in large part by the composition of the fatty acids that are produced by de novo biosynthesis. Almost all of the metabolic energy that is used to produce membrane lipids is expended in the formation of fatty acids, and therefore their production must be precisely controlled to support membrane biogenesis and prevent the wasteful expenditure of ATP. Although the biophysical properties of membranes can be changed by altering the ratio of the polar head groups in membrane phospholipids, bacteria seem to use biochemical and genetic mechanisms to modify the composition of the fatty acids that they synthetize. Also, the membrane composition of cells of a single species can vary according to the environmental conditions to which cells are exposed [1–3]. For example, functions of the cytoplasmic membrane are known to depend critically on the physical state of the lipid bilayer, making it susceptible to changes in environmental temperature. This means that bacteria must process temperature signals to adjust enzyme activities or to activate unique genes necessary to adapt their membranes to the new temperature. Here, the principal genetic and biochemical processes that are responsible for control of membrane lipid homeostasis in bacteria are reviewed.

## General principles of fatty acid biosynthesis

Fatty acids (FA) are the essential building blocks of membrane phospholipids and are the integral part of a diversity of lipids that fulfill a variety of key functions; like energy storage compounds and cell signaling molecules. FA production is an energetically expensive process for cells, and thus, the biosynthetic pathway is subjected to sophisticated genetic and biochemical regulation mechanisms.

In general, all organisms employ a conserved set of chemical reactions that are catalyzed by the fatty acid synthase (FAS), to achieve the de novo FA biosynthesis. FAS works by the sequential extension of the growing carbon chain, two carbons at a time, through a series of decarboxylative condensation reactions. There are two major classes of FAS systems: type I and II [4,5]. These two systems differ on their structural organization, although, the chemical reactions and the catalytic mechanisms for FA biosynthesis are essentially the same. The type II FAS complex (mostly found in bacteria, but also in specialized eukaryotic organelles such as mitochondria and plastids in plants) is constituted by discrete enzymes, each of them encoded by a separate gene and existing as a soluble monofunctional protein. On the other hand, type I FAS (typical of fungi and animals) are huge multifunctional polypeptides that integrate all catalytic steps into large macromolecular assemblies, usually one or two large proteins [5–7]. Interestingly, this scenario of FAS systems distribution (type I and II) is exceptional in bacteria belonging to the group of actinomycetes. The genus *Streptomyces,* for example, contains the classic dissociated type II FAS system [8]; while, mycobacteria contain the two types of FAS found in nature (type I and II) [9] and *Corynebacterium* spp. only harbors a FAS I system [10]. This picture delineates, in part, the increasing complexity of cellular lipid envelope of some members of the actinomycetales group. For instance, while in most bacteria FA synthesized de novo are incorporated mainly into the membrane lipids, being this pathway a major determinant of cell size in *Escherichia coli* and *Bacillus subtilis* [11], in some actinomycetes FA could have other destinies. In mycobacteria, for example, FAS I catalyzes the de novo synthesis of long-chain acyl-CoAs ($C_{16}$- and $C_{18}$-CoA) which are primarily used for the synthesis of membrane phospholipids. Alternatively, this $C_{16}$- and $C_{18}$-CoA can be further elongated by FAS I to produce mostly $C_{24-26}$-CoA in *Mycobacterium tuberculosis;* a characteristic "bimodal" behavior of FAS I [12,13]. The $C_{16-18}$ acyl-CoAs released by FAS I can also be funneled into the FAS II enzyme system where they become elongated iteratively leading to the synthesis of the very long-chain meromycolyl-ACPs. In *M. tuberculosis,* the $C_{24-26}$ fatty acids synthesized by FAS I become carboxylated by a long-chain acyl-CoA carboxylase complex (ACCase 4) to yield a rare β-carboxy-$C_{24-26}$-CoA. This rare acyl-CoA is finally condensated with meromycolyl-AMP, in a reaction catalyzed by the polyketide synthase (PKS) 13, to produce mycolic acids (MA) and their glyco-derivatives, conferring special structural, permeability, and immunopathogenic properties to these bacteria [14]. Furthermore, the acyl-CoAs synthesized by FAS I are also used by diacylglycerol acyl transferases to produce triacylglycerol (TAG) [15], and by different PKSs as important biosynthetic precursors to produce virulence associated polyketide lipids like phtiocerol-dimycoseroic acid (PDIM), poly-acylated trehalose (PATS), and sulfolipids (SL) [14,16].

The following sections will describe in detail the enzymatic steps of phospholipid biosynthesis and the different regulation levels of the pathway, highlighting first the FA biosynthetic process.

## Biochemistry of bacterial fatty acid synthesis

### Acetyl-CoA carboxylase

Malonyl-CoA is the universal elongation unit for the de novo FA biosynthesis. The production of this metabolite takes place by carboxylation of acetyl-CoA by the enzyme acetyl-CoA carboxylase (ACC) (Fig. 1.1). This enzyme belongs to a family of biotin-dependent carboxylases that are widely distributed in nature, being found in animals, fungi, algae, archaea, and bacteria [17,18]. The biotin-dependent carboxylation occurs in two distinct half-reactions [19]; the first one carried out by a biotin carboxylase (BC) component that uses bicarbonate as the $CO_2$ donor catalyzes the MgATP-dependent carboxylation of the biotin cofactor that is covalently linked to a lysine residue of the biotin carboxyl carrier protein (BCCP) component. The second half is performed by a carboxyltransferase (CT) component that catalyzes the transfer of $CO_2$ from carboxybiotin to acetyl-CoA. Besides ACC, within this family of enzymes other carboxylases have distinct substrate preferences, such as propionyl-CoA carboxylase (PCC), 3-methylcrotonyl-CoA carboxylase (MCC), and geranyl-CoA carboxylase (GCC). While ACC and PCC carboxylate at the α carbon of saturated acids, such as acetyl-CoA and propionyl-CoA, respectively; the MCC and GCC enzymes carboxylate the γ carbon of the α,β unsaturated acid, such as 3-methylcrotonyl-CoA or geranyl-CoA. Generically, all these enzymes are called acyl-CoA carboxylases (or YCC) due to their broad substrate preference, mainly for short-chain acyl-CoAs. In actinomycetes, however, the so-called ACC or PCC are also referred as acyl-CoA carboxylases (or ACCases) [9], although in this case, this nomenclature is used because the same enzyme complex can recognize more than one substrate, for example, acetyl-CoA, propionyl-CoA, and even butyryl-CoA, sometimes with very similar specificity

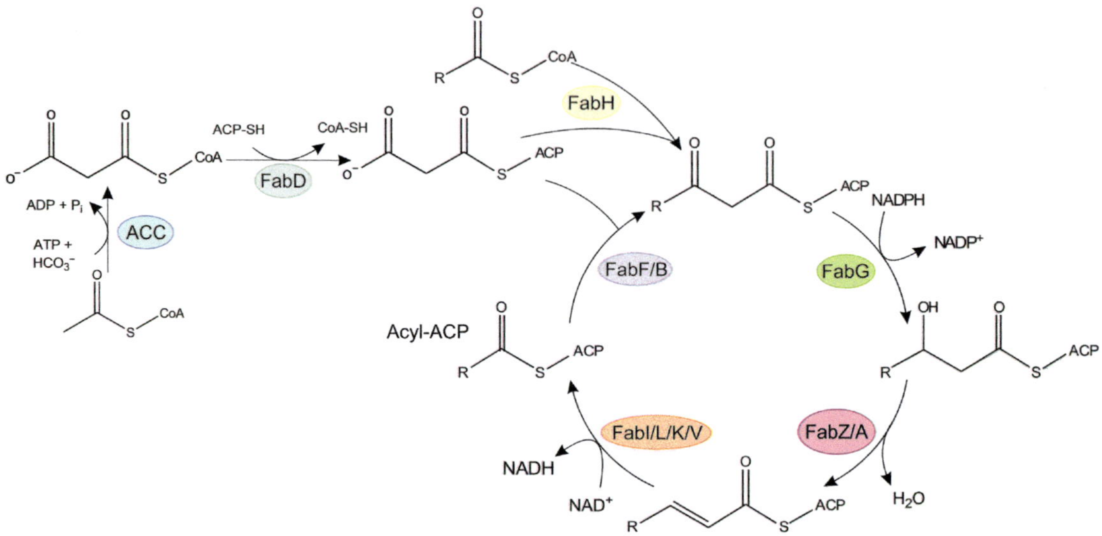


**FIGURE 1.1 Biochemical pathway for the formation of fatty acid in bacteria.** The initiation steps of type II fatty acid synthesis (FAS II) involve the production of malonyl-CoA by the acetyl-CoA carboxylase complex (ACC). Subsequently, the malonyl-CoA-ACP transacylase, FadD, transfers the malonyl groups to the acyl carrier protein (ACP) to produce malonyl-ACP, the elongation unit of the cycle. FabH, the initiating condensing enzyme, utilizes the malonyl-ACP and a priming acyl-CoA substrate to produce the first new C–C bond. FabG, a β-ketoreductase, reduces the β-ketoacyl-ACP to give the corresponding β-hydroxyacyl-ACP, which is then dehydrated to enoyl-ACP by FabA or FabZ. The enoyl-ACP reductase (the multiple isoforms FabI/K/L/V are indicated), performs the final reduction step of each cycle. FabF, and in some bacteria also FabB (required for unsaturated fatty acid production), are the condensing enzymes that start a new round of elongation of the growing acyl-ACP intermediates utilizing malonyl-ACP.

constants. This is the case for the so-called ACC and PCC complexes present in *Streptomyces*, *Mycobacterium*, and *Corynebacterium*, within others [21–22].

Since this chapter focuses on the homeostatic control of membranes lipid biosynthesis, we will mainly discuss the acyl-CoA carboxylases that provide the substrates for the de novo biosynthesis of FA and of other more complex lipids that form part of the outer membrane of some actinomycetes.

The structural organization of the BC, BCCP, and CT components of the ACC/ACCases enzymes varies significantly within bacteria. In *E. coli*, as an example of the first characterized ACC, the BC, BCCP, and CT components are organized in four separate polypeptides, a BC subunit, a BCCP subunit, and two CT subunits. All these subunits interact to constitute the functional enzyme complex [23,18], although the holoenzyme is generally unstable and readily dissociates during purification. Instead, in several actinomycetes ACCases (like the ACC or the PCC complexes) the BC and BCCP components are fused into a single protein, the α subunit, and the two CT subunits are also fused in a single polypeptide (the β subunit) [22,24,25]. A distinctive property of some of these carboxylases is the presence of a small (7–10 kDa) basic peptide, called ε subunit, necessary for maximal enzyme activity [20,21]. In contrast with the *E. coli* ACC, the actinomycetes ACCase complexes are more stable and can be readily purified from the original sources, or reconstituted in vitro from their purified subunits [21,22,24,26]. Interestingly, a new structural organization of a particular multisubunit

ACCase from *M. tuberculosis* has been described recently; this enzyme is called long-chain acyl-CoA carboxylase (LCC). LCC recognizes $C_{24-26}$ acyl-CoAs and it forms a mega-complex with an α and ε subunits and two different β subunits [27]. This LCC complex is also present in other actinomycetes genera, like *Corynebacterium*, *Nocardia*, and *Rhodococus* and their acyl-CoA substrates range between $C_{18}$ and $C_{26}$ acyl-CoAs. Until recently the only known single peptide multidomain ACC had been the eukaryotic one. However, in the last years two multidomains ACCases have been characterized in bacteria, one of them mainly carboxylates long-chain acyl-CoAs [28] while the other, found in *Saccharopolyspora erythraea*, only carboxylates acetyl-CoA and propionyl-CoA [29].

Most bacteria have a unique ACC dedicated to generate malonyl-CoA for de novo FA biosynthesis. However, a much higher complexity exists in relation with the ACCases in actinomycetes that is related with either the complexity of their lipidic content, like in *Mycobacterium*, *Corynebacterium*, or *Rhodococcus*, or with the production of structural reach secondary metabolites like in *Streptomyces* or *Saccharopolyspora* genera. As an example, in *M. tuberculosis* six putative ACCase complexes can be predicted from the genome analysis, and three of them (ACCase 5, ACCase 6, and LCC) are essential for bacterial viability and were found to be involved in lipid metabolism [9]. ACCase 5 and 6 can both recognize the same substrates, acety-CoA and propionyl-CoA, to generate malonyl-CoA and methylmalonyl-CoA, respectively. However, genetics experiments suggest that the physiological role of ACCase 6 is that of an essential ACC, providing the extender unit for the FAS I and FAS II synthases (see below) [30,31], while ACCases 5 most probably provides the methylmalonyl-CoA used for dedicated PKS to synthesize the methyl-branched lipids present in the complex outer membrane of these organisms [21,32,33]. The LCC instead, carboxylates the $C_{24-26}$ acyl-CoAs that become one of the substrate of PKS 13, that after a decarboxylative condensation with the very long-chain meromycolic acids generated by FAS II, will constitute the α branch of the mycolic acids present in the outer membrane of these microorganisms.

## Initiation steps and elongation cycle

Fatty acid biosynthesis proceeds in two stages: initiation and iterative cyclic elongation. As mentioned, the acetyl-CoA carboxylase enzyme complex (ACC) performs the first committed step in bacterial FA synthesis to generate malonyl-CoA through the carboxylation of acetyl-CoA [23,34]. The malonate group from malonyl-CoA is transferred to the acyl carrier protein (ACP) by a malonyl-CoA:ACP transacylase (FabD) [35,36]. The first reaction for the synthesis of the nascent carbon chain comprises the condensation of malonyl-ACP with a short-chain acyl-CoA ($C_2$–$C_5$) catalyzed by a 3-keto-acyl carrier protein synthase III (FabH) (Fig. 1.1). The substrate specificity of FabH plays a determining role in the branched/straight and even/odd characteristics of the diverse fatty acid structures produced by the different bacteria. *E. coli* and *Streptococcus pneumoniae* mostly produce even-number straight-chain saturated and unsaturated fatty acids, since their FabH enzymes utilizes selectively acetyl-CoA as priming unit [37]. On the other hand, many Gram-positive bacteria, such as *Bacillus*, *Staphylococcus*, and *Streptomyces*, produce predominantly odd-numbered branched-chain fatty acids. Biochemical experiments have clearly showed a high selectivity of FabH enzymes from these bacteria for branched chain acyl-CoA primer units (isobutyryl-, isovaleryl-, or anteisovaleryl-CoA) which lead to the production of anteiso- and iso-branched chain FA [38]. It has been demonstrated that *fabH* is not an essential gene in a wild-type *E. coli* strain but it is necessary to adjust cell size. Conversely, *fabH* cannot be disrupted in *E. coli* cells that are defective in the production of the global regulators of the stringent response, guanosine-5′-triphosphate-3′-diphosphate and guanosine-5′-diphosphate-3′-diphosphate [ppGpp and pppGpp, collectively referred as (p)ppGpp] [39]. A very recent study revealed the existence of a new protein, FabY, capable of initiating the biosynthesis of fatty acids in the absence of FabH. Interestingly, it was shown that the expression of *fabY* depends on the presence of the stringent response factors explaining why FabH becomes essential in the absence of these alarmones [40].

The keto-acyl-ACP, product of the FabH condensation, enters the elongation cycle and is reduced by the NADPH dependent β-ketoacyl-ACP reductase (FabG) to give β-hydroxy-acyl-ACP. The β-hydroxyacyl-ACP intermediate is then dehydrated to trans-2-enoyl-ACP by a 3-hydroxyacyl-ACP dehydratase; FabA or FabZ in *E. coli* (Fig. 1.1). The distinction between these two enzymes lies in the dual role of FabA which can also catalyze the isomerization of the *trans*-2-decenoyl-ACP to *cis*-3-decenoyl-ACP intermediate; the point of divergence needed to synthesize unsaturated FA (see below). The monofunctional FabZ is active on all chain lengths of saturated and unsaturated intermediates, and it is the most widely distributed dehydratase. FabA is restricted to α- and γ-proteobacteria and its product, *cis*-3-decenoyl-ACP, requires a 3-ketoacyl-ACP synthase I, FabB, to skip the next enoyl reduction step and lead to the unsaturated FA (UFA) biosynthesis (see below).

The cycle is driven to completion by an enoyl-ACP reductase which reduces the double bond in *trans*-2-enoyl-ACP to form acyl-ACP. The enoyl-ACP reductase Fab I of *E. coli* was the first described, however, this enzyme shows remarkable diversity among different bacteria [1]. For example; FabI of *E. coli* has a preference for NADH over NADPH, while FabI of *S. aureus* prefers NADPH over NADH. *B. subtilis* possesses two enoyl-ACP reductases (FabI and FadL) with opposite cofactor preferences [41]. There is also a widespread enoyl-ACP reductase, first identified in *Vibrio cholerae* and denominated FabV, which is structurally related to FabI. However, many Gram-positive bacteria use an unrelated flavoenzyme, FabK, to perform this step of the FA elongation cycle [42].

In all the successive steps of FA elongation, the produced acyl-ACP is used by the condensing enzyme FabF (3-oxoacyl-ACP-synthase II) or FabB (3-oxoacyl-ACP-synthase I) to initiate a new round of elongation. Both enzymes catalyze a Claisen condensation reaction using malonyl-ACP to elongate the growing acyl chain. Almost all bacteria have the essential condensing enzyme 3-ketoacyl-ACP synthase II (FabF); and γ-proteobacteria that produce unsaturated FA (UFA) also contain the 3-ketoacyl-ACP synthase I (FabB) isoform (Fig. 1.2A). As mentioned, FabB plays a key role in feeding the *cis*-decenoyl-ACP product into the elongation cycle. FabF is not essential for growth in *E. coli*, since FabB efficiently catalyzes the condensation of short to medium chain acyl-ACPs with malonyl-ACP. However, FabB is absolutely required for UFA synthesis, as supported by the fact that *fabB* mutants are UFA auxotrophs and need exogenous oleic acid for growth [43]. Furthermore, in *E. coli*, FabF functions in vivo to elongate palmitoleic acid (16:1Δ9) to *cis*-vaccenic acid (18:1Δ11), a reaction that FabB performs less efficiently [44].

Genomic analyses suggest that the FabA/FabB pathway for UFA biosynthesis might be restricted to α- and γ-proteobacteria. In Gram-positive bacteria, like *S. pneumonia*, the monofunctional enzyme FabM catalyzes the isomerization of the FabZ product, *trans*-2-decenoyl-ACP to *cis*-decenoyl-ACP, and then FabF elongates the UFA [1] (Fig. 1.2B). Many different bacteria such as *Helicobacter*, *Clostridium*, *Campylobacter*, and *Burkholderia* which do not contain

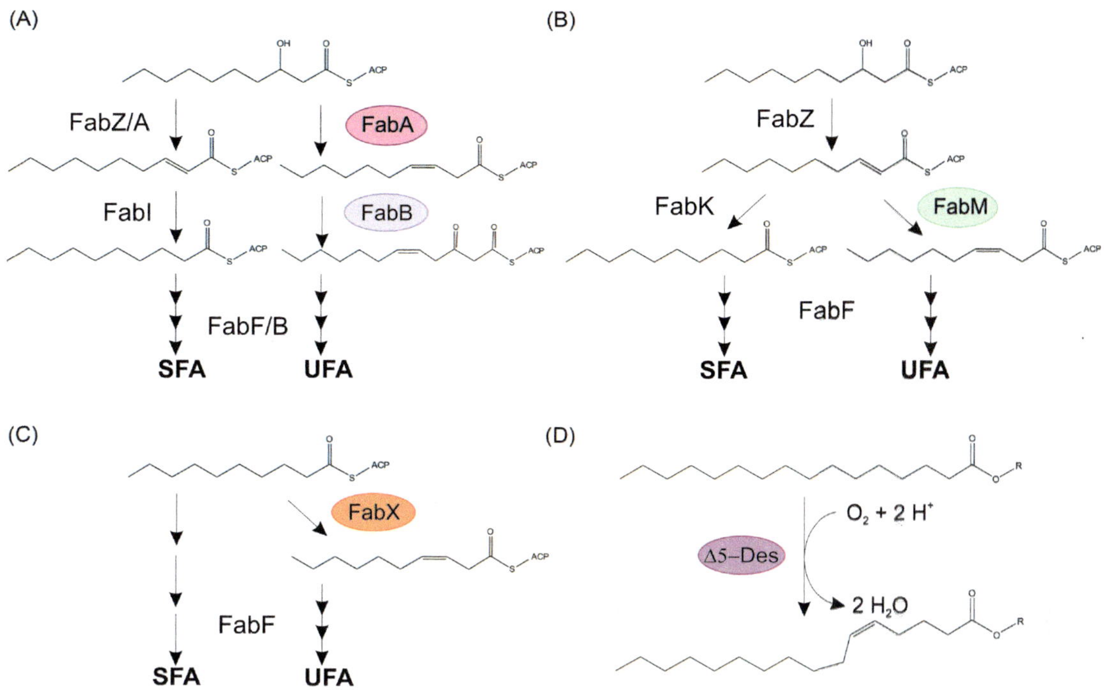


FIGURE 1.2 **Bacterial strategies for the synthesis of unsaturated fatty acids.** (A) *E. coli* contains the bifunctional dehydratase/isomerase FabA enzyme. This enzyme can specifically act on β-C10:0-ACP intermediate to catalyze the *trans*-2-C10:1-ACP dehydratation/isomerization to generate *cis*-3-C10:1-ACP. This product is the substrate of a FabB dependent condensation which channels this intermediate to the synthesis of unsaturated fatty acids. Alternatively, the *trans*-2-C10:1-ACP can be reduced by FabI to continue through the elongation cycles resulting in the production of saturated fatty acids. (B) Members of the genera *Streptococci* and *Clostridia* do not have a FabA homolog in their genomes and have a subtype of FabF as elongation condensing enzyme. Particularly, *Streptococcus* sp. posses a specific isomerase FabM that competes for the substrate with the FabK (the enoyl-ACP reductase), being *fabM* mutants auxotrophs for UFA. Then, FabF elongates the corresponding intermediates leading to the production of both, unsaturated fatty acids and SFA, in these organisms. (C) C10:1-ACP represents the branching point for UFA synthesis in *H. pylori*. FabX enzyme catalyzes a dehydrogenation/isomerization reaction to produce *cis*-3-C10:1-ACP. FabF is involved in the further elongation cycles resulting in UFA and SFA production. (D) Some bacteria employ a post-synthetic mechanism for the production of UFAs which involves the action of a desaturase. These enzymes utilize molecular oxygen as the oxidizing agent and introduce a double bond in the *cis* configuration at specific positions of the acyl chains of membrane phospholipids, acyl-ACPs or acyl-CoAs. The figure depicts the action of Δ5-Des, a Δ5-desaturase from *B. subtilis*, as an example.

homologues to FabA/FabB, utilize FabX for the introduction of a carbon–carbon double bond in the $C_{10}$ intermediate during FA biosynthesis. *H. pylori* FabX, dehydrogenates decanoyl-acyl carrier protein (ACP) and isomerizes *trans*-2-decenoyl-ACP to *cis*-3-decenoyl-ACP, as mentioned the key UFA synthetic intermediate (Fig. 1.2C). Thus, this dual enzyme allows UFA synthesis by diverting and transforming the decanoyl-ACP intermediate that is then elongated by FabF [45]. Alternatively, UFA biosynthesis in some bacteria can be achieved after the FA elongation cycle. The generation of a carbon–carbon double bond is catalyzed by fatty acid desaturases [46] (Fig. 1.2D). For example, *B. subtilis* and *Pseudomonas aeruginosa* fatty acid desaturase, denominated Δ5-Des and DesA, can insert a double bond in the phospholipid acyl chain specifically into the Δ5 position (Δ5-Des) and into the Δ9 position (DesA), respectively [47–49]. Further, an acyl-CoA desaturase (DesBC system) has been described in *P. aeruginosa*, which is responsible for introducing double bonds at the Δ9-position into acyl-CoAs derived from exogenous fatty acids [49].

Finally, the acyl-ACPs of the proper chain length are substrates for acyltransferases involved in membrane phospholipid synthesis; since the longer chain acyl-ACPs are poor substrates for FabB/FabF and become ideal substrates for the acyltransferases (see below).

## Biochemistry of phospholipid biosynthesis

Since phospholipids are the main lipid components of the cell membrane, their synthesis is a key aspect of bacterial physiology. Phospholipids are composed of two fatty acyl chains bound via an ester bond to a glycerol moiety, a phosphate group and a variable head group that defines their properties. Examples are phosphatidic acid (PA), phosphatidylglycerol (PG), phosphatidylethanolamine (PE), phosphatidylserine (PS), cardiolipin (CL), lysyl-phosphatidylglycerol (LPG), and phosphatidylinositol (PI). The biosynthesis of the different phospholipids has been extensively studied in Gram-positive and Gram-negative bacteria (for recent excellent revisions see Refs. [1,2,50]). In this chapter, we will overview the biosynthesis of PA, the key and universal precursor in phospholipid production (for a detailed revision see Ref. [51]).

### Phosphatidic acid biosynthesis

PA is synthetized by esterification of two fatty acyl-chains onto the two hydroxyl groups of *sn*-glycerol-3-phosphate (G3P). In bacteria, G3P can be obtained in three different ways: (i) by de novo synthesis, (ii) by direct transport from the surrounding medium, and (iii) by phosphorylation of uptaken glycerol. The only de novo pathway for the synthesis of G3P in bacteria implies the reduction of the dihydroxyacetone phosphate produced during glycolysis, by the G3P synthase (GpsA) [51]. In *E. coli*, GpsA is a soluble enzyme that is strongly feedback inhibited by G3P. In spite of this, it has been demonstrated that the intracellular G3P concentration does not have a regulatory role on phospholipid production [52]. In the case of *B. subtilis*, GpsA is not inhibited by G3P. It has been hypothesized that this could be related to the fact that Gram-positive bacteria require higher amounts of G3P units for cell wall biosynthesis than for phospholipid formation [53]. Indeed, lipoteichoic acid, a major component of Gram-positive cell wall, contains 14–33 G3P units [54,55]. The increased metabolic demand for G3P possibly requires a less strict regulation of GpsA in Gram-positive bacteria [51].

To obtain G3P directly from the environment bacteria employ the GlpT secondary transporter. Also, they can uptake glycerol and subsequently phosphorylate it by GlpF and GlpK, respectively. *E. coli* and *B. subtilis* can use G3P and glycerol as the sole carbon source thank to a set of genes encoded in the *glp* regulon, but this is not a general case. Certain bacteria, like *S. pneumoniae*, lack the genes for glycerol or G3P metabolism and therefore do not use exogenous G3P or glycerol at all [51].

## Acyltransferases

### *The PlsB/PlsC system*

PA synthesis proceeds by acylation at position 1 of G3P by the G3P acyltransferases [56]. The first G3P acyltransferase to be identified and characterized in bacteria was PlsB from *E. coli* [57,58]. Subsequently, PlsC acylates the 2-position of 1-acyl-glycerol-3-phosphate (lysophosphatidic acid) producing PA [59,60]. PlsB is almost restricted to γ-proteobacteria, while PlsC is universally distributed in bacteria and is an essential protein. PlsB and PlsC use either acyl-ACP, coming from FAS II, or acyl-CoA, derived from exogenous FA, as the acyl donors [61]. In *E. coli* the acyl-CoA synthetase FadD ligates exogenous fatty acids to CoA [62]. The acyl-CoAs are not substrates for further elongation by FAS II but can be directly used by PlsB or PlsC in the transacylation reactions or as carbon source through degradation by β-oxidation [63]. PlsB and PlsC are integral inner membrane proteins (Fig. 1.3A). In *E. coli* PlsB prefers saturated fatty acids, while PlsC has a preference for unsaturated acyl chains. However, not all PlsC homologs have the same substrate preference for acyl chains or acyl donors. For example, in *S. aureus* PlsC prefers branched-chain fatty acids of 15 carbon atoms and only uses acyl-ACP thioesters. Because of the substrate specificity of the two acyltransferases, bacterial phospholipids usually have an asymmetric distribution of fatty acids between the 1- and 2-position of the glycerol phosphate backbone [1].

### *The PlsX/PlsY/PlsC system*

Most bacteria use the PlsX/PlsY system to synthesize PA [61]. PlsX catalyzes the conversion of the long chain acyl-ACPs, end products of FAS II, into acylphosphate. Next, PlsY an integral membrane acylphosphate-G3P acyltransferase, transfers the acyl chain of these intermediates to the 1-position of G3P. Finally, PA synthesis is achieved by the action of PlsC that, as in the PlsB/PlsC pathway, acylates position-2 of lysophosphatidic acid (Fig. 1.3B). Usually, in bacteria employing the PlsX/PlsY/PlsC pathway, PlsC uses only acyl-ACP thioesters as substrates. In these bacteria fatty acids are also asymmetrically distributed between positions

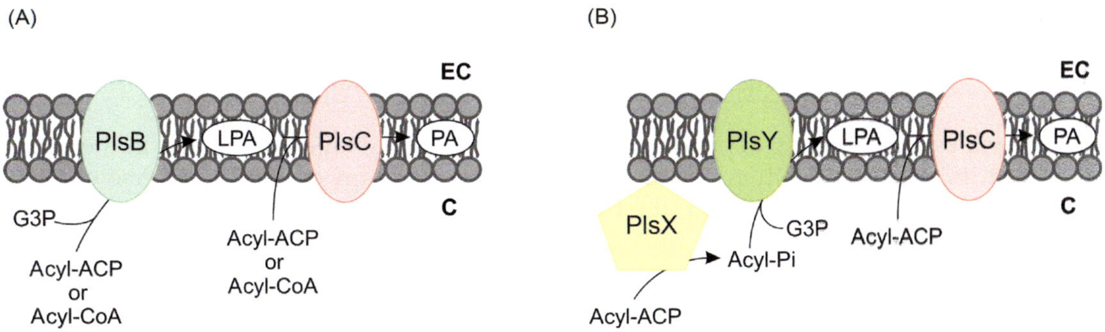


FIGURE 1.3 **Acyltransferase systems in bacteria.** (A) In γ-proteobacteria, PlsB acylates position 1 of glycerol-3-phosphate (G3P) to give 1-acyl-glycerol-3-phosphate (lysophosphatidic acid, LPA). Subsequently, PlsC acylates the 2-position of LPA to give phosphatidic acid (PA). These two enzymes can use either acyl-CoA or acyl-ACP as the acyl donor. (B) In most bacteria, PlsX first converts the acyl-ACPs, end products of FAS II, to the corresponding acyl-phosphate intermediates, which are then used by PlsY to acylate the 1-position of glycerol-phosphate to give LPA. PA is finally synthetized by PlsC. In this system, the enzymes do not use acyl-CoA substrates. *C*, cytosol; *EC*, extracellular.

1 and 2 of G3P due to the specificity of the acyltransferases, making positional asymmetry a common feature of bacterial phospholipids.

The role of the PlsX/PlsY/PlsC system in coupling FAS II with phospholipid synthesis was investigated in *B. subtilis* by Paoletti et al. [64] using conditional mutants in *plsX*, *plsY*, and *plsC*. It was found that a *B. subtilis plsY* conditional mutant strain, upon depletion of PlsY, accumulates free FA coming from the hydrolysis of the physiologically unstable acyl-phosphate intermediates and stops phospholipid synthesis. However, the FA synthesis rate is not affected [64]. Depletion of PlsC resulted in over 200% increase in FA synthesis and, as also observed in the absence of PlsY, the accumulation of high amounts of free FA [64]. It was hypothesized that FA accumulation was due to degradation of the monoglycerides, products of the step catalyzed by PlsY, by an esterase, although this enzyme has not been identified yet. Notably, upon *plsX* depletion not only phospholipid biosynthesis is arrested but also the rate of FA synthesis becomes almost null [64]. This result highlighted PlsX as a key regulatory point that couples FAS II and phospholipid synthesis.

Based on sequence analysis and fractionation, PlsX was originally thought to be a soluble enzyme but subsequent work showed that it is a peripheral membrane protein [64–66]. One of these studies [66] reported that PlsX localization on membranes followed that of the bacterial cell division apparatus, but Sastre et al. [65] has challenged this conclusion by showing that PlsX is evenly distributed on the membrane of proliferating cells, and that its localization was independent of cell division proteins. Sastre et al. [65] also showed that PlsX association with the membrane was independent of PlsY, and that both inhibition of phospholipid synthesis and changes in membrane potential, perturbed PlsX localization. These data suggested that PlsX associates directly with the phospholipid bilayer, but this possibility has not been tested experimentally. Interestingly, PlsX and PlsY are also present in *E. coli*, which has the PlsB acyltransferase. PlsX and PlsY can be individually deleted with no detrimental effect on growth. However, a *plsX plsY* double mutant could not be obtained indicating an essential role for the enzymatic pair in *E. coli* [67]. Why PlsX and PlsY are retained in bacteria with a PlsB/PlsC pathway remains a mystery that will require further research to clarify.

## Control of lipid biosynthesis in bacteria

Bacteria survival largely depends on the integrity of its cell membrane and in the ability to adjust its lipid composition in order to optimize and adapt growth in diverse environments. As mentioned above, since the biophysical properties of the membranes are determined mostly by the composition of FAs, which are the most energetically expensive membrane lipid components; bacteria have evolved sophisticated mechanisms to tightly control the expression of the genes and the activity of the enzymes responsible for the biosynthesis and modification of the fatty acyl chains. In organisms possessing a fatty acid β-oxidation pathway, the expression of the degradation machinery is balanced with fatty acid synthesis enzymes. Furthermore, the expression levels of different fatty acids biosynthetic genes are also coordinated with growth rate, nutrient availability, and environmental stimuli [68,69]. In this section we describe some of the key regulatory components and mechanisms directly related to lipid metabolism.

## Biochemical regulation of fatty acid and phospholipid biosynthesis

### *Regulation at the Initiation steps*

The initial steps are key regulatory points in membrane lipid biosynthesis in order to save cellular resources and energy. The most significant enzymes in this stage are ACC and FabH (Fig. 1.1). As described above, ACC performs the first committed step in FAS I and II systems catalyzing the synthesis of malonyl-CoA, while every intervention of FabH starts the formation of a new fatty-acyl chain. In vivo experimental data indicated that long chain acyl-ACPs, end-products of FAS II, are negative feedback regulators of fatty acid biosynthesis in *E. coli*. Accumulation of long chain acyl-ACPs leads to FAS II arrest in this bacterium. In addition, overexpression of a soluble thioesterase that cleaves the acyl-ACP species resulted in relief of the inhibition and reactivation of FAS II [70]. Later, it was shown that FabH and ACC activities are inhibited in vitro by acyl-ACP [71,72]. In the case of FabH, the longer the length of the acyl chain the stronger the inhibitory effect. This response ensures that fatty acids of the required length are synthesized by tuning the activity of the initiation step to the elongation and acyl transfer stages. On the other hand, ACC is effectively feedback inhibited in vitro by acyl-ACP regardless of the acyl chain length, while its activity is unaffected by unacylated ACP [71]. The individual biotin carboxylase (AccC) and carboxyltransferase (AccAD) reactions are not inhibited by acyl-ACP, suggesting that regulation occurs in the ACC multiprotein complex. Inhibition of ACC by acyl-ACP is a logical feedback system that links the end product of the pathway with the initial reaction and is likely to be widespread in bacteria (Fig. 1.4).

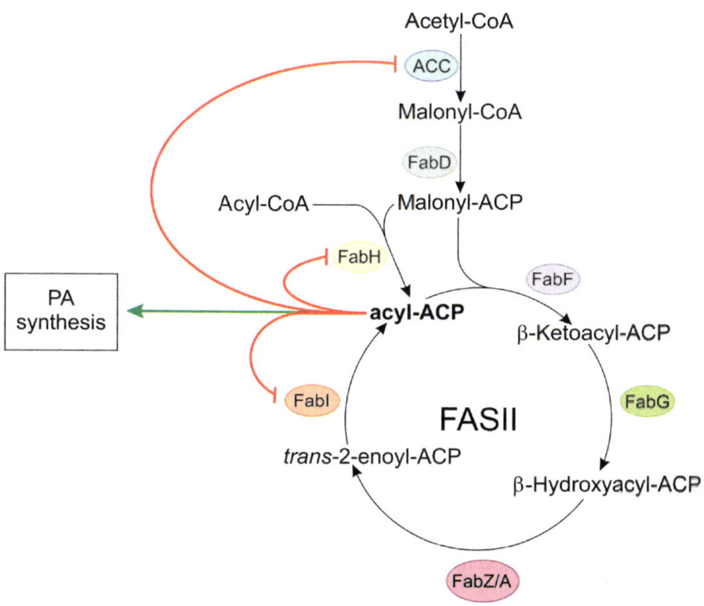


FIGURE 1.4 **Feedback inhibition of fatty acid biosynthesis.** In *E. coli*, long chain acyl-ACPs, end-products of FAS II, are negative feedback regulators of different steps of fatty acid biosynthesis. They inhibit the activity of ACC and FabH (initiation steps) as well as of FabI (elongation step). At the same time, long chain acyl-ACPs are poor substrates for the condensing enzymes (such as FabB and FabF) but good ones for the acyltransferases. Similar control mechanisms might exist in Gram-positive bacteria.

The combined regulation of ACC and FabH by acyl-ACP, the product of the elongation phase, allows for synergistic feedback regulation of the initiation cycle to control the quantity of fatty acid produced. Although the inhibition of FabH by acyl-ACP has only been characterized in *E. coli* it may also exist in other organisms. As mentioned above, the absence of PlsX in *B. subtilis* promotes the arrest of not only phospholipid synthesis but also of fatty acids [64]. Moreover, it has been established that a block at the PlsX reaction promotes accumulation of long chain acyl-ACP from FAS II in *S. aureus* [73]. Hence, it is likely that a similar feedback regulation mechanism mediated by acyl-ACPs works in Gram-positive bacteria. However, up to date, little is known about the biochemical regulation of FAS II in Gram-positive bacteria.

### *Regulation at the elongation steps*

The elongation enzymes are determinants of the length of the final fatty acids and also of the ratio of unsaturated/saturated fatty acids. This implies that their activities and substrate specificities have to be finely balanced to achieve the correct membrane lipid composition. As ACC and FabH, Fab I, the *trans*-2-enoyl-ACP reductase, is feedback inhibited by acyl-ACPs what slows down the rate at which the elongation cycle turns. Moreover, the acyl chain length of the fatty acids generated by FAS II is determined by the competition between the glycerolphosphate acyltransferases (such as PlsB) and the elongation condensing enzymes (such as FabB and FabF) for the acyl-ACPs. FabB and FabF cannot use long chain acyl-ACPs efficiently, but these molecules are ideal substrates for the acyltransferases (Fig. 1.4). In fact, abnormally long chain fatty acids in the plasma membrane of *E. coli* can be obtained by overexpressing FabB or the inhibition of PlsB [74]. A similar accumulation of long-chain fatty acids is observed in *B. subtilis* upon depletion of PlsC [64,75]. Different FabB/FabF crystal structures show that the substrate binding tunnel, which is adjacent to the active site, can accommodate acyl-chains up to 16 carbons, what prevents the enzymes to produce fatty acids of more than 18- or 20-carbon long [76–80]. In summary, biochemical regulation of the elongation steps works in concert to tightly control the rate of FAS II. In addition, the fatty acid chain length is determined by the interplay between the elongation and acyltransfer enzymes.

## Transcriptional regulation of lipid metabolism

Different bacteria possess specific transcriptional regulators that control most or a subset of FA biosynthesis genes. Some transcription factors, like FadR (mostly studied in *E. coli*) are widespread among several different Gram-positive and Gram-negative bacteria. In general, a feedback regulation system by end products, as it is the sensing of FA derivatives by a transcriptional factor, might be a conserved mechanism for transcriptional regulation of FA biosynthetic genes. A remarkable difference occurs with the global regulator FapR of *B. subtilis* (highly conserved in many Gram-positive bacteria) which is an excellent example of feed-forward transcriptional regulation of membrane lipid homeostasis. In this section we will describe some paradigmatic transcriptional regulation examples in model bacteria and its molecular mechanisms.

### *Coordination of fatty acid metabolism in E. coli*

*E. coli* can uptake external FA and use them as sole carbon and energy sources as well as utilize these molecules as precursors for membrane phospholipids biosynthesis. The genes involved in fatty acid degradation (β-oxidation) are regulated by the transcription factor $FadR_{Ec}$, a member of the GntR family of transcription factors. When long-chain fatty acids ($C_{14}$-$C_{18}$) are present in the growth medium they are transported and converted to acyl-CoAs [81]. The binding of FadR to long-chain acyl-CoAs results in a conformational change that promotes releasing of the protein from the regulatory operators and derepression of gene transcription [82–84]. Thus, FadR senses these molecules and derepresses the expression of the genes involved in consuming FA as energy source (*fad* regulon). In addition, earlier studies demonstrated that FadR is also an activator of two genes involved in unsaturated fatty acid synthesis (*fabA* and *fabB*) and of *iclR*. The IclR protein is a repressor of genes of the glyoxylate shunt pathway (*aceAB*), which is required to use the acetyl-CoA derived from β-oxidation. More recently, it has been shown that FadR also acts as an activator of most of the genes involved in FA biosynthesis, including those encoding for the AccB and AccC subunits of the ACC complex [85]. This activation is abolished upon binding of FadR to long chain saturated or unsaturated acyl-CoAs, which, as mentioned above, triggers its dissociation from the operators sites. Thus, there exists a simultaneous activation of FA degradation and a repression of FA synthesis in response to exogenous fatty acids to coordinate these two pathways.

A second transcriptional regulator, FabR, is also involved in fine-tuning the FA composition of the lipid membrane in *E. coli*. Deletion of FabR causes an increase in the unsaturated to saturated fatty acid ratio [86]; and it was demonstrated that FabR represses *fabA* and *fabB* gene expression by binding to their promoter region (downstream of the FadR recognition sequence) [86–89]. Both FadR and FabR can simultaneously bind to the *fabB* promoter sequence, demonstrating combined transcriptional regulation. The regulatory ligands of FabR are long chain acyl-CoAs and acyl-ACP thioesters; and binding of the FabR-18:1Δ9 CoA complex to DNA was antagonized by the presence of saturated 16:0-CoA/16:0-ACP [89]. Overall, it has been proposed that FabR essentially measures the unsaturated/saturated fatty acid ratio in the cell allowing tuning gene expression to achieve an optimal lipid membrane acyl chain composition.

### *Control of unsaturated fatty acid synthesis in P. aeruginosa*

*P. aeruginosa* not only employs the FabA/FabB system for the synthesis of UFAs, but also possess oxygen-dependent Δ9-desaturases that insert a double bond into the acyl-chains of membrane phospholipids (DesA) or saturated 16:0- or 18:0-CoAs (DesBC). The expression of *fabAB* and *desBC* is coordinately controlled by the DesT repressor [90]. DesT is a TetR family transcription factor that binds saturated and unsaturated acyl-CoAs with equal affinity. As *E. coli, P. aeruginosa* can incorporate extracellular FA converting them to their respective CoA thioesters and use them for membrane phospholipids production [49]. Binding of SFA-CoAs promotes releasing of DesT from DNA inducing *desCB* expression. On the other hand, UFA-CoAs enhance DesT binding to its cognate DNA repressing desaturase expression [91]. A structural study of DesT bound to oleoyl- and palmitoyl-CoA revealed that the repressor adopts a "relaxed" conformation when bound to an unsaturated acyl-CoA that facilitates binding to the operator DNA sequence and a "tense" conformation when bound to a saturated acyl-CoA species that is not favorable for DNA binding [92]. Hence, DesT senses the overall fatty acid composition of the acyl-CoA pool allowing *P. aeruginosa* to respond to the availability of exogenous saturated and unsaturated FA, and adjust gene expression to maintain membrane fluidity accordingly.

## *Global regulation of lipid synthesis in B. subtilis*

FapR from *B. subtilis* was the first global transcriptional regulator of lipid biosynthesis discovered in bacteria [93]. It belongs to the DeoR family of transcriptional regulators and negatively controls the expression of at least 10 genes, organized in 6 operons, encoding proteins involved in FA and phospholipid biosynthesis (the *fap* regulon). Notably, the expression of the genes encoding the ACC subunits are not under the control of FapR [93].

FapR is present and highly conserved in all the species of the *Bacillus, Listeria*, and *Staphylococcus genera* (all including important human pathogens like *B. anthracis, B. cereus, Listeria monocytogenes*, and *S. aureus*) as well as in the pathogen *Clostridium difficile* and other related genera. However, FapR is not present in Gram-negative bacteria or other Gram-positive genera [93]. FapR binds to a conserved 17 bp inverted repeat located within or immediately downstream of the predicted *fap* promoters [93]. Interestingly, the activity of FapR is controlled by a feed-forward mechanism. Malonyl-CoA and malonyl-ACP, key metabolites of FA synthesis, can specifically bind to FapR promoting its release from the DNA target sequences and induce the transcription of the *fap* regulon [93–95]. FapR is a two-domain protein with an N-terminal DNA-binding domain connected through a linker α-helix to a larger C-terminal effector-binding domain and works as a homodimer [95]. The effector binding domain exhibits the hot-dog fold also observed in many homodimeric acyl-CoA-binding enzymes [95]. Interestingly, FapR is the first protein to recruit this fold for a no-enzymatic function. The structural snapshots of full-length FapR from *S. aureus* along its regulation cycle revealed distinct quaternary conformations for the DNA-bound (relaxed) and the malonyl-CoA-bound (tense) forms of the repressor. Binding of malonyl-CoA induces a large-scale structural rearrangement that leads to disruption of the repressor-operator complex [96].

## *Control of unsaturated fatty acid synthesis in B. subtilis: The plasma membrane as a signaling structure*

The functions of the cytoplasmic membrane are known to depend critically on the physical state of the lipid bilayer, making it susceptible to changes in environmental temperature. In particular, it has been established that normal cell function requires membrane lipid bilayers that are largely fluid; indeed, the bilayers of most organisms are entirely or mostly fluid at physiological temperatures. However, at low temperature membrane bilayers undergo a reversible change of state from a fluid (disordered) to a nonfluid (ordered) array of the fatty acyl chains. The temperature at the midpoint of this transition is called the transition temperature ($T_m$), and the change of state accompanying an increase in temperature is called the liquid–gel transition. The $T_m$ is a function of the membrane lipid composition and, in organisms deficient in cholesterol, mainly depends on the fatty acid composition of the membrane lipids. A disordered state is imparted by the presence of either unsaturated or terminally branched fatty acids both of which act to offset the closely packed ordered arrangement of the lipid bilayer acyl chains that is conferred by straight chain-saturated acyl chains [97,98].

From these considerations, it seems clear that bacteria and most (if not all) organisms unable to maintain thermal homeostasis must regulate their plasma membrane phase transition in response to temperature [99]. Without regulation, an organism shifted from a high to a low temperature would have membrane lipids with suboptimal fluidity, resulting in subnormal membrane function. This membrane lipid homeostasis that maintains the biophysical properties of membranes is referred to as homeoviscous adaptation [100] and is interpreted as a mechanism that modifies the permeability of the phospholipid bilayer to minimize energy expenditure and optimize growth [101].

The mechanism of homeoviscous adaptation in all cases examined seems to occur via the incorporation of proportionally more UFAs (or fatty acids of analogous properties, such as *anteiso*-branched chain fatty acids) as the temperature decreases. This regulatory mechanism system seems to be a universally conserved adaptation response allowing cells to maintain the appropriate fluidity of membrane lipids regardless of the ambient temperature. This means that cells must process temperature signals to adjust enzyme activities or to activate unique genes necessary to adapt the membranes to the new temperature [102,103].

*B. subtilis* has developed regulatory mechanisms to detect environmental changes in temperature and adjust the expression of a lipid desaturase enzyme accordingly. The DesKR two-component system of this organism is responsible for transcription of the *des* gene, coding for an integral membrane lipid Δ5-desaturase, called Δ5-Des. DesK functions as a kinase when temperature decreases below ~30°C, autophosphorylating a conserved histidine residue within its kinase domain. Then, the phosphoryl group is transferred to the conserved aspartic acid residue of the DNA-binding response regulator DesR. Phosphorylation of DesR promotes its dimerization and binding to the operator DNA sequences where it tetramerizes and activates *des* transcription. Once this lipid desaturase is synthetized it introduces *cis*-double bonds into the acyl chains of membrane phospholipids reestablishing the viscosity of the lipid bilayer. In the context of a fluid membrane, DesK behaves as a phosphatase and dephosphorylates DesR shutting down *des* expression [3]. DesK is an integral membrane protein possessing five transmembrane (TM) helices. It works as a dimer implying that its sensing domain is constituted by 10 TM segments arranged in a yet unknown fashion within the membrane. Genetic, biochemical, structural, and computational studies on wild type, mutant, and truncated versions of the protein indicate that the DesK's sensing mechanism is rooted in temperature-dependent membrane properties. It has been stablished that the reversible formation of a two-helix coiled coil in the fifth TM segment and the N-terminus of the cytoplasmic domain is essential for DesK's sensing and signal transduction mechanisms. It has been proposed that membrane thickening, known to take place upon cooling, is the main driving force for signal sensing and that it acts by inducing helix stretching and rotation prompting an asymmetric kinase-competent state [104].

Some Gram-positive bacteria adjust membrane fluidity by increasing the proportion of the lower melting-point *anteiso*-branched chain fatty acids at lower growth temperatures. The FabH enzyme of bacteria that form branched-chain fatty acids uses isobutyryl-CoA, isovaleryl-CoA, and 2-methylbutyryl-CoA as precursors. 2-methylbutyryl-CoA leads to formation of *anteiso*-fatty acids while the other two lead to the synthesis of *iso*-branched chain fatty acids [105]. In *Listeria monocytogenes* the FabH enzyme possesses an augmented selectivity for the 2-methylbutyryl-CoA primer at 10°C, leading to an increase in the synthesis of anteiso branched-chain fatty acids. This selectivity seems to be an intrinsic property of the FabH enzyme and have direct consequences on the physical properties of the lipid bilayer [106].

### *Control of lipid metabolism in actinomycetes*

As mentioned earlier, the expanded lipid metabolism in actinomycetes brings a higher level of sophistication to the regulation of lipid homeostasis in these bacteria. In fact, the mechanisms involved in this global coordination are only recently beginning to be deciphered.

Initial genomics analyses of different actinomycetes genera indicated that these microorganisms lack genes orthologs to the characterized transcription factors present in other model bacteria. However, the alignment of several actinomycetes genome regions containing the main set of *fasII* genes, highlighted a remarkable synteny, not only along the *fab* genes, but also in the first gene upstream of *fabD* in all *fab* clusters analyzed. This conserved ORF presented a relatively conserved C-terminal domain of approximately 50 amino acids with homology to helix–turn–helix DNA-binding domains of bacterial transcriptional regulators [107]. Biochemical and genetics studies demonstrated that this ORF was the transcriptional regulator of at least part of the FAS II system in *Streptomyces* and in *Mycobacterium* [108,109]. In *Streptomyces* the transcriptional regulator was named FasR, while its ortologue in *Mycobacterium* was named MabR. FasR of *S. coelicolor* ($FasR_{sc}$) functions as an activator of the *fabD–fabH–acpP–fabF* gene cluster, and its disruption resulted in a clear defect in cell growth and in a significant reduction in the relative mRNA abundance of all genes in the cluster. The *fasR*$_{sc}$ knockout strain also showed a strong reduction in the rate of lipid synthesis from acetate, and in the amount of total lipid per cell [107]. In a sharp difference with $FasR_{sc}$, and most other regulators of FA biosynthesis, MabR is essential for cell viability; at least this is true for *Mycobacterium smegmatis*, although this has not been confirmed yet for *M. tuberculosis*. Although the first studies had suggested that MabR was a repressor of the *fasII* operon of mycobacteria, recent experiments carried out with a knock down mutant confirmed that, like its $FasR_{sc}$ ortholog, it is also a transcriptional activator of the *fasII* operon (*fabD-acpM-kasA-kasB-accD6*) [110,111]. These results were consistent with $FasR_{sc}$ and MabR sharing high degree of sequence similarity within the helix–turn–helix DNA binding motif, and also maintaining a similar location of the conserved inverted repeat sequences to which both regulators bind; around −70 bp of the proposed overlapping transcription/translation start sites of the respective *fabD* genes. Phylogenetic analysis of several $FasR_{sc}$ and MabR orthologs show that they segregate into two groups: mycolic acid producers (MabR) and nonproducers ($FasR_{sc}$). The amino-terminal ligand binding domains of these regulators are more dissimilar than their DNA binding domains, suggesting that they interact with different ligands. Although the ligand of $FasR_{sc}$ could not be identified yet, it was clearly demonstrated that the products of the *Mycobacterium* FAS I, acyl-CoAs $\geq C_{16}$, are the ligands that bind MabR to positively modulate its binding to the operator sequences [109]. Other none essential regulatory components of the main *fasII* operon, such as the transcriptional repressor FadRc [112], found to be induced upon starvation [113], add to the complexity of the *fasII* operon regulation in mycobacteria.

A crosstalk between the two FAS systems was demonstrated in a *mabR* conditional mutant strain that showed alterations in both mycolic acids and in de novo FA biosynthesis [111]. However, the cross regulation between these two systems does not occur by a direct binding of MabR to the *fas-acpS* promoter. In fact, the FAS I encoding genes are directly regulated by a different transcriptional regulator called $FasR_{MT}$ that belongs to the TetR family of transcriptional regulators [114]. This protein functions as a transcriptional activator of *fas-acpS* genes expression, and as MabR, it is also essential for *M. smegmatis* viability, a distinctive feature of these transcriptional regulators. Like MabR, $FasR_{MT}$ of mycobacteria also responds to long-chain acyl-CoAs as effector molecules, however, in the case of $FasR_{MT}$ these ligands inhibit the binding of this activator to its cognate DNA sequences. Recently, the crystal structure of $FasR_{MT}$, $FasR_{MT}$ bound to acyl-CoA, and $FasR_{MT}$-DNA complex have been obtained and their analyses gave important insights about the molecular mechanisms that govern the regulation of the de novo fatty acid biosynthesis in *Mycobacterium* [115].

Most genera of the oleaginous bacteria rhodococci present a FAS system similar to *Mycobacterium*. For example, some of the most studied species of this genera, like *Rhodococcus jostii* RHA1 and *Rhodococcus opacus* PD630, present the two FAS systems, I and II, and they also contain orthologs to the MabR and $FasR_{MT}$ transcriptional regulators that control the expression of the FAS II and FAS I components, respectively. Interestingly, a new transcriptional regulator called NlpR (for nitrogen lipid regulator), which contributes to the modulation of nitrogen metabolism, also regulate lipogenesis and triacylglycerol accumulation, linking the N and the C metabolisms in these microorganisms [116]. NlpR acts as an activator of several genes involved in the uptake and assimilation of N species, and remarkably, in genes related with FA and TAG biosynthesis. So far, the ligand sensed by this regulator has not yet been identified. NlpR ortologs are also present in other actinomycetes belonging to *Nocardia, Gordonia, Mycobacterium, Streptomyces,* and *Amycolatopsis* genera; however, the role of this protein in the regulation of C metabolism in these organisms has not been studied yet.

The genus *Corynebacterium* is also quite unique in terms of the FAS system that they use for the de novo fatty acid biosynthesis. FAS II genes are absent from corynebacteria and most species harbor two type I FAS genes (FasA and FasB for FA and MA biosynthesis) [10]. The exceptions are the pathogenic species *C. diphtheriae* that only has one type I FAS, and *C. jeikeium* and *C. urealyticum* that show strict dependence for growth on the presence of exogenous FAs [117,118] because they do not have fatty acid biosynthesis machinery. The regulation of the FAS I systems at the transcriptional level is driven by a unique transcriptional repressor, $FasR_{CG}$, with no homology to either FasR or MabR [119,120].

## Perspectives

Membrane homeostasis in bacteria is robust because many sensory and response mechanisms monitor membrane status and adjust its lipidic composition accordingly. A particularly outstanding question is how the sensor domain of *B. subtilis* DesK operates at the molecular level to sense membrane viscosity. The discovery of this mechanism will impact other fields. For example, molecular machineries responsible for sensing membrane properties also exists in yeast and animals. Thus, the principle of TM signaling to integrate membrane composition with lipid metabolism is ancient and essential for life. Furthermore, there is a sophisticated network of different transcriptional regulators that work in concert with biochemical regulation to control lipid metabolism in bacteria. An understanding of the transcriptional regulation in Actinomycetes has begun and it is expected to guide the progression of therapeutic compounds targeting *Mycobacterium* FAS I and FAS II regulators. There are still huge gaps in our understanding of the regulation of FASII in many bacteria even in the model organisms *E. coli* and *B. subtilis*. For example, the alarmone (p)ppGpp is emerging as a central player in survival of bacteria during antibiotic inhibition of fatty acid synthesis. Nevertheless, the burning question of how fatty acid starvation triggers the (p)ppGpp-mediated "stringent response" has not been yet convincingly answered. It remains unknown the mechanism by which inhibition of the PlsX reaction in *B. subtilis* results in the shutdown of fatty acid synthesis by FAS II. It has not been yet tested if this key regulatory enzymatic point that synchronizes FAS II with phospholipid biosynthesis is also present in other bacteria using the PlsX/PlsY/PlsC pathway. Examining the regulation of these steps and their integration with the other major branches of cellular metabolism will be a major focus for future research.

# References


[1] Parsons JB, Rock CO. Bacterial lipids: metabolism and membrane homeostasis. Prog Lipid Res 2013;52:249–76.
[2] Sohlenkamp C, Geiger O. Bacterial membrane lipids: diversity in structures and pathways. FEMS Microbiol Rev 2016;40:133–59.
[3] de Mendoza D. Temperature sensing by membranes. Annu Rev Microbiol 2014;68:101–16.
[4] Schweizer E, Hofmann J. Microbial type I fatty acid synthases (FAS): major players in a network of cellular FAS systems. Microbiol Mol Biol Rev 2004;68:501–17.
[5] White SW, Zheng J, Zhang YM, Rock CO. The structural biology of type II fatty acid biosynthesis. Annu Rev Biochem 2005;74:791–831.
[6] Leibundgut M, Maier T, Jenni S, Ban N. The multienzyme architecture of eukaryotic fatty acid synthases. Curr Opin Struct Biol 2008;18:714–25.
[7] Rock CO, Jackowski S. Forty years of bacterial fatty acid synthesis. Biochem Biophys Res Commun 2002;292:1155–66.
[8] Revill WP, Bibb MJ, Scheu AK, Kieser HJ, Hopwood DA. Beta-ketoacyl acyl carrier protein synthase III (FabH) is essential for fatty acid biosynthesis in *Streptomyces coelicolor* A3(2). J Bacteriol 2001;183:3526–30.
[9] Gago G, Diacovich L, Arabolaza A, Tsai SC, Gramajo H. Fatty acid biosynthesis in actinomycetes. FEMS Microbiol Rev 2011;35:475–97.
[10] Radmacher E, Alderwick LJ, Besra GS, Brown AK, Gibson KJ, Sahm H, et al. Two functional FAS-I type fatty acid synthases in *Corynebacterium glutamicum*. Microbiology 2005;151:2421–7.
[11] Vadia S, Tse JL, Lucena R, Yang Z, Kellogg DR, Wang JD, et al. Fatty acid availability sets cell envelope capacity and dictates microbial cell size. Curr Biol 2017;27. 1757.e1755-1767.e1755.
[12] Fernandes ND, Kolattukudy PE. Cloning, sequencing and characterization of a fatty acid synthase-encoding gene from *Mycobacterium tuberculosis* var. *bovis* BCG. Gene 1996;170:95–9.
[13] Zimhony O, Vilcheze C, Jacobs WR Jr. Characterization of *Mycobacterium smegmatis* expressing the *Mycobacterium tuberculosis* fatty acid synthase I (fas1) gene. J Bacteriol 2004;186:4051–5.
[14] Marrakchi H, Laneelle MA, Daffe M. Mycolic acids: structures, biosynthesis, and beyond. Chem Biol 2014;21:67–85.
[15] Daniel J, Deb C, Dubey VS, Sirakova TD, Abomoelak B, Morbidoni HR, et al. Induction of a novel class of diacylglycerol acyltransferases and triacylglycerol accumulation in *Mycobacterium tuberculosis* as it goes into a dormancy-like state in culture. J Bacteriol 2004;186:5017–30.
[16] Quadri LE. Biosynthesis of mycobacterial lipids by polyketide synthases and beyond. Crit Rev Biochem Mol Biol 2014;49:179–211.
[17] Salie MJ, Thelen JJ. Regulation and structure of the heteromeric acetyl-CoA carboxylase. Biochim Biophys Acta 2016;1861:1207–13.
[18] Tong L. Structure and function of biotin-dependent carboxylases. Cell Mol Life Sci 2013;70:863–91.
[19] Knowles JR. The mechanism of biotin-dependent enzymes. Annu Rev Biochem 1989;58:195–221.
[20] Diacovich L, Peiru S, Kurth D, Rodriguez E, Podesta F, Khosla C, et al. Kinetic and structural analysis of a new group of Acyl-CoA carboxylases found in *Streptomyces coelicolor* A3(2). J Biol Chem 2002;277:31228–36.
[21] Gago G, Kurth D, Diacovich L, Tsai SC, Gramajo H. Biochemical and structural characterization of an essential acyl coenzyme A carboxylase from *Mycobacterium tuberculosis*. J Bacteriol 2006;188:477–86.
[22] Hunaiti AR, Kolattukudy PE. Isolation and characterization of an acyl-coenzyme A carboxylase from an erythromycin-producing *Streptomyces erythreus*. Arch Biochem Biophys 1982;216:362–71.
[23] Cronan JE Jr, Waldrop GL. Multi-subunit acetyl-CoA carboxylases. Prog Lipid Res 2002;41:407–35.
[24] Henrikson KP, Allen SH. Purification and subunit structure of propionyl coenzyme A carboxylase of *Mycobacterium smegmatis*. J Biol Chem 1979;254:5888–91.
[25] Rodriguez E, Gramajo H. Genetic and biochemical characterization of the alpha and beta components of a propionyl-CoA carboxylase complex of *Streptomyces coelicolor* A3(2). Microbiology 1999;145(Pt. 11):3109–19.
[26] Haase FC, Henrikson KP, Treble DH, Allen SH. The subunit structure and function of the propionyl coenzyme A carboxylase of *Mycobacterium smegmatis*. J Biol Chem 1982;257:11994–9.
[27] Bazet Lyonnet B, Diacovich L, Gago G, Spina L, Bardou F, Lemassu A, et al. Functional reconstitution of the *Mycobacterium tuberculosis* long-chain acyl-CoA carboxylase from multiple acyl-CoA subunits. FEBS J 2017;284:1110–25.
[28] Tran TH, Hsiao YS, Jo J, Chou CY, Dietrich LE, Walz T, et al. Structure and function of a single-chain, multi-domain long-chain acyl-CoA carboxylase. Nature 2015;518:120–4.

[29] Livieri AL, Navone L, Marcellin E, Gramajo H, Rodriguez E. A novel multidomain acyl-CoA carboxylase in *Saccharopolyspora erythraea* provides malonyl-CoA for de novo fatty acid biosynthesis. Sci Rep 2019;9: 6725.

[30] Kurth DG, Gago GM, de la Iglesia A, Bazet Lyonnet B, Lin TW, Morbidoni HR, et al. ACCase 6 is the essential acetyl-CoA carboxylase involved in fatty acid and mycolic acid biosynthesis in mycobacteria. Microbiology 2009;155:2664–75.

[31] Pawelczyk J, Viljoen A, Kremer L, Dziadek J. The influence of AccD5 on AccD6 carboxyltransferase essentiality in pathogenic and non-pathogenic *Mycobacterium*. Sci Rep 2017;7:42692.

[32] Bazet Lyonnet B, Diacovich L, Cabruja M, Bardou F, Quemard A, Gago G, et al. Pleiotropic effect of AccD5 and AccE5 depletion in acyl-coenzyme A carboxylase activity and in lipid biosynthesis in mycobacteria. PLoS ONE 2014;9:e99853.

[33] Lin TW, Melgar MM, Kurth D, Swamidass SJ, Purdon J, Tseng T, et al. Structure-based inhibitor design of AccD5, an essential acyl-CoA carboxylase carboxyltransferase domain of *Mycobacterium tuberculosis*. Proc Natl Acad Sci USA 2006;103:3072–7.

[34] Li SJ, Cronan JE Jr. The genes encoding the two carboxyltransferase subunits of *Escherichia coli* acetyl-CoA carboxylase. J Biol Chem 1992;267:16841–7.

[35] Serre L, Swenson L, Green R, Wei Y, Verwoert II, Verbree EC, et al. Crystallization of the malonyl coenzyme A-acyl carrier protein transacylase from *Escherichia coli*. J Mol Biol 1994;242:99–102.

[36] Serre L, Verbree EC, Dauter Z, Stuitje AR, Derewenda ZS. The *Escherichia coli* malonyl-CoA:acyl carrier protein transacylase at 1.5-A resolution. Crystal structure of a fatty acid synthase component. J Biol Chem 1995;270:12961–4.

[37] Khandekar SS, Gentry DR, Van Aller GS, Warren P, Xiang H, Silverman C, et al. Identification, substrate specificity, and inhibition of the *Streptococcus pneumoniae* beta-ketoacyl-acyl carrier protein synthase III (FabH). J Biol Chem 2001;276:30024–30.

[38] Choi KH, Heath RJ, Rock CO. beta-ketoacyl-acyl carrier protein synthase III (FabH) is a determining factor in branched-chain fatty acid biosynthesis. J Bacteriol 2000;182:365–70.

[39] Yao Z, Davis RM, Kishony R, Kahne D, Ruiz N. Regulation of cell size in response to nutrient availability by fatty acid biosynthesis in *Escherichia coli*. Proc Natl Acad Sci USA 2012;109:E2561–2568.

[40] Sanyal R, Singh V, Harinarayanan R. A novel gene contributing to the initiation of fatty acid biosynthesis in *Escherichia coli*. J Bacteriol 2019;201:e00354–19.

[41] Heath RJ, Su N, Murphy CK, Rock CO. The enoyl-[acyl-carrier-protein] reductases FabI and FabL from *Bacillus subtilis*. J Biol Chem 2000;275:40128–33.

[42] Heath RJ, Rock CO. A triclosan-resistant bacterial enzyme. Nature 2000;406:145–6.

[43] Edwards P, Nelsen JS, Metz JG, Dehesh K. Cloning of the *fabF* gene in an expression vector and in vitro characterization of recombinant *fabF* and *fabB* encoded enzymes from *Escherichia coli*. FEBS Lett 1997;402:62–6.

[44] Garwin JL, Klages AL, Cronan JE Jr. Beta-ketoacyl-acyl carrier protein synthase II of *Escherichia coli*. Evidence for function in the thermal regulation of fatty acid synthesis. J Biol Chem 1980;255:3263–5.

[45] Bi H, Zhu L, Jia J, Zeng L, Cronan JE. Unsaturated fatty acid synthesis in the gastric pathogen *Helicobacter pylori* proceeds via a backtracking mechanism. Cell Chem Biol 2016;23:1480–9.

[46] Altabe SG, Mansilla MC, de Mendoza D. Remodeling of membrane phospholipids by bacterial desaturases. In: Ntambi PDJ, editor. Stearoyl-CoA desaturase genes in lipid metabolism. New York, NY: New York: Springer; 2013. p. 209–31.

[47] Aguilar PS, Cronan JE Jr, de Mendoza D. A *Bacillus subtilis* gene induced by cold shock encodes a membrane phospholipid desaturase. J Bacteriol 1998;180:2194–200.

[48] Altabe SG, Aguilar P, Caballero GM, de Mendoza D. The *Bacillus subtilis* acyl lipid desaturase is a delta5 desaturase. J Bacteriol 2003;185:3228–31.

[49] Zhu K, Choi KH, Schweizer HP, Rock CO, Zhang YM. Two aerobic pathways for the formation of unsaturated fatty acids in *Pseudomonas aeruginosa*. Mol Microbiol 2006;60:260–73.

[50] Zhang Y, Rock CO. Fatty acid and phospholipid biosynthesis in prokaryotes. In: Ridgway ND, Mcleod RS, editors. Biochemistry of lipids, lipoproteins and membranes. 6th ed. UK: Elsevier; 2015. p. 73–112.

[51] Yao J, Rock CO. Phosphatidic acid synthesis in bacteria. Biochim Biophys Acta 2013;1831:495–502.

[52] Goelz SE, Cronan JE Jr. The positional distribution of fatty acids in *Escherichia coli* phospholipids is not regulated by sn-glycerol 3-phosphate levels. J Bacteriol 1980;144:462–4.

[53] Morbidoni HR, de Mendoza D, Cronan JE Jr. Synthesis of sn-glycerol 3-phosphate, a key precursor of membrane lipids, in *Bacillus subtilis*. J Bacteriol 1995;177:5899–905.

[54] Koch HU, Haas R, Fischer W. The role of lipoteichoic acid biosynthesis in membrane lipid metabolism of growing *Staphylococcus aureus*. Eur J Biochem 1984;138:357–63.
[55] Taron DJ, Childs WC 3rd, Neuhaus FC. Biosynthesis of D-alanyl-lipoteichoic acid: role of diglyceride kinase in the synthesis of phosphatidylglycerol for chain elongation. J Bacteriol 1983;154:1110–6.
[56] Zhang YM, Rock CO. Thematic review series: glycerolipids. Acyltransferases in bacterial glycerophospholipid synthesis. J Lipid Res 2008;49:1867–74.
[57] Bell RM. Mutants of *Escherichia coli* defective in membrane phospholipid synthesis. Properties of wild type and Km defective sn-glycerol-3-phosphate acyltransferase activities. J Biol Chem 1975;250:7147–52.
[58] Bell RM, Cronan JE Jr. Mutants of *Escherichia coli* defective in membrane phospholipid synthesis. Phenotypic suppression of sn-glycerol-3-phosphate acyltransferase Km mutants by loss of feedback inhibition of the biosynthetic sn-glycerol-3-phosphate dehydrogenase. J Biol Chem 1975;250:7153–8.
[59] Coleman J. Characterization of *Escherichia coli* cells deficient in 1-acyl-sn-glycerol-3- phosphate acyltransferase activity. J Biol Chem 1990;265:17215–21.
[60] Coleman J. Characterization of the *Escherichia coli* gene for 1-acyl-sn-glycerol-3-phosphate acyltransferase (plsC). Mol Gen Genet 1992;232:295–303.
[61] Lu YJ, Zhang YM, Grimes KD, Qi J, Lee RE, Rock CO. Acyl-phosphates initiate membrane phospholipid synthesis in Gram-positive pathogens. Mol Cell 2006;23:765–72.
[62] Greenway DL, Silbert DF. Altered acyltransferase activity in *Escherichia coli* associated with mutations in acyl coenzyme A synthetase. J Biol Chem 1983;258:13034–42.
[63] Polacco ML, Cronan JE Jr. Mechanism of the apparent regulation of *Escherichia coli* unsaturated fatty acid synthesis by exogenous oleic acid. J Biol Chem 1977;252:5488–90.
[64] Paoletti L, Lu YJ, Schujman GE, de Mendoza D, Rock CO. Coupling of fatty acid and phospholipid synthesis in *Bacillus subtilis*. J Bacteriol 2007;189:5816–24.
[65] Sastre DE, Bisson-Filho A, de Mendoza D, Gueiros-Filho FJ. Revisiting the cell biology of the acyl-ACP:phosphate transacylase PlsX suggests that the phospholipid synthesis and cell division machineries are not coupled in *Bacillus subtilis*. Mol Microbiol 2016;100:621–34.
[66] Takada H, Fukushima-Tanaka S, Morita M, Kasahara Y, Watanabe S, Chibazakura T, et al. An essential enzyme for phospholipid synthesis associates with the *Bacillus subtilis* divisome. Mol Microbiol 2014;91:242–55.
[67] Yoshimura M, Oshima T, Ogasawara N. Involvement of the YneS/YgiH and PlsX proteins in phospholipid biosynthesis in both *Bacillus subtilis* and *Escherichia coli*. BMC Microbiol 2007;7:69.
[68] Bremer HD. P.P. Modulation of chemical composition and other parameters of the cell by growth rate. In: Neidhardt FC, editor. Escherichia coli and Salmonella: cellular and molecular biology. Washington, DC: ASM Press; 1996. p. 1553–69.
[69] Srivatsan A, Wang JD. Control of bacterial transcription, translation and replication by (p)ppGpp. Curr Opin Microbiol 2008;11:100–5.
[70] Jiang P, Cronan JE Jr. Inhibition of fatty acid synthesis in *Escherichia coli* in the absence of phospholipid synthesis and release of inhibition by thioesterase action. J Bacteriol 1994;176:2814–21.
[71] Davis MS, Cronan JE Jr. Inhibition of *Escherichia coli* acetyl coenzyme A carboxylase by acyl-acyl carrier protein. J Bacteriol 2001;183:1499–503.
[72] Heath RJ, Rock CO. Inhibition of beta-ketoacyl-acyl carrier protein synthase III (FabH) by acyl-acyl carrier protein in *Escherichia coli*. J Biol Chem 1996;271:10996–1000.
[73] Parsons JB, Broussard TC, Bose JL, Rosch JW, Jackson P, Subramanian C, et al. Identification of a two-component fatty acid kinase responsible for host fatty acid incorporation by *Staphylococcus aureus*. Proc Natl Acad Sci USA 2014;111:10532–7.
[74] Cronan JE Jr, Weisberg LJ, Allen RG. Regulation of membrane lipid synthesis in *Escherichia coli*. Accumulation of free fatty acids of abnormal length during inhibition of phospholipid synthesis. J Biol Chem 1975;250: 5835–40.
[75] Porrini L, Cybulski LE, Altabe SG, Mansilla MC, de Mendoza D. Cerulenin inhibits unsaturated fatty acids synthesis in *Bacillus subtilis* by modifying the input signal of DesK thermosensor. Microbiologyopen 2014;3: 213–24.
[76] Huang W, Jia J, Edwards P, Dehesh K, Schneider G, Lindqvist Y. Crystal structure of beta-ketoacyl-acyl carrier protein synthase II from *E.coli* reveals the molecular architecture of condensing enzymes. EMBO J 1998;17: 1183–91.

[77] Moche M, Schneider G, Edwards P, Dehesh K, Lindqvist Y. Structure of the complex between the antibiotic cerulenin and its target, beta-ketoacyl-acyl carrier protein synthase. J Biol Chem 1999;274:6031–4.
[78] Olsen JG, Kadziola A, von Wettstein-Knowles P, Siggaard-Andersen M, Larsen S. Structures of beta-ketoacyl-acyl carrier protein synthase I complexed with fatty acids elucidate its catalytic machinery. Structure 2001;9:233–43.
[79] Price AC, Choi KH, Heath RJ, Li Z, White SW, Rock CO. Inhibition of beta-ketoacyl-acyl carrier protein synthases by thiolactomycin and cerulenin. Structure and mechanism. J Biol Chem 2001;276:6551–9.
[80] Price AC, Zhang YM, Rock CO, White SW. Structure of beta-ketoacyl-[acyl carrier protein] reductase from *Escherichia coli*: negative cooperativity and its structural basis. Biochemistry 2001;40:12772–81.
[81] Nunn WD, Colburn RW, Black PN. Transport of long-chain fatty acids in *Escherichia coli*. Evidence for role of fadL gene product as long-chain fatty acid receptor. J Biol Chem 1986;261:167–71.
[82] DiRusso CC, Heimert TL, Metzger AK. Characterization of FadR, a global transcriptional regulator of fatty acid metabolism in *Escherichia coli*. Interaction with the fadB promoter is prevented by long chain fatty acyl coenzyme A. J Biol Chem 1992;267:8685–91.
[83] Henry MF, Cronan JE Jr. *Escherichia coli* transcription factor that both activates fatty acid synthesis and represses fatty acid degradation. J Mol Biol 1991;222:843–9.
[84] Henry MF, Cronan JE Jr. A new mechanism of transcriptional regulation: release of an activator triggered by small molecule binding. Cell 1992;70:671–9.
[85] My L, Ghandour Achkar N, Viala JP, Bouveret E. Reassessment of the genetic regulation of fatty acid synthesis in *Escherichia coli*: global positive control by the dual functional regulator FadR. J Bacteriol 2015;197:1862–72.
[86] Zhang YM, Marrakchi H, Rock CO. The FabR (YijC) transcription factor regulates unsaturated fatty acid biosynthesis in *Escherichia coli*. J Biol Chem 2002;277:15558–65.
[87] Feng Y, Cronan JE. *Escherichia coli* unsaturated fatty acid synthesis: complex transcription of the *fabA* gene and in vivo identification of the essential reaction catalyzed by FabB. J Biol Chem 2009;284:29526–35.
[88] Feng Y, Cronan JE. Complex binding of the FabR repressor of bacterial unsaturated fatty acid biosynthesis to its cognate promoters. Mol Microbiol 2011;80:195–218.
[89] Zhu K, Zhang YM, Rock CO. Transcriptional regulation of membrane lipid homeostasis in *Escherichia coli*. J Biol Chem 2009;284:34880–8.
[90] Subramanian C, Rock CO, Zhang YM. DesT coordinates the expression of anaerobic and aerobic pathways for unsaturated fatty acid biosynthesis in *Pseudomonas aeruginosa*. J Bacteriol 2010;192:280–5.
[91] Zhang YM, Zhu K, Frank MW, Rock CO. A *Pseudomonas aeruginosa* transcription factor that senses fatty acid structure. Mol Microbiol 2007;66:622–32.
[92] Miller DJ, Zhang YM, Subramanian C, Rock CO, White SW. Structural basis for the transcriptional regulation of membrane lipid homeostasis. Nat Struct Mol Biol 2010;17:971–5.
[93] Schujman GE, Paoletti L, Grossman AD, de Mendoza D. FapR, a bacterial transcription factor involved in global regulation of membrane lipid biosynthesis. Dev Cell 2003;4:663–72.
[94] Martinez MA, Zaballa ME, Schaeffer F, Bellinzoni M, Albanesi D, Schujman GE, et al. A novel role of malonyl-ACP in lipid homeostasis. Biochemistry 2010;49:3161–7.
[95] Schujman GE, Guerin M, Buschiazzo A, Schaeffer F, Llarrull LI, Reh G, et al. Structural basis of lipid biosynthesis regulation in Gram-positive bacteria. EMBO J 2006;25:4074–83.
[96] Albanesi D, Reh G, Guerin ME, Schaeffer F, Debarbouille M, Buschiazzo A, et al. Structural basis for feed-forward transcriptional regulation of membrane lipid homeostasis in *Staphylococcus aureus*. PLoS Pathog 2013;9:e1003108.
[97] de Mendoza D, Pilon M. Control of membrane lipid homeostasis by lipid-bilayer associated sensors: a mechanism conserved from bacteria to humans. Prog Lipid Res 2019;76:100996.
[98] Saita E, Albanesi D, de Mendoza D. Sensing membrane thickness: lessons learned from cold stress. Biochim Biophys Acta 2016;1861:837–46.
[99] Sengupta P, Garrity P. Sensing temperature. Curr Biol 2013;23:R304–7.
[100] Sinensky M. Homeoviscous adaptation—a homeostatic process that regulates the viscosity of membrane lipids in *Escherichia coli*. Proc Natl Acad Sci USA 1974;71:522–5.
[101] Zhang YM, Rock CO. Membrane lipid homeostasis in bacteria. Nat Rev Microbiol 2008;6:222–33.
[102] Mansilla MC, Cybulski LE, Albanesi D, de Mendoza D. Control of membrane lipid fluidity by molecular thermosensors. J Bacteriol 2004;186:6681–8.

[103] Marr AG, Ingraham JL. Effect of temperature on the composition of fatty acids in *Escherichia coli*. J Bacteriol 1962;84:1260–7.

[104] Saita E, Abriata LA, Tsai YT, Trajtenberg F, Lemmin T, Buschiazzo A, et al. A coiled coil switch mediates cold sensing by the thermosensory protein DesK. Mol Microbiol 2015;98:258–71.

[105] Kaneda T. Iso- and anteiso-fatty acids in bacteria: biosynthesis, function, and taxonomic significance. Microbiol Rev 1991;55:288–302.

[106] Singh AK, Zhang YM, Zhu K, Subramanian C, Li Z, Jayaswal RK, et al. FabH selectivity for anteiso branched-chain fatty acid precursors in low-temperature adaptation in Listeria monocytogenes. FEMS Microbiol Lett 2009;301:188–92.

[107] Arabolaza A, D'Angelo M, Comba S, Gramajo H. FasR, a novel class of transcriptional regulator, governs the activation of fatty acid biosynthesis genes in *Streptomyces coelicolor*. Mol Microbiol 2010;78:47–63.

[108] Arabolaza A, Shillito ME, Lin TW, Diacovich L, Melgar M, Pham H, et al. Crystal structures and mutational analyses of acyl-CoA carboxylase beta subunit of *Streptomyces coelicolor*. Biochemistry 2010;49:7367–76.

[109] Salzman V, Mondino S, Sala C, Cole ST, Gago G, Gramajo H. Transcriptional regulation of lipid homeostasis in mycobacteria. Mol Microbiol 2010;78:64–77.

[110] Gago G, Diacovich L, Gramajo H. Lipid metabolism and its implication in mycobacteria-host interaction. Curr Opin Microbiol 2018;41:36–42.

[111] Tsai YT, Salzman V, Cabruja M, Gago G, Gramajo H. Role of long-chain acyl-CoAs in the regulation of mycolic acid biosynthesis in mycobacteria. Open Biol 2017;7:170087.

[112] Biswas RK, Dutta D, Tripathi A, Feng Y, Banerjee M, Singh BN. Identification and characterization of Rv0494: a fatty acid-responsive protein of the GntR/FadR family from *Mycobacterium tuberculosis*. Microbiology 2013;159:913–23.

[113] Yousuf S, Angara R, Vindal V, Ranjan A. Rv0494 is a starvation-inducible, auto-regulatory FadR-like regulator from *Mycobacterium tuberculosis*. Microbiology 2015;161:463–76.

[114] Mondino S, Gago G, Gramajo H. Transcriptional regulation of fatty acid biosynthesis in mycobacteria. Mol Microbiol 2013;89:372–87.

[115] Lara J, Diacovich L, Trajtenberg F, Larrieux N, Malchiodi EL, Fernandez MM, Gago G, Gramajo H, Buschiazzo A. Mycobacterium tuberculosis FasR senses long fatty acyl-CoA through a tunnel, inducing DNA-dissociation via a transmission spine.. bioRxiv Preprint 2010;doi: 10.1101/2020.03.05.978833.

[116] Hernandez MA, Lara J, Gago G, Gramajo H, Alvarez HM. The pleiotropic transcriptional regulator NlpR contributes to the modulation of nitrogen metabolism, lipogenesis and triacylglycerol accumulation in oleaginous rhodococci. Mol Microbiol 2017;103:366–85.

[117] Tauch A, Kaiser O, Hain T, Goesmann A, Weisshaar B, Albersmeier A, et al. Complete genome sequence and analysis of the multiresistant nosocomial pathogen *Corynebacterium jeikeium* K411, a lipid-requiring bacterium of the human skin flora. J Bacteriol 2005;187:4671–82.

[118] Tauch A, Trost E, Tilker A, Ludewig U, Schneiker S, Goesmann A, et al. The lifestyle of *Corynebacterium urealyticum* derived from its complete genome sequence established by pyrosequencing. J Biotechnol 2008;136:11–21.

[119] Irzik K, van Ooyen J, Gatgens J, Krumbach K, Bott M, Eggeling L. Acyl-CoA sensing by FasR to adjust fatty acid synthesis in *Corynebacterium glutamicum*. J Biotechnol 2014;192(Pt. A):96–101.

[120] Nickel J, Irzik K, van Ooyen J, Eggeling L. The TetR-type transcriptional regulator FasR of *Corynebacterium glutamicum* controls genes of lipid synthesis during growth on acetate. Mol Microbiol 2010;78:253–65.

CHAPTER

# 2

# Lipid trafficking and signaling in plants

*Amanda M. Koenig*[a], *Christoph Benning*[b,c], *Susanne Hoffmann-Benning*[a,c]

[a]Genetics and Genome Sciences, Michigan State University, East Lansing, MI, United States; [b]DOE-Plant Research Laboratory, Michigan State University, East Lansing, MI, United States; [c]Biochemistry and Molecular Biology, Michigan State University, East Lansing, MI, United States

OUTLINE



Lipid Signaling and Metabolism. http://dx.doi.org/10.1016/B978-0-12-819404-1.00002-6

# Introduction

## General comparison between animal and plant lipids

Plant and animal lipids show not only remarkable similarities but also significant differences. While phospholipids are prevalent in eukaryotic plasma membranes and endomembranes overall, the plant-specific membranes in the chloroplast are composed mostly of nonphosphorus galactolipids and sulfolipids in addition to the phospholipid phosphatidylglycerol. The presence of the chloroplast as the location of fatty acid biosynthesis and the existence of additional lipids in plant membranes prompted a relocalization and expansion of biosynthetic steps, which necessitate additional trafficking between compartments and across multiple membranes. In addition, even within phospholipids, large variations exist. This has a major impact on lipid signaling as well. For example, animal systems have a phosphoinositide/Phospholipase C signaling system: in response to a receptor-binding signal, Phospholipase C (PLC) is activated and hydrolyzes phosphatidylinositol 4,5-bisphosphate ($PIP_2$) into diacylglycerol (DAG) and inositol 1,4,5-trisphosphate (InsP3). Both are secondary messengers that trigger a reprogramming of the animal cell [1]. At the same time, $PIP_2$ is also an important regulator of cellular dynamics, endo-and exocytosis, and of ion channel activity [2]. Plants, on the other hand, have much smaller amounts of $PIP_2$ and essentially no InsP3. Different plant phosphoinositides (PIs) play similar roles in membrane signaling (through regulating ion channels and pumps), vesicle trafficking and possibly, gene expression as well [3–6]. They are also important in signaling abiotic and biotic stresses (for a review, see Barbaglia and Hoffmann-Benning, 2016 [6]). Plant PLCs have evolved the ability to hydrolyze phosphatidylinositol-4-phosphate but can also use phosphatidylinositol; instead of InsP3, InsP6 appears to be the relevant second messenger [7]. PLC-derived DAG can be converted to phosphatidic acid (PA). This minor lipid is emerging as an important signaling molecule in the plant field and will be discussed below.

## Synthesis of membrane lipids

Lipid metabolism is critical for membrane maintenance and composition, energy storage, and signaling in animals and plants. Yet, the photosynthetic membrane in the chloroplast necessitates additional unique lipid metabolic pathways. As such, there are significant differences in plant lipid biosynthesis compared to animal systems. Animal cells assemble fatty acids in the cytoplasm, whereas, in plants, fatty acid synthesis occurs in the chloroplast stroma. The chloroplast-localized fatty acid synthase complex (FAS) generates fatty acids, which supply acyl groups to two separate but parallel lipid synthesis pathways: the eukaryotic pathway through the ER present in both plants and animals, and the plastid or "prokaryotic" pathway unique to plants [8]. The eukaryotic pathway functions similarly in animals and plants. In the ER, acyl-transferases draw from an acyl-CoA pool, derived from free fatty acids exported from the chloroplast into the cytoplasm, to assemble first lysophosphatidic acid (L-PA) from glycerol-3-phosphate (G3P), then convert L-PA to PA [9,10]. PA is subsequently dephosphorylated to DAG, which is further converted to other lipids such as phosphatidylcholine (PC) and phosphatidylethanolamine (PE) or derivatized to CDP-DAG, which is converted to phosphatidylglycerol (PG) and phosphatidylinositol (PI). In plants, the

PC that is produced in the ER can be trafficked back into the chloroplast where it is converted to galactolipids and sulfolipids [8]. The plastid pathway operates in parallel in the chloroplast but uses an acyl-ACP pool instead of acyl-CoA. Plastid-produced DAG is further metabolized to the galactoglycerolipids monogalactosyldiacylglycerol (MGDG) and digalactosyldiacylglycerol (DGDG), which are major components of the chloroplast membrane. The origin of lipid molecular species from each pathway can be distinguished based on their fatty acid composition as a result of differences in acyltransferase substrate specificity for the ER and plastid isoforms. ER-localized acyltransferases add 18-carbon fatty acids to the *sn-2* position of the glycerol backbone, whereas plastid acyltransferases add 16-carbon chains to this position [11,12]. Chloroplast membrane galactolipids in many plants, including the model plant Arabidopsis, comprise both 18-C and 16-C species, indicating that precursors generated in the eukaryotic pathway are utilized in galactolipid synthesis [8]. Therefore, the multiorganelle, two-pathway system necessitates lipid exchange between the ER and the chloroplast and among the chloroplast outer envelope, inner envelope, and thylakoid membranes.

## Lipid trafficking

Lipid trafficking between the ER and the plastid is an active area of plant lipid research. Here, we provide a broad overview of some of the key proteins involved in lipid trafficking and the transported intermediary lipid species. For a more in-depth review, please refer to Lavell and Benning [13].

The tightly regulated interplay between the two lipid biosynthesis pathways in the ER and chloroplast requires the export of fatty acids from the chloroplast across two membranes: the chloroplast inner envelope and the outer envelope membranes (OEM). While the mechanism for export is not definitively known, a role for ER-chloroplast contact sites as well as the tissue-specific involvement of the Arabidopsis inner envelope membrane (IEM) protein Fatty Acid Exporter 1 (FAX1) has been suggested [9,10,14].

Similarly, some of the lipids assembled in the ER require retranslocation to the chloroplast. Genetic mutant screens in Arabidopsis and analysis of lipid profiles have led to the identification of several chloroplast envelope membrane-localized proteins that facilitate lipid trafficking. The TRIGALACTOSYLDIACYLGLYCEROL (TGD) complex has been implicated in the unidirectional import of lipids to the plastid from the ER [15,16]. TGD 1, 2, and 3 form an ABC transporter in the IEM and function in coordination with TGD4 and 5 in the OEM to import DAG-derived lipid species from the ER into the plastid and its thylakoid membrane [17–22]. It should be noted that the imported lipid species has not been unambiguously determined. As both TGD4 in the OEM and TGD2 in the IEM have been characterized as PA-binding proteins [18,23,24], PA has been proposed as a transported lipid, but its binding to TGD2 and TGD4 may also serve regulatory functions. A summary of lipid synthesis and trafficking in plants is shown in Fig. 2.1.

## Phosphatidic acid as a biosynthetic intermediate

As a precursor for all glycerolipid synthesis, PA is a central intermediate. The de novo synthesis of PA through the successive acylation of G3P precedes phospholipid and galactolipid synthesis in both the eukaryotic and plastid pathways and contributes to the diversity

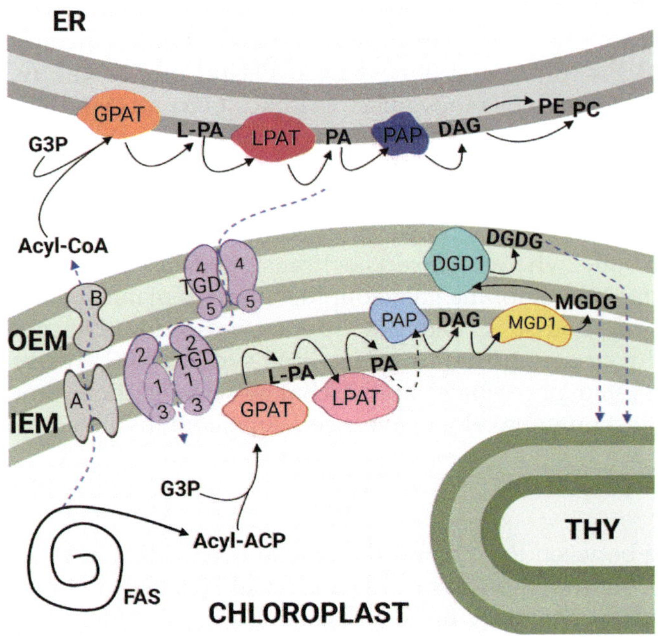


**FIGURE 2.1 Plant membrane lipid biosynthesis and trafficking.** The eukaryotic and plastid pathways for plant lipid biosynthesis are shown, including the localization of enzymes and transporters involved. Black arrows indicate biosynthetic steps; blue arrows show trafficking pathways; dotted arrows signify a mechanism with unknown protein components. *G3P*, glycerol-3-phosphate; *L-PA*, lyso-phosphatidic acid; *PA*, phosphatidic acid; *DAG*, diacylglycerol; *PC*, phosphatidylcholine; *PE*, phosphatidylethanolamine; *MGDG*, monogalactosyldiacylglycerol; *DGDG*, digalactosyldiacylglycerol; *FAS*, fatty acid synthesis; *A*, unknown protein to transport free fatty acids (FFA) across the inner envelope membrane (IEM), possibly Fatty Acid Exporter 1 (FAX1); *B*, unknown protein to transport FFA through the outer envelope membrane (OEM); *ER*, endoplasmic reticulum; *THY*, thylakoids; *GPAT*, G3P acyltransferase; *LPAT*, L-PA acyltransferase; *PAP*, phosphatidic acid phosphatase; *MGD1*, MGDG Synthase 1; *DGD1*, DGDG Synthase 1; *TGD 1, 2, 3, 4, 5*, TRIGALACTOSYLDIACYLGLYCEROL transport complex 1, 2, 3, 4, 5.

of fatty acid composition in lipid molecular species. Phosphatidic acid phosphatases (PAPs) are responsible for the dephosphorylation of PA to DAG, which is a key step in membrane lipid synthesis. The ER-localized phosphatidic acid phosphohydrolases PAH1 and PAH2 influence galactolipid and phospholipid metabolism either directly or through PA or DAG [25, 26]. A lipid phosphate phosphatase (LPP) in cyanobacteria is homologous to several PAPs in Arabidopsis, three of which were characterized as plastid PAP candidates: LPPγ, LPPε1, and LPPε2 [27]. LPPγ, LPPε1, and LPPε2 are all functional and have been localized to the chloroplast [27]. Given that PAP activity is associated with the inner leaflet of the IEM and the MGDG synthase MGD1 is in the outer leaflet, a mechanism by which substrates PA and DAG are made accessible must be considered [28–30]. A predicted rhomboid protease RBL10 has been identified in Arabidopsis to play a role in PA transport across the IEM; *rbl10* mutants show impaired utilization of plastid PA in galactolipid synthesis [31]. Therefore, RBL10 seems to contribute to the transport of the galactolipid precursor to MGD1. While PA is not the direct precursor for MGDG synthesis but rather DAG is the substrate for MGD1, evidence suggests that PA contributes to MGD1 activation [32]. So, not only is PA an essential precursor and intermediate in glycerolipid synthesis, but it also acts as an important regulatory factor in galactolipid metabolism.

## Key signaling lipids in plants

Like animals, plants perceive signals or invaders and respond both on local and organismal levels. Unlike animals, plants do not have a central nervous system for transmission of long-distance signals and coordinated responses nor do they have a blood circulatory system. However, they have a noncircular yet interconnected vasculature for long-distance movement of nutrients, energy carriers and chemical signals, including proteins, nucleic acids, sugars, or lipids (for a review, see [33]).

To cope with environmental factors, plants employ a large variety of lipophilic signaling molecules. Some of these can be complex lipids such as phospholipids, predominantly PIs and PA, ceramides and sphingolipids, and DAG; others include small lipophilic molecules such as many plant hormones or pathogen-response molecules like the dicarboxylic acid azelaic acid (for an in-depth review see [6]). Several of these lipids including PA, L-PA, DAG, and PIs can function as second messengers within plant cells. Some participate in long-distance signaling. One group of important signaling lipids are oxylipins. Oxylipins are the plant equivalent to Eicosanoids. They are oxygenated polyunsaturated fatty acids that are derived from acyl groups of the galactolipids in the chloroplast membrane. They are important for regulating aspects of plant growth and development as well as the response to pathogens and abiotic stresses [34, 35]. As such, they affect the balance between growth and defense [36]. Ceramides and sphingolipids are an integral structural component of plant membranes, particularly for the formation of lipid raft microdomains. Moreover, they participate in developmental and stress signaling such as programmed cell death, temperature stress, and salicylic acid signaling for biotic stress defense. Sphingolipids contribute to these mechanisms as membrane structural components, they are involved in stress perception and they serve as signaling compounds [37]. DAG has proven to be an important signaling lipid in animals through its interaction with protein kinases. Although fluctuations in DAG levels have been observed in correlation with developmental and environmental responses in plants, protein targets for DAG in plants have yet to be uncovered or characterized. Rather, these fluctuations in DAG levels could be indicative of PA production and turnover, while PA per se serves as a lipid signal in plants [38].

PA is emerging as a central regulatory lipid in plant signaling. It acts not only directly by mediating plant responses to both biotic and abiotic stressors but also indirectly by being the central precursor for other signaling lipids such as PIs and DAG. In this role, it influences plant development and local and systemic stress responses. It executes this function directly through imposing changes in membrane curvature, or indirectly by binding of receptor proteins in a signaling cascade, and through interaction with transcription factors. PA is generated rapidly and transiently in response to environmental cues. The expression and activity of PA-generating enzymes are often promoted by hormones and stress cues. Various environmental factors activate different phospholipases, which results in the production of distinct PA pools from different origins, thereby supplying PA as a lipid mediator in stress-specific mechanisms [1,39–45]. For example, Phospholipase Dα1 (PLDα1) and Phospholipase Dδ (PLDδ) are two isoforms of PLD that mediate the production of PA in response to drought/ABA-mediated pathways while the Phospholipase C/Diacylglycerol Kinase (PLC/DGK) pathway is used in cold response [46–48]. Below, we are discussing the role of phospholipids with an emphasis on PA in plant development and signaling.

## Development

Like in animals, hormones play an important role in signaling, differentiating, and coordinating environmental responses in plants. Some key phytohormones are ethylene, abscisic acid (ABA), and auxin. These hormones govern normal plant development as well as adaptations to environmental factors. They bind to specific receptors and act on gene expression, protein modification, and through modifications of phospholipids, mostly PA and PI.

### Plant growth and seedling development

Lipids play an important part in multiple aspects of plant growth and development: (1) as membrane components they are critical for cell and organelle structure as well as for vesicle trafficking. Examples include galactolipids, which fundamentally affect chloroplast development and function [49] or PIs, which are essential for the secretory pathway [7,50]; (2) as signaling molecules, which will be discussed later; and (3) as energy-dense storage compounds.

The less abundant yet important lipid, PA promotes plant growth and biomass accumulation in Arabidopsis. More specifically, PA derived from the hydrolysis of PE by Phospholipase Dε (PLDε) is likely responsible for the observed increase in biomass between *PLDε* overexpression plants compared to the wild-type [51]. Enhanced root development and elongation was also observed in *PLDε* overexpression lines, which may be attributed to PA produced by the action of PLDε and its involvement in nitrogen signaling and acquisition [51]. Lipidomics revealed fluctuations in phospholipid content during different seedling developmental stages, suggesting a role for various phospholipids; in particular, PA and PE exhibit inverse fluctuation patterns in seedling development [52]. For example, PA levels are lowest during imbibition (water uptake into the seed to initiate germination) and increase during germination followed by a decrease when cotyledons fully open, whereas PE levels are highest at imbibition and decrease in subsequent stages. This possible substrate–product relationship suggests that hydrolysis of PE is a major contributor of PA accumulation during radicle emergence, consistent with findings that PLDε-derived PA in roots is likely generated from PE hydrolysis and promotes growth [51, 52]. The plant hormone ABA is involved in the regulation of seed dormancy and germination; PLDα1-derived PA has been implicated in ABA signaling [53]. Acyl-CoA Binding Protein 1 (ACBP1) overexpression increases *PLDα1* expression. It binds both PC and PA and promotes seed germination, potentially by facilitating PC/PA exchange as a result of PLDα1 activity [54]. Thus, it appears that the generation of PA from either PC or PE is fundamental in seed germination.

### Root development

Phospholipids influence root architecture and development under abiotic stress conditions such as water, nitrogen, and/or phosphorus deprivation. As is discussed later in this chapter,

PA influences the localization and activity of PIN-FORMED (PIN) and PINOID (PID), proteins that are critical in auxin transport and signaling and, as a result, governs directional plant and root growth [55, 56]. While plants lack InsP3 targets present in animals, PLC enzymes, along with their substrates and products, influence primary and secondary root growth [57]. Diacylglycerol Kinase 1 (DGK1) affects root architecture by modulating DAG and PA levels in inverse directions and with opposite functions: PA suppresses lateral root formation and enhances root diameter, while DAG stimulates formation of lateral roots and suppresses seminal root growth [58].

## Cytoskeleton

The cytoskeleton is a dynamic system responsible for cell structure and shape, organelle transport, and involved in responses to environmental stimuli. The predominant components of the cytoskeleton are polymeric actin filaments and microtubules, often closely associated with the plasma membrane. Critical for plant growth and development, the rapid reorganization of the cytoskeleton is tightly regulated by proteins and small molecules, including phospholipids. Changes in cytoskeleton organization in response to PIs and PA have been observed in several plant species, including Arabidopsis, *Nicotiana tabacum,* soybean, and *Zea mays* [59–62]. Here, we will highlight the role of PA in cytoskeleton rearrangement; for a review of PI signaling, see [5,63].

In addition to supporting cell structure, polarity, and morphology, actin filaments contribute to organelle movement and signal transduction in plant cells. Interestingly, actin seems to be responsive to PA; an increase in actin polymers has been observed in response to application of PA in soybean, Arabidopsis, and poppy [60, 64]. Actin-binding proteins, many of which have been shown to bind phospholipids, coordinate actin reorganization and function. Arabidopsis Capping Protein (CP) regulates actin polymerization and elongation by binding barbed ends of actin filaments. Arabidopsis CP binds PA, which inhibits its barbed end capping activity, thereby promoting actin elongation and increasing actin filament density [64, 65]. The promotion of actin polymerization by PLD-derived PA, an interaction first observed in mammalian fibroblasts, may contribute to PA's influence on tip growth observed in pollen tube germination and elongation [66–68].

In the plant cell, microtubules function in the deposition and synthesis of cellulose in the cell wall and are essential to cell expansion and morphogenesis during growth and development [69, 70]. The identification of tobacco p90, a PLD enzyme, likely PLDδ, which interacts with microtubules in tobacco, established a link between plant lipid signaling and microtubule architecture [71, 72]. Inhibition of PLD through *n*-butanol modifies PA levels and, consequently, germination and root growth (see above) as well as microtubule organization [71, 73, 74]. While data using BODIPY-PA suggest that PA does not induce microtubule reorganization, Zhang et al. observed that PLDα1-derived PA can reverse microtubule depolymerization and seems to stabilize microtubules under salt-stress conditions [73,75]. More specifically, MAP65-1, a protein responsible for microtubule bundling and stabilization, binds 16:0/18:2 PA (FAs with number of carbons: number of double bonds on *sn*-1/*sn*-2 position of the glyceryl backbone), which promotes its microtubule-bundling activity and enhances tubulin polymerization [75]. This suggests the possibility that, while PA is generally important for all aspects of growth, the PA species determines the specificity of its function.

## Circadian clock

Just as in animal systems, light, or more specifically, the circadian clock and the dark-light cycle determine development and physiological responses. Diurnal cycling and the circadian clock influence glycerolipid metabolism and fatty acid composition in plants [76–78]. For example, PA accumulates during dark periods, which could be due to a reduction in membrane lipid synthesis as some biosynthetic steps may be light-dependent [78,79]. The expression of glycerolipid metabolic enzyme-encoding genes, including genes encoding PA-producing proteins like Lysophosphatidyl Acyltransferase 5 (LPAT5), PLDδ, and PLDζ, may be influenced by the circadian clock and/or diurnal cues [77, 78]. PG (36:4), phosphatidylserine (PS) (38:5), and PA (34:4 and 36:6) levels exhibit daily cycling in wild-type Arabidopsis seedlings, and this cycling pattern is lost in mutants affected in the circadian clock (*lhy-20; cca1-1;* [80,81]), suggesting that phospholipid levels are partially modulated by the circadian clock [82]. Moreover, not only is phospholipid metabolism regulated by the clock but the inverse may also be true: phospholipids modulate the circadian clock. PA, more specifically 16:0-containing PA species, alters circadian clock regulation in Arabidopsis. Clock components LHY and CCA1 bind PA, which inhibits their DNA-binding activity and prevents the suppression of *TIMING OF CAB EXPRESSION 1* (*TOC1*) expression, a key circadian rhythm regulator in plants. Genetic and pharmacological manipulations of PA levels caused shifts in *TOC1* expression oscillations and vertical leaf movement, further confirming that PA regulates the circadian clock [82]. PA, a known stress signal, may act as a mediator between environmental stimuli and the circadian clock. Again, only certain PA species are affected by and interact with the circadian clock.

## Flowering

Phospholipids contribute to the careful coordination of flowering and reproduction in plants. Nakamura et al. observed high PA content in Petunia stamen and pistils compared to petals or leaves as well as enriched PA levels in Arabidopsis reproductive organs [83,84]. Many plants employ self-incompatibility (SI) mechanisms to prevent inbreeding and promote genetic diversity. *PLDα1* is expressed in flowers and is targeted by ARC1, an E3 ligase, during the SI response, suggesting that PA may support compatibility during pollination and fertilization. In fact, PA overcomes incompatibility and enhances pollen attachment. When *PLDα1* is overexpressed, multivesicular bodies exocytose at the plasma membrane even during self-incompatible fertilization, whereas these vesicles are sequestered in the cytosol or vacuole in the wild-type [85]. Thus, PLDα1-derived PA may contribute to membrane-reshaping and thereby facilitates an exocytosis mechanism that supports compatible pollination.

While some data indicate a possible role for PA in flowering, PIs and PC seem to be the predominant phospholipids that affect pollination and flowering [78, 86]. Often, these lipids integrate multiple signals and developmental stages. The switch from vegetative to generative growth is triggered in many plants by a specific light-dark cycle (diurnal rhythm). It is sensed in the leaves and transmitted to the shoot apical meristem in the form of a mobile long-distance, flowering signal, the protein Flowering Locus T (FT) or a homologue. Once there, FT interacts with transcription factors and switches the developmental pattern from leaf formation to flower formation. Nakamura et al. have shown not only that FT binds PC but also that specific PCs display a circadian rhythm in the leaf [78]. This suggests that developmental regulation through lipids is not only dependent on lipid class but also on specific lipid species.

## Abiotic stress

As mentioned above, plants use phytohormones to regulate not only development but also environmental responses. In some cases, these responses are mediated by phospholipids. Here we focus on the role of PA in several hormone-mediated signaling pathways and abiotic stress response. The central role of PA in lipid metabolism, development, and signaling is illustrated in Fig. 2.2.

### Ethylene and stress response

Ethylene is a plant hormone involved in developmental signaling, including seed germination, root hair development, root nodulation, flower senescence, abscission, and fruit ripening. Furthermore, ethylene has been shown to participate in plant stress responses to wounding, hypoxia, ozone, chilling, and freezing [87]. Ethylene mutants of Arabidopsis exhibit a characteristic "triple response" phenotype, including shortened hypocotyls and roots, an exaggerated curvature of the apical hook, and radial swelling of the hypocotyl

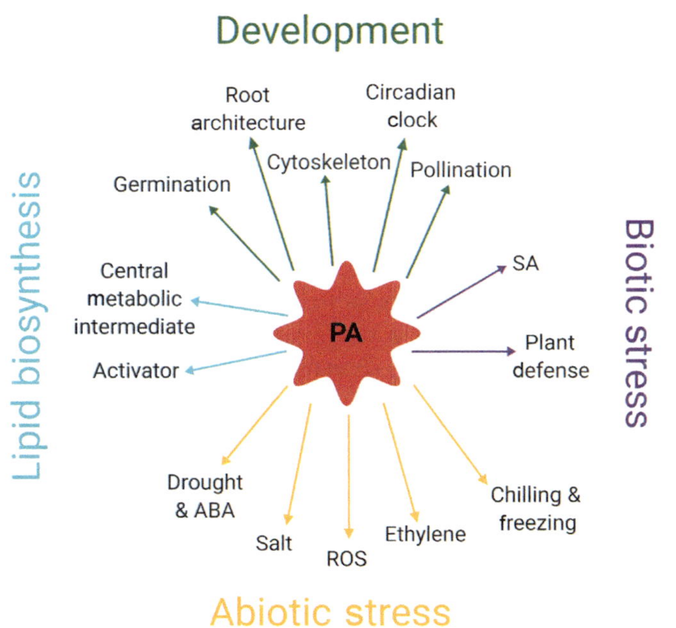


FIGURE 2.2 **Summary of phosphatidic acid functions in plants.** Phosphatidic acid (PA) is generated by several, often stress-dependent pathways, and it functions in various capacities in many developmental, environmental, and metabolic processes in plants. *SA*, salicylic acid signaling; *ROS*, reactive oxygen species; *ABA*, abscisic acid signaling.

in etiolated seedlings. Constitutive triple response 1 (CTR1) is a homolog of mammalian MAPKKK Raf-1 and a negative regulator of ethylene signaling. Ethylene binding deactivates its receptors (ETR1, ETR2, EIN4, ERS1, and ERS2) which, in turn, deactivate the ethylene response repressor CTR1. EIN2 is a positive regulator of ethylene signaling that functions downstream of CTR1 and modulates EIN3 and EIL1 transcription factors by inhibiting their degradation [88, 89]. CTR1 deactivation involves the interaction between ethylene-bound receptors and CTR1, but the specific mechanism is still not known. Testerink et al. showed that CTR1 binds PA, which inhibits CTR1 kinase activity, thereby promoting ethylene signaling [90,91]. The study found that PA interacts preferentially with the kinase domain, with longer chain PA showing a higher inhibitory effect than short chain PA. Furthermore, PA seems to interfere with CTR1-ETR1 interaction. While an increase in PA is not observed in response to ethylene, PA does accumulate in response to ethylene-associated stresses such as wounding or salt stress, and Arabidopsis knockout mutants *pld$\alpha$* and *pld$\delta$* that show a reduced PA content also display a delay in ethylene-induced senescence [92,93]. PA could act in parallel to ethylene signaling to complement or expedite deactivation of CTR1 under wounding, pathogen, or salt stresses. Alternatively, PA may act in the maintenance of long-term ethylene responses.

## Auxin and salt stress response

An overabundance of salt can have adverse effects on plant growth and development by causing water stress through both ionic and osmotic stresses. Thus, salt stress contains some components of water stress, yet the regulation of the response can be vastly different. Plants respond to salt stress by meticulously controlling net influx and transport of $Na^+$ ions and by sequestering $Na^+$ into the vacuole, in order to maintain a higher $K^+$:$Na^+$ ratio in the cytosol. Phospholipase mutants *pld$\alpha$1* and *pld$\delta$* have heightened salt sensitivity, indicating a role for phospholipases and PA in salt tolerance mechanisms [40, 94]. PLD$\alpha$1-derived PA is a proposed activator of MAP kinase MPK6, which phosphorylates SALT OVERLY SENSITIVE 1 (SOS1), a $Na^+/H^+$ antiporter localized in the plasma membrane [95]. SOS1, when activated, maintains ionic homeostasis by exporting $Na^+$ out of the cytoplasm and into the apoplast [96].

In addition to regulating intracellular salt content, plants also adjust their root system architecture to avoid high salinity soil, a process directed by the phytohormone auxin. PIN and PID proteins direct auxin transport, dependent on asymmetric localization of PIN and regulated by reversible phosphorylation [97, 98]. Lipids, including PIs, sterols, and PA, facilitate PIN and PID localization and activation, and thereby play a role in the regulation of salt-responsive auxin transport. PLD$\alpha$/PLD$\delta$-derived PA interacts with PID kinase and promotes the phosphorylation of PIN2, which allows the directional transport of auxin, thus regulating primary root elongation in response to salt stress [56]. PLD$\zeta$-derived PA, on the other hand, seems to affect root gravitropism and halotropism (directional growth in response to gravity and salinity, respectively) in an auxin-dependent manner by regulating Protein Phosphatase 2A (PP2A) activity and PIN1 polar localization [55]. PA accumulates rapidly within minutes of salt-stress and contributes to recruitment of proteins to the membrane, including clathrin assembly proteins, potassium channel subunits, and glyceraldehyde 3P dehydrogenases (GAPDH) shown to interact with PLD$\delta$ [99–102]. The PA-associated protein targets suggest a role for PA in vesicle trafficking, endocytosis, ion exchange, and reactive oxygen species (ROS) signaling in response to salt stress [102]. More broadly, these findings emphasize PA's contribution to protein membrane recruitment and protein localization during plant stress response, in addition to its role as a secondary messenger and regulator.

## Cold stress

Cold stress encompasses both chilling and freezing conditions, two distinct stresses that elicit different plant responses. Cold stress can severely impede plant growth by disrupting normal cell structure and function. For example, chilling stress disrupts normal water uptake, and freezing stress causes cell dehydration and ruptures the plasma membrane as a result of ice accumulation in the intercellular space [103]. Tolerance to freezing stress involves the stabilization of the plasma membrane, which requires adjusting membrane lipid content. Cold acclimated plants shift membrane lipid composition toward more polyunsaturated lipids [104]. Furthermore, in Arabidopsis chloroplasts, monogalactolipids are converted to oligogalactolipids by the enzyme SENSITIVE TO FREEZING 2 (SFR2), resulting in an increase in triacylglycerol (TAG) [105–107]. This leads to a change in the ratio of bilayer- to nonbilayer-forming lipids, prevents water accumulation between membrane bilayers and increases tolerance against freezing. Additionally, freezing conditions cause a decrease in PC, PE, and PG and an increase in PA species [104]. PLD enzymes have been implicated in freezing sensitivity and tolerance. PLDα1-deficient lines are more freezing tolerant than wild-type lines, possibly attributable to higher accumulation of osmolytes under both nonacclimated and cold-acclimated conditions or a reduction in lamellar-to-hexagonal II phase transitions [104, 108]. Conversely, *pldδ* mutants exhibit increased freezing sensitivity and *PLDδ* overexpression confers freezing tolerance after cold acclimation. While it seems PLDs may be involved in freezing tolerance over chilling stress responses, researchers do observe a significant increase in PA within 5 min of cold treatment in both seedlings and mature leaves [46]. Based on $^{32}$P-radiolabelling and transphosphatidylation assays, it was determined that the PLC/DGK pathway, not PLDs, was likely responsible for the accumulation of PA under chilling conditions [46].

Thus, PA plays a role in both cold chilling and freezing tolerance, though it is generated through distinct pathways depending on the temperature stress. Contrary to its role in other environmental signaling, where it serves as a signal or membrane tether, in cold responses, it functions by modifying the physical properties of membranes. The molecular shape of PA can influence lipid phases and membrane curvature in response to changes in physiological conditions [109]. Moreover, shifts in the membrane's biophysical properties affect protein binding and PA's accessibility to its protein targets [110]. Overall, PA may adjust the stability of the membrane in response to the cell's physiological status during cold stress, rather than or, perhaps in addition to, participating as a signal molecule.

## Abscisic acid and drought

ABA is a phytohormone important in growth and environmental stress response, including desiccation, seed development (discussed above), and drought stress response. Drought and osmotic stress cause an accumulation of ABA, which is then sensed by its receptor

PYR/PYL/RCAR. Under normal conditions, Protein Phosphatase 2Cs (PP2C) act as negative regulators, dephosphorylating SNF1-Related Kinase 2 (SnRK2), thereby repressing the ABA response. In the presence of ABA, PP2Cs are sequestered away from SnRK2s, which are then activated to phosphorylate downstream AREB/ABF transcription factors that bind to ABA response elements (ABREs) and promote expression of drought-responsive genes [111]. This ABA-dependent mechanism is important for regulation of stomatal opening and closing, for modulating transpiration and reducing water loss. Importantly, it is mediated by PA. In response to ABA, PLDα1 activity increases, which leads to increased synthesis of PA. PA then acts as an intermediary messenger by interacting with Abscisic Acid Insensitive 1 (ABI1), a PP2C negative regulator of ABA signaling. Thus, PA is likely an essential factor for sequestering PP2Cs away from the ABA/SnRK2 signaling path [53]. PA promotes ABA-induced stomatal closure by decreasing ABI1 activity, likely through a membrane tethering mechanism [53,112]. Furthermore, PLDα1 and PA have been implicated in the inhibition of stomatal opening by regulating GTP-binding protein GPA1 activity [112]. In addition to regulating ABI1 activity, PLDα1-derived PA also promotes stomatal closure by mediating ROS accumulation due to the activity of NADPH oxidases RbohD and RbohF, and by affecting subsequent nitric oxide (NO) signaling [113]. More specifically, 16:0/18:2 PA generated by PLDα1 activates the NADPH oxidase, which results in an accumulation of ROS, followed by downstream NO signal transduction. However, some studies report NO production is upstream of PA and promotes further PA production, perhaps derived from another phospholipase, PLDδ [114,115].

While PLDα1 is the predominant source of PA among PLD isoforms, other PLDs also contribute to various ABA-dependent stress responses: PLDα1 and PLDδ cooperatively function in stomatal closure mechanisms, including ROS production as well as cytosolic alkalization of guard cells [47]. During the early stages of the drought stress response, PLDδ contributes to regulation of stomatal conductance; however, *pldδ* knockout mutants have higher drought tolerance compared to wild-type plants [116]. The accumulation of PLDδ-derived PA in severe drought conditions may cause membrane destabilization and impair membrane integrity [116]. While the two PLDs have overlapping functions in stomatal closure, PLDα1 may play a role in inhibiting light-induced stomatal opening [47]. These results are consistent with a study in *Vicia faba* that showed PA inhibits the catalytic activity of Protein Phosphatase 1C (PP1C) and reduces blue-light dependent stomatal opening [117].

ABA-induced drought stress responses in plants are dependent on PA as part of several regulatory mechanisms, for example through modification of sphingolipids or by binding to MYB transcription factors, like WEREWOLF (WER), affecting its nuclear localization and thereby gene expression [118,119]. Regulation of gene expression through direct interaction between transcription factors and lipids is well studied in animal systems but only emerging in plant research. Overall, the specific role of PA in ABA signaling seems to rely on its derivation from different PLD isoforms, perhaps due to different expression patterns, substrate preference, or cellular localization. This is supported by the fact that PLDα, a PLD that prefers PC over PE, is constitutively present and found in the cytosol and the plasma membrane, while PLDδ, which prefers PE over PC, is induced by dehydration and salt stress [47,120–122]. Further studies will elucidate if these differences extend to other PLDs and other stresses and if regulation depends on the specific lipid species.

## Biotic stress

Plant disease responses rely on pathogen-associated molecular patterns (PAMPS)-triggered immunity (PTI) and effector-triggered immunity (ETI) to combat pathogen attack. Responses are initiated by the recognition of elicitors: molecules produced by pathogens including microbes, fungi, and viruses that trigger plant defense. Phytotoxins, PAMPS, and protein effectors can all stimulate plant defense and immune responses, often involving oxidative bursts and production of secondary messengers like $Ca^+$, NO, and PA. A diversity of elicitors triggers the accumulation of PA including fungal xylanases, chitotetraose, chitosan, *N*-acetylchitooligosaccharides and botrydial; cryptogein from oomycetes; and bacterial flg22, AvrRpm1, and AvrRpt2 [123–129].

Studies have explored the causes and effects of PA accumulation, more specifically the link between NO production, generation of PA, and accumulation of ROS in plant defense responses. NO induces PA accumulation predominantly through PLC/DGK-based mechanisms, however PLDβ1 activity has also been implicated in plant defense responses [43,125,127,128,130–132]. PLDβ1-derived PA can support or suppress pathogen resistance; for example, *pldβ1* knockout lines exhibit increased susceptibility to the fungal pathogen *Botrytis cinerea* but are more resistant to the bacterial pathogen *Pseudemonas syringae* [132]. It has been proposed that the mode of infection—necrotrophic (pathogens kill their hosts to feed) vs. biotrophic (pathogens feed on living hosts)—may influence whether PLD-derived PA acts as a negative or positive regulator in plant defense, a distinction possibly related to the role of salicylic acid (SA) signaling in response to biotrophic and hemibiotrophic pathogens [133,134].

Once generated, PA can participate in plant–pathogen defense in various capacities, including membrane adjustments and protein–lipid interactions. PA binds several proteins involved in ROS-related signaling for plant defense, including 3-Phosphoinositide Dependent Kinase 1 (PDK1), cytosolic glyceraldehyde-3-phosphate dehydrogenases (GAPC), Wheat Kinase Start 1 (WKS1), and NADPH oxidases [128,135–138]. These pathways are reviewed further by Li and Wang [133].

Evidence suggests that PA interacts with not only ROS-associated proteins but also proteins involved in SA signaling as well as antifungal pathways [139, 140]. SA signaling is critical for plant defense responses, predominantly to biotrophic infection. Nonexpressor of Pathogenesis Related 1 (NPR1) is an integral component in the SA signaling cascade; NPR1 oligomers in the cytosol are monomerized during SA signaling and are translocated to the nucleus where NPR1 monomers regulate transcription factors for SA defense genes [139,141,142]. The PLD inhibitor *n*-butanol reduces NPR1 translocation to the nucleus as well as the expression of SA markers [139]. These data suggest that PLD activity influences NPR1 nuclear localization and, since PA accumulation has been observed in response to SA, it is possible that PA plays a role in NPR1 localization directly [139,143]. The antifungal defensin protein MtDef4 from *Medicago truncatula* invades fungal cells, disrupts the plasma membrane, and contributes to cell death. MtDef4 binds PA, and more specifically, the same loop required for PA binding is also necessary for entry into fungal cells, suggesting that PA mediates MtDef4 invasion into fungal cells [140]. Furthermore, genetic evidence suggests that the fungal PLD1 in *Neurospora crassa* may contribute to the antifungal activity of MtDef4, whereas PLD1 in *Fusarium graminearum* did not affect sensitivity to MtDef4 [144].

As becomes very clear, PA is an important signaling lipid in plants. The different triggers appear to affect different PA-biosynthetic mechanisms, each of them somewhat specific for the environmental factor that is causing the response. It would not be surprising if, just as in abiotic stresses, the PA species that is generated were very specific for other signaling paths as well. This possibility of regulating distinct responses through a single lipid class is an exciting field of study.

On a second note, responses to biotic stress contain a local and a systemic component, meaning that some aspects of the response occur directly in the infected cell while others lead to a long-distance signal that prepares other, noninfected parts of the plant for a defense against the pathogen—systemic acquired resistance (SAR). Those latter signals are often small hydrophobic compounds or simple lipids such as oxylipins that are moved through the plant vasculature. Long-distance movement of PA itself is still an untested concept.

## Systemic phospholipid signaling: a new area of lipid research in plants

Environmental and developmental responses not only occur in single affected cells but also throughout the entire organism. In animals, signals are perceived and conducted by neurons, integrated in the brain, and a response is mediated through a second set of signals that, for example trigger muscle movement. Plants do not possess a central nervous system; hence, they largely use chemical signals. These signals transverse the plant through its vasculature, or predominantly the plant phloem, that is directed through a sugar gradient/pressure flow. The signaling molecules used by plants consist of proteins, micro RNAs, mRNAs, sugars, and small lipophilic molecules. The latter are frequently solubilized or associated with amino acids or proteins. In the past, much of this research in plants has focused on the lipophilic oxylipins, predominantly jasmonic acid (JA). Genetic and grafting approaches support the hypothesis that JA, JA conjugates, or an intermediate in the JA pathway, like the oxylipin 12-oxo-phytodienoic acid (OPDA), may act as the long-distance signal in SAR to prime distal tissues following local pathogenic infection [145,146]. Moreover, OPDA was identified by LC/MS in phloem exudate, along with several other lipophilic compounds, fatty acids, triacyl- and diacylglycerols, and phospholipids, including PA [147]. These findings suggested that phospholipids, in particular PA, might serve in long-distance signaling in plants as well, a possibility that had previously been ignored.

Long-distance transport of lipids in aqueous environments like the phloem is not without precedent in nature: in animals, lipids that are bound to proteins move within the bloodstream [148,149]. Several of these mobile lipids have been shown to affect gene expression by binding to transcription factors [150,151]. While these lipid–protein mechanisms are key to mammalian development and health, their possible importance in plants is virtually unexplored. The hypothesis that phospholipids may serve as long-distance signals has been further strengthened by the identification of lipid-binding proteins in phloem exudates. It suggests a mechanism by which small transport proteins may solubilize hydrophobic ligands for translocation through the plant vascular system [152]. Many diverse proteins and small peptides have been identified in the phloem, some of which are known signaling proteins [147]. FT, ACBPs, and Defective in Induced Resistance 1 (DIR1) are examples of mobile lipid-interacting proteins involved in signaling reproductive plant development and biotic stress responses, respectively [54,78,153–155]. Phloem Lipid-Associated Family Protein (PLAFP) is

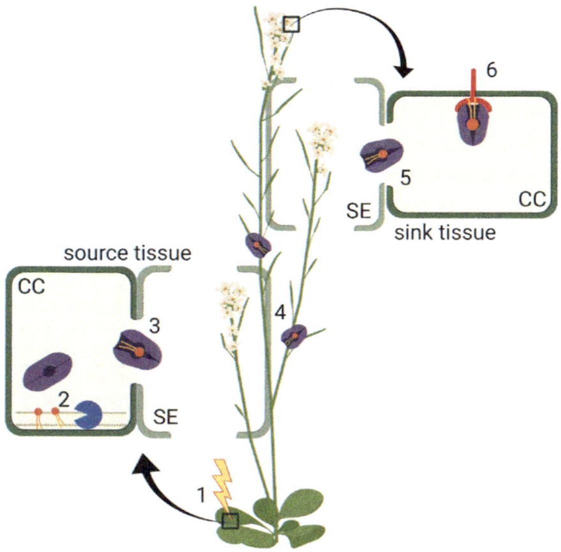


FIGURE 2.3 **Proposed model for long-distance lipid-mediated stress signaling in plants.** (1) Stress sensed in the source tissue (photosynthetically active, energy exporters—i.e., mature leaves) stimulates localized, intracellular stress response mechanisms; (2) Phospholipases generate the signal lipid, for example PA, through hydrolysis of membrane structural lipids like PC and PE, and lipid transport proteins are expressed and accumulate; (3) A lipid transport protein binds and solubilizes the lipid, and the protein-lipid complex moves through plasmodesmata from the companion cells (CC) into the sieve element (SE) of the phloem, where (4) the complex is systemically translocated to (5) sink tissues (photosynthetically inactive, energy importers—i.e., fruits, roots, and flowers). Here, the complex is either unloaded from the phloem through plasmodesmata, and/or (6) interacts with receptors that transmit the stress signal to coordinate plant stress response.

a promising addition to this mobile signal arsenal as a regulator of abiotic stress response and warrants further investigation. The small 20kd PLAT/LH2 (Polycystin-1, Lipoxygenase, Alpha-Toxin domain/Lipoxygenase homology) protein with no predicted catalytic domain was identified in Arabidopsis phloem sap [147]. Data from in vitro lipid overlay and liposome binding assays show that PLAFP binds PA specifically [156]. GUS-reporter studies indicated that the *PLAFP* promoter is active in leaf and root vascular tissue, and qPCR expression studies indicate that *PLAFP* responds to the phytohormone ABA and the drought mimic polyethylene glycol (PEG) [156]. The simultaneous production of PA and PLAFP in response to drought and the drought signal ABA and their joint presence in the phloem suggest a possible role for PLAFP-PA as a mobile signal for lipid-mediated systemic stress responses. Similarly, FT-PC and DIR1-PC or DIR1-Dehydroabietinal, all of which were found in the phloem as well, may function in the signaling of flowering and systemic acquired resistance, respectively [78,157,158].

The current proposed model being investigated is as follows: an external stress locally stimulates an increase in *PLAFP* expression while its lipid ligand PA is generated by phospholipase activity. The increased availability of PLAFP allows for complex formation with PA, solubilizing the lipid for transport out of companion cells and into sieve elements of the phloem. The PLAFP-PA complex can then move to sink tissues where it may interact with receptor proteins, perhaps in a lipid-dependent manner not unlike the Frz-Wnt mechanism [159]. The receptors can then transmit the signal to affect gene expression, and thereby plant development. Alternatively, the lipid could be taken up into the cell and interact with transcription factors similar to PPARα [150,151]. In fact, PA was recently shown to interact with the plant transcription factor WER [119]. Furthermore, PC appears to play a role in the FT-FD-DNA interaction during the regulation of flowering [160].This novel mechanism has the potential to efficiently coordinate locally sensed stimuli and intracellular signals and may open up an entirely new field of plant lipid research (Fig. 2.3).

## Conclusion

Plants contain unique lipids in their chloroplasts, galactolipids, and sulfolipids. The assembly of these lipids requires synthetic pathways that are distributed across distinct organelles, the chloroplast, and the ER, thus necessitating lipid trafficking across multiple membranes. While some of the transmembrane transporters, such as the TGD complex, are known, other aspects such as FA movement from the chloroplast to the ER as well as the identity of the mobile lipids and the mechanism of transport remain the subject of intense investigation. The central lipid in many of these lipid assembly pathways appears to be phosphatidic acid, which is also emerging as an important intracellular and long-distance signal in plants. Distinguishing the distinct functions of PA as lipid precursor and lipid signal will be key to better understanding the integration of lipid metabolism and other cellular functions and may require novel ways of tracking this lipid and its molecular species in vivo.

## Acknowledgments

Work on lipid signaling in plant development and stress response in the Hoffmann-Benning lab is supported by NSF grant 1841251 and USDA grant MICL04147 to SHB, and USDA-NIFA NNF 2015-38420-23697 to AK. Work on lipid biosynthesis, trafficking, and function in the Benning lab is supported by the Division of Chemical Sciences, Geosciences and Biosciences, Office of Basic Energy Sciences of the United States Department of Energy Grant DE-FG02-91ER20021 and MSU AgBioResearch. Figures were created with BioRender (Biorender.com).

## References

[1] Munnik T, Testerink C. Plant phospholipid signaling: "in a nutshell". J Lipid Res 2009;50(Suppl):S260–5.
[2] Balla T. Putting G protein-coupled receptor-mediated activation of phospholipase C in the limelight. J Gen Physiol 2010;135(2):77–80.
[3] Ischebeck T, et al. Phosphatidylinositol 4,5-bisphosphate influences PIN polarization by controlling clathrin-mediated membrane trafficking in Arabidopsis. Plant Cell 2013;25(12):4894–911.
[4] Lin F, et al. A dual role for cell plate-associated PI4Kbeta in endocytosis and phragmoplast dynamics during plant somatic cytokinesis. EMBO J 2019;38(4).
[5] Heilmann I. Phosphoinositide signaling in plant development. Development 2016;143(12):2044–55.
[6] Barbaglia AM, Hoffmann-Benning S. Long-distance lipid signaling and its role in plant development and stress response. Subcell Biochem 2016;86:339–61.
[7] Gillaspy GE. The role of phosphoinositides and inositol phosphates in plant cell signaling. Adv Exp Med Biol 2013;991:141–57.
[8] Ohlrogge J, Browse J. Lipid biosynthesis. Plant Cell 1995;7(7):957–70.

[9] Koo AJ, Ohlrogge JB, Pollard M. On the export of fatty acids from the chloroplast. J Biol Chem 2004;279(16):16101–10.
[10] Li N, et al. FAX1, a novel membrane protein mediating plastid fatty acid export. PLoS Biol 2015;13(2):e1002053.
[11] Heinz E, Roughan PG. Similarities and differences in lipid metabolism of chloroplasts isolated from 18:3 and 16:3 plants. Plant Physiol 1983;72(2):273–9.
[12] Frentzen M, et al. Specificities and selectivities of glycerol-3-phosphate acyltransferase and monoacylglycerol-3-phosphate acyltransferase from pea and spinach chloroplasts. Eur J Biochem 1983;129(3):629–36.
[13] Lavell AA, Benning C. Cellular organization and regulation of plant glycerolipid metabolism. Plant Cell Physiol 2019;60(6):1176–83.
[14] Andersson MX, Goksor M, Sandelius AS. Optical manipulation reveals strong attracting forces at membrane contact sites between endoplasmic reticulum and chloroplasts. J Biol Chem 2007;282(2):1170–4.
[15] Xu C, et al. A permease-like protein involved in ER to thylakoid lipid transfer in Arabidopsis. EMBO J 2003;22(10):2370–9.
[16] Xu C, et al. Mutation of the TGD1 chloroplast envelope protein affects phosphatidate metabolism in Arabidopsis. Plant Cell 2005;17(11):3094–110.
[17] Roston RL, et al. TGD1, -2, and -3 proteins involved in lipid trafficking form ATP-binding cassette (ABC) transporter with multiple substrate-binding proteins. J Biol Chem 2012;287(25):21406–15.
[18] Awai K, et al. A phosphatidic acid-binding protein of the chloroplast inner envelope membrane involved in lipid trafficking. Proc Natl Acad Sci U S A 2006;103(28):10817–22.
[19] Lu B, et al. A small ATPase protein of Arabidopsis, TGD3, involved in chloroplast lipid import. J Biol Chem 2007;282(49):35945–53.
[20] Xu C, et al. Lipid transport mediated by Arabidopsis TGD proteins is unidirectional from the endoplasmic reticulum to the plastid. Plant Cell Physiol 2010;51(6):1019–28.
[21] Roston R, et al. Arabidopsis chloroplast lipid transport protein TGD2 disrupts membranes and is part of a large complex. Plant J 2011;66(5):759–69.
[22] Fan J, et al. Arabidopsis TRIGALACTOSYLDIACYLGLYCEROL5 interacts with TGD1, TGD2, and TGD4 to facilitate lipid transfer from the endoplasmic reticulum to plastids. Plant Cell 2015;27(10):2941–55.
[23] Lu B, Benning C. A 25-amino acid sequence of the Arabidopsis TGD2 protein is sufficient for specific binding of phosphatidic acid. J Biol Chem 2009;284(26):17420–7.
[24] Wang Z, Anderson NS, Benning C. The phosphatidic acid binding site of the Arabidopsis trigalactosyldiacylglycerol 4 (TGD4) protein required for lipid import into chloroplasts. J Biol Chem 2013;288(7):4763–71.
[25] Nakamura Y, et al. Arabidopsis lipins mediate eukaryotic pathway of lipid metabolism and cope critically with phosphate starvation. Proc Natl Acad Sci U S A 2009;106(49):20978–83.
[26] Eastmond PJ, et al. Phosphatidic acid phosphohydrolase 1 and 2 regulate phospholipid synthesis at the endoplasmic reticulum in Arabidopsis. Plant Cell 2010;22(8):2796–811.
[27] Nakamura Y, Tsuchiya M, Ohta H. Plastidic phosphatidic acid phosphatases identified in a distinct subfamily of lipid phosphate phosphatases with prokaryotic origin. J Biol Chem 2007;282(39):29013–21.
[28] Block MA, et al. Preparation and characterization of membrane fractions enriched in outer and inner envelope membranes from spinach chloroplasts. J Biol Chem 1983;258(21):13281–6.
[29] Malherbe A, et al. Feedback inhibition of phosphatidate phosphatase from spinach chloroplast envelope membranes by diacylglycerol. J Biol Chem 1992;267(33):23546–53.
[30] Miege C, et al. Biochemical and topological properties of type A MGDG synthase, a spinach chloroplast envelope enzyme catalyzing the synthesis of both prokaryotic and eukaryotic MGDG. Eur J Biochem 1999;265(3):990–1001.
[31] Lavell A, et al. A predicted plastid rhomboid protease affects phosphatidic acid metabolism in *Arabidopsis thaliana*. Plant J 2019;99(5):978–87.
[32] Dubots E, et al. Activation of the chloroplast monogalactosyldiacylglycerol synthase MGD1 by phosphatidic acid and phosphatidylglycerol. J Biol Chem 2010;285(9):6003–11.
[33] Lucas WJ, et al. The plant vascular system: evolution, development and functions. J Integr Plant Biol 2013;55(4):294–388.
[34] Eckardt NA. Oxylipin signaling in plant stress responses. Plant Cell 2008;20(3):495–7.
[35] Wasternack C, Feussner I. The oxylipin pathways: biochemistry and function. Annu Rev Plant Biol 2018;69:363–86.
[36] Howe GA, Major IT, Koo AJ. Modularity in jasmonate signaling for multistress resilience. Annu Rev Plant Biol 2018;69:387–415.

[37] Huby E, et al. Sphingolipids: towards an integrated view of metabolism during the plant stress response. New Phytol 2019;225(2):659–70.

[38] Dong, W, et al., Does diacylglycerol serve as a signaling molecule in plants? Plant Signal Behav 2012; 472–5.

[39] Xue HW, Chen X, Mei Y. Function and regulation of phospholipid signalling in plants. Biochem J 2009;421(2):145–56.

[40] Hong Y, Zhang W, Wang X. Phospholipase D and phosphatidic acid signalling in plant response to drought and salinity. Plant Cell Environ 2010;33(4):627–35.

[41] Testerink C, Munnik T. Molecular, cellular, and physiological responses to phosphatidic acid formation in plants. J Exp Bot 2011;62(7):2349–61.

[42] Julkowska MM, Testerink C. Tuning plant signaling and growth to survive salt. Trends Plant Sci 2015;20(9):586–94.

[43] Bargmann BO, Munnik T. The role of phospholipase D in plant stress responses. Curr Opin Plant Biol 2006;9(5):515–22.

[44] Yao HY, Xue HW. Phosphatidic acid plays key roles regulating plant development and stress responses. J Integr Plant Biol 2018;60(9):851–63.

[45] Wang X, et al. Phosphatidic acid as lipid messenger and growth regulators in plants,. In: Wang X, editor. Phospholipases in plant signaling,. Springer, Berlin, Heidelberg: Signaling and Communication in Plants; 2014. p. 69–92.

[46] Arisz SA, et al. Rapid phosphatidic acid accumulation in response to low temperature stress in Arabidopsis is generated through diacylglycerol kinase. Front Plant Sci 2013;4(1):1–15.

[47] Uraji M, et al. Cooperative function of PLDdelta and PLDalpha1 in abscisic acid-induced stomatal closure in Arabidopsis. Plant Physiol 2012;159(1):450–60.

[48] Hong Y, et al. Plant phospholipases D and C and their diverse functions in stress responses. Prog Lipid Res 2016;62:55–74.

[49] Doermann P, et al. Isolation and characterization of an Arabidopsis mutant deficient in the thylakoid lipid digalactosyl diacylglycerol. Plant Cell 1995;7(11):1801–10.

[50] Thole JM, Nielsen E. Phosphoinositides in plants: novel functions in membrane trafficking. Curr Opin Plant Biol 2008;11(6):620–31.

[51] Hong Y, et al. Phospholipase D epsilon and phosphatidic acid enhance Arabidopsis nitrogen signaling and growth. Plant J 2009;58(3):376–87.

[52] Wang YS, Yao HY, Xue HW. Lipidomic profiling analysis reveals the dynamics of phospholipid molecules in *Arabidopsis thaliana* seedling growth. J Integr Plant Biol 2016;58(11):890–902.

[53] Zhang W, et al. Phospholipase D alpha 1-derived phosphatidic acid interacts with ABI1 phosphatase 2C and regulates abscisic acid signaling. Proc Natl Acad Sci U S A 2004;101(25):9508–13.

[54] Du ZY, et al. Arabidopsis acyl-CoA-binding protein ACBP1 participates in the regulation of seed germination and seedling development. Plant J 2013;74(2):294–309.

[55] Gao HB, Chu YJ, Xue HW. Phosphatidic acid (PA) binds PP2AA1 to regulate PP2A activity and PIN1 polar localization. Mol Plant 2013;6(5):1692–702.

[56] Wang P, et al. Phosphatidic acid directly regulates PINOID-dependent phosphorylation and activation of the PIN-FORMED2 auxin efflux transporter in response to salt stress. Plant Cell 2019;31(1):250–71.

[57] Zhang Q, et al. Knock-down of Arabidopsis PLC5 reduces primary root growth and secondary root formation while overexpression improves drought tolerance and causes stunted root hair growth. Plant Cell Physiol 2018;59(10):2004–19.

[58] Yuan S, et al. Diacylglycerol kinase and associated lipid mediators modulate rice root architecture. New Phytol 2019;223(1):261–76.

[59] Motes CM, et al. Differential effects of two phospholipase D inhibitors, 1-butanol and N-acylethanolamine, on in vivo cytoskeletal organization and Arabidopsis seedling growth. Protoplasma 2005;226(3–4):109–23.

[60] Lee S, Park J, Lee Y. Phosphatidic acid induces actin polymerization by activating protein kinases in soybean cells. Mol Cells 2003;15(3):313–9.

[61] Apostolakos P, Panteris E, Galatis B. The involvement of phospholipases C and D in the asymmetric division of subsidiary cell mother cells of *Zea mays*. Cell Motil Cytoskeleton 2008;65(11):863–75.

[62] Pleskot R, et al. Turnover of phosphatidic acid through distinct signaling pathways affects multiple aspects of pollen tube growth in tobacco. Front Plant Sci 2012;3(54):1–13.

[63] Heilmann M, Heilmann I. Plant phosphoinositides-complex networks controlling growth and adaptation. Biochim Biophys Acta 2015;1851(6):759–69.

[64] Huang S, et al. Heterodimeric capping protein from Arabidopsis is regulated by phosphatidic acid. Mol Biol Cell 2006;17(4):1946–58.
[65] Li J, et al. Capping protein modulates the dynamic behavior of actin filaments in response to phosphatidic acid in Arabidopsis. Plant Cell 2012;24(9):3742–54.
[66] Ha KS, Exton JH. Activation of actin polymerization by phosphatidic acid derived from phosphatidylcholine in IIC9 fibroblasts. J Cell Biol 1993;123(6 Pt 2):1789–96.
[67] Potocký M, et al. Phosphatidic acid produced by phospholipase D is required for tobacco pollen tube growth. Planta 2003;217(1):122–30.
[68] Pleskot R, et al. Mutual regulation of plant phospholipase D and the actin cytoskeletonS the actin F cytoskeleton. Plant J 2010;62(3):494–507.
[69] Gutierrez R, et al. Arabidopsis cortical microtubules position cellulose synthase delivery to the plasma membrane and interact with cellulose synthase trafficking compartments. Nat Cell Biol 2009;11(7):797–806.
[70] Paredez AR, Somerville CR, Ehrhardt DW. Visualization of cellulose synthase demonstrates functional association with microtubules. Science 2006;312(5779):1491–5.
[71] Gardiner J, et al. The effects of the phospholipase D-antagonist 1-butanol on seedling development and microtubule organisation in Arabidopsis. Plant Cell Physiol 2003;44(7):687–96.
[72] Gardiner JC, et al. A 90-kD phospholipase D from tobacco binds to microtubules and the plasma membrane. Plant Cell 2001;13(9):2143–58.
[73] Dhonukshe P, Gadella TW Jr. Alteration of microtubule dynamic instability during preprophase band formation revealed by yellow fluorescent protein-CLIP170 microtubule plus-end labeling. Plant Cell 2003;15(3):597–611.
[74] Hirase A, et al. n-Butanol induces depolymerization of microtubules in vivo and in vitro. Plant Cell Physiol 2006;47(7):1004–9.
[75] Zhang Q, et al. Phosphatidic acid regulates microtubule organization by interacting with MAP65-1 in response to salt stress in Arabidopsis. Plant Cell 2012;24(11):4555–76.
[76] Ekman A, Bulow L, Stymne S. Elevated atmospheric CO(2) concentration and diurnal cycle induce changes in lipid composition in *Arabidopsis thaliana*. New Phytol 2007;174(3):591–9.
[77] Maatta S, et al. Levels of *Arabidopsis thaliana* leaf phosphatidic acids, phosphatidylserines, and most trienoate-containing polar lipid molecular species increase during the dark period of the diurnal cycle. Front Plant Sci 2012;3:49.
[78] Nakamura Y, et al. Arabidopsis florigen FT binds to diurnally oscillating phospholipids that accelerate flowering. Nat Commun 2014;5:3553.
[79] Browse J, Roughan PG, Slack CR. Light control of fatty acid synthesis and diurnal fluctuations of fatty acid composition in leaves. Biochem J 1981;196(1):347–54.
[80] Schaffer R, et al. The late elongated hypocotyl mutation of Arabidopsis disrupts circadian rhythms and the photoperiodic control of flowering. Cell 1998;93(7):1219–29.
[81] Green RM, Tobin EM. Loss of the circadian clock-associated protein 1 in Arabidopsis results in altered clock-regulated gene expression. Proc Natl Acad Sci U S A 1999;96(7):4176–9.
[82] Kim SC, et al. Interaction and regulation between lipid mediator phosphatidic acid and circadian clock regulators. Plant Cell 2019;31(2):399–416.
[83] Nakamura Y, Ohta H. The diacylglycerol forming pathways differ among floral organs of Petunia hybrida. FEBS Lett 2007;581(28):5475–9.
[84] Yunus IS, et al. Phosphatidic acid is a major phospholipid class in reproductive organs of *Arabidopsis thaliana*. Plant Signal Behav 2015;10(8):e1049790 1–4.
[85] Scandola S, Samuel MA. A flower-specific phospholipase D is a stigmatic compatibility factor targeted by the self-incompatibility response in *Brassica napus*. Curr Biol 2019;29(3). 506-512.e4.
[86] Heilmann I, Ischebeck T. Male functions and malfunctions: the impact of phosphoinositides on pollen development and pollen tube growth. Plant Reprod 2016;29(1–2):3–20.
[87] Dubois M, Van den Broeck L, Inzé D. The pivotal role of ethylene in plant growth. Trends Plant Sci 2018;23(4):311–23.
[88] Wang KL, Li H, Ecker JR. Ethylene biosynthesis and signaling networks. Plant Cell 2002;14(Suppl):S131–51.
[89] Zheng Y, Zhu Z. Relaying the ethylene signal: new roles for EIN2. Trends Plant Sci 2016;21(1):2–4.
[90] Testerink C, et al. Phosphatidic acid binds to and inhibits the activity of Arabidopsis CTR1. J Exp Bot 2007;58(14):3905–14.

[91] Testerink C, et al. PA, a stress-induced short cut to switch-on ethylene signalling by switching-off CTR1? Plant Signal Behav. 2008;681–3.
[92] Fan L, Zheng S, Wang X. Antisense suppression of phospholipase D alpha retards abscisic acid- and ethylene-promoted senescence of postharvest Arabidopsis leaves. Plant Cell 1997;9(12):2183–96.
[93] Jia Y, Li W. Characterisation of lipid changes in ethylene-promoted senescence and its retardation by suppression of phospholipase Dδ in Arabidopsis leaves. Front Plant Sci 2015;6(1045):1–6.
[94] Ji T, et al. Overexpression of cucumber phospholipase D alpha Gene (CsPLDalpha) in tobacco enhanced salinity stress tolerance by regulating Na(+)-K(+) balance and lipid peroxidation. Front Plant Sci. 2017;8:499.
[95] Yu L, et al. Phosphatidic acid mediates salt stress response by regulation of MPK6 in *Arabidopsis thaliana*. New Phytol 2010;188(3):762–73.
[96] Yang Y, Guo Y. Unraveling salt stress signaling in plants. J Integr Plant Biol 2018;60(9):796–804.
[97] Weller B, et al. Dynamic PIN-FORMED auxin efflux carrier phosphorylation at the plasma membrane controls auxin efflux-dependent growth. Proc Natl Acad Sci U S A 2017;114(5):E887–96.
[98] Barbosa ICR, Hammes UZ, Schwechheimer C. Activation and polarity control of PIN-FORMED auxin transporters by phosphorylation. Trends Plant Sci 2018;23(6):523–38.
[99] Munnik T, et al. Hyperosmotic stress stimulates phospholipase D activity and elevates the levels of phosphatidic acid and diacylglycerol pyrophosphate. Plant J 2000;22(2):147–54.
[100] Bargmann BO, et al. Multiple PLDs required for high salinity and water deficit tolerance in plants. Plant Cell Physiol 2009;50(1):78–89.
[101] Guo L, et al. Cytosolic glyceraldehyde-3-phosphate dehydrogenases interact with phospholipase ddelta to transduce hydrogen peroxide signals in the Arabidopsis response to stress. Plant Cell 2012;24(5): 2200–12.
[102] McLoughlin F, et al. Identification of novel candidate phosphatidic acid-binding proteins involved in the salt-stress response of *Arabidopsis thaliana* roots. Biochem J 2013;450(3):573–81.
[103] Thomashow MF. Role of cold-responsive genes in plant freezing tolerance. Plant Physiol 1998;118(1):1–8.
[104] Welti R, et al. Profiling membrane lipids in plant stress responses. Role of phospholipase D alpha in freezing-induced lipid changes in Arabidopsis. J Biol Chem 2002;277(35):31994–2002.
[105] Moellering ER, Muthan B, Benning C. Freezing tolerance in plants requires lipid remodeling at the outer chloroplast membrane. Science 2010;330(6001):226–8.
[106] Roston RL, et al. Structural determinants allowing transferase activity in SENSITIVE TO FREEZING 2, classified as a family I glycosyl hydrolase. J Biol Chem 2014;289(38):26089–106.
[107] Barnes AC, Benning C, Roston RL. Chloroplast membrane remodeling during freezing stress is accompanied by cytoplasmic acidification activating SENSITIVE TO FREEZING2. Plant Physiol 2016;171(3):2140–9.
[108] Rajashekar CB, et al. Suppression of phospholipase Dalpha1 induces freezing tolerance in Arabidopsis: response of cold-responsive genes and osmolyte accumulation. J Plant Physiol 2006;163(9):916–26.
[109] Kooijman EE, et al. Modulation of membrane curvature by phosphatidic acid and lysophosphatidic acid. Traffic 2003;4(3):162–74.
[110] Putta P, et al. Phosphatidic acid binding proteins display differential binding as a function of membrane curvature stress and chemical properties. Biochim Biophys Acta 2016;1858(11):2709–16.
[111] Fujita Y, et al. ABA-mediated transcriptional regulation in response to osmotic stress in plants. J Plant Res 2011;124(4):509–25.
[112] Mishra G, et al. A bifurcating pathway directs abscisic acid effects on stomatal closure and opening in Arabidopsis. Science 2006;312(5771):264–6.
[113] Zhang Y, et al. Phospholipase dalpha1 and phosphatidic acid regulate NADPH oxidase activity and production of reactive oxygen species in ABA-mediated stomatal closure in Arabidopsis. Plant Cell 2009;21(8):2357–77.
[114] Distéfano AM, et al. Nitric oxide-induced phosphatidic acid accumulation: a role for phospholipases C and D in stomatal closure. Plant Cell Environ 2008;31(2):187–94.
[115] Distéfano AM, et al. Phospholipase Ddelta is involved in nitric oxide-induced stomatal closure. Planta 2012;236(6):1899–907.
[116] Distéfano AM, et al. Phospholipase D delta knock-out mutants are tolerant to severe drought stress. Plant Signal Behav 2015;10(11):e1089371.
[117] Takemiya A, Shimazaki K. Phosphatidic acid inhibits blue light-induced stomatal opening via inhibition of protein phosphatase 1. Plant Physiol 2010;153(4):1555–62.
[118] Guo L, et al. Phosphatidic acid binds and stimulates Arabidopsis sphingosine kinases. J Biol Chem 2011;286(15):13336–45.

[119] Yao H, et al. Phosphatidic acid interacts with a MYB transcription factor and regulates its nuclear localization and function in Arabidopsis. Plant Cell 2013;25(12):5030–42.

[120] Fan L, et al. Subcellular distribution and tissue expression of phospholipase Dalpha, Dbeta, and Dgamma in Arabidopsis. Plant Physiol 1999;119(4):1371–8.

[121] Qin C, Wang C, Wang X. Kinetic analysis of Arabidopsis phospholipase Ddelta. Substrate preference and mechanism of activation by Ca2+ and phosphatidylinositol 4,5-biphosphate. J Biol Chem 2002;277(51):49685–90.

[122] Li M, Hong Y, Wang X. Phospholipase D- and phosphatidic acid-mediated signaling in plants. Biochim Biophys Acta 2009;1791(9):927–35.

[123] van der Luit AH, et al. Elicitation of suspension-cultured tomato cells triggers the formation of phosphatidic acid and diacylglycerol pyrophosphate. Plant Physiol 2000;123(4):1507–16.

[124] den Hartog M, Verhoef N, Munnik T. Nod factor and elicitors activate different phospholipid signaling pathways in suspension-cultured alfalfa cells. Plant Physiol 2003;132(1):311–7.

[125] Raho N, et al. Phosphatidic acid production in chitosan-elicited tomato cells, via both phospholipase D and phospholipase C/diacylglycerol kinase, requires nitric oxide. J Plant Physiol 2011;168(6):534–9.

[126] Yamaguchi T, et al. Elicitor-induced activation of phospholipases plays an important role for the induction of defense responses in suspension-cultured rice cells. Plant Cell Physiol 2005;46(4):579–87.

[127] D'Ambrosio JM, et al. The sesquiterpene botrydial from *Botrytis cinerea* induces phosphatidic acid production in tomato cell suspensions. Planta 2018;247(4):1001–9.

[128] Cacas JL, et al. Diacylglycerol kinases activate tobacco NADPH oxidase-dependent oxidative burst in response to cryptogein. Plant Cell Environ 2017;40(4):585–98.

[129] Andersson MX, et al. Phospholipase-dependent signalling during the AvrRpm1- and AvrRpt2-induced disease resistance responses in *Arabidopsis thaliana*. Plant J 2006;47(6):947–59.

[130] Laxalt AM, et al. Nitric oxide is critical for inducing phosphatidic acid accumulation in xylanase-elicited tomato cells. J Biol Chem 2007;282(29):21160–8.

[131] Gonorazky G, et al. Silencing of the tomato phosphatidylinositol-phospholipase C2 (SlPLC2) reduces plant susceptibility to *Botrytis cinerea*. Mol Plant Pathol 2016;17(9):1354–63.

[132] Zhao J, et al. Arabidopsis phospholipase Dbeta1 modulates defense responses to bacterial and fungal pathogens. New Phytol 2013;199(1):228–40.

[133] Li J, Wang X. Phospholipase D and phosphatidic acid in plant immunity. Plant Sci 2019;279:45–50.

[134] Zhang Q, Xiao S. Lipids in salicylic acid-mediated defense in plants: focusing on the roles of phosphatidic acid and phosphatidylinositol 4-phosphate. Front Plant Sci 2015;6:387.

[135] Anthony RG, et al. The Arabidopsis protein kinase PTI1-2 Is activated by convergent phosphatidic acid and oxidative stress signaling pathways downstream of PDK1 and OXI1. J Biol Chem 2006;281(49):37536–46.

[136] Han S, et al. Cytoplastic glyceraldehyde-3-phosphate dehydrogenases interact with ATG3 to negatively regulate autophagy and immunity in *Nicotiana benthamiana*. Plant Cell 2015;27(4):1316–31.

[137] Gou JY, et al. Wheat stripe rust resistance protein WKS1 reduces the ability of the thylakoid-associated ascorbate peroxidase to detoxify reactive oxygen species. Plant Cell 2015;27(6):1755–70.

[138] Kadota Y, et al. Direct regulation of the NADPH oxidase RBOHD by the PRR-associated kinase BIK1 during plant immunity. Mol Cell 2014;54(1):43–55.

[139] Janda M, et al. Phospholipase D affects translocation of NPR1 to the nucleus in Arabidopsis thaliana. Front Plant Sci 2015;6(59):1–11.

[140] Sagaram US, et al. Structural and functional studies of a phosphatidic acid-binding antifungal plant defensin MtDef4: identification of an RGFRRR motif governing fungal cell entry. PLoS One 2013;8(12):e82485.

[141] Mou Z, Fan W, Dong X. Inducers of plant systemic acquired resistance regulate NPR1 function through redox changes. Cell 2003;113(7):935–44.

[142] Lee HJ, et al. Systemic immunity requires SnRK2.8-mediated nuclear import of NPR1 in Arabidopsis. Plant Cell 2015;27(12):3425–38.

[143] Kalachova T, et al. Involvement of phospholipase D and NADPH-oxidase in salicylic acid signaling cascade. Plant Physiol Biochem 2013;66:127–33.

[144] El-Mounadi K, et al. Antifungal mechanisms of a plant defensin MtDef4 are not conserved between the ascomycete fungi *Neurospora crassa* and *Fusarium graminearum*. Mol Microbiol 2016;100(3):542–59.

[145] Strassner J, et al. Characterization and cDNA-microarray expression analysis of 12-oxophytodienoate reductases reveals differential roles for octadecanoid biosynthesis in the local versus the systemic wound response. Plant J 2002;32(4):585–601.

[146] Li L, et al. Distinct roles for jasmonate synthesis and action in the systemic wound response of tomato. Proc Natl Acad Sci U S A 2002;6416–21.
[147] Guelette BS, Benning UF, Hoffmann-Benning S. Identification of lipids and lipid-binding proteins in phloem exudates from *Arabidopsis thaliana*. J Exp Bot 2012;63(10):3603–16.
[148] Charbonneau D, Beauregard M, Tajmir-Riahi HA. Structural analysis of human serum albumin complexes with cationic lipids. J Phys Chem B 2009;113(6):1777–84.
[149] Blaner WS. Retinol-binding protein: the serum transport protein for vitamin A. Endocr Rev 1989;10(3):308–16.
[150] Musille PM, Kohn JA, Ortlund EA. Phospholipid-driven gene regulation. FEBS Lett 2013;587(8):1238–46.
[151] Wahli W, Michalik L. PPARs at the crossroads of lipid signaling and inflammation. Trends Endocrinol Metab 2012;23(7):351–63.
[152] Benning U, et al. New aspects of phloem-mediated long-distance lipid signaling in plants. Front Plant Sci 2012;3:53.
[153] Andres F, Coupland G. The genetic basis of flowering responses to seasonal cues. Nat Rev Genet 2012;13(9):627–39.
[154] Maldonado AM, et al. A putative lipid transfer protein involved in systemic resistance signalling in Arabidopsis. Nature 2002;419(6905):399–403.
[155] Champigny MJ, et al. Long distance movement of DIR1 and investigation of the role of DIR1-like during systemic acquired resistance in Arabidopsis. Front Plant Sci 2013;4:230.
[156] Barbaglia AM, et al. Phloem proteomics reveals new lipid-binding proteins with a putative role in lipid-mediated signaling. Front Plant Sci 2016;7:563.
[157] Lascombe MB, et al. The structure of "defective in induced resistance" protein of *Arabidopsis thaliana*, DIR1, reveals a new type of lipid transfer protein. Protein Sci 2008;17(9):1522–30.
[158] Shah J, et al. Signaling by small metabolites in systemic acquired resistance. Plant J 2014;79(4):645–58.
[159] Janda CY, et al. Structural basis of Wnt recognition by Frizzled. Science 2012;337(6090):59–64.
[160] Nakamura Y, et al. High-resolution crystal structure of Arabidopsis FLOWERING LOCUS T illuminates its phospholipid-binding site in flowering.. iScience 2019;21:577–86.

CHAPTER

# 3

# Sex as a modulator of lipid metabolism and metabolic disease

*Laurent Vergnes*[a], *Carrie Riestenberg*[a], *Karen Reue*[a,b]

[a]Department of Human Genetics, David Geffen School of Medicine at UCLA, Los Angeles, CA, United States; [b]Molecular Biology Institute, University of California, Los Angeles, CA, United States

OUTLINE



## Rationale for the study of sex differences in lipid metabolism

Men and women differ in many aspects of lipid metabolism and in the development of related conditions such as dyslipidemia, obesity, and cardiovascular diseases. While researchers and physicians have been aware of such sex differences for decades, an understanding of the underlying mechanisms has lagged behind. Some sex differences in metabolism are

Lipid Signaling and Metabolism. http://dx.doi.org/10.1016/B978-0-12-819404-1.00003-8

linked to the fundamental role of females in childbearing. For example, females are equipped with mechanisms that promote adipose tissue storage, with increased storage in subcutaneous depots; conversely, males tend to be more efficient at mobilization of stored fat [1–3]. In addition to differences in fat storage, many other aspects of lipid metabolism exhibit sex differences. It is critical to better understand the basis for sex differences in lipid metabolism in order to optimize the diagnosis and treatment of cardiometabolic disease in both sexes. Fortunately, recent increased emphasis on the inclusion of "sex as a biological variable" in basic and clinical research studies [4–7] has begun to inspire greater focus on how components of sex influence lipid metabolism and related disease processes. Here, we briefly describe components of sex and how they are studied, and then review some recent findings regarding sex differences in lipid physiology and related diseases.

## Biological sex versus gender

In considering the effects of sex on lipid metabolism, it is useful to define the components of sex. Biological sex in mammals is determined by the inheritance of XX or XY chromosomes at fertilization, which subsequently determines the development of ovaries or testes, and the classification of an individual as female or male [8,9]. By contrast, gender refers to social or cultural designations of feminine or masculine behavior or roles, and may not correspond to an individuals' biological sex (Fig. 3.1). The studies referred to in the following sections involved humans or experimental models that were grouped into categories based strictly on biological sex, and the effects observed are most likely due to factors such as gonadal hormones and/or sex chromosome effects. However, it is possible that some of the traits discussed could be impacted by gender as well. A good example is food intake in humans, which is likely influenced by both biologically determined sex differences, and social expectations that impact women differently than men. However, gender effects on lipid metabolism have not been well documented in humans, and cannot be approached in standard experimental animal models. Thus, due to a lack of robust studies on the impact of gender on lipid metabolism, the studies discussed here are based on the classification of subjects by biological sex.

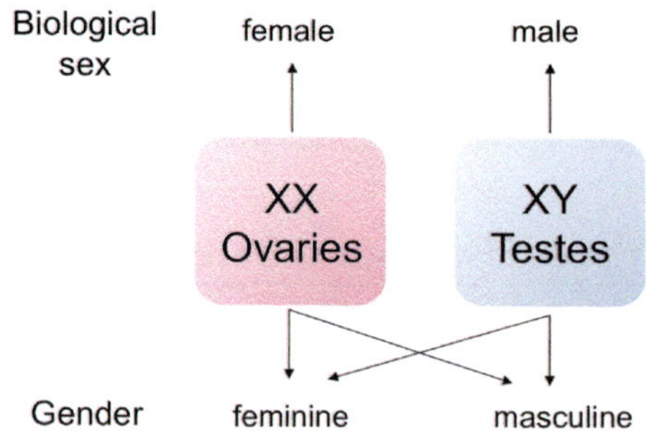


FIGURE 3.1 **Biological sex and gender.** The XX or XY sex chromosome complement inherited at conception specifies the development of ovaries or testes, that is, biological sex. Gender is not equivalent to biological sex, but refers to the sociological norms that are associated with feminine or masculine identity or behavior.

## Components of biological sex—gonadal hormones and sex chromosomes

Biological sex includes a genetic component (XX vs. XY sex chromosomes) and a hormonal component (secretions of the ovaries or testes). These two components may act together or independently to influence lipid metabolism. It is well appreciated that gonadal hormones, particularly estrogens and androgens, influence many aspects of lipid homeostasis. The contribution of gonadal hormones to a particular trait is often addressed in human studies by comparing women at pre- and postmenopausal ages. The reduction in levels of estrogens and other gonadal hormones that occurs following menopause is associated with increased fat storage in visceral adipose depots, and increased risk for hyperlipidemia, insulin resistance, hypertension, and cardiovascular disease [10–13]. Gonadal hormones act primarily through estrogen and androgen receptors. Global ablation of estrogen receptor α in the mouse results in disturbed lipoprotein profile and increased adiposity in both male and female mice [14,15], and ablation specifically in hypothalamic neurons leads to dysregulation of energy balance, and an accumulation of abdominal fat [16,17]. Thus, gonadal hormones are important regulators of lipid metabolism in the brain and in peripheral tissues.

Beyond the well-established roles of gonadal hormones in metabolic homeostasis, sex chromosome complement may independently influence lipid metabolism. Prior to the development of ovaries or testes in an embryo, differences between XX and XY individuals are apparent. For example, in both humans and mice, male fetuses are larger than female fetuses prior to gonadal differentiation [18,19]. At birth, male babies typically weigh more than female babies [20]. Furthermore, male/female differences persist in childhood prior to puberty, including differences in weight, height, and total body lean and fat mass [21]. The important role of the sex chromosome complement in metabolic homeostasis and other processes is also highlighted by abnormalities observed in individuals with altered sex chromosome copy number, the most common being XXY (Klinefelter syndrome) and XO (Turner syndrome). Even as children, Klinefelter individuals have increased risk for abdominal obesity; the occurrence prior to puberty suggests that this is likely attributable to the sex chromosome copy number rather than to abnormal gonadal hormone levels [22,23]. As adults, XXY men have higher incidence than XY men of the Metabolic Syndrome (defined as the presence of at least three of the following risk factors: abdominal obesity, reduced high-density lipoprotein levels, elevated triglyceride levels, elevated fasting glucose levels, and hypertension) [24]. This may relate to both sex chromosome complement and altered gonadal hormone levels. Turner syndrome women have several developmental abnormalities (short stature, gonadal dysgenesis, congenital heart defects). Both pediatric and adult XO females have increased risk for impaired lipid homeostasis, which may include elevated total and low-density lipoprotein cholesterol and triglyceride levels, and diminished high-density lipoprotein levels [25]. The mechanisms responsible for increased prevalence of dyslipidemia in XO individuals are not known, but occurrence prior to puberty suggests that abnormal X-chromosome gene dosage could be a primary factor.

The assessment of sex chromosome effects on lipid metabolism is challenging in humans due to the limited size of research cohorts with altered sex chromosome numbers, as well as confounding effects of abnormal gonadal hormone levels in these individuals. Studies in mouse models overcome these limitations and have provided further evidence for primary effects of sex chromosome complement on lipid metabolism. In particular, studies with the

Four Core Genotypes mouse model have revealed independent actions of the two components of biological sex—gonadal type and sex chromosome complement—on traits such as plasma lipid levels, obesity, and atherosclerosis. This model allows the decoupling of gonadal sex determination from the sex chromosomes by relocating the *Sry* gene (required to initiate testes development) from the Y chromosome to mouse chromosome 3 [9,26]. Rather than just two sexes (XX/ovaries or XY/testes), this model has four combinations (XX/ovaries, XX/testes, XY/ovaries, and XY/testes) (Fig. 3.2A). The analysis of these four mouse genotypes can reveal whether gonadal type, chromosome complement, or both play a role in determining a trait of interest (Fig. 3.2B). Studies with the Four Core Genotypes model have provided evidence for important roles of sex chromosomes, in addition to gonadal hormones, in lipid homeostasis and disease as discussed in subsequent sections. A proposed mechanism by which the dose of the X chromosome (one copy in XY vs. two copies in XX, for example) may influence physiological traits is through continued expression of ~10%–20% of the genes on the inactivated X chromosome in XX cells [27]. Escape from X-inactivation occurs in both humans and mice, and leads to higher expression levels for these genes in tissues of females compared to males. Genes that escape X-inactivation include genes that encode proteins that regulate gene transcription and protein translation, and could therefore influence a network of other genes on the autosomes that regulate metabolism.

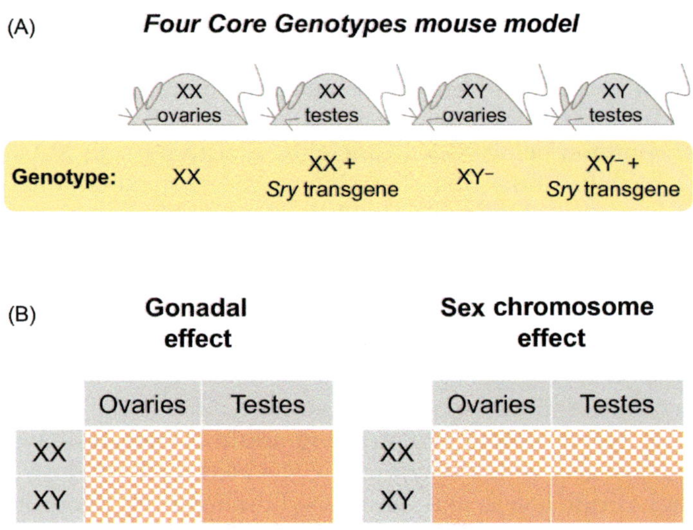


**FIGURE 3.2 Analysis of gonadal and sex chromosome contributions to lipid metabolism.** (A) The Four Core Genotypes mouse model includes four genotypes that separate sex chromosome complement from gonadal type to determine the contribution of each to traits of interest. (B) Characterization of the Four Core Genotypes mouse model for a trait of interest can reveal whether observed sex differences are influenced more by gonadal type (left) or by sex chromosome type (right). Gonadal effects will be evident if XX and XY animals with ovaries are similar, and they differ from XX and XY animals with testes. On the other hand, sex chromosome effects lead to similar phenotypes in mice with XX chromosomes regardless of presence of ovaries or testes, and differ compared to mice with XY chromosomes and ovaries or testes. Not shown here, gonadal and chromosomal sex also may interact to determine specific traits.

## Sex differences in lipoprotein metabolism

Lipids are synthesized in virtually every cell of the body, but only the small intestine and liver synthesize and secrete lipoproteins, which distribute lipids to other tissues [28]. Following food ingestion, dietary fats are hydrolyzed by lipases in the small intestine to release fatty acids and monoacylglycerol. These components are taken up by enterocytes and used to resynthesize triacylglycerol for assembly into intestinal lipoproteins (chylomicrons), which are secreted into the lymphatics. Triglyceride-rich very low density lipoproteins (VLDLs) are synthesized by and secreted from the liver. Once in the circulation, the lipids carried in the lipid-rich core of chylomicrons and VLDL are hydrolyzed by lipoprotein lipase that is present on endothelial cells of tissues such as skeletal muscle, heart, and adipose tissue. Remodeling of lipoprotein particles in the circulation alters the lipid and protein content and gives rise to additional lipoprotein classes, most notably low-density and high-density lipoproteins (LDLs and HDLs, respectively). The fatty acids released are taken up into tissues and oxidized for energy production or stored in lipid droplets. Sex differences have been identified in the lipoprotein profiles and in the kinetics of lipoprotein production and clearance [10,29,30]. Given the strong association between serum lipoprotein profile and cardiovascular disease, it is of interest to understand the contributing mechanisms for these sex differences.

It has been known for decades that premenopausal women typically have a more atheroprotective plasma lipoprotein profile than age-matched men. That is, women tend to have lower levels of LDL cholesterol, lower VLDL cholesterol and VLDL triglycerides, and higher HDL cholesterol levels [31]. Sex differences in lipoprotein particle size and particle concentrations may contribute to sex differences in atherogenicity of the lipoproteins [32–34]. For example, the higher HDL cholesterol levels in women are associated with larger HDL particles without a major difference in the concentration of HDL particles. In addition, women tend to have smaller size and concentration of VLDL particles, and larger, but fewer, LDL particles. Sex differences also occur in lipoprotein kinetics [30]. In nonobese individuals, women exhibit higher rates of VLDL triglyceride secretion from liver coupled with more rapid clearance by peripheral tissues, leading to an overall lower VLDL triglyceride concentration in the circulation compared to men [35,36]. The increased efficiency of VLDL clearance in women may be associated with enhanced lipoprotein lipase action on larger triglyceride cores in nascent VLDL secreted by women [36]. Higher HDL concentrations in women are associated with increased rates of synthesis of the key HDL proteins, apolipoproteins A-I and A-II [37].

The more cardioprotective lipoprotein profile typically seen in premenopausal women has traditionally been attributed to the differential gonadal hormonal milieu. However, this simple explanation is not fully supported by available data. For example, exogenous estrogen replacement in postmenopausal women or in other groups with low estrogen levels (such as women with polycystic ovary syndrome) does not replicate the beneficial effects on lipoprotein profile that are experienced normally by premenopausal women [38,39]. Although estrogens have a role, it is likely the interaction of multiple factors that are responsible for differential lipoprotein profiles in premenopausal women and men, including gonadal hormone profile, genetic sex (XX chromosome complement), levels and distribution of fat deposition, and environmental factors (food intake and food choice). The reduction in gonadal hormone levels at menopause leads to increased total body fat and intra-abdominal adipose tissue, which may promote impaired insulin action. Insulin is an important regulator of lipid synthesis and metabolism, and better insulin sensitivity in premenopausal women compared to men may be lost following menopause.

Studies in the Four Core Genotypes mouse model have demonstrated specific roles for sex chromosomes and for gonadal sex in determining lipoprotein levels. HDL cholesterol levels are influenced by sex chromosomes as a predominant factor, and by gonadal sex as an additional factor [40]. Chow-fed XX mice had higher HDL cholesterol levels than XY mice regardless of gonad type; mice with male gonads also had slightly higher levels than mice with female gonads (Fig. 3.3A). The XX chromosome effect on HDL cholesterol levels was accentuated by feeding mice a cholesterol-rich diet (Fig. 3.3A). Furthermore, XX and XXY mice had higher HDL cholesterol levels than XY mice, indicating that it is the presence of two

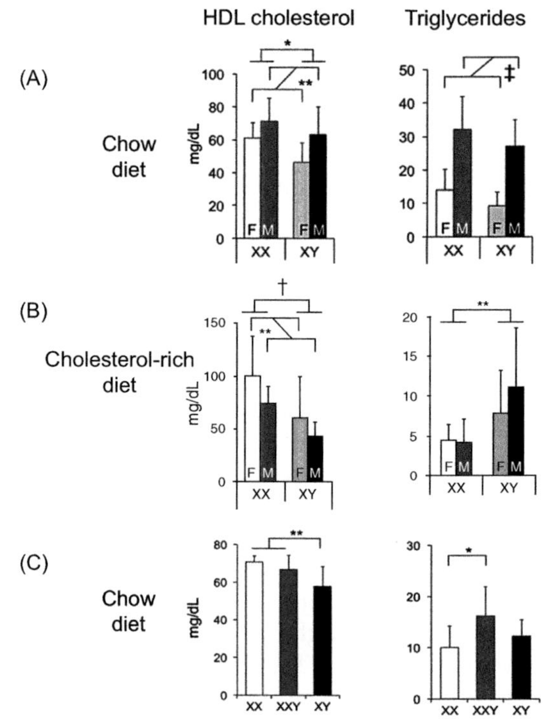


**FIGURE 3.3 Plasma lipid levels are influenced by sex chromosome and gonad types.** (A) In mice fed a chow diet, the majority of lipoprotein cholesterol is in the form of high density lipoproteins (HDL). In C57BL/6 Four Core Genotypes mice, HDL cholesterol levels are higher in XX than XY mice, and in mice with male compared to female gonads. Fasting plasma triglyceride levels are higher in mice with testes (M, for male gonads) than with ovaries (F, for female gonads). (B) In mice fed a cholesterol-enriched diet, HDL cholesterol levels are still highly regulated by sex chromosome type, but presence of ovaries leads to higher levels compared to those with testes. On this diet, triglyceride levels are higher in mice with XY compared to XX chromosomes, regardless of gonad type. (C) The sex chromosome effect on HDL cholesterol levels is due to the dose of the X chromosome, with higher levels in the presence of two X compared to one X chromosome; the presence of a Y chromosome does not influence the HDL cholesterol levels. Triglyceride levels are not as clearly influenced by sex chromosome number as HDL cholesterol levels. Statistical analyses performed by two-way ANOVA. $*P < 0.05$; $**P < 0.01$; $\dagger P < 0.0001$; $\ddagger P < 0.00001$. Source: *Data adapted from reference [40], reproduced with permission.*

X chromosomes in females, rather than the absence of a Y chromosome, that influences HDL cholesterol concentrations (Fig. 3.3C). Sex differences in circulating triglyceride levels were regulated independently of cholesterol levels and segregated with gonadal sex in chow-fed mice, and with chromosomal sex when fed a cholesterol-enriched diet (Fig. 3.3A,B). These studies suggest that chromosomal and gonadal sex interact with diet and other factors to influence plasma lipid levels, and that they are not dictated by a single difference between males and females.

# Sex differences in fat storage and adipose tissue function

## Characteristics of male versus female fat

Males and females differ in many aspects of fat storage (Fig. 3.4). Most apparent are differences between men and women in fat distribution among anatomical depots [41–43]. In general, men have more visceral adipose tissue (within the abdominal-pelvic cavity) than premenopausal women, whereas women store more subcutaneous fat (underlying the dermis). Epidemiologic studies have often linked visceral adipose tissue to Metabolic Syndrome,

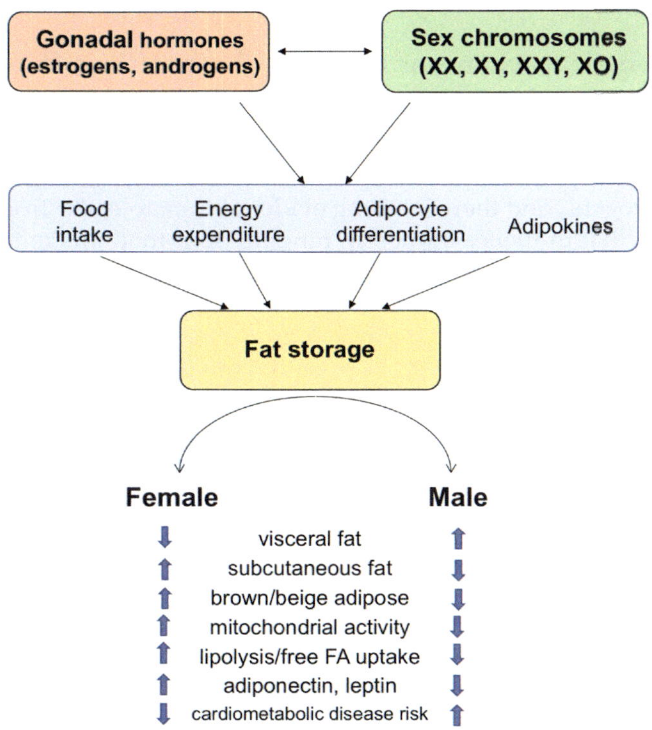


FIGURE 3.4 **Sex differences in adiposity are influenced by the interactions between sex and numerous physiological processes, including food intake, energy expenditure, adipocyte differentiation.** Together, these determine differences in fat storage that are often observed between males and females.

and have provided some evidence that subcutaneous adipose tissue may have protective effects in maintaining glucose homeostasis [44–46]. These fat depots differ in several properties that may influence systemic metabolism, as evidenced by distinct gene expression patterns between visceral and subcutaneous adipose tissues in both humans and mice [42,47]. A particularly notable finding was the differential expression of several homeobox genes in fat depots from upper and lower body in humans, suggesting that fundamental differences exist during development of different fat depots, which may contribute to sex differences in adipose tissue functions [42].

Mature adipocyte function differs in males and females. Adipose tissue is a true endocrine organ, and secretes numerous factors that act as signaling mediators to influence metabolism, including proteins, lipids, metabolites, and noncoding RNAs [48]. The best characterized adipokines are leptin and adiponectin, which have critical roles in energy balance, reproduction, inflammation, and insulin sensitivity. Sex differences in adiponectin and leptin levels have been shown in numerous populations, with unadjusted levels typically being higher in women; however, several additional factors (body fat content, ethnicity, age) also impact the levels of these adipokines [49]. In the mouse, sex differences in leptin levels are altered by removal of gonadal hormones, but adiponectin levels are not, suggesting differential regulation of these adipokines [50]. At present, little is known about whether other signaling mediators produced by adipocytes differ by sex.

## Hormonal and genetic mechanisms that contribute to sex differences in adiposity

The observation that obesity, and particularly central adiposity, increases in women following menopause is consistent with an important role for estrogens in determining the differences between men and premenopausal women [51]. Historically, this has been attributed to the reduction in estrogens, and there is strong evidence for a role of estrogen in the regulation of several processes that influence adiposity, ranging from food intake to regulation of basal metabolic rate and the differentiation of adipocyte progenitor cells. Consistent with a key role for estrogen, deletion of ERα in male and female mice leads to central obesity and insulin resistance [14,15]. There are several physiological mechanisms by which estrogens likely impact adiposity. Estrogens exert effects in several regions of the hypothalamus to modulate food intake as well as energy expenditure [52–54]. Local effects in adipose tissue may also occur, as preadipocytes express ERα, and mature adipocytes in both visceral and subcutaneous fat express ERα and β. In vitro, estrogens are found to promote differentiation of adipocyte progenitors, providing an additional mechanism by which reduced estrogen levels following menopause could influence adipose tissue expansion [55,56].

Genetic studies implicate several genetic loci in the regulation of sex differences in adiposity and related traits. The heritability of obesity is high (estimated at 50%–70%), indicating that genetic factors strongly influence fat accumulation [57]. Genome-wide association studies, in which positions spanning the 22 autosomal chromosomes were interrogated to determine their effect on adiposity, have identified dozens of genetic loci that contribute to variations across the population in adipose tissue accumulation (often measured as body mass index or waist-to-hip ratio). Significantly, many of the genetic loci associated with measures of adiposity are sex-specific. For example, of 49 loci identified in a genetic association study meta-analysis of more than 200,000 individuals, 20 had larger effects in women compared to men [58]. Numerous other genome-wide association studies have similarly found that genetic loci exhibit differential effects in women compared to men [59–61]. An exciting locus that exhibits effects on adiposity in women but not men is *KLF14*, encoding the transcription factor Kruppel-like factor 14. Genetic variants at this locus are associated with body-fat distribution, lipid levels, cardiovascular disease, and type 2 diabetes in a female- and maternal-specific manner [62]. The *KLF14* variants act specifically in adipose tissue to influence expression of ~400 genes in adipocytes with effects on adipocyte differentiation and cell size. Unfortunately, many genome-wide association studies in humans have neglected to include the sex chromosomes, and it is not clear whether there are strong associations with sex chromosome variants in human obesity, but studies in the mouse have implicated X chromosome dosage in the regulation of adipose tissue accumulation, as described below.

The Four Core Genotypes mouse model has been used to investigate the role of gonadal hormones and sex chromosome complement on adiposity and susceptibility to diet-induced obesity [63,64]. These studies revealed that when mice are fed a chow diet, mice with male gonads (on both XX and XY background) have greater body weight and body fat than mice with female gonads (XX or XY backgrounds) [65]. However, when gonads are removed in adult mice, XX mice that previously had male or female gonads gained weight more rapidly than XY mice. The difference between XX and XY mice was accentuated further when fed a high fat diet [65]. XX mice were found to have similar energy expenditure as XY mice, but had slightly increased food intake during the inactive part of the circadian cycle [65,66]. These results indicate that gonadal hormones and sex chromosomes influence adiposity independently.

To better understand sex differences in adiposity, it would be valuable to determine the mechanism by which XX chromosome complement promotes adiposity. Toward this end, body weight and body fat of XX, XXY, XO, and XY mice were compared [65]. These genotypes broke down into two distinct categories in terms of weight gain: XX and XXY mice gained weight more rapidly than XY and XO mice, indicating that the presence of two X chromosomes promotes fat accumulation over presence of a single X chromosome, regardless of the presence of a Y chromosome. These studies set the stage for the identification of mechanisms by which X chromosome dosage influences adiposity, including the involvement of genes that escape X-inactivation.

## Sex differences in adipose tissue energetics

As described above, gonadal hormones act through nuclear hormone receptors in the brain to modulate energy homeostasis. Sex differences also occur in local adipose tissue energy metabolism. These include differences in the lipolytic capacity and energetic activity within white adipose tissue and brown adipose tissue, as well as a differential capacity for browning of white adipose tissue. White adipocytes are a key source of fatty acids for energy in other peripheral tissues during conditions such as fasting and exercise. In general, adipose tissue from women has higher rates of lipolysis (by approximately 40%), but also more efficient utilization of the fatty acids liberated from adipose tissue [67,68]. Differences also occur in lipolytic activity of adipose tissue from different depots, and may depend on the levels of β-adrenergic receptors that are present in various depots [68].

Recent studies in both humans and mice suggest that females have enhanced mitochondrial activity in white adipocytes compared to males. In whole body measurements, women had higher resting metabolic rate per kilogram of adipose tissue than men [69]. This was reflected in increased expression of genes associated with mitochondrial oxidative phosphorylation in women's adipose tissue, but not in liver or muscle [69]. Consistent with these findings, preadipocytes from obese women cultured in vitro had higher mitochondrial to glycolytic activity [70]. The sex difference in white adipose tissue mitochondrial function persists across species, as a systems genetics study in 100 inbred mouse strains found greater mitochondrial activity and mitochondrial DNA content in female compared to males on some genetic backgrounds [71]. The greater mitochondrial activity in females was associated with reduced susceptibility to obesity and insulin resistance. Gonadectomy revealed that the sex difference in adipose tissue mitochondria cannot be attributed to gonadal hormones.

Female human and mouse white adipose tissues also appear to have enhanced capacity for the expression of uncoupling protein 1 (UCP1), which may contribute to enhanced energy dissipation in females. UCP1 allows the uncoupling of mitochondrial fatty acid oxidation from ATP synthesis and results in heat production, known as nonshivering thermogenesis [72]. UCP1 is most abundant in brown adipose tissue, which has an important role in thermogenesis in rodents and in newborn human babies. It is well accepted that many adult humans also have small amounts of brown adipose tissue, typically in supraclavicular, perirenal, and other depots [73]. The best available technique to detect brown adipose tissue activity is the injection of [18F]-fluorodeoxyglucose (FDG) followed by a positron emission tomography (PET)/X-ray computed tomography scan (CT). Using this technique, in individuals with detectable brown adipose tissue, women were observed to have higher mean brown adipose tissue activity than men (i.e., greater FDG uptake in identified brown adipose tissue) [74]. Additionally, in a retrospective analysis of FDG-PET/CT scans from more than 4800 individuals, women were more likely than men to have metabolically active brown adipose tissue and to show an association between increased brown adipose tissue and lower body mass index [75]. Women also have enhanced appearance within the perirenal white adipose tissue depot of beige adipocytes, which express UCP1 and have the capacity for energy dissipation as heat [76].

## Sex differences in atherosclerosis

Men and women differ in the age of onset and pathology of cardiovascular diseases, such as atherosclerosis. In general, women have delayed onset compared to men, and whereas men tend to experience atherosclerotic plaque rupture, women more often experience thrombus formation at intact, nonruptured coronary artery plaques [77]. Key risk factors for atherosclerosis include obesity and hyperlipidemia, and sex differences in these traits (discussed in preceding sections) likely contribute to observed sex differences in the development of cardiovascular disease. In women, estrogens appear to have a protective effect on atherogenesis, whereas in men, androgens appear to have a protective effect; waning of gonadal hormone levels with age may contribute to atherogenesis in both sexes [10]. Estrogens also reduce atherosclerosis in mice, including robust models of atherogenesis such as LDL receptor-deficient and apolipoprotein E-deficient mice [78]. However, with more advanced age (>85 years), women tend to surpass men in the prevalence of atherosclerosis [30,79], suggesting that additional factors beyond gonadal hormones contribute. Sex chromosome complement may be one such factor. Men with XYY sex chromosomes have increased cardiovascular mortality [80], and a common European Y chromosome haplotype promotes a 50% increase in coronary artery disease risk, which is associated with altered immune response [81]. Cardiovascular disease is a common cause of morbidity and mortality in Turner syndrome (XO), but it is not possible to show direct effects of the chromosome complement, as women experience many developmental abnormalities and altered gonadal hormones levels that could contribute to increased risk [82].

Studies in the Four Core Genotypes mouse model have provided more insight into the contributions of gonadal and chromosomal sex to atherosclerosis risk. The removal of gonads in adult mice led to greater atherosclerotic lesion areas in XX compared to XY mice fed an atherogenic diet, regardless of the original male or female gonad type [83]. Similar studies performed in mice that lack either the LDL receptor or apolipoprotein E, both of which increase atherogenesis, confirmed that XX sex chromosome complement promotes atherosclerosis in multiple genetic backgrounds [83]. As described previously [40,65], the XX chromosome complement also led to increased adiposity and elevated LDL cholesterol levels, risk factors for atherosclerosis, making it difficult to determine whether XX chromosome complement directly promotes atherogenesis.

## Sex differences in the gut microbiota influence metabolism

The gut microbiome, estimated to comprise more than 100 trillion bacteria, plays a critical role in the absorption and metabolism of nutrients in the intestine [84]. Sex differences in the gut microbiome composition have been associated with several aspects of lipid metabolism, including dyslipidemia and obesity [85]. The composition of the gut microbiome is determined by sequencing 16S microbial ribosomal RNA extracted from fecal samples. The gut microbiota impact host physiology by production of metabolites that enter the systemic circulation, including short-chain fatty acids (acetate, propionate, and butyrate) and trimethylamine (TMA) [86]. TMA is derived from bacterial metabolism of choline or phosphatidylcholine, which are found in eggs, dairy, red meat, and salt-water fish. TMA is converted to trimethylamine N-oxide (TMAO) through the action of the hepatic enzyme flavin monooxygenase 3 (FMO3), which is expressed at about 1000-fold higher levels in liver of female compared to male mice, leading to about four-fold elevations in TMAO in the circulation of female mice [87,88]. Sex differences in the FMO3/TMA/TMAO axis is one mechanism by which the gut microbiota may contribute to male/female differences in lipid metabolism, including plasma triglyceride and HDL-cholesterol levels, and propensity for browning of white adipose tissue [89,90].

Sex differences in the gut microbiome composition have been observed in children and in adults and found to influence adiposity. Interestingly, in children as young as 2–3 years of age (prior to large changes in gonadal hormone levels at puberty), sex differences in the gut microbiome were associated with fat-free mass index in boys, but not girls [91]. In both adult humans and mice, gut microbiota are associated with the development of obesity, and in particular, with the ratio of Firmicutes-to-Bacteroidetes. A sex-specific relationship was detected between microbiome sequence data and measures of fat distribution in adults, classified as "android fat ratio" and "gynoid fat ratio" [92]. In these studies, the presence of the Ruminococcacaea family had a positive association with gynoid fat ratio in females and a negative association in males. However, not all studies in humans have reached the same conclusions, possibly due to confounding effects on microbiota by diet, age, and ethnicity.

Studies in mouse models have allowed control of variables that may influence microbiome composition in order to assess how sex influences its composition. Studies in different mouse strains reveal that the host genetic background influences the gut microbiome [93]. The microbiome was shown to contribute to sex differences in mouse obesity and dyslipidemia through antibiotic-induced microbiota depletion followed by fecal microbiota transplantation in mice of the opposite sex [94]. Transplantation of female mice with male microbiota induced obesity and elevated plasma triglycerides in female mice above levels normally present in females with their endogenous microbiome. This was associated with an increase in Firmicutes-to-Bacteroidetes ratio. Gonadal hormones were responsible, at least in part, for some sex differences in gut microbiota, as gonadectomy (with and without subsequent hormone replacement) influenced the gut microbiome in mice [93]. There is also evidence for gonadal hormone effects on human gut microbiota. The gut microbiome of pre- and postmenopausal women and age-matched men were compared [95]. When matched for body mass index and nutritional background, postmenopausal women had a higher proportion of Firmicutes than premenopausal women, and a higher Firmicutes/Bacteroidetes ratio than age-matched men. Thus, sex differences in the gut microbiome may be influenced by gonadal hormones, diet, and other factors, and may contribute to sex differences in fat distribution and fat accumulation in obesity.

## Future perspectives

For decades, researchers have cataloged differences between males and females in many aspects of lipid metabolism and storage, both in humans and in preclinical animal models. A challenge is to identify the underlying mechanisms for the observed sex differences. As described herein, tools such as genetically engineered mouse strains have been developed to parse genetic from hormonal bases for sex differences. In many cases, sex effects on a particular trait may be the composite of interactions between gonadal hormones, sex chromosomes, and additional factors, including genetic variants across the genome (sex-by-gene interactions), commensal gut microbiota, and diet (Fig. 3.5). An additional layer of complexity is the intersection between biological sex and gender. Here, we focused primarily on data based on biological sex, since most studies have divided subjects on these criteria and less rigorous data are available for gender effects on lipid metabolism. Nevertheless, it is clear that gender and other cultural influences on lifestyle also impact lipid metabolism and related diseases, for example, through effects on dietary habits and eating behavior. As healthcare moves toward a more individualized approach to achieve "precision medicine," a better understanding of drivers of sex differences in dyslipidemias, obesity, and other risk factors for cardiovascular diseases will promote more effective prevention and treatments for both sexes.

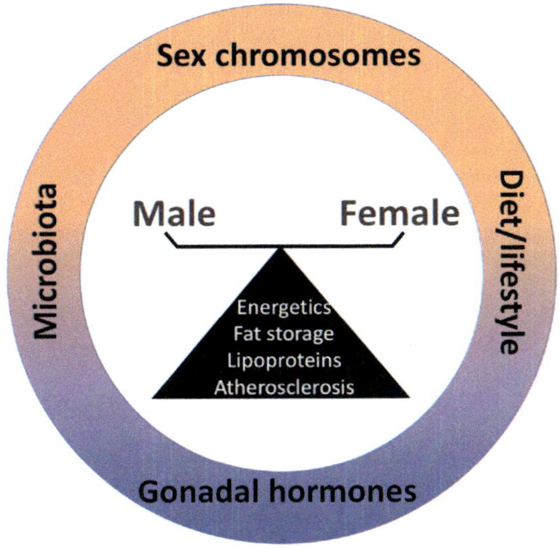


FIGURE 3.5 **Sex differences exist in numerous aspects of lipid homeostasis, including energy storage and utilization, lipoprotein levels, and related disease processes, such as atherogenesis.** Males and females are born with intrinsic differences (sex chromosome complement and gonadal type), which interact with environmental components to bring about differences observed between males and females.

# References


[1] Karastergiou K, Smith SR, Greenberg AS, Fried SK. Sex differences in human adipose tissues—the biology of pear shape. Biol Sex Differ 2012;3:13.

[2] Karastergiou K, Fried SK. . Adv Exp Med Biol 2017;1043:29–51.

[3] Palmer BF, Clegg DJ. The sexual dimorphism of obesity. Mol Cell Endocrinol 2015;402:113–9.

[4] Clayton JA, Collins FS. Policy: NIH to balance sex in cell and animal studies. Nature 2014;509:282–3.

[5] Danska JS. Sex matters for mechanism. Sci Transl Med 2014;6:258fs40.

[6] Miller LR, Marks C, Becker JB, Hurn PD, Chen W-J, Woodruff T, et al. Considering sex as a biological variable in preclinical research. FASEB J 2017;31:29–34.

[7] Sandberg K, Umans JG, Georgetown Consensus Conference Work Group. Recommendations concerning the new U.S. National Institutes of Health initiative to balance the sex of cells and animals in preclinical research. FASEB J 2015;29:1646–52.

[8] Arnold AP, Chen X, Link JC, Itoh Y, Reue K. Cell-autonomous sex determination outside of the gonad. Dev Dyn 2013;242:371–9.

[9] Mauvais-Jarvis F, Arnold AP, Reue K. A guide for the design of pre-clinical studies on sex differences in metabolism. Cell Metab 2017;25:1216–30.

[10] Arnold AP, Cassis LA, Eghbali M, Reue K, Sandberg K. Sex hormones and sex chromosomes cause sex differences in the development of cardiovascular diseases. Arterioscler Thromb Vasc Biol 2017;37:746–56.

[11] Garaulet M, Pérez-Llamas F, Baraza JC, Garcia-Prieto MD, Fardy PS, Tébar FJ, et al. Body fat distribution in pre- and post-menopausal women: metabolic and anthropometric variables. J Nutr Health Aging 2002;6:123–6.

[12] Lima R, Wofford M, Reckelhoff JF. Hypertension in postmenopausal women. Curr Hypertens Rep 2012;14:254–60.

[13] Maas AHEM, Appelman YEA. Gender differences in coronary heart disease. Neth Heart J 2010;18:598–602.

[14] Ohlsson C, Hellberg N, Parini P, Vidal O, Bohlooly M, Bohlooly M, et al. Obesity and disturbed lipoprotein profile in estrogen receptor-alpha-deficient male mice. Biochem Biophys Res Commun 2000;278:640–5.

[15] Heine PA, Taylor JA, Iwamoto GA, Lubahn DB, Cooke PS. Increased adipose tissue in male and female estrogen receptor-alpha knockout mice. Proc Natl Acad Sci U S A 2000;97:12729–34.

[16] Musatov S, Chen W, Pfaff DW, Mobbs CV, Yang X-J, Clegg DJ, et al. Silencing of estrogen receptor alpha in the ventromedial nucleus of hypothalamus leads to metabolic syndrome. Proc Natl Acad Sci U S A 2007;104:2501–6.
[17] Xu Y, Nedungadi TP, Zhu L, Sobhani N, Irani BG, Davis KE, et al. Distinct hypothalamic neurons mediate estrogenic effects on energy homeostasis and reproduction. Cell Metab 2011;14:453–65.
[18] Bukowski R, Smith GCS, Malone FD, Ball RH, Nyberg DA, Comstock CH, et al. Human sexual size dimorphism in early pregnancy. Am J Epidemiol 2007;165:1216–8.
[19] Burgoyne PS, Thornhill AR, Boudrean SK, Darling SM, Bishop CE, Evans EP, et al. The genetic basis of XX-XY differences present before gonadal sex differentiation in the mouse [and discussion]. Philos Trans R Soc B Biol Sci 1995;350:253–61.
[20] Wells JCK. Sexual dimorphism of body composition. Best Pract Res Clin Endocrinol Metab 2007;21:415–30.
[21] Taylor RW, Grant AM, Williams SM, Goulding A. Sex differences in regional body fat distribution from pre- to postpuberty. Obesity 2010;18:1410–6.
[22] Aksglaede L, Molgaard C, Skakkebaek NE, Juul A. Normal bone mineral content but unfavourable muscle/fat ratio in Klinefelter syndrome. Arch Dis Child 2008;93:30–4.
[23] Bardsley MZ, Falkner B, Kowal K, Ross JL. Insulin resistance and metabolic syndrome in prepubertal boys with Klinefelter syndrome. Acta Paediatr 2011;100:866–70.
[24] Bojesen A, Kristensen K, Birkebaek NH, Fedder J, Mosekilde L, Bennett P, et al. The metabolic syndrome is frequent in Klinefelter's syndrome and is associated with abdominal obesity and hypogonadism. Diabetes Care 2006;29:1591–8.
[25] Mavinkurve M, O'Gorman CS. Cardiometabolic and vascular risks in young and adolescent girls with Turner syndrome. BBA Clin. 2015;3:304–9.
[26] Burgoyne PS, Arnold AP. A primer on the use of mouse models for identifying direct sex chromosome effects that cause sex differences in non-gonadal tissues. Biol Sex Differ 2016;7:68.
[27] Zore T, Palafox M, Reue K. Sex differences in obesity, lipid metabolism, and inflammation—a role for the sex chromosomes? Mol Metab 2018;15:35–44.
[28] Abumrad NA, Davidson NO. Role of the gut in lipid homeostasis. Physiol Rev 2012;92:1061–85.
[29] Link JC, Reue K. The genetic basis for sex differences in obesity and lipid metabolism. Annu Rev Nutr 2017;37:225–45.
[30] Wang X, Magkos F, Mittendorfer B. Sex differences in lipid and lipoprotein metabolism: it's not just about sex hormones. J Clin Endocrinol Metab 2011;96:885–93.
[31] Hazzard WR. Atherogenesis: why women live longer than men. Geriatrics 1985;40:42–51. 54.
[32] Freedman DS, Otvos JD, Jeyarajah EJ, Shalaurova I, Cupples LA, Parise H, D'Agostino RB, Wilson PWF, Schaefer EJ, et al. Sex and age differences in lipoprotein subclasses measured by nuclear magnetic resonance spectroscopy: The Framingham Study. Clin Chem 2004;50:1189–200.
[33] Vaidya D, Dobs A, Gapstur SM, Golden SH, Hankinson A, Liu K, et al. The association of endogenous sex hormones with lipoprotein subfraction profile in the Multi-Ethnic Study of Atherosclerosis. Metabolism 2008;57:782–90.
[34] Johnson JL, Slentz CA, Duscha BD, Samsa GP, McCartney JS, Houmard JA, et al. Gender and racial differences in lipoprotein subclass distributions: the STRRIDE study. Atherosclerosis 2004;176:371–7.
[35] Matthan NR, Jalbert SM, Barrett PHR, Dolnikowski GG, Schaefer EJ, Lichtenstein AH. Gender-specific differences in the kinetics of nonfasting TRL, IDL, and LDL apolipoprotein B-100 in men and premenopausal women. Arterioscler Thromb Vasc Biol 2008;28:1838–43.
[36] Magkos F, Patterson BW, Mohammed BS, Klein S, Mittendorfer B. Women produce fewer but triglyceride-richer very low-density lipoproteins than men. J Clin Endocrinol Metab 2007;92:1311–8.
[37] Schaefer EJ, Zech LA, Jenkins LL, Bronzert TJ, Rubalcaba EA, Lindgren FT, et al. Human apolipoprotein A-I and A-II metabolism. J Lipid Res 1982;23:850–62.
[38] Godsland IF. Effects of postmenopausal hormone replacement therapy on lipid, lipoprotein, and apolipoprotein (a) concentrations: analysis of studies published from 1974-2000. Fertil Steril 2001;75:898–915.
[39] Soares GM, Vieira CS, de Paula Martins W, dos Reis RM, de Sá MFS, Ferriani RA. Metabolic and cardiovascular impact of oral contraceptives in polycystic ovary syndrome. Int J Clin Pract 2009;63:160–9.
[40] Link JC, Chen X, Prien C, Borja MS, Hammerson B, Oda MN, et al. Increased high-density lipoprotein cholesterol levels in mice with XX versus XY sex chromosomes. Arterioscler Thromb Vasc Biol 2015;35:1778–86.
[41] Karastergiou K, Fried SK. 2017 Advances in experimental medicine and biology; 2017, p. 29–51.

[42] Karastergiou K, Fried SK, Xie H, Lee M-J, Divoux A, Rosencrantz MA, et al. Distinct developmental signatures of human abdominal and gluteal subcutaneous adipose tissue depots. J Clin Endocrinol Metab 2013;98:362–71.
[43] White UA, Tchoukalova YD. Sex dimorphism and depot differences in adipose tissue function. Biochim Biophys Acta - Mol Basis Dis 2014;1842:377–92.
[44] Fox CS, Massaro JM, Hoffmann U, Pou KM, Maurovich-Horvat P, Liu C-Y, et al. Abdominal visceral and subcutaneous adipose tissue compartments: association with metabolic risk factors in the Framingham Heart Study. Circulation 2007;116:39–48.
[45] Liu J, Fox CS, Hickson DA, May WD, Hairston KG, Carr JJ, et al. Impact of abdominal visceral and subcutaneous adipose tissue on cardiometabolic risk factors: the Jackson Heart Study. J Clin Endocrinol Metab 2010;95:5419–26.
[46] Tran TT, Yamamoto Y, Gesta S, Kahn CR. Beneficial effects of subcutaneous fat transplantation on metabolism. Cell Metab 2008;7:410–20.
[47] Grove KL, Fried SK, Greenberg AS, Xiao XQ, Clegg DJ. A microarray analysis of sexual dimorphism of adipose tissues in high-fat-diet-induced obese mice. Int J Obes (Lond) 2010;34:989–1000.
[48] Funcke J-B, Scherer PE. Beyond adiponectin and leptin: adipose tissue-derived mediators of inter-organ communication. J Lipid Res 2019;R094060.
[49] Lew J, Sanghavi M, Ayers CR, McGuire DK, Omland T, Atzler D, et al. Sex-based differences in cardiometabolic biomarkers. Circulation 2017;135:544–55.
[50] Gui Y, Silha JV, Murphy LJ. Sexual dimorphism and regulation of resistin, adiponectin, and leptin expression in the mouse. Obes Res 2004;12:1481–91.
[51] Flegal KM, Kruszon-Moran D, Carroll MD, Fryar CD, Ogden CL. Trends in obesity among adults in the United States, 2005–2014. JAMA 2016;315:2284–91.
[52] Butera PC. Estradiol and the control of food intake. Physiol Behav 2010;99:175–80.
[53] Mauvais-Jarvis F, Clegg DJ, Hevener AL. The role of estrogens in control of energy balance and glucose homeostasis. Endocr Rev 2013;34:309–38.
[54] Lizcano F, Guzmán G. Estrogen deficiency and the origin of obesity during menopause. Biomed Res Int 2014;2014:1–11.
[55] Wada T, Ihunnah CA, Gao J, Chai X, Zeng S, Philips BJ, et al. Estrogen sulfotransferase inhibits adipocyte differentiation. Mol Endocrinol 2011;25:1612–23.
[56] Dieudonne MN, Pecquery R, Leneveu MC, Giudicelli Y. Opposite effects of androgens and estrogens on adipogenesis in rat preadipocytes: evidence for sex and site-related specificities and possible involvement of insulin-like growth factor 1 receptor and peroxisome proliferator-activated receptor gamma2. Endocrinology 2000;141:649–56.
[57] Herrera BM, Lindgren CM. The genetics of obesity. Curr Diab Rep 2010;10:498–505.
[58] Shungin D, Winkler TW, Croteau-Chonka DC, Ferreira T, Locke AE, Mägi R, et al. New genetic loci link adipose and insulin biology to body fat distribution. Nature 2015;518:187–96.
[59] Heid IM, Jackson AU, Randall JC, Winkler TW, Qi L, Steinthorsdottir V, et al. Meta-analysis identifies 13 new loci associated with waist-hip ratio and reveals sexual dimorphism in the genetic basis of fat distribution. Nat Genet 2010;42:949–60.
[60] Fox CS, Liu Y, White CC, Feitosa M, Smith AV, Heard-Costa N, et al. Genome-wide association for abdominal subcutaneous and visceral adipose reveals a novel locus for visceral fat in women. PLoS Genet 2012;8:e1002695.
[61] Sung YJ, Pérusse L, Sarzynski MA, Fornage M, Sidney S, Sternfeld B, et al. Genome-wide association studies suggest sex-specific loci associated with abdominal and visceral fat. Int J Obes (Lond) 2016;40:662–74.
[62] Small KS, Todorčević M, Civelek M, El-Sayed Moustafa JS, Wang X, Simon MM, et al. Regulatory variants at KLF14 influence type 2 diabetes risk via a female-specific effect on adipocyte size and body composition. Nat Genet 2018;50:572–80.
[63] Link JC, Chen X, Arnold AP, Reue K. Metabolic impact of sex chromosomes. Adipocyte 2013;2:74–9.
[64] Reue K. Sex differences in obesity: X chromosome dosage as a risk factor for increased food intake, adiposity and co-morbidities. Physiol Behav 2017;176:174–82.
[65] Chen X, McClusky R, Chen J, Beaven SW, Tontonoz P, Arnold AP, et al. The number of x chromosomes causes sex differences in adiposity in mice. PLoS Genet 2012;8:e1002709.
[66] Chen X, Wang L, Loh DH, Colwell CS, Taché Y, Reue K, et al. Sex differences in diurnal rhythms of food intake in mice caused by gonadal hormones and complement of sex chromosomes. Horm Behav 2015;75:55–63.

[67] Nielsen S, Guo Z, Albu JB, Klein S, O'Brien PC, Jensen MD. Energy expenditure, sex, and endogenous fuel availability in humans. J Clin Invest 2003;111:981–8.
[68] Karpe F, Dickmann JR, Frayn KN. Fatty acids, obesity, and insulin resistance: time for a reevaluation. Diabetes 2011;60:2441–9.
[69] Nookaew I, Svensson P-A, Jacobson P, Jernås M, Taube M, Larsson I, et al. Adipose tissue resting energy expenditure and expression of genes involved in mitochondrial function are higher in women than in men. J Clin Endocrinol Metab 2013;98:E370–8.
[70] Keuper M, Berti L, Raedle B, Sachs S, Böhm A, Fritsche L, et al. Preadipocytes of obese humans display gender-specific bioenergetic responses to glucose and insulin. Mol Metab 2019;20:28–37.
[71] Norheim F, Hasin-Brumshtein Y, Vergnes L, Chella Krishnan K, Pan C, Seldin MM, et al. Gene-by-sex interactions in mitochondrial functions and cardio-metabolic traits. Cell Metab 2019;29:932–49. e4.
[72] Bartelt A, Heeren J. Adipose tissue browning and metabolic health. Nat Rev Endocrinol 2014;10:24–36.
[73] Cypess AM, Kahn CR. The role and importance of brown adipose tissue in energy homeostasis. Curr Opin Pediatr 2010;22:478–84.
[74] Martinez-Tellez B, Sanchez-Delgado G, Garcia-Rivero Y, Alcantara JMA, Martinez-Avila WD, Muñoz-Hernandez MV, et al. A new personalized cooling protocol to activate brown adipose tissue in young adults. Front Physiol 2017;8:863.
[75] Brendle C, Werner MK, Schmadl M, la Fougère C, Nikolaou K, Stefan N, et al. Correlation of brown adipose tissue with other body fat compartments and patient characteristics. Acad Radiol 2018;25:102–10.
[76] van den Beukel JC, Grefhorst A, Hoogduijn MJ, Steenbergen J, Mastroberardino PG, Dor FJMF, et al. Women have more potential to induce browning of perirenal adipose tissue than men. Obesity 2015;23:1671–9.
[77] de Jager SCA, Meeuwsen JAL, van Pijpen FM, Zoet GA, Barendrecht AD, Franx A, et al. Preeclampsia and coronary plaque erosion: manifestations of endothelial dysfunction resulting in cardiovascular events in women. Eur J Pharmacol 2017;816:129–37.
[78] Hodgin JB, Maeda N. Minireview: estrogen and mouse models of atherosclerosis. Endocrinology 2002;143:4495–501.
[79] Schubert CM, Rogers NL, Remsberg KE, Sun SS, Chumlea WC, Demerath EW, et al. Lipids, lipoproteins, lifestyle, adiposity and fat-free mass during middle age: the Fels Longitudinal Study. Int J Obes (Lond) 2006;30:251–60.
[80] Higgins CD, Swerdlow AJ, Schoemaker MJ, Wright AF, Jacobs PA, UK Clinical Cytogenetics Group. Mortality and cancer incidence in males with Y polysomy in Britain: a cohort study. Hum Genet 2007;121:691–6.
[81] Charchar FJ, Bloomer LD, Barnes TA, Cowley MJ, Nelson CP, Wang Y, et al. Inheritance of coronary artery disease in men: an analysis of the role of the Y chromosome. Lancet (London, England) 2012;379:915–22.
[82] Lichiardopol C, Mota M. Cardiovascular risk factors in Turner syndrome. Rom J Intern Med 2004;42:371–9.
[83] AlSiraj Y, Chen X, Thatcher SE, Temel RE, Cai L, Blalock E, et al. XX sex chromosome complement promotes atherosclerosis in mice. Nat Commun 2019;10:2631.
[84] Yadav M, Verma MK, Chauhan NS. A review of metabolic potential of human gut microbiome in human nutrition. Arch Microbiol 2018;200:203–17.
[85] Razavi AC, Potts KS, Kelly TN, Bazzano LA. Sex, gut microbiome, and cardiovascular disease risk. Biol Sex Differ 2019;10:29.
[86] Canfora EE, Meex RCR, Venema K, Blaak EE. Gut microbial metabolites in obesity, NAFLD and T2DM. Nat Rev Endocrinol 2019;15:261–73.
[87] Wang Z, Klipfell E, Bennett BJ, Koeth R, Levison BS, DuGar B, et al. Gut flora metabolism of phosphatidylcholine promotes cardiovascular disease. Nature 2011;472:57–63.
[88] Veeravalli S, Karu K, Scott F, Fennema D, Phillips IR, Shephard EA. Effect of flavin-containing monooxygenase genotype, mouse strain, and gender on trimethylamine n-oxide production, plasma cholesterol concentration, and an index of atherosclerosis. Drug Metab Dispos 2018;46:20–5.
[89] Fu J, Bonder MJ, Cenit MC, Tigchelaar EF, Maatman A, Dekens JAM, et al. The gut microbiome contributes to a substantial proportion of the variation in blood lipids. Circ Res 2015;117:817–24.
[90] Schugar RC, Shih DM, Warrier M, Helsley RN, Burrows A, Ferguson D, et al. The TMAO-producing enzyme flavin-containing monooxygenase 3 regulates obesity and the beiging of white adipose tissue. Cell Rep 2017;20:279.
[91] Smith-Brown P, Morrison M, Krause L, Davies PSW. Male-specific association between fat-free mass index and fecal microbiota in 2- to 3-year-old Australian children. J Pediatr Gastroenterol Nutr 2018;66:147–51.

[92] Min Y, Ma X, Sankaran K, Ru Y, Chen L, Baiocchi M, et al. Sex-specific association between gut microbiome and fat distribution. Nat Commun 2019;10:2408.

[93] Org E, Mehrabian M, Parks BW, Shipkova P, Liu X, Drake TA, et al. Sex differences and hormonal effects on gut microbiota composition in mice. Gut Microbes 2016;7:313–22.

[94] Kaliannan K, Robertson RC, Murphy K, Stanton C, Kang C, Wang B, et al. Estrogen-mediated gut microbiome alterations influence sexual dimorphism in metabolic syndrome in mice. Microbiome 2018;6:205.

[95] Santos-Marcos JA, Rangel-Zuñiga OA, Jimenez-Lucena R, Quintana-Navarro GM, Garcia-Carpintero S, Malagon MM, et al. Influence of gender and menopausal status on gut microbiota. Maturitas 2018;116:43–53.

CHAPTER

# 4

# Local interactions in the bone marrow microenvironment and their contributions to systemic metabolic processes

*Kenneth T. Lewis*[a], *Ormond A. MacDougald*[a,b]

[a]Department of Molecular and Integrative Physiology, University of Michigan Medical School, Ann Arbor, MI, United States; [b]Department of Internal Medicine, University of Michigan Medical School, Ann Arbor, MI, United States

OUTLINE



Lipid Signaling and Metabolism. http://dx.doi.org/10.1016/B978-0-12-819404-1.00004-X

## Bone marrow adipose: a brief history

The marrow lies completely encapsulated within bone, and in many ways early scientific investigation of marrow was similarly constrained, with most intellectuals attributing the role of marrow to supplying nutrients or drawing waste products from the bone [1]. Franz Ernst Christian Neumann was first to recognize the importance of marrow beyond the bone when he made the observation in 1869 that red blood cells appear to be formed within the marrow and correctly postulated that both red and white blood cells develop from a common hematopoietic precursor marrow cell [1,2]. Neumann later noted the heterogeneity of marrow when he observed that the proximal region of long bones contains hematopoietic red marrow, whereas yellow fatty marrow is located in more distal locations. He further observed that the marrow is almost entirely red at birth but gradually becomes more yellow with age [1,3]. Refinement in microscopic techniques revealed that marrow fat is stored in adipocytes, which are densely packed in distal yellow marrow but sparsely distributed in red marrow [4]. Despite these early advances characterizing the distribution of bone marrow adiposity, most investigators considered marrow fat as simply a space filler derived from connective tissue [5]. Whether marrow fat was a functionally distinct adipose tissue or simply ectopic lipid accumulation remained contested throughout the mid-20th century [5,6]. It was not until 1967 that BMAs were determined to comprise a functionally distinct adipose tissue depot, which was later coined bone marrow adipose tissue (BMAT) [7,8]. Shortly thereafter, Tavassoli suggested that BMAT may be further divided into two groups, based on his observation that BMA lipid composition of yellow marrow differed from that of red marrow [9,10]. The BMAT of red and yellow marrow were coined labile and stable, respectively, because BMAT in proximal marrow sites (red) is far more sensitive to external stimuli. Recently, Scheller et al. confirmed these results and further described differences between these two distinct BMAT types, renaming them regulated (rBMAT) and constitutive (cBMAT) BMAT [11]. In addition to its labile nature, the distinction of rBMAT from cBMAT is further marked by smaller adipocyte volume, higher proportion of saturated lipids, and reduced expression of the adipogenic *cebpa* and *cebpb* genes [11].

## BMAT is distinct from peripheral adipose depots

The study of BMAT has gradually gained interest in the past two decades but is still far understudied compared to white and brown adipose tissue depots, WAT and BAT, respectively. Despite comprising 50%–70% of marrow volume and 10% or more of total adipose mass in adult humans, only a fraction of a percent of adipose publications have investigated BMAT to date [12–14]. While individual BMAs closely resemble white adipocytes in morphology, there are myriad differences in their composition, function, and metabolic properties; thus, physiological discoveries in WAT do not necessarily translate to BMAT biology. The work by Liu et al. highlights the stark difference between WAT and BMAT, reporting that nearly 4000 genes are differentially regulated between isolated BMAs and adipocytes from the epididymal WAT depot [15]. When compared to WAT, BMAs have lower expression of several hallmark adipocyte genes, including fatty acid binding protein 4 (FABP4) and peroxisomal proliferation activated receptor gamma (PPAR-γ), the latter of which has widely been recognized as a master regulator of adipogenesis [16]. Interestingly, BMAs demonstrate high expression of genes involved in pro-inflammatory cytokine signaling, such as interleukin-6 (IL-6) and tumor necrosis factor-alpha (TNF-$\alpha$), compared to white adipocytes, despite a similar degree of contamination from $CD11b^+$ macrophages and monocytes (10%–17%) [15]. However, isolating individual cells from their surrounding tissue via collagenase digestion can cause a pro-inflammatory shift in gene expression, so these results must be viewed with caution, especially since collagenase was used to digest WAT but not BMAT [13,15,17].

The unique gene expression of BMAs is mirrored by functional differences from peripheral adipocytes and the extent of these differences remains incompletely understood. In histological sections, BMAs are morphologically similar to white adipocytes with a large unilocular lipid droplet, although BMAs are smaller than white adipocytes [14,18,19]. The BMAs of cBMAT have approximately one-fifth the volume of white adipocytes, and rBMAT adipocytes are an additional 50% smaller in volume compared to cBMAT [11,14,18,19]. Like WAT, BMAs also express genes involved in creating and maintaining adipocyte characteristics, including *Adipoq, Lep, Fabp4, Pparg, Cebpa,* and more [11,15,20]. Furthermore, glucose uptake is blunted in the femoral marrow of euglycemic hyperinsulinemic humans with an AKT2 mutation, suggesting bone marrow undergoes insulin-mediated glucose uptake in an AKT-dependent manner, similar to that of WAT [21]. However, a recent study in mice indicates that insulin does not stimulate glucose uptake in marrow or BMAT from femur or tibia, and even decreases glucose uptake in humeral marrow [preprint] [22]. BMAT further differs from WAT in its response to nutrient deprivation. Whereas WAT potently mobilizes its lipid stores to fuel systemic metabolic demands, BMAT instead expands lipid uptake dramatically [23–25]. During these periods, the expanding BMAT potently increases secretion of the adipokine adiponectin [20], which sensitizes skeletal muscle, liver, and adipose to insulin [26–28].

BMAs also possess a dense mitochondrial network surrounding the lipid droplet [29], and low levels of BAT gene transcripts have been identified in whole bones [30]. Observations like these have led some to suggest that BMAT may functionally lie between WAT and BAT [30]. The well-established BAT marker genes *Prdm16, Ppargc1a, Foxc2,* and *Dio2* are highly upregulated in whole tibiae compared to WAT [30], but only *Dio2* is significantly upregulated in isolated BMAT versus WAT [15,31]. *Dio2* encodes the type 2 iodothyronine deiodinase enzyme responsible for converting $T_4$ thyroid hormone (TH) to the 10-fold more active $T_3$ TH, which is critical for thermogenic activation of BAT [31,32]. Furthermore, agonism of TH receptor $\beta$ in TH-deficient mice increases both marrow adiposity and *Ucp1* expression dramatically in femoral and tibial epiphyseal tissues, indicating a possible mechanistic link between BMAT and BAT [33]. However, *Ucp1* expression is not necessarily indicative of UCP-1 protein expression or active thermogenesis [34]. Additionally, BMAT is strongly resistant to $\beta$-adrenergic-stimulated lipolysis, an important attribute of BAT which is required for thermogenesis [35]. Thus, BMAT is a truly independent adipose depot whose function is not entirely understood but is clearly distinct from that of either WAT or BAT.

## Bone marrow niche cells arise from multipotent progenitor cells

To understand the factors by which BMAT differs from peripheral adipose depots, we must consider the unusual nature of the bone marrow microenvironment and its resident cells. Conceptually, the cells of the bone marrow can be divided into two major categories:

hematopoietic cells, which perform the primary marrow function of creating new blood cells, and mesenchymal (stromal) cells, which provide structural, nutritional, and signaling support to the hematopoietic cells [36]. These marrow cells can quickly expand to meet immune, hemostatic, or metabolic demands but they are volumetrically constrained by the surrounding ossified matrix of the bone. As a result, expansion of any individual marrow cell type is a zero-sum game, in which other marrow cells must reduce their number or volume to meet the space constraint of the medullary cavity. The balance between hematopoietic and mesenchymal lineage cells within the marrow is often a contributing factor to pathogenesis in marrow, and therefore represents a critical regulatory checkpoint for healthy marrow function.

On the one hand, hematopoietic stem cells (HSCs) are proliferating multipotent cells that reside in specialized endosteal and perivascular niches and differentiate to common myeloid and lymphoid precursor cells, which in turn give rise to erythrocytes, lymphocytes, thrombocytes, and osteoclasts [37]. On the other hand, mesenchymal cells arise from self-renewing multipotent cells called mesenchymal stem cells (MSCs), which differentiate into osteoblasts, adipocytes, chondrocytes, stromal, and vascular cells. MSCs give rise to the bone structure during embryonic development, first through cartilage-producing chondrocytes and then by deposition of osseous tissue by osteoblasts [38]. Mesenchymal cells also give rise to the vascular endothelium, which allows for the rapid vascularization of newly made bone tissue, allowing circulating embryonic HSCs to populate within the bone through proliferation and differentiation [38]. At birth, the marrow of mammals appears red throughout due to high hematopoietic cellularity. The marrow grows more yellow with age as MSCs increasingly differentiate into adipocytes and accumulate lipid, first in distal skeletal sites then expanding proximally, giving rise to cBMAT and rBMAT, respectively [31]. In addition to age, the accumulation of BMAT is also stimulated by, or associated with, obesity, type 2 diabetes mellitus (T2DM), osteoporosis, anorexia nervosa, and exposure to ionizing radiation [12,23,31,39,40]. The wide range of stimuli that influence marrow adiposity, including metabolic disease, make BMAT an intriguing area of study that requires consideration of interactions within the marrow that lead to cell lineage allocation and development.

## Factors from adipose and bone influence the fate of MSCs

Perhaps the most striking example of the close relationship between BMAs and another MSC-derived cell is demonstrated by osteoblasts. Longstanding observations have noted an inverse relationship between marrow adiposity and bone volume and mineralization, where high adiposity associates with diminished bone formation due to reduced osteoblast number and activity [39,41]. Importantly, BMAs and osteoblasts share a common mesenchymal progenitor cell [42]; thus, promoting differentiation of MSCs into BMAs (adipogenesis) may deplete the precursor cell population needed for osteoblast differentiation (osteogenesis), and vice versa. Additionally, paracrine factors from multiple marrow cell types, including adipocytes, osteocytes, and endothelial cells, can influence the fate of the common progenitor by directing MSCs toward adipogenesis or osteogenesis (Fig. 4.1).

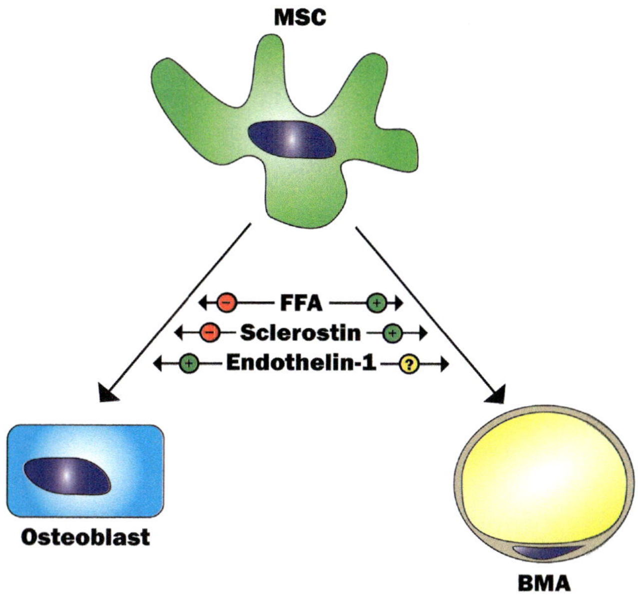


FIGURE 4.1 **Osteo-adipogenic marrow progenitor fate is regulated by paracrine factors from multiple marrow cell types.** Mesenchymal stem cells (*MSC*) serve as an osteo-adipogenic precursor in the bone marrow. MSCs are directed toward adipogenesis by free fatty acids (*FFA*) released by adipocytes, and sclerostin from osteocytes. MSCs can also be directed toward osteogenesis by endothelin-1 secreted by marrow vascular endothelial cells.

## Marrow adiposity and bone formation are not always inversely related

While the inverse relationship between bone density and BMAT is well documented, there are also many states, in which bone formation and marrow adiposity appear directly proportional to one another, suggesting that additional inputs drive the bone-adipose relationship. For example, in humans, distal marrow sites display a sharp increase in adiposity during puberty, which is concurrent with high rates of skeletal acquisition [24]. With aging, however, marrow adiposity continues to increase while bone mass drops precipitously. Further, mouse models do not always demonstrate a clear inverse relationship between marrow adiposity and bone formation. For example, C57BL/6J mice display relatively low marrow adiposity with normal bone mass, whereas C3H/HeJ mice have very high marrow adiposity and bone mass [43]. Other mouse models show direct, inverse, or no relationship at all between bone mass and adiposity [43]. Furthermore, mice with a loss-of-function mutation in the KIT receptor (CD117) have profound depletion of BMAT, with no BMAs present in tibiae or lumbar vertebrae, yet the mice are not protected from ovariectomy-induced osteopenia. This finding suggests that BMAT expansion is not required for negative bone phenotypes, and argues against the "space filling" model of BMAT expansion whereby BMAs simply fill any voids resulting from hypocellularity [44]. Of note, since KIT signaling is utilized in multiple bone marrow progenitors and interruption of KIT signaling can lead to apoptosis, it is possible that widespread depletion of mesenchymal precursors leads to the observed phenotype [45,46]. To understand how the relationship between BMAT and bone can display such heterogeneity, we must consider other signaling pathways that can alter adipocytes, osteoblasts, or their precursor cells.

## PPAR-γ directs MSC fate toward adipogenesis

Activation of the nuclear receptor and transcription factor PPAR-γ, a master regulator of WAT adipogenesis, has emerged as an important factor in controlling the inverse relationship of BMAs and osteoblasts by directing the lineage commitment of mesenchymal precursors toward adipogenesis [47,48]. Inactivating PPAR-γ in mice through genetic haploinsufficiency increases bone formation and decreases BMA abundance in the lower limbs of aged mice [49]. Furthermore, isolated bone marrow cultures from PPAR-γ haploinsufficient mice are more osteogenic and less adipogenic compared to wildtype marrow, suggesting that PPAR-γ insufficiency stimulates osteogenesis and inhibits adipogenesis of mesenchymal precursors [49]. Similarly, pharmacologic inhibition of PPAR-γ in a type 1 diabetes mellitus mouse model causes a marked reduction in marrow adiposity, though no rescue of bone loss was noted [50]. Agreeing with these results in respect to adiposity, activation of PPAR-γ by thiazolidinedione-related compounds leads to dramatic expansion of BMAT; however, bone volume, density, and strength are markedly reduced [51,52].

Since PPAR-γ has numerous endogenous ligands, including free fatty acids, metabolites of lipoproteins and lipid mediators, and platelet activating factor, systemic or local ligands may exert control over MSC lineage fate decisions affecting bone marrow adiposity [53]. The action of free fatty acids as PPAR ligands has been particularly well-studied and has revealed that polyunsaturated fatty acids (PUFA), especially omega-3 species, are potent endogenous activators of PPAR-γ [54,55]. Interestingly, BMAT triacylglycerol (TAG) stores are enriched in long-chain PUFAs compared to WAT; even within bone marrow there is a gradient of increasing lipid unsaturation from rBMAT to cBMAT [9,11]. This observation opens the possibility that adipose lipid stores may act as a pool of PPAR-γ ligands that can be mobilized by lipolysis to influence the lineage commitment of MSCs. If true, BMAT may have an especially potent effect due to its high proportion of long-chain PUFA and its close proximity to MSCs. This could, in part, explain the paradoxical expansion of BMAT during periods of nutrient deprivation when fatty acid mobilization is high [23–25]. It is also possible that systemically high fatty acid levels from diet or WAT lipolysis could influence PPAR-γ activation to alter the bone marrow environment. For example, plasma free fatty acids are greatly increased in mice fed a high fat diet (HFD), which also demonstrates increased marrow adipogenesis and depletion of mesenchymal progenitors resulting in low bone mass [56]. This may allow sequestration of excess free fatty acids through esterification in BMA TAG stores, which could protect peripheral tissues from deleterious effects of fatty acids at the expense of bone formation.

## Bone cells regulate MSC fate

In addition to PPAR-γ-mediated mechanisms, MSCs can also be directed toward adipogenesis via signaling from bone cells. Osteocytes are now recognized as important regulators of the marrow niche. For example, osteocytes secrete the Wnt inhibitor sclerostin, which acts through paracrine mechanisms to inhibit bone formation and induce marrow adipogenesis (Fig. 4.1) [57,58]. Osteocyte-derived sclerostin travels to the marrow space through the canaliculi of the ossified matrix, where it acts to inhibit Wnt signaling in endochondral MSCs [58,59]. Since Wnt signaling in osteoblasts and their precursors increases the expression of Runx2 and other osteogenic genes to induce osteoblast differentiation and proliferation, inhibition of Wnt signaling by sclerostin inhibits bone formation [58,60]. Interestingly, Runx2 also binds to the sclerostin promoter, along with osterix, to increase sclerostin expression [61]. Additionally, Wnt signaling has been shown to inhibit adipogenic transcription factors PPAR-γ and CCAAT/enhancer binding protein alpha (C/EBPα) [59,62,63]. Thus, sclerostin secreted from osteocytes can direct endosteal MSCs toward adipogenesis by inhibiting Wnt-activated osteogenic programming and alleviating Wnt repression of adipogenic transcription factors.

### Osteocalcin promotes peripheral insulin sensitization

As discussed, much of the regulation between bone formation and marrow adiposity is mediated by reprogramming progenitor cells; however, the skeleton can also directly regulate adipose function broadly via paracrine and endocrine signaling to promote metabolic homeostasis. A clear example of the metabolic-modulatory effect of the skeleton is demonstrated by osteocalcin, a peptide secreted from osteoblasts that has been implicated in multiple pathways of insulin sensitization [64]. Insulin receptor-mediated signaling in osteoblasts induces production and secretion of osteocalcin through the transcription factors Twist2 and Runx2 [65]. In wildtype mice, secreted osteocalcin sensitizes white and brown adipocytes to insulin by increasing expression of adiponectin and PPAR-γ coactivator 1-alpha (PGC1α), respectively, and by reducing obesity-induced endoplasmic reticulum stress through an NF-κB-dependent mechanism [64,66]. Furthermore, central adeno-associated virus overexpression of leptin increases serum osteocalcin and improves metabolic parameters in wildtype, obese, and insulinopenic mice. This opens the possibility of a positive feedback loop in which osteocalcin supports adipogenesis and adipocyte production of leptin drives further osteocalcin release via hypothalamic relay circuits [67]. In agreement, leptin has previously been shown to ameliorate calorie restriction-induced reduction of osteocalcin [68]. These findings are consistent with observations in humans with T2DM that increased serum osteocalcin correlates with improved insulin sensitivity, decreased fasting glucose concentration, and increased serum adiponectin [69,70]. The circuitry whereby osteoblasts regulate insulin sensitivity of adipocytes is somewhat clouded by evidence that human adipose tissue and cultured primary adipocytes also express osteocalcin at the gene and protein level [71]. Whether or not physiologically meaningful amounts of osteocalcin are secreted from WAT in vivo remains unknown. What is clear concerning the bone and BMAT is that the relationship between these two tissues is highly regulated by local and systemic factors to influence bone formation, metabolism, and the fate of their common mesenchymal progenitor.

## CXCL12-expressing stromal cells serve as an osteo-adipogenic progenitor and are required to support hematopoiesis

Early anatomists divided organs into two components: the parenchyma, or functional unit, and the stroma, meaning "bed covering," which provides structural support for the functional components of an organ [72]. The stroma of bone marrow includes mesenchymal precursors, blood vessels, neurons, reticulocytes, and more. These cells have proven vital to the development, maintenance, and signaling of the bone marrow, making them much more than simply a "bed" for hematopoietic and bone-forming cells. A subset of mesenchymal derived stromal cells marked by high expression of C-X-C motif chemokine ligand 12 (CXCL12), also known as stromal-derived factor 1α, have emerged as an important regulator of bone formation, adipogenesis, and hematopoiesis in the marrow [73,74]. Often referred to as CXCL12-abundant reticular cells (CAR) cells, these MSCs have reticular morphology and support and recruit surrounding cells through secretion of various chemokines, including CXCL12, interleukin-7 (IL-7), and stem cell factor (SCF) [73,75]. CAR cells are prevalent at the endosteum and perivascular niches and are theorized to play a major role in recruitment and maintenance of HSCs [76]. Supporting this theory, deletion of the CXCL12 receptor CXCR4 in mice results in a severe reduction of HSC due to depletion of the quiescent HSC pool [76]. While CAR cells have an important role in supporting the HSC niche, they are not terminally differentiated cells and may serve as progenitors for multiple mesenchymal-lineage marrow cells, including adipocytes and osteoblasts [73]. Therefore, the decision point that regulates fate determination of CAR cells has important implications for bone marrow composition and metabolism. For example, the transcription factor early B-cell 3 (Ebf3) is preferentially expressed in a self-renewing subset of CAR cells and when knocked out in mice, along with Ebf1, CAR cell self-renewal is severely diminished, resulting in impairment of HSC niche function [77]. With impaired self-renewal abilities, CAR cells differentiate into osteoblasts, leading to profound osteosclerosis and ultimately trabecularization of the entire medullary cavity [77].

CXCL12 is also highly expressed by leptin-receptor positive stromal cells ($LepR^+$), an extensively studied BMAT and osteoblast precursor, and there is likely a large overlap between CAR cells and $LepR^+$ cells [74,78–80]. $LepR^+$ cells arise postnatally and are the source of most BMAT and bone formed in adulthood [74]. Like CAR cells, CXCL12 and SCF secreted from $LepR^+$ cells are vital for maintenance of HSC self-renewal and hematopoiesis [78,81]. Approximately 5% of $LepR^+$ stromal cells are traced by adiponectin-Cre/ER and act as progenitors for SCF-expressing BMAs [81]. When SCF is deleted from adipocytes and adiponectin positive $LepR^+$ cells, post-irradiation hematopoietic recovery is severely diminished, indicating a role for BMAT and/or its progenitors in supporting hematopoiesis [81]. However, while multipotent MSCs have been shown to support hematopoiesis [78], there is evidence that both mature BMAs and adipocyte-committed MSCs negatively regulate HSCs [82,83]. Thus, it is likely that the observed hematopoietic support from SCF can largely be attributed to MSCs rather than BMAs.

A recent study by Mukohira et al. discovered that over 99% of CXCL12-expressing MSCs are efficiently targeted by the adiponectin-Cre transgene, highlighting the close relationship of these cells to BMAs [75]. Furthermore, approximately 85% of all marrow cells targeted by adiponectin-Cre also express CXCL12 [75]. Deletion of IL-7 in adiponectin-Cre marrow cells diminishes B lymphopoiesis [75], consistent with IL-7 conditional deletion driven by LepR-Cre [80], highlighting the importance of CXCL12-expressing stromal cells in regulating hematopoiesis [75]. These results also indicate a functional overlap between the adiponectin-Cre cells studied by Mukohira et al. and the $LepR^+$ cells studied by Zhou et al. [75,81]. The apparent discrepancy between the percentage of $CXCL12^+$ cells targeted by adiponectin-Cre and the percentage of $LepR^+$ cells [81], which are overwhelmingly $CXCL12^+$, targeted by adiponectin-Cre/ER is likely due to differences in Cre expression between the two transgenes [75].

Together, these findings position CXCL12-expressing stromal cells, including CAR cells, at an important intersection of bone marrow cell fate determination; differentiation into osteoblasts or adipocytes could benefit skeletal strength or convey a protective metabolic effect through adipokine secretion, respectively [20,73]. However, either of these cell fate decisions would negatively affect the HSC niche by diminishing CAR-cell secreted CXCL12 and SCF that is required to maintain HSC quiescence [73,75,80]. Fully characterizing the population of CXCL12-expressing MSCs, their contributions to the HSC niche, and their role as osteo-adipogenic progenitors are important areas of study that are currently being addressed.

## Bone marrow vascular endothelial cells regulate MSC development and peripheral endothelial dysfunction

The bone marrow has a dense network of arterioles and sinusoids that supply nutrients to the marrow and serve as highways to move nascent hematopoietic cells and progenitors throughout the body. Each of these vessels is lined by specialized endothelial cells that regulate vasodilation, permeability, and cell-cell signaling within the marrow niche [84]. A growing body of evidence suggests that vascular endothelial cells also help regulate the composition of the marrow microenvironment and are involved in numerous metabolic disorders. For example, the low permeability of bone marrow arterioles creates a hypoxic extravascular environment that is low in reactive oxygen species and helps maintain HSC quiescence, whereas leaky sinusoids encourage differentiation of HSCs and act as the entry point of leukocytes into circulation [85]. Bone marrow vascular endothelial cells also secrete numerous factors that can influence surrounding cells, such as the vasoconstrictor endothelin-1 and the notch-ligand DLL-4 [86,87]. Endothelin-1 directs cultured MSCs toward osteogenesis in an endothelin-receptor-dependent manner but causes increased adipogenesis in adipose-derived stem cells, suggesting that endothelin-1 paracrine signaling from the vascular endothelium can direct cell lineage commitment differentially based on tissue type (Fig. 4.1) [86,88]. Furthermore, vascular endothelial cells modulate their secretome in response to stressors within the marrow microenvironment. Single cell RNA-sequencing of mouse bone marrow cells stressed with exposure to the chemotherapy drug 5-fluorouracil revealed downregulation of vascular endothelial DLL-4 leading to myeloid transcriptional reprograming of HSCs [87]. The vascular endothelial secretome is quite complex, containing over 100 proteins depending on the tissue of origin and many of these proteins are glycosylated or contain transmembrane domains [89,90]. Understanding the autocrine, paracrine, and endocrine roles of these secreted factors will be essential to understanding numerous cell-fate decisions that regulate the striking cellular plasticity of the bone marrow microenvironment.

### Bone marrow vascular endothelial cells mobilize to repair dysfunctional peripheral endothelium

Increased incidence of cardiovascular disease and corresponding vascular endothelial dysfunction is a significant comorbidity of diabetes mellitus and is the primary cause of death among all diabetes patients [91]. Patients suffering from T2DM have markedly fewer circulating endothelial progenitor cells (EPC), which may explain why the endothelium is dysfunctional in diabetes [92]. EPCs are bone-marrow derived progenitor cells that closely resemble circulating hematopoietic progenitors but are utilized for angiogenesis and vascular endothelial repair in response to injury. The contribution of EPCs to endothelial dysfunction in metabolic disease has been reviewed by Yie and Tse and highlights the importance of cellular interactions within the bone marrow for maintenance of metabolic homeostasis [93]. One mechanism by which diabetes reduces circulating EPCs is by interfering with EPC mobilization from the bone marrow. EPCs are drawn to ischemic tissues from the bone marrow via cytokine signaling from stromal-derived CXCL12, among others [93]. Bone marrow CXCL12 is increased in diabetes, which reduces the chemoattractant gradient between the marrow and ischemic tissue bed, thus preventing migration of EPCs from the marrow. Degradation of CXCL12 in the marrow by MMP-9 is stimulated by granulocyte colony stimulating factor (G-CSF) to increase the outward chemoattractant gradient, promoting EPC migration [94]. G-CSF has also been implicated in depletion of BMAs following bariatric surgery [95]. Together, these findings opening the possibility that promoting EPC migration could potentially negatively regulate marrow adiposity. Interestingly, the insulin-sensitizing frontline pharmacological treatment for T2DM, metformin, increases angiogenesis, wound healing, and circulating EPCs, suggesting that its benefit may be due in part to reestablishing proper endothelial function [96].

## Sympathetic nervous system activation fails to induce "browning" of BMAT but induces caloric restriction-induced BMAT expansion

The nervous system plays an important role in coordinating the behavior of adipose depots, particularly with respect to lipolysis, as demonstrated by lipolysis-dependent non-shivering thermogenesis in response to cold exposure [97]. Axons from sympathetic neurons innervate BAT, WAT, and BMAT depots, and their effects on adipose metabolism have been extensively studied, particularly in BAT. Norepinephrine released from sympathetic nerve endings activates β3-adrenergic receptors (β3AR) of brown adipocytes to activate PGC1α, which promotes transcription of uncoupling protein 1 (UCP1); UCP1 in turn decouples the mitochondrial electron transport chain from ATP generation, causing production of heat [97]. Hormone sensitive lipase (HSL) activity is also increased by β3AR-mediated signaling to mobilize free fatty acids from esterified stores and fuel the increased β-oxidation of decoupled mitochondria. Brown adipocytes are particularly well-suited for heat generation due to the huge number of mitochondria relative to other adipose depots; indeed, the distinctive brown color of BAT is caused by the adipocyte's small multilocular lipid droplets surrounded by a vast network of mitochondria. Cold exposure increases lipid mobilization and utilization of thermogenic BAT, which confers a protective metabolic phenotype by removing excess fatty acids from the blood of animals on a nutrient-excessive diet [98].

Unlike BAT, white adipocytes possess one massive unilocular lipid droplet and have all organelles and cytoplasm pushed toward the cell surface. However, some WAT depots also respond to sympathetic activation by increasing mitochondrial activity and lipolysis [99]. This process, called beiging, changes the morphology of the cell and also confers a similar, though less marked, metabolically protective effect as BAT activation. The similarities in thermogenic programing between BAT and WAT beg the question as to what role BMAT may play in sympathetic activation of thermogenesis. As mentioned earlier, BMAs express low levels of some hallmark BAT genes and have tight associations between their lipid droplets and mitochondria, suggesting the necessary hardware is in place for thermogenesis [15,29–31]. In vivo, gene expression of both *Ucp1* and *Prdm16* appears elevated in human bone marrow samples compared to white adipocytes, and murine vertebral BMAT contains detectable levels of UCP-1 protein by immunofluorescence/immunohistochemistry [100]. Additionally, cold stimulation increases fluorodeoxyglucose (18-FDG) uptake, a measure of glucose uptake, in human vertebral bone marrow [100]. However, vertebral marrow contains few BMAs, so it is unlikely that the cold-induced glucose uptake can be attributed to BMAT [12]. A recent study performed under similar conditions demonstrated that cold exposure increases glucose uptake in bone and red marrow, but not in BMAT [preprint] [22]. Together, these data suggest that bone marrow may respond to sympathetic activation of thermogenesis independently of BMAT.

## Sympathetic activation promotes MSC mobilization and adipogenesis

Despite expressing some BAT genes at higher levels than WAT, BMAT is resistant to βAR-stimulated lipolysis when compared to WAT depots [35]. BMAT displays profound resistance to the pan-βAR agonist isoproterenol as well as the β3AR-specific agonist CL316,243(CL), as measured by HSL phosphorylation to indicate active lipolysis. Furthermore, BMAT resistance to β3AR agonism displays a site-specific effect in which rBMAT (red marrow) glycerol release from lipolysis is slightly increased (1.5× vs 15× in WAT) but cBMAT (yellow marrow) glycerol release is unchanged [35]. However, CL treatment causes lipid droplet remodeling of distal femoral BMAs, indicating that BMAT is somewhat sensitive to β3AR signaling [35]. Thus, while BMAT gene and protein expression of BAT markers may be higher than that of WAT, it is functionally less responsive to βAR-stimulated lipolysis than white adipocytes. That said, recent evidence illuminates a potential role for bone marrow progenitor cells in facilitating βAR-induced browning of white adipocytes. Work by Yoneshiro et al. demonstrates a complete failure of β3AR-stimulated inguinal WAT (iWAT) beiging in mice with irradiation-depleted bone marrow, yet BAT activation remains unaffected [101]. Furthermore, transplantation of wildtype marrow progenitors to repopulate the bone marrow restores CL-stimulated beiging of the iWAT. When $GFP^+$ bone marrow cells are transplanted into irradiated mice, CL increases $GFP^+$ cell migration to iWAT and these cells give rise to $UCP1^+$ multilocular adipocytes [101]. These results suggest that β3AR signaling mobilizes bone marrow cells to iWAT, via circulation, to facilitate thermogenic beiging of that tissue (Fig. 4.2) [101]. These findings may help explain why BMAT expresses thermogenic genes and proteins despite its strong resistance to βAR-induced thermogenesis. Additionally, peripheral development of bone marrow-derived precursors presents an intriguing pathway in which the marrow can exert control over systemic metabolism.

Although the sympathetic nervous system does not activate browning or beiging of BMAT, there appear to be multiple processes in marrow adipocytes that depend on direct βAR signaling. For example, βAR blockers diminish BMAT expansion and ameliorate bone loss and hypoleptinemia in calorie-restricted Sprague-Dawley rats compared to *ad libitum* controls [102]. In agreement, C57BL/6 mice treated with the β-blocker propranolol are protected from BMAT expansion induced by either caloric restriction or overfeeding [103]. Furthermore, βAR agonism via isoproterenol stimulates adipogenic differentiation of cultured MSCs and diminishes the anti-adipogenic effect of cultured osteoblasts on undifferentiated 3T3-L1 cells [103]. These findings suggest that βAR signaling contributes to lineage allocation of osteo-adipogenic progenitor cells to favor BMAT expansion when nutrient availability is limited. Sympathetic signaling can also act directly on mature osteoblasts via β2AR-mediated pathways which support osteoblast proliferation to promote bone formation [104].

### Sympathetic activation alters adipocyte function

Beyond directing marrow adipogenesis, sympathetic activation may directly act on mature BMAs to alter their function. Cultured 3T3-L1 and primary mouse white adipocytes treated with CL increase adiponectin exocytosis dependent on Epac (exchange protein directly activated by cAMP), but adiponectin secretion could not be stimulated in primary adipocytes isolated from obese and T2DM mice (Fig. 4.2) [105,106]. Although these experiments utilized white adipocytes and thus may not reflect the response of BMAs, the results are consistent with longstanding in vivo observations that both adipose sympathetic activity and serum adiponectin concentration are increased by caloric restriction [107,108]. These findings present a possible mechanism to explain the paradoxical increase of adiponectin accompanying WAT atrophy during caloric restriction, by potentiating adiponectin secretion through β-adrenergic signaling. However, since BMAs are strongly resistant to β3AR-induced lipolysis, further work is needed to determine the extent to which BMAs are sensitive to other β3AR-activated pathways.

## BMAs and their progenitors support bone marrow malignancies

Just as interactions between marrow cells work to affect the development, proliferation, and behavior of neighboring marrow cells, they also regulate cancerous cells that develop in the bone marrow or are deposited by circulating malignancies. Common bone marrow cancers include multiple myeloma, leukemia, and skeletal metastasis from distant tumors. The properties that make the bone marrow microenvironment uniquely well-suited to care for its self-renewing MSC and HSC pools may also make it fertile soil for the growth and development of cancer cells. The specific attributes of the marrow niche that favor cancerous growth have previously been reviewed in detail [109]. Like BMAT expansion, tumor growth in the confined space of the marrow requires depletion of other marrow cells. Tumors often accomplish this feat by stimulating bone resorption. In turn, factors from the broken down bone (extracellular matrix, growth factors, calcium, etc.) stimulate further growth of the tumor, forming a so-called "viscous cycle" that encourages tumor growth [109]. Bone marrow cancers often invade the same hypoxic niche that is populated by HSCs and surrounded by BMAs, leading some to speculate that adipocytes may play a role in the progression of marrow tumors. Such tumor-adipose interactions are not unprecedented in the literature; metastatic ovarian cancer cells migrate to the omental adipose tissue depot and mobilize adipocyte lipid stores to fuel their own metabolic needs [110]. Furthermore, diet-induced BMAT expansion is associated with increased growth and invasiveness of prostate cancer skeletal tumors, partially by upregulating FABP4 and IL-1β in the cancer cells [111]. Further work is needed to determine if similar adipocyte-tumor interdependencies occur in other marrow cancers like multiple myeloma and leukemia.

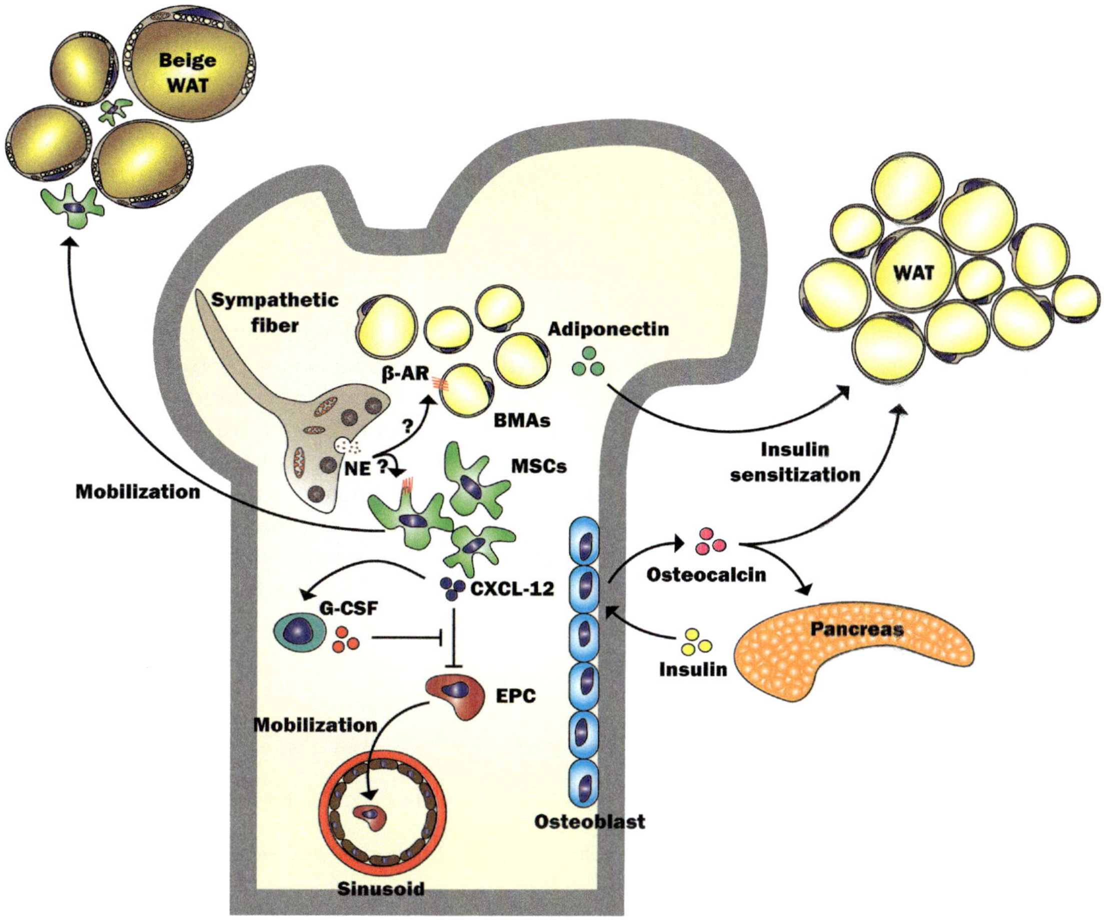


FIGURE 4.2 **Mesenchymal stem cells (*MSCs*) and their daughter cells coordinate peripheral metabolism.** MSCs directly influence peripheral metabolism by mobilizing to white adipose depots to facilitate WAT beiging process, in response to β-adrenergic receptor(β-AR)-mediated signaling, possibly by norepinephrine (*NE*) from innervating sympathetic fibers. MSCs also promote hematopoiesis and inhibit endothelial progenitor cell (*EPC*) migration through the paracrine action of CXCL12. Furthermore, MSCs differentiate into bone marrow adipocytes (*BMAs*) or osteoblasts, and both of which promote insulin sensitization through the endocrine action of adiponectin and osteocalcin, respectively.

Multiple myeloma (MM) is a common bone marrow cancer caused by B-cell malignancy that leads to uncontrolled proliferation of plasma cells. The growth and survival of MM cells are highly dependent on signals from neighboring cells, and the nature of these complex interactions have been reviewed extensively [109,112,113]. As with other bone cancers, MM favors osteolytic conditions caused by simultaneous stimulation of osteoclasts and inhibition of osteoblasts, which further support tumor growth in a viscous cycle. Furthermore, in vitro experiments show that differentiated osteoblasts suppress MM growth and survival, suggesting that inhibition of osteoblasts may further potentiate MM development [114]. There is also ample evidence to suggest that MSCs and adipocytes may support the progression of MM [112,113]. MSCs isolated from MM patients display enhanced expression of genes involved in angiogenesis and tumor growth compared to normal MSCs, and correspondingly, MM MSCs promote growth of myeloma cells in culture while suppressing osteoblastic differentiation [115]. The role for adipocytes in supporting MM progression is further supported by epidemiological observations that measures of increasing adiposity are correlated with greater risk of MM mortality [116]. Taken together, these data highlight the importance of cell-cell interactions in the bone marrow microenvironment, particularly with adipocytes, in regulating the metabolism and progression of marrow cancers.

In summary, the bone marrow microenvironment represents a complex and dynamic arrangement of cellular interactions that work together to maintain skeletal, metabolic, and immune regulation. When the balance of any one of these functions is perturbed, the repercussions to the host can be devastating and far reaching. A deeper understanding of the development of these diverse cells from an assortment of multipotent progenitor cells will improve our ability to treat diseases that arise from their dysfunction. Ever-advancing techniques in the fields of metabolomics, single-cell transcriptomics, and lineage tracing are revealing the complexities of the bone marrow microenvironment in more detail than possibly could have been imagined just a short time ago. These techniques will be essential to fully characterize the development of mesenchymal and hematopoietic cells from their progenitors, and uncover how that development is altered in metabolic disease.

## References

[1] Cooper B. The origins of bone marrow as the seedbed of our blood: from antiquity to the time of Osler. Proc (Bayl Univ Med Cent) 2011;24(2):115–8.

[2] Neumann FEC. Du rôle de la moelle des os dansla formations du sang. Comptes rendus de l'Académie des Sciences 1869;68:1112–3.

[3] Neumann FEC. Das Gesetz Verbreitung des gelben und roten Markes in den Extremitatenknochen. Centralblatt für die medicinischen Wissenschaft 1882;20:321–3.

[4] Dickson WEC. The bone marrow. New York: Longman, Green & Co; 1908.

[5] Wells HG. Adipose tissue, A neglected subject. JAMA 1940;114(23):2284–9.

[6] Newlin HE, Mccay CM. Bone marrow for fat storage in rabbits. Arch Biochem 1948;17(1):125–8.

[7] Zakaria E, Shafrir E. Yellow bone marrow as adipose tissue. Proc Soc Exp Biol Med 1967;124(4):1265–8.

[8] Zach E, Shafrir E. Composition of bone marrow adipose tissue in relation to body fat depots in various species. Isr J Med Sci 1974;10(12):1541–50.

[9] Tavassoli M. Marrow adipose cells. Histochemical identification of labile and stable components. Arch Pathol Lab Med 1976;100(1):16–8.

[10] Tavassoli M. Marrow adipose cells. Ultrastructural and histochemical characterization. Arch Pathol 1974;98(3):189–92.

[11] Scheller EL, Doucette CR, Learman BS, Cawthorn WP, Khandaker S, Schell B, et al. Region-specific variation in the properties of skeletal adipocytes reveals regulated and constitutive marrow adipose tissues. Nat Commun 2015;6:7808.

[12] Scheller EL, Rosen CJ. What's the matter with MAT? Marrow adipose tissue, metabolism, and skeletal health. Ann N Y Acad Sci 2014;1311:14–30.

[13] Scheller EL, Burr AA, MacDougald OA, Cawthorn WP. Inside out: bone marrow adipose tissue as a source of circulating adiponectin. Adipocyte 2016;5(3):251–69.

[14] Hardouin P, Rharass T, Lucas S. Bone marrow adipose tissue: to be or not to be a typical adipose tissue? Front Endocrinol (Lausanne) 2016;7:85.

[15] Liu LF, Shen WJ, Ueno M, Patel S, Kraemer FB. Characterization of age-related gene expression profiling in bone marrow and epididymal adipocytes. BMC Genomics 2011;12:212.

[16] Shao X, Wang M, Wei X, Deng S, Fu N, Peng Q, et al. Peroxisome proliferator-activated receptor-γ: master regulator of adipogenesis and obesity. Curr Stem Cell Res Ther 2016;11(3):282–9.

[17] Hayman DM, Blumberg TJ, Scott CC, Athanasiou KA. The effects of isolation on chondrocyte gene expression. Tissue Eng 2006;12(9):2573–81.

[18] Parlee SD, Lentz SI, Mori H, MacDougald OA. Quantifying size and number of adipocytes in adipose tissue. Methods Enzymol 2014;537:93–122.

[19] Cawthorn WP, Scheller EL, Parlee SD, Pham HA, Learman BS, Redshaw CM, et al. Expansion of bone marrow adipose tissue during caloric restriction is associated with increased circulating glucocorticoids and not with hypoleptinemia. Endocrinology 2016;157(2):508–21.

[20] Cawthorn WP, Scheller EL, Learman BS, Parlee SD, Simon BR, Mori H, et al. Bone marrow adipose tissue is an endocrine organ that contributes to increased circulating adiponectin during caloric restriction. Cell Metab 2014;20(2):368–75.

[21] Latva-Rasku A, Honka MJ, Stancˇáková A, Koistinen HA, Kuusisto J, Guan L, et al. A Partial loss-of-function variant in AKT2 is associated with reduced insulin-mediated glucose uptake in multiple insulin-sensitive tissues: a genotype-based callback positron emission tomography study. Diabetes 2018;67(2):334–42.

[22] Suchacki KJ, Tavares AAS, Mattiucci D, Scheller EL, Papanastasiou G, Gray C, et al. . Bone marrow adipose tissue is a unique adipose subtype with distinct roles in systemic glucose homeostasis 2019;. [Preprint].

[23] Abella E, Feliu E, Granada I, Millá F, Oriol A, Ribera JM, et al. Bone marrow changes in anorexia nervosa are correlated with the amount of weight loss and not with other clinical findings. Am J Clin Pathol 2002;118(4):582–8.

[24] Devlin MJ, Cloutier AM, Thomas NA, Panus DA, Lotinun S, Pinz I, et al. Caloric restriction leads to high marrow adiposity and low bone mass in growing mice. J Bone Miner Res 2010;25(9):2078–88.

[25] Devlin MJ. Why does starvation make bones fat? Am J Hum Biol 2011;23(5):577–85.

[26] Fu Y, Luo N, Klein RL, Garvey WT. Adiponectin promotes adipocyte differentiation, insulin sensitivity, and lipid accumulation. J Lipid Res 2005;46(7):1369–79.

[27] Kim JY, van de Wall E, Laplante M, Azzara A, Trujillo ME, Hofmann SM, et al. Obesity-associated improvements in metabolic profile through expansion of adipose tissue. J Clin Invest 2007;117(9):2621–37.

[28] Stern JH, Rutkowski JM, Scherer PE. Adiponectin, leptin, and fatty acids in the maintenance of metabolic homeostasis through adipose tissue crosstalk. Cell Metab 2016;23(5):770–84.

[29] Robles H, Park S, Joens MS, Fitzpatrick JAJ, Craft CS, Scheller EL. Characterization of the bone marrow adipocyte niche with three-dimensional electron microscopy. Bone 2019;118:89–98.

[30] Krings A, Rahman S, Huang S, Lu Y, Czernik PJ, Lecka-Czernik B. Bone marrow fat has brown adipose tissue characteristics, which are attenuated with aging and diabetes. Bone 2012;50(2):546–52.

[31] Scheller EL, Cawthorn WP, Burr AA, Horowitz MC, MacDougald OA. Marrow adipose tissue: trimming the fat. Trends Endocrinol Metab 2016;27(6):392–403.

[32] de Jesus LA, Carvalho SD, Ribeiro MO, Schneider M, Kim SW, Harney JW, et al. The type 2 iodothyronine deiodinase is essential for adaptive thermogenesis in brown adipose tissue. J Clin Invest 2001;108(9):1379–85.

[33] Lindsey RC, Mohan S. Thyroid hormone acting via TRβ induces expression of browning genes in mouse bone marrow adipose tissue. Endocrine 2017;56(1):109–20.

[34] Nedergaard J, Cannon B. UCP1 mRNA does not produce heat. Biochim Biophys Acta 2013;1831(5):943–9.

[35] Scheller EL, Khandaker S, Learman BS, Cawthorn WP, Anderson LM, Pham HA, et al. Bone marrow adipocytes resist lipolysis and remodeling in response to β-adrenergic stimulation. Bone 2019;118:32–41.

[36] Caplan AI, Dennis JE. Mesenchymal stem cells as trophic mediators. J Cell Biochem 2006;98(5):1076–84.

[37] Cuminetti V, Arranz L. Bone marrow adipocytes: the enigmatic components of the hematopoietic stem cell niche. J Clin Med 2019;8(5):707.

[38] Mikkola HK, Orkin SH. The journey of developing hematopoietic stem cells. Development 2006;133(19):3733–44.

[39] Devlin MJ, Rosen CJ. The bone-fat interface: basic and clinical implications of marrow adiposity. Lancet Diabetes Endocrinol 2015;3(2):141–7.

[40] Meunier P, Aaron J, Edouard C, Vignon G. Osteoporosis and the replacement of cell populations of the marrow by adipose tissue. A quantitative study of 84 iliac bone biopsies. Clin Orthop Relat Res 1971;80:147–54.

[41] Shen W, Chen J, Punyanitya M, Shapses S, Heshka S, Heymsfield SB. MRI-measured bone marrow adipose tissue is inversely related to DXA-measured bone mineral in Caucasian women. Osteoporos Int 2007;18(5):641–7.

[42] Lin YF, Jing W, Wu L, Li XY, Wu Y, Liu L, et al. Identification of osteo-adipo progenitor cells in fat tissue. Cell Prolif 2008;41(5):803–12.
[43] Fazeli PK, Horowitz MC, MacDougald OA, Scheller EL, Rodeheffer MS, Rosen CJ, et al. Marrow fat and bone--new perspectives. J Clin Endocrinol Metab 2013;98(3):935–45.
[44] Iwaniec UT, Turner RT. Failure to generate bone marrow adipocytes does not protect mice from ovariectomy-induced osteopenia. Bone 2013;53(1):145–53.
[45] Huss R, Moosmann S. The co-expression of CD117 (c-kit) and osteocalcin in activated bone marrow stem cells in different diseases. Br J Haematol 2002;118(1):305–12.
[46] Domen J, Weissman IL. Hematopoietic stem cells need two signals to prevent apoptosis, BCL-2 can provide one of these, Kitl/c-Kit signaling the other. J Exp Med 2000;192(12):1707–18.
[47] Tontonoz P, Hu E, Spiegelman BM. Stimulation of adipogenesis in fibroblasts by PPAR gamma 2, a lipid-activated transcription factor. Cell 1994;79(7):1147–56.
[48] Lecka-Czernik B, Gubrij I, Moerman EJ, Kajkenova O, Lipschitz DA, Manolagas SC, et al. Inhibition of Osf2/Cbfa1 expression and terminal osteoblast differentiation by PPARgamma2. J Cell Biochem 1999;74(3):357–71.
[49] Akune T, Ohba S, Kamekura S, Yamaguchi M, Chung UI, Kubota N, et al. PPARgamma insufficiency enhances osteogenesis through osteoblast formation from bone marrow progenitors. J Clin Invest 2004;113(6):846–55.
[50] Botolin S, McCabe LR. Inhibition of PPARgamma prevents type I diabetic bone marrow adiposity but not bone loss. J Cell Physiol 2006;209(3):967–76.
[51] Lazarenko OP, Rzonca SO, Suva LJ, Lecka-Czernik B. Netoglitazone is a PPAR-gamma ligand with selective effects on bone and fat. Bone 2006;38(1):74–84.
[52] Rzonca SO, Suva LJ, Gaddy D, Montague DC, Lecka-Czernik B. Bone is a target for the antidiabetic compound rosiglitazone. Endocrinology 2004;145(1):401–6.
[53] Villacorta L, Schopfer FJ, Zhang J, Freeman BA, Chen YE. PPARgamma and its ligands: therapeutic implications in cardiovascular disease. Clin Sci (Lond) 2009;116(3):205–18.
[54] Varga T, Czimmerer Z, Nagy L. PPARs are a unique set of fatty acid regulated transcription factors controlling both lipid metabolism and inflammation. Biochim Biophys Acta 2011;1812(8):1007–22.
[55] Marion-Letellier R, Savoye G, Ghosh S. Fatty acids, eicosanoids and PPAR gamma. Eur J Pharmacol 2016;785:44–9.
[56] Tencerova M, Figeac F, Ditzel N, Taipaleenmäki H, Nielsen TK, Kassem M. High-fat diet-induced obesity promotes expansion of bone marrow adipose tissue and impairs skeletal stem cell functions in mice. J Bone Miner Res 2018;33(6):1154–65.
[57] Kim SW, Lu Y, Williams EA, Lai F, Lee JY, Enishi T, et al. Sclerostin antibody administration converts bone lining cells into active osteoblasts. J Bone Miner Res 2017;32(5):892–901.
[58] Fairfield H, Falank C, Harris E, Demambro V, McDonald M, Pettitt JA, et al. The skeletal cell-derived molecule sclerostin drives bone marrow adipogenesis. J Cell Physiol 2018;233(2):1156–67.
[59] Kang S, Bennett CN, Gerin I, Rapp LA, Hankenson KD, Macdougald OA. Wnt signaling stimulates osteoblastogenesis of mesenchymal precursors by suppressing CCAAT/enhancer-binding protein alpha and peroxisome proliferator-activated receptor gamma. J Biol Chem 2007;282(19):14515–24.
[60] Gaur T, Lengner CJ, Hovhannisyan H, Bhat RA, Bodine PV, Komm BS, et al. Canonical WNT signaling promotes osteogenesis by directly stimulating Runx2 gene expression. J Biol Chem 2005;280(39):33132–40.
[61] Pérez-Campo FM, Santurtún A, García-Ibarbia C, Pascual MA, Valero C, Garcés C, et al. Osterix and RUNX2 are transcriptional regulators of sclerostin in human bone. Calcif Tissue Int 2016;99(3):302–9.
[62] Ross SE, Hemati N, Longo KA, Bennett CN, Lucas PC, Erickson RL, et al. Inhibition of adipogenesis by Wnt signaling. Science 2000;289(5481):950–3.
[63] Kennell JA, MacDougald OA. Wnt signaling inhibits adipogenesis through beta-catenin-dependent and -independent mechanisms. J Biol Chem 2005;280(25):24004–10.
[64] Zhou B, Li H, Xu L, Zang W, Wu S, Sun H. Osteocalcin reverses endoplasmic reticulum stress and improves impaired insulin sensitivity secondary to diet-induced obesity through nuclear factor-κB signaling pathway. Endocrinology 2013;154(3):1055–68.
[65] Liu JM, Rosen CJ, Ducy P, Kousteni S, Karsenty G. Regulation of glucose handling by the skeleton: insights from mouse and human studies. Diabetes 2016;65(11):3225–32.
[66] Ferron M, Hinoi E, Karsenty G, Ducy P. Osteocalcin differentially regulates beta cell and adipocyte gene expression and affects the development of metabolic diseases in wild-type mice. Proc Natl Acad Sci U S A 2008;105(13):5266–70.

[67] Kalra SP, Dube MG, Iwaniec UT. Leptin increases osteoblast-specific osteocalcin release through a hypothalamic relay. Peptides 2009;30(5):967–73.
[68] Goldstone AP, Howard JK, Lord GM, Ghatei MA, Gardiner JV, Wang ZL, et al. Leptin prevents the fall in plasma osteocalcin during starvation in male mice. Biochem Biophys Res Commun 2002;295(2):475–81.
[69] Kanazawa I, Yamaguchi T, Tada Y, Yamauchi M, Yano S, Sugimoto T. Serum osteocalcin level is positively associated with insulin sensitivity and secretion in patients with type 2 diabetes. Bone 2011;48(4):720–5.
[70] Lee NK, Sowa H, Hinoi E, Ferron M, Ahn JD, Confavreux C, et al. Endocrine regulation of energy metabolism by the skeleton. Cell 2007;130(3):456–69.
[71] Foresta C, Strapazzon G, De Toni L, Gianesello L, Calcagno A, Pilon C, et al. Evidence for osteocalcin production by adipose tissue and its role in human metabolism. J Clin Endocrinol Metab 2010;95(7):3502–6.
[72] Shostak S. Histology's nomenclature: past, present and future. *Biol Syst Open Access* 2013;2(4):122.
[73] Omatsu Y, Sugiyama T, Kohara H, Kondoh G, Fujii N, Kohno K, et al. The essential functions of adipo-osteogenic progenitors as the hematopoietic stem and progenitor cell niche. Immunity 2010;33(3):387–99.
[74] Zhou BO, Yue R, Murphy MM, Peyer JG, Morrison SJ. Leptin-receptor-expressing mesenchymal stromal cells represent the main source of bone formed by adult bone marrow. Cell Stem Cell 2014;15(2):154–68.
[75] Mukohira H, Hara T, Abe S, Tani-Ichi S, Sehara-Fujisawa A, Nagasawa T, et al. Mesenchymal stromal cells in bone marrow express adiponectin and are efficiently targeted by an adiponectin promoter-driven Cre transgene. Int Immunol 2019;32(11):729–42.
[76] Sugiyama T, Kohara H, Noda M, Nagasawa T. Maintenance of the hematopoietic stem cell pool by CXCL12-CXCR4 chemokine signaling in bone marrow stromal cell niches. Immunity 2006;25(6):977–88.
[77] Seike M, Omatsu Y, Watanabe H, Kondoh G, Nagasawa T. Stem cell niche-specific Ebf3 maintains the bone marrow cavity. Genes Dev 2018;32(5–6):359–72.
[78] Ding L, Saunders TL, Enikolopov G, Morrison SJ. Endothelial and perivascular cells maintain haematopoietic stem cells. Nature 2012;481(7382):457–62.
[79] Nagasawa T. CXCL12/SDF-1 and CXCR4. Front Immunol 2015;6:301.
[80] Cordeiro Gomes A, Hara T, Lim VY, Herndler-Brandstetter D, Nevius E, Sugiyama T, et al. Hematopoietic stem cell niches produce lineage-instructive signals to control multipotent progenitor differentiation. Immunity 2016;45(6):1219–31.
[81] Zhou BO, Yu H, Yue R, Zhao Z, Rios JJ, Naveiras O, et al. Bone marrow adipocytes promote the regeneration of stem cells and haematopoiesis by secreting SCF. Nat Cell Biol 2017;19(8):891–903.
[82] Ambrosi TH, Scialdone A, Graja A, Gohlke S, Jank AM, Bocian C, et al. Adipocyte accumulation in the bone marrow during obesity and aging impairs stem cell-based hematopoietic and bone regeneration. Cell Stem Cell 2017;20(6):771–784.e6.
[83] Naveiras O, Nardi V, Wenzel PL, Hauschka PV, Fahey F, Daley GQ. Bone-marrow adipocytes as negative regulators of the haematopoietic microenvironment. Nature 2009;460(7252):259–63.
[84] Sivaraj KK, Adams RH. Blood vessel formation and function in bone. Development 2016;143(15):2706–15.
[85] Itkin T, Gur-Cohen S, Spencer JA, Schajnovitz A, Ramasamy SK, Kusumbe AP, et al. Distinct bone marrow blood vessels differentially regulate haematopoiesis. Nature 2016;532(7599):323–8.
[86] Lee MS, Wang J, Yuan H, Jiao H, Tsai TL, Squire MW, et al. Endothelin-1 differentially directs lineage specification of adipose- and bone marrow-derived mesenchymal stem cells. FASEB J 2019;33(1):996–1007.
[87] Tikhonova AN, Dolgalev I, Hu H, Sivaraj KK, Hoxha E, Cuesta-Domínguez Á, et al. The bone marrow microenvironment at single-cell resolution. Nature 2019;569(7755):222–8.
[88] Nelson J, Bagnato A, Battistini B, Nisen P. The endothelin axis: emerging role in cancer. Nat Rev Cancer 2003;3(2):110–6.
[89] Hocking SL, Wu LE, Guilhaus M, Chisholm DJ, James DE. Intrinsic depot-specific differences in the secretome of adipose tissue, preadipocytes, and adipose tissue-derived microvascular endothelial cells. Diabetes 2010;59(12):3008–16.
[90] Yin X, Bern M, Xing Q, Ho J, Viner R, Mayr M. Glycoproteomic analysis of the secretome of human endothelial cells. Mol Cell Proteomics 2013;12(4):956–78.
[91] Rosamond W, Flegal K, Furie K, Go A, Greenlund K, Haase N, et al. Heart disease and stroke statistics--2008 update: a report from the American Heart Association Statistics Committee and Stroke Statistics Subcommittee. Circulation 2008;117(4):e25–e146.
[92] Avogaro A, Fadini GP, Gallo A, Pagnin E, de Kreutzenberg S. Endothelial dysfunction in type 2 diabetes mellitus. Nutr Metab Cardiovasc Dis 2006;16(Suppl 1):S39–45.

[93] Yiu KH, Tse HF. Specific role of impaired glucose metabolism and diabetes mellitus in endothelial progenitor cell characteristics and function. Arterioscler Thromb Vasc Biol 2014;34(6):1136–43.
[94] Jin F, Zhai Q, Qiu L, Meng H, Zou D, Wang Y, et al. Degradation of BM SDF-1 by MMP-9: the role in G-CSF-induced hematopoietic stem/progenitor cell mobilization. Bone Marrow Transplant 2008;42(9):581–8.
[95] Li Z, Hardij J, Evers SS, Hutch CR, Choi SM, Shao Y, et al. G-CSF partially mediates effects of sleeve gastrectomy on the bone marrow niche. J Clin Invest 2019;129(6):2404–16.
[96] Yu JW, Deng YP, Han X, Ren GF, Cai J, Jiang GJ. Metformin improves the angiogenic functions of endothelial progenitor cells via activating AMPK/eNOS pathway in diabetic mice. Cardiovasc Diabetol 2016;15:88.
[97] Collins S, Surwit RS. The beta-adrenergic receptors and the control of adipose tissue metabolism and thermogenesis. Recent Prog Horm Res 2001;56:309–28.
[98] Bartelt A, Bruns OT, Reimer R, Hohenberg H, Ittrich H, Peldschus K, et al. Brown adipose tissue activity controls triglyceride clearance. Nat Med 2011;17(2):200–5.
[99] Seale P, Conroe HM, Estall J, Kajimura S, Frontini A, Ishibashi J, et al. Prdm16 determines the thermogenic program of subcutaneous white adipose tissue in mice. J Clin Invest 2011;121(1):96–105.
[100] Nishio M, Yoneshiro T, Nakahara M, Suzuki S, Saeki K, Hasegawa M, et al. Production of functional classical brown adipocytes from human pluripotent stem cells using specific hemopoietin cocktail without gene transfer. Cell Metab 2012;16(3):394–406.
[101] Yoneshiro T, Shin W, Machida K, Fukano K, Tsubota A, Chen Y, et al. Differentiation of bone marrow-derived cells toward thermogenic adipocytes in white adipose tissue induced by the β3 adrenergic stimulation. FASEB J 2019;33(4):5196–207.
[102] Baek K, Bloomfield SA. Blocking β-adrenergic signaling attenuates reductions in circulating leptin, cancellous bone mass, and marrow adiposity seen with dietary energy restriction. J Appl Physiol (1985) 2012;113(11):1792–801.
[103] Baek K, Park HJ, Hwang HR, Baek JH. Propranolol attenuates calorie restriction- and high calorie diet-induced bone marrow adiposity. BMB Rep 2014;47(10):587–92.
[104] Takeda S, Elefteriou F, Levasseur R, Liu X, Zhao L, Parker KL, et al. Leptin regulates bone formation via the sympathetic nervous system. Cell 2002;111(3):305–17.
[105] Komai AM, Musovic S, Peris E, Alrifaiy A, El Hachmane MF, Johansson M, et al. White adipocyte adiponectin exocytosis is stimulated via β3-adrenergic signaling and activation of Epac1: catecholamine resistance in obesity and type 2 diabetes. Diabetes 2016;65(11):3301–13.
[106] Musovic S, Olofsson CS. Adrenergic stimulation of adiponectin secretion in visceral mouse adipocytes is blunted in high-fat diet induced obesity. Sci Rep 2019;9(1):10680.
[107] Migliorini RH, Garofalo MA, Kettelhut IC. Increased sympathetic activity in rat white adipose tissue during prolonged fasting. Am J Physiol 1997;272(2 Pt 2):R656–61.
[108] Duszka K, Picard A, Ellero-Simatos S, Chen J, Defernez M, Paramalingam E, et al. Intestinal PPARγ signalling is required for sympathetic nervous system activation in response to caloric restriction. Sci Rep 2016;6:36937.
[109] Reagan MR, Rosen CJ. Navigating the bone marrow niche: translational insights and cancer-driven dysfunction. Nat Rev Rheumatol 2016;12(3):154–68.
[110] Clark R, Krishnan V, Schoof M, Rodriguez I, Theriault B, Chekmareva M, et al. Milky spots promote ovarian cancer metastatic colonization of peritoneal adipose in experimental models. Am J Pathol 2013;183(2):576–91.
[111] Herroon MK, Rajagurubandara E, Hardaway AL, Powell K, Turchick A, Feldmann D, et al. Bone marrow adipocytes promote tumor growth in bone via FABP4-dependent mechanisms. Oncotarget 2013;4(11):2108–23.
[112] Falank C, Fairfield H, Reagan MR. Signaling interplay between bone marrow adipose tissue and multiple myeloma cells. Front Endocrinol (Lausanne) 2016;7:67.
[113] Reagan MR, Liaw L, Rosen CJ, Ghobrial IM. Dynamic interplay between bone and multiple myeloma: emerging roles of the osteoblast. Bone 2015;75:161–9.
[114] Takeuchi K, Abe M, Hiasa M, Oda A, Amou H, Kido S, et al. Tgf-Beta inhibition restores terminal osteoblast differentiation to suppress myeloma growth. PLoS One 2010;5(3):e9870.
[115] Corre J, Mahtouk K, Attal M, Gadelorge M, Huynh A, Fleury-Cappellesso S, et al. Bone marrow mesenchymal stem cells are abnormal in multiple myeloma. Leukemia 2007;21(5):1079–88.
[116] Teras LR, Kitahara CM, Birmann BM, Hartge PA, Wang SS, Robien K, et al. Body size and multiple myeloma mortality: a pooled analysis of 20 prospective studies. Br J Haematol 2014;166(5):667–76.

CHAPTER

# 5

# Lipids in the transcriptional regulation of adipocyte differentiation and metabolism

*Shuwen Qian, Liang Guo, Qiqun Tang*

The Key Laboratory of Metabolism and Molecular Medicine of the Ministry of Education, Institute of Stem Cell Research and Regenerative Medicine of Institutes of Biomedical Sciences, Department of Biochemistry and Molecular Biology of School of Basic Medical Sciences, Fudan University, Shanghai, China

OUTLINE



Lipid Signaling and Metabolism. http://dx.doi.org/10.1016/B978-0-12-819404-1.00005-1

## Introduction

Adipose tissues are special organs distributing in various positions of body, and function to regulate energy homeostasis. This function is fulfilled coordinately by different types of adipose tissues, mainly white adipose tissue (WAT) and brown adipose tissue (BAT) [1]. WAT stores energy in the form of lipid, which is hydrated to provide fuel in case of energy scarcity (need) [2]. In contrast, BAT can dissipate energy by thermogenesis instead of generating ATP due to the high levels of uncoupling protein 1 (UCP1). UCP1 is located in the inner mitochondrial membrane and translocates protons from the intermembrane space to the mitochondrial matrix, by passing the ATP synthase [3,4]. Another type of adipocytes, known as beige cell, was identified about a decade ago. They are recruited in WAT depot by stimuli such as cold and exercise and have thermogenic activity. Both WAT and BAT act as endocrine organs, responsible for the synthesis and secretion of cytokines. These cytokines take part in a range of processes, such as nutritional intake (leptin) [5], inflammation (tumor necrosis factor α (TNFα), interleukin 6 (IL6), adiponectin) [6–8], and biological process for glucose and lipid metabolism in the liver [9,10]. All these processes have direct or indirect influences on whole body metabolic homeostasis.

Adipose tissue expands by hypertrophy and/or hyperplasia [11]. Hypotrophy is the enlargement of preexisting adipocytes, caused by the accumulation of more lipids in adipocytes to accommodate excess dietary nutrients. Hyperplasia is the proliferation of the adipocyte precursor cells (APCs) and the subsequent differentiation into mature adipocytes. The process of adipocyte differentiation is known to participate in the development of adipose tissue in late stage of fetus and after birth. Even in adulthood, adipocyte differentiation continues and maintains an annual renewal rate of about 10% [12]. This means adipocyte hyperplasia occurs in nonobese individuals in physiological context. Adipocyte hyperplasia is also triggered when the size and lipid content of adipocyte increase to a certain level [13]. It is noted that hyperplasia of white adipocyte provides space to store excessive energy, while hyperplasia of beige or brown adipocyte consumes energy. The hyperplasia of both types of adipocytes with opposite functions reflects the body's adaption to environmental cues.

## Transcriptional regulation of adipocyte differentiation

Adipocytes derive from multipotent stem cells. One source, known as pericytes, are those with endothelial origin, residing in the wall of small vasculatures [14–16]. The other group, so-called mesenchymal stem cell, is of hematopoietic origins and recruited from circulation, which is evidenced by the presence of labeled adipocytes differentiated from xenotransplant bone marrow [17,18]. Upon the stimulation by hormones such as BMP4 [19], the multipotent stem cells proceed into preadipocyte stage, which is termed as commitment. Preadipocytes then differentiate into mature adipocyte with the establishment of adipocyte-specific phenotype and metabolic function.

During the past decade, lots of studies have been carried out to define the regulatory events during adipocyte differentiation. The majority of the studies were performed in cultured cell lines, among which 3T3-L1 and 3T3-F442A are the most commonly used preadipocyte cell lines [20,21]. C3H10T1/2 is the most commonly used pluripotent stem cell line [22]. They were all established from mouse embryonic fibroblasts, although with different potentials of adipogenesis. There are also several in vivo mouse models developed to study adipocyte differentiation. With those endeavors embarked by investigators, the network regulating adipocyte differentiation on transcriptional level has been depicted in detail.

As shown in Fig. 5.1, confluent preadipocytes are induced to differentiation upon exposure to the adipogenic cocktail of inducers—fetal bovine serum (FBS), dexamethasone (Dex), isobutylmethylxanthine (IBMX), and insulin. The adipogenic program occurs in two phases. In the first phase, the growth arrest cells synchronously reenter cell cycle and undergo at least two rounds of division known as mitotic clonal expansion (MCE) [23]. During this time, endocellular cAMP level is elevated by Dex and IBMX, thereby activating protein kinase A, which phosphorylates and activates cAMP response element-binding protein (CREB) [24], leading to the expression of CCAAT-enhancer-binding protein β (C/EBPβ) [25,26]. C/EBPβ together with cell-cycle regulators such as retinoblastoma (RB) family of proteins [27] and

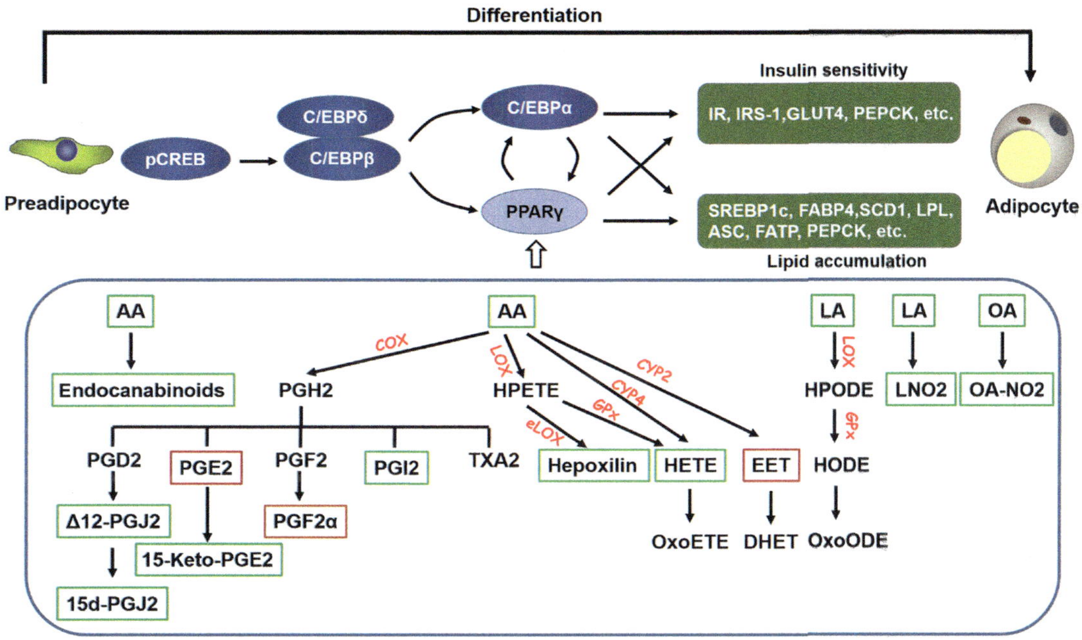


**FIGURE 5.1** **Transcriptional regulation of adipocyte differentiation and fatty acids with regulating effects.** C/EBPs and PPARγ are master transcriptional regulators for adipocyte differentiation. Fatty acids within squares are those have reported role in adipogenesis. *Green line squares* indicate pro-adipogenic effect and *red line squares* indicate inhibitive effect. *CREB*, cAMP response element-binding protein; *C/EBPs*, CCAAT-enhancer-binding proteins; *PPARγ*, peroxisome proliferator-activated receptor γ; *IR*, insulin receptor; *IRS-1*, insulin receptor substrate 1; *PEPCK*, phosphoenol pyruvate carboxykinase; *SREBP1c*, sterol regulatory element-binding proteins 1c; *FABP4*, fatty acid binding protein 4; *SCD1*, stearoyl CoA desaturase 1; *LPL*, lipoprotein lipase; *ASC*, acyl-CoA synthase; *FATP*, fatty acid transport protein; *AA*, arachidonic acid; *LA*, linoleic acid; *OA*, oleic acid; *PGs*, prostaglandins; *HPETE*, hydroperoxyeicosatetraenoic acids; *HETE*, hydroxyeicosatetraenoic acids; *EET*, epoxyeicosatrienoic acid; *OxoETE*, oxo-eicosatetraenoic acids; *DHET*, dihydroxyeicosatrienoic acid; *HPODE*, hydroperoxyoctadecadienoic acid; *HODE*, hydroxyoctadecadienoic acid; *OxoODE*, oxo-octadecadienoic acid, *COX*, cyclooxygenase; *LOX*, lipoxygenase; *CYP*, cytochrome P450 monooxygenase; *GPx*, glutathione peroxidase.

E2Fs [28], facilitates the expression of C/EBPα and peroxisome proliferator-activated receptor γ (PPARγ). After that, the cells exit cell cycle and enter the phase of terminal differentiation. C/EBPα transactivates the expression of adipocyte-specific genes including stearoyl CoA desaturase-1 (SCD1), phosphoenolpyruvate carboxykinase (PEPCK), fatty acid binding protein 4 (FABP4) and glucose transporter 4 (GLUT4), as well as sterol regulatory element-binding proteins 1c (SREBP1c), which is an important transcriptional regulator of genes associated with lipogenesis [29,30]. On the other hand, PPARγ promotes the transactivation of FABP4, lipoprotein lipase (LPL), acyl-CoA synthase (ACS), PEPCK, fatty acid transport protein (FATP) and adipsin [31,32]. Activated by C/EBPα and PPARγ, these genes above help build up the mature adipocyte phenotype and metabolic function.

C/EBPα and PPARγ are generally regarded as master regulators for adipogenesis, which is supported by both gain-of-function and loss-of-function experiments. Ectopical expression of PPARγ in nonadipogenic mouse fibroblasts can initiate the entire adipogenic program, giving rise to the adipocytes that are capable of many functions of adipocyte [33]. Studies performed in vivo shows specific knockout of PPARγ in adipose tissue reduces the mass of adipose tissue and insulin sensitivity [34]. Ectopic expression of C/EBPα in a variety of fibroblasts induces adipogenesis [35], while ablation of C/EBPα in all tissues except the liver reveals that C/EBPα is required for formation of WAT. It is very interesting that although indispensable for white adipogenesis, C/EBPα is not required for the formation of BAT [36].

In addition to the C/EBPs and PPARγ associated network, other transactivators such as Krox20, krüppel-like factors (KLFs), signal transducer and activator of transcription 5A (STAT5A) and wnt signaling are reported to regulate adipogenic differentiation. Krox20 acts early in the adipogenic program and contributes to the induction of C/EBPβ expression [37]. STAT5A induces preadipocyte differentiation, probably by activation of PPARγ [38]. KLF5 and KLF6 promote adipogenesis [39], while KLF2 Inhibits adipogenesis [40,41]. Wnt activates lineage commitment of MSC through the canonical pathway [42], however, in late stage of the program, wnt inhibits adipocyte differentiation [43].

## Lipids in adipocyte differentiation

One of the most characteristic features of adipocytes is high levels of lipids content, showing dynamic changes including synthesis, breakdown and modification. Lipids are accumulated in adipocytes via fatty acid uptake or lipogenesis. De novo lipogenesis begins with fatty acid biosynthesis, with ATP-citrate lyase (ACL), acetyl-CoA carboxylase (ACC), the fatty acid synthase (FAS) complex, fatty acid elongases (Elovl) and stearoyl-CoA desaturases (SCD) as the key enzymes [44,45]. Fatty acids are esterified together with glycerol, cholesterol form triacylglycerol (TAG), cholesterol esters respectively. If there is a phosphate group bonded to carbon-3 of the glycerol backbone, the lipid is called phospholipids [46]. The esters store in lipid droplets (LDs) or construct cellular and organelle membrane. TAG in LDs is hydrolyzed by adipose triglyceride lipase (ATGL), hormone-sensitive lipase (HSL), and monoacylglycerol lipase (MGL) into free fatty acids [47–49]. Those enzymes have prior activities depending on the number of fatty acid chains remaining on the backbone of glycerol [50]. Fatty acid can be oxidized in mitochondria, a process called beta-oxidation [51]. Peroxisome also participates in beta-oxidation for some type of very long-chain fatty acids [52]. Phospholipids embedded in cellular membranes release fatty acids by phospholipase (PLs) act [53]. Those fatty acids are found to act as signaling molecules in biological process including adipocytes differentiation.

## Lipids as PPARγ ligands

To date, almost all the lipids that have been identified to regulate adipocyte differentiation function through PPARγ (Fig. 5.1). PPARγ has a large ligand-binding pocket so that it binds to a variety of ligands with high affinity including fatty acids [54–56]. Most of the fatty acids can activate PPARγ at micromolar concentration. In general, polyunsaturated fatty acids (PUFAs) are better activators than monounsaturated fatty acids (MUFAs) and saturated fatty acids [57–59]. PUFAs have *n*-6 series and *n*-3 series defined by the position of the first double bond from the methyl end. Linoleic acid (LA) (C18:2*n*-6) and α-linolenic acid (ALA) (C18:3*n*-3) belong to *n*-6 and *n*-3 types respectively. Both of them are essential fatty acids because they cannot be endogenously synthesized and thus must be absorbed from food. They can be converted into long-chain PUFA by elongation and/or desaturation. For example, eicosadienoic acid (C20:2*n*-6) and arachidonic acid (AA) (C20:4*n*-6) can be generated from LA [60]. Eicosapentaenoic acid (EPA), docosahexaenoic acid (DHA) are *n*-3 series fatty acid rich in fish oil. While LA, EPA and DHA are all natural ligands of PPARγ [61,62], AA was reported to be the most potent binder and activator of PPARγ, especially when AA were modified with oxidation or nitration. There are three families of enzymes for fatty acids oxidation—cyclooxygenases (COXs), lipoxygenases (LOXs), and cytochrome P450 monooxygenases (CYPs) (Fig. 5.1).

### COXs and their derivatives related to adipocyte differentiation

The COX enzymes in adipocytes, mediating the synthesis of prostaglandin (PG)H2 from AA, exist as two isoforms, COX1 and COX2 [63,64]. PGH2 is the precursor of PGD2, PGE2, PGF2, PGI2, and thromboxane A2. These derivatives are synthesized by PGD2 synthase, PGE2 synthase, PGF2 synthase, PGI2 synthase, and thromboxane A2 synthase, respectively [65]. PGD2 metabolite, 15-deoxy-Δ12,14-prostaglandin J2 (15d-PGJ2), was shown to act as a PPARγ agonist to promote adipocyte differentiation [66]. However, in 3T3-L1 preadipocyte differentiation system, endogenous 15d-PGJ2 is generated by cells during the late phase of differentiation, that is, the maturation phase [67], lagging behind the time point when PPARγ is activated. Moreover, the concentration of 15d-PGJ2 in media of differentiating 3T3-L1 cells is much lower than exogenous concentration required to induce adipogenesis [68]. Therefore, the possibility that 15d-PGJ2 is the endogenous PPARγ ligand functioning during initiation of adipocyte differentiation remains ambiguous. The intermediate metabolite of PGD2, Δ12-PGJ2 seems more likely to be endogenous PPARγ ligand to induce adipogenesis, as its levels increase gradually during adipocyte [68]. In contrast to PGD2, PGE2 is more abundant in 3T3-L1 cells. Its α,β-unsaturated form, 15-keto-PGE2 is a ligand of PPARγ, thus activating PPARγ-mediated transcription and enhancing adipogenesis of 3T3-L1 cells [69]. More supporting data include that its level increases upon induction and peaks at day 2 [70], and expression of prostaglandin reductase that delete α,β-unsaturation converting 15-keto-PGE2 to PPARγ-binding deficient derivative, 13,14-dihydro-15-keto-PGE2, abolishes adipogenesis [69,70].

Other studies showed that preadipocytes 3T3-L1 endogenously synthesize PGI2 in culture condition. After differentiation, the level of prostaglandin remains upregulated in the mature adipose tissues for 4–6 days followed by a slow decline [71]. PGI2 plays a role in elevating intracellular cyclic adenosine monophosphate (cAMP) through PGI2 receptor, which suggests the ability of PGI2 to trigger adipocyte differentiation [72,73].

It is noted that knockdown or inhibition of COX2 promotes adipocyte differentiation [74,75]. Consistently, sustained expression of COX2 represses adipogenesis [76]. These results suggest that COX-2 more preferentially generates endogenous antiadipogenic prostanoids. One of the reported PGs with negative effects on adipocyte differentiation is PGF2α, the possible mechanisms involved are promoting DNA synthesis, activating mitogen-activated protein kinase (MAPK) and finally downregulating the activity of PPARγ by initiating phosphorylation [77–81]. PGF2α may help maintain the undifferentiated state of preadipocytes, as endogenous PGF2α production is lower in differentiating cells compared to unstimulated preadipocytes [77]. Another PGs that can suppress adipocyte differentiation is PGE2. Through binding to its receptor EP4, PGE2 is shown to suppress differentiation in an adipocyte differentiation system from mouse embryonic fibroblasts (MEF) [82,83].

## LOXs and their derivatives related to adipocyte differentiation

LOXs catalyze AA to hydroperoxyeicosatetraenoic acids (HPETEs). These HPETEs may be converted further to various bioactive lipid mediators including hydroxyeicosatetraenoic acids (HETEs), leukotrienes, lipoxins, and hepoxilins [84]. LOXs can also metabolize LA to produce hydroperoxyoctadecadienoic acids (HPODEs) and further yield hydroxyoctadecadienoic acids (HODEs) [85]. It has been documented in macrophage that PPARγ is activated by 13-hydroxyoctadecadienoic acid (13-HODE) and 15-hydroxyeicosatetraenoic acid (15-HETE), which are generated from LA and AA, respectively, by a 12/15-lipoxygenase and sequential peroxidases [86,87]. The products of HETEs and HODEs generated by LOX activity are further metabolized by dehydrogenase, resulting in the formation of oxo fatty acids, which are also reported to bind PPARγ [88].

Reports about the role of LOXs in adipocyte differentiation are rare. Chemical inhibition of LOX decreases the rate of adipocyte differentiation from 3T3-L1 preadipocytes [89,90]. This inhibition is effective only during the first 4 days of the differentiation process, when 3T3-L1 cells synthesize endogenous PPARγ ligand(s) [89]. Although conventional platelet- and leucocyte-type 12(s)-LOXs expresses in white and brown adipose tissue, that expressed in 3T3-L1 cells is epidermis lipoxygenase type 3 (eLOX3) [89]. ELOX3 appears essential for differentiation of 3T3-L1 preadipocytes as knockdown of eLOX3 by short hairpin RNA (shRNA) blocks differentiation. In addition, forced expression of eLOX3 or addition of eLOX3 product, hepoxilins, stimulates adipogenesis of fibroblasts under conditions that normally require an exogenous PPARγ ligand for differentiation [91]. Hepoxilins have the ability to bind to and activate PPARγ [92]. Consistently, the production of hepoxilins increased during the initial stages of adipogenesis, implying its potential for PPARγ ligand [91]. Since eLOX3 uses fatty acid hydroperoxides as substrates, there should be enzymes to produce these substrate [93,94]. To date, the only reported LOX expressing in 3T3-L1 preadipocytes is 5-LOX. However, 5-LOX expresses at 5 days post initiation of differentiation [95]. This time window for 5-LOX expression seems to lag behind that for PPARγ expression. Thus, it is unlikely for 5-LOX to initiate differentiation.

Other than the LOXs, ROS system also regulates adipocyte differentiation. The overexpression of H2O2-scavenging enzyme peroxiredoxin II (Prx II) inhibits adipocyte differentiation from the mesenchymal stem cell. Consistently, inhibition of NAD(P)H oxidase (NOX), which catalyzes the formation of ROS, impairs adipocyte differentiation [96]. NOX4-deficient mice display adipose tissue accumulation resulting from both accelerated adipocyte differentiation and hypertrophy [97]. However, the mechanism of NOX4-derived reactive oxygen species (ROS) in mediating adipocyte differentiation seems rather complicated. In addition to producing fatty acid hydroperoxides as candidates for acting as PPARγ ligands [98], ROS directly or indirectly regulates the expression or activity of enzymes involving synthesis of prostanoids from arachidonic acid such as phospholipase A2 (cPLA2), cyclooxygenase-2 (COX-2) and membrane-bound prostaglandin E synthase 1, and prostaglandin I synthase (mPGES-1) [99].

## CYPs and their derivatives related to adipocyte differentiation

The CYP monooxygenases, mainly CYP2 epoxygenases and CYP4 omega-hydroxylases, add one oxygen to AA generating either epoxyeicosatrienoic (EETs) or ω and ω-1 hydroxyeicosatetraenoic acids (e.g., 20-HETEs), respectively [100]. EET can be further transformed to produce dihydroxyeicosatrienoic acids (DHETs) by the soluble epoxide hydrolase (sEH). 20-HETE, EETs, and DHETs are reported to be the activators of PPARγ [101,102]. However, the functions of those products in adipocyte differentiation have not come into unanimous conclusion. Exogenous EET analog administration to 3T3-L1 elicits antiadipogenic effects via inhibition of the early phase of adipogenesis [103]. The effect is similar for human mesenchymal stem cells (MSCs), in which addition of EETs or EET-agonist decreases the adipogenesis [104,105]. In contrast, exogenous administration of 20-HETE increased adipogenesis of human MSCs, which may be resulted from the pro-adipogenic effects of both 20-HETE and its cyclooxygenase 2 (COX2)-derived metabolite, 20-OH-$PGE_2$ [106].

## Nitrolinoleic acid

Nitroalkene derivatives of linoleic acid (LNO2) and oleic acid (OA-NO2) are present in human blood and urine. Two nitroalkene derivatives of oleic acid (OA-NO2), 9- and 10-nitro-9-*cis*-octadecaenoic acids, are 50% greater than that of nitrated linoleic acid. OA-NO2 is a potent ligand for PPARs at physiological concentrations [107]. As expected, OA-NO2 is able to induce PPARγ-dependent adipogenesis in 3T3-L1 preadipocytes [107]. LNO2 also acts as a PPARγ ligand as well as an inducer of adipocyte differentiation [108,109]. The binding activity of LNO2 for PPARγ is more robust than other reported endogenous PPARγ ligands such as 15d-PGJ2, lysophosphatidic acid (16:0 and 18:1) [109]. Fatty acid nitration is formed by addition of nitrogen dioxide (•NO2) to the double bond, yielding stereoisomers [110]. Importantly, the production of nitric oxide increases during the early stage of adipocyte differentiation, and inhibition of nitric oxide synthase activity abrogated the differentiation process [111].

## Endocannabinoids

Endocannabinoids are arachidonic acid-derived endogenous lipids that activate the cannabinoid (CB) receptors, with a range of effects on the nervous system. It has also been reported that endocannabinoids, 2-arachidonoylglycerol and anandamide (*N*-arachidonoylethanolamine) are produced by human white subcutaneous adipocytes [112]. Anandamide is able to bind directly to PPAR-ligand binding domain and induce transcriptional activation of PPARγ. Anandamide treatment induced 3T3-L1 fibroblast differentiation into adipocytes [113].

# Lipids in brown and beige adipocyte development

## Regulation of brown/beige adipocyte development

BAT is present in human in infancy and predominantly located in the interscapular depot, and gradually diminishes with growth. In normal adult humans, BAT is proportionally smaller and was used to be believed functionally less important. However, the beige adipocytes that arise in WAT depot under certain circumstances such as cold and exercise have BAT-like thermogenic activity, and play an important role in maintaining energy homeostasis. Brown adipocytes derive from myogenic factor 5 (myf5) positive progenitors, while beige adipocytes derive from myf5 negative progenitors [114]. Although brown and beige adipocytes have different origins, both of them are thought to be controlled by the similar transcriptional regulation (Fig. 5.2), in which PR domain containing 16 (PRDM16) plays an important role. PRDM16 coactivates PPARG coactivator 1 (PGC1α); and PPARγ, which then drives biogenesis of mitochondria along with expression of UCP-1 [115]. On the other hand, PPAR agonists induce a white-to-brown fat conversion through stabilization of PRDM16 protein [116]. From this aspect, lipids that activate PPARγ are potent candidates to regulate the biogenesis of brown or beige adipocyte in addition to inducing adipocyte differentiation.

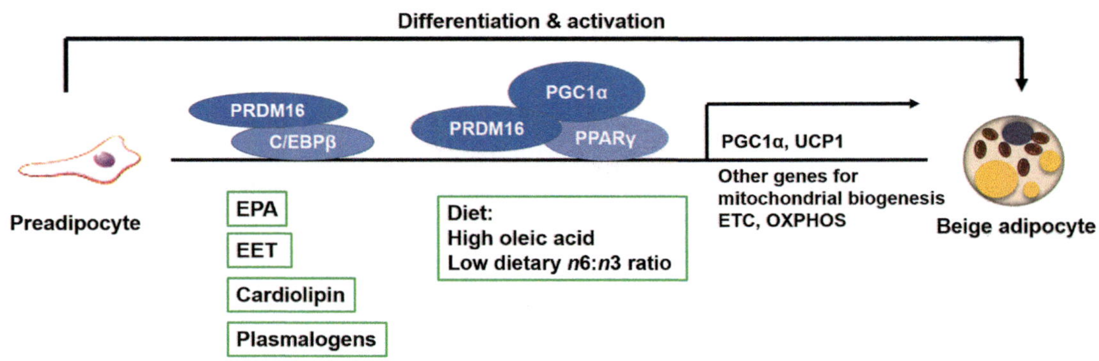


FIGURE 5.2 **Differentiation and activation of beige adipocyte and fatty acids with regulating effects.** Fatty acids within green line squares are those have reported role in promote beige adipocyte biogenesis and activation. *PRDM16*, PR domain containing 16; *PPARγ*, peroxisome proliferator-activated receptor γ; *PGC1α*, PPARG coactivator 1; *ETC*, electron transport chain; *OXPHOS*, oxidative phosphorylation; *EPA*, eicosapentaenoic acid; *EET*, epoxyeicosatrienoic acid.

## Lipidomics related to brown or beige adipocyte

Lipidomics of BAT comparing cold stimuli to room temperature condition show that although the total concentration of TAG remains the same, the composition of acyl chain species changed. Cold exposure increases the levels of the C18:0, C20:0, and C22:0 acyl chains in TAG. Cold-induced changes in glycerophospholipid acyl chain composition were associated with distinct subclasses. Levels of phosphatidylethanolamine (PE), phosphatidylcholine (PC), and lysophosphatidylcholine (LPC) species that contained C16:1 acyl chains were reduced. The level of C18:0 was increased in both PC and PE, and C18:2 was increased in both the PE and phosphatidylserine (PS) subclasses [117]. As studies have demonstrated that the lipidomic profile of BAT and WAT are distinct [118], the lipidomics of WAT undergoing browning are suggested to be distinguished from that in BAT. However there are few reports on comprehensive lipidomic change in cold exposure-induced WAT. Data about effect of exercise on subcutaneous WAT has reported the decrease of major phospholipid molecular species of PA, PC, PE, and PS. Further analysis based on the chain length and saturation state reveal a significant change of the chain length and saturation of PE, PS, and LPS species, i.e., exercise increased the total monounsaturated PEs (PE_O-36:1), and decreased polyunsaturated PS/LPS (18:2- and 20:4) species [119].

As collection of blood samples is a minimally invasive procedure as compared with collection of tissue samples by biopsy, most lipidomic research associated with BAT activity in human was analysis of serum/plasma. One study of human circulating lipidome has revealed positive association between cold-stimulated BAT activity and serum lysophosphatidylcholine (LPC) species, LPC (16:0) and LPC (16:1), as well as phosphatidylcholine (PC), PC (32:1) [120]. Another study evaluated 26 specific plasma FAs associated with cold-induced BAT activity in human, revealing the increased concentrations of lignoceric acid (C24:0), along with the decreased levels of eicosenoic acid (C20:1n9), nervonic acid (C24:1n9), and behenic acid (C22:0) in circulation [121]. A global lipidomic analysis of a panel of more (88) lipids species in the human plasma identifies 12,13-dihydroxy-9Z-octadecenoic acid (12,13-diHOME) as the only species that increased by cold exposure in all individuals measured [122]. 12,13-diHOME is a linoleic acid metabolite, biosynthesized through the formation of 12,13-epoxyoctadecenoic acids (12,13-epOME) catalyzed by cytochrome P450 (Cyp) oxidases, followed by hydrolysis catalyzed by soluble epoxide hydrolases (sEHs) to form the diols 12,13-diHOME [123]. Exercise also causes a pronounced increase in the circulating 12,13-diHOME in human subjects regardless of the gender and age. The same alteration was present in rodents, with data showing that exercise training increased circulating 12,13-diHOME. It was postulated that 12,13-diHOME was derived from BAT as surgical removal of brown adipose tissue (BAT) abolished the increase in 12,13-diHOME [124].

## Lipids promoting biogenesis of brown/beige adipocyte

### *PUFAs*

Studies have tested the effects of *n*-3 PUFAs on WAT browning properties. The addition of EPA (200 mM) during adipocyte differentiation enhanced thermogenic and mitochondrial gene expression in cells from mouse inguinal adipose tissue–derived stem cells [125]. Notably, the conversion from mature white adipocytes to beige-like adipocytes was achieved by the treatment of human adipocytes with 100–200 μM EPA for 24 h, with increased mitochondrial content and an up-regulation of PRDM16, CIDEA (cell death inducing DFFA like effector A), UCP1, and fatty acid oxidation genes [126].

Besides EPA, EETs are potent in inducing the development of beige adipocyte. Although it is reported that EET agonist treatment leads to decreased adipogenesis, the results were controversial as other studies show that EETs actually contributes to browning of white adipocyte in vitro and ameliorate obesity and insulin resistance in vivo [127–130]. The mechanisms involve the activation of PGC1α and the increase of fatty acid oxidation. The discordant conclusion may due to the data based on to interpret adipogenesis. Most studies use lipid content quantitated by the amount of bound oil red O to show the efficiency of differentiation. However, thermogenic adipocytes are active in lipolysis and fatty acid beta-oxidation, leading to decreased lipid content in these adipocytes, which render the phenotype similar to inhibition of adipocyte differentiation. Moreover, the concentration of exogenous EET or EET agonist may also affect the outcomes. EET induces browning at concentration of 1 μM [129], but decreased the rate of differentiation at concentration higher than 1 μM [103]. However, whether EETs at the high concentration is toxic was not evaluated.

### *Plasmalogens*

Peroxisome is an organelle for lipid metabolism, including fatty acid beta-oxidation and ether phospholipid biosynthesis [131]. Plasmalogens are a major subgroup of ether phospholipids. Biosynthesis of plasmalogens involves two essential and sequential intra-peroxisomal steps catalyzed by glyceronephosphate *O*-acyl transferase (GNPAT) and alkylglycerone phosphate synthase (AGPS), which introduce the characteristic ether bond [132]. Knocking out peroxisomal biogenesis factor Per16 specifically in adipocytes impaired cold tolerance, decreased energy expenditure, and increased diet-induced obesity. Meanwhile, plasmalogen present in mitochondria decreased with Per16 inactivation. The effects of Pex16 inactivation can be recapitulated by knockdown of the plasmalogen synthetic enzyme GNPAT. Moreover, dietary supplementation with plasmalogens increased mitochondrial copy number, improved mitochondrial function, and rescued thermogenesis in Pex16 knockout mice [133]. The data suggest that plasmalogen might be an inducer of thermogenic adipocyte. Another study on weight-discordant twins shows the elevation of PUFA-containing phospholipids. Interestingly, most abundant PUFA-containing ether lipids were plasmalogens [134]. The increasing of plasmalogens may represent the adaptive response to counteract obesity. There is also possibility that plasmalogens with different fatty acids esterified in may differ in function related to energy metabolism, which is indicated by the result that the fatty acid dominantly esterified in the sn-2 position is arachidonic acid but not DHA in the elevated plasmalogens of the obese individuals [134].

### *Cardiolipin*

Recently, it has been reported that cardiolipin (CL) biosynthesis is induced in brown and beige adipose tissue upon cold exposure [135]. CL is a kind of diphosphatidylglycerol lipid. It is comprised of four fatty acyl chains, formed as two PA molecules attached to each other by a central glycerol backbone [147]. CLs are synthesized and predominantly located in the mitochondrial inner membrane [136]. CL synthase (CLS) transferring a phosphatidyl group from CDP-diacylglycerol to phosphatidylglycerol (PG) [137–139]. Crls1 deficiency in thermogenic adipocytes revealed a virtual ablation of all CL species, along with the decreased thermogenesis and mitochondrial respiration activity. The function of CL on thermogenesis may be through the regulation of mitochondria function rather than the differentiation of adipocytes, which was also supported by the previously described roles of CLs in mitochondrial biology, including respiratory chain super complex formation, cytochrome c sequestration, OPA1-mediated fusion, and mitochondrial carrier activation [137,139].

## Lipid as dietary supplementation that can activate brown activity

### *Oleic acid*

The Mediterranean diet, rich in oleic acid from olive oil is associated with numerous health benefits including weight reduction as well as improvement in clinical lipid profile and glycemia [140,141]. A study with mice showed that feeding mice with diets rich in oleic acid, linoleic acid, or α-linolenic acid for 12 weeks was associated with reduced weight gain, higher oxygen consumption and greater heat production compared to diets rich in stearic acid. The oleic acid diets-fed group had the highest level of thermogenesis [142]. In a human study, comparing a single meal of oleic acid with the same mix of PUFAs, also demonstrated that the oleic acid was superior in increasing energy expenditure in healthy men [143].

### *PUFAs*

The potential of PUFA to regulate thermogenic function of adipocytes seems dependent on *n*6:*n*3 ratio. In vivo, a low dietary *n*6:*n*3 ratio improved the thermogenic response of brown and white adipose tissues to β3-adrenergic stimulation. This effect was recapitulated in vitro by PUFA treatment of hMADS adipocytes [144]. Fish oil, which is rich in the *n*-3 PUFAs EPA and DHA, shows beneficial effects on metabolism. Fish oil intake for 10 weeks has been shown to increase oxygen consumption and core body temperature in mice. The diet with fish oil supplementation (containing 1.2% pure DHA and 2.4% pure EPA) led to upregulation of UCP1 and the β3 adrenergic receptor in both interscapular BAT and inguinal WAT of mice [145]. However, the effects of WAT browning caused by EPA supplementation are controversial. When mice were fed an HF diet supplemented with pure EPA (3.6% as EPA ethyl ester), the brown marker genes were only elevated in iBAT, but not in inguinal WAT [146].

## Concluding remarks

The adipocyte differentiation program is regulated by an elaborate network of transcriptional factors, among which C/EBPs and PPARγ are master regulators. Lipids in adipocytes not only serve as an energy source for cellular activities, but also act as signal factors to regulate adipocyte development and metabolism. Some PUFAs and they oxidized or nitrated derivatives are PPARγ ligands and able to promote adipocyte differentiation. One of the evidences for their pro-adipogenic effect is that administration of those lipids to cultured cells induces differentiation. However, some lipids don't fit the profile of their endogenous ones during differentiation, or upon applying to in vivo system, they even show adverse event profiles, especially when relating to clinical index. The discrepancy might be attributed to the different nature among cell types, as well as the multiple functions of PPARγ. As in addition to regulating adipocytes differentiation, PPARγ also mediates the beige cell activity and activation of inflammatory cells. So interpreting functions of lipids on adipocyte differentiation in vivo should incorporate in the systematical effects of multiple cells and organs.

Beiging of adipose tissue is a state favoring energy dissipating, so deciphering the lipid composition and identifying the lipid signaling in activation of beige or brown cells may help develop new ways to prevent and treat obesity and related metabolic disorder. So far oleic acid, EPA, EETs, plasmalogens and cardiolipin have been demonstrated to stimulate brown/beige activity (Fig. 5.2). With the help of the lipidomic analytic methods, more active lipids will be identified.

While many studies have quantified the concentration of certain lipids within the adipose tissue by the lipidomic analysis, limited measurements have been performed for analyzing the global changes in lipid composition related to adipocyte differentiation and metabolism. Lipids in adipocytes dynamically change based on the tissue type, physiologic or pathologic condition, which requires more comprehensive experimental settings by taking several key factors into account, such as genotyping, environmental cues, and comparison between different types of adipocytes. Moreover, in addition to the diversiform, lipids in adipocyte have versatile roles such as constructing membrane, regulating differentiation, providing energy fuels, and acting as signals intracellularly or transcellularly. Therefore, it is important to decipher the panorama of lipid changes according to tissue origin, age, sex, differentiation stage or a physiological/pathological state, thus help identify specific bio-active lipid. The improvement in lipidomics along with some advanced technology may help achieve the purpose in the near future.

## References

[1] Cinti S. The adipose organ: morphological perspectives of adipose tissues. Proc Nutr Soc 2001;60(3):319–28.

[2] Spiegelman BM, Flier JS. Obesity and the regulation of energy balance. Cell 2001;104(4):531–43.

[3] Aquila H, Link TA, Klingenberg M. The uncoupling protein from brown fat mitochondria is related to the mitochondrial ADP/ATP carrier. Analysis of sequence homologies and of folding of the protein in the membrane. EMBO J 1985;4(9):2369–76.

[4] Cannon B, Nedergaard J. Brown adipose tissue: function and physiological significance. Physiol Rev 2004;84(1):277–359.

[5] Zhang Y, Proenca R, Maffei M, Barone M, Leopold L, Friedman JM. Positional cloning of the mouse obese gene and its human homologue. Nature 1994;372(6505):425–32.

[6] Bastard JP, Jardel C, Bruckert E, Blondy P, Capeau J, Laville M, et al. Elevated levels of interleukin 6 are reduced in serum and subcutaneous adipose tissue of obese women after weight loss. J Clin Endocrinol Metab 2000;85(9):3338–42.

[7] Kern PA, Saghizadeh M, Ong JM, Bosch RJ, Deem R, Simsolo RB. The expression of tumor necrosis factor in human adipose tissue. Regulation by obesity, weight loss, and relationship to lipoprotein lipase. J Clin Invest 1995;95(5):2111–9.

[8] Scherer PE, Williams S, Fogliano M, Baldini G, Lodish HF. A novel serum protein similar to C1q, produced exclusively in adipocytes. J Biol Chem 1995;270(45):26746–9.

[9] Leal Vde O, Mafra D. Adipokines in obesity. Clin Chim Acta 2013;419:87–94.

[10] Wang GX, Zhao XY, Lin JD. The brown fat secretome: metabolic functions beyond thermogenesis. Trends Endocrinol Metab 2015;26(5):231–7.

[11] Rosen ED, MacDougald OA. Adipocyte differentiation from the inside out. Nat Rev Mol Cell Biol 2006;7(12):885–96.

[12] Spalding KL, Arner E, Westermark PO, Bernard S, Buchholz BA, Bergmann O, et al. Dynamics of fat cell turnover in humans. Nature 2008;453(7196):783–7.

[13] Faust IM, Johnson PR, Stern JS, Hirsch J. Diet-induced adipocyte number increase in adult rats: a new model of obesity. Am J Physiol 1978;235(3):E279–86.

[14] Crisan M, Yap S, Casteilla L, Chen CW, Corselli M, Park TS, et al. A perivascular origin for mesenchymal stem cells in multiple human organs. Cell Stem Cell 2008;3(3):301–13.

[15] Tang W, Zeve D, Suh JM, Bosnakovski D, Kyba M, Hammer RE, et al. White fat progenitor cells reside in the adipose vasculature. Science 2008;322(5901):583–6.

[16] Tran KV, Gealekman O, Frontini A, Zingaretti MC, Morroni M, Giordano A, et al. The vascular endothelium of the adipose tissue gives rise to both white and brown fat cells. Cell Metab 2012;15(2):222–9.

[17] Crossno JT Jr, Majka SM, Grazia T, Gill RG, Klemm DJ. Rosiglitazone promotes development of a novel adipocyte population from bone marrow-derived circulating progenitor cells. J Clin Invest 2006;116(12):3220–8.

[18] Sera Y, LaRue AC, Moussa O, Mehrotra M, Duncan JD, Williams CR, et al. Hematopoietic stem cell origin of adipocytes. Exp Hematol 2009;37(9):1108–20. 1120-1121.

[19] Tang QQ, Otto TC, Lane MD. Commitment of C3H10T1/2 pluripotent stem cells to the adipocyte lineage. Proc Natl Acad Sci U S A 2004;101(26):9607–11.

[20] Green H, Meuth M. An established pre-adipose cell line and its differentiation in culture. Cell 1974;3(2):127–33.

[21] Green H, Kehinde O. Spontaneous heritable changes leading to increased adipose conversion in 3T3 cells. Cell 1976;7(1):105–13.

[22] Reznikoff CA, Brankow DW, Heidelberger C. Establishment and characterization of a cloned line of C3H mouse embryo cells sensitive to postconfluence inhibition of division. Cancer Res 1973;33(12):3231–8.

[23] Tang QQ, Otto TC, Lane MD. Mitotic clonal expansion: a synchronous process required for adipogenesis. Proc Natl Acad Sci U S A 2003;100(1):44–9.

[24] Fox KE, Colton LA, Erickson PF, Friedman JE, Cha HC, Keller P, et al. Regulation of cyclin D1 and Wnt10b gene expression by cAMP-responsive element-binding protein during early adipogenesis involves differential promoter methylation. J Biol Chem 2008;283(50):35096–105.

[25] Niehof M, Manns MP, Trautwein C. CREB controls LAP/C/EBP beta transcription. Mol Cell Biol 1997;17(7):3600–13.

[26] Zhang JW, Klemm DJ, Vinson C, Lane MD. Role of CREB in transcriptional regulation of CCAAT/enhancer-binding protein beta gene during adipogenesis. J Biol Chem 2004;279(6):4471–8.

[27] Classon M, Kennedy BK, Mulloy R, Harlow E. Opposing roles of pRB and p107 in adipocyte differentiation. Proc Natl Acad Sci U S A 2000;97(20):10826–31.

[28] Fajas L, Landsberg RL, Huss-Garcia Y, Sardet C, Lees JA, Auwerx J. E2Fs regulate adipocyte differentiation. Dev Cell 2002;3(1):39–49.

[29] Christy RJ, Yang VW, Ntambi JM, Geiman DE, Landschulz WH, Friedman AD, et al. Differentiation-induced gene expression in 3T3-L1 preadipocytes: CCAAT/enhancer binding protein interacts with and activates the promoters of two adipocyte-specific genes. Genes Dev 1989;3(9):1323–35.

[30] Yeh WC, Cao Z, Classon M, McKnight SL. Cascade regulation of terminal adipocyte differentiation by three members of the C/EBP family of leucine zipper proteins. Genes Dev 1995;9(2):168–81.

[31] Rosen ED, Sarraf P, Troy AE, Bradwin G, Moore K, Milstone DS, et al. PPAR gamma is required for the differentiation of adipose tissue in vivo and in vitro. Mol Cell 1999;4(4):611–7.

[32] Tontonoz P, Hu E, Spiegelman BM. Stimulation of adipogenesis in fibroblasts by PPAR gamma 2, a lipid-activated transcription factor. Cell 1994;79(7):1147–56.

[33] Tontonoz P, Hu E, Graves RA, Budavari AI, Spiegelman BM. mPPAR gamma 2: tissue-specific regulator of an adipocyte enhancer. Genes Dev 1994;8(10):1224–34.

[34] Zhang J, Fu M, Cui T, Xiong C, Xu K, Zhong W, et al. Selective disruption of PPARgamma 2 impairs the development of adipose tissue and insulin sensitivity. Proc Natl Acad Sci U S A 2004;101(29):10703–8.

[35] Freytag SO, Paielli DL, Gilbert JD. Ectopic expression of the CCAAT/enhancer-binding protein alpha promotes the adipogenic program in a variety of mouse fibroblastic cells. Genes Dev 1994;8(14):1654–63.

[36] Linhart HG, Ishimura-Oka K, DeMayo F, Kibe T, Repka D, Poindexter B, et al. C/EBPalpha is required for differentiation of white, but not brown, adipose tissue. Proc Natl Acad Sci U S A 2001;98(22):12532–7.

[37] Chen Z, Torrens JI, Anand A, Spiegelman BM, Friedman JM. Krox20 stimulates adipogenesis via C/EBPbeta-dependent and -independent mechanisms. Cell Metab 2005;1(2):93–106.

[38] Floyd ZE, Stephens JM. STAT5A promotes adipogenesis in nonprecursor cells and associates with the glucocorticoid receptor during adipocyte differentiation. Diabetes 2003;52(2):308–14.

[39] Oishi Y, Manabe I, Tobe K, Tsushima K, Shindo T, Fujiu K, et al. Kruppel-like transcription factor KLF5 is a key regulator of adipocyte differentiation. Cell Metab 2005;1(1):27–39.

[40] Banerjee SS, Feinberg MW, Watanabe M, Gray S, Haspel RL, Denkinger DJ, et al. The Krüppel-like factor KLF2 inhibits peroxisome proliferator-activated receptor-gamma expression and adipogenesis. J Biol Chem 2003;278(4):2581–4.

[41] Wu J, Srinivasan SV, Neumann JC, Lingrel JB. The KLF2 transcription factor does not affect the formation of preadipocytes but inhibits their differentiation into adipocytes. Biochemistry 2005;44(33):11098–105.
[42] Bowers Robert R, Lane Daniel M. Wnt signaling and adipocyte lineage commitment. Cell Cycle 2008;7(9):1191–6.
[43] Ross SE, Hemati N, Longo KA, Bennett CN, Lucas PC, Erickson RL, et al. Inhibition of adipogenesis by Wnt signaling. Science 2000;289(5481):950–3.
[44] Ameer F, Scandiuzzi L, Hasnain S, Kalbacher H, Zaidi N. De novo lipogenesis in health and disease. Metab Clin Exp 2014;63(7):895–902.
[45] Lodhi IJ, Wei X, Semenkovich CF. Lipoexpediency: de novo lipogenesis as a metabolic signal transmitter. Trends Endocrinol Metab 2010;22(1):1–8.
[46] Kent C. Eukaryotic phospholipid biosynthesis. Annu Rev Biochem 1995;64:315–43.
[47] Karlsson M, Contreras JA, Hellman U, Tornqvist H, Holm C. cDNA cloning, tissue distribution, and identification of the catalytic triad of monoglyceride lipase. Evolutionary relationship to esterases, lysophospholipases, and haloperoxidases. J Biol Chem 1997;272(43):27218–23.
[48] Kraemer FB, Shen WJ. Hormone-sensitive lipase: control of intracellular tri-(di-)acylglycerol and cholesteryl ester hydrolysis. J Lipid Res 2002;43(10):1585–94.
[49] Zimmermann R, Strauss JG, Haemmerle G, Schoiswohl G, Birner-Gruenberger R, Riederer M, et al. Fat mobilization in adipose tissue is promoted by adipose triglyceride lipase. Science 2004;306(5700):1383–6.
[50] Thompson BR, Lobo S, Bernlohr DA. Fatty acid flux in adipocytes: the in's and out's of fat cell lipid trafficking. Mol Cell Endocrinol 2010;318(1–2):24–33.
[51] Zhang L, Keung W, Samokhvalov V, Wang W, Lopaschuk GD. Role of fatty acid uptake and fatty acid β-oxidation in mediating insulin resistance in heart and skeletal muscle. BBA Mol Cell Biol Lipids 2010;1801(1):1–22.
[52] Poirier Y, Antonenkov VD, Glumoff T, Hiltunen JK. Peroxisomal β-oxidation—a metabolic pathway with multiple functions. Biochim Biophys Acta Mol Cell Res 2006;1763(12):1413–26.
[53] Chap H. Forty five years with membrane phospholipids, phospholipases and lipid mediators: a historical perspective. Biochimie 2016;125:234–49.
[54] Johnson BA, Wilson EM, Li Y, Moller DE, Smith RG, Zhou G. Ligand-induced stabilization of PPARgamma monitored by NMR spectroscopy: implications for nuclear receptor activation. J Mol Biol 2000;298(2):187–94.
[55] Nolte RT, Wisely GB, Westin S, Cobb JE, Lambert MH, Kurokawa R, et al. Ligand binding and co-activator assembly of the peroxisome proliferator-activated receptor-gamma. Nature 1998;395(6698):137–43.
[56] Uppenberg J, Svensson C, Jaki M, Bertilsson G, Jendeberg L, Berkenstam A. Crystal structure of the ligand binding domain of the human nuclear receptor PPARgamma. J Biol Chem 1998;273(47):31108–12.
[57] Kliewer SA, Sundseth SS, Jones SA, Brown PJ, Wisely GB, Koble CS, et al. Fatty acids and eicosanoids regulate gene expression through direct interactions with peroxisome proliferator-activated receptors α and γ. Proc Natl Acad Sci U S A 1997;94(9):4318–23.
[58] Krey G, Braissant O, L'Horset F, Kalkhoven E, Perroud M, Parker MG, et al. Fatty acids, eicosanoids, and hypolipidemic agents identified as ligands of peroxisome proliferator-activated receptors by coactivator-dependent receptor ligand assay. Mol Endocrinol 1997;11(6):779–91.
[59] Yu K, Bayona W, Kallen CB, Harding HP, Ravera CP, McMahon G, et al. Differential activation of peroxisome proliferator-activated receptors by eicosanoids. J Biol Chem 1995;270(41):23975–83.
[60] Berdeaux O, Blond JP, Brétillon L, Chardigny JM, Mairot T, Vatèle JM, et al. In vitro desaturation or elongation of monotrans isomers of linoleic acid by rat liver microsomes. Mol Cell Biochem 1998;185(1):17–25.
[61] Marion-Letellier R, Butler M, Dechelotte P, Playford RJ, Ghosh S. Comparison of cytokine modulation by natural peroxisome proliferator-activated receptor gamma ligands with synthetic ligands in intestinal-like Caco-2 cells and human dendritic cells—potential for dietary modulation of peroxisome proliferator-activated receptor gamma in intestinal inflammation. Am J Clin Nutr 2008;87(4):939–48.
[62] Xu HE, Lambert MH, Montana VG, Parks DJ, Blanchard SG, Brown PJ. Molecular recognition of fatty acids by peroxisome proliferator–activated receptors. Mol Cell 1999;3(3):397–403.
[63] Rahman MS, Syeda PK, Khan F, Nishimura K, Jisaka M, Nagaya T, et al. Cultured preadipocytes undergoing stable transfection with cyclooxygenase-1 in the antisense direction accelerate adipogenesis during the maturation phase of adipocytes. Appl Biochem Biotechnol 2013;171(1):128–44.
[64] Simmons DL, Botting RM, Hla T. Cyclooxygenase Isozymes: the biology of prostaglandin synthesis and inhibition. Pharmacol Rev 2004;56(3):387–437.
[65] Cipollone F, Cicolini G, Bucci M. Cyclooxygenase and prostaglandin synthases in atherosclerosis: recent insights and future perspectives. Pharmacol Ther 2008;118(2):161–80.

[66] Fujimori K, Maruyama T, Kamauchi S, Urade Y. Activation of adipogenesis by lipocalin-type prostaglandin D synthase-generated Δ12-PGJ2 acting through PPARγ-dependent and independent pathways. Gene 2012;505(1):46–52.

[67] Mazid MA, Chowdhury AA, Nagao K, Nishimura K, Jisaka M, Nagaya T, et al. Endogenous 15-deoxy-delta(12,14)-prostaglandin J(2) synthesized by adipocytes during maturation phase contributes to upregulation of fat storage.". FEBS Lett 2006;580(30):6885–90.

[68] Hossain MS, Chowdhury AA, Rahman MS, Nishimura K, Jisaka M, Nagaya T, et al. Development of enzyme-linked immunosorbent assay for delta12-prostaglandin J2 and its application to the measurement of the endogenous product generated by cultured adipocytes during the maturation phase. Prostaglandins Other Lipid Mediat 2011;94(3–4):73–80.

[69] Chou W-L, Chuang L-M, Chou C-C, Wang AH-J, Lawson JA, FitzGerald GA, et al. Identification of a novel prostaglandin reductase reveals the involvement of prostaglandin E2 catabolism in regulation of peroxisome proliferator-activated receptor gamma activation. J Biol Chem 2007;282(25):18162–72.

[70] Yu Y-H, Chang Y-C, Su T-H, Nong J-Y, Li C-C, Chuang L-M. Prostaglandin reductase-3 negatively modulates adipogenesis through regulation of PPARγ activity. J Lipid Res 2013;54(9):2391–9.

[71] Rahman MS, Khan F, Syeda PK, Nishimura K, Jisaka M, Nagaya T, et al. Endogenous synthesis of prostacyclin was positively regulated during the maturation phase of cultured adipocytes. Cytotechnology 2014;66(4):635–46.

[72] Dorris SL, Peebles SR. PGI2 as a regulator of inflammatory diseases. Mediators Inflamm 2012;2012:926968.

[73] Khan F, Syeda PK, Nartey MNN, Rahman MS, Islam MS, Nishimura K, et al. Pretreatment of cultured preadipocytes with arachidonic acid during the differentiation phase without a cAMP-elevating agent enhances fat storage after the maturation phase. Prostaglandins Other Lipid Mediat 2016;123:16–27.

[74] Chu X, Nishimura K, Jisaka M, Nagaya T, Shono F, Yokota K. Up-regulation of adipogenesis in adipocytes expressing stably cyclooxygenase-2 in the antisense direction. Prostaglandins Other Lipid Mediat 2010;91(1):1–9.

[75] Yan H, Kermouni A, Abdel-Hafez M, Lau DCW. Role of cyclooxygenases COX-1 and COX-2 in modulating adipogenesis in 3T3-L1 cells. J Lipid Res 2003;44(2):424–9.

[76] Cha M-H, Kim I-C, Lee B-H, Yoon Y. Baicalein inhibits adipocyte differentiation by enhancing COX-2 expression. J Med Food 2006;9(2):145.

[77] Casimir DA, Miller CW, Ntambi JM. Preadipocyte differentiation blocked by prostaglandin stimulation of prostanoid FP2 receptor in murine 3T3-L1 cells. Differentiation 1996;60(4):203–10.

[78] Liu L, Clipstone NA. Prostaglandin F2alpha inhibits adipocyte differentiation via a G alpha q-calcium-calcineurin-dependent signaling pathway. J Cell Biochem 2007;100(1):161–73.

[79] Miller CW, Casimir DA, Ntambi JM. The mechanism of inhibition of 3T3-L1 preadipocyte differentiation by prostaglandin F2alpha. Endocrinology 1996;137(12):5641–50.

[80] Reginato MJ, Krakow SL, Bailey ST, Lazar MA. Prostaglandins promote and block adipogenesis through opposing effects on peroxisome proliferator-activated receptor gamma. J Biol Chem 1998;273(4):1855–8.

[81] Serrero G, Lepak NM. Prostaglandin F2alpha receptor (FP receptor) agonists are potent adipose differentiation inhibitors for primary culture of adipocyte precursors in defined medium. Biochem Biophys Res Commun 1997;233(1):200–2.

[82] Inazumi S T., Shirata N, Morimoto K, Takano H, Segi-Nishida E, Sugimoto Y. Prostaglandin $E_2$-EP4 signaling suppresses adipocyte differentiation in mouse embryonic fibroblasts via an autocrine mechanism. J Lipid Res 2011;52(8):1500–8.

[83] Tsuboi H, Sugimoto Y, Kainoh T, Ichikawa A. Prostanoid EP4 receptor is involved in suppression of 3T3-L1 adipocyte differentiation. Biochem Biophys Res Commun 2004;322(3):1066–72.

[84] Pace-Asciak CR, Asotra S. Biosynthesis, catabolism, and biological properties of HPETEs, hydroperoxide derivatives of arachidonic acid. Free Radic Biol Med 1989;7(4):409.

[85] Kuhn H. Biosynthesis, metabolization and biological importance of the primary 15-lipoxygenase metabolites 15-hydro(pero)XY-5Z,8Z,11Z,13E-eicosatetraenoic acid and 13-hydro(pero)XY-9Z,11E-octadecadienoic acid. Prog Lipid Res 1996;35(3):203–26.

[86] Conrad DJ, Kuhn H, Mulkins M, Highland E, Sigal E. Specific inflammatory cytokines regulate the expression of human monocyte 15-lipoxygenase. Proc Natl Acad Sci U S A 1992;89(1):217–21.

[87] Huang JT, Welch JS, Ricote M, Binder CJ, Willson TM, Kelly C, et al. Interleukin-4-dependent production of PPAR-gamma ligands in macrophages by 12/15-lipoxygenase. Nature 1999;400(6742):378–82.

[88] Itoh T, Fairall L, Amin K, Inaba Y, Szanto A, Balint BL, Nagy L, Yamamoto K, Schwabe JWR. Structural basis for the activation of PPARgamma by oxidized fatty acids. Nat Struct Mol Biol 2008;15(9):924.

[89] Madsen L, Petersen RK, Sørensen MB, Jørgensen C, Hallenborg P, Pridal L, et al. Adipocyte differentiation of 3T3-L1 preadipocytes is dependent on lipoxygenase activity during the initial stages of the differentiation process. Biochem J 2003;375(Pt 3):539–49.
[90] Shillabeer G, Kumar V, Tibbo E, Lau DC. Arachidonic acid metabolites of the lipoxygenase as well as the cyclooxygenase pathway may be involved in regulating preadipocyte differentiation. Metab Clin Exp 1998;47(4):461–6.
[91] Hallenborg P, Jørgensen C, Petersen RK, Feddersen S, Araujo P, Markt P, et al. Epidermis-type lipoxygenase 3 regulates adipocyte differentiation and peroxisome proliferator-activated receptor γ activity. Mol Cell Biol 2010;30(16):4077–91.
[92] Brunnstrom A, Hamberg M, Griffiths WJ, Mannervik B, Claesson HE. Biosynthesis of 14,15-hepoxilins in human l1236 Hodgkin lymphoma cells and eosinophils. Lipids 2011;46(1):69–79.
[93] Yu Z, Schneider C, Boeglin WE, Marnett LJ, Brash AR. The lipoxygenase gene ALOXE3 implicated in skin differentiation encodes a hydroperoxide isomerase. Proc Natl Acad Sci U S A 2003;100(16):9162–7.
[94] Yu Z, Schneider C, Boeglin WE, Brash AR. Human and mouse eLOX3 have distinct substrate specificities: implications for their linkage with lipoxygenases in skin. Arch Biochem Biophys 2006;455(2):188–96.
[95] Dzitoyeva S, Chen H, Manev H. 5-Lipoxygenase-activating protein as a modulator of olanzapine-induced lipid accumulation in adipocyte. J Lipids 2013;2013:864593–6.
[96] Kanda Y, Hinata T, Kang SW, Watanabe Y. Reactive oxygen species mediate adipocyte differentiation in mesenchymal stem cells. Life Sci 2011;89(7–8):250–8.
[97] Li Y, Mouche S, Sajic T, Veyrat-Durebex C, Supale R, Pierroz D, et al. Deficiency in the NADPH oxidase 4 predisposes towards diet-induced obesity. Int J Obes 2012;36(12):1503–13.
[98] Girotti AW. Lipid hydroperoxide generation, turnover, and effector action in biological systems. J Lipid Res 1998;39(8):1529–42.
[99] Korbecki J, Baranowska-Bosiacka I, Gutowska I, Chlubek D. The effect of reactive oxygen species on the synthesis of prostanoids from arachidonic acid. J Physiol Pharmacol 2013;64(4):409.
[100] Capdevila JH, Falck JR. Biochemical and molecular properties of the cytochrome P450 arachidonic acid monooxygenases. Prostaglandins Other Lipid Mediat 2002;68–69:325–44.
[101] Fang X, Dillon JS, Hu S, Harmon SD, Yao J, Anjaiah S, et al. 20-Carboxy-arachidonic acid is a dual activator of peroxisome proliferator-activated receptors alpha and gamma. Prostaglandins Other Lipid Mediat 2007;82(1–4):175–84.
[102] Liu Y, Zhang Y, Schmelzer K, Lee TS, Fang X, Zhu Y, et al. The antiinflammatory effect of laminar flow: the role of PPARgamma, epoxyeicosatrienoic acids, and soluble epoxide hydrolase. Proc Natl Acad Sci U S A 2005;102(46):16747–52.
[103] Zha W, Edin ML, Vendrov KC, Schuck RN, Lih FB, Jat JL, et al. Functional characterization of cytochrome P450-derived epoxyeicosatrienoic acids in adipogenesis and obesity. J Lipid Res 2014;55(10):2124–36.
[104] Kim DH, Vanella L, Inoue K, Burgess A, Gotlinger K, Manthati VL, et al. Epoxyeicosatrienoic acid agonist regulates human mesenchymal stem cell-derived adipocytes through activation of HO-1-pAKT signaling and a decrease in PPARγ. Stem Cells Dev 2010;19(12):1863–73.
[105] Vanella L, Kim DH, Sodhi K, Barbagallo I, Burgess AP, Falck JR, et al. Crosstalk between EET and HO-1 downregulates Bach1 and adipogenic marker expression in mesenchymal stem cell derived adipocytes. Prostaglandins Other Lipid Mediat 2011;96(1–4):54–62.
[106] Kim DH, Puri N, Sodhi K, Falck JR, Abraham NG, Shapiro J, et al. Cyclooxygenase-2 dependent metabolism of 20-HETE increases adiposity and adipocyte enlargement in mesenchymal stem cell-derived adipocytes. J Lipid Res 2013;54(3):786–93.
[107] Baker PR, Lin Y, Schopfer FJ, Woodcock SR, Groeger AL, Batthyany C, et al. Fatty acid transduction of nitric oxide signaling: multiple nitrated unsaturated fatty acid derivatives exist in human blood and urine and serve as endogenous peroxisome proliferator-activated receptor ligands. J Biol Chem 2005;280(51):42464–75.
[108] Li Y, Zhang J, Schopfer FJ, Martynowski D, Garcia-Barrio MT, Kovach A, et al. Molecular recognition of nitrated fatty acids by PPAR gamma. Nat Struct Mol Biol 2008;15(8):865–7.
[109] Schopfer FJ, Lin Y, Baker PR, Cui T, Garcia-Barrio M, Zhang J, et al. Nitrolinoleic acid: an endogenous peroxisome proliferator-activated receptor gamma ligand. Proc Natl Acad Sci U S A 2005;102(7):2340–5.
[110] Freeman BA, Baker PR, Schopfer FJ, Woodcock SR, Napolitano A, D'Ischia M. Nitro-fatty acid formation and signaling. J Biol Chem 2008;283(23):15515–9.
[111] Yan H, Aziz E, Shillabeer G, Wong A, Shanghavi D, Kermouni A, et al. Nitric oxide promotes differentiation of rat white preadipocytes in culture. J Lipid Res 2002;43(12):2123–9.

[112] Gonthier MP, Hoareau L, Festy F, Matias I, Valenti M, Bes-Houtmann S, Rouch C, et al. Identification of endocannabinoids and related compounds in human fat cells. Obesity (Silver Spring) 2007;15(4):837–45.

[113] Bouaboula M, Hilairet S, Marchand J, Fajas L, Le Fur G, Casellas P. Anandamide induced PPARgamma transcriptional activation and 3T3-L1 preadipocyte differentiation. Eur J Pharmacol 2005;517(3):174–81.

[114] Seale P, Bjork B, Yang W, Kajimura S, Chin S, Kuang S, et al. PRDM16 controls a brown fat/skeletal muscle switch. Nature 2008;454(7207):961–7.

[115] Puigserver P, Wu Z, Park CW, Graves R, Wright M, Spiegelman BM. A cold-inducible coactivator of nuclear receptors linked to adaptive thermogenesis. Cell 1998;92(6):829–39.

[116] Ohno H, Shinoda K, Spiegelman BM, Kajimura S. PPARγ agonists induce a white-to-brown fat conversion through stabilization of PRDM16 protein. Cell Metab 2012;15(3):395–404.

[117] Marcher AB, Loft A, Nielsen R, Vihervaara T, Madsen JG, Sysi-Aho M, et al. RNA-Seq and mass-spectrometry-based lipidomics reveal extensive changes of glycerolipid pathways in brown adipose tissue in response to cold. Cell Rep 2015;13(9):2000–13.

[118] Hoene M, Li J, Haring HU, Weigert C, Xu G, Lehmann R. The lipid profile of brown adipose tissue is sex-specific in mice. Biochim Biophys Acta 2014;1842(10):1563–70.

[119] May FJ, Baer LA, Lehnig AC, So K, Chen EY, Gao F, et al. Lipidomic adaptations in white and brown adipose tissue in response to exercise demonstrate molecular species-specific remodeling. Cell Rep 2017;18(6):1558–72.

[120] Boon MR, Bakker LEH, Prehn C, Adamski J, Vosselman MJ, Jazet IM, et al. LysoPC-acyl C16:0 is associated with brown adipose tissue activity in men. Metabolomics 2017;13(5):48.

[121] Iwen KA, Backhaus J, Cassens M, Waltl M, Hedesan OC, Merkel M, et al. Cold-induced brown adipose tissue activity alters plasma fatty acids and improves glucose metabolism in men. J Clin Endocrinol Metab 2017;102(11):4226–34.

[122] Lynes MD, Leiria LO, Lundh M, Bartelt A, Shamsi F, Huang TL, et al. The cold-induced lipokine 12,13-diHOME promotes fatty acid transport into brown adipose tissue. Nat Med 2017;23(5):631–7.

[123] Sisignano M, Angioni C, Park CK, Meyer DSS, Jordan H, Kuzikov M, Liu D, et al. Targeting CYP2J to reduce paclitaxel-induced peripheral neuropathic pain. Proc Natl Acad Sci U S A 2016;113(44):12544–9.

[124] Stanford KI, Lynes MD, Takahashi H, Baer LA, Arts PJ, May FJ, et al. 12,13-diHOME: an exercise-induced lipokine that increases skeletal muscle fatty acid uptake. Cell Metab 2018;27(5):1111–20.

[125] Zhao M, Chen X. Eicosapentaenoic acid promotes thermogenic and fatty acid storage capacity in mouse subcutaneous adipocytes. Biochem Biophys Res Commun 2014;450(4):1446–51.

[126] Laiglesia LM, Lorente-Cebrian S, Prieto-Hontoria PL, Fernandez-Galilea M, Ribeiro SM, Sainz N, et al. Eicosapentaenoic acid promotes mitochondrial biogenesis and beige-like features in subcutaneous adipocytes from overweight subjects. J Nutr Biochem 2016;37:76–82.

[127] Abraham NG, Sodhi K, Silvis AM, Vanella L, Favero G, Rezzani R, et al. CYP2J2 targeting to endothelial cells attenuates adiposity and vascular dysfunction in mice fed a high-fat diet by reprogramming adipocyte phenotype. Hypertension 2014;64(6):1352–61.

[128] Burgess A, Li M, Vanella L, Kim DH, Rezzani R, Rodella L, et al. Adipocyte heme oxygenase-1 induction attenuates metabolic syndrome in both male and female obese mice. Hypertension 2010;56(6):1124–30.

[129] Waldman M, Bellner L, Vanella L, Schragenheim J, Sodhi K, Singh SP, Lin D, et al. Epoxyeicosatrienoic acids regulate adipocyte differentiation of mouse 3T3 cells, via PGC-1alpha activation, which is required for HO-1 expression and increased mitochondrial function. Stem Cells Dev 2016;25(14):1084–94.

[130] Zhang S, Chen G, Li N, Dai M, Chen C, Wang P, et al. CYP2J2 overexpression ameliorates hyperlipidemia via increased fatty acid oxidation mediated by the AMPK pathway. Obesity (Silver Spring) 2015;23(7):1401–13.

[131] Lodhi IJ, Semenkovich CF. Peroxisomes: a nexus for lipid metabolism and cellular signaling. Cell Metab 2014;19(3):380–92.

[132] Brites P, Waterham HR, Wanders RJA. Functions and biosynthesis of plasmalogens in health and disease. Biochim Biophys Acta– Mol Cell Biol Lipids 2004;1636(2):219–31.

[133] Park H, He A, Tan M, Johnson JM, Dean JM, Pietka TA, et al. Peroxisome-derived lipids regulate adipose thermogenesis by mediating cold-induced mitochondrial fission. J Clin Invest 2019;129(2):694–711.

[134] Pietiläinen KH, Róg T, Seppänen-Laakso T, Virtue S, Gopalacharyulu P, Tang J, et al. Association of lipidome remodeling in the adipocyte membrane with acquired obesity in humans. PLoS Biol 2011;9(6):e1000623.

[135] Sustarsic EG, Ma T, Lynes MD, Larsen M, Karavaeva I, Havelund JF, et al. Cardiolipin synthesis in brown and beige fat mitochondria is essential for systemic energy homeostasis. Cell Metab 2018;28(1):159–74.

[136] Paradies V, Paradies G, De Benedictis V, Ruggiero FM, Petrosillo G. Functional role of cardiolipin in mitochondrial bioenergetics. BBA Bioenergetics 2014;1837(4):408–17.

[137] Houtkooper RH, Vaz FM. Cardiolipin, the heart of mitochondrial metabolism. Cell Mol Life Sci 2008;65(16):2493–506.

[138] Schlame M, Hostetler KY. Cardiolipin synthase from mammalian mitochondria. Biochim Biophys Acta 1997;1348(1–2):207–13.

[139] Shi Y. Emerging roles of cardiolipin remodeling in mitochondrial dysfunction associated with diabetes, obesity, and cardiovascular diseases. J Biomed Res 2010;24(1):6–15.

[140] Esposito K, Kastorini C-M, Panagiotakos DB, Giugliano D. Mediterranean diet and weight loss: meta-analysis of randomized controlled trials. Metab Syndr Relat Disord 2011;9(1):1–12.

[141] Kastorini CM, Milionis HJ, Esposito K, Giugliano D, Goudevenos JA, Panagiotakos DB. The effect of Mediterranean diet on metabolic syndrome and its components: a meta-analysis of 50 studies and 534,906 individuals. Journal Am Coll Cardiol 2011;57(11):1299–313.

[142] Shin S, Ajuwon KM. Effects of diets differing in composition of 18-C fatty acids on adipose tissue thermogenic gene expression in mice fed high-fat diets. Nutrients 2018;10(2):256.

[143] Jones PJH, Jew S, AbuMweis S. The effect of dietary oleic, linoleic, and linolenic acids on fat oxidation and energy expenditure in healthy men. Metab Clin Exp 2008;57(9):1198–203.

[144] Ghandour RA, Colson C, Giroud M, Maurer S, Rekima S, Ailhaud G, et al. Impact of dietary ω3 polyunsaturated fatty acid supplementation on brown and brite adipocyte function. J Lipid Res 2018;59(3):452–61.

[145] Kim M, Goto T, Yu R, Uchida K, Tominaga M, Kano Y, et al. Fish oil intake induces UCP1 upregulation in brown and white adipose tissue via the sympathetic nervous system. Sci Rep 2015;5(1):18013.

[146] Pahlavani M, Razafimanjato F, Ramalingam L, Kalupahana NS, Moussa H, Scoggin S, et al. Eicosapentaenoic acid regulates brown adipose tissue metabolism in high-fat-fed mice and in clonal brown adipocytes. J Nutr Biochem 2017;39:101–9.

[147] Schlame M. Cardiolipin synthesis for the assembly of bacterial and mitochondrial membranes. J Lipid Res 2008;49(8):1607–20.

CHAPTER

# 6

# Lipid receptors and signaling in adipose tissue

*Ann V. Hertzel*[a], *Timothy D. O'Connell*[b], *David A. Bernlohr*[a]

[a]Department of Biochemistry, Molecular Biology and Biophysics, The University of Minnesota, Minneapolis, MN, United States; [b]Department of Integrative Biology and Physiology, The University of Minnesota, Minneapolis, MN, United States

OUTLINE



## Introduction

Fatty acids as macronutrients and structural elements dominated the biochemical landscape during the latter half of the 20th century with numerous key discoveries linked to their role in energy homeostasis (oxidation and storage) and membrane biogenesis. However, the past 20 years have revealed a wealth of new biology linked to the role of fatty acids as bioregulators implicated in a variety of complex processes such as immunity and inflammation, development and cellular communication. Due to their high chemical solubility in low dielectric environments, protonated fatty acids diffuse rapidly across biological membranes and serve either directly, or indirectly, as allosteric ligands for a variety of proteins controlling diverse processes such as gene expression, mitochondrial uncoupling or lipogenesis. However, fatty acids may also bind with high affinity and specificity to cell-surface receptors and activate/repress signaling pathways thereby serving as multipurposed homeostatic

Lipid Signaling and Metabolism. http://dx.doi.org/10.1016/B978-0-12-819404-1.00006-3

regulators. The focus of this chapter will be on such cell surface fatty acid receptors and their signaling processes.

Expression cloning methodologies coupled with chemical genomic approaches have revealed insight into cell surface receptors for fatty acids and their derivatives (eicosanoids). Adipocytes are rich sources for such receptors and integrate both the nutrient and regulatory nature of fatty acid biology. Adipocytes internalize large amounts of fatty acids for both biosynthesis of membranes/organelles and energy storage/oxidation. Over the past decade, our awareness that adipocytes express cell surface receptors for fatty acids that control signaling pathways toward diverse processes such as calcium homeostasis, gene expression, and insulin action have greatly expanded our understanding of how such molecules can serve as bioregulators. This chapter will focus on the major fatty acid receptors of adipocytes and include a discussion of CD36/SR-B2, eicosanoid receptors, and free fatty acid receptors. The reader is referred to other chapters in this volume for excellent presentations on toll-like receptors found in adipose tissue, particularly toll-like receptor 4.

## Receptor signaling systems in adipocytes

### CD36/SR-B2

CD36 is a complex, cell surface glycoprotein of 88 kDa implicated in fatty acid transport via a combination of biochemical, molecular, and genetic determinants [1–3]. Officially renamed as scavenger receptor B2 (SR-B2), CD36/SR-B2 was identified by Abumrad and colleagues in the mid-1990s as an adipocyte protein covalently modified by sulfo-*N*-succinimidyl acyl ligands [4]. Whereas protonated fatty acids can readily diffuse across biological membranes, CD36/SR-B2 increased the rate of such movement by mechanisms likely linked to increasing the local concentration of fatty acids in the outer leaflet of the plasma membrane. The protein is found in a variety of cells beyond adipocytes including skeletal and cardiac myocytes, hypothalamic neurons, intestinal absorptive cells, and tongue epithelial cells [3]. Molecular disruption of CD36/SR-B2 in murine fat cells or 3T3-L1 adipocytes reduces basal and insulin-stimulated fatty acid uptake by approximately 50% suggesting that this protein is a major determinant in fatty acid influx [5]. As a cell surface protein, CD36/SR-B2 undergoes rapid recycling between the outer membrane and endosomes in an AMPK-dependent manner and high plasma membrane occupancy and lipid hyperaccumulation is obesogenic [6]. Gain or loss of function strategies revealed that CD36/DR-B2 likely functions in fatty acid transport and lipogenesis. Consistent with this, the expression of CD36/SR-B2 in adipocytes is positively regulated by PPAR$\gamma$ and animal models lacking CD36/SR-B2 exhibit reduced triglyceride accumulation [7,8].

CD36/SR-B2 is predicted to have two transmembrane domains located at the N- and C-terminal regions leaving the majority of the polypeptide on the exofacial leaflet of the plasma membrane [2,9]. Biochemical analysis of CD36/SR-B2 suggests that the N and C-terminal regions of the protein are anchored in the inner leaflet of the plasma membrane by acylation/palmitoylation of pairs of cysteine residues [10]. The extracellular domain of the protein is subject to numerous covalent modifications including phosphorylation, glycosylation, and oxidation (disulfide bond formation) and is considered to be the fatty acid-binding domain, particularly the region from amino acids 120 to 280 [2]. The binding specificity for

CD36/SR-B2 is very broad and the protein binds a variety of hydrophobic ligands including advanced glycation end products, oxidized low-density lipoprotein and bacterial fragments along with long-chain fatty acids. By virtue of the two transmembrane domains and covalent acylation, CD36/SR-B2 is mainly localized to lipid raft regions, potentially to facilitate fatty acid flip-flop across the hydrophobic membrane.

The C-terminal intracellular domain serves as a binding site for proteins of lipid metabolism (cytoplasmic fatty acid-binding proteins) and kinase signaling (Syk and Fyn kinases) [11]. Samovski et al., have reported that CD36 promotes tyrosine phosphorylation of the insulin receptor by the Fyn kinase, a process that enhances receptor recruitment of p85 adaptor protein and downstream signaling [12]. Saturated fatty acids suppress CD36/SR-B2-Fyn-dependent insulin receptor phosphorylation, whereas unsaturated fatty acids stimulate the process. The metabolic teleology of this finding is complex given that both unsaturated and saturated fatty acids require a glucose-derived 3-carbon backbone for esterification into triglyceride. Regardless, it is clear that CD36/SR-B2 plays a dual role in metabolism—as both a fatty acid transport protein and as a lipid-dependent signaling sensor regulating such broad phenotypes as glucose uptake and metabolism.

## Leukotriene receptors

Eicosanoids are 20 carbon molecules derived from arachidonic acid that via receptor binding, mediate signal transduction to exert profound physiological consequences. Currently, there are five identified prostanoid receptors, allowing cellular response to prostaglandins, thromboxanes, and prostacyclins, and at least six receptors respond to leukotrienes. Eicosanoid receptors are members of the seven transmembrane G-protein-coupled receptor family, are coupled to one or more signal transduction pathways, and are expressed in a cell type(s) specific manner (Fig. 6.1). These receptors play fundamental roles in normal physiology and in a variety of pathological states. There has been significant interest in these receptors in adipose tissue since immune cells recruited to adipose tissue produce eicosanoids as well as adipocytes themselves that produce eicosanoids through the lipolytic process [13]. Leukotrienes are synthesized from the *n*-6 fatty acid, arachidonic acid, into the dihydroxy acid, LTB4, or the group of cysteinyl leukotrienes, LTC4, LTD4, and LTE4. Thus far there are at least six receptors that bind leukotrienes. Two receptors bind the dihydroxy-leukotriene, LTB4 (BLT receptors) and four binds to the cysteinyl leukotrienes, LTC4, LTD4, and LTE4 (CysLTR1, CysLTR2, GPR17, and GPR99) [14].

### *LTB4 and BLT receptors*

This family of G-protein-coupled receptors include BLT1R and BLT2R. Based on similarity, these two receptors are structurally related to each other, but also display a closeness to the prostaglandin receptor DP2. Twenty years after the discovery of LTB4, the first receptor was cloned [15]. BLT1R displays high affinity for LTB4, that results in stimulated chemotaxis and inflammation of immune cells. However, initial studies, as well as studies on a BLT1R knockout mouse, uncovered a low-affinity response to LTB4 that led to the discovery of BLT2R [16]. In addition to LTB4, this receptor has also been shown to be capable of binding other lipoxygenase products, including some HETEs and HpETEs, as well as the thromboxane product, 12-hydroxyheptadecatrienoic acid (12-HHT) [17]. Unexpectantly, the effect of inflammatory

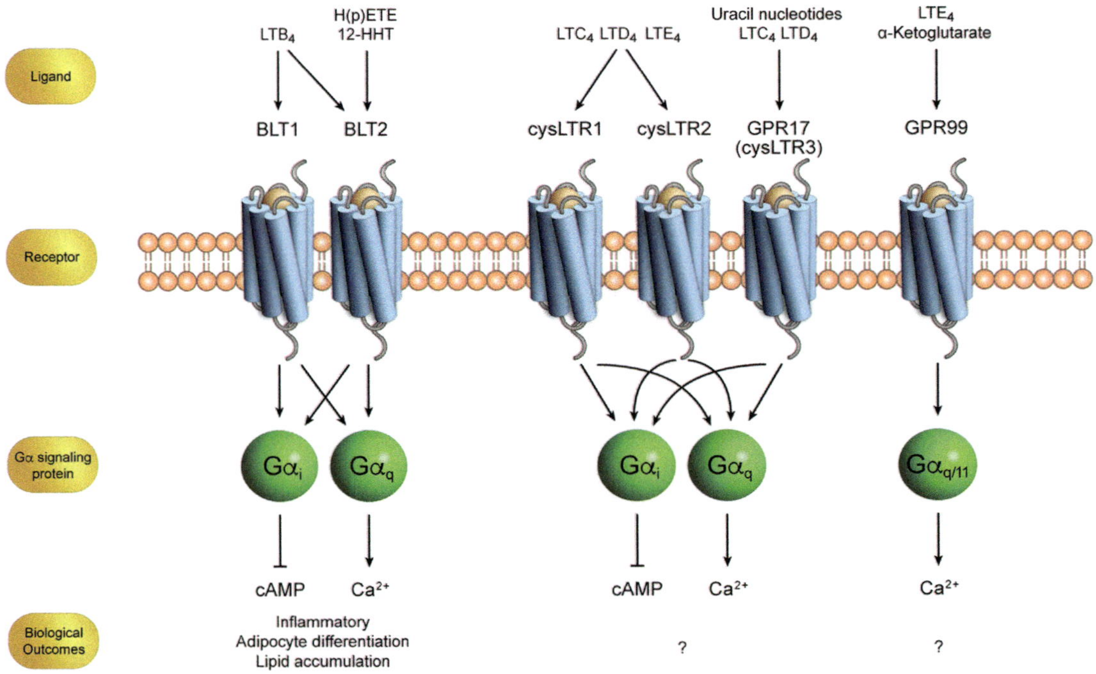


**FIGURE 6.1 Leukotriene signaling via CysLTR1, CysLTR2, GPR17(CysLTR3), and GPR99 receptors.** Leukotriene receptors are members of the G-protein-coupled receptor family. These receptors couple to various Gα proteins, typically resulting in alteration of cAMP or calcium levels. BLT1 and BLT2 bind to LTB4, with additional binding capability to H(p)ETE and 12-HHT, for BLT2. CysLTR1 and CysLTR2 bind LTC4, LTD4, and LTE4, although with varying affinities. GPR17 (CysLTR3) binds LTC4, LTD4, and uracil nucleotides, whereas GPR99 binds LTE4 and α-ketoglutarate. BLT1 and BLT2 are inflammatory in adipose as well as increasing lipid accumulation. The effects of CysLTR1, CysLTR2, GPR17, and GPR99 on adipose function require additional analysis.

factors on BLT1R expression is cell type specific; in some cases, inflammatory factors cause increased BLT1R expression, whereas in others, the same inflammatory factors decrease BLT1R expression [14] (Fig. 6.1).

BLT1R and BLT2R mainly couple to the inhibitory $G_{\alpha i}$ protein, thus reducing cAMP levels, but also to $G_{\alpha q}$, thus activating phospholipase C to increase intracellular calcium [18]. Utilizing either a BLT receptor antagonist or siRNA knockdown of BLT1R and BLT2R resulted in inhibition of adipocyte differentiation and decreased lipid accumulation [19] (Fig. 6.1). BLT1R knockout mice fed a high-fat diet were protected from glucose and insulin intolerance. These knockout mice had decreased inflammation in adipose tissue and liver, with a decrease in hepatic triglyceride accumulation. Furthermore, deletion of BLT1R prevented the high fat diet-induced loss of insulin signaling in liver and skeletal muscle [20]. BLT1R deficiency also led to an alteration of macrophage accumulation in adipose tissue, with a decrease in the adipose tissue expression of CCL2. In a feed-forward cycle, activation of BLT1R by LTB4 induces production of CCL2 in monocytes and, similarly, CCL2 stimulates the production of LTB4 [20]. An additional study found that deficiency of BLT1R by genetic or pharmacologic methods led to an antiinflammatory state and protected the mice from insulin resistance and hepatic steatosis [21]. Altogether, these data demonstrate the major role that BLT1R plays in the chronic inflammation associated with obesity that can lead to insulin resistance.

### *LTC4, LTD4, LTE4 and the CysLT, GPR17, GPR99 receptors*

Over time, several receptors have been found capable of binding to one or more of the cysteinyl leukotrienes. Two of these, CysLT1R and CysLT2R, couple to $G_{\alpha q}$ or $G_{\alpha i}$ in a cell type-specific manner, resulting in activation of phospholipase C and the mobilization of calcium, or the inhibition of cAMP, respectively [22] (Fig. 6.1). CysLT1R has the highest affinity for LTD4, followed by LTC4, then LTE4, whereas CysLT2R binds LTC4 equally to LTD4, with lower affinity for LTE4 [22]. The primary sequences of these two receptors are fairly divergent with only 38% amino acid identity. Additionally, it was discovered that GPR17 can also bind LTC4 and LTD4, in addition to uracil nucleotides; thus this receptor is sometimes referred to as CysLTR3 [23]. However, at least in specific contexts, GPR17 has more recently been shown to not be responsive to leukotrienes or uracil nucleotides [24].

An additional G-protein-coupled receptor, GPR99, can also bind cysteinyl leukotrienes, especially LTE4, in addition to binding α-ketoglutarate. GPR99 typically couples to $G_{\alpha q/11}$ resulting in an increase in calcium. The complexity of cysteinyl receptor signaling arises from the unique cell-specific expression of the cysteinyl receptor(s) and adaptor proteins, in addition to the fact that these receptors can dimerize or oligomerize with self as well as with other members of the family, that is, CysLT1R and CysLT2R [25]. These results in sum point to the need for additional work to assess the effects of cysteinyl leukotrienes on adipose biology.

## Prostanoid receptors

### *$PGD_2$ and DP receptors*

There are two known G-protein-coupled receptors that avidly bind prostaglandin $D_2$ ($PGD_2$): DP1 and DP2/CRTH2 (chemoattractant receptor-homologous molecule expressed on T-helper type 2 cells; formerly called GPR44). These receptors have greatly reduced ability to bind to other prostanoids ($PGD_2 >> PGF_{2\alpha} = PGE_2 > PGI_2$ = thromboxane A2). DP1 is a member of the prostanoid receptor family that couples to pertussis toxin (PTX)-resistant $G_{\alpha s}$ to stimulate adenylate cyclase, ultimately elevating cAMP. In contrast, DP2/CRTH2 is structurally distinct from the prostanoid family of receptors, being part of the formyl-methionyl-leucyl-phenylalanine receptor subfamily with homology to chemokine receptors. Upon $PGD_2$ binding, DP2/CRTH2 couples with PTX-sensitive $G_{\alpha i}$ proteins that suppress adenylate cyclase and intracellular cAMP, while inducing intracellular $Ca^{2+}$ mobilization [26] (Fig. 6.2). An X-ray crystal structure of human DP2/CRTH2 revealed that the carboxylate group of $PGD_2$ extends deep into the ligand-binding pocket following the positive charge gradient, whereas the hydrocarbon chain is stabilized by many aromatic residues in the ligand-binding pocket [27]. Whereas DP1 has not found in adipose tissue, DP2/CRTH2 is expressed in both immune cells and in adipocytes in adipose tissue [28]. DP2 activation suppresses lipolysis in adipocytes, whereas a knockout of DP2 in adipocytes differentiated from mouse embryonic fibroblasts enhanced lipolysis [29,30]. Although $PGD_2$ has been reported to promote adipocyte differentiation, this occurs via PPARγ and not through a DP receptor [31].

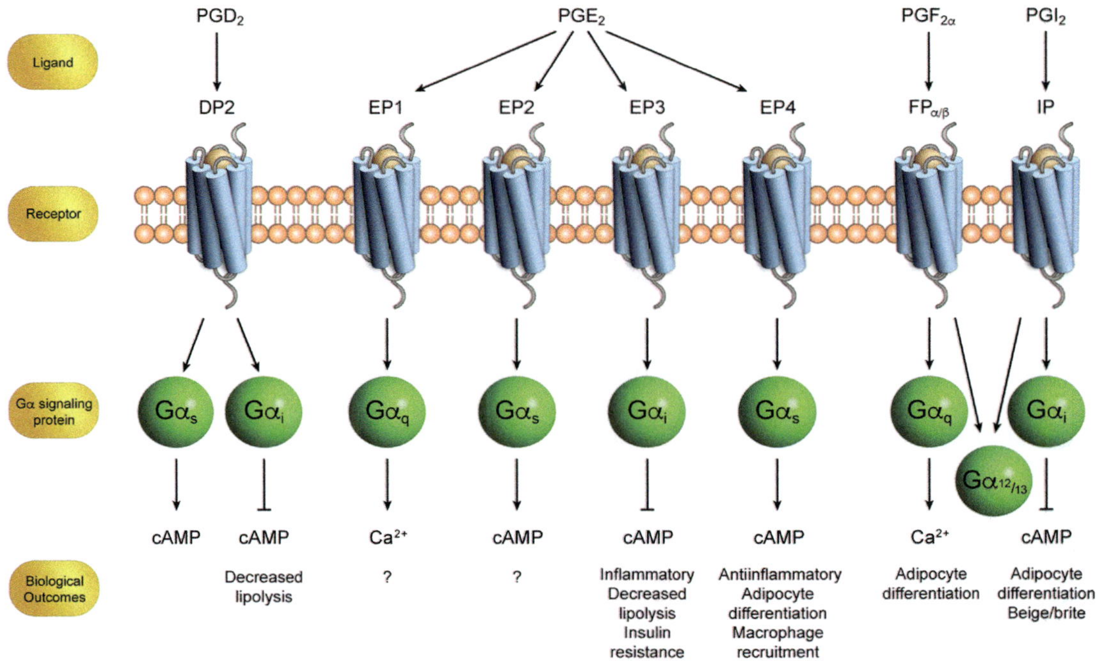


**FIGURE 6.2** **Prostaglandin signaling via DP2, EP1–4, FP, and IP receptors.** Prostaglandin receptors are also members of the G-protein-coupled receptor family. These receptors couple to various Gα proteins, typically resulting in alteration of cAMP or calcium levels. DP2 binds $PGD_2$ that can result in decreased adipocyte lipolysis. EP1, EP2, EP3, and EP4 all bind $PGE_2$. Due to the various receptors, the biological outcome of EP receptor signaling is varied, but can involve inflammation, lipolysis, insulin resistance and macrophage recruitment. The FP and IP receptors bind to $PGF_2$ and $PGI_2$. Less is known about the functional consequences of these prostaglandin receptors, but current data suggests involvement in adipocyte differentiation.

## *PGE₂ and EP receptors*

There are four main receptors that bind to prostaglandin $E_2$ ($PGE_2$) in the low nM range: EP1, EP2, EP3, and EP4, with the ancestral prostanoid receptor thought to be an EP [32,33]. Gene expression of all EP receptor subtypes have been found in mouse epididymal adipose tissue [34]. Of these, EP3 is the most abundant EP receptor in adipose, specifically in adipocytes [35]. Analysis of alternatively spliced EP3 transcripts has currently identified eight isoforms, all with distinctive C-terminal tails [36,37]. Mouse adipose tissue has at least three of these EP3 isoforms (α, β, γ) [35]. Each EP subtype has distinct signal transduction properties, since EP1 couples to $G_{\alpha q}$ to increase calcium, EP3 couples to $G_i$ to decrease cAMP, whereas EP2 and EP4 both couple to $G_{\alpha s}$ resulting in elevated cAMP [38] (Fig. 6.2). This versatility in receptor response to $PGE_2$ allows a collection of diverse responses to $PGE_2$ in a cell- and receptor-dependent manner [38], such as conditions that result in $PGE_2$ stimulated pro-inflammatory versus antiinflammatory pathways [37].

Examination of the effect of obesity on the expression of EP receptors in adipose tissue, via genetic modification (*db/db*) or high-fat diet has produced mixed results. The mRNA levels of EP2, EP3, and EP4 were found to be increased in mouse epididymal adipose tissue from a HFD fed mouse [34]. Furthermore, only EP3 expression (and not other EP receptors) was elevated in primary adipocytes isolated from the HFD-induced obese rats as well as in obese human subjects. This was also demonstrated in cell culture adipocyte models, as EP3 expression was elevated in mouse 3T3-L1 and human SGBS adipocytes treated with palmitate. Consistent with this, human visceral adipocyte EP3 mRNA levels positively correlate with BMI [39]. In contrast to these results, expression of the EP3 isoforms was found to be downregulated in *db/db* or HFD fed mouse white adipose tissue [35]. In mice, rats and human subjects, $PGE_2$ has been shown to inhibit lipolysis [40], especially catecholamine stimulated lipolysis [41,42], in an EP3 receptor-dependent manner [35]. Additionally, EP3 $\alpha$ and $\gamma$ isoforms have been shown to inhibit adipogenesis, food intake, and serum triglyceride levels [35].

Activation of adipocyte $PGE_2$/EP3 signaling during hypertrophy and hypoxia resulted in an increase in proinflammatory adipokine production as well as a decrease in adiponectin production mainly via the activation of the NF-κB-mediated inflammatory pathway, playing a crucial role in the development of obesity-associated inflammation and insulin resistance [39]. Similarly, EP3 inhibitors significantly reversed adipose inflammatory gene/protein expression as well as impairing glucose and insulin tolerance in *db/db* mice. Thus $PGE_2$/EP3 signaling plays a significant role in obesity-induced adipose tissue inflammation and insulin resistance [39]. In contrast to these previous studies, knockout of all EP3 receptors has been demonstrated to cause insulin resistance [35]. Furthermore, $PGE_2$/EP4 signaling in adipocytes differentiated from MEFs resulted in reduced PPARγ2 expression and subsequent limited differentiation, via autocrine effect of $PGE_2$ [43].

Deletion of EP4 increased β3-adrenergic receptor-dependent white adipose remodeling as seen by a reduction in triglyceride storage and thus a decrease in the size of lipid droplets and an increase in mitochondrial biogenesis [44]. Furthermore, $PGE_2$-EP4 signaling modulates inflammatory gene expression in adipose tissue, as evidenced via EP4 knockout displaying increased inflammatory chemokine expression, seen especially upon HFD [34]. Stimulation of adipocyte lipolysis increased $PGE_2$ production that is involved in the recruitment of macrophages into adipose tissue via a mechanism involving macrophage EP4 signaling [45].

### *$PGF_{2\alpha}$ and FP receptor*

The FP receptor has been cloned from human and mouse, and two alternatively spliced isoforms, FPα and FPβ, have been identified in which FPβ lacks 46 amino acids on the C-terminus as compared to FPα [46]. The FP receptors signal via coupling to $G_{\alpha q}$ to mobilize calcium, but also couple to $G_{\alpha 12/13}$, and in a cell-type-specific manner to $G_{\alpha i}$ [32] (Fig. 6.2).

The FP receptors are present throughout the adipocyte differentiation process [47] and $PGF_{2\alpha}$ has been shown to inhibit adipocyte differentiation [47–49]. Consistent with $G_{\alpha q}$ coupling, stimulation of the FP receptor results in a transient calcium increase. These data are consistent with other calcium mobilizing agents that also have been shown to inhibit differentiation [47]. Indeed, the inhibition of adipocyte differentiation by $PGF_{2\alpha}$ involves a FP-$G_{\alpha q}$-calcium-calcineurin-dependent signaling pathway, that leads to reduced expression of PPARγ and C/EBPα involving a HDAC-sensitive mechanism [49]. In abdominally obese women, there is an increase in $PGF_{2\alpha}$ release by omental mature adipocytes and the women with the highest $PGF_{2\alpha}$ release form omental adipocytes had higher BMI and HOMA-IR [50].

### *$PGI_2$ (prostacyclin) and IP receptors*

$PGI_2$, also called prostacyclin, binds to the IP receptor, that couples primarily couples to $G_{\alpha s}$ but can also couple to $G_{\alpha q}$ and $G_{\alpha i}$. Biologically, the $PGI_2$/IP receptor mediates opposite functions as to thromboxane/TP receptors. Originally described as to their involvement in cardiovascular system (vasodilatory and antithrombotic), $PGI_2$/IP has also been shown to be involved in the regulation of adipose tissue development. $PGI_2$ through IP can upregulate C/EBPβ and C/EBPδ, thus promoting adipocyte differentiation. Furthermore, $PGI_2$/IP activates PPARγ leading to the terminal differentiation of adipocytes [30,48]. $PGI_2$ has been also shown to induce thermogenic beige/brite adipocytes partially through IP-dependent, but also IP-independent pathways [40,51].

### *$TXA_2$ (thromboxane) and TP receptor*

The human thromboxane ($TXA_2$) receptor, TP, was the first prostanoid receptor to be cloned, and has been demonstrated to play a role in cardiovascular disease. This receptor has alternative splicing, resulting in two variants (α and β) that differ in their C-terminal tail region. Biologically, the $TXA_2$/TP receptor mediates opposite functions as to $PGI_2$/IP [37].

## Free fatty acid receptors (Ffar) in fat cells

The discovery of GPCRs that bind to and are activated by endogenous free fatty acids establishes an entirely novel paradigm whereby fatty acids function as signaling molecules, not simply as an energy source. Originally identified as a group of related orphan G-protein-coupled receptors (GPRs), GPR40, GPR41, GPR42, and GPR43 were found clustered in a locus in the human genome at 19q13.1 [52]. GPR40, GPR41, and GPR43 were later deorphanized and recognized as receptors for endogenous free fatty acids [53–58], whereas GPR42, although structurally related to GPR41, does not bind fatty acids and appears to have risen through gene duplication [56]. In a similar manner, GPR120 was identified at 10q23.33 in the human genome [59], and was also later identified as a receptor for endogenous free fatty acids [60]. Subsequently, these receptors were grouped into a larger family and reclassified as free fatty acid receptors 1–4 (Ffar1–4) by the International Union of Basic and Clinical Pharmacology (IUPHAR). Ffar2 and Ffar3, formerly GPR43 and GPR41 respectively, are receptors for short-chain fatty acids containing 1–6 carbons (C1–C6) [56–58]. Although sharing only 40% sequence identity, Ffar1 and Ffar4, formerly GPR40 and GPR120 respectively, are receptors for medium- and long-chain fatty acids containing ≥10 carbons (C10–C22) [53–55,60]. All members of the Ffar family have been implicated in regulating metabolism and immune responses [61], although only Ffar2, Ffar3, and Ffar4 are expressed in adipose and will be the focus here (Fig. 6.3).

### *Ffar2 and Ffar3 in adipose*

Ffar2 and Ffar3 share similarities in expression patterns, with both receptors expressed in enteroendocrine cells, including L and I cells, whereas K cells express only Ffar3 [62–67]. Both receptors are also expressed in pancreatic α- and β-cells [68–70]. Conversely, Ffar2 is highly expressed in immune cells [56], including monocytes, neutrophils [71], and regulatory T-cells [72], whereas both Ffar2 and Ffar3 are expressed in dendritic cells [73]. Ffar2 and Ffar3 both

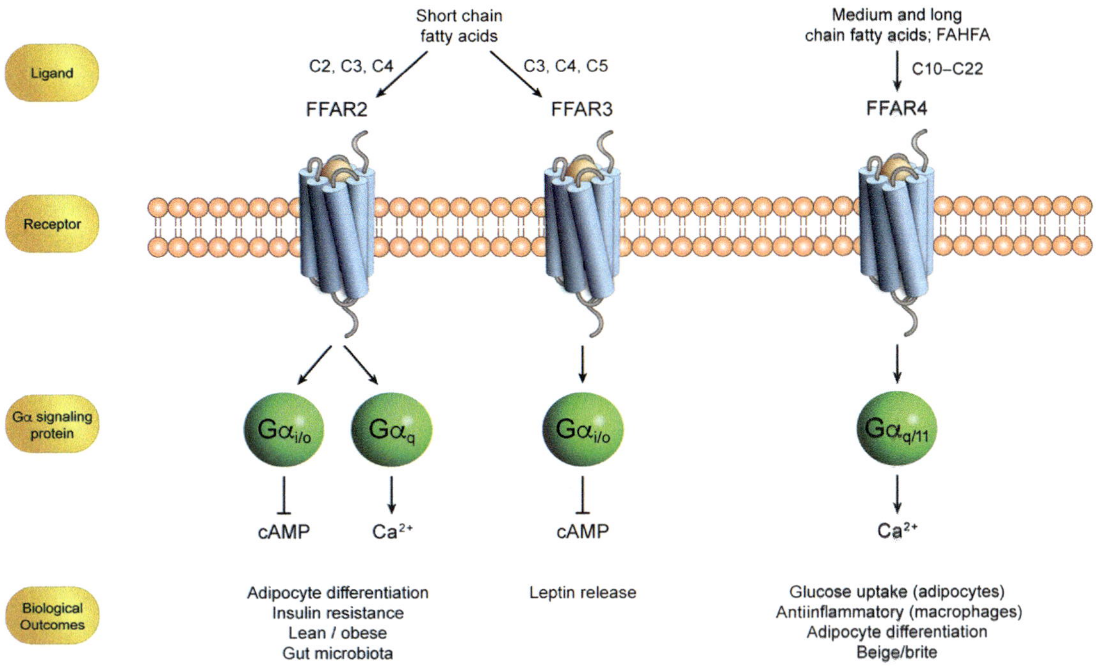


**FIGURE 6.3 Fatty acid signaling via Ffar receptors.** Ffar receptors are also members of the G-protein-coupled receptor family. These receptors couple to various Gα proteins, typically resulting in alteration of cAMP or calcium levels. FFAR2 and FFAR3 bind to short fatty acids, with carbon (C) lengths of 2–4 for FFAR2 and 3–5 carbons for FFAR3. Biological outcomes of FFAR2 ligand binding includes adipocyte differentiation, insulin resistance, and effects on gut microbiota. Less is certain about FFAR3, but it has been implicated in leptin release. FFAR4 binds to medium to long-chain fatty acids, including 10–22 carbon lengths, resulting in glucose uptake, antiinflammation, as well as impacting the occurrence of beige/brite adipocytes.

bind short-chain fatty acids, C1–C6. However, there are differences in potency between receptors with Ffar2 showing the highest affinity for acetate (C2), propionate (C3), and butyrate (C4), whereas Ffar3 shows the highest affinity for propionate (C3), butyrate (C4), and isobutyrate (C5) [56,57]. Original studies indicated that Ffar2 might activate both $G_{\alpha i/o}$ and $G_{\alpha q}$ pathways, and later studies have confirmed Ffar2 signals through $G_{\alpha i/o}$ to inhibit cAMP production [56,57], with the demonstration of $G_{\alpha q}$-mediated GLP-1 secretion in intestinal L-cells [67]. Conversely Ffar3 signals primarily through $G_{\alpha i/o}$ [56,57] (Fig. 6.3). In adipose, some studies indicate that Ffar2 and Ffar3 are both expressed in 3T3-L1, mouse and human adipocytes [57,74,75], while others have failed to confirm Ffar3 expression, suggesting that Ffar2 is the predominant receptor in adipose [76,77]. Ffar2 expression is also increased in adipose tissue of mice fed a high-fat diet [76]. Prior studies have found that short-chain fatty acids regulate adipocyte function, including adipocyte differentiation, lipolysis, and leptin secretion [78,79], suggesting the potential for Ffar2/3 regulation of adipocyte function.

In 3T3-L1 adipocytes, acetate (C2) and propionate (C3) promoted lipolysis and propionate (C3) induced Ffar2 expression during differentiation, while differentiation was attenuated by siRNA-mediated knockdown of Ffar2 [76]. In support of these findings, systemic deletion of Ffar2 (Ffar2 KO) attenuated high-fat diet-induced weight gain and increased lean body mass, with improved insulin sensitivity and increased energy expenditure [80]. Although prior studies suggested Ffar3 was expressed in adipocytes and regulated leptin secretion [75], both of these studies failed to detect Ffar3 expression in adipocytes [76,80]. Conversely, in a different Ffar2 KO model, mice were obese and there was no effect on adipocyte differentiation or circulating leptin levels, while in mice overexpressing Ffar2 in adipose were lean [81]. Interestingly, these phenotypes were lost when mice were raised under germ-free conditions or after treatment with antibiotics, suggesting a significant role for gut microbiota in regulating Ffar2 effects on obesity [81]. A study in primary human adipocytes also found no link between Ffar2 and adipogenesis [82]. Another study in mice with systemic deletion of Ffar3 (Ffar3 KO) demonstrated a loss of acetate (C2) and propionate (C3) induced leptin release in adipocytes, suggesting a requirement for Ffar3, but Ffar2 expression was also decreased in adipocytes in this model, clouding interpretation of these results [77]. In summary, these seemingly conflicting results have failed to define the exact role Ffar2 and Ffar3 adipocytes. Although strain differences might explain some of the discrepancies, the underlying basis for these differences remains unclear.

### *Ffar4 in adipose*

Ffar4 is most highly expressed in lung, but also shows high levels of expression in the GI tract, with lower levels of expression in several tissues and cell-types including, brain, heart, pancreas, adipose, taste buds, and immune cells, including macrophages [60,83–85]. Ffar4 binds to a large range of medium- and long-chain saturated, mono-unsaturated, and polyunsaturated fatty acids, C10–C22, with potency in the low micromolar range. Interestingly, polyunsaturated fatty acids (PUFAs) tend to show higher efficacy to activate signaling responses than saturated FAs (SFAs) [84,86,87]. Recently, a novel class of branched fatty acid esters of hydroxy fatty acids (FAHFAs) were also described as ligands for Ffar4 [88].

Initial reports indicated that Ffar4 activates $G_{\alpha q/11}$ mediated release of intracellular $Ca^{2+}$ and activation of ERK [60], that was subsequently confirmed by others [89,90]. Ffar4 also signals through β-arrestin-2 (β-Arr2) [84,89]. However, Ffar4 signaling through $G_{\alpha q/11}$ or β-Arr2 appears to be cell-type specific, with Ffar4 signaling through $G_{\alpha q/11}$ to regulate glucose uptake in 3T3-L1 and primary mouse adipocytes, whereas Ffar4 signals through β-Arr2 in RAW264.7 and primary peritoneal mouse macrophages to regulate antiinflammatory effects [84]. In humans, there are two Ffar4 isoforms, short and long, differentiated by a 16 amino acid insertion in the third intracellular loop of the long-isoform, whereas other species express only one isoform, homologous to the short-isoform in humans [90]. Interestingly, the long-isoform only signals through β-Arr2 and is unable to activate $G_{\alpha q/11}$-mediated signaling [90] (Fig. 6.3).

In adipose, Ffar4 is expressed in white adipose tissue (WAT) and detected in subcutaneous, perinephric, mesenteric, and epididymal fat pads in mice, and Ffar4 expression is increased in mice on a high-fat diet [91]. In mice with systemic deletion of Ffar4 (Ffar4KO mice), adipocyte size is increased in epididymal and subcutaneous fat in mice on a normal or high-fat diet, but exaggerated weight gain was only observed in Ffar4KO mice on the high-fat diet [92]. In humans, Ffar4 is also expressed in adipose, and Ffar4 expression is higher in obese versus lean individuals [92]. Interestingly, Ffar4 is also expressed in brown adipose tissue (BAT), and Ffar4 induces browning of (WAT) [93]. Furthermore, activation of Ffar4 in BAT increases fatty acid oxidation and reduces weight and fat mass in mice [94].

During differentiation of 3T3-L1 and human adipocytes, Ffar4 expression is increased and differentiation is prevented by knockdown of Ffar4 [91,95]. In mice with systemic deletion of Ffar4 (Ffar4KO mice), adipogenesis is also inhibited [92]. Additionally, the increase in Ffar4 expression during adipocyte differentiation is accompanied by increases in expression of peroxisome proliferator-activated receptor γ (PPARγ) and fatty acid-binding protein 4 (FABP4) [91]. Alternatively, troglitazone, a PPARγ agonist, increases Ffar4 expression [91]. Of note, PPARγ ligands include many fatty acids and eicosanoids that are also ligands for Ffar4, and show some general overlap in biologic function in adipose [96]. This indicates the potential for significant interaction between Ffar4 and PPARγ in the regulation of adipocyte differentiation and function, which at this point remains to be defined.

In adipose, resident adipose tissue macrophages (ATMs) maintain homeostasis, but during obesity, ATMs help drive tissue inflammation exacerbating the condition [97]. Ffar4 is expressed in macrophages, and Ffar4 signals through β-Arr2 to produce an antiinflammatory macrophage phenotype and induce and insulin-sensitizing effect in adipocytes [84]. Furthermore, Ffar4 activates cytoplasmic phospholipase A2 in macrophages, inducing production of pro- and antiinflammatory eicosanoids that could affect adipocyte function [98].

In obesity models, some studies indicate that Ffar4 prevents obesity. In one study, Ffar4KO mice exhibited mild glucose intolerance and insulin resistance but were not obese, while 60% high-fat diet-induced a greater degree of obesity, that was partially reversed by ω3-polyunsaturated fatty acid supplementation [84]. These findings were largely confirmed in a separate Ffar4KO mouse line, where a 60% high-fat diet-induced obesity glucose intolerance, insulin resistance, and decreased adipocyte differentiation and lipogenesis [92]. Furthermore, the R270H polymorphism in human Ffar4, that inactivates receptor signaling, is associated with morbid obesity, also suggesting Ffar4 prevents obesity. However, in another different Ffar4KO mouse, a 45% high-fat diet did not induce more obesity, nor was there any significant effect of ω3-PUFA supplementation [99], that was supported by another study indicating the Ffar4 is not required for ω3-mediated antiobesity effects [100]. A more recent report indicates that Ffar4 activates lipolysis in adipocytes and synergizes with Ffar1, that is known to induce insulin release in pancreatic β-cells [53–55], to improve glycemic control in *db/db* mice and insulin sensitivity in *ob/ob* mice [101]. The suggestion that dual activation of Ffar1 and Ffar4 might attenuate obesity is an intriguing new direction.

## Concluding remarks

Fatty acids represent a biological and biophysical challenge to study. Their chemical insolubility and chemical diversity pose problems for direct analysis and much of our understanding of fatty acid receptors and signaling systems comes from gain or loss of function molecular strategies developed in cells or animal models. The expression of different fatty acid receptors that signal to diverse pathways in fat cells introduces a complex metabolic conundrum. With multiple receptors for the same ligand, what regulates receptor expression and specificity of receptor signaling? Given the diversity of receptors for fatty acids, defining a unique molecular signature toward lipid signaling seems unlikely. Moreover, since each receptor is under exquisite control at the transcriptional and posttranscriptional level, orchestrating the repertoire of fatty acid-derived signals seems daunting.

An additional conundrum for fatty acid receptors and their signaling systems in fat cells derives from questions surrounding presentation of the ligand. That is, since the eicosanoid receptors and the Ffars are all G-protein-coupled receptors, and likely bind their cognate ligand(s) in a membrane-associated cleft in the receptor transmembrane region, how is the ligand delivered to the protein? That is, does the ligand diffuse laterally in the plane of the membrane until encountering a particular receptor or are ligands vectorally directed toward a particular receptor via an accessory binding protein (CD36/SR-B2?). If so, how is specificity of receptor interaction defined or is it stochastically driven? Additionally, what is the physical organization of the receptors in the membrane relative to each other and the potential initial fatty acid-binding protein CD36/SR-B2?

Given the breadth of questions and physiological impact of fatty acid receptors and their signaling processes, much is left to be learned. The opportunity for advanced analysis using appropriate gene editing and/or chemical probes will likely bring many answers to the most vexing questions but as usual, introduce new questions to be studied.

## Acknowledgments

The authors would like to thank Anthony Hertzel for graphic artwork. Supported by NIH HL R01130099 to TDO and NIH DK053189 to DAB.

## References

[1] Glatz JFC, Luiken JJFP. From fat to FAT (CD36/SR-B2): understanding the regulation of cellular fatty acid uptake. Biochimie 2017;136:21–6.
[2] Martin C, Chevrot M, Poirier H, Passilly-Degrace P, Niot I, Besnard P. CD36 as a lipid sensor. Physiol Behav 2011;105(1):36–42.
[3] Niculite C-M, Enciu A-M, Hinescu ME. CD 36: focus on epigenetic and post-transcriptional regulation. Front Genet 2019;10:680.
[4] Abumrad NA, el-Maghrabi MR, Amri EZ, Lopez E, Grimaldi PA. Cloning of a rat adipocyte membrane protein implicated in binding or transport of long-chain fatty acids that is induced during preadipocyte differentiation. Homology with human CD36. J Biol Chem 1993;268(24):17665–8.
[5] Lobo S, Wiczer BM, Bernlohr DA. Functional analysis of long-chain acyl-CoA synthetase 1 in 3T3-L1 adipocytes. J Biol Chem 2009;284(27):18347–56.
[6] Glatz JFC, Luiken JJFP, Bonen A. Membrane fatty acid transporters as regulators of lipid metabolism: implications for metabolic disease. Physiol Rev 2010;90(1):367–417.
[7] Sato O, Takanashi N, Motojima K. Third promoter and differential regulation of mouse and human fatty acid translocase/CD36 genes. Mol Cell Biochem 2007;299(1–2):37–43.
[8] Coburn CT, Knapp FF, Febbraio M, Beets AL, Silverstein RL, Abumrad NA. Defective uptake and utilization of long chain fatty acids in muscle and adipose tissues of CD36 knockout mice. J Biol Chem 2000;275(42):32523–9.
[9] Baillie AG, Coburn CT, Abumrad NA. Reversible binding of long-chain fatty acids to purified FAT, the adipose CD36 homolog. J Membr Biol 1996;153(1):75–81.
[10] Tao N, Wagner SJ, Lublin DM. CD36 is palmitoylated on both N- and C-terminal cytoplasmic tails. J Biol Chem 1996;271(37):22315–20.
[11]Huang MM, Bolen JB, Barnwell JW, Shattil SJ, Brugge JS. Membrane glycoprotein IV (CD36) is physically associated with the Fyn, Lyn, and Yes protein-tyrosine kinases in human platelets. Proc Natl Acad Sci U S A 1991;88(17):7844–8.
[12] Samovski D, Dhule P, Pietka T, Jacome-Sosa M, Penrose E, et al. Regulation of insulin receptor pathway and glucose metabolism by CD36 signaling. Diabetes 2018;67(7):1272–84.
[13] Gartung A, Zhao J, Chen S, Mottillo E, VanHecke GC, et al. Characterization of eicosanoids produced by adipocyte lipolysis: implication of cyclooxygenase-2 in adipose inflammation. J Biol Chem 2016;291(31):16001–10.
[14] Bäck M, Hansson GK. Leukotriene receptors in atherosclerosis. Ann Med 2006;38(7):493–502.

[15] Yokomizo T, Izumi T, Chang K, Takuwa Y, Shimizu T. A G-protein-coupled receptor for leukotriene B4 that mediates chemotaxis. Nature 1997;387(6633):620–4.

[16] Yokomizo T, Kato K, Terawaki K, Izumi T, Shimizu T. A second leukotriene B4receptor, BLT2: a new therapeutic target in inflammation and immunological disorders. J Exp Med 2000;192(3):421–31.

[17] Yokomizo T, Kato K, Hagiya H, Izumi T, Shimizu T. Hydroxyeicosanoids bind to and activate the low affinity leukotriene B 4 receptor, BLT2. J Biol Chem 2001;276(15):12454–9.

[18] Yokomizo T. Leukotriene B4 receptors: novel roles in immunological regulations. Adv Enzyme Regul 2011;51(1):59–64.

[19] Hirata K, Wada K, Murata Y, Nakajima A, Yamashiro T, Kamisaki Y. Critical role of leukotriene B4 receptor signaling in mouse 3T3-L1 preadipocyte differentiation. Lipids Health Dis 2013;12(1):1.

[20] Spite M, Hellmann J, Tang Y, Mathis SP, Kosuri M, et al. Deficiency of the leukotriene B 4 receptor, BLT-1, protects against systemic insulin resistance in diet-induced obesity. J Immunol 2011;187(4):1942–9.

[21] Li P, Oh DY, Bandyopadhyay G, Lagakos WS, Talukdar S, et al. LTB4 promotes insulin resistance in obese mice by acting on macrophages, hepatocytes and myocytes. Nat Med 2015;21(3):239–47.

[22] Yokomizo T, Nakamura M, Shimizu T. Leukotriene receptors as potential therapeutic targets. J Clin Invest 2018;128(7):2691–701.

[23] Ciana P, Fumagalli M, Trincavelli ML, Verderio C, Rosa P, et al. The orphan receptor GPR17 identified as a new dual uracil nucleotides/cysteinyl-leukotrienes receptor. EMBO J 2006;25(19):4615–27.

[24] Simon K, Merten N, Schröder R, Hennen S, Preis P, et al. The orphan receptor GPR17 is unresponsive to uracil nucleotides and cysteinyl leukotrienes. Mol Pharmacol 2017;91(5):518–32.

[25] Singh RK, Gupta S, Dastidar S, Ray A. Cysteinyl leukotrienes and their receptors: molecular and functional characteristics. Pharmacology 2010;85(6):336–49.

[26] Moon TC, Campos-Alberto E, Yoshimura T, Bredo G, Rieger AM, et al. Expression of DP2 (CRTh2), a prostaglandin D2 receptor, in human mast cells. PLoS One 2014;9(9).

[27] Wang L, Yao D, Deepak RNVK, Liu H, Xiao Q, et al. Structures of the human PGD 2 receptor CRTH2 reveal novel mechanisms for ligand recognition. Mol Cell 2018;72(1):48–59. e4.

[28] Virtue S, Masoodi M, De Weijer BAM, Van Eijk M, Mok CYL, et al. Prostaglandin profiling reveals a role for haematopoietic prostaglandin D synthase in adipose tissue macrophage polarisation in mice and humans. Int J Obes 2015;39(7):1151–60.

[29] Wakai E, Aritake K, Urade Y, Fujimori K. Prostaglandin D2 enhances lipid accumulation through suppression of lipolysis via DP2 (CRTH2) receptors in adipocytes. Biochem Biophys Res Commun 2017;490(2):393–9.

[30] Rahman MS. Prostacyclin: a major prostaglandin in the regulation of adipose tissue development. J Cell Physiol 2019;234(4):3254–62.

[31] Mazid MA, Chowdhury AA, Nagao K, Nishimura K, Jisaka M, et al. Endogenous 15-deoxy-Δ12,14-prostaglandin J2 synthesized by adipocytes during maturation phase contributes to upregulation of fat storage. FEBS Lett 2006;580(30):6885–90.

[32] Hirata T, Narumiya S. Prostanoid receptors. Chem Rev 2011;111(10):6209–30.

[33] Yokoyama U, Iwatsubo K, Umemura M, Fujita T, Ishikawa Y. The prostanoid EP4 receptor and its signaling pathway. Pharmacol Rev 2013;65(3):1010–52.

[34] Tang EHC, Cai Y, Wong CK, Rocha VZ, Sukhova GK, et al. Activation of prostaglandin e2-ep4 signaling reduces chemokine production in adipose tissue. J Lipid Res 2015;56(2):358–68.

[35] Xu H, Fu JL, Miao YF, Wang CJ, Han QF, et al. Prostaglandin E2 receptor EP3 regulates both adipogenesis and lipolysis in mouse white adipose tissue. J Mol Cell Biol 2016;8(6):518–29.

[36] Breyer RM, Bagdassarian CK, Myers SA, Breyer MD. Prostanoid receptors: subtypes and signaling. Annu Rev Pharmacol Toxicol 2001;41(1):661–90.

[37] Ricciotti E, Fitzgerald AG. [Eicosanoid neuroinflammation] prostaglandins and inflammation. Arter Thromb Vasc Biol 2011;31(5):986–1000.

[38] Kawahara K, Hohjoh H, Inazumi T, Tsuchiya S, Sugimoto Y. Prostaglandin E2-induced inflammation: relevance of prostaglandin E receptors. Biochim Biophys Acta Mol Cell Biol Lipids 2015;1851(4):414–21.

[39] Chan PC, Hsiao FC, Chang HM, Wabitsch M, Shiuan Hsieh P. Importance of adipocyte cyclooxygenase-2 and prostaglandin E2-prostaglandin E receptor 3 signaling in the development of obesity-induced adipose tissue inflammation and insulin resistance. FASEB J 2016;30(6):2282–97.

[40] Chatzipanteli K, Rudolph S, Axelrod L. Coordinate control of lipolysis by prostaglandin E2 and prostacyclin in rat adipose tissue. Diabetes 1992;41(8):927–35.

[41] Fain JN, Leffler CW, Bahouth SW, Rice AM, Rivkees SA. Regulation of leptin release and lipolysis by PGE2 in rat adipose tissue. Prostaglandins Other Lipid Mediat 2000;62(4):343–50.

[42] García-Alonso V, Titos E, Alcaraz-Quiles J, Rius B, Lopategi A, et al. Prostaglandin E2 exerts multiple regulatory actions on human obese adipose tissue remodeling, inflammation, adaptive thermogenesis and lipolysis. PLoS One 2016;11(4):1–17.

[43] Inazumi T, Shirata N, Morimoto K, Takano H, Segi-Nishida E, Sugimoto Y. Prostaglandin E2-EP4 signaling suppresses adipocyte differentiation in mouse embryonic fibroblasts via an autocrine mechanism. J Lipid Res 2011;52(8):1500–8.

[44] Ying F, Cai Y, Cai Y, Wang Y, Tang EHC. Prostaglandin E receptor subtype 4 regulates lipid droplet size and mitochondrial activity in murine subcutaneous white adipose tissue. FASEB J 2017;31(9):4023–36.

[45] Hu X, Cifarelli V, Sun S, Kuda O, Abumrad NA, Su X. Major role of adipocyte prostaglandin E2 in lipolysis-induced macrophage recruitment. J Lipid Res 2016;57(4):663–73.

[46] Fujino H, Srinivasan D, Pierce KL, Regan JW. Differential regulation of prostaglandin receptor isoforms by protein kinase C. Mol Pharmacol 2000;57(2):353–8.

[47] Miller CW, Casimir DA, Ntambi JM. The mechanism of inhibition of 3T3-L1 preadipocyte differentiation by prostaglandin F2alpha. Endocrinology 1996;137(12):5641–50.

[48] Fujimori K, Ueno T, Nagata N, Kashiwagi K, Aritake K, et al. Suppression of adipocyte differentiation by aldo-keto reductase 1B3 acting as prostaglandin F2α synthase. J Biol Chem 2010;285(12):8880–6.

[49] Liu L, Clipstone NA. Prostaglandin F2α inhibits adipocyte differentiation via a Gαq-calcium-calcineurin-dependent signaling pathway. J Cell Biochem 2007;100(1):161–73.

[50] Michaud A, Lacroix-Pépin N, Pelletier M, Veilleux A, Noël S, et al. Prostaglandin (PG) F2 alpha synthesis in human subcutaneous and omental adipose tissue: modulation by inflammatory cytokines and role of the human aldose reductase AKR1B1. PLoS One 2014;9(3):e90861.

[51] Bayindir I, Babaeikelishomi R, Kocanova S, Sousa IS, Lerch S, et al. Transcriptional pathways in cPGI2-induced adipocyte progenitor activation for browning. Front Endocrinol (Lausanne) 2015;6(August):1–14.

[52] Sawzdargo M, George SR, Nguyen T, Xu S, Kolakowski LF, et al. A cluster of four novel human G protein-coupled receptor genes occurring in close proximity to CD22 gene on chromosome 19q13.1. Biochem. Biophys. Res. Commun 1997;547(239):543–7.

[53] Briscoe CP, Tadayyon M, Andrews JL, Benson WG, Chambers JK, et al. The orphan G protein-coupled receptor GPR40 is activated by medium and long chain fatty acids. J Biol Chem 2003;278(13):11303–11.

[54] Itoh Y, Kawamata Y, Harada M, Kobayashi M, Fujii R, et al. Free fatty acids regulate insulin secretion from pancreatic β cells through GPR40. Nature 2003;422(6928):173–6.

[55] Kotarsky K, Nilsson NE, Flodgren E, Owman C, Olde B. A human cell surface receptor activated by free fatty acids and thiazolidinedione drugs. Biochem Biophys Res Commun 2003;301(2):406–10.

[56] Brown AJ, Goldsworthy SM, Barnes AA, Eilert MM, Tcheang L, et al. The orphan G protein-coupled receptors GPR41 and GPR43 are activated by propionate and other short chain carboxylic acids. J Biol Chem 2003;278(13):11312–9.

[57] Le Poul E, Struyf S, Springael JY, Lannoy V, Decobecq ME, et al. Functional characterization of human receptors for short chain fatty acids and their role in polymorphonuclear cell activation. J Biol Chem 2003;278(28):25481–9.

[58] Nilsson NE, Kotarsky K, Owman C, Olde B. Identification of a free fatty acid receptor, FFA2R, expressed on leukocytes and activated by short-chain fatty acids. Biochem Biophys Res Commun 2003;303(4):1047–52.

[59] Fredriksson R, Höglund PJ, Gloriam DEII, Lagerström MC, Schiöth HB, Lagerstro MC. Seven evolutionarily conserved human rhodopsin G protein-coupled receptors lacking close relatives. FEBS Lett 2003;554(3):381–8.

[60] Hirasawa A, Tsumaya K, Awaji T, Katsuma S, Adachi T, et al. Free fatty acids regulate gut incretin glucagon-like peptide-1 secretion through GPR120. Nat Med 2005;11(1):90–4.

[61] Milligan G, Shimpukade B, Ulven T, Hudson BD. Complex pharmacology of free fatty acid receptors. Chem Rev 2017;117(1):67–110.

[62] Karaki SI, Mitsui R, Hayashi H, Kato I, Sugiya H, et al. Short-chain fatty acid receptor, GPR43, is expressed by enteroendocrine cells and mucosal mast cells in rat intestine. Cell Tissue Res 2006;324(3):353–60.

[63] Karaki S, Tazoe H, Hayashi H, Kashiwabara H, Tooyama K, et al. Expression of the short-chain fatty acid receptor, GPR43, in the human colon. J Mol Histol 2008;39(2):135–42.

[64] Nøhr MK, Pedersen MH, Gille A, Egerod KL, Engelstoft MS, et al. GPR41/FFAR3 and GPR43/FFAR2 as cosensors for short-chain fatty acids in enteroendocrine cells vs FFAR3 in enteric neurons and FFAR2 in enteric leukocytes. Endocrinology 2013;154(10):3552–64.

[65] Tazoe H, Otomo Y, Kaji I, Tanaka R, Karaki S-I, Kuwahara A. Roles of short-chain fatty acids receptors, GPR41 and GPR43 on colonic functions. J Physiol Pharmacol 2008;59(Suppl. 2):251–62.
[66] Tazoe H, Otomo Y, Karaki S-I, Kato I, Fukami Y, et al. Expression of short-chain fatty acid receptor GPR41 in the human colon. Biomed Res 2009;30(3):149–56.
[67] Tolhurst G, Heffron H, Lam YS, Parker HE, Habib AM, et al. Short-chain fatty acids stimulate glucagon-like peptide-1 secretion via the G-protein-coupled receptor FFAR2. Diabetes 2012;61(2):364–71.
[68] McNelis JC, Lee YS, Mayoral R, van der Kant R, Johnson AMF, et al. GPR43 potentiates β-cell function in obesity. Diabetes 2015;64(9):3203–17.
[69] Priyadarshini M, Villa SR, Fuller M, Wicksteed B, Mackay CR, et al. An acetate-specific GPCR, FFAR2, regulates insulin secretion. Mol Endocrinol 2015;29(7):1055–66.
[70] Tang C, Ahmed K, Gille A, Lu S, Gröne H-J, et al. Loss of FFA2 and FFA3 increases insulin secretion and improves glucose tolerance in type 2 diabetes. Nat Med 2015;21(2):173–7.
[71] Maslowski KM, Vieira AT, Ng A, Kranich J, Sierro F, et al. Regulation of inflammatory responses by gut microbiota and chemoattractant receptor GPR43. Nature 2009;461(7268):1282–6.
[72] Smith PM, Howitt MR, Panikov N, Michaud M, Gallini CA, et al. The microbial metabolites, short-chain fatty acids, regulate colonic Treg cell homeostasis. Science 2013;341(6145):569–73.
[73] Trompette A, Gollwitzer ES, Yadava K, Sichelstiel AK, Sprenger N, et al. Gut microbiota metabolism of dietary fiber influences allergic airway disease and hematopoiesis. Nat Med 2014;20(2):159–66.
[74] Han J-H, Kim I-S, Jung S-H, Lee S-G, Son H-Y, Myung C-S. The effects of propionate and valerate on insulin responsiveness for glucose uptake in 3T3-L1 adipocytes and C2C12 myotubes via G protein-coupled receptor 41. PLoS One 2014;9(4):e95268.
[75] Xiong Y, Miyamoto N, Shibata K, Valasek MA, Motoike T, et al. Short-chain fatty acids stimulate leptin production in adipocytes through the G protein-coupled receptor GPR41. Proc Natl Acad Sci U S A 2004;101(4):1045–50.
[76] Hong Y-H, Nishimura Y, Hishikawa D, Tsuzuki H, Miyahara H, et al. Acetate and propionate short chain fatty acids stimulate adipogenesis via GPCR43. Endocrinology 2005;146(12):5092–9.
[77] Zaibi MS, Stocker CJ, O'Dowd J, Davies A, Bellahcene M, et al. Roles of GPR41 and GPR43 in leptin secretory responses of murine adipocytes to short chain fatty acids. FEBS Lett 2010;584(11):2381–6.
[78] Canfora EE, Jocken JW, Blaak EE. Short-chain fatty acids in control of body weight and insulin sensitivity. Nat Rev Endocrinol 2015;11(10):577–91.
[79] Miyamoto J, Kasubuchi M, Nakajima A, Kimura I. Anti-inflammatory and insulin-sensitizing effects of free fatty acid receptors. Handb Exp Pharmacol 2017;236:221–31.
[80] Bjursell M, Admyre T, Göransson M, Marley AE, Smith DM, et al. Improved glucose control and reduced body fat mass in free fatty acid receptor 2-deficient mice fed a high-fat diet. Am J Physiol Endocrinol Metab 2011;300(1):E211–20.
[81] Kimura I, Ozawa K, Inoue D, Imamura T, Kimura K, et al. The gut microbiota suppresses insulin-mediated fat accumulation via the short-chain fatty acid receptor GPR43. Nat Commun 2013;4(1):1829.
[82] Dewulf EM, Ge Q, Bindels LB, Sohet FM, Cani PD, et al. Evaluation of the relationship between GPR43 and adiposity in human. Nutr Metab (Lond) 2013;10(1):11.
[83] Eclov JA, Qian Q, Redetzke R, Chen Q, Wu SC, et al. EPA, not DHA, prevents fibrosis in pressure overload-induced heart failure: potential role of free fatty acid receptor 4. J Lipid Res 2015;56(12):2297–308.
[84] Oh DY, Talukdar S, Bae EJ, Imamura T, Morinaga H, et al. GPR120 is an omega-3 fatty acid receptor mediating potent anti-inflammatory and insulin-sensitizing effects. Cell 2010;142(5):687–98.
[85] Tanaka T, Yano T, Adachi T, Koshimizu T, Hirasawa A, Tsujimoto G. Cloning and characterization of the rat free fatty acid receptor GPR120: in vivo effect of the natural ligand on GLP-1 secretion and proliferation of pancreatic beta cells. Naunyn Schmiedebergs Arch Pharmacol 2008;377(4–6):515–22.
[86] Christiansen E, Watterson KR, Stocker CJ, Sokol E, Jenkins L, et al. Activity of dietary fatty acids on FFA1 and FFA4 and characterisation of pinolenic acid as a dual FFA1/FFA4 agonist with potential effect against metabolic diseases. Br J Nutr 2015;113(11):1677–88.
[87] Suckow AT, Polidori D, Yan W, Chon S, Ma JY, et al. Alteration of the glucagon axis in GPR120 (FFAR4) knockout mice: a role for GPR120 in glucagon secretion. J Biol Chem 2014;289(22):15751–63.
[88] Yore MM, Syed I, Moraes-Vieira PM, Zhang T, Herman MA, et al. Discovery of a class of endogenous mammalian lipids with anti-diabetic and anti-inflammatory effects. Cell 2014;159(2):318–32.
[89] Hudson BD, Shimpukade B, Mackenzie AE, Butcher AJ, Pediani JD, et al. The pharmacology of TUG-891, a potent and selective agonist of the free fatty acid receptor 4 (FFA4/GPR120), demonstrates both potential opportunity and possible challenges to therapeutic agonism. Mol Pharmacol 2013;84(5):710–25.

[90] Watson S-J, Brown AJH, Holliday ND. Differential signaling by splice variants of the human free fatty acid receptor GPR120. Mol Pharmacol 2012;81(5):631–42.
[91] Gotoh C, Hong Y-H, Iga T, Hishikawa D, Suzuki Y, et al. The regulation of adipogenesis through GPR120. Biochem Biophys Res Commun 2007;354(2):591–7.
[92] Ichimura A, Hirasawa A, Poulain-Godefroy O, Bonnefond A, Hara T, et al. Dysfunction of lipid sensor GPR120 leads to obesity in both mouse and human. Nature 2012;483(7389):350–4.
[93] Quesada-López T, Cereijo R, Turatsinze J-V, Planavila A, Cairó M, et al. The lipid sensor GPR120 promotes brown fat activation and FGF21 release from adipocytes. Nat Commun 2016;7(1):13479.
[94] Schilperoort M, van Dam AD, Hoeke G, Shabalina IG, Okolo A, et al. The GPR120 agonist TUG-891 promotes metabolic health by stimulating mitochondrial respiration in brown fat. EMBO Mol Med 2018;10(3):e8047.
[95] Song T, Zhou Y, Peng J, Tao Y-X, Yang Y, et al. GPR120 promotes adipogenesis through intracellular calcium and extracellular signal-regulated kinase 1/2 signal pathway. Mol Cell Endocrinol 2016;434:1–13.
[96] Marion-Letellier R, Savoye G, Ghosh S. Fatty acids, eicosanoids and PPAR gamma. Eur J Pharmacol 2016;785:44–9.
[97] Catrysse L, van Loo G. Adipose tissue macrophages and their polarization in health and obesity. Cell Immunol 2018;330:114–9.
[98] Liu Y, Chen L-Y, Sokolowska M, Eberlein M, Alsaaty S, et al. The fish oil ingredient, docosahexaenoic acid, activates cytosolic phospholipase $A_2$ via GPR120 receptor to produce prostaglandin $E_2$ and plays an anti-inflammatory role in macrophages. Immunology 2014;143(1):81–95.
[99] Bjursell M, Xu X, Admyre T, Böttcher G, Lundin S, et al. The beneficial effects of n-3 polyunsaturated fatty acids on diet induced obesity and impaired glucose control do not require Gpr120. PLoS One 2014;9(12):e114942.
[100] Pærregaard SI, Agerholm M, Serup AK, Ma T, Kiens B, et al. FFAR4 (GPR120) signaling is not required for anti-inflammatory and insulin-sensitizing effects of omega-3 fatty acids. Mediators Inflamm 2016;2016:1–12.
[101] Satapati S, Qian Y, Wu MS, Petrov A, Dai G, et al. GPR120 suppresses adipose tissue lipolysis and synergizes with GPR40 in antidiabetic efficacy. J Lipid Res 2017;58(8):1561–78.

CHAPTER

# 7

# Adipocyte lipolysis and lipid-derived metabolite signaling

*Charlie Kirsh*[a], *Abigail M. Harris*[b], *Judith Simcox*[a]

[a]Biochemistry, University of Wisconsin Madison, United States; [b]Nutritional Sciences, University of Wisconsin Madison, United States

OUTLINE



## Dysfunction of adipocyte lipolysis is central to metabolic disease

The major energy reserves of mammals are stored as triglycerides in white adipose tissue. The abundance of these stored energy reserves is communicated to other organs through endocrine signaling from white adipose tissue mediated by secreted proteins or lipids. Multiple signaling proteins have been identified to be secreted from white adipocytes including leptin, which conveys satiety and controls energy expenditure, or adiponectin, a known regulator of insulin sensitivity [1,2]. Adipocytes also regulate systemic energy metabolism through lipid signaling and lipid-derived metabolites. Secretion of lipids from adipocytes regulates the adaptive response to environmental stress such as the switch from the fed to

Lipid Signaling and Metabolism. http://dx.doi.org/10.1016/B978-0-12-819404-1.00007-5

the fasted state and plays an important role in the etiology of metabolic diseases such as type 2 diabetes.

The abundance of the triglyceride stores in white adipocytes is maintained by a balance of lipid synthesis through lipogenesis and breakdown with lipolysis. Lipids in the adipose tissue are constantly turning over, with the contents of fat cells being entirely replaced every 2–3 weeks in humans [3]. Lipogenesis is activated by insulin secretion in response to elevated glucose as well as by circulating lipids packaged in chylomicrons and very low-density lipoproteins (LDLs). Adipocytes have incredible plasticity and are able to respond to energy excess through expanding their mass; more than 80% of white adipocyte volume is occupied by stored triglycerides [4,5]. Conversely, lipolysis mobilizes stored lipids during energy replete conditions such as fasting or during energetically demanding states such as exercise or cold exposure. These lipids are secreted either as free fatty acids, by-products of free fatty acid processing, or as protein lipid complexes that enter the circulation. The surge of lipids in the circulation from increased adipocyte lipolysis signals for a shift in fuel utilization, impairs insulin signaling, and leads to an accumulation of lipids in distal tissues such as the liver, and in chronic conditions signals for cell death. This multitude of changes in response to lipolysis highlights the importance of dynamic response to energy availability. Adipocytes are central to this nutritional regulation with their ability for energy storage and regulated release of these stores.

Much of the research focus has been on lipid signaling during the fed state because obesity is caused by overnutrition and is linked with metabolic diseases such as type 2 diabetes, cardiovascular disease, and certain types of cancer. However, obesity is a chronic state that is complex in nature because it utilizes nutritional signals from both the fed and fasted state. In obesity, basal lipolysis is elevated while glucose remains abundantly available [6]. Correcting the increased lipolysis through either reducing adipocyte lipolysis specifically or whole body lipolysis through pharmacological intervention led to improved insulin responsiveness, decreased hepatic steatosis, and improved glycemic control [7,8]. Understanding the signals that regulate lipolysis and the complex interplay of cellular communication can inform on the development of metabolic diseases associated with obesity.

During white adipocyte lipolysis, free fatty acids are mobilized from triglycerides stored in lipid droplets. Some of these free fatty acids are directly released into the circulation, while others are further processed in adipocytes before release. More work is needed to understand how adipocytes signal to other cells and organs through lipid-mediated signals during lipolysis. In this chapter, we will discuss the signals that regulate lipolysis, the mobilization of lipolysis derived lipids, and the signaling of free fatty acids in other tissues. Although there are a multitude of other lipids that signal in the adipose tissue, we will focus on intercellular signaling discussing both the function in normal physiology and how these signals impact a disease state.

## Regulation of lipolysis

In adipocytes, intracellular lipolysis leads to the break-down of triglycerides stored in lipid droplets. The lipid droplets are composed of an external phospholipid bilayer surrounding sterol esters that wrap around the core of triglycerides. The release of free fatty acids from this storage depot is regulated by sequential hydrolysis to produce three free fatty acids and a glycerol. This process is catalyzed by three cytosolic enzymes that when activated associate with the phospholipid bilayer of lipid droplets including adipose-triglyceride lipase (ATGL), hormone-sensitive lipase (HS)L, and monoacylglycerol lipase. The release of free fatty acids from adipocyte lipolysis supplies energy to tissues such as the liver, skeletal muscle, and heart during fasting to preserve glucose stores for the brain and red blood cells.

Lipolysis in white adipose tissue is regulated at multiple levels including activation of cell surface receptors to communicate energy demand as well as direct stimulation by shifts in cellular metabolism. At the cell surface lipolysis can be induced by activation of the glucagon receptor in fasting or by catecholamine stimulation in cold exposure and cachexia.

Although lipolysis is activated by multiple cellular receptors and intracellular nutrient signaling, most of the pathways are protein kinase A (PKA) mediate. For example, during fasting, glucagon is secreted from α cells in the pancreatic islets to interact with glucagon receptors in the liver and adipose tissue. Glucagon receptor is a class B G-protein-coupled receptor (GPCR), which undergoes a conformational change upon glucagon binding that alters the interaction with $G_{\alpha s}$- and $G_q$-proteins, which activate adenylyl cyclase and phospholipase C activities [9,10]. Adenyl cyclase activation increases cAMP, which binds to PKA and causes a dissociation of the regulator and catalytic subunits to phosphorylate HSL on multiple sites [11]. HSL is then able to translocate from the cytosol to the lipid droplet.

Catecholamine-stimulated lipolysis is activated similarly through a PKA-mediated pathway. With external stimuli such as cold exposure, the catecholamine norepinephrine is released by sympathetic neurons that innervate adipose tissue depots to activate β3 adrenergic receptors. In brown adipose tissue, this activation leads to the production of heat through mitochondrial uncoupling but in white adipose tissue β3-adrenergic receptor, stimulation leads to activation of lipolysis to support energy demand. Adrenergic receptors interact with $G_s$-proteins to activate adenylyl cyclase, which increases cAMP and activate PKA [12].

PKA signaling can also activate lipolysis through HSL independent mechanisms. One example of this receptor independent signaling is seen with α-β hydrolase domain containing protein 5 (ABHD5), which regulates ATGL localization. ABHD5 is typically sequestered through binding to perilipin 1 or 5 that resides on the phospholipid bilayer of the lipid droplet, but once perilipin is phosphorylated by PKA, binding is disrupted and ABHD5 is released to the cytoplasm. Once in the cytoplasm, ABHD5 binds ATGL, which translocates ATGL to the surface of lipid droplets where it activates lipolysis [13].

Other cell nonautonomous activators of lipolysis include thyroxine, cortisol, adrenocorticotropic hormone, and growth hormones [14]. More work is needed to establish the mechanism of these lipolytic activators and their differential regulation between species. An example of this is seen in thyroxine, which has differential regulation of lipolysis in humans and rodents. In humans, hyperthyroid patients have increased catecholamine-stimulated lipolysis but normal basal lipolysis [15]. Conversely in rodents, the regulation of lipolysis by thyroxine was more dependent upon the fasted or fed state rather than overarching hyperthyroidism [16]. The complex regulation highlights the differences in regulation between basal and catecholamine-stimulated lipolysis as well as the interplay from multiple signals that activate lipolysis.

As highlighted in the examples above, basal lipolysis is differentially regulated from catecholamine stimulated lipolysis. Basal lipolysis is known to regulate adipose tissue hypertrophy and circulating free fatty acids [17,18]. Basal lipolysis is increased with obesity while catecholamine stimulated lipolysis is decreased, suggesting that chronic conditions impact signaling in these pathways [6]. The molecular mechanisms regulating basal lipolysis have been difficult to study but have similarly been shown to be controlled by PKA regulated lipolysis through ATGL activation [19]. More recent studies have pointed to regulation of lipolysis by purinergic receptor 2Y (P2Y), a family of G protein coupled receptor that are activated by ATP, adenosine diphosphate, uridine-5′-triphosphate, and uridine-5′-diphosphate. Ali et al. showed that stimulation of $P2Y_2$ led to activation of phospholipase C to regulate production of diacylglycerol (DAG) and inositol triphosphate to increased levels of $Ca^{2+}$ into the cytosol leading inhibition of PKA [20]. This would imply that the regulation of basal lipolysis is through inhibition of PKA signaling while catecholamine regulation is through activation of PKA .

Other pathways of inhibition of lipolysis can be receptor mediated such as insulin, which is released by pancreatic β-cells in response to feeding. Insulin functions by binding to insulin receptor, which triggers an AKT-dependent signaling cascade that activates phosphodiesterase 3B, which hydrolyzes cAMP into the inactive 5′AMP diminishing PKA activation. The regulation of insulin-mediated inhibition of lipolysis is complex and can be impacted by other signals including circulating lactate levels, which activate G-protein coupled receptor 81 (GPCR81) [21]. Parathyroid hormone also inhibits lipolysis through active conversion of 25 hydroxycholecalciferol to vitamin $D_3$, the rise in vitamin $D_3$ leads to elevated adipocyte calcium levels [22,23]. These high calcium levels decrease cAMP, which leads to decreased HSL translocation to the lipid droplet. Other signals inhibit lipolysis through nonreceptor-mediated pathways such as increased circulating ethanol. Elevated ethanol leads to activation of phosphodiesterase 4 and subsequent decreased PKA activation [24]. This inhibited lipolysis associated with alcohol consumption causes white adipose tissue dysfunction that occurs in alcoholism, which then leads to increased lipid deposition in the liver and alcohol associated hepatic steatosis.

## Increased lipolysis in disease

Lipolysis regulation and rates are different between fat depots. Anatomically, there are two compartments of white adipose tissue: subcutaneous and visceral, which are found beneath the skin in the hypodermis and in the body cavity, respectively. In humans, catecholamine-stimulated lipolysis is highest in visceral adipose tissue while basal lipolysis is highest in subcutaneous [3,25]. In obese subjects with metabolic syndrome, the visceral white adipose tissue exhibits higher expansion and lipolysis [26]. White adipose tissue lipolysis regulates hepatic insulin responsiveness, with increased lipolysis causing insulin resistance [7]. Moreover, impaired lipolysis is directly associated with future weight gain and the development of metabolic disease. Despite the unknown regulation of basal lipolysis in obesity, blocking lipolysis in obesity has been shown to be protective. Loss of HSL in mouse models leads to increased adipocytes hypertrophy while also being protective again high fat diet induced obesity and impaired glucose handling [27].

Besides free fatty acids, there are a host of other lipids released by white adipocytes during lipolysis. An example of this is found in prostaglandin $E_2$ and prostaglandin $D_2$, which are known to regulate macrophage infiltration and inflammatory state in adipose tissue [28]. These lipids are important signals for immune response and have been linked mechanistically to the detrimental inflammatory state of adipose tissue during obesity. Advances in lipid quantification driven my mass spectrometry have shown that there is a diverse array of lipids released in response to catecholamine induced lipolysis. In one study using LC/MS, it was observed that β3-adrenergic receptor agonist treatment increased the secretion of more than fifty lipid species including cyclooxygenases, lipoxygenases, and epoxygenases as well as metabolites of linoleic acid, eicosapentaenoic acid, and docosahexaenoic acid [29].

## Lipid mobilization during lipolysis

There are two major mechanisms through which lipids can be released from the cell during lipolysis. These two mechanisms are through protein-mediated transport of free fatty acids into the circulation, and through exocytosis of extracellular vesicles. It has been shown that free fatty acids require a transporter to pass through the cell membrane and the kinetics of mobility suggests facilitated diffusion [30]. Multiple proteins have been identified that facilitate fatty acid transport including CD36, plasma membrane-associated fatty acid binding protein ($FABP_{pm}$), and a family of fatty acid transport proteins (FATP1-6) [31]. The actual function of these proteins remains contested, and more research is needed to elucidate their regulation and role in free fatty acid mobilization. These studies are difficult because of the high affinity of albumin and potential transporters for fatty acid as well as the rapid entry of free fatty acids into the cell. Although it is clear, these proteins are required to facilitate fatty acid transport. For example, structural characterization of CD36 shows a free fatty acid binding tunnel that could facilitate transport, and loss of CD36 leads to impaired free fatty acid uptake in C2C12 muscle cells and over expression increases fatty acid uptake [32,33]. Moreover, in human mutations of CD36, there is impaired free fatty acid uptake into the muscle, adipose tissue, and heart but CD36 function was not necessary for uptake of free fatty acids in the liver [34].

Once the free fatty acids enter the plasma, the majority—99% of the non-esterified fatty acids—are noncovalently bound to albumin. Albumin has ten free fatty acid binding sites with three of these sites having high affinity. Lipolysis can also be mediated by extracellular lipases such as lipoprotein lipase, which hydrolyze free fatty acids from chylomicrons and very LDLs in the capillary bed.

Another mechanism of lipid mobilization during lipolysis is the release of extracellular vesicles. Extracellular Vesicles are small vesicles that have been secreted from a cell that carry an assortment of signaling molecules such as microRNAs, proteins, and various lipid species. The components of extracellular vesicles are derived from the cell of origin; however, the relative enrichment of a species may be drastically different. This suggests that extracellular vesicles are preferentially loaded with certain molecules over others [35]. Proteomics data have identified around 13,000 eukaryotic extracellular vesicle proteins with 7806 extracellular vesicle proteins found in humans. Of the human extracellular vesicle proteins, nearly 800 proteins have been categorized as common, representing proteins such as tetraspanins, integrins, heat shock proteins, ribosomal proteins, and metabolic enzymes [36]. Once the components of an extracellular are released into a recipient cell, they have the potential to impact cellular processes. An example of this impact on cellular processes can be seen in treatment of a cell culture model of melanoma with adipocyte-derived extracellular vesicles, which contain lipolysis regulating enzymes such as HSL and ABHD5. Treatment of melanoma cells with adipocytes-secreted extracellular vesicles induces fatty acid oxidation and lipolysis [37].

Extracellular vesicles are categorized based on size and biogenesis and are commonly grouped into two species: microvesicles and exosomes. Microvesicles originate through budding of a cell's plasma membrane to form extracellular vesicles, which are typically 50–1000 nm in size [38]. Microvesicle biogenesis is the result of multiple lipids and protein pathways. Two important lipids, phosphatidylserine (PS) and phosphatidylethanolamine, translocate to the outer leaflet of a cell membrane through phospholipid translocators. Through these translocator, flippases, floppases, and scramblases, the asymmetry of the plasma membrane is lost [39]. The loss of asymmetry in combination with an influx of $Ca^{2+}$ ions allows for the presentation of PS on the plasma membrane, helping to promote shedding of the microvesicles.

The other class of extracellular vesicles are collectively known as exosomes, which are slightly smaller than macrovesicles, around 50–150 nm in size, and they are a product of the late endosome. Exosomes begin as small, intraluminal vesicles confined inside a multivesicular body (MVB) that fuses with the plasma membrane. After fusion, the intraluminal vesicles are released to the extracellular matrix as exosomes [40]. Adipocytes produce both microvesicles and exosomes.

The biogenesis of exosomes is thought to be the result of two pathways. One pathway is dependent on the endosomal sorting complexes required for transport (ESCRT), a four subunit complex that is involved in the recognition of ubiquitinated cargo and formation of a MVB. The ESCRT complex interacts with associated proteins and the endosome to promote budding of intraluminal vesicles and eventual membrane shedding. An alternative pathway, coined ESCRT-independent, centers around signaling from lipids and other proteins. Ceramides, created from sphingomyelin, help to package exosome contents such as miRNA and proteins, like tetraspanins [41]. The packaging and subsequent localization of tetraspanin, CD81, promotes and regulates the biogenesis of exosomes. For exosomes to be released from the MVB, protein classes such as Rab GTPase and SNARE interact with actin and microtubules to move the MVB toward the membrane (Fig. 7.1) [42].

In adipocytes, a common marker for classification is the fatty acid-binding protein 4 (FABP-4) with its concentration dependent on the maturation of the adipocyte [43]. Extracellular vesicles derived from adipocytes are typically classified by adiponectin enrichment. Adipocyte extracellular vesicles can produce both adiponectin-positive and adiponectin-negative vesicles; however, the adiponectin-positive vesicles are thought to be produced by stromal cells in the adipose tissue, rather than the adipocytes. Adipokines are also enriched in adipocyte extracellular vesicles. Macrophage migration inhibitory factor, tumor necrosis factor alpha, and macrophage colony-stimulating factor-1 (MCSF) are all present in adipocyte extracellular vesicles [44]. The inflammatory adipokine prevalence in adipocyte extracellular vesicles is thought to contribute to the pathogenesis of diabetes and obesity [44,45].

Lipid enrichment of the extracellular vesicle membrane through phospholipids, cholesterol, and sphingomyelins provide stability and structure to the extracellular vesicles, whereas other lipids from the extracellular vesicle also convey information to a recipient cell. Prostaglandins and lysophosphatidylcholine associated with extracellular vesicles are able to impact their respective pathways, such as activating the nuclear receptor PPARγ, once uptake has occurred by a recipient cell. Researchers have shown that the activity of these lipids may be increased while bound to the extracellular vesicle membrane [46]. Enzymes involved in lipid metabolism, such as fatty acid synthase and phospholipase, have also been found in extracellular vesicles [35].

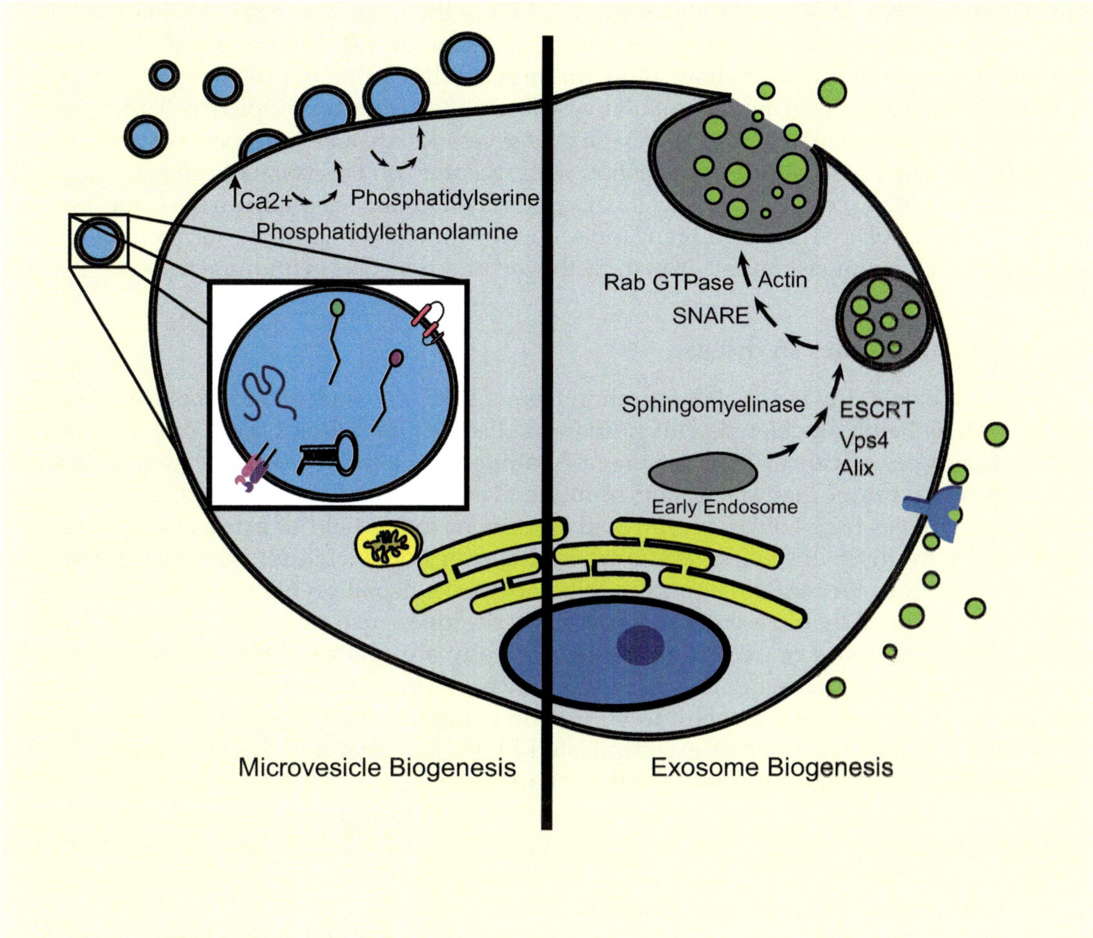


FIGURE 7.1 **Biogenesis of Microvesicles (left) and Exosomes (right).** The biogenesis of extracellular vesicles from a generic cell is shown above. Microvesicles (*blue in web version*) bud from the plasma membrane after signaling from, but not limited to, $Ca^{2+}$, phosphatidylserine, phosphatidylethanolamine, and scramblases. The inset shows a typical microvesicle containing various fatty acids, miRNA, mRNA, and transmembrane proteins. Exosomes (*green in web version*) begin as an early endosome. Signaling from ESCRT and associated proteins or through lipid signaling from sphingomyelinase results in the creation and sorting of a MVB. Actin and microtubules interact with various Rab and SNARE proteins to fuse with the plasma membrane and release the exosomes. *Credit: Charlie Kirsh.*

Multiple types of RNA have been observed in extracellular vesicles, including coding mRNA, noncoding miRNA, siRNA, and tRNA [47,48]. RNA is stored inside of extracellular vesicles, which offer protection of RNAs from the various RNase species present in many bodily fluids. By avoiding degradation, mRNA, miRNA, and siRNA from one cell are able to impact cellular processes in another cell. The horizontal transfer of mRNA through extracellular vesicle-mediated processes allows for the production of different proteins in multiple cell types. The transfer of mRNA, however, is less common when compared to that of miRNA and siRNA.

miRNA and siRNA are typically the most enriched of the nucleic acid species in extracellular vesicles. These RNA molecules are used for the silencing of mRNA throughout the body and the delivery through exosomes may allow for targeted disruption of protein synthesis [49].

Once the components of an extracellular vesicle are released into a recipient cell, they have the potential to impact cellular processes. Extracellular vesicles are able to impact the recipient cell through two major mechanisms: (1) reaction with receptors on the recipient cell and (2) absorption into the recipient cell where extracellular vesicle contents are released into the cytoplasm. The ability of extracellular vesicle membrane molecules to induce a change in the recipient cell's activity has far-reaching impact on cell growth, pathogen infection, and in immune response [50–52].

### Extracellular vesicles in disease

Increased extracellular vesicle's secretion from adipocytes is observed in obesity and fasting as well as being associated with numerous diseases including type 2 diabetes, obesity, and certain types of cancer [53]. A range of pathogenic cellular processes such as immune cell infiltration, tumor metastasis, and immune evasion can all be regulated, in part, by extracellular vesicles [52,54,55]. Currently, advancement in the field of extracellular vesicles is impeded by multiple obstacles, including issues with nomenclature, isolation methods, as well as techniques for targeting specific cells for uptake. Isolation of extracellular vesicles is typically performed by ultracentrifugation, but techniques such as density gradient centrifugation, filtration, and size exclusion chromatography are also used [56]. These methods of isolation have been successful; however, differences in protocols, controls, and characterization methods have made it difficult to verify and compare results [57]. Moreover, validation of isolated extracellular vesicles is complicated by the fact that microvesicles and exosomes share biochemical properties [58]. Another challenge is to determine methods by which an extracellular vesicle is targeted for internalization by a cell. This will likely depend on the specific extracellular vesicle, and its composition of membrane bound ligands. Ultimately, the inability to classify extracellular vesicles reliably puts a limitation on the knowledge and progress surrounding targeting and uptake [59].

The role of extracellular vesicles has transformed from a method of waste disposal to an integral communication device for cell to cell signaling. The ability to harness the targeted signaling of extracellular vesicles for use in therapeutics and biotechnology would have broad implications. An extracellular vesicle engineered with a molecule of choice and a delivery system to specifically target a tissue could allow for drug delivery even for agents that are easily degraded or unable to be absorbed, as is the case with the blood brain barrier [55,60]. The future of extra cellular vesicles is likely to be one of important discoveries and, with such areas of interest, there remains much work to be done.

## Free fatty acid signaling to peripheral tissues

Free fatty acids are taken up and used directly as an energy substrate by tissues including the skeletal muscle, liver, and heart. The release of free fatty acids allows for the preservation of glucose stores for tissues that are unable to use lipid stores such as the brain. After these free fatty acids are taken up into cells, they can either be broken down in β-oxidation or they can be processed into other lipids. Free fatty acids are the precursors for multipole lipid species including: acylcarnitines, phospholipids, sphingolipids, eicosanoids, and sterols as well as being incorporated in lipids like triglyceride and DAG species [61,62]. They can also be processed into triglyceride-rich lipoproteins including LDL and high-density lipoprotein (HDL). Besides being incorporated into other lipid species, free fatty acids also play important signaling roles including serving as a ligand to nuclear receptors and regulating cellular sensitivity to insulin.

## Free fatty acids activate transcriptional regulation

Free fatty acids can serve as ligands to activate nuclear receptors, steroid superfamily of intracellular receptors, and G-protein-coupled receptors (GPCRs). The most highly studied of these is the activation of the nuclear receptors peroxisome proliferator-activated receptors (PPARs) by free fatty acid binding [62]. Activation of nuclear receptors requires binding of a lipid ligand, heterodimerization with another nuclear receptor, and coregulator binding. There are multiple PPARs including PPARα, PPARγ1 and 2, and PPARδ; these various PPARs exhibit tissue-specific expression and have differential mechanisms of activation. PPARα is abundantly expressed in the liver and is activated by lipolysis; knockdown of ATGL in mice led to decreased PPARα signaling in the liver [63]. This decreased PPARα signaling leads to lower oxidative phosphorylation, cardiomyopathy, and peripheral lipid accumulation in the heart. The cardiomyopathy and lipid accumulation in the heart are likely dependent on PPARα since they were ameliorated by treatment with PPARα activators [64].The dependence of this lipolysis-driven activation of PPARα upon free fatty acid binding has not been entirely explored; established ligands of PPARα were not examined such as 1-palmitoyl-2-oleoyl-*sn*-glycerol-3-phosphocholine [65]. Moreover, other lipids that are increased with lipolysis have also been shown to bind PPARs including eicosanoids and glycerophospholipids [29].

## Free fatty acids reduce insulin sensitivity

Free fatty acids are known to cause insulin resistance in multiple tissues including the liver, skeletal muscle, and pancreas. The mechanisms for this increased insulin resistance are likely diverse and stem from multiple pathways of free fatty acid incorporation and processing. Free fatty acids alone are able to cause insulin resistance through increasing the ratio of Acetyl-CoA to free CoA, which inhibits pyruvate dehydrogenase and increases phosphofructokinase [66]. This is separate in humans where glucose-6-phosphate and glucose are both decreased with elevated lipolysis and fatty acids are able to inhibit insulin-stimulated PI3-kinase activity in human muscle cells [67]. Fatty acids can also serve as structural elements for other lipid species, which allows them to be processed into these lipid classes including acylcarnitines, phospholipids, sphingolipids, neutral lipids, and lipoprotein complexes. Many of these lipids are appreciated markers of metabolic disease with ceramides, DAGs, triglycerides, and acylcarnitines all being increased in the circulation with obesity and type 2 diabetes.

There are two major mechanisms that are proposed to mediate this insulin sensitivity including increased DAG levels and increased ceramides. Both of these lipids have increased production in the liver in response to elevated lipolysis in fasting, as well as chronic states of increased basal lipolysis observed in fasting. Both hepatic DAG and ceramides are elevated in obesity and are correlated with insulin resistance in human studies on obese non-diabetic subjects [68].

Diversity in DAG species and their regulation are mediated by acyl chains and cellular localization. Acyl-chains can be attached to DAGs through the sn1, sn2, or sn3 position with preference being shown for the sn1, sn2 position occupancy. The cellular localization of DAGs is also varied with metabolic state with significant pools in the endoplasmic reticulum, mitochondria, plasma membrane, lipid droplets, and the cytosol. In the development of hepatic insulin resistance observed in lipolysis associated with a 48-hour fast, the largest changes in DAGs are observed in the plasma membrane pool [68,69].

The mechanisms of DAG-mediated insulin resistance are controlled by sn-1,2-DAG activation of Protein kinase C *epsilon* type (PKCε) which leads to translocation to the membrane. Once at the plasma membrane, PKCε phosphorylates insulin receptor on threonine 1160 which destabilizes the active conformation of insulin receptor kinase leading to impaired tyrosine kinase activity. This inhibition of insulin receptor kinase function impacts all downstream pathways regulated by insulin signaling including decreased glycogen synthesis, increased gluconeogenesis, and lower de novo lipogenesis. The negative effects of high fat diet on insulin resistance can be ablated by liver specific knockdown of PKCε or mutation of the insulin receptor threonine 1160 [70,71]. Moreover, elevating the DAG levels in the liver by overexpressing diacylglycerol O-acyltransferase 2 (DGAT2) in mice leads to elevated PKCε translocation and worsened hepatic glucose tolerance and preventing the buildup of DAGs by liver-specific knockout of DGAT2 is protective against high fat diet induced insulin resistance.

Increased flux of free fatty acids during lipolysis also increases production of sphingolipids in the liver and skeletal muscle which regulate insulin sensitivity. Sphingolipids are a class of lipids that are major components of biological membranes. Beyond their structural role in membranes, the diverse structure and function of sphingolipids allow them to have a bioactive role in metabolic and inflammatory signaling. All sphingolipids contain a base unit called sphingosine, which results from the condensation of an amino acid and a fatty acid—specifically serine and palmitoyl-CoA. De novo synthesis of sphingolipids begins in the endoplasmic reticulum and is completed in the Golgi where complex sphingolipids are generated by the addition of fatty acids and other metabolites. Sphingolipids can be further modified in the membrane compartment in which they reside, and intermediates in sphingolipid metabolism can be recycled [72].

Serine palmitoyl transferase catalyzes the first step in de novo sphingolipid synthesis in which serine and palmitoyl CoA are condensed to form 3-keto-dihydrosphingosine. The reduction of 3-keto-dihydrosphingosine generates dihydro-sphingosine, which is acylated by (dihydro)-ceramide synthase (CerS) and desaturated to produce a ceramide. Ceramides can be further modified by phosphorylation or through modification of their terminal hydroxyl group, ceramides can be processed into the 5 major classes of sphingolipids—sphingomyelins, cerebrosides, sulfatides, globosides, and gangliosides. Not only do ceramides occupy a central position in the synthesis and catabolism of sphingolipids, they have also been shown to regulate lipid metabolism through inhibiting lipolysis, controlling mitochondrial function, and regulating the transport of free fatty acids in the cell [73].

Because sphingolipids are hydrophobic, their metabolism is restricted to membrane compartments. In cells, sphingolipids require transport across and between membranes to different cellular compartments by vesicle transport or specialized transport proteins like ceramide transfer protein (CERT) [74]. Interestingly, ceramides transported from the ER to the Golgi via CERT are acted on by sphingomyelin synthase 1 to produce sphingomyelin, whereas ceramides trafficked to the Golgi by vesicle transport are fated to be glycosylated [75]. The membrane localization and transport of sphingolipids has been shown to be functionally relevant to cellular metabolism since loss of CERT function leads to insulin resistance [76]. Ceramides specifically have been linked to selective insulin resistance through inhibition of AKT disrupting insulin signaling or through inhibition of protein kinase B a known repressor of gluconeogenesis [73,77].

In obesity, there are increased sphingolipid in the circulation where they are transported on lipoproteins, albumin, or as part of an extracellular vesicle. Of the sphingolipids present in the plasma 87% is sphingomyelin, about 10% is a diverse array of more than 50 species of glycosphingolipids, and the remaining 3% is ceramides. Additionally, S1P is the only sphingolipid that is more abundant in the plasma than in tissue, and it is transported on HDL and albumin [75]. The tissue of origin for these circulating sphingolipids and changes in sphingolipid abundance in tissues is dependent upon the metabolic stress or diseased state. For example, sphingolipids have been shown to be increased in adipose tissue in obese women with metabolic syndrome but not in obese women without metabolic syndrome [78]. This suggests that sphingolipids do not simply accumulate with increasing body fat, but rather signify metabolic health status. Additionally, adipose dysfunction leads to an imbalance in essential metabolic hormones, like leptin and adiponectin [79]. In turn, these changes decrease levels of the adiponectin receptor, which can catabolize ceramides [80].

## Perspectives on lipolysis-mediated lipid signals

Adipocyte lipolysis acts to signal the metabolic state of energy demand and induces a shift from glucose to lipid utilization. Activation of lipolysis by glucocorticoids and glucagon leads to the release of free fatty acids from lipid droplet triglycerides. Free fatty acid release into the circulation signals for insulin resistance, lipid processing in distal tissues, and a dynamic switch to lipid utilization to preserve glucose stores for the brain. These free fatty acids can also be the substrate for many lipid species including acylcarnitines, sphingolipids, and LDLs. More questions remain in the field of lipolytic signals as we begin to untangle the changes in lipolysis that occur with obesity, the various lipid species that are altered with metabolic disease, and as we develop technology to answer these questions.

White adipocytes are known to secrete a number of other lipids that are not discussed in this chapter because their release with lipolysis has yet to be characterized. These lipids include lysophosphatidic acid and fatty acid esters of hydroxyl fatty acids (FAHFAs), which regulate adipocyte development and insulin sensitivity respectively [2,81,82]. The diversity of lipids from adipose tissue highlights the importance of these lipids for dynamic regulation of metabolism and response to environmental stress.

There are many questions remaining associated with lipolysis both from a biological and in lipid characterization perspective. Biologically, the identity and specificity of lipid transporters, the mechanisms through which elevated free fatty acids impact multiple tissues, and regulation of basal lipolysis in obesity all remain elusive. Central to all of these questions is the shift in response from acute to chronic lipolysis and understanding the dynamics of this shift.

The burgeoning use of mass spectrometry-based lipidomics has uncovered a multitude of questions that deserve further exploration. Tantalizingly, there are several hundred uncharacterized peaks that appear in mass spectrometry-based lipidomic analysis as well as peaks that are difficult to assess in the neutral lipids range [83]. Further investigation of these lipids, identification, and functional characterization will yield novel regulators of whole-body energy expenditure and metabolism. Many questions remain on the nature of well-studied lipids and their role in metabolic disease including questions on lipid species chain length and structure. The use of ion mobility spectrometry to assess lipid modifications, double bond positions, and acyl-chain position will allow us to uncover the role of these lipid signatures as signaling molecules and add depth to our current view [84]. The use of matrix-assisted laser desorption/ionization mass spectrometric imaging will allow for spatial profiling of lipids in tissues; this technology could answer questions on tissue heterogeneity of lipids, changes in lipid droplet composition, and the spatial regulation of lipolysis in white adipose tissue [85]. These new lipid quantification tools allow us to carve new frontiers in the metabolic map and re-explore longstanding questions with a better vantage point.

Adipose tissue lipolysis has been shown to regulate systemic insulin sensitivity, cardiovascular function, and mediate weight loss. Treatment of these diseases by targeting lipolysis is promising, and several recent discoveries in technology and lipid quantification hold therapeutic potential. One of these technological innovations is a luciferase reporter system for nuclear receptors that quantifies lipolysis and the available free fatty acid pool in real time [86]. Other advances have been made in quantification of lipids by mass spectrometry which has led to a revolution in profiling circulating lipids as diagnostic markers of disease such as hepatic steatosis and type 2 diabetes [87]. The final focus should be on pushing these discoveries for therapeutic potential by targeting adipocyte lipolysis or pathways of sphingolipid synthesis [8]. As we face an ever-expanding obesity epidemic, it is important to determine the basic mechanisms of lipolytic regulation, signaling, and how this information can be harnessed for the treatment of metabolic disease.

## References

[1] Friedman JM. Leptin and the endocrine control of energy balance. Nat Metab 2019;1(8):754–64.

[2] Funcke JB, Scherer PE. Beyond adiponectin and leptin: adipose tissue-derived mediators of inter-organ communication. J Lipid Res 2019;doi: 10.1194/jlr.R094060.

[3] Arner P. Human fat cell lipolysis: biochemistry, regulation and clinical role. Best practice & research. Clin Endocrinol Metab 2005;19(4):471–82.

[4] Body DR. The lipid composition of adipose tissue. Prog Lipid Res 1988;27(1):39–60.

[5] Brasaemle DL. Thematic review series: adipocyte biology. The perilipin family of structural lipid droplet proteins: stabilization of lipid droplets and control of lipolysis. J Lipid Res 2007;48(12):2547–59.

[6] Arner P, Andersson DP, Backdahl J, Dahlman I, Ryden M. Weight gain and impaired glucose metabolism in women are predicted by inefficient subcutaneous fat cell lipolysis. Cell Metab 2018;28(1):45–54.e3.

[7] Schoiswohl G, Stefanovic-Racic M, Menke MN, et al. Impact of reduced ATGL-mediated adipocyte lipolysis on obesity-associated insulin resistance and inflammation in male mice. Endocrinology 2015;156(10):3610–24.

[8] Schweiger M, Romauch M, Schreiber R, et al. Pharmacological inhibition of adipose triglyceride lipase corrects high-fat diet-induced insulin resistance and hepatosteatosis in mice. Nat Commun 2017;8:14859.

[9] Zhang H, Qiao A, Yang L, et al. Structure of the glucagon receptor in complex with a glucagon analogue. Nature 2018;553(7686):106–10.

[10] Galsgaard KD, Pedersen J, Knop FK, Holst JJ, Wewer Albrechtsen NJ. . Glucagon receptor signaling and lipid metabolism 2019;10(413)doi: 10.3389/fphys.2019.00413.

[11] Kim C, Xuong NH, Taylor SS. Crystal structure of a complex between the catalytic and regulatory (RIalpha) subunits of PKA. Science (New York, N. Y.) 2005;307(5710):690–6.
[12] Duncan RE, Ahmadian M, Jaworski K, Sarkadi-Nagy E, Sul HS. Regulation of lipolysis in adipocytes. Ann Rev Nutr 2007;27:79–101.
[13] Sanders MA, Madoux F, Mladenovic L, et al. Endogenous and synthetic ABHD5 ligands regulate ABHD5-perilipin interactions and lipolysis in fat and muscle. Cell Metab 2015;22(5):851–60.
[14] Xu C, He J, Jiang H, et al. Direct effect of glucocorticoids on lipolysis in adipocytes. Mol Endocrinol 2009;23(8):1161–70.
[15] Wahrenberg H, Wennlund A, Arner P. Adrenergic regulation of lipolysis in fat cells from hyperthyroid and hypothyroid patients. J Clin Endocrinol Metab 1994;78(4):898–903.
[16] Ben Cheikh R, Chomard P, Beltramo JL, Autissier N, Dumas P. Influence of prolonged fasting and thyroid hormone on the alpha 2-adrenergic response in isolated epididymal adipocytes of Wistar rats. Eur J Endocrinol 1997;136(2):223–9.
[17] Heckmann BL, Zhang X, Xie X, et al. Defective adipose lipolysis and altered global energy metabolism in mice with adipose overexpression of the lipolytic inhibitor G0/G1 switch gene 2 (G0S2). J Biol Chem 2014;289(4):1905–16.
[18] Ryden M, Arner P. Subcutaneous adipocyte lipolysis contributes to circulating lipid levels. Arteriosclerosis, thrombosis, and vascular biology 2017;37(9):1782–7.
[19] Miyoshi H, Perfield 2nd JW, Obin MS, Greenberg AS. Adipose triglyceride lipase regulates basal lipolysis and lipid droplet size in adipocytes. J Cell Biochem 2008;105(6):1430–6.
[20] Ali SB, Turner JJO, Fountain SJ. Constitutive P2Y2 receptor activity regulates basal lipolysis in human adipocytes. J Cell Sci 2018;131(22).
[21] Ahmed K, Tunaru S, Tang C, et al. An autocrine lactate loop mediates insulin-dependent inhibition of lipolysis through GPR81. Cell Metab 2010;11(4):311–9.
[22] Shi H, Dirienzo D, Zemel MB. Effects of dietary calcium on adipocyte lipid metabolism and body weight regulation in energy-restricted aP2-agouti transgenic mice. FASEB J 2001;15(2):291–3.
[23] Xue B, Greenberg AG, Kraemer FB, Zemel MB. Mechanism of intracellular calcium ([Ca2+]i) inhibition of lipolysis in human adipocytes. FASEB J 2001;15(13):2527–9.
[24] Kang L, Nagy LE. Chronic ethanol feeding suppresses beta-adrenergic receptor-stimulated lipolysis in adipocytes isolated from epididymal fat. Endocrinology 2006;147(9):4330–8.
[25] Tchernof A, Belanger C, Morisset AS, et al. Regional differences in adipose tissue metabolism in women: minor effect of obesity and body fat distribution. Diabetes 2006;55(5):1353–60.
[26] Booth A, Magnuson A, Foster M. Detrimental and protective fat: body fat distribution and its relation to metabolic disease. Horm Mol Biol Clin Investig 2014;17(1):13–27.
[27] Harada K, Shen WJ, Patel S, et al. Resistance to high-fat diet-induced obesity and altered expression of adipose-specific genes in HSL-deficient mice. Am J Physiol Endocrinol Metab 2003;285(6):E1182–95.
[28] Hu X, Cifarelli V, Sun S, Kuda O, Abumrad NA, Su X. Major role of adipocyte prostaglandin E2 in lipolysis-induced macrophage recruitment. J Lipid Res 2016;57(4):663–73.
[29] Gartung A, Zhao J, Chen S, et al. Characterization of eicosanoids produced by adipocyte lipolysis: implication of cyclooxygenase-2 in adipose inflammation. J Biol Chem 2016;291(31):16001–10.
[30] Abumrad NA, Park JH, Park CR. Permeation of long-chain fatty acid into adipocytes. Kinetics, specificity, and evidence for involvement of a membrane protein. J Biol Chem 1984;259(14):8945–53.
[31] Schwenk RW, Holloway GP, Luiken JJ, Bonen A, Glatz JF. Fatty acid transport across the cell membrane: regulation by fatty acid transporters. Prostaglandins Leukot Essent Fatty Acids 2010;82(4–6):149–54.
[32] Nahle Z, Hsieh M, Pietka T, et al. CD36-dependent regulation of muscle FoxO1 and PDK4 in the PPAR delta/beta-mediated adaptation to metabolic stress. J Biol Chem 2008;283(21):14317–26.
[33] Pepino MY, Kuda O, Samovski D, Abumrad NA. Structure-function of CD36 and importance of fatty acid signal transduction in fat metabolism. Ann Rev Nutr 2014;34:281–303.
[34] Hames KC, Vella A, Kemp BJ, Jensen MD. Free fatty acid uptake in humans with CD36 deficiency. Diabetes 2014;63(11):3606–14.
[35] Zaborowski MP, Balaj L, Breakefield XO, Lai CP. Extracellular vesicles: composition, biological relevance, and methods of study. Bioscience 2015;65(8):783–97.
[36] Choi D-S, Kim D-K, Kim Y-K, Gho YS. Proteomics, transcriptomics and lipidomics of exosomes and ectosomes. Proteomics 2013;13(10–11):1554–71.

[37] Lazar I, Clement E, Dauvillier S, et al. Adipocyte exosomes promote melanoma aggressiveness through fatty acid oxidation: a novel mechanism linking obesity and cancer. Cancer Res 2016;76(14):4051–7.
[38] van Niel G, D'Angelo G, Raposo G. Shedding light on the cell biology of extracellular vesicles. Nat Rev Mol Cell Biol 2018;19(4):213–28.
[39] Daleke DL. Regulation of transbilayer plasma membrane phospholipid asymmetry. J Lipid Res 2003;44(2):233–42.
[40] Record M, Carayon K, Poirot M, Silvente-Poirot S. Exosomes as new vesicular lipid transporters involved in cell-cell communication and various pathophysiologies. Biochim Biophys Acta 2014;1841(1):108–20.
[41] Trajkovic K, Hsu C, Chiantia S, et al. Ceramide triggers budding of exosome vesicles into multivesicular endosomes. Science (New York, N. Y.) 2008;319(5867):1244–7.
[42] Hessvik NP, Llorente A. Current knowledge on exosome biogenesis and release. Cell Mol Life Sci 2018;75(2):193–208.
[43] Connolly KD, Guschina IA, Yeung V, et al. Characterisation of adipocyte-derived extracellular vesicles released pre- and post-adipogenesis. J Extracell Vesicles 2015;4:29159.
[44] Gao X, Salomon C, Freeman DJ. Extracellular vesicles from adipose tissue-A potential role in obesity and type 2 diabetes? Front Endocrinol 2017;8:202.
[45] Kranendonk MEG, Visseren FLJ, van Herwaarden JA, et al. Effect of extracellular vesicles of human adipose tissue on insulin signaling in liver and muscle cells. Obesity 2014;22(10):2216–23.
[46] Subra C, Grand D, Laulagnier K, et al. Exosomes account for vesicle-mediated transcellular transport of activatable phospholipases and prostaglandins. J Lipid Res 2010;51(8):2105–20.
[47] Kim KM, Abdelmohsen K, Mustapic M, Kapogiannis D, Gorospe M. RNA in extracellular vesicles. Wiley Interdiscip Rev RNA 2017;8(4)doi: 10.1002/wrna.1413.
[48] Turchinovich A, Drapkina O, Tonevitsky A. Transcriptome of extracellular vesicles: state-of-the-art. Front Immunol 2019;10:202.
[49] Kanada M, Bachmann MH, Hardy JW, et al. Differential fates of biomolecules delivered to target cells via extracellular vesicles. Proc Natl Acad Sci 2015;112(12):E1433–42.
[50] Desrochers LM, Bordeleau F, Reinhart-King CA, Cerione RA, Antonyak MA. Microvesicles provide a mechanism for intercellular communication by embryonic stem cells during embryo implantation. Nat Commun 2016;7:11958.
[51] Antonyak MA, Li B, Boroughs LK, et al. Cancer cell-derived microvesicles induce transformation by transferring tissue transglutaminase and fibronectin to recipient cells. Proc Natl Acad Sci U S A 2011;108(12):4852–7.
[52] Robbins PD, Morelli AE. Regulation of immune responses by extracellular vesicles. Nat Rev Immunol 2014;14:195. https://www.nature.com/articles/nri3622#supplementary-information.
[53] Lazar I, Clement E, Attane C, Muller C, Nieto L. A new role for extracellular vesicles: how small vesicles can feed tumors' big appetite. J Lipid Res 2018;59(10):1793–804.
[54] Ricklefs FL, Alayo Q, Krenzlin H, et al. Immune evasion mediated by PD-L1 on glioblastoma-derived extracellular vesicles. Sci Adv 2018;4(3):eaar2766.
[55] Wiklander OPB, Brennan MÁ, Lötvall J, Breakefield XO, El Andaloussi S. Advances in therapeutic applications of extracellular vesicles. Sci Transl Med 2019;11(492):eaav8521.
[56] Gardiner C, Di Vizio D, Sahoo S, et al. Techniques used for the isolation and characterization of extracellular vesicles: results of a worldwide survey. J Extracell Vesicles 2016;5:32945.
[57] Witwer KW, Buzas EI, Bemis LT, et al. Standardization of sample collection, isolation and analysis methods in extracellular vesicle research. J Extracell Vesicles 2013;2.
[58] Skotland T, Sandvig K, Llorente A. Lipids in exosomes: current knowledge and the way forward. Prog Lipid Res 2017;66:30–41.
[59] Mulcahy LA, Pink RC, Carter DR. Routes and mechanisms of extracellular vesicle uptake. J Extracell Vesicles 2014;3doi: 10.3402/jev.v3.24641.
[60] Sancho-Albero M, Rubio-Ruiz B, Pérez-López AM, et al. Cancer-derived exosomes loaded with ultrathin palladium nanosheets for targeted bioorthogonal catalysis. Nat Catal 2019;doi: 10.1038/s41929-019-0333-4.
[61] Simcox J, Geoghegan G, Maschek JA, et al. Global analysis of plasma lipids identifies liver-derived acylcarnitines as a fuel source for brown fat thermogenesis. Cell Metab 2017;26(3):509–522.e6.
[62] Zechner R, Zimmermann R, Eichmann TO, et al. FAT SIGNALS--lipases and lipolysis in lipid metabolism and signaling. Cell Metab 2012;15(3):279–91.
[63] Ong KT, Mashek MT, Bu SY, Greenberg AS, Mashek DG. Adipose triglyceride lipase is a major hepatic lipase that regulates triacylglycerol turnover and fatty acid signaling and partitioning. Hepatology (Baltimore, Md.) 2011;53(1):116–26.

[64] Haemmerle G, Moustafa T, Woelkart G, et al. ATGL-mediated fat catabolism regulates cardiac mitochondrial function via PPAR-alpha and PGC-1. Nat Med 2011;17(9):1076–85.
[65] Chakravarthy MV, Lodhi IJ, Yin L, et al. Identification of a physiologically relevant endogenous ligand for PPA-Ralpha in liver. Cell 2009;138(3):476–88.
[66] Randle PJ, Garland PB, Hales CN, Newsholme EA. The glucose fatty-acid cycle. Its role in insulin sensitivity and the metabolic disturbances of diabetes mellitus. Lancet (London, England) 1963;1(7285):785–9.
[67] Dresner A, Laurent D, Marcucci M, et al. Effects of free fatty acids on glucose transport and IRS-1–associated phosphatidylinositol 3-kinase activity. J Clin Invest 1999;103(2):253–9.
[68] Kumashiro N, Erion DM, Zhang D, et al. Cellular mechanism of insulin resistance in nonalcoholic fatty liver disease. Proc Natl Acad Sci 2011;108(39):16381.
[69] Perry RJ, Wang Y, Cline GW, et al. Leptin mediates a glucose-fatty acid cycle to maintain glucose homeostasis in starvation. Cell 2018;172(1–2). 234-48.e17.
[70] Samuel VT, Shulman GI. The pathogenesis of insulin resistance: integrating signaling pathways and substrate flux. J Clin Invest 2016;126(1):12–22.
[71] Samuel VT, Shulman GI. Mechanisms for insulin resistance: common threads and missing links. Cell 2012;148(5):852–71.
[72] Olson DK, Frohlich F, Farese RV, Walther TC. Taming the sphinx: mechanisms of cellular sphingolipid homeostasis. Biochim Biophys Acta-Mol Cell Biol Lipids 2016;1861(8):784–92.
[73] Chaurasia B, Tippetts TS, Mayoral Monibas R, et al. Targeting a ceramide double bond improves insulin resistance and hepatic steatosis. Science (New York, N. Y.) 2019;365(6451):386–92.
[74] Hanada K, Kumagai K, Tomishige N, Yamaji T. CERT-mediated trafficking of ceramide. Biochim Biophys Acta-Mol Cell Biol Lipids 2009;1791(7):684–91.
[75] Iqbal J, Walsh MT, Hammad SM, Hussain MM. Sphingolipids and lipoproteins in health and metabolic disorders. Trends Endocrinol Metab 2017;28(7):506–18.
[76] Bandet CL, Mahfouz R, Veret J, Ceramide Transporter CERT, et al. Is Involved in muscle insulin signaling defects under lipotoxic conditions. Diabetes 2018;67(7):1258–71.
[77] Chaurasia B, Summers SA. Ceramides - Lipotoxic Inducers of Metabolic Disorders: (Trends in Endocrinology and Metabolism 26 538-550; 2015). Trends Endocrinol Metab 2018;29(1):66–7.
[78] Choromanska B, Mysliwiec P, Hady HR, et al. Metabolic syndrome is associated with ceramide accumulation in visceral adipose tissue of women with morbid obesity. Obesity 2019;27(3):444–53.
[79] Lopategi A, Lopez-Vicario C, Alcaraz-Quiles J, et al. Role of bioactive lipid mediators in obese adipose tissue inflammation and endocrine dysfunction. Mol Cell Endocrinol 2016;419(C):44–59.
[80] Holland WL, Miller RA, Wang ZV, et al. Receptor-mediated activation of ceramidase activity initiates the pleiotropic actions of adiponectin. Na Med 2011;17(1). 55-U226.
[81] Moraes-Vieira PM, Saghatelian A, Kahn BB. GLUT4 Expression in adipocytes regulates de novo lipogenesis and levels of a novel class of lipids with antidiabetic and anti-inflammatory effects. Diabetes 2016;65(7):1808–15.
[82] Yung YC, Stoddard NC, Chun J. LPA receptor signaling: pharmacology, physiology, and pathophysiology. J Lipid Res 2014;55(7):1192–214.
[83] Murphy RC. Challenges in mass spectrometry-based lipidomics of neutral lipids. TrAC Trends Analyt Chem 2018;107:91–8.
[84] Paglia G, Astarita G. Metabolomics and lipidomics using traveling-wave ion mobility mass spectrometry. Nat Protoc 2017;12(4):797–813.
[85] Shi Y, Johnson J, Wang B, et al. Mass spectrometric imaging reveals temporal and spatial dynamics of bioactive lipids in arteries undergoing restenosis. J Proteome Res 2019;18(4):1669–78.
[86] Mottillo EP, Zhang H, Yang A, Zhou L, Granneman JG. Genetically-encoded sensors to detect fatty acid production and trafficking. Mol Metab 2019;29:55–64.
[87] Jha P, McDevitt MT, Gupta R, et al. Systems analyses reveal physiological roles and genetic regulators of liver lipid species. Cell Syst 2018;6(6). 722-733.e6.

CHAPTER

# 8

# Regulation of intracellular lipid storage and utilization

*Alyssa S. Zembroski, Kimberly K. Buhman*

Purdue University, West Lafayette, IN, United States

OUTLINE



## Abbreviations

**ACAT** acyl-CoA:cholesterol acyltransferase
**AGPAT** 1-acyl-glycerol-3-phosphate acyltransferase
**ApoB-100** apolipoproteinB-100
**ATGL** adipose triglyceride lipase
**ATP** adenosine triphosphate
**CCT1** CTP:phosphocholine cytidylyltransferase
**CE** cholesteryl ester
**CGI-58** comparative gene identification-58
**CIDE** cell death-inducing DFF45-like effector
**CLD** cytoplasmic lipid droplet
**CM** chylomicron
**COX** cyclooxygenase
**CPT** carnitine palmitoyltransferase

Lipid Signaling and Metabolism. http://dx.doi.org/10.1016/B978-0-12-819404-1.00008-7

| | |
|---|---|
| **DAG** | diacylglycerol |
| **DGAT** | diacylglycerol acyltransferase |
| **ER** | endoplasmic reticulum |
| **ERAD** | endoplasmic reticulum-associated degradation |
| **FA** | fatty acid(s) |
| **FAO** | fatty acid oxidation |
| **G3P** | glycerol-3-phosphate |
| **GPAT** | glycerol-3-phosphate acyltransferase |
| **FIT** | fat storage-inducing transmembrane protein |
| **HSL** | hormone sensitive lipase |
| **LAL** | lysosomal acid lipase |
| **LDL** | low density lipoprotein |
| **MGAT** | acyl-CoA:monoacylglycerol acyltransferase |
| **MAG** | monoacylglycerol |
| **MGL** | monoacylglycerol lipase |
| **NAFLD** | non-alcoholic fatty liver disease |
| **NEFA** | nonesterified fatty acid |
| **PAP** | phosphatidic acid phosphatases |
| **PC** | phosphatidylcholine |
| **PL** | phospholipid |
| **Plin** | perilipin |
| **PPAR** | peroxisome proliferator activated receptor |
| **TAG** | triacylglycerol |
| **TCA** | tricarboxylic acid |
| **VLDL** | very low density lipoprotein |

## Introduction

Cytoplasmic lipid droplets (CLDs) are dynamically regulated lipid-storage organelles present in almost every cell type and conserved across organisms [1,2]. The major physiological function of CLDs is to store fatty acids (FA) in the form of neutral lipid, mostly triacylglycerol (TAG) and cholesteryl esters (CE) [3,4]. These neutral lipids are stored in the core of CLDs and emulsified in the cell cytosol by a phospholipid (PL) monolayer coat and associated proteins [5,6]. Generally, CLDs form in the presence of excess cellular lipid and are broken down when lipid substrate is needed, helping to control cellular FA levels and protect from lipotoxicity. The flexibility of CLD metabolism makes them major cellular metabolic hubs that respond to fluctuating systemic and cellular signals [7]. Because these organelles have important roles in supporting cellular and physiological homeostasis, complex regulatory systems are in place to control the flux of FA into storage, oxidation, or complex lipid synthesis.

CLDs within specific cell types contribute to the distinct systemic and cellular functions of those cell types in regulating and maintaining systemic nutrient and energy needs. For example, CLDs in adipocytes serve as a long-term TAG storage depot to provide FA to the rest of the body during periods of fasting [8], whereas CLDs in myocytes provide FA needed locally for oxidation and energy production in contracting muscle [9]. Further, CLDs in enterocytes [10] and hepatocytes [11] play diverse roles including serving as a cellular energy source as well as providing substrate for lipoprotein synthesis and secretion to deliver FA throughout the body. Therefore, the regulation of CLD formation, breakdown, and use as substrate for cellular and systemic processes must be tightly controlled to avoid abnormal cellular and systemic lipid levels.

Interestingly, CLDs in diverse cell types contribute to both health and disease, and it is not always clear if they are a cause or consequence of a disease state. A consequence of altered cellular and systemic lipid levels is metabolic abnormalities, such as those present in obesity and insulin resistance [12,13]. Disruptions in the metabolic regulation of lipid metabolism contribute to the development of secondary diseases such as nonalcoholic fatty liver disease (NAFLD) [14], diabetic cardiomyopathy [15], atherosclerosis [16], and cancer [17]. Therefore, understanding the regulation of intracellular lipid storage and utilization in a cell-specific manner is important in both normal physiology and pathophysiological conditions.

## Cytoplasmic lipid droplet composition and formation

CLD formation is determined by substrate supply as well as energy and building needs of the cell depending on physiological state. Generally, CLDs form when there is an excess of intracellular lipid. In doing so, they not only protect the cell from lipotoxicity but also store energy dense FA, cholesterol substrate necessary for synthesis of cell membranes and steroid hormones, and fat-soluble vitamins used in metabolic pathways. The majority of FA are stored in the form of TAG but can also be esterified to cholesterol. The level of FA in a cell at any given time depends on the rate of FA uptake, rate of FA synthesis, rate of FA oxidation (FAO), and FA use in complex lipid synthesis. Excess cellular FA can accumulate during different periods of metabolic flux, for example in energy excess. An abnormally high level of cellular FA in the absence of energy demand initiates CLD formation, which has the potential to contribute to negative health consequences in cell types that do not typically store lipids. Although CLDs are present for different reasons in physiological and pathophysiological states, the general process of CLD formation is similar [5].

CLD formation first requires the synthesis of neutral lipids, including TAG and CE. TAG and CE are synthesized in the endoplasmic reticulum (ER) by their respective enzymatic pathways. The collection of TAG and CE in the ER initiates the early steps of CLD formation, eventually leading to the emergence of CLDs from the cytosolic ER membrane. Once formed, CLDs can increase in size by the addition of TAG and PL. Finally, the protein coat of CLDs reflects the physiological state of the cell and plays a role in the maintenance and metabolism of CLDs.

### CLD composition

#### *Source of lipid for CLD formation*

TAG and CE stored in the core of CLDs must first be synthesized from FA and either glycerol or cholesterol, respectively, and these substrates may originate from sources either exogenous or endogenous to the cell. Exogenous sources of these substrates are those from circulation, including nonesterified fatty acids (NEFAs) and glycerol primarily originating from adipose tissue, or FA and cholesterol carried in circulating lipoproteins such as very low density lipoproteins (VLDLs), chylomicrons (CMs) and low density lipoproteins (LDLs). Endogenous sources of these substrates are lipids synthesized through intracellular de novo lipogenesis and glycerol from glycolysis. Either or both sources provide lipids that are used for CLD synthesis (Fig. 8.1).

*For Section II: CLD composition and formation*

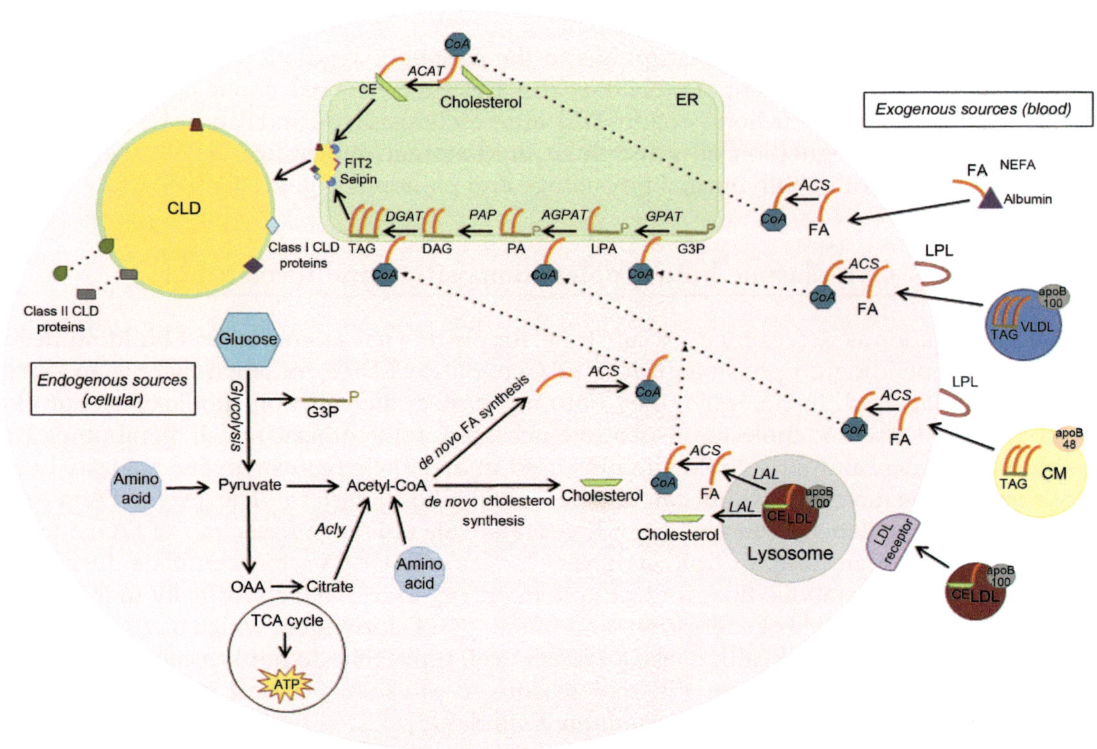


**FIGURE 8.1 CLD composition and formation.** CLDs form in the presence of excess cellular lipid from exogenous and/or endogenous sources. Exogenous sources are those from the blood, which includes NEFAs from adipose tissue, TAG transported in VLDL and CM, and cholesterol transported in LDL. LPL cleaves FA from TAG in VLDL and CM. LDL is endocytosed and broken down in the lysosome by LAL to release FA and cholesterol. Once FA enter the cell, they are activated by ACS enzymes before they can be further utilized. Endogenous sources of lipid include those from de novo FA and cholesterol synthesis, and glycerol (G3P) from glycolysis which are used for TAG and CE synthesis at the ER membrane. TAG is synthesized in most cell types by the G3P pathway composed of GPAT, AGPAT, PAP, and DGAT enzymes. CEs are synthesized by ACAT. Both TAG and CE collect in the ER bilayer forming a nascent CLD, and as it grows it is stabilized by seipin and FIT2 at the ER membrane. Class I proteins in the ER relocate to growing CLDs. The CLD eventually buds from the ER into the cytosol. CLDs can reversibly acquire Class II proteins from the cytosol. *ACAT*, acyl-CoA:cholesterol acyltransferase; *ACS*, acyl-CoA synthetase; *Acly*, ATP citrate lyase; *ATP*, adenosine triphosphate; *AGPAT*, 1-acyl-glycerol-3-phosphate acyltransferase; *apoB48*, apolipoprotein B48; *apoB100*, apolipoprotein B100; *CE*, cholesteryl ester; *CM*, chylomicron; *CLD*, cytoplasmic lipid droplet; *DAG*, diacylglycerol; *DGAT*, diacylglycerol acyltransferase; *ER*, endoplasmic reticulum; *FA*, fatty acid(s); *FIT2*, fat storage-inducing transmembrane protein; *G3P*, glycerol-3-phosphate; *GPAT*, glycerol-3-phosphate acyltransferase; *LAL*, lysosomal lipase; *LDL*, low density lipoprotein; *LPA*, lysophosphatidic acid; *LPL*, lipoprotein lipase; *NEFA*, nonesterified fatty acid; *OAA*, oxaloacetate; *PA*, phosphatidic acid; *PAP*, phosphatidic acid phosphatase; *TAG*, triacylglycerol; *TCA*, tricarboxylic acid; *VLDL*, very low density lipoprotein.

### Exogenous sources

Cells acquire FA and cholesterol from exogenous sources including circulating lipoproteins (CM, VLDL, and LDL) and NEFAs. CMs circulate after a meal and carry dietary TAG from the small intestine. Between meals, VLDL carries endogenous TAG originating from the liver [18]. TAG carried in circulating lipoproteins is cleared by the action of endothelial lipoprotein lipase, which hydrolyzes lipoprotein TAG to release FA for tissue uptake [19,20]. Circulating LDL delivers cholesterol to cells by endocytosis and subsequent lysosomal breakdown [21]. Another exogenous source of FA is circulating NEFAs, which are released from adipocytes (along with glycerol) by cytoplasmic lipolysis in response to fasting or certain other stimuli [22]. During long-term fasting or insulin resistance, adipocyte lipolysis is increased leading to elevated levels of circulating NEFAs, which stimulate lipid accumulation and CLD formation in cell types such as hepatocytes and myocytes [23].

### Endogenous sources

Cells endogenously acquire FA and cholesterol through de novo lipogenesis and glycerol from glycolysis. De novo FA and cholesterol synthesis is activated during times of nutrient excess, utilizing acetyl-CoA derived from excess carbohydrates or amino acids to build the FA palmitate [24], or cholesterol through the mevalonate pathway [25,26]. De novo FA synthesis is often upregulated during disease states, for example it contributes to the lipid accumulation that occurs in NAFLD [27] and certain types of cancer [28]. Glycerol that is used for TAG synthesis is produced as an intermediate of the glycolytic pathway, glycerol-3-phosphate (G3P) [29]. Ultimately, cellular FA and cholesterol acquired from either exogenous or endogenous sources are used as substrates to build TAG or CE before being stored in CLDs.

## *TAG and CE synthesis pathways*

TAG is synthesized by either two major pathways, the G3P pathway or the acyl-CoA:monoacylglycerol acyltransferase (MGAT) pathway. Both of these pathways include multiple sequential enzymatic steps that take place predominantly at the ER membrane. The G3P pathway is the major TAG synthesis pathway in most cell types, which catalyzes the addition of three fatty acyl-CoAs to a G3P backbone [30]. The enzymes involved in this reaction include glycerol-3-phosphate acyltransferases (GPAT), 1-acyl-glycerol-3-phosphate acyltransferases (AGPAT), phosphatidic acid phosphatases/lipins (PAP), and diacylglycerol acyltransferases (DGAT), and different isoforms of each enzyme show cell type and substrate specificity [30]. For example, DGAT1 is more active toward exogenous FA while DGAT2 preferentially utilizes endogenously synthesized FA [31–33]. Alternative to the G3P pathway, TAG may be synthesized by the MGAT pathway. The MGAT pathway is most active in small intestinal enterocytes during the process of dietary fat absorption due to the high concentration of digestion product sn-2 monoacylglycerol [10]. In this pathway, two fatty acyl-CoAs are successively esterified to a sn-2 monoacylglycerol backbone by MGAT and DGAT enzymes [34]. Both the G3P and MGAT pathways share the same last committed step in TAG synthesis with the addition of FA to diacylglycerol (DAG) by either DGAT1 or DGAT2, demonstrating the importance of these enzymes in TAG synthesis and lipid homeostasis [31].

Intracellular CE are synthesized at the ER membrane by either acyl-CoA:cholesterol acyltransferase (ACAT) 1 or 2, which esterify fatty acyl-CoA to cholesterol forming CE [35]. CE is stored in CLDs in cell types active in steroid hormone synthesis, including ovarian and testicular cells as well as adrenocortical cells [36]. Once TAG and CE are synthesized by their respective pathways, the neutral lipid collects in the ER membrane and facilitates the initial steps of CLD formation.

## CLD formation

### *CLD formation and budding from the ER membrane*

The most widely accepted mechanism of CLD formation is budding from the ER membrane, which is facilitated by multiple proteins [5,37]. The mechanism of CLD budding is favored by biophysical principles and includes phases of nucleation, growth, and eventual emergence from the ER membrane into the cytosol [38]. The first step of CLD formation begins when TAG accumulates and condenses at sites within the ER bilayer forming an oil lens, such as those observed in *Saccharomyces cerevisiae* after stimulation of TAG synthesis [39]. Certain nucleation sites may exist in the ER membrane that favor CLD formation due to specificities in PL and protein composition, which induce changes in surface tension to facilitate ER membrane alterations and encourage budding [40–42]. Conversely, the intrinsic properties of the accumulating lipid itself may be sufficient to stimulate budding [43]. The nascent CLD then emerges from the ER toward the cytosol by a dewetting process [38]. Once the contact angle between the cytosolic ER membrane and nascent CLD reaches a sufficient value, the CLD buds from the ER membrane, taking the outer cytosolic ER membrane to serve as its PL monolayer [5].

Multiple proteins have been implicated in mediating CLD formation and budding by stabilizing CLD growth, and ER–CLD contacts are important for this process [3,37,42,44,45]. One such protein is seipin, an ER protein mutated in the lipid storage disorder Berardinelli-Seip Congenital Lipodystrophy [46]. Seipin has a specific structure that allows it to localize to sites of CLD formation by detecting lipid-packing defects in the ER membrane, where it then stabilizes nascent CLDs for growth to occur [37,47–50]. The ER–CLD contacts facilitated by seipin have recently been found to be membrane "necks," which allow for the transfer of lipid from the ER to growing CLDs, preventing CLD ripening in the ER membrane and in turn regulating CLD size [45]. However, since loss of seipin results in various extremes in CLD sizes [45,47,49] and has many protein interacting partners [42,44], its exact molecular function has proven difficult to unravel.

Another protein family that mediates CLD formation is the fat storage-inducing transmembrane (FIT) proteins, most notably FIT2 [3,51]. Deletion of FIT2 in *S. cerevisiae* and 3T3-L1 cells, an adipocyte cell model, inhibits LD budding and results in TAG accumulation in the ER [39]. Similarly, CLD accumulation is decreased upon FIT2 knockdown in 3T3-L1 adipocytes and zebrafish [51]. Conversely, overexpression of FIT2 increases the packaging of TAG in CLDs in human embryonic kidney HEK293 cells [51]. How FIT2 regulates CLD budding is not clear, however, it may involve its ability to bind TAG and DAG [52] or it may guide the directionality of CLD budding toward the cytosol [39]. Multiple other proteins have been shown to maintain connections between the ER and CLDs during initial CLD formation to facilitate CLD growth and budding, including Rab18, DFCP1, and Ldo proteins [53–57].

### *CLD expansion*

Once budded from the ER membrane, CLDs can grow by reestablishing ER-CLD connections, local TAG and PL synthesis on the CLD itself, or lipid transfer between neighboring

CLDs. CLDs can either separate from or remain attached to the ER membrane [58]. CLDs dissociated from the ER membrane can grow after reconnecting back to the ER by membrane bridges formed by mechanisms of the Arf1/Cop1 complex, allowing for the relocalization of enzymes in the G3P pathway to CLDs [58–60]. Relocalization of these enzymes to the CLD surface facilitates CLD growth by local TAG synthesis, denoting this subpopulation of CLDs as "expanding lipid droplets" [5,58]. However, a recent study investigating the structure of DGAT2, an enzyme catalyzing the last committed step in TAG synthesis, proposed a different mechanism for local TAG synthesis on CLDs. In Cos-7 fibroblast-like cells, DGAT2 interacts with growing CLDs through its C-terminal domain while remaining embedded in the ER membrane to catalyze TAG synthesis [61]. Therefore, complete relocalization of the enzyme from the ER to the CLD surface itself may not be required. On the other hand, enzymes in the G3P pathway of TAG synthesis were identified on all sizes of CLD subpopulations isolated from CHO K2 cells, and the smallest CLDs were able to synthesize TAG even in the absence of ER [62]. These results suggest the presence of these enzymes on CLDs alone is sufficient for TAG synthesis. Although the interaction of TAG synthesis enzymes with CLDs appears to be regulated, the exact signal that stimulates CLD growth and localization of TAG synthesis enzymes to CLDs is unclear.

In addition to the generation of ER-CLD connections and local TAG synthesis, CLDs can expand by lipid transfer between neighboring CLDs. Lipid transfer is mediated by the cell death-inducing DFF45-like effector (CIDE) family of proteins, including Cidea, Cideb, and Cidec/Fsp27 [63]. These proteins congregate at CLD-CLD contact sites and with the addition of other proteins create a pore between two CLDs, allowing for the transfer of lipid from smaller to larger CLDs [63,64]. The role of Cidec/Fsp27 in mediating lipid droplet growth is most apparent in adipocytes, where it is required for the formation of large unilocular lipid droplets [65–67].

Expanding CLDs requires an increase in PL levels to keep their growing neutral lipid core emulsified in the aqueous cell cytosol, and the amount of PL surrounding CLDs is regulated by the rate-limiting enzyme in phosphatidylcholine (PC) synthesis, CTP:phosphocholine cytidylyltransferase (CCT1) [68]. Insufficient levels of PC, the major PL surrounding CLDs, during CLD growth results in CLD instability and coalescence [69]. Therefore, maintenance of adequate PC levels by CCT1 is required to prevent CLD fusion. Upon oleate loading in *Drosophila S2* cells, PC deficiency on expanding CLDs is detected by CCT1, initiating its translocation from the nucleus to the CLD surface [70]. Binding and activation at the CLD surface initiates PC synthesis to match the increase in neutral lipid deposition of growing CLDs, preventing CLD fusion [69,70]. The importance of CCT in regulating CLD size and PC levels is also apparent in mammalian cells [71]. However, CCT may not localize to CLDs directly in these cell types due to differences in the PL monolayer composition of CLDs between mammals and *Drosophila S2* cells, which influences the binding ability of CCT [71]. Whether CCT directly localizes to CLDs or synthesizes PC at an alternative location, adequate levels of PC are required to maintain CLD stability during expansion.

## CLD protein association, removal, and role of Perilipins

The formation of a complete CLD is established by the presence of proteins associated with its PL monolayer [72], which not only help to emulsify CLDs in the cell cytosol but are also thought to regulate various aspects of CLD metabolism including stabilization, expansion, and catabolism [2,73]. How proteins exactly target to CLDs is unclear, as no specific CLD-targeting sequence or unique structure in CLD-associated proteins has been discovered to date. However, similarities in the structures of proteins that associate with CLDs have led to their designation into one of either two groups: Class I or Class II [72]. The structure of Class I proteins includes a hairpin loop, which is a hydrophobic sequence containing a proline residue that inserts into membranes [74]. Conversely, Class II proteins contain amphipathic helices that associate with the surface of membranes [72]. Due to their structural distinctions, the association and removal of Class I and Class II proteins from the CLD surface occurs by different mechanisms.

### *CLD protein association*

Proteins that associate with CLDs can originate from either the ER or from the cytosol during CLD formation or after CLDs have budded from the ER membrane. Proteins originating from the ER are Class I proteins, which relocalize to CLDs when CLDs are present. Therefore, due to the hydrophobic nature of these proteins, their association with CLDs depends on connections with the ER during different stages of CLD maturation. Diffusion of proteins from the ER to CLDs can occur during CLD formation [72,75] or through membrane bridges formed between budded CLDs and the ER membrane [58]. Proteins originating from the cytosol are Class II proteins, which are translated in the cytosol and subsequently associate with the PL monolayer of CLDs [74]. Class II proteins recognize packing defects common to CLD monolayers, and upon binding form an $\alpha$-helix that enables their reversible association [76]. A list of proteins harboring either Class I or Class II features that have been validated as CLD-associated proteins is listed in [2].

### *CLD protein removal*

Proteins are removed from CLDs by moving back to the ER, degradation, or molecular crowding. Proteins that move back to the ER from the CLD surface are Class I proteins, relying again on contacts between the two organelles. Class I proteins diffuse back to the ER through membrane connections, where they can either reside in the ER or be degraded by the ER-associated degradation (ERAD) system. One example of a Class I protein that translocates back and forth from the ER to CLDs is Dga1, the yeast homolog of the TAG synthesis enzyme DGAT1. In *S. cerevisiae*, Dga1 relocalizes from the ER to the CLD surface upon stimulation of CLD formation [75]. However, under lipolytic conditions Dga1 diffuses back to the ER from the shrinking CLD surface [75]. These results suggest that connections between the ER and CLDs are not only important for CLD growth but also for exchange of proteins between the two organelles under fluctuating metabolic conditions.

Class I proteins that translocate back to the ER through ER-CLD connections can be degraded by ERAD, which is a mechanism to control the movement of proteins back to the CLD surface [74]. In yeast cells, the hydrophobic hairpin structure in certain proteins that allows for their CLD association is also required for their degradation by the ER-localized ubiquitin ligase Doa10, preventing their localization to CLDs from the ER [77]. In addition to controlling the localization of CLD proteins in yeast, ERAD also modulates the association of a high-confidence CLD protein identified in human osteosarcoma U2OS cells, c18orf32, to the CLD surface by degradation in the ER [78]. Another potential mechanism of protein removal from

CLDs is the ubiquitin-proteasome system, which has been implicated in the degradation of multiple CLD proteins including certain Plins and lipolytic enzymes [74].

Proteins that are removed from CLDs by molecular crowding are Class II proteins. Since Class II proteins are not as securely integrated into the CLD PL monolayer like Class I proteins are, Class II proteins dissociate from CLDs when the area of the CLD surface decreases as in CLD breakdown. This forces proteins to compete with each other for association. Loss of binding area and protein crowding at the CLD surface initiates protein release and unfolding into the cell cytosol due to their amphipathic nature [76,79]. Although there are multiple mechanisms in place controlling the association and removal of proteins from CLDs, what exactly determines the composition of the CLD proteome is unclear. However, since the presence of CLDs is determined by fluctuating metabolic conditions, the proteome likely reflects this and is regulated accordingly.

### *Perilipin proteins regulate CLD dynamics*

Certain proteins that associate with CLDs contribute to CLD dynamics and cellular lipid homeostasis, for example the perilipin (Plin) family, which are well-established structural CLD proteins that include Plin1-Plin5 [80]. Plins are differentially present depending on cell type, for example Plin1 and Plin4 are expressed exclusively in adipocytes, Plin5 is expressed mostly in highly oxidative tissue such as the heart, while Plin2 and Plin3 are ubiquitous [80]. The general function of Plins is to protect CLDs from lipolysis, thereby promoting TAG storage [81]. Each Plin has a unique way of regulating CLD lipolysis. For example Plin1 regulates the recruitment of lipolytic machinery to CLDs upon phosphorylation in adipocytes [82], Plin2 forms a barrier between the lipid core of CLDs and lipolytic enzymes [83], and Plin5 regulates the use of FA released by lipolysis for mitochondrial FAO [84]. Despite their differences in maintaining CLD integrity, controlling CLD lipolysis makes Plins regulators of cellular lipid homeostasis and important residents of the CLD proteome.

While Plins are present on all CLDs, the specific Plin that associates depends on competition with other Plin species during different stages of CLD maturation in certain cell types. In adipocytes, Plin1 replaces Plin2 when CLDs increase in size during adipocyte differentiation [85]. In enterocytes, Plin2 and Plin3 differentially associate with CLDs upon chronic and acute high fat feeding, respectively, as well as in an obese state [86,87]. These observations suggest that the different species of Plins have specific yet unresolved roles in CLD metabolism. One reason Plins may replace each other on CLDs is their ability to stabilize CLDs and protect from lipolysis [88]. A recent study found that the differential association of Plin1-3 is due to differences in cooperative binding affinity of the amino- and carboxy-terminal regions of the different proteins to the CLD surface, accounting for their differences in CLD stabilization [88]. Therefore, the specific species of Plin present on CLDs in certain cell types plays a major role in controlling CLD metabolism.

## CLD breakdown and fates of released lipids

### CLD breakdown

TAG and CE stored in CLDs are broken down when there is a cellular or systemic need for FA or cholesterol. The enzymatic processes that catalyze CLD breakdown are TAG lipolysis, which releases FA and glycerol, and CE lipolysis, which releases FA and cholesterol. The two types of TAG lipolysis are cytoplasmic TAG lipolysis and lysosomal TAG lipolysis, termed lipophagy. Cytoplasmic TAG lipolysis is the major enzymatic pathway for the breakdown of CLDs in most cell types, while lipophagy is a newly recognized form of CLD breakdown active in hepatocytes [89,90]. Each lipolytic process consists of different enzymes and different mechanisms of CLD breakdown (Fig. 8.2).

### *Cytoplasmic TAG lipolysis*

The classical pathway of CLD breakdown is cytoplasmic TAG lipolysis. This lipolytic pathway is composed of three enzymes: adipose triglyceride lipase (ATGL), hormone-sensitive lipase (HSL), and monoacylglycerol lipase (MGL), which successively hydrolyze TAG releasing FA and glycerol [20]. These enzymes are highly active in adipocytes; therefore, their role in the lipolytic cascade has been studied extensively in this cell type. However, these enzymes are also expressed in most other cell types. The lipolytic cascade in adipocytes is stimulated by activation of the β-adrenergic signaling pathway. Activation of the β-adrenergic signaling pathway results in the recruitment and activation of the lipolytic machinery to CLDs from their basal cellular locations [20,22]. Briefly, elevated levels of cAMP activate protein kinase A, which phosphorylates Plin1 on CLDs and HSL in the cytosol. Phosphorylation of Plin1 allows ATGL to interact with its coactivator, comparative gene identification-58 (CGI-58) on the CLD surface, which initiates the first TAG hydrolysis reaction releasing DAG [20]. Phosphorylation of HSL in the cytosol results in its translocation to the CLD where it catalyzes the hydrolysis of DAG, releasing monoacylglycerol (MAG). Lastly, MAG is hydrolyzed by MGL, releasing FA and glycerol. FA released from stimulated cytoplasmic TAG lipolysis in adipocytes provides systemic tissues with substrate for oxidation and energy production. In addition to FA provided by adipocyte lipolysis, other cell types can break down their own CLDs by cytoplasmic TAG lipolysis and use the FA released for oxidation and energy production, as well as for other cellular functions (discussed below).

Due to the potential negative consequences of elevated lipolytic activity and excess FA release, the activity of these enzymes is tightly regulated. While catecholamines stimulate lipolysis through β-adrenergic signaling to release FA during times of energy demand, insulin potently inhibits lipolysis through the insulin signaling pathway to promote TAG storage during times of nutrient excess [22]. Altered lipolytic activity indicates metabolic dysregulation and its consequences contribute to the development of metabolic disease. For example, loss of insulin signaling during insulin resistance results in an elevated level of adipocyte lipolysis and circulating FA levels, which can contribute to lipotoxicity and ectopic fat deposition [23]. Therefore, proper storage of TAG and control of CLD breakdown by specific signals is an important regulatory node maintaining cellular and systemic lipid homeostasis.

### *Lysosomal TAG lipolysis*

Another mechanism of CLD breakdown has recently come to light, called lipophagy, or lysosomal TAG hydrolysis. Lysosomal TAG hydrolysis entails engulfment of whole or a part of CLDs by autophagic vesicles, which are then targeted to the lysosome for enzymatic breakdown of TAG and CE by lysosomal acid lipase (LAL) [89]. Autophagic breakdown of CLDs was first discovered in hepatocytes in response to starvation as well as to oleic acid loading, providing FA for oxidation [91]. Since then, lipophagy has been shown to catalyze CLD

*For Section III: CLD breakdown and fates of released lipids*

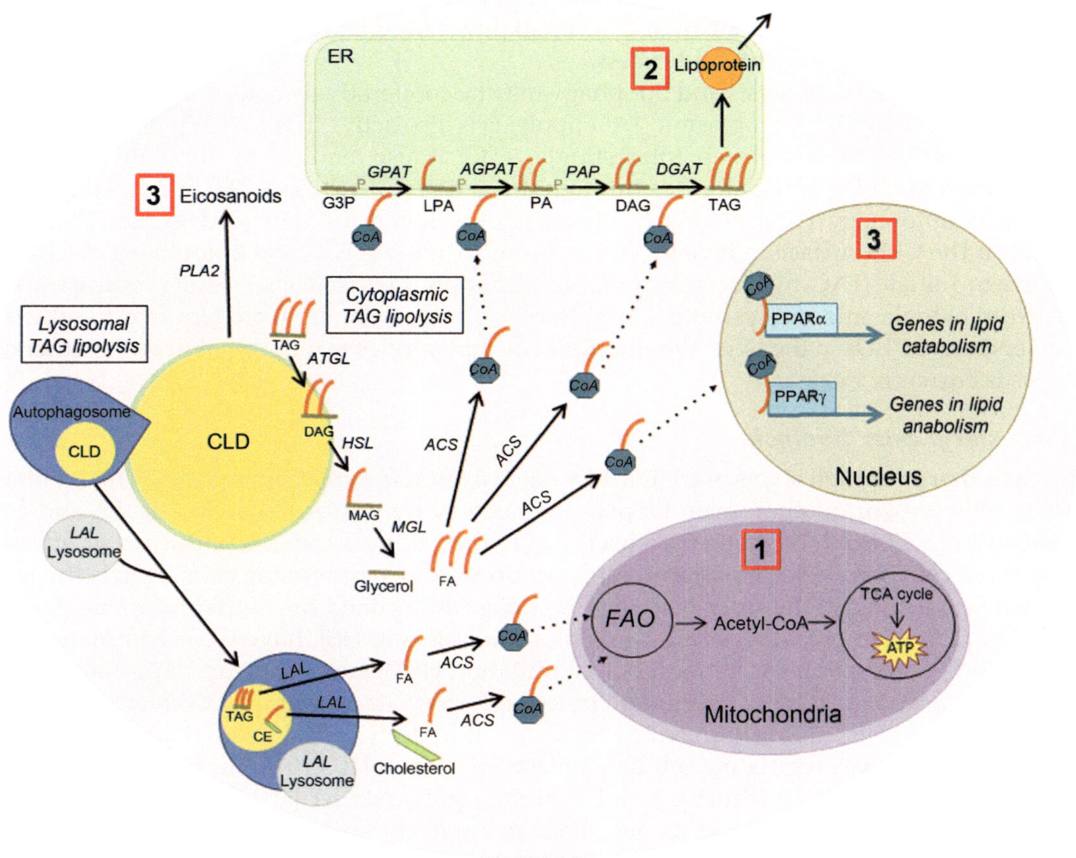


**FIGURE 8.2 CLD breakdown and fates of released lipids.** CLDs are broken down by either cytoplasmic TAG lipolysis or lysosomal TAG lipolysis. Cytoplasmic TAG lipolysis consists of the sequential breakdown of TAG by ATGL, HSL, and MGL enzymes releasing glycerol and FA. Lysosomal TAG lipolysis consists of engulfment of a whole or part of CLD by autophagosomes, which is then targeted to the lysosome for enzymatic breakdown of TAG and CE by LAL. FA released by either cytoplasmic or lysosomal TAG lipolysis must be first activated by ACS enzymes before utilization in the cell. FA released from CLDs can be (1) oxidized in the mitochondria, producing acetyl-CoA to be shuttled into the TCA cycle for ATP synthesis, (2) resynthesized into TAG at the ER membrane and packaged into lipoproteins such as VLDL and CM in hepatocytes and enterocytes, respectively, or (3) used for signaling purposes, including binding to and activating PPAR transcription factors in the nucleus to stimulate the transcription of genes involved in lipid metabolism, and also eicosanoid production to be used in cellular signaling pathways. *ACS*, acyl-CoA synthetase; *ATGL*, adipose triglyceride lipase; *ATP*, adenosine triphosphate; *AGPAT*, 1-acyl-glycerol-3-phosphate acyltransferase; *CE*, cholesteryl ester; *CLD*, cytoplasmic lipid droplet; *DAG*, diacylglycerol; *DGAT*, diacylglycerol acyltransferase; *ER*, endoplasmic reticulum; *FA*, fatty acid(s); *FAO*, fatty acid oxidation; *G3P*, glycerol-3-phosphate; *GPAT*, glycerol-3-phosphate acyltransferase; *HSL*, hormone sensitive lipase; *LAL*, lysosomal lipase; *LPA*, lysophosphatidic acid; *MAG*, monoacylglycerol; *MGL*, monoacylglycerol lipase; *PA*, phosphatidic acid; *PAP*, phosphatidic acid phosphatase; *PLA2*, phospholipase A2; *PPAR*, peroxisome proliferator activated receptor; *TAG*, triacylglycerol; *TCA*, tricarboxylic acid.

breakdown in multiple cell types [92] and its dysfunction has been implicated in various diseases [93,94]. Lipophagy is an important regulator of cellular lipid levels in cell types that often handle large amounts of lipid, including hepatocytes [89] and enterocytes [95,96]. This allows for an efficient mechanism of CLD breakdown in order to maintain intracellular lipid homeostasis during times of lipid excess.

Cytoplasmic TAG lipolysis and lipophagy may be cooperatively and complementarily regulated in the liver [97]. Cytoplasmic TAG lipolysis is also active in the liver [98], and relies on ATGL to catalyze the initial step. Interestingly, ATGL is also required to stimulate autophagic breakdown of CLDs in hepatocytes through the activity of sirtuin 1 [99], a deacetylase that regulates autophagy [100]. Similarly, chaperone-mediated autophagy degrades Plin2 and Plin3 on the CLD surface in hepatocytes in order to allow ATGL and autophagic machinery access to initiate TAG hydrolysis and CLD breakdown [101]. These results indicate a link between cytoplasmic and lysosomal TAG lipolysis, suggesting both are required for efficient catabolism of CLDs in the liver. Whether this concept applies to CLD breakdown in other cell types is currently unclear.

### *Cholesteryl ester lipolysis*

Cells that store high levels of CE in CLDs, such as steroid synthesizing cells or macrophage foam cells present in atherosclerotic plaques, actively hydrolyze CE to release FA and cholesterol from CLDs that can then be used inside the cell for synthesis pathways or effluxed out of the cell by reverse cholesterol transport. In steroid synthesizing cells, HSL is the main enzyme catalyzing CE hydrolysis to release cholesterol needed for steroid hormone production [36]. For example, adrenocortical cells of mice lacking HSL have CE accumulation and decreased corticosterone secretion upon stimulation [102], suggesting an important role for this enzyme in regulating intracellular cholesterol substrate availability. Conversely, in macrophage foam cells, multiple enzymes catalyze CE hydrolysis to initiate reverse cholesterol transport, a pathway regulating cellular cholesterol content [103,104]. These enzymes include neutral cholesterol ester hydrolase 1, carboxylesterase 1, and LAL [103,104]; the latter is active in multiple other cell types and its deficiency is implicated in both cholesterol ester storage disease and Wolman disease resulting in defective breakdown of CE [105]. Further, defective CLD breakdown and CE accumulation in macrophage foam cells can contribute to the development of atherosclerosis [106]. Therefore, the hydrolysis of CE from storage in CLDs is a central regulatory node of cholesterol homeostasis.

## Fates of FA released from CLD breakdown

The FA released from CLD breakdown have catabolic or anabolic fates depending on cell type and metabolic status [4]. In cardiomyocytes for example, FA released from CLDs by cytoplasmic TAG lipolysis undergo oxidation to provide the majority of resting cellular energy for this cell type. On the other hand, during a fasted state, FA released from CLD breakdown in hepatocytes are used for TAG synthesis and VLDL formation for systemic lipid delivery. In addition to energy production and lipoprotein formation, FA released from CLD breakdown can also serve as signaling molecules that regulate gene transcription and downstream metabolic processes. Due to the multiple different fates of FA stored in CLDs, CLD breakdown must be tightly regulated to control systemic and cellular FA levels and their associated metabolic pathways.

### *Fatty acid oxidation*

A major function of CLDs is to store excess cellular lipid during times of energy surplus; during energy deprivation, the FA released from CLD breakdown can be used for mitochondrial FAO and subsequent adenosine triphosphate (ATP) production [107]. FAO takes place in the inner mitochondrial membrane, requiring the transport of FA into this organelle. FA are shuttled into the inner mitochondrial membrane by carnitine conjugation reactions catalyzed by carnitine palmitoyltransferase (CPT) 1 and CPT2. A series of enzymatic reactions then catabolize FA into acetyl-CoA, which enters the tricarboxylic acid (TCA) cycle to produce ATP [107]. Therefore, CLDs provide a readily available source of FA to generate cellular energy upon demand.

FA that are used for oxidation can originate from exogenous sources, including circulating NEFAs from adipocyte lipolysis or TAG carried in lipoproteins, or endogenous sources such as intracellular CLD breakdown. Adipocyte lipolysis is stimulated during fasting to release FA into circulation, providing systemic tissues with substrate to generate cellular energy by FAO [8]. In the liver, these FA can also be partially oxidized to ketones and released into circulation, which provide the brain with an alternative source of energy when glucose is limited [108]. FA taken up from circulation originating from either adipocyte lipolysis or lipoprotein-mediated delivery are often first packaged in TAG and stored in CLDs, which serves to protect cells such as cardiomyocytes from lipotoxicity [109]. Using FA stored in CLDs for oxidation not only maintains adequate cellular energy levels but is also a mechanism to control cellular lipid content. Highly oxidative cell types such as cardiac and skeletal myocytes use FA to generate a large amount of their cellular energy, making CLDs an important component of their cell metabolism [9,109]. However, excessive CLD accumulation in these cell types can interfere with cellular function [109,110], requiring a tight balance between lipid storage and FAO. Therefore, the flexibility of CLD metabolism in these cell types is critical for the storage or catabolism of FA in response to fluctuating physiological states.

Genes involved in the mitochondrial FAO machinery and lipid catabolism are regulated by the transcription factor peroxisome proliferator activated receptor (PPAR)-α, the major PPAR isoform expressed in oxidative tissues [111,112]. PPAR-α is activated by FA ligands, such as those released from TAG lipolysis during CLD breakdown. Activation of PPAR-α is required for functional mitochondrial FAO and therefore contributes to regulation of intracellular lipid levels. PPAR-α regulates lipid accumulation and FAO in the heart, which is a central process in this tissue due to its high energy demands. PPAR-α in cardiomyocytes is specifically activated by FA released from CLDs by ATGL [113]. PPAR-α activation stimulates the transcription of genes involved in lipid catabolism and FAO, allowing for efficient FA breakdown and ATP generation. Therefore, ATGL in cardiomyocytes is necessary for cardiac function by releasing FA from CLDs to activate PPAR-α and FAO pathways. Consistently, ATGL null mice develop TAG accumulation, mitochondrial dysfunction, and heart failure due to loss of PPAR-α-regulated oxidative gene expression in cardiomyocytes [113]. These results demonstrate that CLD-derived FA act as both signaling molecules and substrate for FAO in the heart, which is necessary for maintaining efficient FA catabolism and cardiac function.

In humans, CLD accumulation in cardiomyocytes occurs during diabetes and obesity, as well as in patients with inborn errors of FAO. This is due in part to decreased intracellular TAG turnover and reduced or defective FAO [109]. A failure to compensate for accumulating intracellular lipid contributes to defective cardiac function in these conditions [109], making the regulation of FA catabolism and balance of TAG storage an important aspect of human cardiac lipid homeostasis and health.

CLDs in skeletal muscle fibers provide substrate for FAO during exercise [9,110,114]. There are two types of muscle fibers: slow-twitch Type I and fast-twitch Type II [115]. Type I muscle fibers use FA provided by intramyocellular lipids stored in CLDs to power their oxidative metabolism during moderate intensity exercise [116–119], while Type II muscle fibers primarily use glucose stored in glycogen for rapid high-intensity exercise [120]. ATGL is also involved in CLD breakdown and FA release in skeletal muscle fibers, and ATGL protein levels increase during exercise training in rat and human skeletal muscle [121,122]. ATGL-mediated TAG lipolysis also regulates the expression of FAO genes in skeletal muscle by releasing FA that activate PPAR-$\alpha$ and stimulate gene transcription [113,123].

After a bout of moderate intensity exercise, CLD size and density decrease in Type I muscle fibers [118], indicating CLD breakdown through TAG lipolysis and subsequent FAO to power muscle contraction during exercise. A rapid demand for FA substrate and catabolism during exercise requires CLDs to be in close proximity to the location of FA degradation. Consistently, CLDs in Type 1 fibers are often closely associated with mitochondria for direct shuttling of FA into the oxidation pathway [116,124]. These CLD-mitochondrial interactions are facilitated by Plin5 [125], similar to the role of Plin5 in regulating FA release from CLDs for oxidation in other cell types (discussed below). Therefore, CLD breakdown in skeletal muscle provides important substrate to maintain adequate cellular energy and muscle performance during moderate intensity exercise.

In sedentary individuals, excess lipid accumulation and the presence of detrimental lipid metabolites such as DAG and ceramide interfere with the insulin sensitivity of skeletal muscle fibers [126]. Exercise training can decrease CLD accumulation and recover muscle insulin sensitivity by restoring intracellular lipid and energy homeostasis [127]. Interestingly, however, trained athletes have an increase in CLD accumulation while maintaining insulin sensitivity [128], which may instead be a positive adaptation to chronic exercise for muscle fuel provision [127,129]. This contradictory phenomenon is called the "athlete's paradox" and demonstrates the ability of skeletal muscle to adapt to changing metabolic demands and conditions [127]. In addition, this example of CLD dynamics provides evidence of the differential role CLDs may play depending on physiological or disease state.

### *Lipoprotein synthesis*

CLDs in hepatocytes and enterocytes provide a regulated pool of lipid substrate to be used for lipoprotein assembly and secretion, helping to maintain adequate blood lipid levels during different physiological states. For example, hepatocytes package lipid stored in intracellular CLDs into VLDL particles between meals or during fasting, while enterocytes package dietary fat into either CLDs or CMs after a meal depending on the amount of fat consumed [10,130]. FA provided to systemic tissues by VLDL or CM can be oxidized for energy or used as cellular building blocks to synthesize necessary cell components such as PL membranes. Therefore, CLDs in hepatocytes and enterocytes help to maintain systemic lipid homeostasis by serving as a storage pool of lipid for lipoprotein assembly.

#### Lipoprotein synthesis in enterocytes

CLDs form in enterocytes when enterocytes are exposed to a large amount of dietary fat [130], helping to prevent cellular lipotoxicity that could result from a major acute influx of lipid during a high fat meal. These CLDs are thought to play a role in the dynamic process of dietary fat absorption by serving as a regulated pool of substrate for CM assembly and secretion, controlling the amount of dietary lipid secreted into the body at one time [130,131]. Although the exact fate of lipid stored in enterocyte CLDs and the molecular mechanisms of CLD mobilization is unclear, evidence of an early triglyceridemic response following certain physiological stimuli posits them as substrate for CM secretion. Factors such as a subsequent meal [132], glucose [133,134], sham fat feeding [135], and the enteroendocrine hormone GLP-2 [136] stimulate postprandial intestinal TAG secretion in humans, suggesting mobilization of intestinal TAG stores. For example, in human subjects, consumption of a mid-day meal containing primarily oleic acid results in the rapid initial secretion of CM containing linoleic acid consumed during a morning meal [132]. Secreted CM containing linoleic acid from the morning meal is followed by the secretion of CM containing oleic acid from the mid-day meal but at a slower rate [132]. These results demonstrate that lipid consumed previously can be stored in human intestinal cells up to 6 hours before it is used for CM assembly and secretion, confirming a role for the intestine in temporary lipid storage.

The idea that enterocyte CLDs serve as a temporary TAG storage pool is also demonstrated in the triglyceridemic response to glucose consumption. Consumption of a glucose drink 5 hours after the consumption of a mixed meal or fat emulsion in human subjects results in a rise in plasma TAG, with most of this TAG carried in CM originating from the intestine [133,134]. Further, quantitative CLD analysis of human intestinal biopsy samples revealed a fewer total number of CLDs per cell and an increased number of smaller CLDs after glucose ingestion compared to water ingestion [133]. These results suggest that lipid stored in enterocyte CLDs is mobilized and secreted as CM in response to oral glucose. Although the molecular mechanisms of this effect are currently unclear, initial proteomic analysis of the intestinal biopsy samples has identified potential molecular players that may mediate the enterocyte response to glucose [133]. Interestingly, both enteral [137] and intravenous [138] glucose administration also stimulate the secretion of CM from intestinal lipid stores. These results suggest that glucose may have both direct and indirect effects on enterocyte lipid metabolism, and may contribute to an increase in CM secretion that occurs when blood glucose levels are elevated during insulin resistance [139]. However, future studies are required to uncover the exact target of glucose-mediated intestinal TAG mobilization.

The mobilization of lipid stored in enterocytes requires CLDs to be broken down by either cytoplasmic or lysosomal TAG lipolysis. However, the exact lipase that hydrolyzes TAG in enterocyte CLDs to be used for CM synthesis and secretion is unclear. Enzymes involved in the cytoplasmic TAG hydrolysis pathway, including ATGL [140], HSL [141,142], and MGL [143], are present and active in enterocytes. However, loss of these enzymes results in either an unchanged [140,142] or decreased [143] TAG secretion rate after administration of olive oil gavage, suggesting cytoplasmic TAG hydrolysis, may not quantitatively contribute to mobilizing CLDs for CM assembly and secretion. Instead, lysosomal TAG hydrolysis may preferentially direct mobilized CLD substrate to lipoprotein assembly.

A role for lipophagy in CLD mobilization and CM secretion is apparent in DGAT1$^{-/-}$ mice, whose intestinal phenotype is critical for their resistance to diet-induced obesity [144]. DGAT1$^{-/-}$ mice have a lower intestinal TAG secretion rate and massive CLD accumulation in enterocytes compared to wild-type mice 2 h after an olive oil gavage [145]. The decrease in CM secretion is thought to be due in part to defective lipophagy, as demonstrated by TAG accumulation in autophagic vesicles and decreased lysosomal acidification in enterocytes of DGAT1$^{-/-}$ mice [96]. In support of a role for autophagic breakdown of enterocyte CLDs, treating differentiated Caco-2 cells, a cell model of enterocytes, with lipid micelles initiates an autophagic response and targets CLDs for lysosomal breakdown [95]. These results suggest that CLD breakdown by lysosomal TAG hydrolysis in enterocytes is a cellular response to help mitigate a large influx of lipid, and by doing so provides substrates for CM assembly. Future studies are required to determine the exact mechanism of CLD targeting to autophagosomes and its quantitative contribution to CM secretion.

#### Lipoprotein synthesis in hepatocytes

Lipid stored in hepatocyte CLDs is used for the assembly and secretion of VLDL through a cycle of lipolysis and reesterification [11,146,147]. Hepatocyte lipid stores originate from the clearance of remnant lipoproteins, NEFA flux to the liver downstream of adipocyte lipolysis, or de novo FA synthesis [148]. The CLDs formed from these lipids are broken down via lipolysis, and the FA released are reesterified back into TAG at the ER membrane and are either secreted on VLDL or repackaged into CLDs [146,147]. In fact, about 70% of lipid used for VLDL assembly and secretion originates from the lipolysis and reesterification of hepatocyte TAG stores [147]. Therefore, hepatocyte CLDs serve as a readily available regulated storage pool of lipid to provide the rest of the body with FA in between meals.

Factors that regulate the lipolysis and reesterification cycle in hepatocytes include insulin and Plin2. For example, in primary rat hepatocytes, insulin promotes the reesterification of TAG into storage instead of the packaging of hydrolyzed TAG into VLDL, overall decreasing the rate of VLDL secretion [147]. However, when insulin signaling in hepatocytes is lost, VLDL secretion is increased due in part to the channeling of excess lipid that accumulates during insulin resistance into VLDL assembly instead of storage [148]. Excess VLDL secretion can contribute to elevated blood lipid levels and negative health consequences that often occur during insulin resistance. Therefore, modulating the release or storage of lipid into CLDs can influence hepatic VLDL secretion and affect blood lipid levels.

Another factor directing the storage versus secretion of lipid stored in hepatocyte CLDs is Plin2. Plin2 overexpression in McA-RH7777 hepatoma cells increases CLD accumulation and decreases TAG turnover [149], supporting its role in inhibiting CLD lipolysis [81]. Further, VLDL secretion is decreased when Plin2 is overexpressed [149] and increased when Plin2 expression is lost [150]. This suggests that preventing TAG hydrolysis and promoting TAG storage in CLDs reduces the availability of FA to be channeled to VLDL assembly and vice versa. The involvement of Plin2 in CLD breakdown and formation in hepatocytes is thus an important mechanism to maintain a normal VLDL secretion rate and blood lipid levels.

The lipolytic enzymes directing FA released from CLD TAG hydrolysis for VLDL secretion in hepatocytes have yet to be identified. Overexpression of ATGL and HSL in primary mouse hepatocytes [98] and McArdle RH777 rat hepatoma cells [98,151] partitions FA to oxidation instead of to VLDL assembly and secretion, similar to its action in enterocytes [140]. An enzyme separate from the classic lipolysis pathway may instead play a role. Evidence for a liver-specific enzyme in catalyzing CLD hydrolysis for VLDL assembly is liver triacylglycerol hydrolase (TGH) [152,153], also known as carboxylesterase 1d/3 [154]. TGH promotes the turnover of stored TAG and secretion of VLDL in McArdle RH7777 rat hepatoma cells [152] and liver-specific TGH transgenic mice have increased TAG and apolipoproteinB-100 (apoB-100) secretion [155], which is the structural protein of VLDL. On the other hand, when TGH is inhibited in primary rat hepatocytes, TAG mobilization and VLDL secretion are decreased [156]. These results support a role for TGH in the lipolysis and reesterification of stored TAG for VLDL assembly hepatocytes. Further, TGH has been identified on ER luminal lipid droplets by proteomic analysis [157], suggesting TGH may promote VLDL secretion by hydrolyzing TAG stored in ER luminal lipid droplets in preparation for VLDL assembly [157,158]. Further investigation is required to determine exactly how TGH is involved in the turnover of hepatic TAG stores and its effect on VLDL secretion.

### *Signaling molecules*

FA stored in CLDs can activate gene transcription and serve as substrates or intermediates in signaling pathways upon CLD breakdown. Gene transcription is activated by FA binding to and activating the PPAR family of nuclear receptors, which are transcription factors that regulate the expression of certain genes involved in lipid metabolism [112]. For example, PPAR-α controls the expression of genes involved in catabolic lipid pathways such as FAO [111], while PPAR-γ controls genes involved in anabolic lipid pathways such as lipid synthesis and storage [112]. Therefore, stimulated CLD breakdown can provide a FA signal that initiates the transcription of genes involved in specific metabolic pathways. For example, as discussed earlier, FA released from CLDs by ATGL-catalyzed TAG lipolysis in oxidative tissues activate PPAR-α and stimulate the transcription of genes involved in FAO to generate cellular energy [113,123].

CLDs can also store FA that are used for lipid signaling pathways. A major group of lipid signaling molecules is the eicosanoids, which mediate various inflammatory responses [159,160]. Eicosanoids are derived from polyunsaturated FA, specifically arachidonic acid, which is commonly present in membrane PL and cleaved by phospholipase A2. Specific enzymes that are involved in intermediate reactions of arachidonic acid metabolism, including the cytochrome P450s, cyclooxygenases (COX), and lipoxygenases, determine the type of eicosanoid produced [160]. Eicosanoid synthesis enzymes have been found to localize to CLDs in different cell types, suggesting CLDs store arachidonic acid that can be metabolized into different inflammatory signaling molecules. For example, phospholipase A2 is present and active in the lipid droplet fraction of human monocytic leukemia U937 cells [161], and COX-2 localizes to CLDs in skeletal muscle C2C12 cells treated with the saturated FA palmitate [162]. Further, COX-2 expression was associated with a decrease in insulin signaling in cells treated with palmitate [162], suggesting that eicosanoids potentially derived from lipid stored in CLDs can affect cellular metabolism. In fact, CLDs formed in response to oxidized LDL in mouse macrophages actively synthesize leukotrienes, which may contribute to the inflammation present in atherosclerotic lesions [163]. Therefore, CLDs are signaling hubs that can contribute to mediating cellular inflammatory responses. Future studies are required to determine whether disruption of CLD storage directly influences the inflammation response.

## Other functions of CLD proteins

Proteins that associate with CLDs also have roles outside of CLD synthesis and catabolism, including enabling connections to other organelles and regulating protein degradation and storage. These other functions of CLD proteins confirm that CLDs play multiple other roles outside of mediating lipid metabolism and contribute to total physiological cell function [4]. Lastly, the functional relevance of CLD proteins identified by proteomic studies is an important consideration for future studies in the field of CLD biology.

### *CLD proteins mediate connections to organelles*

The identification of proteins specific to certain organelles on the CLD surface supports the concept of CLD–organelle interactions. CLDs interact with multiple cellular organelles depending on metabolic state, allowing for lipid transfer between organelles [164,165]. For example, CLDs interact with the ER during times of lipid excess for their formation and growth, and interact with lysosomes and mitochondria during times of energy deprivation and CLD breakdown [89,166]. The flexibility of CLD–organelle interactions contributes to the dynamic nature of CLDs. Recently, a novel multispectral imaging technique was used to study the dynamics of organelle interactions in live COS-7 fibroblast-like cells [164]. CLD contact with ER was the most common interaction; however, CLDs also interact with mitochondria, Golgi, lysosomes, and peroxisomes. Changes in cellular metabolic homeostasis by starvation or FA treatment alter the frequency of CLD interactions with certain organelles, demonstrating the ability of CLDs to adapt to changing metabolic signals [164]. Although these interactions are visible by imaging techniques, the molecular players that facilitate the anchoring or attachment of CLDs to organelles are currently under investigation.

Recent efforts have ventured into the study of Plin5 and its involvement in CLD–mitochondrial interactions [167]. Plin5 localizes to CLD–mitochondrial contact sites in highly oxidative tissues and is thought to regulate CLD lipolysis and FA transfer into the mitochondria for FA β-oxidation [84,168,169]. Control of mitochondrial interactions with CLDs and FA entry into the mitochondria by Plin5 depends on metabolic state. In a basal state, Plin5 promotes TAG storage in CLDs by reducing CLD lipolysis and FAO [84]. When lipolysis is stimulated, mitochondria are recruited to CLDs by Plin5 to allow for FA transfer and FAO to occur [84,169]. In addition to providing FA for their oxidation during stimulated CLD breakdown, CLD–mitochondrial interactions by Plin5 also facilitate CLD growth. Mitochondria interacting with CLDs in brown adipose tissue produce ATP to power the energy demands of TAG synthesis, stimulating CLD expansion and growth [170]. These results shed new light on the role CLD–mitochondrial interactions play in lipid metabolism. Overall, Plin5 is an important component of the CLD proteome in oxidative tissues enabling connections to the mitochondria upon metabolic flux and changing energy demand.

### *CLDs as a protein storage reservoir*

While many proteins that associate with CLDs are active in lipid synthesis [58], CLD breakdown [20], or CLD dynamics [81], some proteins that associate with CLDs may have no functional role in CLD metabolism but instead are targeted to the CLD surface for storage or degradation [171]. This unique function of CLDs in protein sequestration allows them to regulate cellular protein levels. An example of a group of proteins that is stored on the CLD surface is histone proteins, which serves as a regulatory mechanism to control gene transcription in growing *Drosophila* embryos. Excess histone proteins H2A, H2Av, and H2B collect on the CLD surface in *Drosophila* embryos and are later released from CLDs and translocate to the nucleus as development progresses [172]. The CLD protein Jabba mediates the localization and binding of histones to the CLD surface [173]. The ability of CLDs to harbor histone proteins during different stages of development suggests that CLDs are able to regulate intracellular protein localization. In addition to *Drosophila*, histone proteins are consistently identified in CLD proteomic studies of numerous cell types [2], suggesting their storage on CLDs may be a common regulatory mechanism.

In addition to a depot for histones, CLDs can also act as reservoir for proteins that are targeted for degradation. Multiple proteins are degraded on CLDs as a mechanism to control the composition of the CLD proteome [74]. However, proteins may also be specifically targeted to the CLD surface before they are degraded. One example of a protein that localizes to CLDs for degradation in hepatocytes is apoB-100, the structural protein of VLDL. ApoB-100 is stabilized by the packaging of TAG and CE onto VLDL itself, and in the absence of adequate lipidation, apoB-100 is targeted for ubiquitination and proteosomal degradation [174]. CLDs are involved in an intermediate step in the degradation of apoB-100. In Huh7 human hepatoma cells, ubiquitinated apoB-100 forms crescent-like structures around CLDs [175,176]. The presence of these apoB-crescents around CLDs is exaggerated when proteosomal or autosomal degradation is inhibited, suggesting localization of apoB-100 to the CLD may prevent apoB-100 accumulation in other cellular locations when its efficient degradation is compromised [175]. The ability of CLDs to harbor proteins for storage and future use or in preparation for their degradation makes them important regulators of protein quality control.

### *Functions of other CLD proteins*

Although the role of Plins as structural CLD proteins has been recognized and the function of certain lipid metabolism proteins in CLD remodeling has been studied, the function of most other CLD-associated proteins remains fairly enigmatic. Advancements in mass spectrometry technology and proteomic software have enabled the identification of numerous CLD proteins in various cell types, continually expanding the list of CLD proteome constituents [2]. However, the tendency for contamination with other organelle proteins during the CLD isolation procedure and lack of functional studies makes the role of these proteins on CLDs difficult to delineate. Recently, a list of high-confidence CLD-associated proteins was developed by Bersuker et al. using a proximity labeling strategy in U2OS, Huh7, and HEK293 cells [78]. This study demonstrates new mechanisms of identifying bona-fide CLD-associated proteins, which can help define a list of functionally relevant players regulating CLD metabolism to study in future experiments. Further, the proteins identified by Bersuker et al. are members of protein functional groups that are repeatedly identified in CLD proteomic studies in different cell types, suggesting that they may play a core role in CLD function. Among the protein groups commonly identified are those involved in lipid turnover, histone proteins, ribosomal proteins, structural proteins, and proteins involved in protein degradation, signaling, membrane trafficking, and CLD dynamics [2,78]. Determining the functional significance of proteins that associate with CLDs is a necessary, albeit challenging, next step in CLD research.

## Conclusions and future perspectives

CLDs are major contributors to the physiological role of certain tissues in maintaining systemic lipid homeostasis. Each step in CLD dynamics from neutral lipid synthesis, formation at the ER, maintenance in the cytosol, and fate of the lipid stored in CLDs is a mechanism to control cellular lipid concentrations. Regulation of CLD metabolism in major metabolic tissues maintains a balance between TAG storage and utilization, in turn maintaining cellular and systemic lipid homeostasis. TAG storage in adipocyte CLDs fluctuates depending on metabolic state to regularly provide cells of the body with FA substrate upon energy demand. CLDs in cardiac and skeletal myocytes provide a readily available source of cellular energy that is essential to power their high energy demands, necessitating a tight balance between TAG storage and utilization. Lastly, CLDs in enterocytes and hepatocytes safely sequester high concentrations of lipid present during opposing physiological states, which can be used for lipoprotein synthesis and secretion at later times to provide an additional source of cellular FA to the body's tissues. Changes in CLD dynamics in these cell types during pathological states such as obesity and diabetes alter the balance between TAG storage and utilization, contributing to abnormal cellular and systemic lipid levels and negative health consequences. Future studies into the regulation of intracellular lipid storage and utilization will hope to define the molecular mechanisms controlling specific aspects of CLD dynamics that can be targeted to prevent and treat metabolic disease.

## References


[1] Cohen S. Lipid droplets as organelles. Int Rev Cell Mol Biol 2018;337:83–110.

[2] Zhang C, Liu P. The new face of the lipid droplet: lipid droplet proteins. Proteomics 2019;19(10):e1700223.

[3] Olzmann JA, Carvalho P. Dynamics and functions of lipid droplets. Nat Rev Mol Cell Biol 2019;20(3):137–55.

[4] Welte MA, Gould AP. Lipid droplet functions beyond energy storage. Biochim Biophys Acta Mol Cell Biol Lipids 2017;1862(10 Pt B):1260–72.

[5] Walther TC, Chung J, Farese RV Jr. Lipid droplet biogenesis. Annu Rev Cell Dev Biol 2017;33:491–510.

[6] Thiam AR, Farese RV Jr, Walther TC. The biophysics and cell biology of lipid droplets. Nat Rev Mol Cell Biol 2013;14(12):775–86.

[7] Krahmer N, Farese RV Jr, Walther TC. Balancing the fat: lipid droplets and human disease. EMBO Mol Med 2013;5(7):973–83.

[8] Duncan RE, Ahmadian M, Jaworski K, Sarkadi-Nagy E, Sul HS. Regulation of lipolysis in adipocytes. Annu Rev Nutr 2007;27:79–101.

[9] Bosma M. Lipid droplet dynamics in skeletal muscle. Exp Cell Res 2016;340(2):180–6.

[10] D'Aquila T, Hung YH, Carreiro A, Buhman KK. Recent discoveries on absorption of dietary fat: presence, synthesis, and metabolism of cytoplasmic lipid droplets within enterocytes. Biochim Biophys Acta 2016;1861(8 Pt A):730–47.

[11] Gibbons GF, Islam K, Pease RJ. Mobilisation of triacylglycerol stores. Biochim Biophys Acta 2000;1483(1):37–57.

[12] Bray GA. Medical consequences of obesity. J Clin Endocrinol Metab 2004;89(6):2583–9.

[13] Petersen MC, Shulman GI. Mechanisms of insulin action and insulin resistance. Physiol Rev 2018;98(4):2133–223.

[14] Kawano Y, Cohen DE. Mechanisms of hepatic triglyceride accumulation in non-alcoholic fatty liver disease. J Gastroenterol 2013;48(4):434–41.

[15] Jia G, Hill MA, Sowers JR. Diabetic cardiomyopathy: an update of mechanisms contributing to this clinical entity. Circ Res 2018;122(4):624–38.

[16] Bergheanu SC, Bodde MC, Jukema JW. Pathophysiology and treatment of atherosclerosis: current view and future perspective on lipoprotein modification treatment. Neth Heart J 2017;25(4):231–42.

[17] Petan T, Jarc E, Jusovic M. Lipid droplets in cancer: guardians of fat in a stressful world. Molecules 2018;23(8):E1941.

[18] Kwiterovich PO Jr. The metabolic pathways of high-density lipoprotein, low-density lipoprotein, and triglycerides: a current review. Am J Cardiol 2000;86(12a):5l–10l.

[19] Kersten S. Physiological regulation of lipoprotein lipase. Biochim Biophys Acta 2014;1841(7):919–33.

[20] Young SG, Zechner R. Biochemistry and pathophysiology of intravascular and intracellular lipolysis. Genes Dev 2013;27(5):459–84.

[21] Brown MS, Goldstein JL. Receptor-mediated endocytosis: insights from the lipoprotein receptor system. Proc Natl Acad Sci U S A 1979;76(7):3330–7.

[22] Fruhbeck G, Mendez-Gimenez L, Fernandez-Formoso JA, Fernandez S, Rodriguez A. Regulation of adipocyte lipolysis. Nutr Res Rev 2014;27(1):63–93.
[23] Morigny P, Houssier M, Mouisel E, Langin D. Adipocyte lipolysis and insulin resistance. Biochimie 2016;125:259–66.
[24] Ameer F, Scandiuzzi L, Hasnain S, Kalbacher H, Zaidi N. De novo lipogenesis in health and disease. Metabolism 2014;63(7):895–902.
[25] Buhaescu I, Izzedine H. Mevalonate pathway: a review of clinical and therapeutical implications. Clin Biochem 2007;40(9–10):575–84.
[26] Ikonen E. Cellular cholesterol trafficking and compartmentalization. Nat Rev Mol Cell Biol 2008;9(2):125–38.
[27] Postic C, Girard J. Contribution of de novo fatty acid synthesis to hepatic steatosis and insulin resistance: lessons from genetically engineered mice. J Clin Invest 2008;118(3):829–38.
[28] Beloribi-Djefaflia S, Vasseur S, Guillaumond F. Lipid metabolic reprogramming in cancer cells. Oncogenesis 2016;5:e189.
[29] Lunt SY, Vander Heiden MG. Aerobic glycolysis: meeting the metabolic requirements of cell proliferation. Annu Rev Cell Dev Biol 2011;27:441–64.
[30] Wang H, Airola MV, Reue K. How lipid droplets "TAG" along: Glycerolipid synthetic enzymes and lipid storage. Biochim Biophys Acta Mol Cell Biol Lipids 2017;1862(10 Pt B):1131–45.
[31] Yen CL, Stone SJ, Koliwad S, Harris C, Farese RV Jr. Thematic review series: glycerolipids. DGAT enzymes and triacylglycerol biosynthesis. J Lipid Res 2008;49(11):2283–301.
[32] Yen CL, Monetti M, Burri BJ, Farese RV Jr. The triacylglycerol synthesis enzyme DGAT1 also catalyzes the synthesis of diacylglycerols, waxes, and retinyl esters. J Lipid Res 2005;46(7):1502–11.
[33] Man WC, Miyazaki M, Chu K, Ntambi J. Colocalization of SCD1 and DGAT2: implying preference for endogenous monounsaturated fatty acids in triglyceride synthesis. J Lipid Res 2006;47(9):1928–39.
[34] Yen CL, Nelson DW, Yen MI. Intestinal triacylglycerol synthesis in fat absorption and systemic energy metabolism. J Lipid Res 2015;56(3):489–501.
[35] Rogers MA, Liu J, Song BL, Li BL, Chang CC, Chang TY. Acyl-CoA: cholesterol acyltransferases (ACATs/SOATs): enzymes with multiple sterols as substrates and as activators. J Steroid Biochem Mol Biol 2015;151:102–7.
[36] Shen WJ, Azhar S, Kraemer FB. Lipid droplets and steroidogenic cells. Exp Cell Res 2016;340(2):209–14.
[37] Jackson CL. Lipid droplet biogenesis. Curr Opin Cell Biol 2019;59:88–96.
[38] Thiam AR, Foret L. The physics of lipid droplet nucleation, growth and budding. Biochim Biophys Acta 2016;1861(8 Pt A):715–22.
[39] Choudhary V, Ojha N, Golden A, Prinz WA. A conserved family of proteins facilitates nascent lipid droplet budding from the ER. J Cell Biol 2015;211(2):261–71.
[40] Choudhary V, Golani G, Joshi AS, Cottier S, Schneiter R, Prinz WA, et al. Architecture of lipid droplets in endoplasmic reticulum is determined by phospholipid intrinsic curvature. Curr Biol 2018;28(6):915–26. e9.
[41] Ben M'barek K, Ajjaji D, Chorlay A, Vanni S, Foret L, Thiam AR. ER membrane phospholipids and surface tension control cellular lipid droplet formation. Dev Cell 2017;41(6):591–604. e7.
[42] Nettebrock NT, Bohnert M. Born this way - Biogenesis of lipid droplets from specialized ER subdomains. Biochim Biophys Acta Mol Cell Biol Lipids 2020;1865(1):S1388–1981. (19)30061-7.
[43] Deslandes F, Thiam AR, Foret L. Lipid droplets can spontaneously bud off from a symmetric bilayer. Biophys J 2017;113(1):15–8.
[44] Chen X, Goodman JM. The collaborative work of droplet assembly. Biochim Biophys Acta Mol Cell Biol Lipids 2017;1862(10 Pt B):1205–11.
[45] Salo VT, Li S, Vihinen H, Holtta-Vuori M, Szkalisity A, Horvath P, et al. Seipin facilitates triglyceride flow to lipid droplet and counteracts droplet ripening via endoplasmic reticulum contact. Dev Cell 2019;50(4):478–93. e9.
[46] Magre J, Delepine M, Khallouf E, Gedde-Dahl T Jr, Van Maldergem L, Sobel E, et al. Identification of the gene altered in Berardinelli-Seip congenital lipodystrophy on chromosome 11q13. Nat Genet 2001;28(4):365–70.
[47] Sui X, Arlt H, Brock KP, Lai ZW, DiMaio F, Marks DS, et al. Cryo-electron microscopy structure of the lipid droplet-formation protein seipin. J Cell Biol 2018;217(12):4080–91.
[48] Yan R, Qian H, Lukmantara I, Gao M, Du X, Yan N, et al. Human SEIPIN: binds anionic phospholipids. Dev Cell 2018;47(2):248–56. e4.
[49] Salo VT, Belevich I, Li S, Karhinen L, Vihinen H, Vigouroux C, et al. Seipin regulates ER-lipid droplet contacts and cargo delivery. EMBO J 2016;35(24):2699–716.

[50] Wang H, Becuwe M, Housden BE, Chitraju C, Porras AJ, Graham MM, et al. Seipin is required for converting nascent to mature lipid droplets. Elife 2016;5:e16582.
[51] Kadereit B, Kumar P, Wang WJ, Miranda D, Snapp EL, Severina N, et al. Evolutionarily conserved gene family important for fat storage. Proc Natl Acad Sci U S A 2008;105(1):94–9.
[52] Gross DA, Zhan C, Silver DL. Direct binding of triglyceride to fat storage-inducing transmembrane proteins 1 and 2 is important for lipid droplet formation. Proc Natl Acad Sci U S A 2011;108(49):19581–6.
[53] Xu D, Li Y, Wu L, Zhao D, Yu J, Huang T, et al. Rab18 promotes lipid droplet (LD) growth by tethering the ER to LDs through SNARE and NRZ interactions. J Cell Biol 2018;217(3):975–95.
[54] Li D, Zhao YG, Zhao H, Huang J, Miao G, Feng D, et al. The ER-localized protein DFCP1 modulates ER-lipid droplet contact formation. Cell Rep 2019;27(2):343–58. e5.
[55] Teixeira V, Johnsen L, Martinez-Montanes F, Grippa A, Buxo L, Idrissi FZ, et al. Regulation of lipid droplets by metabolically controlled Ldo isoforms. J Cell Biol 2018;217(1):127–38.
[56] Nguyen TB, Olzmann JA. Getting a handle on lipid droplets: insights into ER-lipid droplet tethering. J Cell Biol 2019;218(4):1089–91.
[57] Moving out but keeping in touch: contacts between endoplasmic reticulum and lipid droplets. Current Opinion in Cell Biology. 2019;57:64–70.
[58] Wilfling F, Wang H, Haas JT, Krahmer N, Gould TJ, Uchida A, et al. Triacylglycerol synthesis enzymes mediate lipid droplet growth by relocalizing from the ER to lipid droplets. Dev Cell 2013;24(4):384–99.
[59] Thiam AR, Antonny B, Wang J, Delacotte J, Wilfling F, Walther TC, et al. COPI buds 60-nm lipid droplets from reconstituted water-phospholipid-triacylglyceride interfaces, suggesting a tension clamp function. Proc Natl Acad Sci U S A 2013;110(33):13244–9.
[60] Wilfling F, Thiam AR, Olarte MJ, Wang J, Beck R, Gould TJ, et al. Arf1/COPI machinery acts directly on lipid droplets and enables their connection to the ER for protein targeting. Elife 2014;3:e01607.
[61] McFie PJ, Banman SL, Stone SJ. Diacylglycerol acyltransferase-2 contains a c-terminal sequence that interacts with lipid droplets. Biochim Biophys Acta Mol Cell Biol Lipids 2018;1863(9):1068–81.
[62] Zhang S, Wang Y, Cui L, Deng Y, Xu S, Yu J, et al. Morphologically and functionally distinct lipid droplet subpopulations. Sci Rep 2016;6:29539.
[63] Gao G, Chen FJ, Zhou L, Su L, Xu D, Xu L, et al. Control of lipid droplet fusion and growth by CIDE family proteins. Biochim Biophys Acta Mol Cell Biol Lipids 2017;1862(10 Pt B):1197–204.
[64] Gong J, Sun Z, Wu L, Xu W, Schieber N, Xu D, et al. Fsp27 promotes lipid droplet growth by lipid exchange and transfer at lipid droplet contact sites. J Cell Biol 2011;195(6):953–63.
[65] Nishino N, Tamori Y, Tateya S, Kawaguchi T, Shibakusa T, Mizunoya W, et al. FSP27 contributes to efficient energy storage in murine white adipocytes by promoting the formation of unilocular lipid droplets. J Clin Invest 2008;118(8):2808–21.
[66] Sun Z, Gong J, Wu H, Xu W, Wu L, Xu D, et al. Perilipin1 promotes unilocular lipid droplet formation through the activation of Fsp27 in adipocytes. Nat Commun 2013;4:1594.
[67] Toh SY, Gong J, Du G, Li JZ, Yang S, Ye J, et al. Up-regulation of mitochondrial activity and acquirement of brown adipose tissue-like property in the white adipose tissue of fsp27 deficient mice. PLoS One 2008;3(8):e2890.
[68] Vance JE. Phospholipid synthesis and transport in mammalian cells. Traffic 2015;16(1):1–18.
[69] Guo Y, Walther TC, Rao M, Stuurman N, Goshima G, Terayama K, et al. Functional genomic screen reveals genes involved in lipid-droplet formation and utilization. Nature 2008;453(7195):657–61.
[70] Krahmer N, Guo Y, Wilfling F, Hilger M, Lingrell S, Heger K, et al. Phosphatidylcholine synthesis for lipid droplet expansion is mediated by localized activation of CTP:phosphocholine cytidylyltransferase. Cell Metab 2011;14(4):504–15.
[71] Aitchison AJ, Arsenault DJ, Ridgway ND, Nuclear-localized CTP. phosphocholine cytidylyltransferase alpha regulates phosphatidylcholine synthesis required for lipid droplet biogenesis. Mol Biol Cell 2015;26(16):2927–38.
[72] Kory N, Farese RV Jr, Walther TC, Targeting Fat:. Mechanisms of protein localization to lipid droplets. Trends Cell Biol 2016;26(7):535–46.
[73] Xu S, Zhang X, Liu P. Lipid droplet proteins and metabolic diseases. Biochim Biophys Acta Mol Basis Dis 2018;1864(5 Pt B):1968–83.
[74] Bersuker K, Olzmann JA. Establishing the lipid droplet proteome: mechanisms of lipid droplet protein targeting and degradation. Biochim Biophys Acta Mol Cell Biol Lipids 2017;1862(10 Pt B):1166–77.
[75] Jacquier N, Choudhary V, Mari M, Toulmay A, Reggiori F, Schneiter R. Lipid droplets are functionally connected to the endoplasmic reticulum in Saccharomyces cerevisiae. J Cell Sci 2011;124(Pt 14):2424–37.

[76] Prevost C, Sharp ME, Kory N, Lin Q, Voth GA, Farese RV Jr, et al. Mechanism and determinants of amphipathic helix-containing protein targeting to lipid droplets. Dev Cell 2018;44(1):73–86. e4.
[77] Ruggiano A, Mora G, Buxo L, Carvalho P. Spatial control of lipid droplet proteins by the ERAD ubiquitin ligase Doa10. EMBO J 2016;35(15):1644–55.
[78] Bersuker K, Peterson CWH, To M, Sahl SJ, Savikhin V, Grossman EA, et al. A proximity labeling strategy provides insights into the composition and dynamics of lipid droplet proteomes. Dev Cell 2018;44(1):97–112. e7.
[79] Kory N, Thiam AR, Farese RV Jr, Walther TC. Protein crowding is a determinant of lipid droplet protein composition. Dev Cell 2015;34(3):351–63.
[80] Itabe H, Yamaguchi T, Nimura S, Sasabe N, Perilipins:. a diversity of intracellular lipid droplet proteins. Lipids Health Dis 2017;16(1):83.
[81] Sztalryd C, Brasaemle DL. The perilipin family of lipid droplet proteins: gatekeepers of intracellular lipolysis. Biochim Biophys Acta Mol Cell Biol Lipids 2017;1862(10 Pt B):1221–32.
[82] Greenberg AS, Coleman RA, Kraemer FB, McManaman JL, Obin MS, Puri V, et al. The role of lipid droplets in metabolic disease in rodents and humans. J Clin Invest 2011;121(6):2102–10.
[83] Listenberger LL, Ostermeyer-Fay AG, Goldberg EB, Brown WJ, Brown DA. Adipocyte differentiation-related protein reduces the lipid droplet association of adipose triglyceride lipase and slows triacylglycerol turnover. J Lipid Res 2007;48(12):2751–61.
[84] Wang H, Sreenivasan U, Hu H, Saladino A, Polster BM, Lund LM, et al. Perilipin 5, a lipid droplet-associated protein, provides physical and metabolic linkage to mitochondria. J Lipid Res 2011;52(12):2159–68.
[85] Brasaemle DL, Barber T, Wolins NE, Serrero G, Blanchette-Mackie EJ, Londos C. Adipose differentiation-related protein is an ubiquitously expressed lipid storage droplet-associated protein. J Lipid Res 1997;38(11):2249–63.
[86] Lee B, Zhu J, Wolins NE, Cheng JX, Buhman KK. Differential association of adipophilin and TIP47 proteins with cytoplasmic lipid droplets in mouse enterocytes during dietary fat absorption. Biochim Biophys Acta 2009;1791(12):1173–80.
[87] D'Aquila T, Zembroski AS, Buhman KK. Diet induced obesity alters intestinal cytoplasmic lipid droplet morphology and proteome in the postprandial response to dietary fat. Front Physiol 2019;10:180.
[88] Ajjaji D, Ben M'barek K, Mimmack ML, England C, Herscovitz H, Dong L, et al. Dual binding motifs underpin the hierarchical association of perilipins1-3 with lipid droplets. Mol Biol Cell 2019;30(5):703–16.
[89] Schulze RJ, Sathyanarayan A, Mashek DG. Breaking fat: the regulation and mechanisms of lipophagy. Biochim Biophys Acta Mol Cell Biol Lipids 2017;1862(10 Pt B):1178–87.
[90] Schulze RJ, Drizyte K, Casey CA, McNiven MA. Hepatic Lipophagy: New insights into autophagic catabolism of lipid droplets in the liver. Hepatol Commun 2017;1(5):359–69.
[91] Singh R, Kaushik S, Wang Y, Xiang Y, Novak I, Komatsu M, et al. Autophagy regulates lipid metabolism. Nature 2009;458(7242):1131–5.
[92] Martinez-Lopez N, Singh R. Autophagy and lipid droplets in the liver. Annu Rev Nutr 2015;35:215–37.
[93] Martinez-Vicente M, Talloczy Z, Wong E, Tang G, Koga H, Kaushik S, et al. Cargo recognition failure is responsible for inefficient autophagy in Huntington's disease. Nat Neurosci 2010;13(5):567–76.
[94] Zhou K, Yao P, He J, Zhao H. Lipophagy in nonliver tissues and some related diseases: pathogenic and therapeutic implications. J Cell Physiol 2019;234(6):7938–47.
[95] Khaldoun SA, Emond-Boisjoly MA, Chateau D, Carrière V, Lacasa M, Rousset M, et al. Autophagosomes contribute to intracellular lipid distribution in enterocytes. Mol Biol Cell 2014;25(1):118–32.
[96] Hung YH, Buhman KK. DGAT1 deficiency disrupts lysosome function in enterocytes during dietary fat absorption. Biochim Biophys Acta Mol Cell Biol Lipids 2019;1864(4):587–95.
[97] Zechner R, Madeo F, Kratky D. Cytosolic lipolysis and lipophagy: two sides of the same coin. Nat Rev Mol Cell Biol 2017;18(11):671–84.
[98] Ong KT, Mashek MT, Bu SY, Greenberg AS, Mashek DG. Adipose triglyceride lipase is a major hepatic lipase that regulates triacylglycerol turnover and fatty acid signaling and partitioning. Hepatology 2011;53(1):116–26.
[99] Sathyanarayan A, Mashek MT, Mashek DG. ATGL promotes autophagy/lipophagy via SIRT1 to control hepatic lipid droplet catabolism. Cell Rep 2017;19(1):1–9.
[100] Lee IH, Cao L, Mostoslavsky R, Lombard DB, Liu J, Bruns NE, et al. A role for the NAD-dependent deacetylase Sirt1 in the regulation of autophagy. Proc Natl Acad Sci U S A 2008;105(9):3374–9.
[101] Kaushik S, Cuervo AM. Degradation of lipid droplet-associated proteins by chaperone-mediated autophagy facilitates lipolysis. Nat Cell Biol 2015;17(6):759–70.

[102] Kraemer FB, Shen WJ, Harada K, Patel S, Osuga J, Ishibashi S, et al. Hormone-sensitive lipase is required for high-density lipoprotein cholesteryl ester-supported adrenal steroidogenesis. Mol Endocrinol 2004;18(3):549–57.
[103] Sekiya M, Osuga J, Igarashi M, Okazaki H, Ishibashi S. The role of neutral cholesterol ester hydrolysis in macrophage foam cells. J Atheroscler Thromb 2011;18(5):359–64.
[104] Ouimet M, Marcel YL. Regulation of lipid droplet cholesterol efflux from macrophage foam cells. Arterioscler Thromb Vasc Biol 2012;32(3):575–81.
[105] Li F, Zhang H. Lysosomal acid lipase in lipid metabolism and beyond. Arterioscler Thromb Vasc Biol 2019;39(5):850–6.
[106] Sekiya M, Osuga J, Nagashima S, Ohshiro T, Igarashi M, Okazaki H, et al. Ablation of neutral cholesterol ester hydrolase 1 accelerates atherosclerosis. Cell Metab 2009;10(3):219–28.
[107] Houten SM, Violante S, Ventura FV, Wanders RJ. The biochemistry and physiology of mitochondrial fatty acid beta-oxidation and its genetic disorders. Annu Rev Physiol 2016;78:23–44.
[108] Balasse EO, Fery F. Ketone body production and disposal: effects of fasting, diabetes, and exercise. Diabetes Metab Rev 1989;5(3):247–70.
[109] Schulze PC, Drosatos K, Goldberg IJ. Lipid use and misuse by the heart. Circ Res 2016;118(11):1736–51.
[110] Daemen S, van Polanen N, Hesselink MKC. The effect of diet and exercise on lipid droplet dynamics in human muscle tissue. J Exp Biol 2018;221(Pt Suppl 1). jeb167015.
[111] Rakhshandehroo M, Knoch B, Muller M, Kersten S. Peroxisome proliferator-activated receptor alpha target genes. PPAR Res 2010;2010:612089.
[112] Grygiel-Gorniak B. Peroxisome proliferator-activated receptors and their ligands: nutritional and clinical implications—a review. Nutr J 2014;13:17.
[113] Haemmerle G, Moustafa T, Woelkart G, Buttner S, Schmidt A, van de Weijer T, et al. ATGL-mediated fat catabolism regulates cardiac mitochondrial function via PPAR-alpha and PGC-1. Nat Med 2011;17(9):1076–85.
[114] Lundsgaard AM, Fritzen AM, Kiens B. Molecular regulation of fatty acid oxidation in skeletal muscle during aerobic exercise. Trends Endocrinol Metab 2018;29(1):18–30.
[115] Scott W, Stevens J, Binder-Macleod SA. Human skeletal muscle fiber type classifications. Phys Ther 2001;81(11):1810–6.
[116] Shaw CS, Jones DA, Wagenmakers AJ. Network distribution of mitochondria and lipid droplets in human muscle fibres. Histochem Cell Biol 2008;129(1):65–72.
[117] De Bock K, Richter EA, Russell AP, Eijnde BO, Derave W, Ramaekers M, et al. Exercise in the fasted state facilitates fibre type-specific intramyocellular lipid breakdown and stimulates glycogen resynthesis in humans. J Physiol 2005;564(Pt 2):649–60.
[118] van Loon LJ, Koopman R, Stegen JH, Wagenmakers AJ, Keizer HA, Saris WH. Intramyocellular lipids form an important substrate source during moderate intensity exercise in endurance-trained males in a fasted state. J Physiol 2003;553(Pt 2):611–25.
[119] Koh HE, Nielsen J, Saltin B, Holmberg HC, Ortenblad N. Pronounced limb and fibre type differences in subcellular lipid droplet content and distribution in elite skiers before and after exhaustive exercise. J Physiol 2017;595(17):5781–95.
[120] Herbison GJ, Jaweed MM, Ditunno JF. Muscle fiber types. Arch Phys Med Rehabil 1982;63(5):227–30.
[121] Turnbull PC, Longo AB, Ramos SV, Roy BD, Ward WE, Peters SJ. Increases in skeletal muscle ATGL and its inhibitor G0S2 following 8 weeks of endurance training in metabolically different rat skeletal muscles. Am J Physiol Regul Integr Comp Physiol 2016;310(2):R125–33.
[122] Alsted TJ, Nybo L, Schweiger M, Fledelius C, Jacobsen P, Zimmermann R, et al. Adipose triglyceride lipase in human skeletal muscle is upregulated by exercise training. Am J Physiol Endocrinol Metab 2009;296(3):E445–53.
[123] Meex RC, Hoy AJ, Mason RM, Martin SD, McGee SL, Bruce CR, et al. ATGL-mediated triglyceride turnover and the regulation of mitochondrial capacity in skeletal muscle. Am J Physiol Endocrinol Metab 2015;308(11):E960–70.
[124] Tarnopolsky MA, Rennie CD, Robertshaw HA, Fedak-Tarnopolsky SN, Devries MC, Hamadeh MJ. Influence of endurance exercise training and sex on intramyocellular lipid and mitochondrial ultrastructure, substrate use, and mitochondrial enzyme activity. Am J Physiol Regul Integr Comp Physiol 2007;292(3):R1271–8.
[125] Bosma M, Minnaard R, Sparks LM, Schaart G, Losen M, de Baets MH, et al. The lipid droplet coat protein perilipin 5 also localizes to muscle mitochondria. Histochem Cell Biol 2012;137(2):205–16.
[126] Bosma M, Kersten S, Hesselink MK, Schrauwen P. Re-evaluating lipotoxic triggers in skeletal muscle: relating intramyocellular lipid metabolism to insulin sensitivity. Prog Lipid Res 2012;51(1):36–49.

[127] van Loon LJ, Goodpaster BH. Increased intramuscular lipid storage in the insulin-resistant and endurance-trained state. Pflugers Arch 2006;451(5):606–16.
[128] Goodpaster BH, He J, Watkins S, Kelley DE. Skeletal muscle lipid content and insulin resistance: evidence for a paradox in endurance-trained athletes. J Clin Endocrinol Metab 2001;86(12):5755–61.
[129] Coen PM, Goodpaster BH. Role of intramyocellular lipids in human health. Trends Endocrinol Metab 2012;23(8):391–8.
[130] Zhu J, Lee B, Buhman KK, Cheng JX. A dynamic, cytoplasmic triacylglycerol pool in enterocytes revealed by ex vivo and in vivo coherent anti-Stokes Raman scattering imaging. J Lipid Res 2009;50(6):1080–9.
[131] Xiao C, Stahel P, Carreiro AL, Buhman KK, Lewis GF. Recent advances in triacylglycerol mobilization by the gut. Trends Endocrinol Metab 2018;29(3):151–63.
[132] Fielding BA, Callow J, Owen RM, Samra JS, Matthews DR, Frayn KN. Postprandial lipemia: the origin of an early peak studied by specific dietary fatty acid intake during sequential meals. Am J Clin Nutr 1996;63(1):36–41.
[133] Xiao C, Stahel P, Carreiro AL, Hung YH, Dash S, Bookman I, et al. Oral glucose mobilizes triglyceride stores from the human intestine. Cell Mol Gastroenterol Hepatol 2019;7(2):313–37.
[134] Robertson MD, Parkes M, Warren BF, Ferguson DJ, Jackson KG, Jewell DP, et al. Mobilisation of enterocyte fat stores by oral glucose in humans. Gut 2003;52(6):834–9.
[135] Chavez-Jauregui RN, Mattes RD, Parks EJ. Dynamics of fat absorption and effect of sham feeding on postprandial lipema. Gastroenterology 2010;139(5):1538–48.
[136] Dash S, Xiao C, Morgantini C, Connelly PW, Patterson BW, Lewis GF. Glucagon-like peptide-2 regulates release of chylomicrons from the intestine. Gastroenterology 2014;147(6):1275–84. e4.
[137] Xiao C, Dash S, Morgantini C, Lewis GF. Novel role of enteral monosaccharides in intestinal lipoprotein production in healthy humans. Arterioscler Thromb Vasc Biol 2013;33(5):1056–62.
[138] Xiao C, Dash S, Morgantini C, Lewis GF. Intravenous glucose acutely stimulates intestinal lipoprotein secretion in healthy humans. Arterioscler Thromb Vasc Biol 2016;36(7):1457–63.
[139] Adeli K, Lewis GF. Intestinal lipoprotein overproduction in insulin-resistant states. Curr Opin Lipidol 2008;19(3):221–8.
[140] Obrowsky S, Chandak PG, Patankar JV, Povoden S, Schlager S, Kershaw EE, et al. Adipose triglyceride lipase is a TG hydrolase of the small intestine and regulates intestinal PPARalpha signaling. J Lipid Res 2013;54(2):425–35.
[141] Grober J, Lucas S, Sorhede-Winzell M, Zaghini I, Mairal A, Contreras JA, et al. Hormone-sensitive lipase is a cholesterol esterase of the intestinal mucosa. J Biol Chem 2003;278(8):6510–5.
[142] Obrowsky S, Chandak PG, Patankar JV, Pfeifer T, Povoden S, Schreiber R, et al. Cholesteryl ester accumulation and accelerated cholesterol absorption in intestine-specific hormone sensitive lipase-null mice. Biochim Biophys Acta 2012;1821(11):1406–14.
[143] Douglass JD, Zhou YX, Wu A, Zadroga JA, Gajda AM, Lackey AI, et al. Global deletion of MGL in mice delays lipid absorption and alters energy homeostasis and diet-induced obesity. J Lipid Res 2015;56(6):1153–71.
[144] Lee B, Fast AM, Zhu J, Cheng JX, Buhman KK. Intestine-specific expression of acyl CoA:diacylglycerol acyltransferase 1 reverses resistance to diet-induced hepatic steatosis and obesity in Dgat1-/- mice. J Lipid Res 2010;51(7):1770–80.
[145] Hung YH, Carreiro AL, Buhman KK. Dgat1 and Dgat2 regulate enterocyte triacylglycerol distribution and alter proteins associated with cytoplasmic lipid droplets in response to dietary fat. Biochim Biophys Acta 2017;1862(6):600–14.
[146] Gibbons GF, Wiggins D. Intracellular triacylglycerol lipase: its role in the assembly of hepatic very-low-density lipoprotein (VLDL). Adv Enzyme Regul 1995;35:179–98.
[147] Wiggins D, Gibbons GF. The lipolysis/esterification cycle of hepatic triacylglycerol. Its role in the secretion of very-low-density lipoprotein and its response to hormones and sulphonylureas. Biochem J 1992;284(Pt 2):457–62.
[148] Choi SH, Ginsberg HN. Increased very low density lipoprotein (VLDL) secretion, hepatic steatosis, and insulin resistance. Trends Endocrinol Metab 2011;22(9):353–63.
[149] Magnusson B, Asp L, Bostrom P, Ruiz M, Stillemark-Billton P, Linden D, et al. Adipocyte differentiation-related protein promotes fatty acid storage in cytosolic triglycerides and inhibits secretion of very low-density lipoproteins. Arterioscler Thromb Vasc Biol 2006;26(7):1566–71.
[150] Chang BH, Li L, Saha P, Chan L. Absence of adipose differentiation related protein upregulates hepatic VLDL secretion, relieves hepatosteatosis, and improves whole body insulin resistance in leptin-deficient mice. J Lipid Res 2010;51(8):2132–42.

[151] Reid BN, Ables GP, Otlivanchik OA, Schoiswohl G, Zechner R, Blaner WS, et al. Hepatic overexpression of hormone-sensitive lipase and adipose triglyceride lipase promotes fatty acid oxidation, stimulates direct release of free fatty acids, and ameliorates steatosis. J Biol Chem 2008;283(19):13087–99.
[152] Lehner R, Cui Z, Vance DE. Subcellular localization, developmental expression and characterization of a liver triacylglycerol hydrolase. Biochem J 1999;338(Pt 3):761–8.
[153] Alam M, Ho S, Vance DE, Lehner R. Heterologous expression, purification, and characterization of human triacylglycerol hydrolase. Protein Expr Purif 2002;24(1):33–42.
[154] Lian J, Nelson R, Lehner R. Carboxylesterases in lipid metabolism: from mouse to human. Protein Cell 2018;9(2):178–95.
[155] Wei E, Alam M, Sun F, Agellon LB, Vance DE, Lehner R. Apolipoprotein B and triacylglycerol secretion in human triacylglycerol hydrolase transgenic mice. J Lipid Res 2007;48(12):2597–606.
[156] Gilham D, Ho S, Rasouli M, Martres P, Vance DE, Lehner R. Inhibitors of hepatic microsomal triacylglycerol hydrolase decrease very low density lipoprotein secretion. FASEB J 2003;17(12):1685–7.
[157] Wang H, Gilham D, Lehner R. Proteomic and lipid characterization of apolipoprotein B-free luminal lipid droplets from mouse liver microsomes: implications for very low density lipoprotein assembly. J Biol Chem 2007;282(45):33218–26.
[158] Lehner R, Lian J, Quiroga AD. Lumenal lipid metabolism: implications for lipoprotein assembly. Arterioscler Thromb Vasc Biol 2012;32(5):1087–93.
[159] Harizi H, Corcuff JB, Gualde N. Arachidonic-acid-derived eicosanoids: roles in biology and immunopathology. Trends Mol Med 2008;14(10):461–9.
[160] Dennis EA, Norris PC. Eicosanoid storm in infection and inflammation. Nat Rev Immunol 2015;15(8):511–23.
[161] Yu W, Bozza PT, Tzizik DM, Gray JP, Cassara J, Dvorak AM, et al. Co-compartmentalization of MAP kinases and cytosolic phospholipase A2 at cytoplasmic arachidonate-rich lipid bodies. Am J Pathol 1998;152(3):759–69.
[162] Chen X, Xu S, Wei S, Deng Y, Li Y, Yang F, et al. Comparative proteomic study of fatty acid-treated myoblasts reveals role of Cox-2 in palmitate-induced insulin resistance. Sci Rep 2016;6:21454.
[163] Silva AR, Pacheco P, Vieira-de-Abreu A, Maya-Monteiro CM, D'Alegria B, Magalhaes KG, et al. Lipid bodies in oxidized LDL-induced foam cells are leukotriene-synthesizing organelles: a MCP-1/CCL2 regulated phenomenon. Biochim Biophys Acta 2009;1791(11):1066–75.
[164] Valm AM, Cohen S, Legant WR, Melunis J, Hershberg U, Wait E, et al. Applying systems-level spectral imaging and analysis to reveal the organelle interactome. Nature 2017;546(7656):162–7.
[165] Schuldiner M, Bohnert M. A different kind of love - lipid droplet contact sites. Biochim Biophys Acta 2017;1862(10 Pt B):1188–96.
[166] Aon MA, Bhatt N, Cortassa SC. Mitochondrial and cellular mechanisms for managing lipid excess. Front Physiol 2014;5:282.
[167] Benador IY, Veliova M, Liesa M, Shirihai OS. Mitochondria bound to lipid droplets: where mitochondrial dynamics regulate lipid storage and utilization. Cell Metab 2019;29(4):827–35.
[168] Gemmink A, Daemen S, Kuijpers HJH, Schaart G, Duimel H, Lopez-Iglesias C, et al. Super-resolution microscopy localizes perilipin 5 at lipid droplet-mitochondria interaction sites and at lipid droplets juxtaposing to perilipin 2. Biochim Biophys Acta Mol Cell Biol Lipids 2018;1863(11):1423–32.
[169] Varghese M, Kimler VA, Ghazi FR, Rathore GK, Perkins GA, Ellisman MH, et al. Adipocyte lipolysis affects Perilipin 5 and cristae organization at the cardiac lipid droplet-mitochondrial interface. Sci Rep 2019;9(1):4734.
[170] Benador IY, Veliova M, Mahdaviani K, Petcherski A, Wikstrom JD, Assali EA, et al. Mitochondria bound to lipid droplets have unique bioenergetics, composition, and dynamics that support lipid droplet expansion. Cell Metab 2018;27(4):869–85. e6.
[171] Welte MA. Proteins under new management: lipid droplets deliver. Trends Cell Biol 2007;17(8):363–9.
[172] Cermelli S, Guo Y, Gross SP, Welte MA. The lipid-droplet proteome reveals that droplets are a protein-storage depot. Curr Biol 2006;16(18):1783–95.
[173] Li Z, Thiel K, Thul PJ, Beller M, Kuhnlein RP, Welte MA. Lipid droplets control the maternal histone supply of Drosophila embryos. Curr Biol 2012;22(22):2104–13.
[174] Olofsson SO, Asp L, Boren J. The assembly and secretion of apolipoprotein B-containing lipoproteins. Curr Opin Lipidol 1999;10(4):341–6.
[175] Ohsaki Y, Cheng J, Fujita A, Tokumoto T, Fujimoto T. Cytoplasmic lipid droplets are sites of convergence of proteasomal and autophagic degradation of apolipoprotein B. Mol Biol Cell 2006;17(6):2674–83.
[176] Suzuki M, Otsuka T, Ohsaki Y, Cheng J, Taniguchi T, Hashimoto H, et al. Derlin-1 and UBXD8 are engaged win dislocation and degradation of lipidated ApoB-100 at lipid droplets. Mol Biol Cell 2012;23(5):800–10.

CHAPTER

# 9

# The lipid droplet as a signaling node

*Charles P. Najt*[a], *Douglas G. Mashek*[a,b]

[a]Department of Biochemistry, Molecular Biology and Biophysics, University of Minnesota, Minneapolis, MN, United States; [b]Department of Medicine, Division of Diabetes, Endocrinology and Metabolism, University of Minnesota, Minneapolis, MN, United States

OUTLINE



## Lipid droplet composition

### Lipids

To understand how lipid droplets impact cellular signaling pathways, an understanding of the composition of lipid droplets is first required. Lipid droplets are most notably recognized as cellular storage depots for neutral lipids—triacylglycerol (TAG) and cholesterol esters, surrounded by a unique phospholipid monolayer. While numerous studies have analyzed cellular or tissue lipidomes, the lipid droplet lipidome has been less extensively characterized. Using a LC–MS approach, Chiraju and colleagues showed that TAGs represent >90% of the lipid species in hepatic lipid droplets; it should be noted that cholesterol and cholesterol esters were not analyzed due to technical limitations [1,2]. The next most abundant lipid species detected were diacylglycerol and phosphatidylcholine followed by relatively small amounts of other phospholipid species. The composition of lipid species in lipid droplets were also

Lipid Signaling and Metabolism. http://dx.doi.org/10.1016/B978-0-12-819404-1.00009-9

impacted by various nutritional interventions such as fasting, feeding or long-term administration of a high fat diet [1]. NMR spectroscopy of lipid droplets reveals that cholesteryl esters (~34%) and TAG (~44%) are the most abundant lipid species in CHO K2 cells [3]. However, the same analysis also showed that the ether lipid monoalk(en)yl diacylglycerol comprised 10%–20% of lipids, highlighting an important role for lipid droplets in ether lipid metabolism. Over 160 phospholipid species were also identified with phosphatidylcholine being the most abundant class, followed by phosphatidylethanolamine, ether-linked phosphatidylcholine and phosphatidylinositol. In contrast, lipid droplets were relatively deficient in sphingomyelin, phosphatidylserine, and phosphatidic acid. Lipid droplets have also been shown to be a reservoir for diacylglycerols and ceramides, both of which are robustly increased in liver, heart, and muscle of a genetic model of lipodystrophy and insulin resistance [4]; the importance of these lipid metabolites in signaling will be discussed in more depth later in this chapter.

In addition to neutral lipid storage, lipid droplets are also well known to provide storage for fat-soluble vitamins, of which the most studied are retinyl esters, the storage of vitamin A. While lipid droplets in hepatic stellate cells are the primary storage site of vitamin A in the body [5], other cells store vitamin A in lipid droplets albeit at a lower concentration [6]. Vitamin E has also been shown to be present in adipose tissue lipid droplets [7]. To the best of our knowledge, the role of lipid droplets in the storage of vitamin D and K has not been studied. Another understudied area regarding lipophilic molecules that appear to collect in lipid droplets are that of the N-acylethanolamines more commonly known as endocannabinoids [8,9]. These lipid species, involved in various physiological functions including nociception and inflammation, are the most recent lipid species discovered to exist in the lipid droplet neutral lipid core. Collectively, these studies show that lipid droplets are comprised of numerous lipid species that vary depending upon the cell type and nutritional/metabolic state of the organism.

## Proteins

In contrast to the lipid droplet lipidome, the protein composition of lipid droplets has been characterized to a much greater extent. Dozens of studies have reported on the protein composition of lipid droplets, which varies widely depending upon the tissues analyzed, metabolic state, and method of analysis [10]. The 100s of proteins that comprise the lipid droplet proteome serve a wide range of functions that ultimately influence the morphology, function, and signaling of these dynamic organelles. While many of these proteins likely have functions that are simply structural, numerous proteins have been documented to contribute to the regulation of lipid metabolism and link alterations in lipid droplet composition to various cell signaling pathways and, ultimately, cellular function. It should be noted that many proteins found on lipid droplets are also associated with pathways not typically associated with lipid metabolism. The presence of histones, ribosomal proteins, and proteins associated with other organelles on lipid droplets illustrate the diverse nature of the lipid droplet proteome and suggests that lipid droplets play a key role in protein storage [11]. Examples of how some of these select proteins provide communication channels between lipid droplet and various pathways will be discussed in more detail in later sections.

# Lipid droplet signaling

There are several mechanisms through which lipid droplets can influence cellular signaling and function. One is simply providing a reservoir for lipophilic molecules that, if not readily stored in lipid droplets, would be "free" to signal via their intrinsic mechanisms. An additional mechanism is through the supply of substrates (often fatty acids) for the production of various lipid metabolites that serve as signaling molecules. Finally, a third mechanism discussed in this section through which lipid droplets signal is via the sequestration or release of proteins on the droplet surface. These proteins may be involved in lipid transport, have enzymatic functions, or facilitate organelle interactions as a means to govern lipid droplet communication in cells.

## Lipid droplets as lipophilic storage units

Lipid droplets and lipid bilayer membranes constitute the primary lipophilic environments in cells. While the storage capacity of lipid bilayers is limited due to size, lipid droplets are much less stringent in storage of lipophilic molecules given their much larger nonpolar core. As such, lipids or other lipophilic molecules that are incorporated into lipid droplets are readily impacted by alterations in lipid droplet metabolism. If their incorporation is impaired, then the molecules are then available to initiate various modes of cellular signaling unique to each molecule.

### *Lipotoxicity*

Perhaps the most-studied example illustrating the importance of lipid droplets sequestering lipids is lipotoxicity. Fatty acids are readily activated to acyl-CoAs and metabolized through numerous pathways including storage as TAG or phospholipids, or oxidation. If a major metabolic pathway such as incorporation into TAG and storage in lipid droplets is attenuated, then fatty acids or their metabolites can accumulate and cause cellular dysfunction, in a process termed lipotoxicity [12]. Saturated fatty acids, primarily palmitate, are especially prone to trigger endoplasmic reticulum (ER) stress, inflammatory signaling, insulin resistance, and apoptosis when given to cells at high concentrations [13]. Recent studies have detailed the molecular mechanism underlying palmitate-induced lipotoxicity. A lipidomic analysis of cells exposed to palmitate reveals increased levels of saturated glycerolipid intermediates such as lysophosphatidic acid, phosphatidic acid, and diacylglycerol [12]. The accumulation of these intermediates in phospholipid and TAG synthetic pathways is likely due to the substrate specificity of the acyltransferase enzymes involved in this pathway. While the first enzyme in the pathway, glycerol-3-phosphate acyltransferase (GPAT1 or 4), readily use saturated fatty acids as substrates to form lysophosphatidic acid [14,15], subsequent enzymes have reduced activity toward saturated lipids. Indeed, both the lipin and diacylglycerol acyltransferase enzymes, which catalyze the last two steps in TAG synthesis, have reduced activity toward di-saturated substrates [12,16]. As a consequence, flux of palmitate to TAG is attenuated resulting in accumulation of di-saturated intermediates. Consistent with the importance of saturated glycerolipid intermediates playing a key role, inhibition of the ER-localized GPAT3 and 4 reduce saturated intermediates and prevent palmitate-mediated

ER stress and lipotoxicity [12]. These results highlight the importance of the acyltransferases and di-saturated intermediates as mediators of smooth muscle cell lipotoxicity and vascular function [16]. While accumulation of inert metabolic intermediates would not be detrimental, phosphatidic acid and diacylglycerol are potent signaling molecules that can regulate numerous cellular functions involved in the lipotoxic response (see [17,18] for reviews).

In addition to accumulation of metabolites in the glycerolipid pathway, overflow of palmitate into additional biosynthetic pathways such as sphingolipid and ceramide synthesis can also occur [19,20]. It should be noted that while saturated fatty acids are often grouped as a single class, their lipotoxic effects are largely limited to palmitate with fewer studies showing detrimental effects of myristate and stearate [21–23]. Consistent with the benefits of incorporation of fatty acids into TAG, blocking diacylglycerol acyltransferase 1, which is known to esterify exogenous fatty acids, also increases saturated ceramide [24]. Similarly, increasing fatty acid oxidation via a malonyl-CoA insensitive carnitine palmitoyl transferase-1 prevents palmitate-induced insulin resistance and apoptosis in muscle cells concurrent with reduced diacylglycerol and ceramide [25]. If impaired production of TAG is a driving factor in saturated fatty acid-induced lipotoxicity then exposure of cells to unsaturated fatty acids should alleviate the detrimental effects of palmitic acid given alone. Indeed, many studies show that co-incubation of cells with unsaturated fatty acids promotes the esterification of palmitate into TAG and alleviates lipotoxicity [19,26,27]. These data are consistent with an inverse relationship between rates of fatty acid esterification and lipotoxicity [28]. Taken together, this work suggests that lipid droplets play a key role in sequestering saturated fatty acids into TAG as a means to prevent lipotoxicity in numerous cell types.

An additional mechanism through which lipid droplets influence lipotoxicity is through their regulation of oxidative stress. Exposure of cells to oxidative stress triggers lipid droplet accumulation, which may act to sequester fatty acids into lipid droplets in order to prevent reactive oxygen species production [29]. In cancer cells, hypoxia also triggers lipid droplet accumulation via the activation of the hypoxia-induced transcription factor 1$\alpha$. Once activated, it induces HIF-induced gene 2 (HIG2; also known as HILPDA), which directly interacts with and antagonizes adipose triglyceride lipase (ATGL) to prevent the production of fatty acids from lipid droplets and subsequent production of reactive oxygen species by limiting fatty acid oxidation [30–32]. Moreover, hypoxia also triggers the accumulation of saturated fatty acid species in lipid droplets through the inhibition of oxygen-dependent desaturases [33] suggesting that lipid droplets act to buffer the lipid profile under hypoxic conditions.

### *Ceramides*

As mentioned above, lipid droplets are also contain ceramides, potent sphingolipid signaling molecules that impact a myriad of cellular functions [4]. Recent work shows that diacylglycerol acyltransferase 2, known to catalyze the terminal stop in TAG synthesis, also catalyzes the conversion of ceramide to acylceramides, which are subsequently stored within lipid droplets [34]. Moreover, ceramide synthetase 6 interacts with long chain acyl-CoA synthetase 5 and diacylglycerol acyltransferase 2 to traffic fatty acids into ceramide/acylceramides at ER/lipid droplet junctions. These studies highlight a potential role of lipid droplets in the synthesis and storage of ceramides, which may ultimately impact ceramide signaling.

### *Toxins*

Due to their intrinsic physical properties, lipid droplets can also provide a reservoir for other lipophilic molecules. Numerous toxins including polycyclic aromatic hydrocarbons and polychlorinated biphenyls are stored within lipid droplets [35,36]. This sequestration in lipid droplets is likely to influence their pharmacokinetics and downstream effects on cells. As an example, yeast lipid droplets sequester phototoxic perylenequinones produced from fungi as a means to defend against the lethal effects of the toxin [37]. Moreover, yeast deficient in lipid droplets are hypersusceptible to the lipophilic toxins. Together these studies provide a glimpse of the potential role of lipid droplets as a mediator of lipophilic xenobiotics.

### *Drugs*

Many pharmaceutical agents have hydrophobic properties and, therefore, could potentially incorporate into lipid droplets. As an example, the hydrophobic anticancer prodrug CHR2863 is stored in lipid droplets before being metabolized to its active form [38]. Similarly, the antifungal drug ST-669 has also been shown to localize to lipid droplets although the importance of this in regards to efficacy of the drug is not known [39]. Recently it was shown that lipid droplets may mediate the treatment of *Mycobacterium tuberculosis*. Lipid droplets in macrophages infected with *M. tuberculosis* transiently store the antibiotic bedaquiline, which is commonly used to treat the disease, and are directly involved in the transfer of the drug to the bacteria [40]. While studies are limited, this work clearly shows that the potential importance of lipid droplets as mediators of drug storage and metabolism, and will likely be a growing research focus in the future.

## Lipid droplets as a source of lipid signaling molecules

There are two major mechanisms through which lipid droplets are degraded. The most well studied is the classical pathway of enzyme-catalyzed lipolysis, whereas autophagy-mediated degradation, termed lipophagy, has been the focus of more recent studies. The role of lipophagy as a contributor to cellular signaling has not been explored, but undoubtedly will be an emphasis of future studies as this field expands. In contrast, lipolysis of lipid droplets has been well studied and is a major source of signaling lipids that impacts many different intracellular pathways.

### *Lipolytically derived fatty acids*

Recent studies have focused on signaling as a consequence of TAG hydrolysis. Gene expression analysis of tissues from mice lacking either ATGL or hormone-sensitive lipase, which catalyze the first and second reactions of TAG hydrolysis, respectively, reveals divergent effects of these two enzymes on gene expression patterns [41]. ATGL deficiency suppresses oxidative gene expression in numerous tissues such as heart, liver, skeletal muscle, and brown adipose, whereas, hormone-sensitive lipase deficiency only affects oxidative genes in brown fat. Numerous other biological processes, such as cell growth and death, lipid synthesis, and signal transduction, are also differentially modulated in specific tissues in response to ATGL ablation. These data suggest that lipolysis, particularly by ATGL, may play a role in transcriptional control of metabolism. In support, hepatocytes that overexpress ATGL have higher peroxisome proliferator activated receptor-α (PPARα) activity and target gene expression, whereas knockdown of hepatic ATGL in vivo suppresses the expression of PPARα target genes [42,43]. ATGL overexpression in adipocytes results in the expression of genes involved in fatty acid oxidation and oxidative metabolism, including PPARα and -δ targets and results in enhanced fatty oxidation [44]. Because fatty acids are known ligands for PPARα, these data suggest that ATGL-mediated TAG hydrolysis provides an important source of PPARα ligands. Indeed, providing a PPARα agonist partially rescues the cardiomyopathy and impairments in oxidative metabolism that characterizes ATGL knockout mice [45,46]. However, administering the PPARα agonist fenofibrate to mice with suppressed hepatic ATGL does not increase the expression of PPARα target genes to levels in control mice treated with the same drug [47] suggesting a more complicated mechanism may underlay this regulation. Moreover, while PPARα drives the expression of genes involved in fatty acid oxidation, it does not drive genes involved in mitochondrial biogenesis or oxidative phosphorylation [48,49], which are commonly suppressed in models of ATGL ablation or inhibition.

Since peroxisome proliferator activated receptor-gamma coactivator-1α (PGC1α) is a major driver of PPARα signaling and in regulating mitochondrial biogenesis and function, additional studies explored the role of ATGL-catalyzed lipolysis in regulating PGC1α. Indeed, ATGL overexpression activates PGC1α activity and inhibition of ATGL prevents cAMP/PKA-mediated activation of PGC-1α suggesting that lipolysis is key node in mediating hormonal induction of PGC1α and mitochondrial biogenesis/function [50]. This study also revealed that ATGL regulates PGC1α acetylation through the activation of sirtuin 1 (SIRT1), a protein known to deacetylate and activate PGC1α [51]. More recent work has further defined the novel mechanism through which ATGL controls SIRT1 activity. These studies show that monounsaturated fatty acids released from ATGL-catalyzed lipolysis allosterically activate SIRT1 toward PGC1α and several other known SIRT1 peptide substrates [52]. This activation is specific to monounsaturated fatty acids as other long chain saturated and unsatured fatty acids lack the ability to allosterically modulate SIRT1. These data are also consistent with ATGL showing preference for cleaving monounsaturated fatty acids from TAG [53]. Previous work has shown liver fatty acid binding protein, the major intracellular fatty acid carrier in hepatocytes that is thought to transport fatty acids to the nucleus, is not required for ATGL-mediated changes in gene expression [47], thus providing a conundrum on how cytosolic liberated fatty acids are able to activate SIRT1, which is primarily nuclear. Recent evidence has pointed toward an important role of the lipid droplet protein perilipin 5 (PLIN5) in linking lipolysis to SIRT1 activity. In muscle and brown adipocytes, PLIN5 translocates from lipid droplets to the nucleus where it interacts with SIRT1 and PGC-1α to promote oxidative metabolism in response to cAMP/PKA signaling [54]. Similar to what was noted above regarding ATGL, PLIN5 is also required for cAMP/PKA-mediated activation of SIRT1 and the PGC1α/PPARα signaling axis. Additional findings reveal that PLIN5 is a fatty acid binding protein that preferentially binds MUFAs, an effect that is enhanced with PKA-mediate phosphorylation of PLIN5 [52]. In addition, PLIN5 is required for ATGL-mediated activation of SIRT1 and PGC-1α/PPAR-α. Collectively, these data identify a novel signaling node involving PLIN5 trafficking of lipolytically derived MUFAs to the nucleus to activate SIRT1 as a key signaling pathway governing oxidative metabolism. From a physiological viewpoint, this mechanism also couples the supply of substrates (fatty acids produced via ATGL) with the oxidative machinery (increased mitochondria) during times of enhanced lipid catabolism, which likely plays a key role in preventing lipotoxicity.

In addition to regulating gene expression, fatty acids may regulate energy metabolism via effects on AMP-activated kinase (AMPK). Once hydrolyzed, fatty acids must be activated to acyl-CoAs before they can enter reesterification or oxidation pathways. In the activation reaction catalyzed by long chain acyl-CoA synthetases, ATP is hydrolyzed to AMP plus orthophosphate. Thus, during times of enhanced lipolysis, acyl-CoA synthetase activity can modulate cellular levels of ATP and AMP. One example is in 3T3-L1 adipocytes where isoproterenol-stimulated lipolysis decreases cellular ATP levels while increasing AMP concentrations, which subsequently promotes AMPK activity [55]. Inhibiting acyl-CoA synthetase activity attenuates the changes in nucleotide levels and blocks the activation of AMPK further verifying the role of fatty acids as a mediator of the changes in AMPK activity. This study demonstrates a novel mechanism that links fatty acids released from lipolysis to the regulation of cellular energy metabolism mediated by phosphorylation. Additional work shows that lipolysis in adipose tissue inhibits mammalian target of rapamycin (mTOR) 1 and 2 complex formation [56]. Moreover, catecholamine-induced lipolysis and mTOR complex dissociation was blocked in the absence of ATGL. While the contribution of SIRT1 or AMPK, which often work in tandem, was not studied, these data are consistent with their known effects to antagonize mTOR activity. Thus, ATGL-catalyzed lipolysis promotes AMPK and SIRT1, nutrients sensors that respond to low energy (NAD and AMP) while inhibiting mTOR, which is driven by excess nutrients and growth-promoting hormones. Collectively, these data suggest that ATGL is a key signaling node during periods of catabolism that not only provides fatty acid substrates, but also coordinates a host of downstream signaling nodes to promote oxidative metabolism and inhibit mTOR-mediated anabolic pathways.

While genetic manipulations of genes in mouse models and cells have helped tease apart key proteins that regulate lipid droplet catabolism, the most well studied and perhaps significant genetic variant to regulate lipid droplets is patatin-like phospholipase domain-containing protein 3 (PNPLA3). An I148M variant of PNPLA3, which has prevalence as high as 50% in some populations, is the single largest genetic predictor of fatty liver disease [57]. Recent evidence suggests that this variant binds to comparative gene identification-58 (CGI-58), a potent activator of ATGL, and thereby antagonizes lipolysis and lipophagy as a means to promote lipid droplet accumulation [58–60]. Moreover, this PNPLA3 variant accumulates on lipid droplets, which can enhance its antagonistic abilities, as are result of its resistance to ubiquitination and subsequent degradation [61]. These studies highlight the importance of how a genetic variant in lipid droplet proteins can have a profound impact on lipid droplet metabolism, signaling, and disease development.

### *Steroids*

As noted above cholesterol esters are also stored within lipid droplets. Since they are the storage from of free cholesterol, the regulation of cholesterol ester turnover can influence the downstream metabolism of numerous pathways involved in cholesterol metabolism. One such example is in steroidogenic cells where cholesterol derived from lipid droplets is the preferred substrate for steroid hormone synthesis. Whereas most steroid hormone synthesis is thought to occur in the mitochondria, proteomics studies have revealed numerous steroid synthetic enzymes to be present on lipid droplets [62–64]; many have subsequently been confirmed with confocal microscopy as bonafide lipid droplet proteins. These studies reveal the entire cadre of proteins needed for testosterone production exists on lipid droplets suggesting that this organelle may be a major site of testosterone synthesis.

### *Eicosanoids*

Eicosanoids represent a large class of oxidized lipids that act as potent signaling molecules involved in numerous pathways including immune regulation. Historically, it was viewed that polyunsaturated fatty acids stored in plasma membrane phospholipids exclusively provided the substrate for numerous phospholipases that catalyze the hydrolysis of fatty acids in the first step of the eicosanoid synthetic pathway. However, lipid droplets are increasingly recognized as a major contributor to eicosanoid synthesis in numerous cell types. Several enzymes involved in eicosanoid synthesis including cyclooxogenases, lipoxygenases, and more terminal eicosanoid synthetic enzymes are found on lipid droplets [64,65]. In agreement with the presence of these enzymes, the surface of the lipid droplet is an active site of eicosanoid synthesis [66]. Moreover, inflammatory conditions where eicosanoid production is increased also promotes lipid droplet accumulation in numerous cell types [66–68]. Whereas phospholipids have been historically regarded as the sole source of fatty acid substrates for eicosanoids, fatty acids liberated from TAG hydrolysis appear to be a major contributor to eicosanoid production [69]. ATGL appears to be the primary supplier of hydrolyzed fatty acids from TAG-derived polyunsaturated fatty acids in mast and endothelial cells and neutrophils [70–72].

In addition to ATGL, other lipid droplet proteins have been shown to influence inflammatory signaling. Perilipin 2 (PLIN2) is a ubiquitously expressed lipid droplet protein that is known to antagonize ATGL-catalyzed lipolysis [73]. Based on this information, it would seem logical that PLIN2 would prevent eicosanoid production and inflammation. However, the opposite is observed as PLIN2 is required for inflammatory signaling in response to high fat feeding in mice, lipopolysaccharide administration in hepatocytes [74], and oxidized low density lipoprotein treatment in macrophages [75]. Because PLIN2 ablation reduces lipid droplet levels, it is difficult to ascertain if PLIN2 is directly involved in the production of inflammatory lipid metabolites or if it promotes inflammatory signaling simply by enlarging the lipid droplet pool. In contrast, perilipin 1 (PLIN1) deficiency in adipose tissue promotes eicosanoid production and inflammation [76] highlighting distinct effects of lipid droplet proteins, which also likely elicit their effects in a tissue-specific manner.

### *Ceramides*

While lipid droplets appear to be an important storage organelle for ceramides/acylceramides as noted above, they may also provide the fatty acid substrates to generate ceramides. CGI-58 recruits and co-activates PNPLA1 at the surface of lipid droplets to generate fatty acid substrates needed for the synthesis of acylceramides, which are required for proper skin barrier function and are impaired in subjects with mutations in CGI-58 or PNPLA1 [77]. In addition, these effects are independent of ATGL suggesting the presence of other lipases on the lipid droplet surface, including members of the PNPLA family, are involved in the channeling of lipolytic products to specific downstream lipid metabolic pathways [78].

## Proteins that link lipid droplets to cell signaling

Numerous lipid droplet proteins are involved in regulating lipid metabolism and/or trafficking as a means to influence cell signaling. The family of PLIN proteins has important but very different effects on cell signaling depending upon the isoform and cell type. It is unclear if these signaling effects are due directly to their ability to antagonize lipolysis or if separate functions of each protein impact signaling. Perhaps the best example if this type of multifunctional protein is PLIN5. In addition to its role in mediating lipid droplet-nuclear signaling, PLIN5 also acts as a tether between lipid droplets and mitochondria [79]. While this function was originally proposed to serve as a bridge for transfer of lipolytically derived fatty acids to the mitochondria during time of energy demand, more recent work as suggested that mitochondrial and lipid droplet interactions facilitate lipid biogenesis [80]. In this model, glucose oxidized in mitochondria adjacent to lipid droplets provides the ATP to fuel lipid synthesis. This study also highlights the potential importance of lipid droplet interactions with other cellular organelles as a key mediator of lipid droplet signaling. Indeed, lipid droplets are widely recognized to interact with numerous organelles including ER, mitochondria, peroxisomes, ribosomes, autophagosomes, lysosomes, and plasma membrane to name a few. These interactions are facilitated by a variety of proteins depending upon the organelle and serve to coordinate metabolism and signaling.

### *Histones*

It is becoming increasingly recognized that lipid droplets act to sequester proteins, many of which have apparent functions beyond regulation of lipid metabolism. One class of proteins consistently found on lipid droplets, perhaps surprisingly, is histones. Numerous proteomic analysis of lipid droplets have found histones commonly present on lipid droplets across cell types and amongst different organisms. Evidence suggests that histone sequestration on lipid droplets serves different roles including storage of maternal histones to facilitate rapid chromatin assembly during development, and in transient sequestration of newly synthesized histones to provide a buffer if histone production is too high [81]. In addition, histone storage in lipid droplets may also serve a role as antibacterial agents [82] as histones are known to have potent antibacterial properties [83].

### *CIDE*

Several members of the cell death-inducing DNA fragmentation factor α-like effector (CIDE) family or proteins are present on lipid droplets. In adipocytes, CIDEC (also known as FSP27) is well documented to promote lipid transfer between lipid droplets and, thereby, regulate lipid droplet expansion [84]. In addition, CIDEC interacts with nuclear factor of activated T cells 5 (NFAT5), a transcription factor that regulates osmoprotective and inflammatory gene expression [85]. This interaction prevents nuclear translocation of NFAT5 under isotonic conditions as CIDEC overexpression attenuates and knockdown promotes NFAT5 translocation and transcriptional activity.

### *Protein turnover*

Lipid droplets may also influence protein-mediated signaling and function through the regulation of protein turnover. Misfolded proteins in the ER are commonly translocate to the cytosol for degraded through the ubiquitin/proteasome system in a process termed

ER-associated degradation (ERAD). Indeed, proteasomes are commonly found in the vicinity of lipid droplets [86]. Perhaps the best example of lipid droplets involvement in protein degradation is ApoB100. To eliminate lipidated ApoB100 from the ER lumen, it translocates to the lipid droplet surface where it undergoes ubiquitination targeting it for degradation [87]. Consistent with this, inhibition of proteosomal degradation causes accumulation of specific proteins including ApoB100 on the surface of lipid droplets [88]. These data are also supported by studies showing the presence of numerous proteins involved in ubiquitination including ancient ubiquitous protein 1, which recruits various ubiquitin ligases to the surface of lipid droplets [89,90]. Similar to ApoB100, translocation to lipid droplets and proteosomal degradation also occurs for 3-hydroxy-3-methyl-glutaryl-CoA reductase, the rate limiting step in cholesterol biosynthesis [91]. The presence of chaperone proteins including heat shock protein 70 on lipid droplets also suggests that the lipid droplets may be involved in the regulation misfolded proteins as well [92].

## Summary

In the past decade, the number of publications in the lipid droplet field has increased exponentially. A focus of much of this work has been on characterizing the composition (both lipid and protein) of lipid droplets and on the underlying fundamental biology of lipid droplet biogenesis and degradation. However, increasing attention is being placed on characterizing

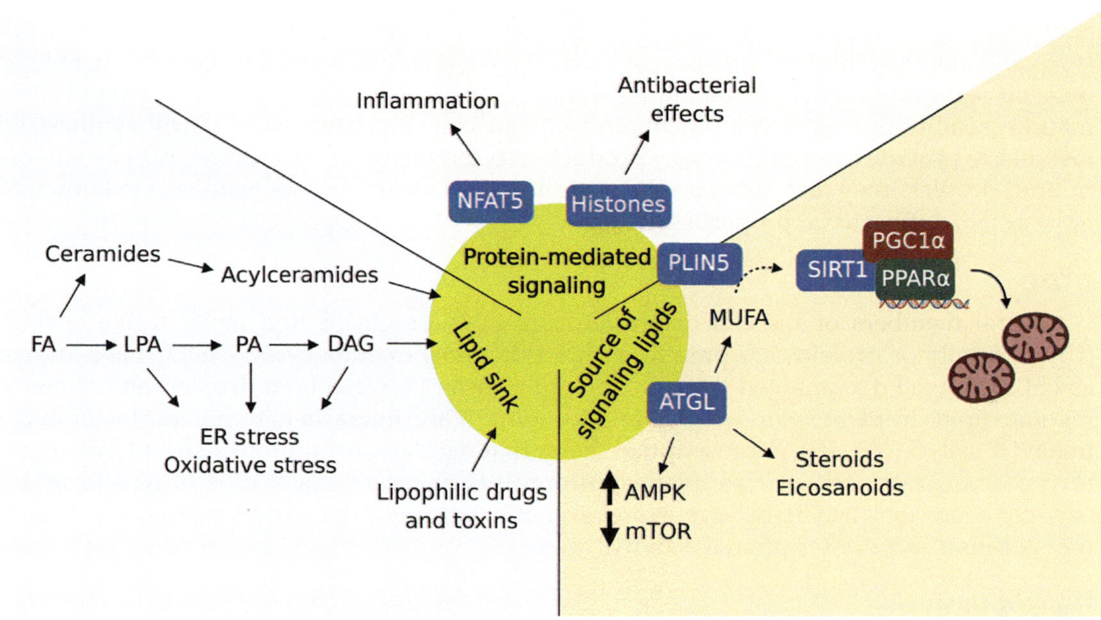


**FIGURE 9.1** **Key mechanisms through which lipid droplets regulate signaling. Signaling events primarily occur through the sequestration of lipids or lipophilic compounds, through the production of lipid signaling molecules, or through protein-mediated signaling.**

on how lipid droplets orchestrate cellular metabolism and signaling as a means to link nutrient storage and cell function. While storage of lipophilic metabolites, production of signaling molecules, and protein-mediated signaling provide avenues for how lipid droplets communicate (Fig. 9.1), our global understanding of lipid droplet signaling is still in its infancy. In addition to novel signaling pathways being unraveled in future studies, how lipid droplets signaling differs amongst cell types and during disease development will be key questions needing to be answered to facilitate development of therapeutic approaches toward diseases influenced by alterations in lipid droplet metabolism and signaling.

## References

[1] Chitraju C, Trötzmüller M, Hartler J, Wolinski H, Thallinger GG, Lass A, et al. Lipidomic analysis of lipid droplets from murine hepatocytes reveals distinct signatures for nutritional stress. J Lipid Res 2012;53(10):2141–52. Available from: http://www.jlr.org/content/53/10/2141.

[2] Chitraju C, Trötzmüller M, Hartler J, Wolinski H, Thallinger GG, Haemmerle G, et al. The impact of genetic stress by ATGL deficiency on the lipidome of lipid droplets from murine hepatocytes. J Lipid Res 2013;54(8):2185–94. Available from: http://www.jlr.org/lookup/doi/10.1194/jlr.M037952.

[3] Bartz R, Li W-H, Venables B, Zehmer JK, Roth MR, Welti R, et al. Lipidomics reveals that adiposomes store ether lipids and mediate phospholipid traffic. J Lipid Res 2007;48(4):837–47. Available from: http://www.jlr.org/lookup/doi/10.1194/jlr.M600413-JLR200.

[4] Preuss C, Jelenik T, Bódis K, Müssig K, Burkart V, Szendroedi J, et al. A new targeted lipidomics approach reveals lipid droplets in liver, muscle and heart as a repository for diacylglycerol and ceramide species in non-alcoholic fatty liver. Cells 2019;8(3):277. Available from: https://www.mdpi.com/2073-4409/8/3/277.

[5] Blaner WS, Li Y, Brun P-J, Yuen JJ, Lee S-A, Clugston RD. Vitamin A absorption storage and mobilization. Subcell Biochem 2016;81:95–125. Available from: http://link.springer.com/10.1007/978-94-024-0945-1_4.

[6] Nagy NE, Holven KB, Roos N, Senoo H, Kojima N, Norum KR, et al. Storage of vitamin A in extrahepatic stellate cells in normal rats. J Lipid Res 1997;38(4):645–58. Available from: http://www.ncbi.nlm.nih.gov/pubmed/9144080.

[7] Traber MG, Kayden HJ. Tocopherol distribution and intracellular localization in human adipose tissue. Am J Clin Nutr 1987;46(3):488–95. Available from: https://academic.oup.com/ajcn/article/46/3/488-495/4694551.

[8] Kaczocha M, Glaser ST, Chae J, Brown DA, Deutsch DG. Lipid droplets are novel sites of N-acylethanolamine inactivation by fatty acid amide hydrolase-2. J Biol Chem 2010;285(4):2796–806. Available from: http://www.jbc.org/lookup/doi/10.1074/jbc.M109.058461.

[9] Tsuboi K, Uyama T, Okamoto Y, Ueda N. Endocannabinoids and related N-acylethanolamines: biological activities and metabolism. Inflamm Regen 2018;38(1):28. Available from: https://inflammregen.biomedcentral.com/articles/10.1186/s41232-018-0086-5.

[10] Bersuker K, Olzmann JA. Establishing the lipid droplet proteome: mechanisms of lipid droplet protein targeting and degradation. Biochim Biophys Acta Mol Cell Biol Lipids 2017;1862(10 Pt B):1166–77. Available from: http://www.sciencedirect.com/science/article/pii/S1388198117301105.

[11] Cermelli S, Guo Y, Gross SP, Welte MA. The lipid-droplet proteome reveals that droplets are a protein-storage depot. Curr Biol 2006;16(18):1783–95. Available from: https://linkinghub.elsevier.com/retrieve/pii/S0960982206019750.

[12] Piccolis M, Bond LM, Kampmann M, Pulimeno P, Chitraju C, Jayson CBK, et al. Probing the global cellular responses to lipotoxicity caused by saturated fatty acids. Mol Cell 2019;74(1). 32–44.e8. Available from: http://www.ncbi.nlm.nih.gov/pubmed/30846318.

[13] Ertunc ME, Hotamisligil GS. Lipid signaling and lipotoxicity in metaflammation: indications for metabolic disease pathogenesis and treatment. J Lipid Res 2016;57(12):2099–114. Available from: http://www.jlr.org/lookup/doi/10.1194/jlr.R066514.

[14] Dircks LK, Sul HS. Mammalian mitochondrial glycerol-3-phosphate acyltransferase. Biochim Biophys Acta 1997;1348(1–2):17–26. Available from: http://www.ncbi.nlm.nih.gov/pubmed/9370312.

[15] Chen YQ, Kuo M-S, Li S, Bui HH, Peake DA, Sanders PE, et al. AGPAT6 is a novel microsomal glycerol-3-phosphate acyltransferase. J Biol Chem 2008;283(15):10048–57. Available from: http://www.jbc.org/lookup/doi/10.1074/jbc.M708151200.

[16] Masuda M, Miyazaki-Anzai S, Keenan AL, Okamura K, Kendrick J, Chonchol M, et al. Saturated phosphatidic acids mediate saturated fatty acid-induced vascular calcification and lipotoxicity. J Clin Invest 2015;125(12):4544–58. Available from: https://www.jci.org/articles/view/82871.

[17] Liu Y, Su Y, Wang X. Phosphatidic acid-mediated signaling. Adv Exp Med Biol 2013;991:159–76. Available from: http://link.springer.com/10.1007/978-94-007-6331-9_9.

[18] Eichmann TO, Lass A. DAG tales: the multiple faces of diacylglycerol–stereochemistry, metabolism, and signaling. Cell Mol Life Sci 2015;72(20):3931–52. Available from: http://link.springer.com/10.1007/s00018-015-1982-3.

[19] Ricchi M, Odoardi MR, Carulli L, Anzivino C, Ballestri S, Pinetti A, et al. Differential effect of oleic and palmitic acid on lipid accumulation and apoptosis in cultured hepatocytes. J Gastroenterol Hepatol 2009;24(5):830–40. Available from: http://doi.wiley.com/10.1111/j.1440-1746.2008.05733.x.

[20] Shih L-M, Tang H-Y, Lynn K-S, Huang C-Y, Ho H-Y, Cheng M-L. Stable isotope-labeled lipidomics to unravel the heterogeneous development lipotoxicity. Molecules 2018;23(11):2862. Available from: http://www.mdpi.com/1420-3049/23/11/2862.

[21] Martínez L, Torres S, Baulies A, Alarcón-Vila C, Elena M, Fabriàs G, et al. Myristic acid potentiates palmitic acid-induced lipotoxicity and steatohepatitis associated with lipodystrophy by sustaning de novo ceramide synthesis. Oncotarget 2015;6(39):41479–96. Available from: http://www.oncotarget.com/fulltext/6286.

[22] Blachnio-Zabielska AU, Chacinska M, Vendelbo MH, Zabielski P. The crucial role of C18-Cer in fat-induced skeletal muscle insulin resistance. Cell Physiol Biochem 2016;40(5):1207–20. Available from: https://www.karger.com/Article/FullText/453174.

[23] Deguil J, Pineau L, Rowland Snyder EC, Dupont S, Beney L, Gil A, et al. Modulation of lipid-induced ER stress by fatty acid shape. Traffic 2011;12(3):349–62. Available from: http://doi.wiley.com/10.1111/j.1600-0854.2010.01150.x.

[24] Nguyen TB, Louie SM, Daniele JR, Tran Q, Dillin A, Zoncu R, et al. DGAT1-dependent lipid droplet biogenesis protects mitochondrial function during starvation-induced autophagy. Dev Cell 2017;42(1).

[25] Henique C, Mansouri A, Fumey G, Lenoir V, Girard J, Bouillaud F, et al. Increased mitochondrial fatty acid oxidation is sufficient to protect skeletal muscle cells from palmitate-induced apoptosis. J Biol Chem 2010;285(47):36818–27. Available from: http://www.jbc.org/content/285/47/36818.abstract.

[26] Listenberger LL, Han X, Lewis SE, Cases S, Farese RV Jr, Ory DS, et al. Triglyceride accumulation protects against fatty acid-induced lipotoxicity. Proc Natl Acad Sci U S A 2003;100(6):3077–82. Available from: http://www.ncbi.nlm.nih.gov/pubmed/12629214.

[27] Maedler K, Spinas GA, Dyntar D, Moritz W, Kaiser N, Donath MY. Distinct effects of saturated and monounsaturated fatty acids on beta-cell turnover and function. Diabetes 2001;50(1):69–76. Available from: http://diabetes.diabetesjournals.org/cgi/doi/10.2337/diabetes.50.1.69.

[28] Kharroubi I, Ladriere L, Cardozo AK, Dogusan Z, Cnop M, Eizirik DL. Free fatty acids and cytokines induce pancreatic beta-cell apoptosis by different mechanisms: role of nuclear factor-kappaB and endoplasmic reticulum stress. Endocrinology 2004;145(11):5087–96.

[29] Liu S, Geng B, Zou L, Wei S, Wang W, Deng J, et al. Development of hypertrophic cardiomyopathy in perilipin-1 null mice with adipose tissue dysfunction. Cardiovasc Res 2015;105(1):20–30. Available from: http://www.ncbi.nlm.nih.gov/pubmed/25416668.

[30] Padmanabha D, Baker KD. Drosophila gains traction as a repurposed tool to investigate metabolism. Trends Endocrinol Metab 2014;25(10):518–27. Available from: http://www.sciencedirect.com/science/article/pii/S1043276014000587.

[31] DiStefano MT, Danai LV, Roth Flach RJ, Chawla A, Pedersen DJ, Guilherme A, et al. The lipid droplet protein hypoxia-inducible gene 2 promotes hepatic triglyceride deposition by inhibiting lipolysis. J Biol Chem 2015;290(24):15175–84. Available from: http://www.ncbi.nlm.nih.gov/pubmed/25922078.

[32] Zhang X, Saarinen AM, Hitosugi T, Wang Z, Wang L, Ho TH, et al. Inhibition of intracellular lipolysis promotes human cancer cell adaptation to hypoxia. Elife 2017;6. Available from: http://www.ncbi.nlm.nih.gov/pubmed/29256392.

[33] Ackerman D, Tumanov S, Qiu B, Michalopoulou E, Spata M, Azzam A, et al. Triglycerides promote lipid homeostasis during hypoxic stress by balancing fatty acid saturation. Cell Rep 2018;24(10):2596–605. e5. Available from: https://linkinghub.elsevier.com/retrieve/pii/S2211124718312695.

[34] Senkal CE, Salama MF, Snider AJ, Allopenna JJ, Rana NA, Koller A, et al. Ceramide is metabolized to acylceramide and stored in lipid droplets. Cell Metab 2017;25(3):686–97. Available from: https://linkinghub.elsevier.com/retrieve/pii/S1550413117301006.

[35] Bourez S, Le Lay S, Van den Daelen C, Louis C, Larondelle Y, Thomé J-P, et al. Accumulation of polychlorinated biphenyls in adipocytes: selective targeting to lipid droplets and role of caveolin-1. PLoS ONE 2012;7(2):e31834. Available from: https://dx.plos.org/10.1371/journal.pone.0031834.

[36] Murphy G, Rouse RL, Polk WW, Henk WG, Barker SA, Boudreaux MJ, et al. Combustion-derived hydrocarbons localize to lipid droplets in respiratory cells. Am J Respir Cell Mol Biol 2008;38(5):532–40. Available from: http://www.atsjournals.org/doi/abs/10.1165/rcmb.2007-0204OC.

[37] Chang W, Zhang M, Zheng S, Li Y, Li X, Li W, et al. Trapping toxins within lipid droplets is a resistance mechanism in fungi. Sci Rep 2015;5(1):15133. Available from: http://www.nature.com/articles/srep15133.

[38] Verbrugge SE, Al M, Assaraf YG, Kammerer S, Chandrupatla DMSH, Honeywell R, et al. Multifactorial resistance to aminopeptidase inhibitor prodrug CHR2863 in myeloid leukemia cells: down-regulation of carboxylesterase 1, drug sequestration in lipid droplets and pro-survival activation ERK/Akt/mTOR. Oncotarget 2016;7(5):5240–57. Available from: http://www.oncotarget.com/fulltext/6169.

[39] Sandoz KM, Valiant WG, Eriksen SG, Hruby DE, Allen RD, Rockey DD. The broad-spectrum antiviral compound ST-669 restricts chlamydial inclusion development and bacterial growth and localizes to host cell lipid droplets within treated cells. Antimicrob Agents Chemother 2014;58(7):3860–6. Available from: http://aac.asm.org/lookup/doi/10.1128/AAC.02064-13.

[40] Greenwood DJ, Dos Santos MS, Huang S, Russell MRG, Collinson LM, MacRae JI, et al. Subcellular antibiotic visualization reveals a dynamic drug reservoir in infected macrophages. Science 2019;364(6447):1279–82. Available from: http://www.ncbi.nlm.nih.gov/pubmed/31249058.

[41] Pinent M, Hackl H, Burkard TR, Prokesch A, Papak C, Scheideler M, et al. Differential transcriptional modulation of biological processes in adipocyte triglyceride lipase and hormone-sensitive lipase-deficient mice. Genomics 2008;92(1):26–32. Available from: http://www.sciencedirect.com/science/article/B6WG1-4ST3VN5-3/1/ce8a4e2f22145fe431dec06b08546671.

[42] Sapiro JMM, Mashek MTT, Greenberg ASS, Mashek DGG. Hepatic triacylglycerol hydrolysis regulates peroxisome proliferator-activated receptor alpha activity. J Lipid Res 2009;50(8):1621–9.

[43] Ong KT, Mashek MT, Bu SY, Greenberg AS, Mashek DG. Adipose triglyceride lipase is a major hepatic lipase that regulates triacylglycerol turnover and fatty acid signaling and partitioning. Hepatology 2011;53(1):116–26.

[44] Ahmadian M, Duncan RE, Varady KA, Frasson D, Hellerstein MK, Birkenfeld AL, et al. Adipose overexpression of desnutrin promotes fatty acid use and attenuates diet-induced obesity. Diabetes 2009;58(4):855–66. Available from: http://diabetes.diabetesjournals.org/content/58/4/855.

[45] Haemmerle G, Moustafa T, Woelkart G, Büttner S, Schmidt A, van de Weijer T, et al. ATGL-mediated fat catabolism regulates cardiac mitochondrial function via PPAR-α and PGC-1. Nat Med 2011;17(9):1076–85. Available from: http://www.nature.com/nm/journal/v17/n9/full/nm.2439.html.

[46] Birse RT, Choi J, Reardon K, Rodriguez J, Graham S, Diop S, et al. High-fat-diet-induced obesity and heart dysfunction are regulated by the TOR pathway in Drosophila. Cell Metab 2010;12(5):533–44. Available from: http://www.sciencedirect.com/science/article/pii/S1550413110003505.

[47] Ong KT, Mashek MT, Davidson NO, Mashek DG, Hepatic ATGL. mediates PPAR-alpha signaling and fatty acid channeling through a L-FABP independent mechanism. J Lipid Res 2014;55(5):808–15. Available from: http://www.ncbi.nlm.nih.gov/pubmed/24610891.

[48] de la Rosa Rodriguez MA, Sugahara G, Hooiveld GJEJ, Ishida Y, Tateno C, Kersten S. The whole transcriptome effects of the PPARα agonist fenofibrate on livers of hepatocyte humanized mice. BMC Genomics 2018;19(1):443. Available from: https://bmcgenomics.biomedcentral.com/articles/10.1186/s12864-018-4834-3.

[49] Rakhshandehroo M, Knoch B, Müller M, Kersten S. Peroxisome proliferator-activated receptor alpha target genes. PPAR Res 2010;2010:1–20. Available from: http://www.hindawi.com/journals/ppar/2010/ 612089/.

[50] Khan SA, Sathyanarayan A, Mashek MT, Ong KT, Wollaston-Hayden EE, Mashek DG. ATGL-catalyzed lipolysis regulates SIRT1 to control PGC-1α/PPAR-α signaling. Diabetes 2015;64(2):418–26.

[51] Rodgers JT, Lerin C, Haas W, Gygi SP, Spiegelman BM, Puigserver P. Nutrient control of glucose homeostasis through a complex of PGC-1α and SIRT1. Nature 2005;434(7029):113–8. Available from: http://www.nature.com/nature/journal/v434/n7029/full/nature03354.html.

[52] Najt CP, Khan SA, Heden TD, Witthuhn BA, Perez M, Mead LE, et al. Lipid droplet-derived monounsaturated fatty acids traffic via PLIN5 to allosterically activate SIRT1. Mol Cell 2019;doi: 10.1016/j.molcel.2019.12.003.

[53] Eichmann TO, Kumari M, Haas JT, Farese RV, Zimmermann R, Lass A, et al. Studies on the substrate and stereo/regioselectivity of adipose triglyceride lipase, hormone-sensitive lipase, and diacylglycerol-O-acyltransferases. J Biol Chem 2012;287(49):41446–57. Available from: http://www.jbc.org/content/287/49/41446.

[54] Gallardo-Montejano VI, Saxena G, Kusminski CM, Yang C, McAfee JL, Hahner L, et al. Nuclear perilipin 5 integrates lipid droplet lipolysis with PGC-1α/SIRT1-dependent transcriptional regulation of mitochondrial function. Nat Commun 2016;7:12723. Available from: http://www.ncbi.nlm.nih.gov/ pubmed/27554864.

[55] Gauthier M-S, Miyoshi H, Souza SC, Cacicedo JM, Saha AK, Greenberg AS, et al. AMP-activated protein kinase is activated as a consequence of lipolysis in the adipocyte: potential mechanism and physiological relevance. J Biol Chem 2008;283(24):16514–24. Available from: http://www.jbc.org.floyd.lib.umn.edu/cgi/content/abstract/283/24/16514.

[56] Mullins GR, Wang L, Raje V, Sherwood SG, Grande RC, Boroda S, et al. Catecholamine-induced lipolysis causes mTOR complex dissociation and inhibits glucose uptake in adipocytes. Proc Natl Acad Sci U S A 2014;111(49):17450–5. Available from: http://www.ncbi.nlm.nih.gov/pubmed/25422441.

[57] Romeo S, Kozlitina J, Xing C, Pertsemlidis A, Cox D, Pennacchio LA, et al. Genetic variation in PNPLA3 confers susceptibility to nonalcoholic fatty liver disease. Nat Genet 2008;40(12):1461–5. Available from: http://www.nature.com/ng/journal/v40/n12/full/ng.257.html.

[58] Negoita F, Blomdahl J, Wasserstrom S, Winberg ME, Osmark P, Larsson S, et al. PNPLA3 variant M148 causes resistance to starvation-mediated lipid droplet autophagy in human hepatocytes. J Cell Biochem 2018;120(1):343–56. Available from: http://www.ncbi.nlm.nih.gov/pubmed/30171718.

[59] Wang Y, Kory N, BasuRay S, Cohen JC, Hobbs HH. PNPLA3, CGI-58, and inhibition of hepatic triglyceride hydrolysis in mice. Hepatology 2019;69(6):2427–41. Available from: http://www.ncbi.nlm.nih.gov/pubmed/30802989.

[60] Yang A, Mottillo EP, Mladenovic-Lucas L, Zhou L, Granneman JG. Dynamic interactions of ABHD5 with PNPLA3 regulate triacylglycerol metabolism in brown adipocytes. Nat Metab 2019;1:560–9.

[61] BasuRay S, Smagris E, Cohen JC, Hobbs HH. The PNPLA3 variant associated with fatty liver disease (I148M) accumulates on lipid droplets by evading ubiquitylation. Hepatology 2017;66(4):1111–24. Available from: http://www.ncbi.nlm.nih.gov/pubmed/28520213.

[62] Khor VK, Ahrends R, Lin Y, Shen W-J, Adams CM, Roseman AN, et al. The proteome of cholesteryl-ester-enriched versus triacylglycerol-enriched lipid droplets. PLoS ONE 2014;9(8):e105047. Available from: http://www.ncbi.nlm.nih.gov/pubmed/25111084.

[63] Yamaguchi T, Fujikawa N, Nimura S, Tokuoka Y, Tsuda S, Aiuchi T, et al. Characterization of lipid droplets in steroidogenic MLTC-1 Leydig cells: protein profiles and the morphological change induced by hormone stimulation. Biochim Biophys Acta 2015;1851(10):1285–95. Available from: https://linkinghub.elsevier.com/retrieve/pii/S1388198115001298.

[64] Wang W, Wei S, Li L, Su X, Du C, Li F, et al. Proteomic analysis of murine testes lipid droplets. Sci Rep 2015;5(1). 12070. Available from: http://www.nature.com/articles/srep12070.

[65] Chen X, Xu S, Wei S, Deng Y, Li Y, Yang F, et al. Comparative proteomic study of fatty acid-treated myoblasts reveals role of Cox-2 in palmitate-induced insulin resistance. Sci Rep 2016;6(1). 21454. Available from: http://www.nature.com/articles/srep21454.

[66] Bozza PT, Yu W, Penrose JF, Morgan ES, Dvorak AM, Weller PF. Eosinophil lipid bodies: specific, inducible intracellular sites for enhanced eicosanoid formation. J Exp Med 1997;186(6):909–20. Available from: http://www.jem.org/lookup/doi/10.1084/jem.186.6.909.

[67] D'Avila H, Melo RCN, Parreira GG, Werneck-Barroso E, Castro-Faria-Neto HC, Bozza PT. Mycobacterium bovis bacillus Calmette-Guérin induces TLR2-mediated formation of lipid bodies: intracellular domains for eicosanoid synthesis in vivo. J Immunol 2006;176(5):3087–97. Available from: http://www.jimmunol.org/lookup/doi/10.4049/jimmunol.176.5.3087.

[68] Pacheco P, Bozza FA, Gomes RN, Bozza M, Weller PF, Castro-Faria-Neto HC, et al. Lipopolysaccharide-induced leukocyte lipid body formation in vivo: innate immunity elicited intracellular Loci involved in eicosanoid metabolism. J Immunol 2002;169(11):6498–506. Available from: http://www.jimmunol.org/lookup/doi/10.4049/jimmunol.169.11.6498.

[69] Dichlberger A, Schlager S, Lappalainen J, Käkelä R, Hattula K, Butcher SJ, et al. Lipid body formation during maturation of human mast cells. J Lipid Res 2011;52(12):2198–208. Available from: http://www.jlr.org/lookup/doi/10.1194/jlr.M019737.

[70] Dichlberger A, Schlager S, Maaninka K, Schneider WJ, Kovanen PT. Adipose triglyceride lipase regulates eicosanoid production in activated human mast cells. J Lipid Res 2014;55(12):2471–8. Available from: http://www.jlr.org/lookup/doi/10.1194/jlr.M048553.

[71] Riederer M, Lechleitner M, Köfeler H, Frank S. Reduced expression of adipose triglyceride lipase decreases arachidonic acid release and prostacyclin secretion in human aortic endothelial cells. Arch Physiol Biochem 2017;123(4):249–53. Available from: https://www.tandfonline.com/doi/full/10.1080/13813455.2017.1309052.

[72] Schlager S, Goeritzer M, Jandl K, Frei R, Vujic N, Kolb D, et al. Adipose triglyceride lipase acts on neutrophil lipid droplets to regulate substrate availability for lipid mediator synthesis. J Leukoc Biol 2015;98(5):837–50. Available from: http://doi.wiley.com/10.1189/jlb.3A0515-206R.

[73] Sztalryd C, Brasaemle DL. The perilipin family of lipid droplet proteins: gatekeepers of intracellular lipolysis. Biochim Biophys acta Mol Cell Biol Lipids 2017;1862(10 Pt B):1221–32. Available from: https://linkinghub.elsevier.com/retrieve/pii/S1388198117301403.

[74] Najt CP, Senthivinayagam S, Aljazi MB, Fader KA, Olenic SD, Brock JR, et al. Liver-specific loss of perilipin 2 alleviates diet-induced hepatic steatosis, inflammation, and fibrosis. Am J Physiol Gastrointest Liver Physiol 2016;310:G726–38. Available from: http://www.ncbi.nlm.nih.gov/pubmed/26968211.

[75] Chen FL, Yang ZH, Wang XC, Liu Y, Yang YH, Li LX, et al. Adipophilin affects the expression of TNF-alpha, MCP-1, and IL-6 in THP-1 macrophages. Mol Cell Biochem 2010;337(1–2):193–9. Available from: http://link.springer.com/10.1007/s11010-009-0299-7.

[76] Sohn JH, Lee YK, Han JS, Jeon YG, Kim JI, Choe SS, et al. Perilipin 1 (Plin1) deficiency promotes inflammatory responses in lean adipose tissue through lipid dysregulation. J Biol Chem 2018;293:13974–88. Available from: http://www.ncbi.nlm.nih.gov/pubmed/30042231.

[77] Kien B, Grond S, Haemmerle G, Lass A, Eichmann TO, Radner FPW. ABHD5 stimulates PNPLA1-mediated omega-O-acylceramide biosynthesis essential for a functional skin permeability barrier. J Lipid Res 2018;59:2360–7. Available from: http://www.ncbi.nlm.nih.gov/pubmed/30361410.

[78] Fischer J, Lefevre C, Morava E, Mussini JM, Laforet P, Negre-Salvayre A, et al. The gene encoding adipose triglyceride lipase (PNPLA2) is mutated in neutral lipid storage disease with myopathy. Nat Genet 2007;39(1):28–30.

[79] Wang H, Sreenivasan U, Hu H, Saladino A, Polster BM, Lund LM, et al. Perilipin 5, a lipid droplet-associated protein, provides physical and metabolic linkage to mitochondria. J Lipid Res 2011;52(12):2159–68. Available from: http://www.ncbi.nlm.nih.gov/pubmed/21885430.

[80] Benador IY, Veliova M, Mahdaviani K, Petcherski A, Wikstrom JD, Assali EA, et al. Mitochondria bound to lipid droplets have unique bioenergetics, composition, and dynamics that support lipid droplet expansion. Cell Metab 2018;27(4):869–85. e6. Available from: http://www.ncbi.nlm.nih.gov/pubmed/29617645.

[81] Mashek DG, Khan SA, Sathyanarayan A, Ploeger JM, Franklin MP. Hepatic lipid droplet biology: getting to the root of fatty liver. Hepatology 2015;62(3):964–7.

[82] Anand P, Cermelli S, Li Z, Kassan A, Bosch M, Sigua R, et al. A novel role for lipid droplets in the organismal antibacterial response. Elife 2012;1. e00003. Available from: https://elifesciences.org/articles/00003.

[83] Miller BF, Abrams R, Dorfman A, Klein M. Antibacterial properties of protamine and histone. Science 1942;96(2497):428–30. Available from: http://www.ncbi.nlm.nih.gov/pubmed/17729719.

[84] Gong J, Sun Z, Wu L, Xu W, Schieber N, Xu D, et al. Fsp27 promotes lipid droplet growth by lipid exchange and transfer at lipid droplet contact sites. J Cell Biol 2011;195(6):953–63. Available from: http://www.ncbi.nlm.nih.gov/pubmed/22144693.

[85] Ueno M, Shen W-J, Patel S, Greenberg AS, Azhar S, Kraemer FB. Fat-specific protein 27 modulates nuclear factor of activated T cells 5 and the cellular response to stress. J Lipid Res 2013;54(3):734–43. Available from: http://www.ncbi.nlm.nih.gov/pubmed/23233732.

[86] Ohsaki Y, Cheng J, Fujita A, Tokumoto T, Fujimoto T. Cytoplasmic lipid droplets are sites of convergence of proteasomal and autophagic degradation of apolipoprotein B. Mol Biol Cell 2006;17(6):2674–83. Available from: http://www.ncbi.nlm.nih.gov/pubmed/16597703.

[87] Suzuki M, Otsuka T, Ohsaki Y, Cheng J, Taniguchi T, Hashimoto H, et al. Derlin-1 and UBXD8 are engaged in dislocation and degradation of lipidated ApoB-100 at lipid droplets. Mol Biol Cell 2012;23(5):800–10. Available from: https://www.molbiolcell.org/doi/10.1091/mbc.e11-11-0950.

[88] Ohsaki Y, Cheng J, Suzuki M, Fujita A, Fujimoto T. Lipid droplets are arrested in the ER membrane by tight binding of lipidated apolipoprotein B-100. J Cell Sci 2008;121(14):2415–22. Available from: http://jcs.biologists.org/cgi/doi/10.1242/jcs.025452.

[89] Spandl J, Lohmann D, Kuerschner L, Moessinger C, Thiele C. Ancient ubiquitous protein 1 (AUP1) localizes to lipid droplets and binds the E2 ubiquitin conjugase G2 (Ube2g2) via its G2 binding region. J Biol Chem 2011;286(7):5599–606. Available from: http://www.ncbi.nlm.nih.gov/pubmed/21127063.

[90] Klemm EJ, Spooner E, Ploegh HL. Dual role of ancient ubiquitous protein 1 (AUP1) in lipid droplet accumulation and endoplasmic reticulum (ER) protein quality control. J Biol Chem 2011;286(43):37602–14. Available from: http://www.ncbi.nlm.nih.gov/pubmed/21857022.

[91] Hartman IZ, Liu P, Zehmer JK, Luby-Phelps K, Jo Y, Anderson RGW, et al. Sterol-induced dislocation of 3-hydroxy-3-methylglutaryl coenzyme A reductase from endoplasmic reticulum membranes into the cytosol through a subcellular compartment resembling lipid droplets. J Biol Chem 2010;285(25):19288–98. Available from: http://www.jbc.org/lookup/doi/10.1074/jbc.M110.134213.

[92] Jiang H, He J, Pu S, Tang C, Xu G. Heat shock protein 70 is translocated to lipid droplets in rat adipocytes upon heat stimulation. Biochim Biophys Acta 2007;1771(1):66–74. Available from: https://linkinghub.elsevier.com/retrieve/pii/S1388198106003167.

CHAPTER

# 10

# Lipid droplets in the immune response and beyond

*Sheetal Gandotra*

Council of Scientific and Industrial Research-Institute of Genomics and Integrative Biology, Academy of Scientific and Innovative Research, New Delhi, India

OUTLINE



Free fatty acids represent less than 2% of the total cellular fatty acid content, with different cell types exhibiting different capacity to store excess fatty acids in the form of triglycerides and cholesterol esters. The extreme hydrophobicity of these storage lipids enables efficient packaging of these lipids into organelles called lipid droplets. These organelles can range in diameter from less than a micron to about 50 μm, depending on the cell type. It is the central role of these organelles in lipid signaling and mobilization, particularly relevant to the

Lipid Signaling and Metabolism. http://dx.doi.org/10.1016/B978-0-12-819404-1.00010-5

immune system, which will form the basis of this chapter. While most of our understanding of lipid droplets comes from nonimmune cells, the newly emerged concept that altering the metabolic state of an immune cell can have important consequences for its polarized state has rekindled interest in the role of lipid droplets in immune cells, particularly the myeloid cells.

## Structure and topology of lipid droplets

Triglycerides and cholesterol esters form bulk of the neutral lipids present in lipid droplets [1]. The neutral lipid core is surrounded by a phospholipid monolayer, which ensures that the droplets can exist as nonaggregating lipid depots within the cell. This also means that the droplets do not aggregate with other hydrophobic biomolecules and can be metabolized in a regulated manner. These phospholipid monolayer encapsulated droplets originate at specified regions of the endoplasmic reticulum membrane; these sites are defined by enzymes that are involved in initiation of lipid droplet synthesis. Free fatty acids must first be charged to an ester with Coenzyme A before they can be utilized for synthesis of neutral lipids. Acyl CoA synthetase long chain family member 3 (ACSL3) is the primary enzyme involved in this process. ACSL3 protein enrichment at sites of new LD initiation in response to exogenous fatty acids has suggested that ACSL3-dependent fatty acyl CoA must make high density of fatty acids locally available for subsequent steps in triglyceride synthesis [2]. Triglyceride synthesis can be initiated from glycerol-3-phosphate (Kennedy pathway) or monoacylglycerol. In case of the former, membrane localized phospholipids serve as intermediates, which upon the action of phosphatidic acid phosphatase, generate diacylglycerol (DAG). In case of the latter, monoacylglycerol is sequentially acylated by monoacylglycerol O-acyl transferase (MGAT) to DAG. Both pathways converge on the diacylglycerol O-acyl transferase (DGAT)-dependent acylation of DAG to TAG [3]. The initial "lens" of triglyceride synthesized within the ER bilayer membrane must at some point egress outward to form a LD bud. These bud points are specified by local phospholipid composition and surface tension [4]. For example, positive curvature induced by lysophosphatidyl choline can spontaneously initiate lipid droplet budding from an artificial ER membrane-triglyceride in vitro model system and in ER membranes in vivo [5,6]. ER-LD contacts regulate the flux of lipids into LDs. Proteins such as Seipin, FATP1, sorting nexin 14 (SNX14), DFCP1, Rab18, and Zw10 are known to localize to the ER-LD contact sites and mediate this interaction [7–9]. Sorting nexin 14 contains an N terminal transmembrane region that localizes it to the ER membrane and a C terminus nexin domain with amphipathic helices with which it latches on to the lipid droplet, thereby acting as a bridge between the ER membrane and the budding lipid droplet [10]. Although Seipin and Snx14 have similar mechanisms of targeting across the ER and LD neither can compensate for the loss of the other, suggesting independent requirement for lipid droplet growth.

In addition to the ER, lipid droplets have been shown to interact with several intracellular organelles, including the mitochondria, peroxisomes, Golgi, lysosomes, and autophagosomes [11] (Fig. 10.1). The LD-mitochondria contact provides the means to traffic fatty acids for beta-oxidation under conditions of starvation [12]. The LD-peroxisome contact maybe required for fatty acid trafficking to and biogenesis of peroxisomes [13]. LD-Golgi contacts are mediated by Arf1, enabling transport of LDs for the secretion of VLDL particles [14]. LD-autophagosome contacts are thought to provide access of lipids for degradation under conditions of starvation [15]. Molecular players involved in these interactions are just

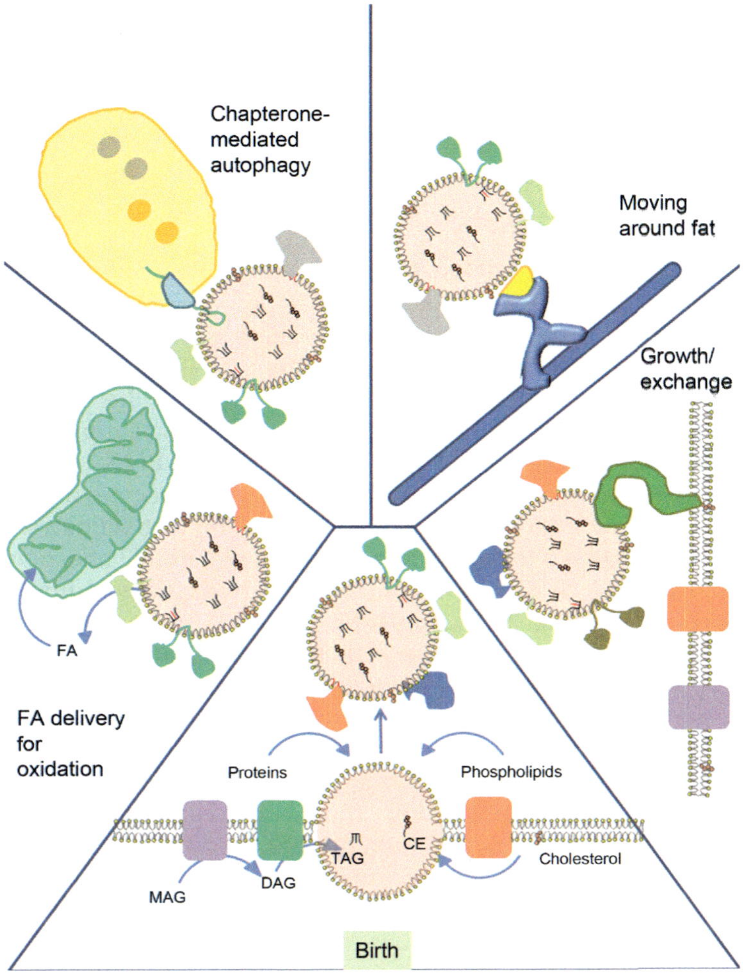


**FIGURE 10.1 Lipid droplet life cycle.** The "birth" of a lipid droplet takes place at the ER membrane where neutral lipid triglycerides and cholesterol esters are synthesize and then packaged as a lens between the two leaflets. Phospholipids and proteins at the cytosolic face of the membrane provide positive curvature resulting in pinching out of the newly formed lipid droplet. Proteins can bind to the lipid droplet in a stable and exchangeable manner. The droplet can interact with many organelles, including the mitochondria where the interaction leads to efficient transfer of fatty acid for oxidation, or the lysosome where the interaction leads to lipophagy or with the ER membrane for exchange of lipids. The droplet can be moved from one location in the cell efficiently and in a vectorial manner by the action of motor proteins kinesins and dyneins.

beginning to be understood. Proteins are thought to play important roles in the interactions and on determining the diverse functions of the LD. However, considering that the core of the LD is nonpolar, and the surface composed of a phospholipid monolayer, the localization of proteins within the core of the droplet seems implausible and also not supported by cryo-electron microscopy. Most data support either ER to LD or cytosol to LD targeting of proteins. The differential targeting of proteins generates a diversity within the pool of lipid droplets. This diversity has been a challenge to investigate within a single cellular system; the diversity across different models has illuminated the many functions of the lipid droplet. For example, stellate cells that contain mostly cholesterol ester as the LD core are likely to recruit different proteins to the monolayer compared to say macrophages which have mainly triglyceride as the LD core. The LD proteome also reflects the kinds of interactions that droplets are likely to be making under the physiological state from which they were isolated.

The presence of specific coat proteins is likely linked to the preferred method of lipolysis in that cell type. Lipid droplet coat proteins play a role in regulating lipolysis. The PAT family of coat proteins, comprising of Perilipin (Perilipin 1), Adipophilin (Perilipin 2), and TIP47 (Perilipin 3) are the major lipid droplet coat proteins. They follow different mechanisms of regulating lipolysis; PLIN1, the major LD coat protein in adipocytes (Fig. 10.2) seems to prevent basal lipolysis by preventing interaction between the lipase adipose tissue triglyceride lipase (ATGL) and its activator CGI58, by scavenging the latter [16–19]. Beta adrenergic stimulation leads to protein kinase A-dependent phosphorylation of PLIN1, thereby relieving CGI58 for activation of ATGL at the droplet surface. PLIN2 and PLIN3 also suppress basal lipolysis but do not follow the same mechanism of preventing lipolysis [20]. PLIN2 was the major LD coat protein identified in human macrophages [21] (Fig. 10.2). Hepatocytes also express PLIN2 on their lipid droplets and much of our knowledge on PLIN2 regulated lipolysis comes from this cell type. One model of lipolysis in hepatocytes proposes that PLIN2 phosphorylation by AMP regulated protein kinase (AMPK) targets it for chaperone-mediated autophagy in liver cells [22]. These findings suggest that PLIN2 must be degraded for lipolysis to proceed via lipophagy. Recently, the stepwise role of cytosolic lipases and lysosomal lipases for lipid droplet turnover was proposed. This alternative model proposes that cytosolic lipases must degrade larger lipid droplets to smaller ones, which can then become substrates of lysosomal lipases [23]. Together, there seems to be a requirement of regulating lipid droplet size prior to lipolysis. This idea is resonant with the fragmentation of lipid droplets in catecholamine stimulated 3T3L1 adipocytes which exhibit lipid droplet fragmentation independent of the presence of activated lipases on the lipid droplets, although phosphorylated HSL tended to localize to shrinking lipid droplets [24,25]. Moreover, inhibition of lipolysis still allows fragmentation of lipid droplets in response to catecholamine stimulation while catecholamine responsive phosphorylation of Perilipin 1 at serine 492 is required for this fragmentation [26]. Lipolysis in macrophages has not been studied with respect to precise mechanisms that are operative at the level of the lipid droplet or its coat proteins but may be crucial to understanding how lipid mediators of inflammation and lipid droplets participate in inflammation and host defense.

## Signaling intermediates and lipid droplets

Fatty acids can appear in the cellular environment from four possible routes, uptake of free fatty acids via scavenger receptors, uptake of lipid-esterified fatty acids that are made available within the lysosome upon degradation of the incoming lipid, degradation of cellular lipids, and de novo fatty acid synthesis. But the steady-state levels of cellular free fatty acids remain largely unchanged with an increase in fatty acids from any of these routes. This is because the free fatty acid is rapidly esterified as a fatty acyl CoA ester which can be used for β-oxidation or for synthesis of new lipids. Free fatty acids, their oxidation products, and lipid intermediates of triglyceride synthesis or hydrolysis such as phosphatidic acid (PA), lysophosphatidic acid (LPA),

DAG, and monoacylglycerol can all function as signaling intermediates. Much of our knowledge about how lipid droplets regulate these comes from studies on cells that do not belong to the immune system but the concepts may translate to cells of the immune system also.

## Fatty acids in ER stress and lipotoxicity: lipid droplets to the rescue

The concern about exogenous free fatty acids is relevant in the context of findings that elevated plasma free fatty acids correlate with risk of metabolic disease and heart disease. This phenomenon is likely to be a greater concern for cells with limited triglyceride storage capacity, such as pancreatic β cells and cardiomyocytes [27,28] under conditions of elevated levels of circulating free fatty acids. The ability to store exogenous fatty acids is indeed a beneficial trait for cardiomyocytes, revealed by the ability of the stored fat to regulate mitochondrial fatty acid beta-oxidation, provide energy, and limit lipotoxicity [29]. Fatty acids mobilized by ATGL in fact regulate PPARα in cardiomyocytes. The absence of ATGL in cardiomyocytes leads to decreased expression of the master regulator of mitochondrial biogenesis and fatty acids oxidation, PGC-1α and PGC-1β [29]. Therefore, storage of circulating fatty acids in cardiomyocyte lipid droplets is a homeostatic response tied in closely with their own metabolism.

Could the knowledge of how cells respond to exogenous fatty acids therefore allow means of increasing the cellular triglyceride storage capacity and mitigate cardiotoxicity? A transient increase in free fatty acids is enough to inhibit the cellular synthesis of fatty acids and enhance their incorporation into triglycerides. But different fatty acids can induce lipotoxicity and triglyceride synthesis to different extents, with saturated fatty acids such as palmitate (16:0) and stearate (18:0) being more adept at causing lipotoxicity while oleic acid (18:1) and linoleic acid (18:2) being more adept at ensuring triglyceride storage and mitigating the saturated fatty acid-mediated lipotoxicity [28,30]. In fact, it is not just that the unsaturated fatty acid can itself be assimilated better but it ensures improved assimilation of other fatty acid species [28]. How does this happen? Firstly, saturated fatty acids such as palmitate induce ER stress-mediated lipotoxicity [30]. This involves incorporation of palmitate to phospholipids and triglycerides in microsomal membranes, associated with dilation of the ER and redistribution of ER resident chaperones to the cytosol. These lipotoxic events can be averted by increasing fatty acid oxidation, which also lead to decreased assimilation of palmitate to phospholipids, suggesting this assimilation in microscoma membranes to be a key driver of palmitate-induced toxicity [30]. It turns out that long chain unsaturated fatty acids can promote triglyceride synthesis by binding to an endoplasmic reticulum localized fatty acid sensor, ubiquitin-like (UBX)-domain-containing protein UBXD8 [31,32]. In the absence of exogenous fatty acids, UBXD8 promotes de novo fatty acid synthesis but prevents triglyceride storage. Exogenous unsaturated fatty acids promote the polymerization of the sensor, rendering it inactive and allowing triglyceride storage [32]. Furthermore, LD localized UBXD8 directly binds to ATGL and prevents its activation by CGI58, thereby promoting growth of the droplet in the presence of exogenous fatty acids [33].

## Free and esterified eicosanoids

Eicosanoids are arachidonic acid-derived lipid mediators of inflammation. They are synthesized by stereo- and regio-specific peroxidation of arachidonic acid by three enzyme families

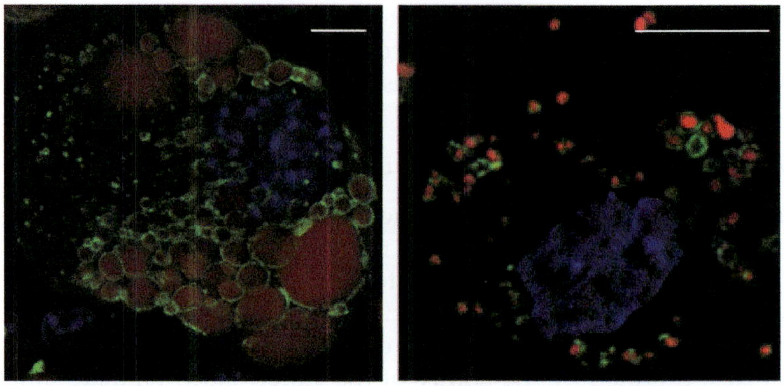

FIGURE 10.2 **Size, density, and coat proteins in adipocytes versus and macrophages.** 3T3L1 adipocytes and THP-1 macrophages immunostained for PLIN1 and PLIN2 (*pseudocolored green* [*black* in print version]), respectively. Neutral lipids stained using BODIPY493/503 (*pseudocolored red* [*light gray* in print version]). DAPI stained nuclei can be seen in *blue* (*dark gray* in print version). Scale bar = 10 μm. Source*: Data are from Menon D, Singh K, Pinto SM, Nandy A, Jaisinghani N, Kutum R, et al. Quantitative lipid droplet proteomics reveals mycobacterium tuberculosis induced alterations in macrophage response to infection.* ACS Infect Dis *2019;***5***: 559–569. https://pubs.acs.org/doi/pdf/10.1021/acsinfecdis.8b00301.*

namely lipoxygenases, cyclooxygenases, and cytochrome P450 (Fig. 10.3). Eicosanoids signal via an autocrine and paracrine effect wherein the eicosanoid can diffuse out of the cell in which it is synthesized to activate G protein coupled receptors on neighboring cells. In addition, eicosanoids can be taken up by cells to be incorporated into phospholipids. For example, 15-hydroxyeicosatetraenoic acid (15-HETE), a product of 15-lipoxygenase activity can be assimilated to phospholipids if provided exogenously. Synthesis of both free and esterified eicosanoids is possible by all three enzymes families. While earlier literature on eicosanoids points toward the role of free arachidonic acid-derived eicosanoids in inflammation, esterified eicosanoids have also been recognized as mediators and regulators of inflammation. The role of lipid droplets in both free and esterified eicosanoids will be discussed in this section and is summarized in Fig. 10.3.

Eicosanoids have been classically linked with the activity of phospholipase A2, which releases arachidonic acid present at the sn-2 position of phospholipids. The evidence that lipid droplets house arachidonic acid dates back to early 1980s, when lipid droplets had barely assumed their organelle status [34]. Mast cells are a rapid source of eicosanoids during infection; the observation that these cells contain abundant lipid droplets in vivo, led to the question of whether these droplets contribute to eicosanoid production. The observation that arachidonic acid fed to mast cells in vitro led to its incorporation primarily in cytosolic lipid droplets was the first clue that lipid droplets could potentially be a source of arachidonic acid in times of inflammatory burst [34]. The obvious question that comes to mind with these findings is, does the accumulation of arachidonic acid into neutral lipids ensure that free arachidonic acid is not available under nonstimulated condition and upon exposure to a danger signal, induction of COX2 will enable conversion of whatever arachidonic acid is available toward eicosanoid production? In order to poise the cells for robust release of arachidonic acid to achieve full inflammatory potential, there must be either induction of the lipase, or its activation at the lipid droplet surface. Some 20 years later, this concept was revisited when the role of triglyceride hyrdolase Patatin-like phospholipase domain-containing protein

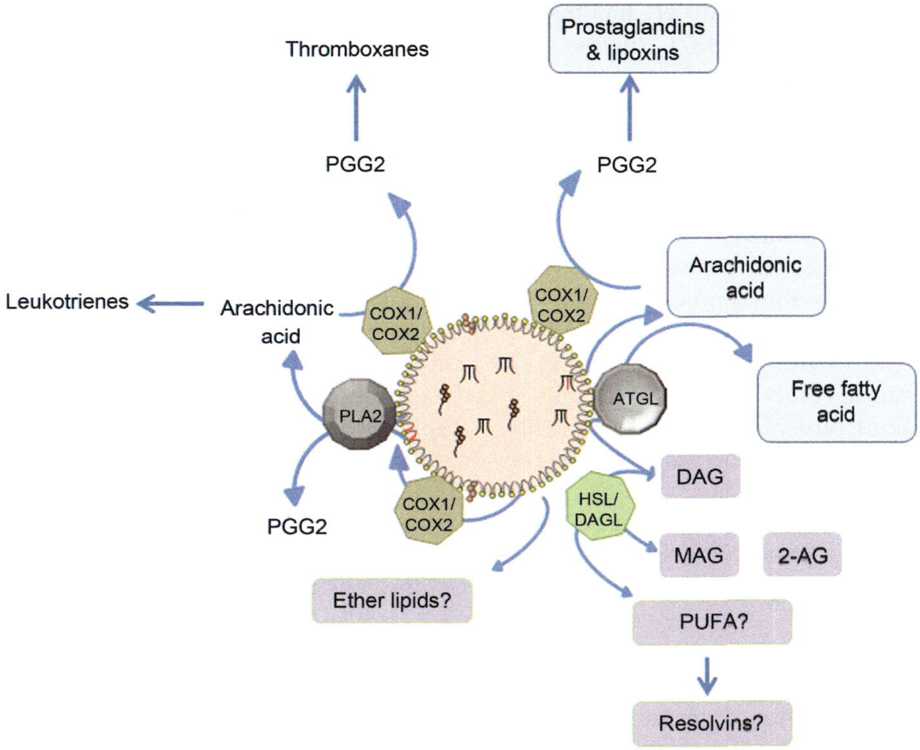


FIGURE 10.3 **Lipid droplets and signaling lipids.** Indicated are possible pathways for generating fatty acid containing mediators of inflammation and inflammation resolution from lipolytic activity at the lipid droplet surface. Production of prostaglandins and the lipoxin LXB4 is regulated by the presence of lipid droplets. Reliance on neutral lipid pools for generating signaling lipids such as MAG, DAG, polar or nonpolar ether lipids, or of PUFA derived resolvins is yet to be related with the abundance of lipid droplet stored pools of neutral lipid precursors.

2 (PNPLA2)/adipose triglyceride lipase (ATGL) in human mast cells was addressed [35]. Enzymes involved in liberating the free eicosanoids from triglycerides and phospholipids, ATGL and cPLA2 were not induced in response to IgE crosslinking, a condition that is known to induce COX-2. Instead, the basal levels of COX-1 were found to coincide with the initial burst of prostaglandin D2 and leukotriene C4 release. Silencing of either ATGL or cPLA2 could reduce this eicosanoid release. Therefore, it is thought that arachidonate released from triglyceride could be either directly utilized by COX1 or esterified to phospholipids, which in turn could be hydrolyzed by PLA2 and then utilized by COX1 (Fig. 10.3).

Besides mast cells, macrophages are also important source of eicosanoids. Several human pathologies are associated with macrophages laden with lipid droplets; this includes atherosclerosis plaques, tuberculosis granulomas, and neoplastic processes. In contrast, normal resident macrophages do not contain abundant lipid droplets. Patricia Bozza's group developed an assay called EicosaCell wherein chemical crosslinking of newly synthesized eicosanoids to nearby proteins could be detected using an antibody against a chosen eicosanoid, followed by a fluorophore tagged secondary antibody [36]. Specifically the COX1 products PGD2 and LTC4 could be localized to lipid droplets of *M. bovis* BCG infected macrophages. Recently, the presence of lipid droplets has been linked to the release of eicosanoids in response to *M. tuberculosis* infection [37]. Release of COX1-dependent prostaglandins (PGD2, PGE2, and PGF2) was found to be triglyceride synthesis dependent. Surprisingly, the release of LXB4 was dependent on triglyceride levels, while 15HETE was not, whereas both are products of 15LOX activity, suggesting that substrate availability itself may not be the sole reason for specific eicosanoids to be generated in a triglyceride dependent manner.

Arachidonic acid is certainly not the only fatty acid within lipid bodies. In fact, besides saturated and unsaturated fatty acids, polyunsaturated fatty acids belonging to n-6 PUFA and n-3 PUFA are also present in triglycerides. The n-3 PUFAs such as eicosapentaenoic acid (EPA) and docosahexaenoic acid (DHA) also regulate synthesis of important lipid mediators of inflammation. EPA being a competitive substrate for COX can also inhibit arachidonic acid metabolism directly [38]. In addition, EPA and DHA metabolism results in production of inflammation resolving lipid mediators called as resolvins [39]. What then governs whether the final outcome of ATGL activity on lipid droplets is going to be proinflammatory or anti-inflammatory? One can speculate that the composition of fatty acids within the triglycerides may regulate the final outcome. But how is ATGL regulated in these cells? Today we understand a great deal about lipid mobilization in adipocytes wherein the role of ATGL has been well described. Our knowledge on regulated lipolysis in immune cells is relatively limited. The nature of the lipid droplet proteome in immune cells may dictate pathways involved.

## Diacylglycerol

DAG, an activator of protein kinase C (PKC), can lead to monocyte to macrophage differentiation. One of the targets is the NADPH oxidase complex which is required for pathogen induced oxidative burst. Phorbol 12-myristate 13-acetate, a DAG homolog is in fact used for differentiating monocyte cell lines to macrophages in vitro. Of the different PKC isoforms, PKCα is the DAG/PMA responsive one present in macrophages [40]. Phospholipase C (PLC) activity on phosphatidylinositol 4,5-bisphosphate results in two secondary messengers- inositol 1,4,5-bisphosphate and DAG. PLC isoforms are linked with several proinflammatory signal transduction pathways involving nuclear factor kappa (NF-κB), mitogen-activated protein kinase (MAPK), and activation of interferon regulatory factors (IRFs) (reviewed in Zhu et al. [40a]). DAG therefore has the potential to regulate key inflammatory processes of macrophages.

## Monoacylglycerol

Monoacylglycerols have more recently emerged as important signaling intermediates that are involved in regulating immune responses, neuronal activity, tumor growth, and cardiovascular function. Lipolysis is the major route of MAG formation through a stepwise degradation of TAG to DAG, followed by activity of HSL or DAGLβ which hydrolyse DAG to MAG [41]. Phospholipid-derived MAG is the alternate route. 1-MAG or 2-MAG can be generated by lysophosphatidic acid phosphatase activity on LPA. 1-MAG can also be derived from DAG which in turn can be derived from PL by the activity of phospholipase C. The localization of enzymes on the outer surface of the cell or ER lumen regulates where MAG synthesis takes place. The levels of MAG can be further regulated by monoacylglycerol lipases, of which MAGL exhibits the highest contribution, followed by α/β hydrolase domain 6(ABHD6) and ABHD12. While MAGL hydrolyzes both 1-MAG and 2-MAG, ABHD6 exhibits preference for 1-MAG. Of all the MAG species, the endocannabinoid 2-arachindoyl glycerol (2-AG) is the best studied. 2-AG exhibits pleiotropic effects on proinflammatory cytokine secretion with an inhibitory effect on LPS stimulated TNF secretion from murine macrophages while inhibition of its production suppresses LPS stimulated TNF secretion in both human dendritic cells and murine models [41–43]. Some of these observed differences could be due to different signaling pathways adopted by externally provided 2-AG versus intracellular 2-AG. 2-AG binds to several cannabinoid receptors 1 and 2 (CB1 and CB2) while other MAG species can bind to the GPCR GPR119, the capsaicin receptor TRPV1, and intracellular receptors PPARα/γ and Munc13-1 [44]. Expression of CB1 receptors is most abundant in the central nervous system while that of CB2 receptors is restricted to peripheral tissues and immune cells, with expression being high in B lymphocytes, natural killer cells, and cells of the myeloid lineage. Role of MAG signaling via the other receptors is largely unstudied in the immune system, probably due to higher expression of these receptors in neurons, cells of the GI tract, and pancreatic beta cells.

## Glyceryl prostaglandins

Glyceryl prostaglandins are esters of prostaglandins generated by the action of COX-2 on glyceride lipids [45]. They exhibit functions distinct from free prostaglandins or their parent glyceride. For example, PGE2-glycerol (PGE2-G) in macrophages triggers calcium mobilization and transient rise in IP3, eventually activating the ERK/MAPK pathway [46]. In contrast, PGE2 activates the ERK/MAPK pathway independent of calcium mobilization and more importantly, PGE2-G conversion to PGE2 is not required for this activity of PGE2-glyceride. Moreover, PGE2-G signals via the G protein-coupled receptor $P2Y_6$ at an EC50 value of ~1 pM, indicating activation even at the physiologically low levels of the agonist. [47]. The presence of enzymes involved in the synthesis and metabolism of prostaglandin-glycerides at the lipid droplet surface (discussed in section "Lipid droplet proteome in immune cells") implicates a role for lipid droplets in regulating processes affected by free and esterified glycerides.

## Ether lipids

Ether lipids differ from other glycerolipids in that the *sn-1* position of the glycerol backbone is attached by an ether bond. Their synthesis initiates in the peroxisome and is completed in the ER. Plasmalogens are a type of etherphospholipid with a vinyl group containing fatty acid at the sn-1 position. It constitutes nearly 50% of the PC pool of neutrophils. De novo synthesized fatty acid is required for neutrophil plasmalogen which is important for neutrophil survival [48]. Ether lipids can bind to PPARγ and promote PPARγ-dependent luciferase reporter activity [49]. Exogenous alkyl glycerols can in fact compensate for loss of de novo ether lipid synthesis in promoting adipogenesis [50]. Ether triglycerides are also present in lipid droplets and their synthesis is also dependent on the enzyme DGAT1. However, the precise function of ether triglycerides and their metabolites derived from have not been addressed so far. A decrease in ether lipid content in pathologies such as neurodegenerative diseases and metabolic disorders and as a biomarker in breast cancer has invigorated interest in this relatively unexplored class of lipids [51].

## Lipid metabolism in polarization of the immune response

Akin to adipocytes, lipid droplet formation for regulated utilization upon hormonal demands, lipid droplets in immune cells could also play a role in changing metabolic programs during the different phases of a disease. For example, immune cell activation at the site of injury, infection, or other pathologies follows a biphasic response with cell types and secreted mediators promoting the recruitment of macrophages and neutrophils to remove the pathogenic insult and then second phase of resolution which removes the dying cells, induces fibrosis and allow reestablishment of the healthy tissue. This homeostatic response if impaired in the second phase leads to chronic inflammation and tissue damage. The involvement of macrophage and T cell polarization in either pro- or antiinflammatory phenotype is noted during most pathologies [52,53]. This polarization seems to exhibit a reciprocal relationship with fatty acid oxidation and fatty acid synthesis. Antigen-stimulated T cells promote aerobic glycolysis, fatty acid synthesis, and amino acid metabolism to support clonal expansion [54]. In contrast, memory cells and naive T cells rely on oxidative phosphorylation to survive in a quiescent state [55]. Macrophages are polarized to M1, or inflammatory state by stimuli such LPS and IFNγ. This polarization is associated with their ability to increase glucose uptake and de novo fatty acid and sterol synthesis in an SREBP-dependent manner, yet they also increase their uptake of exogenous fatty acids [56,57]. In fact, activation of several toll like receptors (TLR1,2,3, and 4) induce human and murine macrophages to store triglycerides in an exogenous fatty acid-dependent manner [57]. M2 macrophages, on the other hand, are associated with wound healing and helminth infections; these are also termed alternatively activated macrophages. Interleukin 4 is a T helper type 2 (Th2) cell-derived cytokine which leads to alternative activation of macrophages which involves shift from glucose oxidation to fatty acid oxidation. For this shift also, macrophages rely on uptake of exogenous lipoproteins via the scavenger receptor CD36, followed by lysosomal lipolysis and metabolism of the incoming fatty acid for elevated oxidative phosphorylation [58]. So while M1 macrophages take up more glucose, they use it for the synthesis of fatty acids while M2 macrophages utilize glucose for oxidative metabolism and while M1 macrophages utilize exogenous and de novo synthesized fatty acid for storage of lipids, M2 macrophages utilize exogenous lipids for fatty acid oxidation [59].

## Adipogenic response to exogenous lipids

In contrast to exogenous fatty acids, esterified fatty acids in the form of exogenous lipids do not induce lipotoxicity in the same manner. In fact, necrotic cells which are a source of exogenous lipids in the context of various infectious and noninfectious pathologies can be taken up by macrophages and assimilated into triglycerides [21]. This involves lysosomal lipase-dependent hydrolysis of the lipids followed by assimilation of the fatty acid to triglycerides, similar to the process of lipid assimilation from lipoproteins. Unlike palmitate, lipoprotein exposure of macrophages does not elicit a proinflammatory response, and the levels of tumor necrosis factor (TNFα) expression seem to be independent of DGAT1 expression [60]. Inhibition of triglyceride synthesis does not lead to increased death of the macrophage under either necrotic cell or lipoprotein exposure, unlike free fatty acid, suggesting the free form of the fatty acid to be more toxic than the esterified form [21, 28, 30, 60]. In any case, both sources of the fatty acid result in increased abundance of lipid droplets. Central to both processes is the ability of the ER resident enzyme diacylglycerol O-acyltransferase 1 to esterify fatty acids with diacylglyceride to generate triglyceride. This is a polytopic transmembrane protein, with the N terminus exposed to the cytosol and the C terminus facing the ER lumen with the active site close to the inner leaflet of the ER membrane [61]. Inhibition, deletion, and silencing of DGAT1 have been shown to play a cell type-dependent effect on the outcome of fatty diet overload. DAG activates protein kinase C epsilon, which leads to lipid induced insulin resistance [62]. Therefore, flux of DAG toward triglyceride is likely to limit activation of this pathway. Paradoxically, silencing of *Dgat2* in rats fed a high fat diet, led to decrease in hepatic triglyceride and improvement in hepatic insulin sensitivity [63]. This was due to the induction of genes involved in fatty acid oxidation and thermogenesis (*Cpt1* and *Ucp2*) and reduction in expression of one of the genes involved in synthesis of DAG (*Gpat*), thereby lowering the levels of DAG itself. Surprisingly, silencing of *Dgat1* in the same model had no effect on hepatic lipid levels.

Because fatty acid biosynthesis is highly regulated and responsive to availability of exogenous fatty acids and esterified lipids, the formation of lipid droplets may have roles additional to circumventing lipotoxicity. It is therefore not surprising that lipid droplets are being recognized as having functions in addition to being "storage depots" of fatty acids. These functions have only begun to be explored which range from regulation of bioactive lipids, immune activation, and antiviral immune response.

## LDs and inflammation

Whether the accumulation of lipid droplets in macrophages and dendritic cells is linked to the proinflammatory state or antiinflammatory state depends on the context of the pathology. The source and type of lipid involved and the underlying metabolic reprogramming of the cell is key to the final outcome. Let us take the example of fatty acids; while stearate, oleate, and linoleate are able to induce LD formation in recipient cells, stearate induces a proinflammatory state while oleate and linoleate suppress it. While these can be regarded as properties of the fatty acids independent of their lipid droplet inducing properties, evidence from dendritic cells in healthy versus cancer tissue are also pointing to the same facet. DCs can be sorted into "high LD" and "low LD" populations from a healthy liver tissue. It was found that the "high LD" population of DCs from normal liver tissue were better able to stimulate T cells, natural killer (NK) cells, and NKT cells and provide enhanced cytotoxic T cell activity when used for immunization [64]. However, LD accumulation in cancer resident DCs is associated with immunogenic dysfunction [65]. In this case, "high LD" DCs from tumor cell-conditioned media were less able to stimulate T cells and exhibited poor MHC class I-restricted antigen presentation [66]. ER stress is a driver for LD biogenesis in DCs yet, it plays very different roles in case of normal liver DCs and tumor resident DCs. In the context of normal liver DCs, ER stress was required for the immunogenic function of "high LD" DCs [64]. In contrast cancer resident DCs exhibited an opposite phenotype; silencing of the ER stress adaptor XBP1 in tumor associated DCs led to decrease in LDs per cell and restored a protective immune response against the tumor via improved antigen presentation [66]. Whether the lipid droplet itself plays a role in antigen presentation or it is simply associated with the phenotypic shift with XBP1 deletion, remains to be elucidated. These studies also collectively point toward the possibility of diversity in lipid droplet types that are generated in response to cancer tissue in comparison with a healthy tissue. Understanding the precise composition of lipids and proteins of the lipid droplet in these scenarios may provide explanations for these differences and provide mechanistic insights into lipid droplet function in health and disease.

A similar dichotomy of lipid droplets and inflammatory status is found in case of macrophages. In case of atherosclerotic plaques, triglyceride and cholesterol ester-rich lipid droplets are the hallmark of foamy macrophages. Much of the literature in this area arises from simplistic models of stimulation of macrophages with isolated low-density lipoproteins and their oxidized forms, resulting in the acquisition of the M1 type proinflammatory phenotype. However, in the plaque, not only the M1 and M2 macrophages, but a spectrum of different macrophage phenotypes exist [67]. Lipid-loaded macrophages are known to release inflammatory cytokines [68], yet the contribution of neutral lipid synthesis and the presence of lipid droplets in this process seems more of a bystander, protective response. Depletion of DGAT1 in the macrophage population did not lead to a decrease in the overall lipid content of atherosclerotic plaques, but led to higher macrophage recruitment, decreased fibrosis, and higher circulating neutrophil counts, indicative of an antiinflammatory role of triglyceride synthesis in atherosclerosis [60]. Cholesterol esters are synthesized by the enzyme acyl-CoA:cholesterol acyltransferase (ACAT). ACAT2 is involved in liver CE storage while ACAT1 is the isoform responsible for CE accumulation in foam cells in atherosclerotic plaques. Therefore, assessment of the role of macrophage CE synthesis requires specific tools against macrophage specific ACAT1 activity. Ablation of *Acat1* gene in the macrophage population in hypercholesteremic mice (LDLR$^{-/-}$ mice on a high fat diet) led to increased lesion area with increased macrophage apoptosis and necrosis, possibly due to accumulation of cholesterol in the plaques [69]. This suggested a protective role for CE synthesis in foam cells. These findings do not, however, undermine the benefits of ACAT inhibition which would decrease cholesterol ester loading of LDL particles, thereby reducing available cholesterol in circulation [70]. Paradoxically, peritoneal macrophages from LDLR KO mice fed a high fat diet exhibit suppression of proinflammatory genes while that in the plaque exhibit a proinflammatory phenotype [71]. The presence of high levels of desmosterol, the product of the penultimate step in cholesterol biosynthesis was found to be responsible for this phenotype via LXR-dependent gene regulation. The presence of desmosterol in foam cells within the plaque yet simultaneous proinflammatory phenotype suggests that macrophage activation in the atherosclerotic plaque may in fact be due to extrinsic proinflammatory signals generated within the artery wall. Parallel pathological consequences of lipid loading such as impaired efferocytosis, impaired lysosomal function, and ER stress may account for much of the pathology than the actual lipid storage capacity, which is rather protective [72].

## Lipid droplet proteome in immune cells

Given the above evidences for the involvement of lipid droplets in the immune response, either as precursors of bioactive lipids or as scaffolds for protein-protein interactions, we are far from understanding the complex mechanisms for lipogenesis and lipolysis as well as the role of lipid droplets in the immune response. Research toward understanding how infection or pathologic insults alter not only the protein composition but also the lipid composition of the droplet is going to increase our ability to address these questions. The most comprehensive lipid droplet proteome of an immune cell was recently published [73]. Phorbol myristate acetate-differentiated THP-1 macrophages, similar to spontaneously differentiated human monocyte-derived macrophages, contain abundant lipid droplets in the basal state. This allows the isolation of lipid droplets from these cells without the requirement of oleic acid treatment. High sensitivity proteomics revealed that lipid droplets of THP1 macrophages contained Perilipin 2 and 3 as the coat proteins but lacked Perilipin 1, the major coat protein of adipocytes. While triglyceride was the major neutral lipid identified, no enzymes involved in triglyceride synthesis were identified. Surprisingly, few lipolytic enzymes were identified, namely alpha beta hydrolase domain-containing 5 (ABHD5), carboxylesterase 1 (CES1), and phospholipase D3. Interestingly, enzymes involved in arachidonic acid to prostanoid synthesis were indeed identified. This included prostaglandin G/H synthase 1 (PTGS1) (also called cyclooxygenase 1) and PTGS2 (COS2) which carry out the committed step in prostanoid synthesis and prostaglandin E synthase 3 (PTGES3) which then converts PGH2 to PGE2. This points toward the possibility that perhaps the arachidonic acid released from PL or TAG hydrolysis is directly converted to prostaglandin H2 and E2 in an efficient manner. Hydrolysis of prostaglandin glycerides can also be achieved by CES1 and palmitoyl protein thioesterase 1 [74]. PPT1 was also identified in our study of the macrophage LD proteome suggesting the possibility of lipid droplet localized degradation of prostaglandin glycerides.

While the presence of proteins belonging to lipid metabolism on lipid droplets seems intuitive, a larger set of proteins with functions unrelated to lipid metabolism have also been found in the macrophage proteome. For example, cytoskeletal proteins such as tubulin and vimentin, and motor proteins such as myosins were found in this study and previous LD proteome datasets. The presence of histones and ribosomal proteins on lipid droplets is also a feature shared between these studies. Several small GTPases involved in phagosome–lysosome pathway and lysosomal proteins were identified in the macrophage LD proteome, pointing toward interactions between the endosomal trafficking machinery with lipid droplets. When comparing these findings to the lipid droplet proteome of other mammalian cell types, one can appreciate the cell type specificity of the lipid droplet proteome. For example, the skeletal muscle lipid droplet proteome identified SNAP23 as a likely mediator of LD–mitochondria interactions that are crucial for energy metabolism from fatty acid oxidation [75]. The high density of LD–endosome/lysosome shared proteins suggest that perhaps these proteins may have distinct roles on lipid droplets or this may be a result of high density of interaction between lysosomes and lipid droplets in macrophages. What could be the purpose of these interactions? One can speculate that these interactions could promote lipid transfer between lipid droplets and lysosomes. In fact, one of the lysosomal proteins identified in this study was lysosomal integral membrane protein 2 (LIMP2/SCARB2) which was recently identified as a cholesterol transporter in lysosomes [76]. LIMP2 was required for vectorial transfer of exogenous cholesterol to LDs; whether this takes place directly or via the ER where ACAT is localized, remains to be understood. Therefore, the macrophage LD proteome is a rich repository to understand lipid trafficking pathway in these cells.

## Lipid droplets in host–pathogen interaction

This section is focused on the strategies that intracellular pathogens such as *mycobacteria* spp., *Chlamydia trachomatis*, and RNA viruses employ in utilizing host lipid droplets toward their pathogenic onslaught (Fig. 10.4).

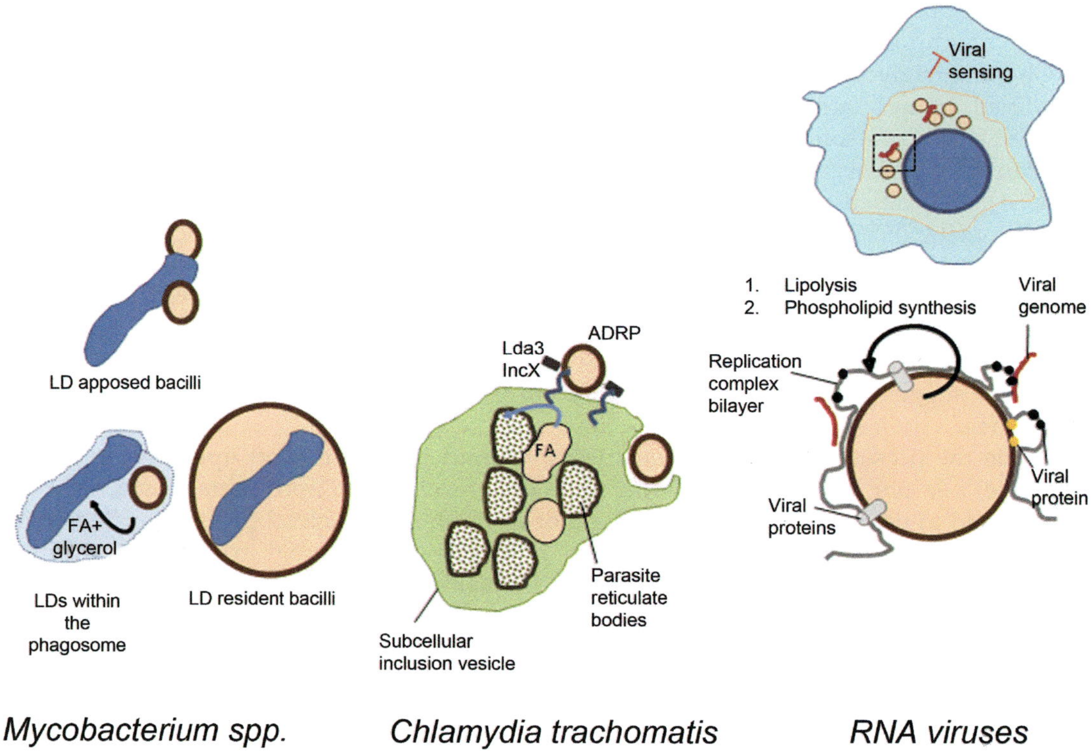


**FIGURE 10.4 Lipid droplets in host–pathogen interaction.** Several pathogenic mycobacteria, including *M. tuberculosis*, *M. leprae*, and *M. marinum*, exhibit intimate association with host lipid droplets; these interactions can range from interaction of lipid droplets with cytosolic bacteria, presence of lipid droplets within the phagosome, or in rare cases, presence of bacteria within lipid droplet-like compartments. *Chlamydia trachomatis* develops into reticulate bodies within a specialized subcellular compartment called an inclusion vesicle. This vesicle derives lipids from several host membranes and this lipid is then vectorialy transferred to the parasite. Shown here is the peripheral association of a PLIN2-coated lipid droplet which is denuded as it enters the inclusion vesicle and then is lipolytically degraded to generate fatty acids that are taken up by the parasite. RNA viruses assemble on specialized membrane compartments elicited upon infection which are called replication complexes. In the case of poliovirus, this perinuclear clustering of host lipid droplets with the replication complexes enables hiding the viral genome from cytosolic RNA sensing pathways of the host thereby preventing an antiviral response to be generated. The intimate association of nonstructural proteins of RNA viruses (*yellow dots* [*light gray* in print version]) with lipid droplets ensures mobilization of fatty acids from lipid droplets, and the synthesis of phospholipids which is required for formation of the replication complex. Structural proteins (*black dots*) and viral RNA genome (*red strands* [*gray* in print version]) assemble at these membrane compartments in close proximity with the lipid droplet.

## Viruses and lipid droplets

In the last decade, several studies have reported the use of lipid droplets as surfaces for replication and assembly of virions of positive strand RNA viruses. These viruses include human and veterinary pathogens, including yellow fever virus, Japanese encephalitis virus, hepatitis C virus, Dengue virus, Zika virus, West Nile virus, and poliovirus. The positive-strand RNA viral genome which is transcribed into a single transcript which is then translated into a single peptide, then processed into 10 proteins co- and posttranslationally. The genome encodes for seven nonstructural proteins, besides the envelope and core proteins. The nonstructural (NS) proteins of the family have been found to interact with lipid droplet coat protein TIP47. Hepatitis C virus is an especially interesting case as the infectious virion itself is a lipoprotein particle, containing apoE. The egress of the virus for secretion requires passage through the ER and the Golgi. However, core, envelope, and nonstructural proteins also localize to the lipid droplet. In case of HCV, the NS5 protein of the virus interacts with the C terminus of DGAT1, at the site of LD budding. In addition, inhibition of DGAT1 leads to diminished virion production, independent of the role of the physical interaction of NS5 and DGAT1. This two-step process of protein–protein interaction and the requirement of the triglyceride ensures that the NS5 is favorably loaded onto the newly formed lipid droplet. It is intriguing that HCV infection is associated with steatohepatitis; the virus induces lipid droplet formation by activation of SREBP1c and reduced LD turnover, and also utilizes the lipid droplets as scaffolds for production of infectious virion [77–79] (Fig. 10.4).

Viruses are adept at generating de novo contact sites between their replication complexes and lipid droplets, a process that is required for their replication. Recent high-resolution microscopy has revealed a lipid droplet associated membranous compartment induced by the HCV [80]. HCV infection induces double membrane extensions of the ER membrane around lipid droplets. It is tempting to speculate that the proximity of the viral RNA replication complex and its proteins in this specialized compartment enables viral particles to enter the ER lumen from where the virions arrive at the Golgi complex. Rab18, a lipid droplet localized small GTPase, is required for the interaction between HCV replication sites and the lipid droplets [81]. This is mediated by a direct interaction between Rab18 and NS5A, in the absence of which viral replication is impaired. Poliovirus, with a positive stranded RNA genome, but lacking a lipid membrane, organizes a replication organelle comprising of single and double membrane compartments (replication complexes) organized in the perinuclear region of its host cells [82,83]. This replication organelle is peculiar in that it encapsulates lipid droplets of the host cells [84]. George Belov's group observed translocation of CCTα from the nucleus to cytoplasm in poliovirus-infected cells, akin to its nuclear to cytoplasmic translocation under conditions of PC synthesis at the growing LD surface [85]. Concomitant with this was the preferential localization of fluorescently labelled fatty acids from lipid droplets to a distinct perinuclear region corresponding to the viral replication complex [85]. Depriving infected cells of either choline or treatment with the lipase inhibitor diethylumbelliferylphosphate led to impairment in formation of the replication organelle, pointing toward the requirement for new phosphatidylcholine synthesis from existing neutral lipid pools. High-resolution microscopy evidence supports these findings and reveals contact sites between lipid droplets and replication complexes [84]. These contact points are enriched in nonstructural viral proteins and also contain the lipases HSL and ATGL. Surprisingly, the inability to mobilize lipids from lipid droplets does not affect the first round of viral replication but does so to subsequent rounds [84,85]. This is partly explained by the finding that the antiviral responses mediated by sensing of dsRNA, a replication intermediate for picornaviruses, is prevented in the absence of the lipid mobilization process. Recognition of dsRNA takes place by sensors such as retinoic acid inducible gene (RIG)-like receptors (RLR) RIG-I and MDA5, and toll-like receptors (TLR) 3 and 7. These pathways converge to induce expression of interferons α/β as well as expression of proinflammatory cytokines. While IFNα/β induction is dependent on phosphorylation of transcription factors IRF3/7, the expression of proinflammatory cytokines is dependent on phosphorylation and degradation of Iκβ, thereby releasing transcriptionally active subunits of NF-κβ. Inhibition of membrane synthesis in infected cells promoted phosphorylation and dimerization of IRF3, suggesting this arm of viral sensing to be dampened by the formation of the replication organelle [85].

While pathogenic viruses seem to rely on lipid droplets for the generation of infectious virions, the lipid droplet also harbors an antiviral protein, Viperin. Viperin is an interferon-stimulated gene which localizes to the lipid droplet via its N-terminal amphipathic alpha helix [86]. Viperin contains a domain associated with generation of the oxidizing deoxyadenosyl radical, and affecting lipid rafts and membrane fluidity. While these features were earlier thought to mediate its antiviral effect, recent studies indicate that it generates 3′-deoxy-3′,4′-didehydro-CTP (ddhCTP), an inhibitor of RNA-dependent RNA polymerases, from CTP in a S-adenosyl methionine-dependent manner [87,88]. Viperin's role as a signal integrator downstream of TLR7 and TLR9 has also emerged recently. Viperin was found to localize the interaction of IRAK1 and TRAF6 to lipid droplets in plasmacytoid dendritic cells. This led to the ubiquitination of IRAK1, a step required for nuclear translocation of IRF7 and type I IFN expression [89]. Further proof that lipid droplets may play a role in antiviral defense comes from observations wherein inhibition of lipid droplet synthesis suppresses the antiviral Type I IFN response generation to dsRNA and Sendai virus [90]. While Viperin seems to be the only direct mediator so far, other players might also be involved.

## Intracellular bacteria hijacking lipid droplets

Intracellular bacterial pathogens are adept at manipulating vesicular transport and membrane lipid trafficking pathways in their favor. Lipid droplets, by virtue of their association with high lipid content as well as proteins involved in lipid mobilization seem to be highly likely to be targets for pathogens. Infection with the successful human pathogen *M. tuberculosis* is associated with the presence of lipid droplet rich foamy macrophages within granuloma structures. Firstly, the biogenesis of lipid droplets in cells infected with *M. tuberculosis* has been highly debatable. While earlier studies attributed these phenomenon to lipids of *M. tuberculosis* [91,92], subsequent studies failed to find a pathogen-induced lipogenic phenomenon at play. Instead, the role of cytokines such as IFNγ, TNFα, and the role of necrosis emerged as likely players in formation of triglyceride-rich foamy macrophages in TB infection [21,37,93]. However, the ability of Mtb to induce lipolysis, and utilize the free fatty acid for its own triglyceride synthesis has emerged across multiple studies [94–96]. As mentioned previously, triglycerides have also been found to be required for eicosanoid production in response to *M. tuberculosis* infection of murine macrophages [37]. Whether and how *M. tuberculosis* is able to induce alterations in the lipid droplet proteome has been a fascinating question, given that it does actively modulate the endocytic system [97]. This was addressed recently by studying the lipid droplet proteome of human THP1 monocyte-derived macrophages infected with *M. tuberculosis*. Using quantitative lipid droplet proteomics, abundance of 86 proteins was found to be actively regulated by Mtb [73]. The depletion of lipid droplets by silencing of *DGAT1* limits the proinflammatory response of THP1 macrophages to *M. tuberculosis* [21]. Whether this involves lipid mediators or a differential proteome-dependent mechanism remains to be understood. Proteins differentially abundant belonged to pathways ranging from protein synthesis, vesicular trafficking, metabolism, protein folding, and signal transduction. Lipases were not found to be differentially abundant, suggesting possibility of noncanonical mechanisms of lipolytic regulation in response to infection, at play in human macrophages. Evidence from in vitro infections with several mycobacterial species points toward an intimate association of host lipid droplets and pathogens such as *M. avium*, *M. leprae*, *M. marinum*, and *M. tuberculosis*; these range from localization of lipid droplets within the mycobacterial phagosome, association of cytosolic-escaped bacteria with lipid droplets, and entry or mycobacteria within lipid droplets [98–101] (Fig. 10.4).

*Chlamydia trachomatis*, another successful human pathogen, also modifies the host cell lipid droplet proteome. Chlamydia is adept at acquiring host cell membranes for the formation of its habitat, also called "inclusion vesicle". The clustering of small lipid droplets within this bacterial inclusion generated the idea that possibly *C. trachomatis* could hijack the host lipid droplets [102] (Fig. 10.4). Moreover, unlike *M. tuberculosis*, *C. trachomatis* lacks biosynthetic pathways that it requires for growth, thereby relying on host lipids. Inhibition of host PLA2 limits replication of the pathogen within its host. Using a label-free approach, lipid droplets from infected cells were found to be enriched in proteins involved in lipid metabolic processes, lipid biosynthetic processes and LD-specific functions, in addition to proteins localized to the inclusion membrane [103]. This not only supported the idea that lipid droplet biogenesis increases upon infection with *Chlamydia*, it also supported the finding of close apposition of lipid droplets with the inclusion membrane. Furthermore, inhibition of lipid droplet biogenesis led to inhibition of intracellular growth of *Chlamydia*.

Intracellular bacterial pathogens can also usurp trafficking of proteins to the lipid droplet to achieve immune evasion. The Type I IFN response, discussed above in the context of viral infection, is also induced in response to several bacteria including *M. tuberculosis* and *C. trachomatis* [104]. Activation of this response leads to induction of several interferon stimulated genes (ISG). IRGM1 and IRGM3 are members of the immunity-related GTPase (IRG) family, which localize to lipid droplets [105,106]. Irgm3 was found to interact with Plin2/Adfp in mouse DCs and depletion of either Irgm3 or Adfp prevented cross presentation of antigens [105]. Absence of these proteins promoted localization of other IRGM oligomeric complexes to pathogen containing vacuoles presumable marking them as "self", rather than "nonself" cargo and thereby preventing their degradation [106].

## Lipid droplets in immune defense of a newborn

Mammalian nutrition post birth depends on milk and milk fat globules are the most important component. The lipid and protein composition of these globules has many a role to play from developing the structure of the newborn's gut, providing essential first line of defense in the gut, and developing the gut microbial community structure [107]. The development and secretion of milk fat globules encompasses the early life of an intracellular lipid droplet which is matured and secreted from mammary epithelial cells through their apical surface. The milk fat globule (also called milk secretion granule) is in fact a lipid droplet (referred as such in this section) encapsulated within a double membrane vesicle derived from the plasma membrane of the epithelial cells. This trilayered membrane structure is also called the milk fat globule membrane (MFGM). Approximately 99% of MFGM lipids are triglycerides; polar lipids comprise of phospholipids and sphingolipids. The surface glycocalyx of the membrane comprising of glycolipids and proteins imparts unique biophysical characteristics compared to infant milk formulae that simply meet the composition of human milk [108]. Ablation of the *dgat1* gene in mice renders females unable to lactate, supporting the essentiality of triglyceride synthesis in the formation of milk fat globules [109]. One of the major proteins contributed from the apical membrane of the mammary epithelial cells is butyrophilin (BTN). BTN belongs to the immunoglobulin family of proteins, which in the bilayer membrane exists as a type 1 transmembrane glycoprotein. Expression of BTN is essential for the secretion of the developing lipid droplet [110]. BTN is found enriched at regions of the surrounding bilayer of the milk fat globule, yet concentrated at ridge-like structures on the inner leaflet of the bilayer, suggesting physical proximity of butyrophilin positive regions of the membrane to the lipid droplet monolayer [111]. Several models exist for BTN-mediated secretion of the lipid droplet: (1) interaction of BTN with xanthine oxidase, a cytoplasmic protein which interacts with PLIN2 on the LD monolayer, (2) direct interaction between BTN and PLIN2, (3) dual localization of BTN on the plasma membrane and lipid droplet monolayer and homotypic interactions at the ridges [111], and (4) transient interaction of BTN with multiple proteins at the ridge structures [112]. Understanding pathways linked with packaging of lipid for secretion during lactation are crucial for finding solutions to tackle early childhood nutrition.

Beige adipose tissue development in humans and mice has been linked with protection against diet-induced obesity. The risk for obesity in adults is highest for those who had a higher BMI between 2 and 6 years of age [113]. In addition, individuals who have not been breast fed as children are at higher risk of obesity in childhood [114]. This raises the question as to whether specific components in breast milk provide "protective" functions in the developing adipose tissue of a newborn. Recently, alkylglycerols in breast milk have been shown to induce synthesis of the bioactive lipid platelet-activating factor (PAF) in the adipose tissue macrophages (ATMs) of newborn mouse pups [115]. ATMs respond to alkylglycerols by also increasing the levels of the PAF receptor PTAF. This autocrine activity of PAF then leads to increased expression of IL6 in the neighboring beige preadipocytes. IL6R engagement on beige preadipocytes leads to phosphorylation and activation of STAT3, leading to beige adipogenesis. Alkylglycerols are less abundant in adult dietary sources. The conversion of alkylglycerols to PAF requires the expression of the enzyme lysophosphocholine acyltransferase 2 (LPCAT2). PAF can also be rapidly degraded by the enzyme alkylglycerol monooxygenase (AGMO). The higher expression of LPCAT2 and decreased expression of AGMO in M1 macrophages compared to M2 macrophages ensures that ATMs of the M1 type are better geared to providing this protective response. As a result, lean mice which harbor M2 ATMs did not exhibit this response to AKGs while obese mice which harbor M1 ATMs did respond. These studies point toward the role of breast milk derived lipids in possibly shaping the adult adipose tissue properties.

## Concluding statement

Cells of the immune system are central to tissue homeostasis and response to pathogenic insults. Their ability to regulate the onset and resolution of the immune response relies on their choice of metabolic state. Lipid utilization vis-a-vis storage is central to much of this decision making. Most of our understanding of how these processes are regulated comes from cells that do not belong to the immune system. However, the role of macrophage and mast cell lipid droplets in host defense and that of dendritic cells in tumor surveillance, points to the need to understand precise mechanisms of lipid trafficking and cellular homeostasis mediated via these specialized organelles. While much of our understanding today is focused on lipid droplets as a source of bioactive lipids, the unique proteome of lipid droplets of macrophages compared to other cell types is an indicator that lipid droplets in immune cells may play distinct roles. Manipulation of lipid droplets by intracellular pathogens provides us insights into how a lipid storage organelle, potentially involved in host defense, can be usurped toward propathogen functions. Proteomics and lipidomics studies, strengthened by high-resolution microscopy, hold the key to unlocking possibilities of the involvement of lipid droplets in the immune system and beyond.

## References

[1] Thiam AR, Farese RV Jr, Walther TC. The biophysics and cell biology of lipid droplets. Nat Rev Mol Cell Biol 2013;14:775–86.
[2] Kassan A, Herms A, Fernandez-Vidal A, Bosch M, Schieber NL, Reddy BJ, et al. Acyl-CoA synthetase 3 promotes lipid droplet biogenesis in ER microdomains. J Cell Biol 2013;203:985–1001.
[3] Walther TC, Chung J, Farese RV Jr. Lipid droplet biogenesis. Annu Rev Cell Dev Biol 2017;33:491–510.
[4] Chorlay A, Thiam AR. An asymmetry in monolayer tension regulates lipid droplet budding direction. Biophys J 2018;114:631–40.
[5] Chorlay A, Monticelli L, Ferreira JV, Ben M'barek K, Ajjaji D, Wang SH, et al. Membrane asymmetry imposes directionality on lipid droplet emergence from the ER. Dev Cell 2019;50:25.
[6] Choudhary V, Golani G, Joshi AS, Cottier S, Schneiter R, Prinz WA, et al. Architecture of lipid droplets in endoplasmic reticulum is determined by phospholipid intrinsic curvature. Curr Biol 2018;28:915–26. e919.
[7] Li D, Zhao YG, Zhao H, Huang J, Miao G, Feng D, et al. The ER-localized protein DFCP1 modulates ER-lipid droplet contact formation. Cell Rep 2019;27:343–58. e345.
[8] Salo VT, Li SQ, Vihinen H, Holtta-Vuori M, Szkalisity A, Horvath P, et al. Seipin facilitates triglyceride flow to lipid droplet and counteracts droplet ripening via endoplasmic reticulum contact. Dev Cell 2019;50:478.
[9] Xu D, Li Y, Wu L, Zhao D, Yu J, Huang T, et al. Rab18 promotes lipid droplet (LD) growth by tethering the ER to LDs through SNARE and NRZ interactions. J Cell Biol 2018;217:975–95.
[10] Datta S, Liu Y, Hariri H, Bowerman J, Henne WM. Cerebellar ataxia disease-associated Snx14 promotes lipid droplet growth at ER-droplet contacts. J Cell Biol 2019;218:1335–51.
[11] Valm AM, Cohen S, Legant WR, Melunis J, Hershberg U, Wait E, et al. Applying systems-level spectral imaging and analysis to reveal the organelle interactome. Nature 2017;546:162–7.
[12] Rambold AS, Cohen S, Lippincott-Schwartz J. Fatty acid trafficking in starved cells: regulation by lipid droplet lipolysis, autophagy, and mitochondrial fusion dynamics. Dev Cell 2015;32:678–92.
[13] Joshi AS, Cohen S. Lipid droplet and peroxisome biogenesis: do they go hand-in-hand? Front Cell Dev Biol 2019;7:92.
[14] Rai P, Kumar M, Sharma G, Barak P, Das S, Kamat SS, et al. Kinesin-dependent mechanism for controlling triglyceride secretion from the liver. Proc Natl Acad Sci USA 2017;114:12958–63.
[15] Kaushik S, Cuervo AM. Degradation of lipid droplet-associated proteins by chaperone-mediated autophagy facilitates lipolysis. Nat Cell Biol 2015;17:759–70.

[16] Gandotra S, Le Dour C, Bottomley W, Cervera P, Giral P, Reznik Y, et al. Perilipin deficiency and autosomal dominant partial lipodystrophy. N Engl J Med 2011;364:740–8.
[17] Gandotra S, Lim K, Girousse A, Saudek V, O'Rahilly S, Savage DB. Human frame shift mutations affecting the carboxyl terminus of perilipin increase lipolysis by failing to sequester the adipose triglyceride lipase (ATGL) coactivator AB-hydrolase-containing 5 (ABHD5). J Biol Chem 2011;286:34998–5006.
[18] Greenberg AS, Egan JJ, Wek SA, Garty NB, Blanchette-Mackie EJ, Londos C. Perilipin, a major hormonally regulated adipocyte-specific phosphoprotein associated with the periphery of lipid storage droplets. J Biol Chem 1991;266:11341–6.
[19] Miyoshi H, Perfield JW 2nd, Souza SC, Shen WJ, Zhang HH, Stancheva ZS, et al. Control of adipose triglyceride lipase action by serine 517 of perilipin A globally regulates protein kinase A-stimulated lipolysis in adipocytes. J Biol Chem 2007;282:996–1002.
[20] Patel S, Yang W, Kozusko K, Saudek V, Savage DB. Perilipins 2 and 3 lack a carboxy-terminal domain present in perilipin 1 involved in sequestering ABHD5 and suppressing basal lipolysis. Proc Natl Acad Sci USA 2014;111:9163–8.
[21] Jaisinghani N, Dawa S, Singh K, Nandy A, Menon D, Bhandari PD, et al. Necrosis driven triglyceride synthesis primes macrophages for inflammation during Mycobacterium tuberculosis Infection. Front Immunol 2018;9:1490.
[22] Kaushik S, Cuervo AM. AMPK-dependent phosphorylation of lipid droplet protein PLIN2 triggers its degradation by CMA. Autophagy 2016;12:432–8.
[23] Schott MB, Weller SG, Schulze RJ, Krueger EW, Drizyte-Miller K, Casey CA, et al. Lipid droplet size directs lipolysis and lipophagy catabolism in hepatocytes. J Cell Biol 2019;218(10):3320–35.
[24] Nagayama M, Shimizu K, Taira T, Uchida T, Gohara K. Shrinking and development of lipid droplets in adipocytes during catecholamine-induced lipolysis. FEBS Lett 2010;584:86–92.
[25] Yamaguchi T, Omatsu N, Morimoto E, Nakashima H, Ueno K, Tanaka T, et al. CGI-58 facilitates lipolysis on lipid droplets but is not involved in the vesiculation of lipid droplets caused by hormonal stimulation. J Lipid Res 2007;48:1078–89.
[26] Marcinkiewicz A, Gauthier D, Garcia A, Brasaemle DL. The phosphorylation of serine 492 of perilipin a directs lipid droplet fragmentation and dispersion. J Biol Chem 2006;281:11901–9.
[27] Chiu HC, Kovacs A, Blanton RM, Han X, Courtois M, Weinheimer CJ, et al. Transgenic expression of fatty acid transport protein 1 in the heart causes lipotoxic cardiomyopathy. Circ Res 2005;96:225–33.
[28] Listenberger LL, Han X, Lewis SE, Cases S, Farese RV Jr, Ory DS, et al. Triglyceride accumulation protects against fatty acid-induced lipotoxicity. Proc Natl Acad Sci USA 2003;100:3077–82.
[29] Haemmerle G, Moustafa T, Woelkart G, Buttner S, Schmidt A, van de Weijer T, et al. ATGL-mediated fat catabolism regulates cardiac mitochondrial function via PPAR-alpha and PGC-1. Nat Med 2011;17:1076–85.
[30] Borradaile NM, Han X, Harp JD, Gale SE, Ory DS, Schaffer JE. Disruption of endoplasmic reticulum structure and integrity in lipotoxic cell death. J Lipid Res 2006;47:2726–37.
[31] Kim H, Zhang H, Meng D, Russell G, Lee JN, Ye J. UAS domain of Ubxd8 and FAF1 polymerizes upon interaction with long-chain unsaturated fatty acids. J Lipid Res 2013;54:2144–52.
[32] Lee JN, Kim H, Yao H, Chen Y, Weng K, Ye J. Identification of Ubxd8 protein as a sensor for unsaturated fatty acids and regulator of triglyceride synthesis. Proc Natl Acad Sci USA 2010;107:21424–9.
[33] Olzmann JA, Richter CM, Kopito RR. Spatial regulation of UBXD8 and p97/VCP controls ATGL-mediated lipid droplet turnover. Proc Natl Acad Sci USA 2013;110:1345–50.
[34] Dvorak AM, Dvorak HF, Peters SP, Shulman ES, MacGlashan DW Jr, Pyne K, et al. Lipid bodies: cytoplasmic organelles important to arachidonate metabolism in macrophages and mast cells. J Immunol 1983;131:2965–76.
[35] Dichlberger A, Schlager S, Maaninka K, Schneider WJ, Kovanen PT. Adipose triglyceride lipase regulates eicosanoid production in activated human mast cells. J Lipid Res 2014;55:2471–8.
[36] Bandeira-Melo C, Weller PF, Bozza PT. EicosaCell—an immunofluorescent-based assay to localize newly synthesized eicosanoid lipid mediators at intracellular sites. Methods Mol Biol 2011;689:163–81.
[37] Knight M, Braverman J, Asfaha K, Gronert K, Stanley S. Lipid droplet formation in Mycobacterium tuberculosis infected macrophages requires IFN-gamma/HIF-1alpha signaling and supports host defense. PLoS Pathog 2018;14:e1006874.
[38] Dong L, Zou H, Yuan C, Hong YH, Kuklev DV, Smith WL. Different fatty acids compete with arachidonic acid for binding to the allosteric or catalytic subunits of cyclooxygenases to regulate prostanoid synthesis. J Biol Chem 2016;291:4069–78.

[39] Weylandt KH, Chiu CY, Gomolka B, Waechter SF, Wiedenmann B. Omega-3 fatty acids and their lipid mediators: towards an understanding of resolvin and protectin formation. Prostaglandins Other Lipid Mediat 2012;97:73–82.

[40] von Knethen A, Tautenhahn A, Link H, Lindemann D, Brune B. Activation-induced depletion of protein kinase C alpha provokes desensitization of monocytes/macrophages in sepsis. J Immunol 2005;174:4960–5.

[40a] L. Zhu, Jones C, Zhang G. The role of phospholipase C signaling in macrophage-mediated inflammatory response. J Immunol Res 2018;2018:5201759.

[41] Shin M, Buckner A, Prince J, Bullock TNJ, Hsu KL. Diacylglycerol lipase-beta is required for TNF-alpha response but not CD8(+) T cell priming capacity of dendritic cells. Cell Chem Biol 2019;26:1036–41. e1033.

[42] Gallily R, Breuer A, Mechoulam R. 2-Arachidonylglycerol, an endogenous cannabinoid, inhibits tumor necrosis factor-alpha production in murine macrophages, and in mice. Eur J Pharmacol 2000;406:R5–7.

[43] Wilkerson JL, Donvito G, Grim TW, Abdullah RA, Ogasawara D, Cravatt BF, et al. Investigation of diacylglycerol lipase alpha inhibition in the mouse lipopolysaccharide inflammatory pain model. J Pharmacol Exp Ther 2017;363:394–401.

[44] Poursharifi P, Madiraju SRM, Prentki M. Monoacylglycerol signalling and ABHD6 in health and disease. Diabetes Obes Metab 2017;19(Suppl. 1):76–89.

[45] Alhouayek M, Masquelier J, Cani PD, Lambert DM, Muccioli GG. Implication of the anti-inflammatory bioactive lipid prostaglandin D2-glycerol ester in the control of macrophage activation and inflammation by ABHD6. Proc Natl Acad Sci USA 2013;110:17558–63.

[46] Nirodi CS, Crews BC, Kozak KR, Morrow JD, Marnett LJ. The glyceryl ester of prostaglandin E2 mobilizes calcium and activates signal transduction in RAW264.7 cells. Proc Natl Acad Sci USA 2004;101:1840–5.

[47] Bruser A, Zimmermann A, Crews BC, Sliwoski G, Meiler J, Konig GM, et al. Prostaglandin E2 glyceryl ester is an endogenous agonist of the nucleotide receptor P2Y6. Sci Rep 2017;7:2380.

[48] Hillebrand M, Gersting SW, Lotz-Havla AS, Schafer A, Rosewich H, Valerius O, et al. Identification of a new fatty acid synthesis-transport machinery at the peroxisomal membrane. J Biol Chem 2012;287:210–21.

[49] Lodhi IJ, Yin L, Jensen-Urstad AP, Funai K, Coleman T, Baird JH, et al. Inhibiting adipose tissue lipogenesis reprograms thermogenesis and PPARgamma activation to decrease diet-induced obesity. Cell Metab 2012;16:189–201.

[50] Hofer DC, Pessentheiner AR, Pelzmann HJ, Schlager S, Madreiter-Sokolowski CT, Kolb D, et al. Critical role of the peroxisomal protein PEX16 in white adipocyte development and lipid homeostasis. Biochim Biophys Acta Mol Cell Biol Lipids 2017;1862:358–68.

[51] Dean JM, Lodhi IJ. Structural and functional roles of ether lipids. Protein Cell 2018;9:196–206.

[52] Phan AT, Goldrath AW, Glass CK. Metabolic and epigenetic coordination of T cell and macrophage immunity. Immunity 2017;46:714–29.

[53] Stienstra R, Netea-Maier RT, Riksen NP, Joosten LAB, Netea MG. Specific and complex reprogramming of cellular metabolism in myeloid cells during innate immune responses. Cell Metab 2017;26:142–56.

[54] Buck MD, O'Sullivan D, Pearce EL. T cell metabolism drives immunity. J Exp Med 2015;212:1345–60.

[55] O'Sullivan D, van der Windt GJ, Huang SC, Curtis JD, Chang CH, Buck MD, et al. Memory CD8(+) T cells use cell-intrinsic lipolysis to support the metabolic programming necessary for development. Immunity 2014;41:75–88.

[56] Feingold KR, Shigenaga JK, Kazemi MR, McDonald CM, Patzek SM, Cross AS, et al. Mechanisms of triglyceride accumulation in activated macrophages. J Leukoc Biol 2012;92:829–39.

[57] Huang YL, Morales-Rosado J, Ray J, Myers TG, Kho T, Lu M, et al. Toll-like receptor agonists promote prolonged triglyceride storage in macrophages. J Biol Chem 2014;289:3001–12.

[58] Huang SC, Everts B, Ivanova Y, O'Sullivan D, Nascimento M, Smith AM, et al. Cell-intrinsic lysosomal lipolysis is essential for alternative activation of macrophages. Nat Immunol 2014;15:846–55.

[59] Remmerie A, Scott CL. Macrophages and lipid metabolism. Cell Immunol 2018;330:27–42.

[60] Vujic N, Porter Abate J, Schlager S, David T, Kratky D, Koliwad SK. Acyl-CoA:Diacylglycerol acyltransferase 1 expression level in the hematopoietic compartment impacts inflammation in the vascular plaques of atherosclerotic mice. PLoS ONE 2016;11:e0156364.

[61] McFie PJ, Stone SL, Banman SL, Stone SJ. Topological orientation of acyl-CoA:diacylglycerol acyltransferase-1 (DGAT1) and identification of a putative active site histidine and the role of the n terminus in dimer/tetramer formation. J Biol Chem 2010;285:37377–87.

[62] Szendroedi J, Yoshimura T, Phielix E, Koliaki C, Marcucci M, Zhang D, et al. Role of diacylglycerol activation of PKCtheta in lipid-induced muscle insulin resistance in humans. Proc Natl Acad Sci USA 2014;111:9597–602.
[63] Choi CS, Savage DB, Kulkarni A, Yu XX, Liu ZX, Morino K, et al. Suppression of diacylglycerol acyltransferase-2 (DGAT2), but not DGAT1, with antisense oligonucleotides reverses diet-induced hepatic steatosis and insulin resistance. J Biol Chem 2007;282:22678–88.
[64] Ibrahim J, Nguyen AH, Rehman A, Ochi A, Jamal M, Graffeo CS, et al. Dendritic cell populations with different concentrations of lipid regulate tolerance and immunity in mouse and human liver. Gastroenterology 2012;143:1061–72.
[65] Herber DL, Cao W, Nefedova Y, Novitskiy SV, Nagaraj S, Tyurin VA, et al. Lipid accumulation and dendritic cell dysfunction in cancer. Nat Med 2010;16:880–6.
[66] Cubillos-Ruiz JR, Silberman PC, Rutkowski MR, Chopra S, Perales-Puchalt A, Song M, et al. ER stress sensor XBP1 controls anti-tumor immunity by disrupting dendritic cell homeostasis. Cell 2015;161:1527–38.
[67] Tabas I, Bornfeldt KE. Macrophage phenotype and function in different stages of atherosclerosis. Circ Res 2016;118:653–67.
[68] Moore KJ, Sheedy FJ, Fisher EA. Macrophages in atherosclerosis: a dynamic balance. Nat Rev Immunol 2013;13:709–21.
[69] Fazio S, Major AS, Swift LL, Gleaves LA, Accad M, Linton MF, et al. Increased atherosclerosis in LDL receptor-null mice lacking ACAT1 in macrophages. J Clin Invest 2001;107:163–71.
[70] Rong JX, Blachford C, Feig JE, Bander I, Mayne J, Kusunoki J, et al. ACAT inhibition reduces the progression of preexisting, advanced atherosclerotic mouse lesions without plaque or systemic toxicity. Arterioscler Thromb Vasc Biol 2013;33:4–12.
[71] Spann NJ, Garmire LX, McDonald JG, Myers DS, Milne SB, Shibata N, et al. Regulated accumulation of desmosterol integrates macrophage lipid metabolism and inflammatory responses. Cell 2012;151:138–52.
[72] Back M, Yurdagul A Jr, Tabas I, Oorni K, Kovanen PT. Inflammation and its resolution in atherosclerosis: mediators and therapeutic opportunities. Nat Rev Cardiol 2019;16:389–406.
[73] Menon D, Singh K, Pinto SM, Nandy A, Jaisinghani N, Kutum R, et al. Quantitative lipid droplet proteomics reveals mycobacterium tuberculosis induced alterations in macrophage response to infection. ACS Infect Dis 2019;5:559–69.
[74] Wang R, Borazjani A, Matthews AT, Mangum LC, Edelmann MJ, Ross MK. Identification of palmitoyl protein thioesterase 1 in human THP1 monocytes and macrophages and characterization of unique biochemical activities for this enzyme. Biochemistry 2013;52:7559–74.
[75] Zhang H, Wang Y, Li J, Yu J, Pu J, Li L, et al. Proteome of skeletal muscle lipid droplet reveals association with mitochondria and apolipoprotein a-I. J Proteome Res 2011;10:4757–68.
[76] Heybrock S, Kanerva K, Meng Y, Ing C, Liang A, Xiong ZJ, et al. Lysosomal integral membrane protein-2 (LIMP-2/SCARB2) is involved in lysosomal cholesterol export. Nat Commun 2019;10:3521.
[77] Harris C, Herker E, Farese RV Jr, Ott M. Hepatitis C virus core protein decreases lipid droplet turnover: a mechanism for core-induced steatosis. J Biol Chem 2011;286:42615–25.
[78] Herker E, Harris C, Hernandez C, Carpentier A, Kaehlcke K, Rosenberg AR, et al. Efficient hepatitis C virus particle formation requires diacylglycerol acyltransferase-1. Nat Med 2010;16:1295–8.
[79] Lerat H, Kammoun HL, Hainault I, Merour E, Higgs MR, Callens C, et al. Hepatitis C virus proteins induce lipogenesis and defective triglyceride secretion in transgenic mice. J Biol Chem 2009;284:33466–74.
[80] Lee JY, Cortese M, Haselmann U, Tabata K, Romero-Brey I, Funaya C, et al. Spatiotemporal coupling of the hepatitis C virus replication cycle by creating a lipid droplet- proximal membranous replication compartment. Cell Rep 2019;27:3602–17. e3605.
[81] Salloum S, Wang H, Ferguson C, Parton RG, Tai AW. Rab18 binds to hepatitis C virus NS5A and promotes interaction between sites of viral replication and lipid droplets. PLoS Pathog 2013;9:e1003513.
[82] Belov GA, Nair V, Hansen BT, Hoyt FH, Fischer ER, Ehrenfeld E. Complex dynamic development of poliovirus membranous replication complexes. J Virol 2012;86:302–12.
[83] Dales S, Eggers HJ, Tamm I, Palade GE. Electron microscopic study of the formation of poliovirus. Virology 1965;26:379–89.
[84] Laufman O, Perrino J, Andino R. Viral generated inter-organelle contacts redirect lipid flux for genome replication. Cell 2019;178:275–89. e216.
[85] Viktorova EG, Nchoutmboube JA, Ford-Siltz LA, Iverson E, Belov GA. Phospholipid synthesis fueled by lipid droplets drives the structural development of poliovirus replication organelles. PLoS Pathog 2018;14:e1007280.

[86] Hinson ER, Cresswell P. The antiviral protein, viperin, localizes to lipid droplets via its N-terminal amphipathic alpha-helix. Proc Natl Acad Sci USA 2009;106:20452–7.
[87] Gizzi AS, Grove TL, Arnold JJ, Jose J, Jangra RK, Garforth SJ, et al. A naturally occurring antiviral ribonucleotide encoded by the human genome. Nature 2018;558:610–4.
[88] Seo JY, Yaneva R, Cresswell P. Viperin: a multifunctional, interferon-inducible protein that regulates virus replication. Cell Host Microbe 2011;10:534–9.
[89] Saitoh T, Satoh T, Yamamoto N, Uematsu S, Takeuchi O, Kawai T, et al. Antiviral protein Viperin promotes Toll-like receptor 7- and Toll-like receptor 9-mediated type I interferon production in plasmacytoid dendritic cells. Immunity 2011;34:352–63.
[90] Monson EA, Crosse KM, Das M, Helbig KJ. Lipid droplet density alters the early innate immune response to viral infection. PLoS ONE 2018;13:e0190597.
[91] Kim MJ, Wainwright HC, Locketz M, Bekker LG, Walther GB, Dittrich C, et al. Caseation of human tuberculosis granulomas correlates with elevated host lipid metabolism. EMBO Mol Med 2010;2:258–74.
[92] Peyron P, Vaubourgeix J, Poquet Y, Levillain F, Botanch C, Bardou F, et al. Foamy macrophages from tuberculous patients' granulomas constitute a nutrient-rich reservoir for M. tuberculosis persistence. PLoS Pathog 2008;4:e1000204.
[93] Guerrini V, Prideaux B, Blanc L, Bruiners N, Arrigucci R, Singh S, et al. Storage lipid studies in tuberculosis reveal that foam cell biogenesis is disease-specific. PLoS Pathog 2018;14:e1007223.
[94] Daniel J, Maamar H, Deb C, Sirakova TD, Kolattukudy PE. Mycobacterium tuberculosis uses host triacylglycerol to accumulate lipid droplets and acquires a dormancy-like phenotype in lipid-loaded macrophages. PLoS Pathog 2011;7:e1002093.
[95] Jaisinghani N, Dawa S, Singh K, Nandy A, Menon D, Bhandari PD, et al. Necrosis triggered by Mycobacterium tuberculosis alters macrophage triglyceride metabolism and inflammatory response in a DGAT1 – dependent manner. BioRXiv 2017;. https://www.biorxiv.org/content/10.1101/187104v1.
[96] Santucci P, Diomande S, Poncin I, Alibaud L, Viljoen A, Kremer L, et al. Delineating the physiological roles of the PE and catalytic domains of LipY in lipid consumption in mycobacterium-infected foamy macrophages. Infect Immun 2018;86. e00394-18.
[97] Bussi C, Gutierrez MG. Mycobacterium tuberculosis infection of host cells in space and time. FEMS Microbiol Rev 2019;43:341–61.
[98] Barisch C, Soldati T. Mycobacterium marinum degrades both triacylglycerols and phospholipids from its dictyostelium host to synthesise its own triacylglycerols and generate lipid inclusions. PLoS Pathog 2017;13:e1006095.
[99] Caire-Brandli I, Papadopoulos A, Malaga W, Marais D, Canaan S, Thilo L, et al. Reversible lipid accumulation and associated division arrest of Mycobacterium avium in lipoprotein-induced foamy macrophages may resemble key events during latency and reactivation of tuberculosis. Infect Immun 2014;82:476–90.
[100] Mattos KA, Lara FA, Oliveira VG, Rodrigues LS, D'Avila H, Melo RC, et al. Modulation of lipid droplets by Mycobacterium leprae in Schwann cells: a putative mechanism for host lipid acquisition and bacterial survival in phagosomes. Cell Microbiol 2011;13:259–73.
[101] Neyrolles O, Hernandez-Pando R, Pietri-Rouxel F, Fornes P, Tailleux L, Barrios Payan JA, et al. Is adipose tissue a place for Mycobacterium tuberculosis persistence? PLoS ONE 2006;1:e43.
[102] Cocchiaro JL, Kumar Y, Fischer ER, Hackstadt T, Valdivia RH. Cytoplasmic lipid droplets are translocated into the lumen of the Chlamydia trachomatis parasitophorous vacuole. Proc Natl Acad Sci USA 2008;105:9379–84.
[103] Saka HA, Thompson JW, Chen YS, Dubois LG, Haas JT, Moseley A, et al. Chlamydia trachomatis infection leads to defined alterations to the lipid droplet proteome in epithelial cells. PLoS ONE 2015;10:e0124630.
[104] Boxx GM, Cheng G. The roles of type I interferon in bacterial infection. Cell Host Microbe 2016;19:760–9.
[105] Bougneres L, Helft J, Tiwari S, Vargas P, Chang BH, Chan L, et al. A role for lipid bodies in the cross-presentation of phagocytosed antigens by MHC class I in dendritic cells. Immunity 2009;31:232–44.
[106] Haldar AK, Saka HA, Piro AS, Dunn JD, Henry SC, Taylor GA, et al. IRG and GBP host resistance factors target aberrant, "non-self" vacuoles characterized by the missing of "self" IRGM proteins. PLoS Pathog 2013;9:e1003414.
[107] Lis-Kuberka J, Orczyk-Pawilowicz M. Sialylated oligosaccharides and glycoconjugates of human milk. The impact on infant and newborn protection, development and well-being. Nutrients 2019;11:E306.
[108] Gallier S, Vocking K, Post JA, Van De Heijning B, Acton D, Van Der Beek EM, et al. A novel infant milk formula concept: mimicking the human milk fat globule structure. Colloids Surf B Biointerfaces 2015;136:329–39.

[109] Cases S, Zhou P, Shillingford JM, Wiseman BS, Fish JD, Angle CS, et al. Development of the mammary gland requires DGAT1 expression in stromal and epithelial tissues. Development 2004;131:3047–55.
[110] Ogg SL, Weldon AK, Dobbie L, Smith AJ, Mather IH. Expression of butyrophilin (Btn1a1) in lactating mammary gland is essential for the regulated secretion of milk-lipid droplets. Proc Natl Acad Sci USA 2004;101:10084–9.
[111] Robenek H, Hofnagel O, Buers I, Lorkowski S, Schnoor M, Robenek MJ, et al. Butyrophilin controls milk fat globule secretion. Proc Natl Acad Sci USA 2006;103:10385–90.
[112] Jeong J, Lisinski I, Kadegowda AK, Shin H, Wooding FB, Daniels BR, et al. A test of current models for the mechanism of milk-lipid droplet secretion. Traffic 2013;14:974–86.
[113] Geserick M, Vogel M, Gausche R, Lipek T, Spielau U, Keller E, et al. Acceleration of BMI in early childhood and risk of sustained obesity. N Engl J Med 2018;379:1303–12.
[114] Arenz S, Ruckerl R, Koletzko B, von Kries R. Breast-feeding and childhood obesity—a systematic review. Int J Obes Relat Metab Disord 2004;28:1247–56.
[115] Yu H, Dilbaz S, Cossmann J, Hoang AC, Diedrich V, Herwig A, et al. Breast milk alkylglycerols sustain beige adipocytes through adipose tissue macrophages. J Clin Invest 2019;129:2485–99.

CHAPTER

# 11

# Fatty acid mediators and the inflammasome

*Frédérik Desmarais[a], Karl-F. Bergeron[a], James M. Ntambi[b], Eric Rassart[a], Catherine Mounier[a]*

[a]Département des Sciences Biologiques, Centre de recherche CERMO-FC, Université du Québec à Montréal, Montreal, QC, Canada; [b]Departments of Biochemistry and Nutritional Sciences, University of Wisconsin-Madison, Madison, WI, United States

OUTLINE



Lipid Signaling and Metabolism. http://dx.doi.org/10.1016/B978-0-12-819404-1.00011-7

# Introduction

## The global obesity crisis

The global obesity crisis has steadily worsened in the last few decades. Thirty-nine percent of the world population can now either be considered overweight or obese. Many American countries (USA, Brazil, Mexico, and Argentina) are leading this trend with average overweight and obesity rates of 64.2% and 28.3%, respectively [1]. While excess fat accumulation in adipose tissues is not by itself a major health issue, obesity's comorbidities are. Type 2 diabetes (T2D), hepatic steatosis, respiratory problems, cardiovascular diseases and cancers are comorbidities responsible for the increased mortality of obese individuals [2].

## Role of inflammation in the genesis and progression of comorbidities

The commonality of most, if not all, of obesity's comorbidities is the presence of chronic inflammation that induces or exacerbates these pathologies [3, 4]. The impact of chronic inflammation on T2D has been studied exhaustively. Evidence shows that inflammation, through the action of cytokines, inhibits insulin signaling and induces insulin resistance in tissues. Reduced insulin sensitivity then results in increased glycemia. This phenomenon is why T2D is often referred to as an inflammatory disease [3, 5, 6].

Tumor necrosis factor-α (TNF-α) and interleukin-1β (IL-1β) are omnipresent cytokines in metabolic diseases. TNF-α and IL-1β are overexpressed in the adipose tissues of obese animals and humans, and disrupt insulin signaling at multiple levels [7, 8]. For example, TNF-α induces the phosphorylation of serine residues on IRS-1 and -2, leading to their inhibition. This then results in the inhibition of GLUT4-mediated glucose uptake in tissues and hyperglycemia, as well as hyperinsulinemia [7]. TNF-α can also inhibit insulin-producing cells by triggering JNK-dependent inhibition of insulin production [9]. Meanwhile, IL-1β reduces the expression of IRS-1 in adipocytes by an ERK-dependent pathway, thus reducing the insulin sensitivity of adipose tissues during prolonged inflammation [8]. Endothelial cells as well as leukocytes can also produce IL-1β. IL-1β induces the expression of COX-2 and the production of prostanoids, a class of lipid mediators which are potent modulators of vasoconstriction/vasodilation as well as effectors of the phenotypic signs of inflammation, including leukocyte recruitment and endothelium hyperpermeability. Therefore, IL-1β is also an important factor in cardiovascular diseases [10].

In addition to cytokine production, an important factor in chronic inflammation induced by obesity is the continued recruitment of macrophages. It is, however, important to distinguish two distinct groups of macrophages in order to understand their role in metabolic diseases. Under normal physiological conditions, the macrophages that patrol tissues (resident macrophages) have a phenotype dedicated to the phagocytosis of dying cells (efferocytosis) and help maintain an antiinflammatory environment by secreting antiinflammatory cytokines (IL-4, IL-10). These macrophages are "polarized" toward an M2 antiinflammatory state [11]. M2 macrophages mainly maintain homeostatic conditions and help in tissue remodeling [12]. However, monocytes/macrophages that invade the tissue in response to different proinflammatory stimuli, such as tissue damage, oxidative stress, and the presence of pathogen, are M1-polarized macrophages. Their phenotype is oriented toward the secretion of proinflammatory cytokines and lipid mediators. M1 macrophages also have reduced efferocytosis capabilities but an increased capacity for phagocytosis of foreign organisms [13]. In the chronic low-grade inflammation found in obesity, the population of M1 macrophages remains high and thus contributes to maintain the tissue in a perpetual state of inflammation [14]. The accumulation of M1 macrophages also correlates with the degree of insulin resistance in rodents fed on a high-fat diet [11, 15].

## Contribution of the western diet to obesity and inflammation

One of the causes of the current obesity and T2D epidemic is the western diet. It is a high fat, high cholesterol, high sugar diet, which is rich in saturated fat, with a high ω-6 to ω-3 fatty acids ratio [16,17] and low proportion of fibers [18]. Chronic consumption of the western diet is considered an important risk factor in the development of all chronic inflammation diseases [19] since it impacts the biosynthesis of important lipid mediators that control the inflammatory environment.

From a signaling standpoint, the high dietary ω-6/ω-3 ratio (or low ω-3/ω-6 ratio) is particularly problematic. Lipids are multipurpose molecules that have structural functions (such as cell membrane composition) and energetic functions (such as production of ATP by oxidation). In addition, lipids also serve as signaling molecules that can modulate inflammation. These lipids can be divided into different classes according to their size, composition and varying effects. These classes are the long chain fatty acids (LCFA), the medium chain fatty acids (MCFA), the short chain fatty acids (SCFA), the ω-3, ω-6, and ω-9 polyunsaturated fatty acids (PUFA), as well as the ceramides. In this review we will focus on the roles of ω-3 and ω-6 PUFA in the control of inflammation as these fatty acids are the main contributors of inflammation in a large majority of metabolic diseases.

The ω-3 fatty acids eicosapentaenoic acid (EPA), docosapentaenoic acid (DPA) and docosahexaenoic acid (DHA) serve as precursors for the biosynthesis of local hormones commonly known as docosanoids, whilst the ω-6 arachidonic acid (ARA) serves as the precursor for the eicosanoids. While docosanoids are essentially antiinflammatory molecules, eicosanoids are generally proinflammatory. However, several antiinflammatory functions have recently been found for a small portion of the eicosanoids and new antiinflammatory members of this family have been discovered. This suggests a complex relationship between ω-6 fatty acid intake and inflammation regulation. Humans evolved in an environment in which the dietary ω-3/ω-6 ratio was 1:1. In modern diets, this ratio has shifted toward a ratio closer to 1:20 [17]. This intake imbalance disrupts the normal inflammatory balance leading to increased synthesis of proinflammatory eicosanoids [20] and lower production of antiinflammatory docosanoids [21].

In humans, and mammalian species in general, the synthesis of the ω-3 precursors starts with α-linolenic acid (ALA, C18:3n3), which is an essential fatty acid. It is abundantly found in vegetal sources, particularly in rapeseeds and walnuts [26]. Once absorbed by the organism, α-linolenic acid can be elongated and unsaturated to produce EPA (C20:5n3), then DPA (C22:5n3), and finally DHA (C22:6n3) (Fig. 11.2) [22]. The ω-6 ARA is produced through elongation and desaturation of its precursor linoleic acid (LA). LA is commonly found in oils derived from plants such as safflower, sunflower, and corn [22, 23]. In most vertebrate species, LA is an essential fatty acid because of the absence of a Δ12 desaturase capable of desaturation at the appropriate position in the hydrocarbon chain of oleic acid (C18:1n9) [22, 24] (Fig. 11.2). ARA can also be acquired directly through consumption of multiple animal products such as chicken, eggs, and beef [25].

The elongation and desaturation of LA (18:2n-6) to produce ARA (C20:4n6) requires many enzymes that are also necessary for the synthesis of ω-3 fatty acids from the ALA precursor. For example, the desaturation of LA to γ-LA (18:3n-6) is catalyzed by the same Δ6 desaturase (Fads2) responsible for the desaturation of ALA to stearidonic acid (18:4n3). Elovl5 and Fads1 are also shared for further elongation and desaturation [22]. However, these enzymes have a higher affinity toward the ω-3 fatty acids. Therefore, the relative abundance of ALA versus LA is an important factor that determines EPA, DPA and DHA output versus ARA output, and their availability for conversion into their respective metabolites. The excessive amount of LA, the ARA precursor, present in the western diet drives a powerful competitive effect that redirects these synthetic enzymes toward the production of ARA and its derived inflammatory metabolites [27, 28]. A significantly decreased ω-3/ω-6 ratio therefore results in excessive inflammatory reactions and delayed inflammation resorption. It can even result in a complete absence of resorption, and thus create a state of chronic low-grade inflammation like the one associated with obesity [16].

Interestingly, the maintenance of an appropriate ω-3/ω-6 ratio seems to preserve an antiinflammatory climate even under extreme conditions. Tissues, including the liver, naturally accumulate fat with age. The result of age-induced lipid accumulation in tissues is very similar to that of obesity-induced lipid accumulation, leading to lipotoxicity, oxidative stress, inflammation, and insulin resistance [29]. This may be explained by a change in the lipid composition of the liver as the proportion of ω-3 fatty acids is reduced with age [30]. Interestingly, our group showed that 12-month-old transgenic mice overexpressing apolipoprotein D (ApoD) maintain the same proportion of PUFA as young mice (3-month-old) [30]. ApoD is thought to participate in PUFA transport. ApoD binds and transports ARA [31], but while it is suspected that ApoD binds to ω-3 fatty acids, no study has yet confirmed it. It appears that under prolonged ApoD overexpression, a greater efflux of PUFA are transported to and accumulated in hepatocytes. This results in a severe hepatic steatosis (averaging a Kleiner score of 2 out of 3) with high PUFA concentrations. In transgenic mice, the liver presents a more favorable ω-3/ω-6 ratio than their wildtype counterparts. Despite this severe lipid accumulation, inflammatory balance is maintained and hepatic steatosis does not progress to nonalcoholic steatohepatitis (NASH), showing that the nature of the lipids accumulated in organs is far more important than their overall quantity for inducing and controlling inflammation [30].

## Regulation of inflammation in obesity

Considerable advancements have been made in the understanding of the initiation and the resolution of the inflammation process. For a long time, inflammation was thought as a process entirely controlled by the production of proinflammatory eicosanoids. Inflammation resorption was understood as passive process that occurs when these proinflammatory mediators stop being produced. It is only through the discovery of multiple antiinflammatory lipid mediators derived from PUFA, named specialized proresolving mediators (SPM), that this concept was laid to rest [32]. Without these mediators, inflammation can continue, unhindered, for the full lifespan of the host. This is the case in obesity where multiple tissues are in a state of constant low-grade inflammation. Interestingly, SPM synthesis is hindered in obesity by multiple possible mechanisms working conjointly and reviewed in later sections [33].

## The NLRP3 inflammasome

In the last decade, the NLR family pyrin domain containing 3 (NLRP3) protein has been linked to inflammatory mechanisms in a variety of metabolic disease contexts. Since many NLRP3 activator signals, such as high concentrations of glucose and palmitate, are present in excessive amount in the western diet, it is not surprising that NLRP3 plays a crucial role in metabolic disorders. NLRP3 is overly expressed and activated in the metabolic tissues of obese individuals exposed to western diet [34]. This results in tissue damage and proinflammatory cytokine secretion that inhibits insulin signaling. In fact, NLRP3 expression is strongly correlated with insulin resistance and metabolic syndrome [35]. This makes NLRP3 a priority target in the fight against metabolic diseases.

Inflammasomes are key modulators of inflammation and play a major role in metabolic diseases. The maturation of important cytokines (IL-1β and IL-18), the onset of pyroptosis and the production of eicosanoid storms are dependent upon their activation. Inflammasomes were first recognized as crucial elements in the defense against pathogens [36] and are now known to be activated in metabolic [37], autoimmune [38] and neurodegenerative diseases [39], as well as in cancer [40–42] and dysbiosis [43, 44].

The study of inflammasomes, in particular the NLRP3 inflammasome, has changed our perspective on how inflammation affects metabolic tissues. Inflammasomes are multiprotein platforms that act as a sort of multistep "arming key" resulting in potent inflammatory responses. In addition, NLRP3 activation has an inhibitory effect on SPM synthesis [45]. In a mouse model of sepsis caused by cecal ligation and puncture, mice lacking *NLRP3* display less severe inflammation-induced symptoms and mortality. Genetic deficiency in NLRP3 results in reduced prostaglandin biosynthesis, inflammation, and increased biosynthesis of lipids mediators such as lipoxins (particularly $LXB_4$, see details in subsequent sections). This suggests a role for the NLRP3 inflammasome in regulating the output of SPM [45]. The proposed mechanism for this regulation is that NLRP3 activation results in the inhibition of lipoxin synthesis in macrophages, through a caspase-7-mediated pathway [45]. Interestingly, inhibition of 12-LOX, an essential enzyme required for the production of many SPM, results in the activation of caspase-7 in cancer cells [46–48], showing a mutual inhibitory relationship between 12-LOX and caspase-7. This interaction could be one of the factors that explain why supplementation with ω-3 fatty acids alone is often unable to ameliorate metabolic diseases in humans [49, 50]. Since, NLRP3 is already activated in the tissues of obese patients, it could prevent the conversion of ω-3 fatty acids into their antiinflammatory metabolites.

Inflammasome assembly occurs when a pathogen-associated molecular pattern (PAMP) or a danger-associated molecular pattern (DAMP) activates one of the cell's pattern-recognition receptors (PRR) such as Toll-like receptors (TLR). While PAMP are easily defined as pathogen components (e.g., LPS) being recognized by the host's cells, DAMP are more complex. DAMP could be defined as homeostatic abnormalities (extracellular ATP, abnormally high concentration of glucose or palmitate, oxidized DNA) and out of place cell components (cytoplasmic host DNA) [43]. In the context of obesity induced by excessive consumption of the western diet, many of these DAMP are present in tissues, resulting in PRR activation.

Upon PRR activation, multiple downstream events occur and result in the activation of an inflammasome scaffolding protein. To date, five inflammasome scaffolding proteins have been exhaustively studied: NLRP1, NLRP3, NLRC4, AIM2, and Pyrin. Each scaffolding protein is activated by a specific set of DAMP and PAMP and has its own upstream signaling pathway, but they operate almost identically once assembled. Scaffolding proteins initiate inflammasome assembly by recruiting multiple copies of the adaptor protein apoptosis-associated speck-like protein containing CARD (ASC). ASC then recruits procaspase-1 dimers which are cleaved into the mature caspase-1. Caspase-1 proceeds to activate multiple proinflammatory pathways. Since the major difference between inflammasome complexes is the scaffolding protein, the inflammasome complex is named according to its scaffolding protein [43]. In the context of metabolic diseases, the most pertinent inflammasome complex is probably the NLRP3 inflammasome due to its specific set of activators [43].

## NLRP3 priming (signal 1)

The individual components of the NLRP3 inflammasome are not constitutively expressed. The first step in the activation of the NLRP3 inflammasome is therefore the transcription and translation of its three major components: the NLRP3 protein itself, pro-IL-1β and pro-IL-18 [51]. Without this first priming step, the amount of NLRP3 inflammasome protein components in the cell is insufficient to generate a robust downstream signaling cascade [51]. TLR ligands such as LPS [52], cytokines such as TNF-α [53], and IL-1R receptor ligands have been shown to efficiently prime NLRP3 (see Fig. 11.1A) [51, 54]. The excessive production of reactive oxygen species (ROS) primes the NLRP3 inflammasome [55]. The exact signaling event responsible for this phenomenon is unclear. However, ROS have often been shown to

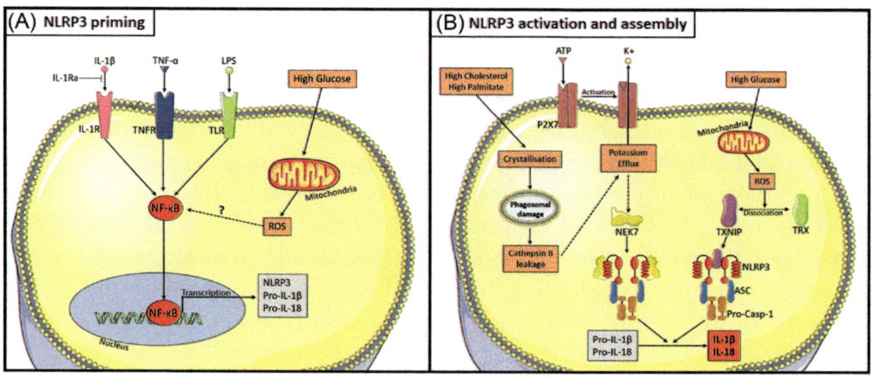


**FIGURE 11.1 NLRP3 inflammasome priming and activation.** Schematic representation of the principal signaling events that modulates the (A) priming and (B) activation of the NLRP3 inflammasome. *ASC*, apoptosis-associated speck-like protein containing a CARD; *IL-1Ra*, interleukin-1 receptor antagonist; *NEK7*, NimA related kinase 7; *NLRP3*, NLR family pyrin domain containing 3; *P2X7*, purinergic type 2 receptor X7; *ROS*, reactive oxygen species; *TRX*, thioredoxin; *TXNIP*, thioredoxin interacting protein.

activate NF-κB through alternative IκBα phosphorylation [56]. Oxidative stress that induces the production of ROS is present in most metabolic diseases such as T2D [57], atherosclerosis [58], and NASH [59]. This priming results in NF-κB activation and the transcription of NLRP3 as well as pro-IL-1β and pro-IL-18 [51].

### NLRP3 activation (signal 2)

Multiple stimuli and pathways have been reported to activate NLRP3. Most of these pathways have a common signaling step that includes a potassium efflux from the cell. Consequently, NEK7 is recruited to NLRP3 and assists in its oligomerization [60–62]. Once oligomerization has been achieved, the complex can proceed to recruit ASC and two procaspases-1. The activated caspase-1 can then cleave pro-IL-1β and pro-IL-18, thus producing the mature form of the cytokines [43]. High extracellular potassium concentrations or addition of glyburide, an inhibitor of ATP-sensitive potassium channels, prevent the activation of NLRP3 and the release of the proinflammatory cytokine IL-1β [63].

Multiple PAMP and DAMP have been shown to activate NLRP3. Cholesterol crystals that are found in atherosclerotic plaques are recognized as DAMP. These crystals can be phagocytosed by macrophages which causes lysosomal destabilization and leakage of cathepsin B [64], activating NLRP3 [65] (Fig. 11.1B). Saturated fatty acids can also be identified as DAMP. Palmitate and stearate, in excessive concentrations, tend to crystalize in macrophages. This induces lysosomal dysfunction and downstream signaling that activates NLRP3 and induces IL-1β release from macrophages. This NLRP3 activation and cytokine secretion is reversed dose dependently by addition of the monounsaturated fatty acid oleate. Addition of oleate allows the appropriate storing of saturated fatty acids into triglycerides, preventing lipid intracellular crystallization. In theory, the stearoyl-CoA desaturase-1 (SCD1) enzyme responsible for the desaturation of palmitate and stearate could mitigate this issue and alleviate crystal formation and inflammation, but failed to do so in one study [66]. SCD1, however, has been described as a crucial element in the formation of autophagosomes [67], which can degrade inflammasome complexes and components. Additionally, palmitate was shown to activate NLRP3 by causing mitochondrial dysfunction and ROS production [68].

Glucose has also been identified as a DAMP. As mentioned before, high glucose concentration can prime NLRP3 via a ROS/NF-κB pathway but it can also activate NLRP3 via a ROS/TXNIP-dependent pathway [69–71]. This means that high glucose could act as both a signal 1 and signal 2 molecule and trigger inflammation by itself [67, 69].

Finally, exogenous ATP can also be classified as a DAMP. When released from necrotic cells or secreted by invading neutrophils, it activates NLRP3 by signaling through the ionotropic P2X7 receptor [72] which then triggers the release of potassium [73].

## The eicosanoid classes

The induction of the classical features of inflammation are produced in response to the biosynthesis of eicosanoids. In the context of obesity, eicosanoids contribute to the metabolic syndrome in numerous ways such as leukocyte recruitment, cytokine expression, vascular endothelium functions, preadipocyte differentiation, and prolipogenic gene expression.

As previously mentioned, the precursor of eicosanoids is the ω-6 fatty acid ARA. Once produced or after uptake by the cell, free ARA is quickly stored as phospholipids [74]. ARA needs to be liberated from phospholipids by a phospholipase to enter eicosanoid biosynthetic pathways. This is usually carried out by the calcium-dependent $cPLA_2$ enzyme, an important rate-limiting step of eicosanoid production [24]. Free ARA can then be transformed further to produce prostanoids, leukotrienes, hydroxyeicosatetraenoic acids (HETE), endocannabinoids, isoprostanes, and lipoxins. The most effective ARA-derived inflammatory modulators are found in the prostanoid, leukotriene, and lipoxin families.

## The prostanoids

Prostanoids are further subdivided into prostaglandin, prostacyclin, and thromboxane subclasses which are synthesized by various pathways. Their precursor ARA is first transformed into prostaglandin $G_2$ ($PGG_2$) and then successively to prostaglandin $H_2$ ($PGH_2$) by either cyclooxygenase-1 or -2 (COX-1/COX-2) (Fig. 11.2) [75].

COX-1 is generally thought to be constitutively expressed while COX-2 expression is highly inducible [76]. COX-2 is at the center of many inflammatory pathways that regulate the output of prostanoids in the inflammatory reaction and in its resorption. Inflammatory cytokines such as IL-1β and TNF-α have been shown to induce robust COX-2 expression [77–80] through the NF-κB and AP-1 transcription factors [81]. Since, the expression of NLRP3 inflammasome components is under the transcriptional regulation of these same transcription

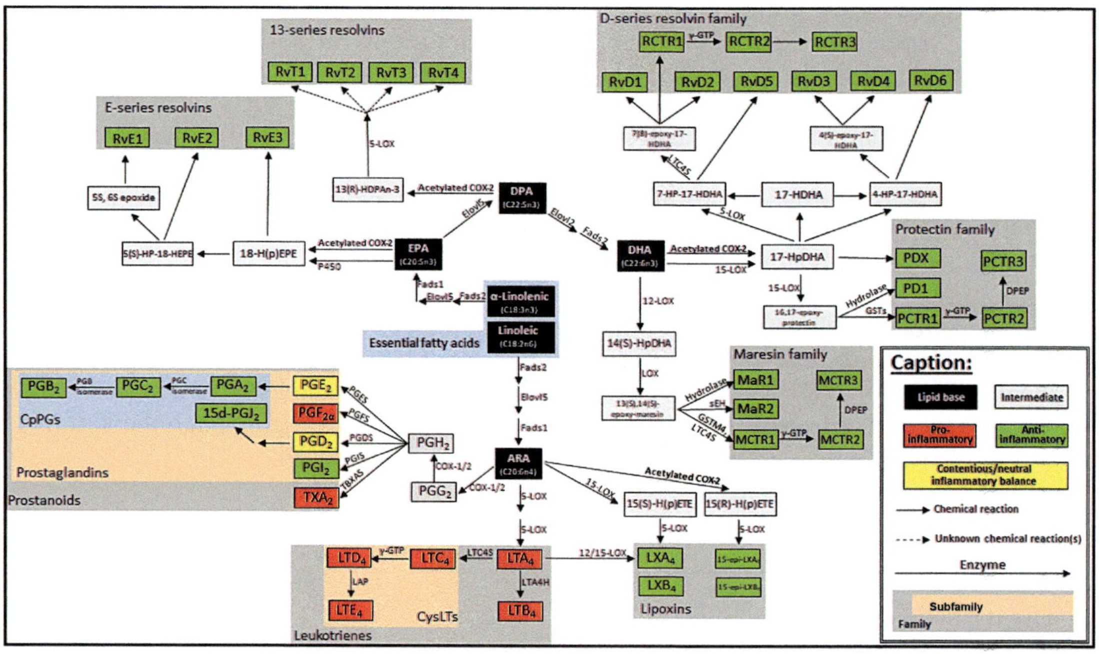


**FIGURE 11.2** **Lipid mediator biosynthetic pathways.** Schematic representation of the biosynthetic pathways that produce the eicosanoids and the docosanoids.

factors, and because the activation of NLRP3 results in the production of IL-1β, COX-2 expression is in general parallel to the expression and activation of elements of the NLRP3 signaling pathway. COX-2 greatly amplifies the NLRP3 inflammasome response by producing ROS in response to signal 2 stimulation (ATP) and by inducing the synthesis of prostaglandins ($PGE_2$) in macrophages [82]. COX-2 activity is also amplified in response to ROS production, creating a positive feedback loop [83].

While there are dozens of different prostanoid species that can be produced from $PGH_2$, not all of them are bioactive or able to produce a potent physiological and inflammatory response. The following section will mainly focus on prostanoids having a direct effect on metabolic disorders.

### *$PGE_2$*

Prostaglandin $E_2$ ($PGE_2$) is produced from $PGH_2$ through the action of one of the prostaglandin E synthases: microsomal prostaglandin E synthase-1 and -2 (mPGES-1 and mPGES-2) or the cytosolic prostaglandin E synthase (cPGES). The signaling pathways activated by $PGE_2$ depend on its concentration as PGE2 binds with various affinity to its four known receptors (EP1-EP4) [84]. $PGE_2$ can regulate numerous biological processes including cell proliferation, apoptosis, angiogenesis, and inflammation in multiple cell types. Most importantly, $PGE_2$ is a key mediator involved in both the initiation and the resolution of inflammation [85].

$PGE_2$ acts as a vasodilator that facilitates the infiltration of leukocytes into the tissue [85]. In macrophages, $PGE_2$ activates NF-κB. This potentiates NLRP3 activation and IL-1β secretion [82]. This also constitutes a positive feedback loop that stimulates COX-2 expression. $PGE_2$ is tough to activate NLRP3 by modulation of cAMP production. EP3 activation has been shown to reduce cAMP production. Since cAMP is able to bind directly to NLRP3 and prevent its activation [86], the activation of the EP3 receptor facilitates NLRP3 activation by diminishing the cAMP pool [87, 88]. On the other hand, activation of EP2 and EP4 stimulates the formation of cAMP [84, 88], producing antiinflammatory effects. Signaling through EP4 has been linked to reduced NF-κB activation and reduced secretion of proinflammatory cytokines [35, 40].

In a normal inflammatory reaction, the end of the initiation phase of inflammation is characterized by a peak in $PGE_2$ concentration. This is followed by a marked increase in anti-inflammatory lipoxin $A_4$ ($LXA_4$) secretion and a reduction in proinflammatory leukotriene $B_4$ ($LTB_4$) secretion. This shift in LXA4 and LTB4 synthesis induces a sudden reduction in the neutrophil population and a concomitant decrease in inflammatory cytokine production [89]. This can be explained by the fact that high concentrations of $PGE_2$ inactivate 5-LOX and activate 15-LOX, the enzymes responsible for the synthesis of LTB4 and LXA4, respectively. Thus, high concentrations of $PGE_2$ divert ARA utilization away from the leukotriene pathway toward the lipoxin pathway [89].

$PGE_2$ also modulates metabolism through its effect on adipocyte differentiation. The EP3 receptor is abundantly expressed in white adipose tissue (WAT). Its expression is decreased in the WAT of hyperphagic db/db mice (an obesity/T2D model) and in mice fed a high-fat diet. Silencing EP3 also produces an obese phenotype associated with increased food intake, reduced insulin sensitivity, elevated serum triglycerides [90] as well as increased macrophage infiltration and inflammation [91]. This might stem from the fact that $PGE_2$ decreases PPARγ expression in adipocytes. PPARγ activation skews preadipocyte differentiation toward a white phenotype. The $PGE_2$ mediated inhibition of PPARγ instead orients preadipocyte differentiation toward a beige phenotype. This is associated with increased expression of uncoupling protein 1 (UCP1) and energy expenditure [92]. $PGE_2$ treatment of human WAT explants and differentiated 3T3 adipocytes also reduces the expression of fibrogenic and inflammatory genes, including IL-6 and MCP-1 [93].

While $PGE_2$ seems to produce beneficial effects in adipose tissue, its role in the liver is more ambiguous. $PGE_2$ biosynthesis is increased in the liver of obese patients [94]. The induction of COX-2 and PGES (mPGES-1 and mPGES-2), responsible for the production of $PGE_2$, in mouse primary hepatocytes enhances de novo lipogenesis while reducing β-oxidation and lipoprotein secretion [95]. $PGE_2$ also dose dependently increases glucose output in rat hepatocytes thus representing an aggravating factor in hyperglycemia [96]. However, mPGES silencing in a mouse model of NASH increases the secretion of IL-1β and TNF-α as well as tissue damage [94]. In addition, COX-2 and mPGES-1 expression are both elevated in renal diseases. $PGE_2$ signaling through EP1 and EP3 leads to vasoconstriction and hypertension which contribute to renal insufficiency and injury, and increases cardiovascular risk [97].

## $TXA_2$

Thromboxane $A_2$ ($TXA_2$) induces multiple physiological responses, including platelet cell adhesion and aggregation, vascular smooth muscle cell contraction, and endothelial inflammatory responses [98]. Additionally, $TXA_2$ induces monocyte and T lymphocyte recruitment through activation of NF-κB and AP1 which results in production of MCP-1, a potent chemoattractant toward leukocytes [99].

One of the greatest physiological challenges that the body experiences in obesity is the proper irrigation of the growing adipose tissue. The angiogenesis process is typically insufficient, and a hypoxic milieu is created in the tissue. This dysregulates secretory functions, promoting inflammation in local blood vessels and secretion of the vasoconstrictive $TXA_2$. Dysfunctional endothelial cells imbedded in adipose tissue express high levels of COX-2. This overexpression results in the overproduction of $TXA_2$, resulting in excessive vasoconstriction mediated by the vascular smooth muscle cells [100–102]. High glucose concentration is also a factor that elevates the expression of COX-2 in endothelial [103] and smooth muscle cells [104], thus promoting hypertension in T2D.

$TXA_2$ also plays an important role in the development of preeclampsia in obese women. The usual treatment for preeclampsia is a daily dose of aspirin. At low dose (between 60 and 150 mg/day), aspirin induces the acetylation of the COX enzymes. Acetylation of COX-1 results in lower production of $TXA_2$, decreasing the risk of preeclampsia. However, obese women produce more $TXA_2$ than lean women, even when medicated, suggesting that aspirin resistance may occur in obese patients [105] (Box 11.1).

## $PGF_{2\alpha}$

Prostaglandin $F_{2\alpha}$ ($PGF_{2\alpha}$) has hypertensive, as well as proinflammatory properties. $PGF_{2\alpha}$ is overproduced in renovascular hypertension in response to increased oxidative stress. This leads to reduced renal blood flow, aggravation of renal damage, and increased mean arterial pressure [83]. Similarly, $PGF_{2\alpha}$ contributes to endothelial dysfunction and hypertension in T2D [108]. $PGF_{2\alpha}$ also represses the progression of the early stage of adipogenesis by increasing intracellular calcium concentrations through activation of a calcium/calmodulin-dependent protein kinase (CaMK) [109].

BOX 11.1

Aspirin covalently modifies COX enzymes by acetylating Ser530 within the active site. This covalent modification leads to complete irreversible loss of COX-1 activity. On COX-2 however, the acetylation results in a shift of reaction specificity. This shift converts COX-2 into a lipoxygenase [106]. Therefore, COX-2 acetylation results in a blockage of prostanoid production in favor of precursors for antiinflammatory compounds such as 15R-HETE (precursor of 15-epi-lipoxins) and 17-HpDHA (precursor of protectins and resolvins) (Fig. 11.2) [106, 107].

## $PGI_2$

Prostaglandin $I_2$ ($PGI_2$), synthesized by the prostaglandin I synthase (PGIS), is primarily produced in the vasculature, particularly in pulmonary arterial segments [110, 111]. In the cardiovascular system, it induces vasodilation and inhibits platelet aggregation [112–115]. $PGI_2$ directly inhibits the vasoconstrictive and proaggregating functions of $TXA_2$ [116–118]. Insufficient synthesis of $PGI_2$ is associated to hypertension in obese individuals [102].

$PGI_2$ was discovered to modulate the expression of specific miRNAs. A study by Mohite et al. used a hybrid enzyme composed of both COX-1 and PGIS linked by a 10-amino acid linker. The resulting enzyme possesses triple catalytic function, converting ARA into $PGG_2$, then into $PGH_2$ and finally into $PGI_2$ in quick succession. The expression of this enzyme into mouse mature adipocytes successfully diverts ARA away from $PGE_2$ to $PGI_2$ synthesis. This results in the differential expression of various miRNAs [118]. One of the most upregulated miRNAs, miR-711 (targeting CCAAT/C/EBP-β and Akt1), reduces adipocyte differentiation via a probable PPARγ-dependent mechanism. In fact, $PGI_2$ is able to directly activate PPARβ/δ [119] and indirectly activate PPARγ by first activating the prostaglandin $I_2$ receptor [120].

## $PGD_2$

Prostaglandin D2 ($PGD_2$) produces either pro- or antiinflammatory effects depending on the context. $PGD_2$ is produced during allergic lung inflammation, where it induces inflammation [121]. However, $PGD_2$ production by Kupffer cells (liver resident macrophages) helps to maintain an antiinflammatory environment in the liver [122].

$PGD_2$ can be produced by either the lipocalin-type prostaglandin D synthase (L-PGDS) or the hematopoietic prostaglandin D synthase (HPGDS). High-fat diet (HFD) selectively induces the expression of L-PGDS in the adipose tissue, increasing the production of $PGD_2$ and stimulating the progression of adipogenesis [123]. However, most of the $PGD_2$ produced in the WAT is synthesized in macrophages, using HPGDS. Thus, $PGD_2$ not only stimulates the production of adipocytes but polarizes their resident macrophages toward the M2 antiinflammatory state [124]. Similarly to $PGI_2$, $PGD_2$ activates adipogenesis via PPARγ [125–127]. *PTGDS*$^{-/-}$ mice (lacking the gene encoding L-PGDS) have increased insulin resistance and reduced glucose tolerance under HFD, developing a more severe T2D than wildtype control mice. Absence of L-PGDS expression also results in larger adipocytes, even in mice fed a low-fat diet [128]. *PTGDS* expression is correlated with brown adipose tissue activity and PPAR activation (γ, 1α, and 1β). Upon cold acclimation, *PTGDS*$^{-/-}$ mice rely more on carbohydrate to fuel thermogenesis than wildtype mice [129]. The increased usage of glucose results in a better outcome in glucose tolerance test in *PTGDS*$^{-/-}$ mice. However, *PTGDS* knockout has no effect on insulin levels, explaining why under normal temperature conditions *PTGDS*$^{-/-}$ mice develop T2D when fed a HFD [129]. Interestingly, the specific knockdown of *PTGDS* in adipocytes results in improved metabolic health when mice are fed with HFD. This is characterized by reduced body weight, increased glucose tolerance and insulin sensitivity as well as reduced macrophage M1 inflammatory markers in the adipose tissue [130].

#### *15d-PGJ$_2$*

15-Deoxy-delta-12-14-prostaglandin $J_2$ (15d-PGJ$_2$), a nonenzymatic derivative of PGD$_2$, is strictly an antiinflammatory metabolite. 15d-PGJ$_2$ acts through multiple pathways to inhibit inflammation. First, it acts directly on NF-κB by reducing its transcriptional activity [131–133]. Additionally, 15d-PGJ$_2$ is a potent activator of PPARγ that in turn further inhibits the transcriptional activity of NF-κB [133]. The direct inhibitory effect of 15d-PGJ$_2$ on NF-κB is possible because of its particular chemical structure [134]. 15d-PGJ$_2$ inhibits NF-κB's DNA-binding capacity by alkylating its RelA (p65) [135, 136] and p50 subunits [137] or IKKα/β, the upstream activators of NF-κB [138].

#### *PGA$_2$*

Prostaglandin $A_2$ (PGA$_2$) is the product of a nonenzymatic dehydration within the cyclopentane ring of PGE$_2$ [139]. Despite very few studies available, PGA$_2$ appears to have mostly antiinflammatory properties. PGA$_2$ suppresses LPS-induced inflammation in pulmonary endothelial cells, microglia and astrocytes by inhibiting NF-κB signaling [140, 141]. Like 15d-PGJ$_2$, PGA$_2$ is a cyclopentenone prostaglandin which inhibits NF-κB signaling in the same manner [139, 142]. Interestingly, EP4 has been identified as a receptor of PGA$_2$ [140]. This means that some of the antiinflammatory properties of PGE$_2$ could be attributable to PGA$_2$ after a spontaneous transformation of PGE$_2$ to PGA$_2$ [143].

Other effects of PGA$_2$ include its capacity to act as a redox signal that modulates metabolism in atherosclerosis. PGA$_2$, as well as its derivatives PGC$_2$ and PGB$_2$, have potent inhibitory effects on the cholesterol metabolism of inflammatory macrophages. In macrophages, PGA$_2$ inhibits lipogenesis and diverts acetyl-CoA units from cholesterogenesis toward phospholipid production. Accordingly PGA$_2$, when delivered in vivo via liposomes, dramatically reduces cholesterol accumulation in foam cells and atherosclerosis lesions [144, 145]. Interestingly, PGA$_2$ has also been reported to activate all three PPAR (γ, α, and β), suggesting again antiinflammatory properties [146].

### Leukotrienes

Leukotrienes are strictly proinflammatory molecules. Leukotriene $A_4$ (LTA$_4$) is first produced from ARA by the 5-lipoxygenase enzyme (5-LOX). Then, LTA$_4$ is either transformed into LTB$_4$ by leukotriene $A_4$ hydrolase or into LTC$_4$ by leukotriene $C_4$ synthase (Fig. 11.2) [147].

The main characteristic of LTB$_4$ is its potent chemotactic effect on leukocytes [148]. In response to HFD, LTB$_4$ production increases in the liver, muscles and adipose tissues. Silencing the LTB$_4$ receptor BLT$_1$ induces an antiinflammatory phenotype and rescues insulin sensitivity in HFD-fed mice [149]. In opposition, $LTB_4$ treatment of primary hepatocytes activates de novo lipogenesis and gluconeogenesis in addition to impairing insulin-mediated suppression of hepatic glucose output [149].

### Lipoxins

Lipoxins are an antiinflammatory eicosanoid family produced from ARA which are potent inhibitors of neutrophil recruitment [89]. Their biosynthesis can occur through two different pathways. The lipoxygenation of ARA by 15-LOX results in the formation of 15(S)-H(p)ETE while the lipoxygenation of ARA by acetylated COX-2 or P450 results in the formation of 15(R)-H(p)ETE. Following this, a second lipoxygenation of these products by 5-LOX results in the production of lipoxins ($LXA_4$ and $LXB_4$) and their epi-lipoxin counterparts (15-epi-$LXA_4$ and 15-epi-$LXB_4$) (Fig. 11.2). $LXA_4$ can also be synthesized through lipoxygenation of $LTA_4$ by either 12-LOX or 15-LOX.

In obese mice, periodic intraperitoneal injection of $LXA_4$ attenuates hepatic steatosis and chronic kidney disease. $LXA_4$ injection also reduces obesity-induced inflammation by reducing the secretion of TNF-α and shifting adipose tissue macrophages toward the M2 polarization [150]. The effect of $LXA_4$ on renal morphology and function is potentially the result of NF-κB and ERK/p38 MAPK signaling inhibition [151].

## The docosanoids

### The specialized proresolving mediators

Much of the antiinflammatory and tissue regenerative properties attributed to ω-3 fatty acids may in fact be attributable to their metabolites, the maresins, protectins and resolvins, but also to their respective conjugates: maresin conjugates in tissue regeneration (MCTR), protectin conjugates in tissue regeneration (PCTR) as well as resolvin conjugates in tissue regeneration (RCTR) [152]. Together, these lipid mediators have been coined as specialized proresolving mediators (SPM). In addition to these ω-3 derived lipid mediators, the ω-6 derived lipoxins are also recognized as members of the SPM superfamily [152]. SPM have a particularly potent antiinflammatory effect, even at nanomolar concentrations. They serve as agonists for specific G protein-coupled receptors (GPCR).

#### *Maresins and MCTR*

Maresins (MaR), also known as macrophage mediators in resolving inflammation, were initially discovered in macrophages [153]. MaR1, MaR2 and their sulfide-conjugate MCTR1 are produced from a 13S,14S-epoxide-maresin intermediate. 13S,14S-epoxide-maresin is first obtained from DHA via two consecutive lipoxygenation reactions (Fig. 11.2) [154–159].

Maresins inhibit the recruitment of polymorphonuclear cells and increase macrophage efferocytosis, the phagocytosis of dying cells. Maresins block NF-κB activation [160] and downregulate many proinflammatory cytokines such as TNF-α, IL-1β and IL-6. It also reduces the production of prostanoids and leukotrienes [32, 159, 161]. Finally, MaR1 improves the insulin sensitivity of *ob/ob* obese mice and stimulates the M2 polarization of macrophages [162].

### *Protectins and PCTR*

Protectins and their sulfide-conjugates (PCTR) are produced from the 17-HpDHA intermediate, a molecule obtained from the transformation of DHA by 15-LOX (or acetylated COX-2 after aspirin treatment) [163]. In a mouse model of carbon tetrachloride-induced liver injury, supplementation of the diet with DHA induces the production of protectin D1 (PD1) and 17-HpDHA [163]. The combined effect of PD1 and 17-HDHA (a reduced form of 17-HpDHA) greatly reduces liver injury in part via activation of PPARγ. This is associated with reduced 5-LOX activity, reduced COX-2 expression and lower secretion of TNF-α [163]. Protectin X (PDX) also displays liver protective effects [164]. Jugular infusions of intralipid emulsions in mice results in rapid systemic inflammation and reduction of insulin sensitivity. However, addition of PDX in the lipid infusion prevents the secretion of proinflammatory cytokines, rescues insulin signaling and increases the plasma concentration of cytokine IL-6 [164] that has, in this context, antiinflammatory properties [165]. The PDX insulin sensitization effects in skeletal muscle are abrogated by IL-6 silencing. Therefore PDX acts through induction of IL-6 production by skeletal muscle to modulate glucose metabolism [164]. In skeletal muscles, IL-6 also increases glucose uptake as well as glycogen synthesis [166, 167].

### *Resolvins and RCTR*

The production of resolvins and their sulfide-conjugates (RCTR) is far more complex than their maresin and protectin counterparts because multiple distinct series of resolvins exist. The E-series is produced from the transformation EPA into the intermediate 18-H(p)EPE by P450 or the acetylated form of COX-2, while the 13-series are produced from the 13(R)-HDPAn-3 intermediate formed by the conversion of DPA using the acetylated form of COX-2. Finally, the D-series of resolvins (RvD) as well as the RCTR are produced from the intermediate 17-HpDHA, which is also the precursor of the protectins [158, 168] (Fig. 11.2).

In a murine model of nonresolving lung inflammation, RvD1 reduces the expression of proinflammatory cytokines (CCL5, CXCL10, CXCL1, IL-1β, IL-17) [169]. This is linked to an increased expression of miR-155 and miR-21 known to target PRR such as TLR as well as the NF-κB transcription factor [169]. RvD1 also upregulates miR-146b and miR-219 that target NF-κB signaling and 5-LOX, respectively [170]. RvD1 can thus interfere with the production of prostaglandins and leukotrienes. It can also limit the expression of proteins dependent on NF-κB such as NLRP3 and many proinflammatory cytokines [171].

## Role of lipid inflammatory mediators in metabolic diseases

Docosanoid functions are specifically antiinflammatory and often overlap. The traditional approach to inflammatory treatment has been the inhibition of the formation of eicosanoids that possess inflammatory properties. However, nonspecific inhibition of eicosanoid production can lead to undesired consequences (e.g., gastrointestinal complications) [172] because the eicosanoid class contains both pro- and antiinflammatory mediators that play important homeostatic functions (see above). Thus, stimulating the production of docosanoid SPM instead of inhibiting the synthesis of eicosanoids appears to be a promising approach to treat inflammation in metabolic disease.

## Lipid mediators in atherosclerosis

High fat feeding of *ApoE*$^{-/-}$ mice (a mouse model of atherosclerosis) leads to the rapid development of atherosclerosis. Under these conditions, the circulating concentration of eicosanoids ($PGE_2$, $PGD_2$, $LTB_4$) increases while the concentrations of RvD2 and MaR1 decrease [173]. This shift toward the accumulation of eicosanoids correlates with the elevation of plaque instability markers, including the expansion of the necrotic core, the accumulation of macrophages, the reduction of the fibrous cap's thickness and the reduction in smooth muscle cell number. Interestingly, injection of mice with RvD2 and MaR1 prevents atheroprogression and promotes M2 proresolving polarization of macrophages [173].

NLRP3 and IL-1β are also thought to be important in the progression of atherosclerosis. Cholesterol deposition occurs early in the formation of the atherosclerotic lesion and results in the formation of cholesterol crystals that can activate NLRP3 in invading macrophages/monocytes [174]. Autophagy is able to inhibit NLRP3 activation either by directly targeting NLRP3 and its complex for degradation in autophagosomes [175] or indirectly by mitophagy which eliminates dysfunctional mitochondria and limits their ROS production [176]. Autophagy-defective *Atg5*$^{-/-}$ mice have higher levels of NLRP3 activation and develop atherosclerosis at a higher rate [177]. Berberine, isolated from the pant *Berberis vulgaris*, increases the autophagic flux of cells [178] and limits the protein expression of NLRP3 components [179]. In *ApoE*$^{-/-}$ mice under HFD, oral administration of berberine significantly reduces the serum levels of proinflammatory cytokines and atherosclerosis markers [180]. A specific NLRP3 inhibitor, IFM-2427, is presently in clinical trials and could be soon available for treatment of atherosclerotic patients [181].

## Lipid mediators in nonalcoholic steatohepatitis

Western diet feeding often induces hepatic steatosis characterized by lipotoxicity, overproduction of ROS, and inflammation. It is also associated with activation of prolipogenic pathways and accelerated lipid accumulation. This condition, known as nonalcoholic steatohepatitis (NASH), can then lead to fibrosis, cirrhosis, and the development of hepatocellular carcinomas. NASH also contributes to insulin resistance and hyperglycemia [182]. *Caenorhabditis elegans* possesses a n3 desaturase named *fat-1* capable of transforming ω-6 into ω-3 fatty acids. Transgenic mice expressing *fat-1* thus have a high ω-3/ω-6 ratio in their tissues [183]. Mice fed with DHA and *fat-1* mice present a higher M2 polarization of liver macrophages and resorb inflammation more efficiently compared to their respective controls [182]. This effect is largely mediated through the activation of the RORα transcription factor by the maresin MaR1 via a signaling loop. DHA is converted to MaR1 by 12-LOX (and other enzymes, see Fig. 11.2) which then activates RORα [182]. RORα then proceeds to activate the transcription of *Alox-12* (the gene transcribing 12-LOX), *Rora* (the gene transcribing RORα), and *Klf4*. The newly produced KLF4 transcription factors then mediate the polarization shift of macrophages toward M2. The increased expression of 12-LOX results in a positive feedback loop increasing MaR1 production and activation of RORα. Unsurprisingly, daily MaR1 intra-peritoneal injection in mice results in the complete abolition of all symptoms associated to NASH [182]. 12-LOX expression is reduced in human hepatic NASH biopsies [182]. As 12-LOX is an important enzyme that enables the production of lipoxins, maresins and resolvins, its reduced expression is likely to play a key role in the failed resolution of inflammation characterizing NASH.

NLRP3 expression and activation is also enhanced in NASH, exacerbating liver damage, lipid accumulation and insulin resistance. In a mouse model where NASH is induced by a methionine- and choline-deficient diet, intraperitoneal administration of berberine, which increases the autophagic flux of cells [178], limits NLRP3 expression and IL-1β secretion. Consequently, mice survival is increased, and NASH symptoms are attenuated [179]. These results were replicated by using MCC950, a selective NLRP3 inhibitor, in hyperphagic *foz/foz* mice fed a methionine/choline-deficient diet. MCC950 was able to block the activation of caspase-1, the maturation of IL-1β, and greatly alleviate liver fibrosis but not lipid accumulation [184]. This suggests that MCC950 treatment blocks the transition of hepatic steatosis towards NASH.

## Lipid mediators in type 2 diabetes

One of the key contributing factors of peripheral insulin resistance is adipose tissue inflammation associated with high production of proinflammatory cytokines (IL-6, TNF-α and MCP-1). These cytokines can then leak from the WAT, enter the circulation and reach other organs, interfering with their insulin sensitivity. Daily MaR1 intraperitoneal injection in overfed *ob/ob* obese mice successfully reverses WAT inflammation and promotes the M2 polarization of macrophages in the tissue. MaR1 treatment also reduces the expression of proinflammatory genes in the WAT and improves insulin sensitivity [162].

The role of NLRP3 in T2D is particularly evident in pancreatic β-cells. In fact, NLRP3-dependent maturation of IL-1β plays a crucial role in β-cell death [185, 186]. Human pancreatic islets secrete IL-1β in response to high glucose concentrations. IL-1β then activates inflammatory pathways by signaling through its receptor IL-1R, resulting in the recruitment of immune cells (mainly macrophages) into the islets. This state of chronic inflammation eventually leads to tissue damage and islet dysfunction. Pancreatic islets also secrete interleukin-1 receptor agonists (IL-1Ra) to block IL-1β signaling. However, after prolonged exposure to high glucose this compensatory mechanism is unable to efficiently block IL-1β signaling [187].

In ex vivo human pancreatic islets, hypoxia or combined LPS/ATP treatment activates NLRP3 expression, resulting in β-cell death. This cell death is inhibited by glyburide, an NLRP3 inhibitor [188]. Similar observations were made in mice where genetic silencing of NLRP3 protected pancreatic islets against T2D induced by either oxidative stress [189] or HFD [185]. *NLRP3*$^{-/-}$ mice display reduced macrophage infiltration in the islets, reduced β-cell death and improved glucose homeostasis [189].

### *Lipid mediators in nonhealing diabetic wounds*

Diabetic patients often suffer from delayed inflammatory resolution as well as delayed wound healing [190]. Initiation of the inflammatory reaction as well as its resorption must occur properly for the wound to heal. In diabetic patients, the reparative functions of macrophages are impaired, leading to an increase in the apoptotic cell burden in the wound [191] as well as a reduction of vascular endothelial growth factor and platelet-derived growth factor secretion [192]. Using a murine splinted excisional wound healing model, it was determined that a DHA metabolite (14S,21R-diHDHA) is underproduced in the macrophages of diabetic *db/db* mice. However, a single treatment with 14S,21R-diHDHA was sufficient to restore macrophage wound healing capabilities [192]. Similarly, 14,22-diHDHA, also named Maresin-like (MaR-L), is able to rescue macrophage wound healing functions in in vitro experiments [193].

NLRP3, caspase-1 and IL-1β are overexpressed in the wounds of diabetic patients compared to nondiabetic controls. The increased expression of NLRP3 colocalizes strikingly with the macrophage marker CD68, suggesting that macrophages are mostly responsible for this phenomenon [194]. In vitro cultures of macrophages under glucose concentrations similar to those found in diabetic patients (30 mM) display high expression of NLRP3, caspase-1 and IL-1β compared to macrophages exposed to normal levels of glucose (5.5 mM) [194]. Furthermore, selective silencing of NLRP3 in the bone marrow of diabetic mice and the topical application of pharmacological inhibitors of NLRP3 (glyburide) and caspase-1 (YVAD) are both able to improve diabetic wound healing [195].

***Lipids mediators in diabetic nephropathy***

Another comorbidity of T2D is diabetic nephropathy. Formation of glomerular mesangial lesions and increased synthesis of extracellular matrix leads to glomerular sclerosis and interstitial fibrosis. The cause of the disease is thought to be inflammatory in nature. Continuous hyperglycemia leads to the overproduction of ROS in renal tissues, leading to the activation of NLRP3 and inflammatory signaling, which contributes to disease progression [196]. Interestingly, ex vivo treatment of mouse glomerular mesangial cells with MaR1 inhibits the production of ROS, the activation of NLRP3 and the expression of proinflammatory cytokines (IL-1β, TGF-β1), thus limiting glomerular inflammation and early fibrosis [197]. The genetic deletion of NLRP3 in diabetic mice attenuates tubular injuries, inflammation as well as fibrosis, partially restoring kidney function. This is associated with a decrease in ROS formation and with the inhibition of TXNIP expression, a potent oxidative stress inducer [198].

## Conclusion

Lipid mediators and NLRP3 are key modulators of inflammation. The activation or inhibition of their respective pathways is what allows the progression of obesity's comorbidities. In this context, the deleterious nature of the western diet cannot be stressed enough. The low proportion of ω-3 compared to ω-6 fatty acids promotes the production of prostanoids and leukotrienes and reduces the production of SPM, a phenomenon that fosters a proinflammatory environment limiting the capacity for inflammation resorption. Additionally, many components of the western diet (high concentration of palmitate, cholesterol and glucose) also promote the priming and the activation of the NLRP3 inflammasome. This second "hit" further promotes inflammation by recruiting leukocytes to the tissue, but also by limiting even further the production of SPM. Together, these two "hits" boost the development of atherosclerosis, NASH, T2D, nonhealing diabetic wounds and diabetic nephropathy.

Fortunately, the comprehension of inflammation initiation and resolution has dramatically evolved in the last decade. The discovery of SPM and inflammasomes as well as the elucidation of their roles in metabolic diseases has given us a way to target inflammation in a new and more efficient way. The prior approach of inhibiting the synthesis of eicosanoids by using nonsteroidal antiinflammatory drugs has shown some results in the fight against obesity-related inflammation and obesity's comorbidities. However, it is apparent that this approach has its limits. Future approaches to treat theses inflammation-driven diseases will have to target pathways that can trigger inflammatory resorption instead of just limiting the synthesis of proinflammatory mediators. Since NLRP3 activation blocks the production of essential proresolving SPM, the combination of diets enriched in ω-3 fatty acids with NLRP3 inhibitor treatments could have synergic effects in promoting the resolution of inflammation in the tissues of obese individuals.

The limitation to the direct therapeutic use of SPM is the requirement of daily injections. Since SPM exert their effect partly via activation of GPCR, pharmacological agonists could be developed for these receptors and administered orally to mimic the effects of SPM. Such drugs may have the potential to become the best method of treating chronic inflammation. Only a handful of receptors have been identified for SPM. The discovery of new receptors could provide additional antiinflammatory drug targets.

## References


[1] Chooi YC, Ding C, Magkos F. The epidemiology of obesity. Metabolism 2019;92:6–10.

[2] Gonzalez-Muniesa P, et al. Obesity. Nat Rev Dis Primers 2017;3:17034.

[3] Saltiel AR, Olefsky JM. Inflammatory mechanisms linking obesity and metabolic disease. J Clin Invest 2017;127(1):1–4.

[4] Yamashita AS, et al. Regulation of metabolic disease-associated inflammation by nutrient sensors. Mediators Inflamm 2018;2018:8261432.

[5] Donath MY, Shoelson SE. Type 2 diabetes as an inflammatory disease. Nat Rev Immunol 2011;11(2):98–107.

[6] Zozulinska D, Wierusz-Wysocka B. Type 2 diabetes mellitus as inflammatory disease. Diab Res Clin Pract 2006;74(2):S12–6.

[7] Nieto-Vazquez I, et al. Insulin resistance associated to obesity: the link TNF-alpha. Arch Physiol Biochem 2008;114(3):183–94.

[8] Jager J, et al. Interleukin-1beta-induced insulin resistance in adipocytes through down-regulation of insulin receptor substrate-1 expression. Endocrinology 2007;148(1):241–51.

[9] Agrawal N, et al. The Drosophila TNF eiger is an adipokine that acts on insulin-producing cells to mediate nutrient response. Cell Metab 2016;23(4):675–84.

[10] Libby P. Interleukin-1 beta as a target for atherosclerosis therapy: biological basis of CANTOS and beyond. J Am Coll Cardiol 2017;70(18):2278–89.

[11] Kraakman MJ, et al. Macrophage polarization in obesity and type 2 diabetes: weighing down our understanding of macrophage function? Front Immunol 2014;5:470.

[12] Jenkins SJ, et al. Local macrophage proliferation, rather than recruitment from the blood, is a signature of TH2 inflammation. Science 2011;332(6035):1284–8.

[13] Korns D, et al. Modulation of macrophage efferocytosis in inflammation. Front Immunol 2011;2:57.

[14] Lumeng CN, Bodzin JL, Saltiel AR. Obesity induces a phenotypic switch in adipose tissue macrophage polarization. J Clin Invest 2007;117(1):175–84.

[15] Fujisaka S, et al. Regulatory mechanisms for adipose tissue M1 and M2 macrophages in diet-induced obese mice. Diabetes 2009;58(11):2574–82.

[16] Simopoulos AP. Evolutionary aspects of diet, the omega-6/omega-3 ratio and genetic variation: nutritional implications for chronic diseases. Biomed Pharmacother 2006;60(9):502–7.

[17] Simopoulos AP. An increase in the omega-6/omega-3 fatty acid ratio increases the risk for obesity. Nutrients 2016;8(3):128.

[18] Statovci D, et al. The impact of western diet and nutrients on the microbiota and immune response at mucosal interfaces. Front Immunol 2017;8:838.

[19] Drake I, et al. A western dietary pattern is prospectively associated with cardio-metabolic traits and incidence of the metabolic syndrome. Br J Nutr 2018;119(10):1168–76.

[20] Kelley DS, et al. Arachidonic acid supplementation enhances synthesis of eicosanoids without suppressing immune functions in young healthy men. Lipids 1998;33(2):125–30.
[21] Katakura M, et al. Chronic arachidonic acid administration decreases docosahexaenoic acid- and eicosapentaenoic acid-derived metabolites in kidneys of aged rats. PLoS ONE 2015;10(10):e0140884.
[22] Guillou H, et al. The key roles of elongases and desaturases in mammalian fatty acid metabolism: insights from transgenic mice. Prog Lipid Res 2010;49(2):186–99.
[23] Lau BY, et al. Investigating the role of polyunsaturated fatty acids in bone development using animal models. Molecules 2013;18(11):14203–27.
[24] Gijon MA, Leslie CC. Regulation of arachidonic acid release and cytosolic phospholipase A2 activation. J Leukoc Biol 1999;65(3):330–6.
[25] Identification of top food sources of various dietary components. Table 4. Food sources of arachidonic acid (PFA 20:4), listed in descending order by percentages of their contribution to intake, based on data from the National Health and Nutrition Examination Survey 2005-2006 2018 2018-04-20; Available from: https://epi.grants.cancer.gov/diet/foodsources/fatty_acids/table4.html.
[26] Blondeau N, et al. Alpha-linolenic acid: an omega-3 fatty acid with neuroprotective properties-ready for use in the stroke clinic? Biomed Res Int 2015;2015:519830.
[27] Husted KS, Bouzinova EV. The importance of n-6/n-3 fatty acids ratio in the major depressive disorder. Medicina (Kaunas) 2016;52(3):139–47.
[28] Emken EA, Adlof RO, Gulley RM. Dietary linoleic acid influences desaturation and acylation of deuterium-labeled linoleic and linolenic acids in young adult males. Biochim Biophys Acta 1994;1213(3):277–88.
[29] Kuk JL, et al. Age-related changes in total and regional fat distribution. Ageing Res Rev 2009;8(4):339–48.
[30] Desmarais F, et al. Apolipoprotein D overexpression alters hepatic prostaglandin and omega fatty acid metabolism during the development of a non-inflammatory hepatic steatosis. Biochim Biophys Acta Mol Cell Biol Lipids 2019;1864(4):522–31.
[31] Morais Cabral JH, et al. Arachidonic acid binds to apolipoprotein D: implications for the protein's function. FEBS Lett 1995;366(1):53–6.
[32] Serhan CN. Treating inflammation and infection in the 21st century: new hints from decoding resolution mediators and mechanisms. FASEB J 2017;31(4):1273–88.
[33] Neuhofer A, et al. Impaired local production of proresolving lipid mediators in obesity and 17-HDHA as a potential treatment for obesity-associated inflammation. Diabetes 2013;62(6):1945–56.
[34] Rheinheimer J, et al. Current role of the NLRP3 inflammasome on obesity and insulin resistance: a systematic review. Metabolism 2017;74:1–9.
[35] Esser N, et al. Obesity phenotype is related to NLRP3 inflammasome activity and immunological profile of visceral adipose tissue. Diabetologia 2013;56(11):2487–97.
[36] von Moltke J, et al. Recognition of bacteria by inflammasomes. Annu Rev Immunol 2013;31:73–106.
[37] Vandanmagsar B, et al. The NLRP3 inflammasome instigates obesity-induced inflammation and insulin resistance. Nat Med 2011;17(2):179–88.
[38] Yang CA, Chiang BL. Inflammasomes and human autoimmunity: a comprehensive review. J Autoimmun 2015;61:1–8.
[39] Lee E, et al. MPTP-driven NLRP3 inflammasome activation in microglia plays a central role in dopaminergic neurodegeneration. Cell Death Differ 2019;26(2):213–28.
[40] Man SM, Karki R, Kanneganti TD. AIM2 inflammasome in infection, cancer, and autoimmunity: role in DNA sensing, inflammation, and innate immunity. Eur J Immunol 2016;46(2):269–80.
[41] Zhong FL, et al. Germline NLRP1 mutations cause skin inflammatory and cancer susceptibility syndromes via inflammasome activation. Cell 2016;167(1). p. 187–202 e17.
[42] Wei Q, et al. Deregulation of the NLRP3 inflammasome in hepatic parenchymal cells during liver cancer progression. Lab Invest 2014;94(1):52–62.
[43] Broz P, Dixit VM. Inflammasomes: mechanism of assembly, regulation and signalling. Nat Rev Immunol 2016;16(7):407–20.
[44] Hu B, et al. Microbiota-induced activation of epithelial IL-6 signaling links inflammasome-driven inflammation with transmissible cancer. Proc Natl Acad Sci U S A 2013;110(24):9862–7.
[45] Lee S, et al. NLRP3 inflammasome deficiency protects against microbial sepsis via increased lipoxin B4 synthesis. Am J Respir Crit Care Med 2017;196(6):713–26.

[46] Wong BC, et al. 12-Lipoxygenase inhibition induced apoptosis in human gastric cancer cells. Carcinogenesis 2001;22(9):1349–54.
[47] Pidgeon GP, et al. Mechanisms controlling cell cycle arrest and induction of apoptosis after 12-lipoxygenase inhibition in prostate cancer cells. Cancer Res 2002;62(9):2721–7.
[48] Li Y, Maher P, Schubert D. A role for 12-lipoxygenase in nerve cell death caused by glutathione depletion. Neuron 1997;19(2):453–63.
[49] Aung T, et al. Associations of omega-3 fatty acid supplement use with cardiovascular disease risks: meta-analysis of 10 trials involving 77917 individuals. JAMA Cardiol 2018;3(3):225–34.
[50] Group ASC, et al. Effects of n-3 fatty acid supplements in diabetes mellitus. N Engl J Med 2018;379(16):1540–50.
[51] Bauernfeind FG, et al. Cutting edge: NF-kappaB activating pattern recognition and cytokine receptors license NLRP3 inflammasome activation by regulating NLRP3 expression. J Immunol 2009;183(2):787–91.
[52] Ghonime MG, et al. Inflammasome priming by lipopolysaccharide is dependent upon ERK signaling and proteasome function. J Immunol 2014;192(8):3881–8.
[53] Franchi L, Eigenbrod T, Nunez G. Cutting edge: TNF-alpha mediates sensitization to ATP and silica via the NLRP3 inflammasome in the absence of microbial stimulation. J Immunol 2009;183(2):792–6.
[54] He Y, Hara H, Nunez G. Mechanism and regulation of NLRP3 inflammasome activation. Trends Biochem Sci 2016;41(12):1012–21.
[55] Groslambert M, Py BF. Spotlight on the NLRP3 inflammasome pathway. J Inflamm Res 2018;11:359–74.
[56] Morgan MJ, Liu ZG. Crosstalk of reactive oxygen species and NF-kappaB signaling. Cell Res 2011;21(1):103–15.
[57] Wang J, Wang H. Oxidative stress in pancreatic beta cell regeneration. Oxid Med Cell Longev 2017;2017:1930261.
[58] Kattoor AJ, et al. Oxidative stress in atherosclerosis. Curr Atheroscler Rep 2017;19(11):42.
[59] Masarone M, et al. Role of oxidative stress in pathophysiology of nonalcoholic fatty liver disease. Oxid Med Cell Longev 2018;2018:9547613.
[60] He Y, et al. NEK7 is an essential mediator of NLRP3 activation downstream of potassium efflux. Nature 2016;530(7590):354–7.
[61] Shi H, et al. NLRP3 activation and mitosis are mutually exclusive events coordinated by NEK7, a new inflammasome component. Nat Immunol 2016;17(3):250–8.
[62] Schmid-Burgk JL, et al. A Genome-wide CRISPR (clustered regularly interspaced short palindromic repeats) screen identifies NEK7 as an essential component of NLRP3 inflammasome activation. J Biol Chem 2016;291(1):103–9.
[63] Cullen SP, et al. Diverse activators of the NLRP3 inflammasome promote IL-1beta secretion by triggering necrosis. Cell Rep 2015;11(10):1535–48.
[64] Rajamaki K, et al. Cholesterol crystals activate the NLRP3 inflammasome in human macrophages: a novel link between cholesterol metabolism and inflammation. PLoS ONE 2010;5(7):e11765.
[65] Bai H, et al. Cathepsin B links oxidative stress to the activation of NLRP3 inflammasome. Exp Cell Res 2018;362(1):180–7.
[66] Karasawa T, et al. Saturated fatty acids undergo intracellular crystallization and activate the NLRP3 inflammasome in macrophages. Arterioscler Thromb Vasc Biol 2018;38(4):744–56.
[67] Ogasawara Y, et al. Stearoyl-CoA desaturase 1 activity is required for autophagosome formation. J Biol Chem 2014;289(34):23938–50.
[68] Wen H, et al. Fatty acid-induced NLRP3-ASC inflammasome activation interferes with insulin signaling. Nat Immunol 2011;12(5):408–15.
[69] Feng H, et al. High glucose and lipopolysaccharide prime NLRP3 inflammasome via ROS/TXNIP pathway in mesangial cells. J Diabetes Res 2016;2016:6973175.
[70] Shi H, et al. Inhibition of autophagy induces IL-1beta release from ARPE-19 cells via ROS mediated NLRP3 inflammasome activation under high glucose stress. Biochem Biophys Res Commun 2015;463(4):1071–6.
[71] Hughes MM, O'Neill LAJ. Metabolic regulation of NLRP3. Immunol Rev 2018;281(1):88–98.
[72] Iyer SS, et al. Necrotic cells trigger a sterile inflammatory response through the Nlrp3 inflammasome. Proc Natl Acad Sci U S A 2009;106(48):20388–93.
[73] Gombault A, Baron L, Couillin I. ATP release and purinergic signaling in NLRP3 inflammasome activation. Front Immunol 2012;3:414.
[74] Brash AR. Arachidonic acid as a bioactive molecule. J Clin Invest 2001;107(11):1339–45.
[75] Kawahara K, et al. Prostaglandin E2-induced inflammation: relevance of prostaglandin E receptors. Biochim Biophys Acta 2015;1851(4):414–21.

[76] de Leval X, et al. Recent advances in inducible cyclooxygenase (COX-2) inhibition. Curr Med Chem 2000;7(10):1041–62.

[77] Dinarello CA. Immunological and inflammatory functions of the interleukin-1 family. Annu Rev Immunol 2009;27:519–50.

[78] Zweers MC, et al. Celecoxib: considerations regarding its potential disease-modifying properties in osteoarthritis. Arthritis Res Ther 2011;13(5):239.

[79] Medeiros R, et al. The role of TNF-alpha signaling pathway on COX-2 upregulation and cognitive decline induced by beta-amyloid peptide. Behav Brain Res 2010;209(1):165–73.

[80] Tamura M, et al. Interleukin-1beta elevates cyclooxygenase-2 protein level and enzyme activity via increasing its mRNA stability in human endometrial stromal cells: an effect mediated by extracellularly regulated kinases 1 and 2. J Clin Endocrinol Metab 2002;87(7):3263–73.

[81] Harper KA, Tyson-Capper AJ. Complexity of COX-2 gene regulation. Biochem Soc Trans 2008;36(Pt 3):543–5.

[82] Hua KF, et al. Cyclooxygenase-2 regulates NLRP3 inflammasome-derived IL-1beta production. J Cell Physiol 2015;230(4):863–74.

[83] Tian XY, et al. Oxidative stress-dependent cyclooxygenase-2-derived prostaglandin f(2alpha) impairs endothelial function in renovascular hypertensive rats. Antioxid Redox Signal 2012;16(4):363–73.

[84] Markovic T, et al. Structural features of subtype-selective EP receptor modulators. Drug Discov Today 2017;22(1):57–71.

[85] Nakanishi M, Rosenberg DW. Multifaceted roles of PGE2 in inflammation and cancer. Semin Immunopathol 2013;35(2):123–37.

[86] Lee GS, et al. The calcium-sensing receptor regulates the NLRP3 inflammasome through Ca2+ and cAMP. Nature 2012;492(7427):123–7.

[87] Sugimoto Y, et al. Cloning and expression of a cDNA for mouse prostaglandin E receptor EP3 subtype. J Biol Chem 1992;267(10):6463–6.

[88] Kalinski P. Regulation of immune responses by prostaglandin E2. J Immunol 2012;188(1):21–8.

[89] Levy BD, et al. Lipid mediator class switching during acute inflammation: signals in resolution. Nat Immunol 2001;2(7):612–9.

[90] Xu H, et al. Prostaglandin E2 receptor EP3 regulates both adipogenesis and lipolysis in mouse white adipose tissue. J Mol Cell Biol 2016;8(6):518–29.

[91] Ceddia RP, et al. The PGE2 EP3 receptor regulates diet-induced adiposity in male mice. Endocrinology 2016;157(1):220–32.

[92] Garcia-Alonso V, et al. Coordinate functional regulation between microsomal prostaglandin E synthase-1 (mPGES-1) and peroxisome proliferator-activated receptor gamma (PPARgamma) in the conversion of white-to-brown adipocytes. J Biol Chem 2013;288(39):28230–42.

[93] Garcia-Alonso V, et al. Prostaglandin E2 exerts multiple regulatory actions on human obese adipose tissue remodeling, inflammation, adaptive thermogenesis and lipolysis. PLoS ONE 2016;11(4):e0153751.

[94] Henkel J, et al. Augmented liver inflammation in a microsomal prostaglandin E synthase 1 (mPGES-1)-deficient diet-induced mouse NASH model. Sci Rep 2018;8(1):16127.

[95] Henkel J, et al. Stimulation of fat accumulation in hepatocytes by PGE(2)-dependent repression of hepatic lipolysis, beta-oxidation and VLDL-synthesis. Lab Invest 2012;92(11):1597–606.

[96] Mine T, Kojima I, Ogata E. Mechanism of prostaglandin E2-induced glucose production in rat hepatocytes. Endocrinology 1990;126(6):2831–6.

[97] Nasrallah R, Hassouneh R, Hebert RL. PGE2, kidney disease, and cardiovascular risk: beyond hypertension and diabetes. J Am Soc Nephrol 2016;27(3):666–76.

[98] Nakahata N. Thromboxane A2: physiology/pathophysiology, cellular signal transduction and pharmacology. Pharmacol Ther 2008;118(1):18–35.

[99] Ishizuka T, et al. Thromboxane A2 (TXA2) receptor blockade suppresses monocyte chemoattractant protein-1 (MCP-1) expression by stimulated vascular endothelial cells. Clin Exp Immunol 2000;120(1):71–8.

[100] Lobato NS, et al. Obesity induced by neonatal treatment with monosodium glutamate impairs microvascular reactivity in adult rats: role of NO and prostanoids. Nutr Metab Cardiovasc Dis 2011;21(10):808–16.

[101] Akamine EH, et al. Decreased endothelium-dependent vasodilation in diabetic female rats: role of prostanoids. J Vasc Res 2006;43(5):401–10.

[102] Lobato NS, et al. Mechanisms of endothelial dysfunction in obesity-associated hypertension. Braz J Med Biol Res 2012;45(5):392–400.

[103] Cosentino F, et al. High glucose causes upregulation of cyclooxygenase-2 and alters prostanoid profile in human endothelial cells: role of protein kinase C and reactive oxygen species. Circulation 2003;107(7):1017–23.
[104] Shi Y, Vanhoutte PM. Oxidative stress and COX cause hyper-responsiveness in vascular smooth muscle of the femoral artery from diabetic rats. Br J Pharmacol 2008;154(3):639–51.
[105] Finneran MM, et al. Obesity and laboratory aspirin resistance in high-risk pregnant women treated with low-dose aspirin. Am J Obstet Gynecol 2019;220(4):385e1–6.
[106] Lucido MJ, et al. Crystal structure of aspirin-acetylated human cyclooxygenase-2: insight into the formation of products with reversed stereochemistry. Biochemistry 2016;55(8):1226–38.
[107] Sharma NP, et al. Asymmetric acetylation of the cyclooxygenase-2 homodimer by aspirin and its effects on the oxygenation of arachidonic, eicosapentaenoic, and docosahexaenoic acids. Mol Pharmacol 2010;77(6):979–86.
[108] Vessieres E, et al. COX-2-derived prostanoids and oxidative stress additionally reduce endothelium-mediated relaxation in old type 2 diabetic rats. PLoS ONE 2013;8(7):e68217.
[109] Miller CW, Casimir DA, Ntambi JM. The mechanism of inhibition of 3T3-L1 preadipocyte differentiation by prostaglandin F2alpha. Endocrinology 1996;137(12):5641–50.
[110] El-Haroun H, et al. IL-1beta, BK, and TGF-beta1 attenuate PGI2-mediated cAMP formation in human pulmonary artery smooth muscle cells by multiple mechanisms involving p38 MAP kinase and PKA. Am J Physiol Lung Cell Mol Physiol 2008;294(3):L553–62.
[111] Dorris SL, Peebles RS Jr. PGI2 as a regulator of inflammatory diseases. Mediators Inflamm 2012;2012:926968.
[112] Kelton JG, Blajchman MA. Prostaglandin I2 (prostacyclin). Can Med Assoc J 1980;122(2):175–9.
[113] Kobzar G, et al. Comparison of anti-aggregatory effects of PGI2, PGI3 and iloprost on human and rabbit platelets. Cell Physiol Biochem 2001;11(5):279–84.
[114] Armstrong JM, et al. Cardiovascular actions of prostacyclin (PCI2) in chloralose anaesthetized dogs [proceedings]. Br J Pharmacol 1977;61(1):136P.
[115] Miller SB. Prostaglandins in health and disease: an overview. Semin Arthritis Rheum 2006;36(1):37–49.
[116] Moncada S, et al. Differential formation of prostacyclin (PGX or PGI2) by layers of the arterial wall. An explanation for the anti-thrombotic properties of vascular endothelium. Thromb Res 1977;11(3):323–44.
[117] Majerus PW. Arachidonate metabolism in vascular disorders. J Clin Invest 1983;72(5):1521–5.
[118] Mohite A, et al. Novel mechanism of the vascular protector prostacyclin: regulating microRNA expression. Biochemistry 2011;50(10):1691–9.
[119] Bishop-Bailey D, Wray J. Peroxisome proliferator-activated receptors: a critical review on endogenous pathways for ligand generation. Prostaglandins Other Lipid Mediat 2003;71(1–2):1–22.
[120] Falcetti E, et al. IP receptor-dependent activation of PPARgamma by stable prostacyclin analogues. Biochem Biophys Res Commun 2007;360(4):821–7.
[121] Moore ML, Peebles RS Jr. Update on the role of prostaglandins in allergic lung inflammation: separating friends from foes, harder than you might think. J Allergy Clin Immunol 2006;117(5):1036–9.
[122] Fujita T, Narumiya S. Roles of hepatic stellate cells in liver inflammation: a new perspective. Inflamm Regen 2016;36:1.
[123] Fujimori K, Aritake K, Urade Y. A novel pathway to enhance adipocyte differentiation of 3T3-L1 cells by up-regulation of lipocalin-type prostaglandin D synthase mediated by liver X receptor-activated sterol regulatory element-binding protein-1c. J Biol Chem 2007;282(25):18458–66.
[124] Virtue S, et al. Prostaglandin profiling reveals a role for haematopoietic prostaglandin D synthase in adipose tissue macrophage polarisation in mice and humans. Int J Obes (Lond) 2015;39(7):1151–60.
[125] Forman BM, et al. 15-Deoxy-Δ12,14-prostaglandin J2 is a ligand for the adipocyte determination factor PPARγ. Cell 1995;83(5):803–12.
[126] Kliewer SA, et al. A prostaglandin J2 metabolite binds peroxisome proliferator-activated receptor γ and promotes adipocyte differentiation. Cell 1995;83(5):813–9.
[127] Fujimori K, et al. Activation of adipogenesis by lipocalin-type prostaglandin D synthase-generated delta(1)(2)-PGJ(2) acting through PPARgamma-dependent and independent pathways. Gene 2012;505(1):46–52.
[128] Ragolia L, et al. Accelerated glucose intolerance, nephropathy, and atherosclerosis in prostaglandin D2 synthase knock-out mice. J Biol Chem 2005;280(33):29946–55.
[129] Virtue S, et al. A new role for lipocalin prostaglandin d synthase in the regulation of brown adipose tissue substrate utilization. Diabetes 2012;61(12):3139–47.
[130] Fujimori K, et al. L-PGDS-produced PGD2 in premature, but not in mature, adipocytes increases obesity and insulin resistance. Sci Rep 2019;9(1):1931.

[131] Straus DS, et al. 15-Deoxy-D12,14-prostaglandin J2 inhibits multiple steps in the NF-kB signaling pathway. Proc Natl Acad Sci U S A 2000;97(9):4844–9.

[132] Suh J, et al. 15-Deoxy-delta(12,14)-prostaglandin J2 activates PI3K-Akt signaling in human breast cancer cells through covalent modification of the tumor suppressor PTEN at cysteine 136. Cancer Lett 2018;424:30–45.

[133] Remels AH, et al. PPARgamma inhibits NF-kappaB-dependent transcriptional activation in skeletal muscle. Am J Physiol Endocrinol Metab 2009;297(1):E174–83.

[134] Talalay P, De Long MJ, Prochaska HJ. Identification of a common chemical signal regulating the induction of enzymes that protect against chemical carcinogenesis. Proc Natl Acad Sci U S A 1988;85(21):8261–5.

[135] Villagomez R, et al. Effect of natural and semisynthetic pseudoguianolides on the stability of NF-kappaB:DNA complex studied by agarose gel electrophoresis. PLoS ONE 2015;10(1):e0115819.

[136] Widen JC, et al. Helenalin analogues targeting NF-kappaB p65: thiol reactivity and cellular potency studies of varied electrophiles. ChemMedChem 2018;13(4):303–11.

[137] Nguyen VS, et al. Specificity and inhibitory mechanism of andrographolide and its analogues as antiasthma agents on NF-kappaB p50. J Nat Prod 2015;78(2):208–17.

[138] Dong T, et al. Ainsliadimer A selectively inhibits IKKalpha/beta by covalently binding a conserved cysteine. Nat Commun 2015;6:6522.

[139] Straus DS, Glass CK. Cyclopentenone prostaglandins: new insights on biological activities and cellular targets. Med Res Rev 2001;21(3):185–210.

[140] Ohmura T, et al. Regulation of lung endothelial permeability and inflammatory responses by prostaglandin A2: role of EP4 receptor. Mol Biol Cell 2017;28(12):1622–35.

[141] Storer PD, et al. Cyclopentenone prostaglandins PGA2 and 15-deoxy-delta12,14 PGJ2 suppress activation of murine microglia and astrocytes: implications for multiple sclerosis. J Neurosci Res 2005;80(1):66–74.

[142] Rossi A, et al. Anti-inflammatory cyclopentenone prostaglandins are direct inhibitors of IkappaB kinase. Nature 2000;403(6765):103–8.

[143] Thurnher M, et al. The cyclopentenone prostaglandin PGA2 costimulates the maturation of human dendritic cells. Exp Hematol 2005;33(2):144–50.

[144] Homem de Bittencourt PI Jr, et al. LipoCardium: endothelium-directed cyclopentenone prostaglandin-based liposome formulation that completely reverses atherosclerotic lesions. Atherosclerosis 2007;193(2):245–58.

[145] Gutierrez LL, et al. Atherosclerosis: a redox-sensitive lipid imbalance suppressible by cyclopentenone prostaglandins. Biochem Pharmacol 2008;75(12):2245–62.

[146] Corton JC, Anderson SP, Stauber A. Central role of peroxisome proliferator-activated receptors in the actions of peroxisome proliferators. Annu Rev Pharmacol Toxicol 2000;40:491–518.

[147] Murphy RC, Gijon MA. Biosynthesis and metabolism of leukotrienes. Biochem J 2007;405(3):379–95.

[148] Yokomizo T, Nakamura M, Shimizu T. Leukotriene receptors as potential therapeutic targets. J Clin Invest 2018;128(7):2691–701.

[149] Li P, et al. LTB4 promotes insulin resistance in obese mice by acting on macrophages, hepatocytes and myocytes. Nat Med 2015;21(3):239–47.

[150] Borgeson E, et al. Lipoxin A4 attenuates obesity-induced adipose inflammation and associated liver and kidney disease. Cell Metab 2015;22(1):125–37.

[151] Guo YP, et al. Lipoxin A4 may attenuate the progression of obesity-related glomerulopathy by inhibiting NF-kappaB and ERK/p38 MAPK-dependent inflammation. Life Sci 2018;198:112–8.

[152] Dalli J, Serhan CN. Immunoresolvents signaling molecules at intersection between the brain and immune system. Curr Opin Immunol 2018;50:48–54.

[153] Serhan CN, et al. Maresins: novel macrophage mediators with potent antiinflammatory and proresolving actions. J Exp Med 2009;206(1):15–23.

[154] Dalli J, et al. The novel 13S,14S-epoxy-maresin is converted by human macrophages to maresin 1 (MaR1), inhibits leukotriene A4 hydrolase (LTA4H), and shifts macrophage phenotype. FASEB J 2013;27(7):2573–83.

[155] Deng B, et al. Maresin biosynthesis and identification of maresin 2, a new anti-inflammatory and pro-resolving mediator from human macrophages. PLoS ONE 2014;9(7):e102362.

[156] Dalli J, Serhan CN. Pro-resolving mediators in regulating and conferring macrophage function. Front Immunol 2017;8:1400.

[157] Dalli J, et al. Maresin conjugates in tissue regeneration biosynthesis enzymes in human macrophages. Proc Natl Acad Sci U S A 2016;113(43):12232–7.

[158] Serhan CN, Levy BD. Resolvins in inflammation: emergence of the pro-resolving superfamily of mediators. J Clin Invest 2018;128(7):2657–69.
[159] Tang S, et al. Maresins: specialized proresolving lipid mediators and their potential role in inflammatory-related diseases. Mediators Inflamm 2018;2018:2380319.
[160] Akagi D, et al. Systemic delivery of proresolving lipid mediators resolvin D2 and maresin 1 attenuates intimal hyperplasia in mice. FASEB J 2015;29(6):2504–13.
[161] Spite M, Claria J, Serhan CN. Resolvins, specialized proresolving lipid mediators, and their potential roles in metabolic diseases. Cell Metab 2014;19(1):21–36.
[162] Martinez-Fernandez L, et al. Maresin 1 improves insulin sensitivity and attenuates adipose tissue inflammation in ob/ob and diet-induced obese mice. FASEB J 2017;31(5):2135–45.
[163] Gonzalez-Periz A, et al. Docosahexaenoic acid (DHA) blunts liver injury by conversion to protective lipid mediators: protectin D1 and 17S-hydroxy-DHA. FASEB J 2006;20(14):2537–9.
[164] White PJ, et al. Protectin DX alleviates insulin resistance by activating a myokine-liver glucoregulatory axis. Nat Med 2014;20(6):664–9.
[165] Scheller J, et al. The pro- and anti-inflammatory properties of the cytokine interleukin-6. Biochim Biophys Acta 2011;1813(5):878–88.
[166] Belizario JE, et al. Skeletal muscle wasting and renewal: a pivotal role of myokine IL-6. Springerplus 2016;5:619.
[167] Weigert C, et al. Interleukin-6 acts as insulin sensitizer on glycogen synthesis in human skeletal muscle cells by phosphorylation of Ser473 of Akt. Am J Physiol Endocrinol Metab 2005;289(2):E251–7.
[168] Dalli J, et al. Novel proresolving and tissue-regenerative resolvin and protectin sulfido-conjugated pathways. FASEB J 2015;29(5):2120–36.
[169] Codagnone M, et al. Resolvin D1 enhances the resolution of lung inflammation caused by long-term Pseudomonas aeruginosa infection. Mucosal Immunol 2018;11(1):35–49.
[170] Recchiuti A, et al. MicroRNAs in resolution of acute inflammation: identification of novel resolvin D1-miRNA circuits. FASEB J 2011;25(2):544–60.
[171] Croasdell A, Sime PJ, Phipps RP. Resolvin D2 decreases TLR4 expression to mediate resolution in human monocytes. FASEB J 2016;30(9):3181–93.
[172] Singh G, Triadafilopoulos G. Epidemiology of NSAID induced gastrointestinal complications. J Rheumatol Suppl 1999;56:18–24.
[173] Viola JR, et al. Resolving lipid mediators maresin 1 and resolvin D2 prevent atheroprogression in mice. Circ Res 2016;119(9):1030–8.
[174] Karasawa T, Takahashi M. Role of NLRP3 inflammasomes in atherosclerosis. J Atheroscler Thromb 2017;24(5):443–51.
[175] Shi CS, et al. Activation of autophagy by inflammatory signals limits IL-1beta production by targeting ubiquitinated inflammasomes for destruction. Nat Immunol 2012;13(3):255–63.
[176] Zhou R, et al. A role for mitochondria in NLRP3 inflammasome activation. Nature 2011;469(7329):221–5.
[177] Razani B, et al. Autophagy links inflammasomes to atherosclerotic progression. Cell Metab 2012;15(4):534–44.
[178] Sun H, et al. Berberine ameliorates blockade of autophagic flux in the liver by regulating cholesterol metabolism and inhibiting COX2-prostaglandin synthesis. Cell Death Dis 2018;9(8):824.
[179] Vivoli E, et al. NLRP3 inflammasome as a target of berberine in experimental murine liver injury: interference with P2X7 signalling. Clin Sci (Lond) 2016;130(20):1793–806.
[180] Wan Q, et al. Suppressive effects of berberine on atherosclerosis via downregulating visfatin expression and attenuating visfatin-induced endothelial dysfunction. Int J Mol Med 2018;41(4):1939–48.
[181] Mullard A. NLRP3 inhibitors stoke anti-inflammatory ambitions. Nat Rev Drug Discov 2019;18(6):405–7.
[182] Han YH, et al. A maresin 1/RORalpha/12-lipoxygenase autoregulatory circuit prevents inflammation and progression of nonalcoholic steatohepatitis. J Clin Invest 2019;130:1684–98.
[183] Kang JX. Fat-1 transgenic mice: a new model for omega-3 research. Prostaglandins Leukot Essent Fatty Acids 2007;77(5–6):263–7.
[184] Mridha AR, et al. NLRP3 inflammasome blockade reduces liver inflammation and fibrosis in experimental NASH in mice. J Hepatol 2017;66(5):1037–46.
[185] Youm YH, et al. Elimination of the NLRP3-ASC inflammasome protects against chronic obesity-induced pancreatic damage. Endocrinology 2011;152(11):4039–45.
[186] Zhao G, et al. Possible role of interleukin-1beta in type 2 diabetes onset and implications for anti-inflammatory therapy strategies. PLoS Comput Biol 2014;10(8):e1003798.

[187] Herder C, et al. The IL-1 pathway in type 2 diabetes and cardiovascular complications. Trends Endocrinol Metab 2015;26(10):551–63.
[188] Lebreton F, et al. NLRP3 inflammasome is expressed and regulated in human islets. Cell Death Dis 2018;9(7):726.
[189] Sokolova M, et al. NLRP3 inflammasome mediates oxidative stress-induced pancreatic islet dysfunction. Am J Physiol Endocrinol Metab 2018;315(5):E912–23.
[190] Pierce GF. Inflammation in nonhealing diabetic wounds. Am J Pathol 2001;159(2):399–403.
[191] Khanna S, et al. Macrophage dysfunction impairs resolution of inflammation in the wounds of diabetic mice. PLoS ONE 2010;5(3):e9539.
[192] Tian H, et al. Autacoid 14S,21R-dihydroxy-docosahexaenoic acid counteracts diabetic impairment of macrophage prohealing functions. Am J Pathol 2011;179(4):1780–91.
[193] Hong S, et al. Maresin-like lipid mediators are produced by leukocytes and platelets and rescue reparative function of diabetes-impaired macrophages. Chem Biol 2014;21(10):1318–29.
[194] Zhang X, et al. NLRP3 inflammasome expression and signaling in human diabetic wounds and in high glucose induced macrophages. J Diabetes Res 2017;2017:5281358.
[195] Mirza RE, et al. Sustained inflammasome activity in macrophages impairs wound healing in type 2 diabetic humans and mice. Diabetes 2014;63(3):1103–14.
[196] Wang C, et al. Quercetin and allopurinol ameliorate kidney injury in STZ-treated rats with regulation of renal NLRP3 inflammasome activation and lipid accumulation. PLoS ONE 2012;7(6):e38285.
[197] Tang S, et al. Maresin 1 mitigates high glucose-induced mouse glomerular mesangial cell injury by inhibiting inflammation and fibrosis. Mediators Inflamm 2017;2017:2438247.
[198] Wu M, et al. NLRP3 deficiency ameliorates renal inflammation and fibrosis in diabetic mice. Mol Cell Endocrinol 2018;478:115–25.

CHAPTER

# 12

# Identification and pathophysiological roles of $LTB_4$ receptors BLT1 and BLT2

*Yumiko Ishii*[a], *Toshiaki Okuno*[b], *Takehiko Yokomizo*[b]

[a]Meakins-Christie Laboratories, Department of Medicine, McGill University, Montreal, QC, Canada; [b]Department of Biochemistry, Juntendo University Graduate School of Medicine, Tokyo, Japan

OUTLINE



Lipid Signaling and Metabolism. http://dx.doi.org/10.1016/B978-0-12-819404-1.00012-9

## Abbreviations

| | |
|---|---|
| **AA** | arachidonic acid |
| **AD** | atopic dermatitis |
| **AHR** | airway hyperresponsiveness |
| **AMD** | age-related macular degeneration |
| **BAL** | bronchoalveolar lavage |
| **CF** | cystic fibrosis |
| **COPD** | chronic obstructive pulmonary disease |
| **CS** | crystalline silica |
| **CYP450** | cytochrome P450 |
| **CysLTs** | cysteinyl leukotrienes |
| **CysLT1R** | CysLT receptor type 1 |
| **DC** | dendritic cells |
| **DSS** | dextran sodium sulfate |
| **EETs** | epoxyeicosatrienic acids |
| **FLAP** | 5-LO-activating protein |
| **GPCR** | G-protein-coupled receptor |
| **HPLC** | high-performance liquid chromatography |
| **LXs** | lipoxins |
| **$LTA_4$** | leukotriene $A_4$ |
| **$LTA_4H$** | leukotriene $A_4$ hydrolase |
| **$LTB_4$** | leukotriene $B_4$ |
| **$LTC_4S$** | leukotriene $C_4$ synthase |
| **LTs** | leukotrienes |
| **MGST** | microsomal glutathione *S*-transferase |
| **MVs** | microvesicles |
| **Nox** | NADPH oxidases |
| **NSAIDs** | nonsteroidal antiinflammatory drugs |
| **OVA** | ovalbumin |
| **oxLDL** | oxidized low-density lipoprotein |
| **PGs** | prostaglandins |
| **$PGH_2$** | prostaglandin $H_2$ |
| **$PLA_2$** | phospholipase $A_2$ |
| **PLC** | phospholipase C |
| **PMNs** | polymorphonuclear neutrophils |
| **ROS** | reactive oxygen species |
| **TNF-α** | tumor necrosis factor-α |
| **TJ** | tight junction |
| **$T_XA_2$** | thromboxane $A_2$ |
| **$T_XA_2S$** | thromboxane $A_2$ synthase |
| **VEGF** | vascular endothelial growth factor |
| **5-HpETE** | 5-hydroxyeicosatetraenoic acid |
| **5-LO** | 5-lipoxigenase |
| **12-HHT** | 12(*S*)-hydroxy-5*Z*,8*E*,10*E*-heptadecatrienoic acid |
| **12(*S*)-HETE** | 12(*S*)-hydroxy-5*Z*,8*Z*,10*E*,14*Z*-eicosatetraenoic acid |

## Introduction

The term "leukotriene" was introduced by Samuelsson in 1979 [1] based on the fact that leukotrienes (LTs) are mainly produced in leukocytes and contain three conjugated double bonds forming a triene. Leukotriene $B_4$ ($LTB_4$, 5(*S*),12(*R*)-dihydroxy-6,14-*cis*-8,10-*trans*-eicosatetraenoic acid) is a bioactive lipid mediator derived from arachidonic acid (AA) and was initially discovered as a potent chemoattractant for polymorphonuclear neutrophils (PMNs) [2]. $LTB_4$ promotes the migration of neutrophils, monocytes, eosinophils, differentiated T-cells, and some subsets of macrophages and dendritic cells (DCs) into inflammatory sites [3–6]. It also enhances phagocytosis in macrophages [7] and promotes antimicrobial defenses [8–11]. We previously identified two $LTB_4$ receptors: the high-affinity receptor BLT1 [12] and the low-affinity receptor BLT2 [13]. BLT1 is mainly expressed in leukocytes. Numerous studied demonstrated that the $LTB_4$–BLT1 axis plays important roles in immune and inflammatory diseases including asthma, chronic obstructive pulmonary disease (COPD), atopic dermatitis (AD), psoriasis, atherosclerosis, arthritis, obesity, cancer, and age-related macular degeneration (AMD) [14–16]. Meanwhile, BLT2 is primarily expressed in epithelial cells. BLT2 exhibits low-affinity binding to $LTB_4$ and can be activated by various fatty acids including 12(*S*)-hydroxy-5*Z*,8*E*,10*E*-heptadecatrienoic acid (12-HHT) [13,17]. Recent studies revealed that the 12-HHT–BLT2 axis is involved in skin and corneal wound healing, acute lung injury, asthma, and chemotherapy resistance [18]. We summarize the biosynthesis, physiological functions, and receptors of $LTB_4$ and 12-HHT, and their roles in human diseases and mouse disease models in this review.

## Biosynthesis and metabolism of $LTB_4$

AA is a 20-carbon fatty acid and an important component of the cell membrane. While intracellular free AA is not available at steady state, AA is released intracellularly by activated phospholipase $A_2$ ($PLA_2$), especially cytosolic $PLA_2$ [19–21], upon appropriate cell stimulation. AA is metabolized to produce lipid mediators via 5-lipoxigenase (5-LO), cyclooxygenase (COX), and cytochrome P450 (CYP450) enzymes to form LTs and lipoxins (LXs), prostaglandins (PGs), and epoxyeicosatrienic acids (EETs), respectively [22]. In resting cells, 5-LO is localized in the cytosol, but when cells are activated, 5-LO is translocated to the nuclear membrane, where 5-LO-activating protein (FLAP), a membrane-spanning protein, transfers AA to 5-LO [23,24]. 5-LO oxygenates AA to form 5-hydroxyeicosatetraenoic acid (5-HpETE) and subsequently leukotriene $A_4$ ($LTA_4$) [25–27]. Interestingly, the localization of 5-LO determines $LTB_4$ synthesis, and inhibition of nuclear localization via mutation of the nuclear localization sequence of 5-LO reduces $LTB_4$ synthesis by more than 60% [28]. Additionally, resolvin D1, a pro-resolving mediator derived from docosahexaenoic acid, translocates 5-LO from the nucleus to the cytosol and decreases the $LTB_4$:lipoxin $A_4$ ratio via calcium-calmodulin-dependent protein kinase II (CaMKII)-p38-mitogen-activated protein kinase-activated protein kinase 2 kinases [29]. Furthermore, phosphorylation of Ser 523 within the nuclear localization sequence and of Ser271 within the nuclear export sequence causes cytosolic localization and inhibition of 5-LO [30,31]. Thus, phosphorylation regulates the localization and activity of 5-LO (Fig. 12.1).

$LTA_4$ is metabolized to $LTB_4$ by leukotriene $A_4$ hydrolase ($LTA_4H$) or cysteinyl leukotrienes (CysLTs) by leukotriene $C_4$ synthase ($LTC_4S$), microsomal glutathione *S*-transferase 2 (MGST2), and microsomal glutathione *S*-transferase 3 (MGST3) [32–42]. FLAP is localized on both the inner and outer nuclear membrane, and $LTA_4H$ is localized in the cytosol and nucleus. Meanwhile, $LTC_4S$ is only present at the outer nuclear membrane. In addition, the

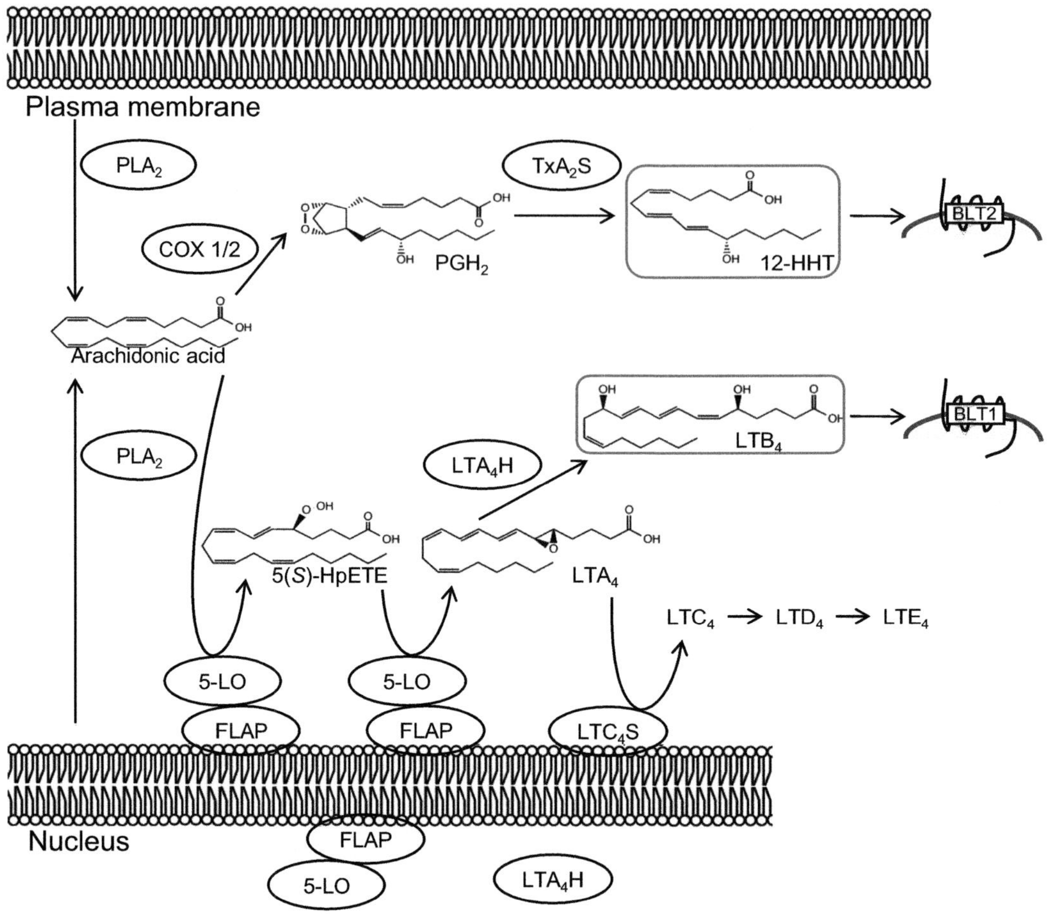


**FIGURE 12.1 Biosynthesis of leukotrienes and 12-HHT.** Arachidonic acid released from membrane phospholipid by PLA2 is converted to leukotrienes, prostanoids, and 12-HHT. LTB4 binds BLT1, a high-affinity receptor for LTB4, and induces chemotaxis and/or various subsets of leukocytes. BLT2, originally identified as a low-affinity receptor for LTB4, is a high-affinity receptor for 12-HHT, enhances epithelial barrier function and promotes wound healing. Enzyme names are abbreviated as follows: *PLA2*, phospholipase A2; *COX 1/2*, cyclooxygenase-1 and cyclooxygenase-2; *TxA2S*, thromboxane A2 synthase; *5-LO*, 5-lipoxygenase; *FLAP*, 5-LO-activating protein; *LTA4H*, leukotriene A4 hydrolase; *LTC4S*, leukotriene C4 synthase.

nuclear localization of 5-LO contributes to $LTB_4$ synthesis. Thus, $LTB_4$ may be preferentially produced on the inner nuclear membrane [43].

In addition to its epoxide hydrolase activity, $LTA_4H$ also has a zinc-dependent peptidase activity [44,45]. Recently, this aminopeptidase activity of $LTA_4H$ was reported to limit inflammation by degrading the neutrophil chemoattractant Pro-Gly-Pro (PGP) [46]. 5-LO is expressed mainly in hematopoietic cells, including PMNs, neutrophils, eosinophils, monocytes/macrophages, DCs, mast cells, and B-lymphocytes [47–52], while $LTA_4H$ is detected in various mammalian cells, even those lacking 5-LO expression [35]. $LTA_4$ is released from human leukocytes, and $LTB_4$ generation is increased by co-culture of activated neutrophils and erythrocytes/alveolar macrophages that do not express 5-LO but do express $LTA_4H$ [53–55]. This coordinated mechanism is referred to as transcellular biosynthesis [56], and transcellular biosynthesis of $LTB_4$ was confirmed in vivo [57,58]. $LTC_4$ is also produced in the same way [57,59].

There are two major pathways of $LTB_4$ inactivation: omega oxidation and $LTB_4$ 12-hydroxydehydrogenase pathways. In granulocytes and hepatocytes, $LTB_4$ is metabolized to a trihydroxy compound by oxidation at C-20 of $LTB_4$ (20-OH-$LTB_4$), followed by transformation to the dicarboxylic acid (20-COOH-$LTB_4$) by several CYP450 enzymes [60–62]. In other cells, the cytosolic enzyme $LTB_4$ 12-hydroxydehydrogenase converts $LTB_4$ into 12-keto-$LTB_4$ [63,64]. $LTB_4$ 12-hydroxydehydrogenase also inactivates other eicosanoids including PGs and lipoxin $A_4$ by reduction of the 13-14 double bond [65,66]. Another possible pathway of $LTB_4$ inactivation is via peroxidase, $H_2O_2$, and halides from eosinophils [67].

## Sex differences in the $LTB_4$ pathway

Biosynthesis of LTs is also regulated by sex hormones, and sex differences are prominent in LT-related diseases including asthma, allergic dermatitis, atherosclerosis, and autoimmune diseases [68–70]. Stimulation of neutrophils with fMLP after priming with LPS/adenosine deaminase showed that the $LTB_4$ level was 4.9-fold higher in females than males, and this was associated with the 5α-dihydrotestosterone-ERKs pathway [71]. Androgens downregulate LT biosynthesis by affecting the localization of 5-LO in human monocytes and murine peritoneal macrophages [72,73]. In fact, 5-LO inhibitors licofelone and sulindac sulfide, and FLAP inhibitor MK886 exhibit higher efficiency in female blood, and the formation of active 5-LO/FLAP complexes is more robust in female leukocytes in human and mouse [74]. Moreover, BLT1 deficiency protects against platelet-activating factor (PAF)-induced shock only in female mice, although the mechanisms are not clear [75]. These reports suggest that LT production is higher in females than males, and this difference might be related to gender-biased inflammatory diseases such as systemic lupus erythematosus and rheumatoid arthritis.

## Identification and characterization of BLT1, a high-affinity receptor of $LTB_4$

A high-affinity binding site for radiolabeled $LTB_4$ on human neutrophils was initially detected by Goldman and colleagues in 1982 [76]. Pharmacological studies showed that the $LTB_4$ receptor is a G-protein-coupled receptor (GPCR). Although numerous studies revealed the presence of a high-affinity $LTB_4$ receptor in granulocytes in human, rat, guinea pig, and other mammalian species, the attempt to identify an $LTB_4$ receptor by conventional strategies was not successful for many years. Based on the finding that $LTB_4$ binding activity is increased during granulocyte-like differentiation of HL-60 cells by retinoic acid, we screened mRNA expression levels in retinoic acid-stimulated HL-60 cells and identified the $LTB_4$ receptor by isolation of cDNA clones encoding GPCRs [12]. We initially designated this receptor as BLTR, but this was subsequently changed to "high-affinity receptor BLT1" by the International Union of Basic and Clinical Pharmacology (IUPHAR) [77]. BLT1 is mainly expressed in leukocytes, such as neutrophils, eosinophils, monocytes, macrophages, DCs, and differentiated $CD4^+$

T-cells, but not in naïve T- or B-cells. Recently, we found that BLT1 is specifically expressed in alternatively activated (M1) macrophages but not in classically activated (M2) macrophages [78]. Membrane fractions of COS-7 and HEK 293 cells transfected with a BLT1 cDNA displayed high-affinity binding to $LTB_4$, with $K_d$ values of 0.15 and 1.2 nM, respectively, sufficient to be competitive with various BLT1 antagonists. $LTB_4$ competes most efficiently with $[^3H]LTB_4$, followed by 20-hydroxy-$LTB_4$, 12-*epi*-$LTB_4$, and 12-oxo-$LTB_4$, but 20-carboxyl-$LTB_4$, 12(*R*)-HETE, all-trans-$LTB_4$, and all-trans-12-epi-$LTB_4$ bind more weakly [13], suggesting that BLT1 recognizes a 6-*cis* configuration, the 12(*R*) hydroxyl moiety, and the C-20 methyl group of $LTB_4$. The human BLT1 gene is located on chromosome 14 (14q11.2-q12) and encodes a single polypeptide chain with 352 amino acids. The BLT1 amino acid sequence reveals a typical seven transmembrane domain structure common to GPCRs [12,79]. Interestingly, BLT1 is able to couple with multiple G-proteins. BLT1 couples mainly with $G\alpha i$ protein(s) and $G\alpha 16$ in granulocytes, and $G\alpha i$ protein(s) and $G\alpha q/11$ in CHO and HEK cells [80]. Activated BLT1 coupling with $G\alpha q/11$ induces phospholipase C (PLC)-β activation and inositol 1,4,5-triphosphate ($IP_3$) accumulation, and a subsequent increase in intracellular calcium concentration. BLT1 coupled with $G\alpha i$ inhibits adenylyl cyclase, leading to a decrease in cAMP. Intracellular loop 3 is important for $G\alpha i$ coupling [80], and helix 8 in the cytoplasmic tail is important in the conformational change to the low-affinity state after G-protein activation [81,82] and desensitization [83–86]. Molecular modeling and mutagenesis studies of BLT1 suggest potent $LTB_4$ binding sites in transmembrane domains III, V, VI, and extracellular loop 2 [87,88], and the crystal structure with the inverse agonist BIIL260 has been determined [89]. However, the precise binding sites of $LTB_4$ are still unknown. Many GPCRs are phosphorylated on the cytoplasmic tail and intracellular loops in response to various extracellular stimuli, and this phosphorylation is important in desensitization, inactivation, internalization, and G-protein-independent signaling of GPCRs [90,91]. BLT1 phosphorylation by GRK6 is involved in desensitization of BLT1 [92]. A hydrophobic dileucine motif in BLT1 helix 8 inhibits BLT1 internalization after $LTB_4$-dependent phosphorylation of BLT1 [84]. In addition to phosphorylation, we found that receptor for advanced glycation end products (RAGE) interacts with BLT1 and modulates its signaling [93].

## BLT1 in allergic diseases

BLT1 is expressed in various types of leukocytes, and studies using BLT1-deficient mice demonstrate that BLT1 is a potent chemoattractant for neutrophils [75], eosinophils [94,95], effector $CD8^+$ T-cells [96,97], and effector $CD4^+$ T-cells [98]. Hence, BLT1 is involved in inflammatory diseases, including allergy.

Large numbers of BLT1-expressing effector memory $CD8^+$ T-cells are present in bronchoalveolar lavage (BAL) samples from patients with asthma [99], and high levels of $LTB_4$ were detected in blood, BAL fluid [100], and exhaled breath condensate [101,102] from children and adolescents with asthma. In vivo studies also demonstrate the importance of BLT1 in asthma. BLT1-deficient mice display reduced accumulation of neutrophils, eosinophils [103,104], lymphocytes [105], and DCs in lung [106], and are resistant to ovalbumin (OVA)-induced allergic airway hyperresponsiveness (AHR). In addition, BLT1 antagonist CP-105,696 inhibits $LTB_4$-mediated neutrophil chemotaxis and upregulation of $CD11b^+$ cells in BAL, accompanied by amelioration of AHR in monkey [107]. However, a clinical trial of the BLT receptor antagonist LY293111 in asthmatic patients failed to improve respiratory function or AHR after allergen challenge, although it reduced the number of neutrophils in BAL fluid [108]. The lack of clinical effects may be because neutrophil levels in the patients were not high before treatment, and the severity of asthma was moderate.

AD is a chronic, pruritic inflammatory skin disease, and $LTB_4$ injection elicits itch-associated responses in mice [109]. Neutrophils from patients with AD exhibit higher $LTA_4H$ activity [110] and produce more $LTB_4$ in response to several stimuli [111]. A high concentration of $LTB_4$ [112–114] and high BLT1 expression [115] were detected in skin lesions of AD patients. In vivo, BLT1-deficient mice failed to develop allergic skin inflammation following repeated epicutaneous sensitization of tape-stripped skin with OVA. Transfer of sensitized immune cells from BLT1-deficient mice revealed that BLT1-expressing neutrophils were the major source of the increased $LTB_4$, and elevated $LTB_4$ recruited BLT1-expressing neutrophils and $CD4^+$ effector T-cells [115]. Besides neutrophils, keratinocytes also produce $LTB_4$ following stimulation with substance P, which elicits an itch sensation [116].

Allergic rhinitis is a chronic inflammatory disease of nasal mucosa, triggered by environmental allergens. Specific allergen challenge induces an increase in $LTB_4$, CysLTs, histamine in nasal lavage fluid, the abundance of neutrophils and eosinophils, protein levels, and nasal airway resistance in patients with allergic rhinitis [117,118]. An OVA-sensitized allergic rhinitis rat model also revealed an increase in $LTB_4$, $LTA_4$, and histamine [119]. Despite this evidence, the role of BLT1 in allergic rhinitis is still unclear.

Activation of Th2 cell-mediated cascades results in acute and chronic ocular allergies, and $LTB_4$ levels are increased in the tears of patients with seasonal allergic conjunctivitis [120]. The contribution of BLT1 to allergic conjunctivitis was demonstrated by inhibiting ragweed pollen-induced ocular scratching behavior by ONO-4057, a BLT1 antagonist [121]. Interestingly, $LTB_4$ is also involved in both late and delayed secondary allergic conjunctivitis [122].

## BLT1 in autoimmune diseases

The immune system normally reacts only to foreign antigens. However, it sometimes fails to distinguish self from nonself and produces autoantibodies, which causes autoimmune diseases. Recent studies demonstrated the role of BLT1 in autoantibody driven-autoimmune diseases. The deposition of IgG autoantibodies in tissues and the subsequent activation of the complement system leads to the accumulation of anaphylatoxin C5a, a potent chemoattractant for granulocytes. C5a and its receptor C5aR1 mediate the arrest of neutrophils on the endothelium by activating $\beta_2$ integrin, and they induce the release of $LTB_4$ from arrested neutrophils. The $LTB_4$–BLT1 axis recruits neutrophils into the tissues in rheumatoid arthritis [123,124] and pemphigoid diseases [125,126]. On the other hand, the CXCL1–2-CXCR2 and $LTB_4$–BLT1 axes play an important role in neutrophil recruitment into skin and in the development of psoriasis [127]. The antagonistic action on BLT1 by resolvin E1 inhibits the migration of cutaneous DCs and $\gamma\delta$ T-cells in murine psoriatic dermatitis [128].

Among autoimmune diseases, the correlation between BLT1 and rheumatoid arthritis is perhaps the most documented. Many studies defined a critical role for $LTB_4$ and BLT1 in the pathogenesis of arthritis using mouse models (Fig. 12.2) [129–131]. In a K/BxN serum

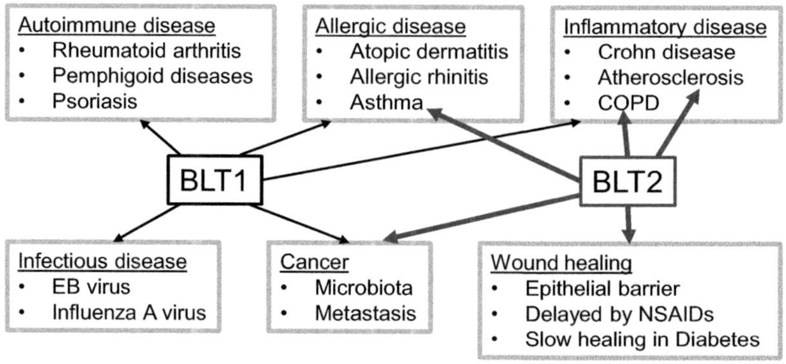


**FIGURE 12.2 The correlation of BLT1 and BLT2 with diseases.** Both of BLT1 and BLT2 play roles in some disease. BLT2, but not BLT1, contributes to wound healing.

transfer arthritis model, BLT1 was required for peripheral neutrophil recruitment and the subsequent induction of IL-1β via immune complex-FcγR interactions [132]. Additionally, $LTB_4$ levels were higher in serum [133,134], synovial fluid [135], and synovial tissue [136] from rheumatoid arthritis patients. BLT1 is also involved in autoimmune uveitis [137] and autoimmune encephalomyelitis [138].

## BLT1 in inflammatory diseases

$LTB_4$ is also increased in colonic mucosa in patients with inflammatory bowel diseases, including ulcerative colitis and Crohn's disease [139]. BLT1 deficiency in DCs reduces pro-inflammatory cytokines including IL-6, TNF-α, and IL-12 via the Gαi βγ subunit, as well as Th1 and Th17 differentiation in a trinitrobenzene sulfonic acid (TNBS)-induced colitis model [140]. Interestingly, in a dextran sodium sulfate (DSS)-induced colitis model, BLT1-deficient mice exhibited a similar severity of colitis to wild-type mice [141]. The DSS-induced colitis model informs on innate immune cells such as neutrophils and macrophages, and mucosal inflammation develops in the absence of T-cells mediating adaptive immunity. Thus, the $LTB_4$–BLT1 axis may modulate intestinal inflammation through T-cells.

Atherosclerosis is a chronic vascular inflammatory disease characterized by infiltration of lipid-laden macrophages and blood vessel narrowing. In vivo, BLT1 deficiency in ApoE knockout mice reduced lesion formation in the aortic valve during the initiation of atherosclerosis by inhibiting macrophage recruitment and vascular smooth muscle cell chemotaxis and proliferation [142]. Two BLT1 antagonists decreased aortic atherosclerosis in ApoE knockout mice [143] and monocyte foam cells in mice [144]. Foam cell formation driven by CD36 regulates the internalization of oxidized low-density lipoprotein (oxLDL), leading to activation of monocytes and the release of microvesicles (MVs) derived from monocytes. MVs are a heterogeneous population of cell-derived membrane-encapsulated particles ranging from 0.1 to 1.0 μm in size, and they are characterized by membrane proteins, receptors, and cytosolic materials including mRNAs and microRNAs from their parental cell of origin [145]. Therefore, MVs are implicated as biomarkers of diseases associated with vascular injury, inflammation, and pro-thrombotic states [146]. Unexpectedly, levels of BLT1$^+$ MVs are reduced in patients with atherosclerosis [147], although $LTB_4$ levels are higher in atherosclerosis patients than in healthy controls [148]. Lower levels of BLT1$^+$ MVs might be attributed to internalization and downregulation by $LTB_4$ binding.

COPD is a common inflammatory disease of the airways characterized by a progressive and irreversible decline in lung function. $LTB_4$ levels in exhaled breath condensate and sputum in patients with COPD are increased [149]. The $LTB_4$–BLT1 axis evokes neutrophil chemoattractants in sputum and exhaled breath condensate [150], and promotes neutrophil survival [151] in COPD patients. There was a negative correlation between $LTB_4$ level in BAL fluid and SOCS-1 mRNA levels in lung, and a positive correlation between SOCS-1 mRNA levels in lung in patients with COPD, accompanied by airflow limitation. In addition, BLT1 expression is increased and SOCS-1 expression is decreased in macrophages from COPD patients [152]. These findings indicate that the $LTB_4$/BLT1/SOCS-1 pathway is crucial for the progression of COPD. Indeed, in vivo studies showed that cigarette smoke extract increases $LTB_4$ levels and SOCS-1 expression in mouse macrophages, and BLT1 antagonist U-75302 inhibits SOCS-1 expression and subsequent TNF-$\alpha$ and IL-6 secretion.

## BLT1 in virus infection

BLT1 plays an important role in both Th2 and Th1 immunological reactions. The $LTB_4$–BLT1 axis enhances Fc$\gamma$R-mediated phagocytosis in macrophages [7,153] and the production of IL-12 in DCs, leading to IFN-$\gamma$ production [154]. In patients with acute Epstein–Barr virus (EBV) infection, BLT1 expression is increased in peripheral blood Tet$^+$ CD8$^+$ T-cells [155]. Recently, $LTB_4$–BLT1 signaling was reported to enhance the activation of type I IFN-$\alpha/\beta$ receptor, followed by IFN-$\alpha$ production and decreased susceptibility to influenza A virus infection in mouse [156]. These findings indicate an antiinflammatory role for the $LTB_4$–BLT1 axis.

## BLT1 in lung disease

Cystic fibrosis (CF), the most common inherited disease, is caused by mutations in the CF transmembrane conductance regulator (*CFTR*) gene, resulting in thick and sticky secretions that clog lungs and other organs. High levels of $LTB_4$ are in sputum and BAL fluid from patients with CF [157]. Tumor necrosis factor-$\alpha$ (TNF-$\alpha$) primes human neutrophils and increases $LTB_4$ levels [158]. In CF patients, TNF-$\alpha$ levels and both $LTB_4$ and CysLTs in sputum are positively correlated, and TNF-$\alpha$ and $LTB_4$ levels are negatively correlated with pulmonary function [159]. Although such evidence supports the potential for clinical efficacy of $LTB_4$ receptor antagonists in CF, one such antagonist (BIIL 284 BS) failed to improve symptoms and pulmonary function in CF patients due to increased severe adverse events including pulmonary infections [160].

## BLT1 in cancer

Chronic inflammation mediated by myeloid cells promotes tumor progression, whereas immune surveillance mediated by cytotoxic T-lymphocytes suppresses tumor growth [161]. LTs including $LTB_4$ are linked to the regulation of the tumor microenvironment and promote inflammation [162]. Recently, the role of BLT1 was revealed in several types of cancer. In a murine spontaneous colon cancer model ($Apc^{Min}$ mice), chemokine-mediated recruitment of mast cells initiated $LTB_4$/BLT1-regulated $CD8^+$ T-cell homing and generation of effective antitumor immunity [163]. Antitumor immunity of $CD8^+$ T-cells was shown to involve the BLT1–CXCR3 axis in a syngeneic mouse model of B16 melanoma [164] and a syngeneic TC-1 cervical cancer model [165]. BLT1 also mediates antitumor effects through memory $CD4^+$ T-cells [166].

Neutrophilic inflammation is related to tumorigenesis [167]. There is a positive correlation between neutrophil numbers within tumors and poor prognosis in patients with pulmonary adenocarcinoma [168]. In vitro and in vivo models of crystalline silica (CS)-induced inflammation in lung demonstrated that mast cell-derived $LTB_4$ initiated macrophage phagocytosis and led to sustained activation of neutrophils via an autocrine loop of $LTB_4$. CS-activated macrophages and epithelial cells produce CXC chemokines, such as CXCL1, 2, 3, and 5, and CC chemokines, such as CCL3 and 4, all of which induce the recruitment of a tumor-promoting subset of neutrophils (N2) [169].

Besides inflammation in cancer tissues, the gut microbiota is related to tumorigenesis [170,171]. $BLT1^{-/-}Apc^{Min/+}$ mice display increased levels of *Akkermansia muciniphila*, a mucin-degrading bacterium present in human intestines [172]. *A. muciniphila* is present at high levels in the colon of cancer patients. Germ-free $BLT1^{-/-}Apc^{Min/+}$ mice do not suffer from colon tumors, but tumors can be induced by fecal transplantation.

These findings suggest that $LTB_4$ could be a therapeutic target for cancer. Indeed, a recent study demonstrated the effect of $LTB_4$ inhibition on colorectal cancer [173]. $LTA_4H$ and BLT1 are overexpressed in human colon adenocarcinoma tissues. Bestatin, an $LTA_4H$ inhibitor, suppresses $LTB_4$ levels, BLT1–ERK1/2 signaling, and subsequent cell proliferation in colon tissues in patients with colorectal cancer and in a patient-derived xenografts mouse model.

The role of BLT1 was investigated in an experimental pulmonary metastasis mouse model [174]. Exposure to diesel exhaust particles (DEPs), the major component of particulate matter 2.5 (PM2.5), elicited neutrophil infiltration by $LTB_4$–BLT1 signaling in lung, and promoted lung metastasis in a mouse model. Administration of BLT1 antagonist U-75302 but not BLT2 antagonist LY255283 before the onset of inflammation reduced DEP-enhanced lung metastasis. Interestingly, BLT1 blockade after lung metastasis had no effect on its magnitude.

## BLT1 in other diseases

BLT1 is also involved in contact dermatitis [154,175], osteoporosis [176], obesity [177–180], endotoxic shock [181], and alveolar bone loss [182]. Recently, we revealed that M2 macrophages expressing BLT1 augment neovascularization in a mouse model of AMD [183]. Damaged retinal pigment epithelial cells release $LTB_4$, triggering recruitment of immune cells such as neutrophils and inflammatory monocytes/macrophages to the injured retina.

Among these cells, M2 macrophages produce pro-angiogenic factors including vascular endothelial growth factor (VEGF)-A and proinflammatory cytokines that accelerate choroidal neovascularization.

## BLT2, a low-affinity receptor of $LTB_4$, and its ligand 12-HHT

In human and mouse, genes encoding BLT1 (*LTB4R1* and *Ltb4r1*) and BLT2 (*LTB4R2* and *Ltb4r2*) are located in close proximity on chromosome 14. Indeed, the BLT2 gene was identified during analysis of the BLT1 promoter [184]. BLT2 is a seven-transmembrane-spanning GPCR, like BLT1. BLT2 is highly conserved between human and mouse (92.7% amino acid identity) [185], and functional BLT2 is also expressed in zebra fish [186]. Because the amino acid sequence is similar between BLT1 and BLT2 in human (45.2% amino acid identity), activation of BLT2 by $LTB_4$ was predicted. Indeed, the membrane fraction of HEK 293 cells transfected with human BLT2 cDNA demonstrated specific and saturable $LTB_4$ binding with a K*d* of 22.7 nM, 20-fold higher than that of human BLT1. The dose–response curve of intracellular calcium mobilization for $LTB_4$ in BLT2-expressing CHO cells was shifted to the right compared with that of BLT1-expressing cells, by two orders of magnitude [185]. BLT2 was also cloned by three other research groups [187–189], and all four groups concluded that BLT2 is a low-affinity $LTB_4$ receptor. Even though, in addition to $LTB_4$, several eicosanoids including 12(*S*)-hydroxy-5*Z*,8*Z*,10*E*,14*Z*-eicosatetraenoic acid (12(*S*)-HETE) and 12(*S*)-hydroperoxy-5*Z*,8*Z*,10*E*,14*Z*-eicosatetraenoic acid (12(*S*)-HpETE) are able to activate BLT2, BLT2 activation requires very high concentrations of these ligands [185]. Thus, we hypothesized that there are other unidentified endogenous BLT2 ligands besides $LTB_4$. To identify these endogenous ligands of BLT2, we extracted lipids from several rat tissues and separated the components by high-performance liquid chromatography (HPLC). Fractions containing strong agonistic activity toward BLT2 were analyzed by mass spectrometry (MS) to elucidate the molecular mass and structure of BLT2 agonistic lipids in the fraction. The combination of exact mass measurement and MS/MS analysis revealed that the BLT2 agonist is a C17 fatty acid, namely 12(*S*)-hydroxyheptadeca-5*Z*,8*E*,10*E*-trienoic acid (12-HHT). Synthetic 12-HHT activates human and mouse BLT2 at lower concentrations than $LTB_4$ in several intracellular signaling assays, and has no activity toward BLT1. $EC_{50}$ values for 12-HHT and $LTB_4$ were found to be 19 and 142 nM, respectively, based on a calcium mobilization assay using BLT2-overexpressing CHO cells, and 10–100 nM 12-HHT induced the migration of mouse bone marrow-derived mast cells via BLT2. These findings demonstrated that 12-HHT is an endogenous ligand for BLT2 [17].

12-HHT was identified in the 1970s, but it was considered to be a mere byproduct of thromboxane $A_2$ ($TxA_2$) production [190,191]. Thromboxane $A_2$ synthase ($TxA_2S$) catalyzes not only the rearrangement of prostaglandin $H_2$ ($PGH_2$) to $TxA_2$, but also the fragmentation of $PGH_2$ into an almost equal amount of 12-HHT and malondialdehyde (MDA) [192]. Approximately 300–700 nM 12-HHT is produced during human blood coagulation [193]. 12-HHT is produced mainly by COX and $TxA_2S$ activities in activated platelets; lipids extracted from small intestines of COX-1-deficient mice display much lower agonistic activity for BLT2 than wild-type mice [17], and 12-HHT production is completely inhibited by the COX inhibitor aspirin, a nonsteroidal antiinflammatory drug (NSAID). In addition to catalysis by $TxA_2$, 12-HHT is synthesized from $PGH_2$ via a nonenzymatic pathway [193]. Recently, 12-HHT synthesis by CYP450 enzyme CYP2S1 in macrophages was reported [194,195], although the contribution of CYP2S1 to 12-HHT production is uncertain. 12-HHT is preferentially metabolized to a 12-keto derivative by 15-hydroxyprostaglandin dehydrogenase (15-PGDH) [196].

Extracellular loop 2 of BLT2 has a potential ligand contact point. The BLT2 SNP (rs1950504) in extracellular loop 2 enhances the ligand-binding affinity of BLT2, Akt activation, and subsequent production of reactive oxygen species (ROS) in CHO-K1 cells [197]. BLT2 is expressed relatively ubiquitously in human tissues [185] and is highly expressed in mouse epithelial cells of the small intestine, colon, skin, lung, and cornea [141,198–201], in contrast to BLT1 that is expressed primarily in leukocytes. Interestingly, BLT2 is only detected at the lateral membrane in epithelial cells, while BLT1 is detected at both the apical and lateral membranes [202]. These findings suggest that although BLT2 shares high amino acid sequence identity with BLT1, its distribution and function are different from BLT1.

## BLT2 in wound healing

The epithelium plays a major role as a physical barrier, and barrier dysfunction causes infection and inflammatory diseases. BLT2 is expressed in the epithelium and localized at the lateral membrane. BLT2-overexpressing Madin-Darby canine kidney (MDCK) II cells recover from the disassembly of cell–cell junctions faster than Mock cells, and 12-HHT stimulation accelerates recovery. Among cell–cell junctions, tight junctions (TJs) are an essential structure for physical barrier function. 12-HHT/BLT2 increases the expression level of claudin-4, a component of TJs that promotes barrier function through the $G\alpha i$ protein-p38 MAPK pathway [202]. This pathway was also confirmed in keratinocytes [203] and bronchial epithelial cells [204]. In addition, the 12-HHT–BLT2 axis enhances cell–cell contacts in low calcium conditions that mimic the basal layer through actin polymerization with phosphorylation of myosin phosphatase target subunit 1 [205]. Enhancement of the epithelial barrier and promotion of wound healing by 12-HHT/BLT2 were reported both in vitro and in vivo. BLT2-deficient mice exhibit higher transepidermal water loss, higher sensitivity to epicutaneous sensitization with OVA [202], and a more severe phenotype in a DSS-induced colitis model [141]. 12-HHT accumulates in wound fluid in mice, and accelerates re-epithelialization and wound closure after skin punching through the production of $TNF\alpha$ and MMP9 [199]. In addition to skin effects, corneal wound healing is also delayed in BLT2-deficient mice [201].

As mentioned above, 12-HHT production is inhibited by NSAIDs. Indeed, NSAIDs delay skin and corneal wound healing in wild-type mice but not in BLT2-deficient mice, suggesting that the clinical side effects of NSAIDs may delay wound healing and/or ulcers might be related to the 12-HHT–BLT2 axis. Additionally, the 12-HHT–BLT2 axis is related to slow wound healing in diabetes [206]. A high fat diet reduces 12-HHT levels and BLT2 expression in mice. Additionally, pro-regenerative mediators (iNOS and nitrites) and TJ-associated proteins claudin-4 and occludin are also downregulated, accompanied by higher MMP9 levels in skin, in mice fed a high fat diet. BLT2-expressing HaCaT cells produce higher levels of nitric oxide and claudin-4 under high glucose conditions and lower levels of MMP9 [203]. The BLT2 antagonist CAY10583 accelerates wound healing in diabetic rats [207]. Thus, BLT2 plays an important role in epithelial homeostasis.

## BLT2 in asthma

In contrast to the well-documented role of BLT1 in asthma, BLT2 function in asthma is controversial. BLT2-deficient mice exhibit enhanced airway eosinophilia and increased IL-13 levels in BAL in an OVA-induced allergic airway model [208]. IL-13-producing $CD4^+$ T-cells are more abundant in the lungs of BLT2-deficient mice after allergen challenge. Knockdown of BLT2 by siRNA in $CD4^+$ T-cells increases IL-13 production. Also, BLT2 mRNA expression in $CD4^+$ T-cells from asthmatic patients is lower than in cells from healthy subjects. These findings suggest that BLT2 may function as an antiinflammatory receptor in asthma. However, in contrast to its protective role, BLT2 antagonist LY255283 and antisense BLT2 reduce OVA-induced airway inflammation in mice [209]. BLT2 mRNA expression in airway fibroblasts is higher in asthmatic patients than in healthy subjects [210]. Aerobic exercise in an OVA-induced asthmatic mouse model reduces AHR and remodeling, accompanied by a reduction in BLT2 expression [211]. In addition, BLT2 studies using the 12(*S*)-HETE ligand revealed the role of BLT2 in mast cells. 12(*S*)-HETE/BLT2 induces the synthesis of IL-4 and IL-13 through ROS generation in antigen-stimulated mast cells [212], and IL-13 is induced by IL-33 [213]. Further investigation is needed to clarify the precise role of BLT2 in asthma.

## BLT2 in cancer

ROS levels are elevated in many cancers, where they promote tumor development and progression, and most chemotherapeutics also elevate intracellular ROS levels [214]. A delicate balance of intracellular ROS levels contributes to cancer cell function. BLT2 increases ROS levels through NADPH oxidases (Nox) and promotes cancer progression [215,216]. The BLT2 antagonist LY255283 and BLT2 siRNA both induce cell cycle arrest and apoptosis in androgen receptor-positive prostate and bladder tissues, and in estrogen receptor-negative breast cancer cells [217–219]. BLT2 increases the invasiveness of ovarian cancer and breast cancer cells through Nox4/ROS/MMP-2 and Nox1/ROS/IL-8 pathways, respectively [220,221]. In addition, 12(*S*)-HETE/BLT2 promotes VEGF-mediated angiogenesis [222]. These findings indicate that BLT2 enhances survival and progressive signaling pathways in cancer cells. Indeed, higher expression levels of BLT2 are associated with a lower disease-free-survival rate in triple-negative breast cancer [223]. BLT2 expression is increased in a colorectal cell line with the KRAS mutation, resulting in high morbidity and mortality rates, and inhibition of BLT2 decreases proliferation [224].

BLT2 was also linked to cancer chemoresistance [225]. Platinum-induced fatty acids (PIFAs) induce chemoresistance through $F4/80^+/CD11b^{low}$ splenocytes. Moreover, 12-HHT, but not 16:4 (n-3), induces chemoresistance, and this is blocked by BLT2 antagonist LY255283 and does not occur in BLT2-deficient mice.

## BLT2 in other diseases

BLT2 is expressed in vascular endothelial cells in lung tissue. BLT2-deficient mice exhibit higher CysLT receptor type 1 (CysLT1R) mRNA expression levels in pulmonary vascular endothelial cells, which causes higher mortality in pneumolysin-dependent acute lung injury [200]. The correlation between BLT2 and CysLT1R indicates tight regulation among LTs. Pharmacological studies demonstrated that BLT2 mediates vascular contractions in the guinea pig lung parenchyma, even though it is induced by $LTB_4$ [226]. BLT2 function was also linked to COPD [227], LPS-induced sepsis [181], and atherosclerosis [228].

## Conclusion

In this review, we summarized the biosynthesis and metabolism of $LTB_4$ and 12-HHT, and the characterization and pathophysiological roles of BLT1 and BLT2. The $LTB_4$–BLT1 and 12-HHT–BLT2 axes are mainly involved in acute and chronic inflammation. Agonists and antagonists are available, and gene-deficient mice have been established, enabling us and others to clarify the novel functions of $LTB_4$ receptors since the identification of $LTB_4$ receptor BLT1 in 1997. Agonists and antagonists may be candidates for therapeutic agents against various diseases including inflammation and cancer. Also, successful BLT1 structural analysis will likely accelerate the development of BLT1-based therapeutics in the future.

## Acknowledgments

We thank numerous collaborators and scientists in our laboratory for experiments, as well as Professor James Ntambi for giving us the opportunity to write this review.

## References

[1] Samuelsson B, Borgeat P, Hammarström S, Murphy RC. Introduction of a nomenclature: leukotrienes. Prostaglandins 1979;17(6):785–7.

[2] Samuelsson B. Leukotrienes: mediators of immediate hypersensitivity reactions and inflammation. Science 1983;220(4597):568–75.

[3] Borgeat P, Naccache PH. Biosynthesis and biological activity of leukotriene B4. Clin Biochem 1990;23(5):459–68.

[4] Colom B, Bodkin JV, Beyrau M, Woodfin A, Ody C, Rourke C, et al. Leukotriene B4-neutrophil elastase axis drives neutrophil reverse transendothelial cell migration in vivo. Immunity 2015;42(6):1075–86.

[5] Lämmermann T, Afonso PV, Angermann BR, Wang JM, Kastenmüller W, Parent CA, et al. Neutrophil swarms require LTB4 and integrins at sites of cell death in vivo. Nature 2013;498(7454):371–5.

[6] Costa MF, de Souza-Martins R, de Souza MC, Benjamim CF, Piva B, Diaz BL, et al. Leukotriene B4 mediates gammadelta T lymphocyte migration in response to diverse stimuli. J Leukoc Biol 2010;87(2):323–32.

[7] Okamoto F, Saeki K, Sumimoto H, Yamasaki S, Yokomizo T. Leukotriene B4 augments and restores Fc gammaRs-dependent phagocytosis in macrophages. J Biol Chem 2010;285(52):41113–21.

[8] Widegren HAM, Borgeat P, Flamand L, Johnston S, Greiff L. LTB4 increases nasal neutrophil activity and conditions neutrophils to exert antiviral effects. Respir Med 2011;105(7):997–1006.

[9] Gaudreault E, Gosselin J. Leukotriene B4 induces release of antimicrobial peptides in lungs of virally infected mice. J Immunol 2008;180(9):6211–21.

[10] Flamand N, Mancuso P, Serezani CH, Brock TG. Leukotrienes: mediators that have been typecast as villains. Cell Mol Life Sci 2007;64(19–20):2657–70.

[11] Wan M, Sabirsh A, Wetterholm A, Agerberth B, Haeggström JZ. Leukotriene B4 triggers release of the cathelicidin LL-37 from human neutrophils: novel lipid-peptide interactions in innate immune responses. FASEB J 2007;21(11):2897–905.

[12] Yokomizo T, Izumi T, Chang K, Takuwa Y, Shimizu T. A G-protein-coupled receptor for leukotriene B4 that mediates chemotaxis. Nature 1997;387(6633):620–4.

[13] Yokomizo T, Kato K, Hagiya H, Izumi T, Shimizu T. Hydroxyeicosanoids bind to and activate the low affinity leukotriene B4 receptor, BLT2. J Biol Chem 2001;276(15):12454–9.

[14] Sasaki F, Yokomizo T. The leukotriene receptors as therapeutic targets of inflammatory diseases. Int Immunol 2019;31:607–15.

[15] Beeh KM, Kornmann O, Buhl R, Culpitt SV, Giembycz MA, Barnes PJ. Neutrophil chemotactic activity of sputum from patients with COPD: role of interleukin 8 and leukotriene B4. Chest 2003;123(4):1240–7.

[16] Corhay JL, Henket M, Nguyen D, Duysinx B, Sele J, Louis R. Leukotriene B4 contributes to exhaled breath condensate and sputum neutrophil chemotaxis in COPD. Chest 2009;136(4):1047–54.

[17] Okuno T, Iizuka Y, Okazaki H, Yokomizo T, Taguchi R, Shimizu T. 12(S)-Hydroxyheptadeca-5Z, 8E, 10E-trienoic acid is a natural ligand for leukotriene B4 receptor 2. J Exp Med 2008;205(4):759–66.

[18] Saeki K, Yokomizo T. Identification, signaling, and functions of LTB. Semin Immunol 2017;33:30–6.

[19] Bonventre JV, Huang Z, Taheri MR, O'Leary E, Li E, Moskowitz MA, et al. Reduced fertility and postischaemic brain injury in mice deficient in cytosolic phospholipase A2. Nature 1997;390(6660):622–5.

[20] Uozumi N, Kume K, Nagase T, Nakatani N, Ishii S, Tashiro F, et al. Role of cytosolic phospholipase A2 in allergic response and parturition. Nature 1997;390(6660):618–22.

[21] Shimizu T. Lipid mediators in health and disease: enzymes and receptors as therapeutic targets for the regulation of immunity and inflammation. Annu Rev Pharmacol Toxicol 2009;49:123–50.

[22] Melanie J, Stables DWG. Old and new generation lipid mediators in acute inflammation and resolution. Prog Lipid Res 2011;50(1):35–51.

[23] Rouzer CA, Kargman S. Translocation of 5-lipoxygenase to the membrane in human leukocytes challenged with ionophore A23187. J Biol Chem 1988;263(22):10980–8.

[24] Woods JW, Coffey MJ, Brock TG, Singer II. Peters-Golden M. 5-Lipoxygenase is located in the euchromatin of the nucleus in resting human alveolar macrophages and translocates to the nuclear envelope upon cell activation. J Clin Invest 1995;95(5):2035–46.

[25] Haeggström J, Meijer J, Rådmark O. Leukotriene A4. Enzymatic conversion into 5,6-dihydroxy-7,9,11,14-eicosatetraenoic acid by mouse liver cytosolic epoxide hydrolase. J Biol Chem 1986;261(14):6332–7.

[26] Rådmark O, Werz O, Steinhilber D, Samuelsson B. 5-Lipoxygenase: regulation of expression and enzyme activity. Trends Biochem Sci 2007;32(7):332–41.

[27] Wan M, Tang X, Stsiapanava A, Haeggström JG. Biosynthesis of leukotriene B. Semin Immunol 2017;33:3–15.

[28] Luo M, Jones SM, Peters-Golden M, Brock TG. Nuclear localization of 5-lipoxygenase as a determinant of leukotriene B4 synthetic capacity. Proc Natl Acad Sci U S A 2003;100(21):12165–70.

[29] Fredman G, Ozcan L, Spolitu S, Hellmann J, Spite M, Backs J, et al. Resolvin D1 limits 5-lipoxygenase nuclear localization and leukotriene B4 synthesis by inhibiting a calcium-activated kinase pathway. Proc Natl Acad Sci U S A 2014;111(40):14530–5.

[30] Luo M, Jones SM, Flamand N, Aronoff DM, Peters-Golden M, Brock TG. Phosphorylation by protein kinase a inhibits nuclear import of 5-lipoxygenase. J Biol Chem 2005;280(49):40609–16.

[31] Flamand N, Luo M, Peters-Golden M, Brock TG. Phosphorylation of serine 271 on 5-lipoxygenase and its role in nuclear export. J Biol Chem 2009;284(1):306–13.

[32] Rådmark O, Shimizu T, Jörnvall H, Samuelsson B. Leukotriene A4 hydrolase in human leukocytes. Purification and properties. J Biol Chem 1984;259(20):12339–45.

[33] Funk CD, Rådmark O, Fu JY, Matsumoto T, Jörnvall H, Shimizu T, et al. Molecular cloning and amino acid sequence of leukotriene A4 hydrolase. Proc Natl Acad Sci U S A 1987;84(19):6677–81.

[34] Minami M, Ohno S, Kawasaki H, Rådmark O, Samuelsson B, Jörnvall H, et al. Molecular cloning of a cDNA coding for human leukotriene A4 hydrolase. Complete primary structure of an enzyme involved in eicosanoid synthesis. J Biol Chem 1987;262(29):13873–6.

[35] Ohishi N, Izumi T, Minami M, Kitamura S, Seyama Y, Ohkawa S, et al. Leukotriene A4 hydrolase in the human lung. Inactivation of the enzyme with leukotriene A4 isomers. J Biol Chem 1987;262(21):10200–5.

[36] Minami M, Minami Y, Emori Y, Kawasaki H, Ohno S, Suzuki K, et al. Expression of human leukotriene A4 hydrolase cDNA in *Escherichia coli*. FEBS Lett 1988;229(2):279–82.

[37] Welsch DJ, Creely DP, Hauser SD, Mathis KJ, Krivi GG, Isakson PC. Molecular cloning and expression of human leukotriene-C4 synthase. Proc Natl Acad Sci U S A 1994;91(21):9745–9.

[38] Lam BK, Penrose JF, Freeman GJ, Austen KF. Expression cloning of a cDNA for human leukotriene C4 synthase, an integral membrane protein conjugating reduced glutathione to leukotriene A4. Proc Natl Acad Sci U S A 1994;91(16):7663–7.

[39] Jakobsson PJ, Mancini JA, Ford-Hutchinson AW. Identification and characterization of a novel human microsomal glutathione S-transferase with leukotriene C4 synthase activity and significant sequence identity to 5-lipoxygenase-activating protein and leukotriene C4 synthase. J Biol Chem 1996;271(36):22203–10.

[40] Ahmad S, Niegowski D, Wetterholm A, Haeggström J, Morgenstern R, Rinaldo-Matthis A. Catalytic characterization of human microsomal glutathione S-transferase 2: identification of rate-limiting steps. Biochemistry (Mosc) 2013;52(10):1755–64.

[41] Jakobsson PJ, Mancini JA, Riendeau D, Ford-Hutchinson AW. Identification and characterization of a novel microsomal enzyme with glutathione-dependent transferase and peroxidase activities. J Biol Chem 1997;272(36):22934–9.

[42] Schröder O, Sjöström M, Qiu H, Stein J, Jakobsson PJ, Haeggström JZ. Molecular and catalytic properties of three rat leukotriene C4 synthase homologs. Biochem Biophys Res Commun 2003;312(2):271–6.

[43] Mandal AK, Jones PB, Bair AM, Christmas P, Miller D, Yamin TT, et al. The nuclear membrane organization of leukotriene synthesis. Proc Natl Acad Sci U S A 2008;105(51):20434–9.

[44] Minami M, Bito H, Ohishi N, Tsuge H, Miyano M, Mori M, et al. Leukotriene A4 hydrolase, a bifunctional enzyme. Distinction of leukotriene A4 hydrolase and aminopeptidase activities by site-directed mutagenesis at Glu-297. FEBS Lett 1992;309(3):353–7.

[45] Wetterholm A, Medina J, Rådmark O, Shapiro R, Harggstrom J, Vallee B, et al. Leukotriene A4 hydrolase: abrogation of the peptidase activity by mutation of glutamic acid-296. Proc Natl Acad Sci U S A 1992;89(19): 9141–5.

[46] Snelgrove RJ, Jackson PL, Hardison MT, Noerager BD, Kinloch A, Gaggar A, et al. A critical role for LTA4H in limiting chronic pulmonary neutrophilic inflammation. Science 2010;330(6000):90–4.

[47] Rouzer CA, Samuelsson B. On the nature of the 5-lipoxygenase reaction in human leukocytes: enzyme purification and requirement for multiple stimulatory factors. Proc Natl Acad Sci U S A 1985;82(18):6040–4.

[48] Brash AR, Murray JJ, Oates JA. The 5-lipoxygenase and 15-lipoxygenase of neutrophils and eosinophils. Prog Clin Biol Res 1985;199:143–52.

[49] Nabe T, Miura M, Kamiki T, Kohno S. Arachidonate 5-lipoxygenase and cyclooxygenase metabolites from guinea pig eosinophils and macrophages. Jpn J Pharmacol 2000;83(3):261–4.

[50] Macchia L, Hamberg M, Kumlin M, Butterfield JH, Haeggström JZ. Arachidonic acid metabolism in the human mast cell line HMC-1: 5-lipoxygenase gene expression and biosynthesis of thromboxane. Biochim Biophys Acta 1995;1257(1):58–74.

[51] Jakobsson PJ, Steinhilber D, Odlander B, Rådmark O, Claesson HE, Samuelsson B. On the expression and regulation of 5-lipoxygenase in human lymphocytes. Proc Natl Acad Sci U S A 1992;89(8):3521–5.

[52] Rådmark O, Werz O, Steinhilber D, Samuelsson B. 5-Lipoxygenase, a key enzyme for leukotriene biosynthesis in health and disease. Biochim Biophys Acta 2015;1851(4):331–9.

[53] Sala A, Bolla M, Zarini S, Müller-Peddinghaus R, Folco G. Release of leukotriene A4 versus leukotriene B4 from human polymorphonuclear leukocytes. J Biol Chem 1996;271(30):17944–8.

[54] McGee JE, Fitzpatrick FA. Erythrocyte-neutrophil interactions: formation of leukotriene B4 by transcellular biosynthesis. Proc Natl Acad Sci U S A 1986;83(5):1349–53.

[55] Grimminger F, Sibelius U, Seeger W. Amplification of LTB4 generation in AM-PMN cocultures: transcellular 5-lipoxygenase metabolism. Am J Physiol 1991;261(2 Pt 1):L195–203.

[56] Capra V, Rovati GE, Mangano P, Buccellati C, Murphy RC, Sala A. Transcellular biosynthesis of eicosanoid lipid mediators. Biochim Biophys Acta 2015;1851(4):377–82.

[57] Zarini S, Gijón M, Ransome A, Murphy R, Sala A. Transcellular biosynthesis of cysteinyl leukotrienes in vivo during mouse peritoneal inflammation. Proc Natl Acad Sci U S A 2009;106(20):8296–301.

[58] Fabre JE, Goulet JL, Riche E, Nguyen M, Coggins K, Offenbacher S, et al. Transcellular biosynthesis contributes to the production of leukotrienes during inflammatory responses in vivo. J Clin Invest 2002;109(10): 1373–80.

[59] Feinmark SJ, Cannon PJ. Endothelial cell leukotriene C4 synthesis results from intercellular transfer of leukotriene A4 synthesized by polymorphonuclear leukocytes. J Biol Chem 1986;261(35):16466–72.

[60] Sumimoto H, Minakami S. Oxidation of 20-hydroxyleukotriene B4 to 20-carboxyleukotriene B4 by human neutrophil microsomes. Role of aldehyde dehydrogenase and leukotriene B4 omega-hydroxylase (cytochrome P-450LTB omega) in leukotriene B4 omega-oxidation. J Biol Chem 1990;265(8):4348–53.

[61] Kikuta Y, Kusunose E, Sumimoto H, Mizukami Y, Takeshige K, Sakaki T, et al. Purification and characterization of recombinant human neutrophil leukotriene B4 omega-hydroxylase (cytochrome P450 4F3). Arch Biochem Biophys 1998;355(2):201–5.

[62] Kikuta Y, Kusunose E, Kusunose M. Characterization of human liver leukotriene B(4) omega-hydroxylase P450 (CYP4F2). J Biochem 2000;127(6):1047–52.

[63] Yokomizo T, Izumi T, Takahashi T, Kasama T, Kobayashi Y, Sato F, et al. Enzymatic inactivation of leukotriene B4 by a novel enzyme found in the porcine kidney. Purification and properties of leukotriene B4 12-hydroxydehydrogenase. J Biol Chem 1993;268(24):18128–35.

[64] Yokomizo T, Ogawa Y, Uozumi N, Kume K, Izumi T, Shimizu T. cDNA cloning, expression, and mutagenesis study of leukotriene B4 12-hydroxydehydrogenase. J Biol Chem 1996;271(5):2844–50.

[65] Ensor CM, Zhang H, Tai HH. Purification, cDNA cloning and expression of 15-oxoprostaglandin 13-reductase from pig lung. Biochem J 1998;330(Pt 1):103–8.

[66] Clish CB, Levy BD, Chiang N, Tai HH, Serhan CN. Oxidoreductases in lipoxin A4 metabolic inactivation: a novel role for 15-onoprostaglandin 13-reductase/leukotriene B4 12-hydroxydehydrogenase in inflammation. J Biol Chem 2000;275(33):25372–80.

[67] Henderson WR, Jörg A, Klebanoff SJ. Eosinophil peroxidase-mediated inactivation of leukotrienes B4, C4, and D4. J Immunol 1982;128(6):2609–13.

[68] Osman M. Therapeutic implications of sex differences in asthma and atopy. Arch Dis Child 2003;88(7):587–90.

[69] McKenzie C, Silverberg J. The prevalence and persistence of atopic dermatitis in urban United States children. Ann Allergy Asthma Immunol 2019;123(2):173–8.

[70] Spence JD, Pilote L. Importance of sex and gender in atherosclerosis and cardiovascular disease. Atherosclerosis 2015;241(1):208–10.

[71] Pergola C, Dodt G, Rossi A, Neunhoeffer E, Lawrenz B, Northoff H, et al. ERK-mediated regulation of leukotriene biosynthesis by androgens: a molecular basis for gender differences in inflammation and asthma. Proc Natl Acad Sci U S A 2008;105(50):19881–6.

[72] Pergola C, Rogge A, Dodt G, Northoff H, Weinigel C, Barz D, et al. Testosterone suppresses phospholipase D, causing sex differences in leukotriene biosynthesis in human monocytes. FASEB J 2011;25(10):3377–87.

[73] Rossi A, Pergola C, Pace S, Rådmark O, Werz O, Sautebin L. In vivo sex differences in leukotriene biosynthesis in zymosan-induced peritonitis. Pharmacol Res 2014;87:1–7.

[74] Pace S, Pergola C, Dehm F, Rossi A, Gerstmeier J, Troisi F, et al. Androgen-mediated sex bias impairs efficiency of leukotriene biosynthesis inhibitors in males. J Clin Invest 2017;127(8):3167–76.

[75] Haribabu B, Verghese MW, Steeber DA, Sellars DD, Bock CB, Snyderman R. Targeted disruption of the leukotriene B(4) receptor in mice reveals its role in inflammation and platelet-activating factor-induced anaphylaxis. J Exp Med 2000;192(3):433–8.

[76] Goldman D, Goetzl E. Specific binding of leukotriene B4 to receptors on human polymorphonuclear leukocytes. J Immunol 1982;129(4):1600–4.

[77] Brink C, Dahlén S-E, Drazen J, Evans JF, Hay DWP, Nicosia S, et al. International Union of Pharmacology XXXVII. Nomenclature for leukotriene and lipoxin receptors. Pharmacol Rev 2003;55(1):195–227.

[78] Sasaki F, Koga T, Ohba M, Saeki K, Okuno T, Ishikawa K, et al. Leukotriene B4 promotes neovascularization and macrophage recruitment in murine wet-type AMD models. JCI Insight 2018;3(18):e96902.

[79] Huang W, Garcia-Zepeda E, Sauty A, Oettgen H, Rothenberg M, Luster A. Molecular and biological characterization of the murine leukotriene B4 receptor expressed on eosinophils. J Exp Med 1998;188(6):1063–74.

[80] Kuniyeda K, Okuno T, Terawaki K, Miyano M, Yokomizo T, Shimizu T. Identification of the intracellular region of the leukotriene B4 receptor type 1 that is specifically involved in Gi activation. J Biol Chem 2007;282(6):3998–4006.

[81] Okuno T, Ago H, Terawaki K, Miyano M, Shimizu T, Yokomizo T. Helix 8 of the leukotriene B4 receptor is required for the conformational change to the low affinity state after G-protein activation. J Biol Chem 2003;278(42):41500–9.

[82] Okuno T, Yokomizo T, Hori T, Miyano M, Shimizu T. Leukotriene B4 receptor and the function of its helix 8. J Biol Chem 2005;280(37):32049–52.

[83] Nakanishi Y, Tan M, Ichiki T, Inoue A, Yoshihara J, Maekawa N, et al. Stepwise phosphorylation of leukotriene B4 receptor 1 defines cellular responses to leukotriene B4. Sci Signal 2018;11(544):eaao5390.

[84] Aratake Y, Okuno T, Matsunobu T, Saeki K, Takayanagi R, Furuya S, et al. Helix 8 of leukotriene B4 receptor 1 inhibits ligand-induced internalization. FASEB J 2012;26(10):4068–78.

[85] Gaudreau R, Le Gouill C, Venne M-H, Stankova J, Rola-Pleszczynski M. Threonine 308 within a putative casein kinase 2 site of the cytoplasmic tail of leukotriene B4 receptor (BLT1) is crucial for ligand-induced, G-protein-coupled receptor-specific kinase 6-mediated desensitization. J Biol Chem 2002;277(35):31567–76.

[86] Chen Z, Gaudreau R, Le Gouill C, Rola-Pleszczynski M, Staňková J. Agonist-induced internalization of leukotriene B4 receptor 1 requires G-protein-coupled receptor kinase 2 but not arrestins. Mol Pharmacol 2004;66(3):377–86.

[87] Sabirsh A, Bywater RP, Bristulf J, Owman C, Haeggström JZ. Residues from transmembrane helices 3 and 5 participate in leukotriene B4 binding to BLT1. Biochemistry (Mosc) 2006;45(18):5733–44.
[88] Basu S, Jala VR, Mathis S, Rajagopal ST, Del Prete A, Maturu P, et al. Critical role for polar residues in coupling leukotriene B4 binding to signal transduction in BLT1. J Biol Chem 2007;282(13):10005–17.
[89] Hori T, Okuno T, Hirata K, Yamashita K, Kawano Y, Yamamoto M, et al. Na+-mimicking ligands stabilize the inactive state of leukotriene B4 receptor BLT1. Nat Chem Biol 2018;14:262.
[90] Luttrell LM, Lefkowitz RJ. The role of β-arrestins in the termination and transduction of G-protein-coupled receptor signals. J Cell Sci 2002;115(3):455–65.
[91] Yang Z, Yang F, Zhang D, Liu Z, Lin A, Liu C, et al. Phosphorylation of G protein-coupled receptors: from the barcode hypothesis to the flute model. Mol Pharmacol 2017;92(3):201–10.
[92] Gaudreau R, Beaulieu M, Chen Z, Gouill C, Lavigne P, Stankova J, et al. Structural determinants regulating expression of the high affinity leukotriene B4 receptor. J Biol Chem 2004;279(11):10338–45.
[93] Ichiki T, Koga T, Okuno T, Saeki K, Yamamoto Y, Yamamoto H, et al. Modulation of leukotriene B4 receptor 1 signaling by receptor for advanced glycation end products (RAGE). FASEB J 2016;30(5):1811–22.
[94] Tager A, Dufour J, Goodarzi K, Bercury S, von Andrian U, Luster A. BLTR mediates leukotriene B(4)-induced chemotaxis and adhesion and plays a dominant role in eosinophil accumulation in a murine model of peritonitis. J Exp Med 2000;192(3):439–46.
[95] Patnode M, Bando J, Krummel M, Locksley R, Rosen S. Leukotriene B4 amplifies eosinophil accumulation in response to nematodes. J Exp Med 2014;211(7):1281–8.
[96] Ott V, Cambier J, Kappler J, Marrack P, Swanson B. Mast cell-dependent migration of effector CD8+ T cells through production of leukotriene B4. Nat Immunol 2003;4(10):974–81.
[97] Goodarzi K, Goodarzi M, Tager A, Luster A, von Andrian U. Leukotriene B4 and BLT1 control cytotoxic effector T cell recruitment to inflamed tissues. Nat Immunol 2003;4(10):965–73.
[98] Tager A, Bromley S, Medoff B, Islam S, Bercury S, Friedrich E, et al. Leukotriene B4 receptor BLT1 mediates early effector T cell recruitment. Nat Immunol 2003;4(10):982–90.
[99] Gelfand E, Dakhama A. CD8+ T lymphocytes and leukotriene B4: novel interactions in the persistence and progression of asthma. J Allergy Clin Immunol 2006;117(3):577–82.
[100] Pal K, Feng X, Steinke J, Burdick M, Shim Y, Sung S, et al. Leukotriene A4 hydrolase activation and leukotriene B4 production by eosinophils in severe asthma. Am J Respir Cell Mol Biol 2019;60(4):413–9.
[101] Kazani S, Planaguma A, Ono E, Bonini M, Zahid M, Marigowda G, et al. Exhaled breath condensate eicosanoid levels associate with asthma and its severity. J Allergy Clin Immunol 2013;132(3):547–53.
[102] Trischler J, Müller C, Könitzer S, Prell E, Korten I, Unverzagt S, et al. Elevated exhaled leukotriene B4 in the small airway compartment in children with asthma. Ann Allergy Asthma Immunol 2001;114(2):111–6.
[103] Medoff B, Tager A, Jackobek R, Means T, Wang L, Luster A. Antibody-antigen interaction in the airway drives early granulocyte recruitment through BLT1. Am J Physiol Lung Cell Mol Physiol 2006;290(1):170–8.
[104] Terawaki K, Yokomizo T, Nagase T, Toda A, Taniguchi M, Hashizume K, et al. Absence of leukotriene B4 receptor 1 confers resistance to airway hyperresponsiveness and Th2-type immune responses. J Immunol 2005;175(7):4217–25.
[105] Miyahara N, Takeda K, Miyahara S, Taube C, Joetham A, Koya T, et al. Leukotriene B4 receptor-1 is essential for allergen-mediated recruitment of CD8+ T cells and airway hyperresponsiveness. J Immunol 2005;174(8):4979–84.
[106] Miyahara Nobuaki N. Leukotriene B4 receptor 1 expression on dendritic cells is required for the development of Th2 responses and allergen-induced airway hyperresponsiveness. J Immunol 2008;181(2):1170–8.
[107] Turner C, Breslow R, Conklyn M, Andresen C, Patterson D, Lopez-Anaya A, et al. In vitro and in vivo effects of leukotriene B4 antagonism in a primate model of asthma. J Clin Invest 1996;97(2):381–7.
[108] Evans D, Barnes P, Spaethe S, van Alstyne E, Mitchell M, O'Connor B. Effect of a leukotriene B4 receptor antagonist, LY293111, on allergen induced responses in asthma. Thorax 1996;51(12):1178–84.
[109] Andoh T, Kuraishi Y. Intradermal leukotriene B4, but not prostaglandin E2, induces itch-associated responses in mice. Eur J Pharmacol 1998;353(1):93–6.
[110] Okano-Mitani H, Ikai K, Imamura S. Leukotriene A4 hydrolase in peripheral leukocytes of patients with atopic dermatitis. Arch Dermatol Res 1996;288(4):168–72.
[111] Hilger R, Neuber K, König W. Conversion of leukotriene A4 by neutrophils and platelets from patients with atopic dermatitis. Immunology 1991;74(4):689–95.
[112] Fogh K, Herlin T, Kragballe K. Eicosanoids in skin of patients with atopic dermatitis: prostaglandin E2 and leukotriene B4 are present in biologically active concentrations. J Allergy Clin Immunol 1989;83(2):450–5.

[113] Koro O, Furutani K, Hide M, Yamada S, Yamamoto S. Chemical mediators in atopic dermatitis: involvement of leukotriene B4 released by a type I allergic reaction in the pathogenesis of atopic dermatitis. J Allergy Clin Immunol 1999;103(4):663–70.
[114] Ruzicka T, Simmet T, Peskar B, Ring J. Skin levels of arachidonic acid-derived inflammatory mediators and histamine in atopic dermatitis and psoriasis. J Invest Dermatol 1986;86(2):105–8.
[115] Oyoshi M, He R, Li Y, Mondal S, Yoon J, Afshar R, et al. Leukotriene B4-driven neutrophil recruitment to the skin is essential for allergic skin inflammation. Immunity 2012;37(4):747–58.
[116] Katsube N, Maruyama M, Andoh T, Kuraishi Y. Involvement of leukotriene B4 in substance P-induced itch-associated response in mice. J Invest Dermatol 2001;117(6):1621–6.
[117] Shaw R, Fitzharris P, Cromwell O, Wardlaw A, Kay A. Allergen-induced release of sulphidopeptide leukotrienes (SRS-A) and LTB4 in allergic rhinitis. Allergy 1985;40(1):1–6.
[118] Meslier N, Braunstein G, Lacronique J, Dessanges J, Rakotosihanaka F, Devillier P, et al. Local cellular and humoral responses to antigenic and distilled water challenge in subjects with allergic rhinitis. Am Rev Respir Dis 1988;137(3):617–24.
[119] Chen S, Chen G, Shu S, Xu Y, Ma X. Metabolomics analysis of baicalin on ovalbumin-sensitized allergic rhinitis rats. R Soc Open Sci 2019;6(2):181081.
[120] Nathan H, Naveh N, Meyer E. Levels of prostaglandin E2 and leukotriene B4 in tears of vernal conjunctivitis patients during a therapeutic trial with indomethacin. Doc Ophthalmol 1994;85(3):247–57.
[121] Andoh T, Sakai K, Urashima M, Kitazawa K, Honma A, Kuraishi Y. Involvement of leukotriene B4 in itching in a mouse model of ocular allergy. Exp Eye Res 2012;98:97–103.
[122] Pelikan Z. Mediator profiles in tears during the conjunctival response induced by allergic reaction in the nasal mucosa. Mol Vis 2013;19:1453–70.
[123] Sadika CD, Miyabe Y, Sezina T, Luster AD. The critical role of C5a as an initiator of neutrophil-mediated autoimmune inflammation of the joint and skin. Semin Immunol 2018;37:21–9.
[124] Miyabe Y, Miyabe C, Murooka TT, Kim EY, Newton GA, Kim ND, et al. Complement C5a receptor is the key initiator of neutrophil adhesion igniting immune complex-induced arthritis. Sci Immunol 2017;2(7): eaaj2195.
[125] Sezin T, Krajewski M, Wutkowski A, Mousavi S, Chakievska L, Bieber K, et al. The leukotriene B4 and its receptor BLT1 act as critical drivers of neutrophil recruitment in murine bullous pemphigoid-like epidermolysis bullosa acquisita. J Invest Dermatol 2017;137(5):1104–13.
[126] Sezin Tanya T. Dual inhibition of complement factor 5 and leukotriene B4 synergistically suppresses murine pemphigoid disease. JCI Insight 2019;4(15):e128239.
[127] Sumida H, Yanagida K, Kita Y, Abe J, Matsushima K, Nakamura M, et al. Interplay between CXCR2 and BLT1 facilitates neutrophil infiltration and resultant keratinocyte activation in a murine model of imiquimod-induced psoriasis. J Immunol 2014;192(9):4361–9.
[128] Sawada Y, Honda T, Nakamizo S, Otsuka A, Ogawa N, Kobayashi Y, et al. Resolvin E1 attenuates murine psoriatic dermatitis. Sci Rep 2018;8(1):11873.
[129] Shao W-H, Del Prete A, Bock CB, Haribabu B. Targeted disruption of leukotriene B4 receptors BLT1 and BLT2: a critical role for BLT1 in collagen-induced arthritis in mice. J Immunol 2006;176(10):6254–61.
[130] Griffiths RJ, Pettipher ER, Koch K, Farrell CA, Breslow R, Conklyn MJ, et al. Leukotriene B4 plays a critical role in the progression of collagen-induced arthritis. Proc Natl Acad Sci U S A 1995;92(2):517–21.
[131] Chen M, Lam BK, Kanaoka Y, Nigrovic PA, Audoly LP, Austen KF, et al. Neutrophil-derived leukotriene B4 is required for inflammatory arthritis. J Exp Med 2006;203(4):837–42.
[132] Kim ND, Chou RC, Seung E, Tager AM, Luster AD. A unique requirement for the leukotriene B4 receptor BLT1 for neutrophil recruitment in inflammatory arthritis. J Exp Med 2006;203(4):829–35.
[133] Xu S, Lin J, LuE H, Chen Z, Jiang D. Regulation of TNFalpha and IL1beta in rheumatoid arthritis synovial fibroblasts by leukotriene B4. Rheumatol Int 2010;30(9):1183–9.
[134] Gürsel T, Firat S, Ercan Z. Increased serum leukotriene B4 level in the active stage of rheumatoid arthritis in children. Prostaglandins Leukot Essent Fatty Acids 1997;56(3):205–7.
[135] Hashimoto A, Endo H, Hayashi I, Murakami Y, Kitasato H, Kono S, et al. Differential expression of leukotriene B4 receptor subtypes (BLT1 and BLT2) in human synovial tissues and synovial fluid leukocytes of patients with rheumatoid arthritis. J Rheumatol 2003;30(8):1712–8.
[136] Ahmadzadeh N, Shingu M, Nobunaga M, Tawara T. Relationship between leukotriene B4 and immunological parameters in rheumatoid synovial fluids. Inflammation 1991;15(6):497–503.

[137] Liao T, Ke Y, Shao W-H, Haribabu B, Kaplan HJ, Sun D, et al. Blockade of the interaction of leukotriene b4 with its receptor prevents development of autoimmune uveitis. Invest Ophthalmol Vis Sci 2006;47(4):1543–9.

[138] Kihara Y, Yokomizo T, Kunita A, Morishita Y, Fukayama M, Ishii S, et al. The leukotriene B4 receptor, BLT1, is required for the induction of experimental autoimmune encephalomyelitis. Biochem Biophys Res Commun 2010;394(3):673–8.

[139] Sharon P, Stenson W. Enhanced synthesis of leukotriene B4 by colonic mucosa in inflammatory bowel disease. Gastroenterology 1984;86(3):453–60.

[140] Zhou J, Lai W, Yang W, Pan J, Shen H, Cai Y, et al. BLT1 in dendritic cells promotes Th1/Th17 differentiation and its deficiency ameliorates TNBS-induced colitis. Cell Mol Immunol 2018;15(12):1047–56.

[141] Iizuka Y, Okuno T, Saeki K, Uozaki H, Okada S, Misaka T, et al. Protective role of the leukotriene B4 receptor BLT2 in murine inflammatory colitis. FASEB J 2010;24(12):4678–90.

[142] Bäck M, Bu DX, Bränström R, Sheikine Y, Yan ZQ, Hansson GK. Leukotriene B4 signaling through NF-kappaB-dependent BLT1 receptors on vascular smooth muscle cells in atherosclerosis and intimal hyperplasia. Proc Natl Acad Sci U S A 2005;102(48):17501–6.

[143] Ketelhuth DFJ, Hermansson A, Hlawaty H, Letourneur D, Yan Z-q, Bäck M. The leukotriene B4 receptor (BLT) antagonist BIIL284 decreases atherosclerosis in ApoE-/- mice. Prostaglandins Other Lipid Mediat 2015;121(Pt):105–9.

[144] Aiello RJ, Bourassa P-A, Lindsey S, Weng W, Freeman A, Showell HJ. Leukotriene B4 receptor antagonism reduces monocytic foam cells in mice. Arterioscler Thromb Vasc Biol 2002;22(3):443–9.

[145] Burger D, Schock S, Thompson CS, Montezano AC, Hakim AM, Touyz RM. Microparticles: biomarkers and beyond. Clin Sci 2013;124(7):423–41.

[146] Mastronardi ML, Mostefai HA, Soleti R, Agouni A, Martínez MC, Andriantsitohaina R. Microparticles from apoptotic monocytes enhance nitrosative stress in human endothelial cells. Fundam Clin Pharmacol 2011;25(6):653–60.

[147] Sanden M, Bothca J, Nielsen MRS, Nielsen MH, Schmidt EB, Handberg A. BLTR1 and CD36 expressing microvesicles in atherosclerotic patients and healthy individuals. Front Cardiovasc Med 2018;5:156.

[148] Bäck M, Hansson GK. Leukotriene receptors in atherosclerosis. Ann Med 2006;38(7):493–502.

[149] Kostikas K, Gaga M, Papatheodorou G, Karamanis T, Orphanidou D, Loukides S. Leukotriene B4 in exhaled breath condensate and sputum supernatant in patients with COPD and asthma. Chest 2005;127(5):1553–9.

[150] Corhay J-L, Henket M, Nguyen D, Duysinx B, Sele J, Louis R. Leukotriene B4 contributes to exhaled breath condensate and sputum neutrophil chemotaxis in COPD. Chest 2009;136(4):1047–54.

[151] Lee E, Lindo T, Jackson N, Meng-Choong L, Reynolds P, Hill A, et al. Reversal of human neutrophil survival by leukotriene b4 receptor blockade and 5-lipoxygenase and 5-lipoxygenase activating protein inhibitors. Am J Respir Crit Care Med 1999;160(6):2079–85.

[152] Dong R, Xie L, Zhao K, Zhang Q, Zhou M, He P. Cigarette smoke-induced lung inflammation in COPD mediated via LTB4/BLT1/SOCS1 pathway. Int J Chron Obstruct Pulmon Dis 2016;11:31–41.

[153] Serezani CH, Aronoff DM, Sitrin RG, Peters-Golden M. FcgammaRI ligation leads to a complex with BLT1 in lipid rafts that enhances rat lung macrophage antimicrobial functions. Blood 2009;114(15):3316–24.

[154] Toda A, Terawaki K, Yamazaki S, Saeki K, Shimizu T, Yokomizo T. Attenuated Th1 induction by dendritic cells from mice deficient in the leukotriene B4 receptor 1. Biochimie 2010;92(6):682–91.

[155] Islam SA, Thomas SY, Hess C, Medoff BD, Means TK, Brander C, et al. The leukotriene B4 lipid chemoattractant receptor BLT1 defines antigen-primed T cells in humans. Blood 2006;107(2):444–53.

[156] Pernet E, Downey J, Vinh DC, Powell WS, Divangahi M. Leukotriene B4-type I interferon axis regulates macrophage-mediated disease tolerance to influenza infection. Nat Microbiol 2019;4(8):1389–400.

[157] Konstan MW, Walenga RW, Hilliard KA, Hilliard JB. Leukotriene B4 markedly elevated in the epithelial lining fluid of patients with cystic fibrosis. Am Rev Respir Dis 1993;148(4):896–901.

[158] Roubin R, Elsas PP, Fiers W, Dessein AJ. Recombinant human tumour necrosis factor (rTNF)2 enhances leukotriene biosynthesis in neutrophils and eosinophils stimulated with the Ca2+ ionophore A23187. Clin Exp Immunol 1987;70(2):484–90.

[159] Greally P, Hussein MJ, Cook AJ, Sampson AP, Piper PJ, Price JF. Sputum tumour necrosis factor-alpha and leukotriene concentrations in cystic fibrosis. Arch Dis Child 1993;68(3):389–92.

[160] Konstan MW, Döring G, Heltshe SL, Lands LC, Hilliard KA, Koker P, et al. A randomized double blind, placebo controlled phase 2 trial of BIIL 284 BS (an LTB4 receptor antagonist) for the treatment of lung disease in children and adults with cystic fibrosis. J Cyst Fibros 2014;13(2):148–55.

[161] Rossi JF, Céballos P, Lu ZY. Immune precision medicine for cancer: a novel insight based on the efficiency of immune effector cells. Cancer Commun 2019;39(1):34.
[162] Wang D, DuBois RN. Eicosanoids and cancer. Nat Rev Cancer 2010;10:181.
[163] Bodduluri SR, Mathis S, Maturu P, Krishnan E, Satpathy SR, Chilton PM, et al. Mast cell–dependent CD8+ T-cell recruitment mediates immune surveillance of intestinal tumors in ApcMin/+ mice. Cancer Immunol Res 2018;6(3):332–47.
[164] Chheda ZS, Sharma RK, Jala VR, Luster AD, Haribabu B. Chemoattractant receptors BLT1 and CXCR3 regulate antitumor immunity by facilitating CD8+ T cell migration into tumors. J Immunol 2016;197(5):2016–26.
[165] Sharma RK, Chheda Z, Jala VR, Haribabu B. Expression of leukotriene B4 receptor-1 on CD8+ T cells is required for their migration into tumors to elicit effective antitumor immunity. J Immunol 2013;191(6):3462–70.
[166] Yokota Y, Inoue H, Matsumura Y, Nabeta H, Narusawa M, Watanabe A, et al. Absence of LTB4/BLT1 axis facilitates generation of mouse GM-CSF-induced long-lasting antitumor immunologic memory by enhancing innate and adaptive immune systems. Blood 2012;120(17):3444–54.
[167] Galdiero MR, Garlanda C, Jaillon S, Marone G, Mantovani A. Tumor associated macrophages and neutrophils in tumor progression. J Cell Physiol 2013;228(7):1404–12.
[168] Wislez M, Rabbe N, Marchal J, Milleron B, Crestani B, Mayaud C, Antoine M, Soler P, Cadranel J. Hepatocyte growth factor production by neutrophils infiltrating bronchioloalveolar subtype pulmonary adenocarcinoma: role in tumor progression and death. Cancer Res 2003;63(6):1405–12.
[169] Satpathy SR, Jala VR, Bodduluri SR, Krishnan E, Hegde B, Hoyle GW, et al. Crystalline silica-induced leukotriene B4-dependent inflammation promotes lung tumour growth. Nat Commun 2015;6:7064.
[170] Candela M, Guidotti M, Fabbri A, Brigidi P, Franceschi C, Fiorentini C. Human intestinal microbiota: cross-talk with the host and its potential role in colorectal cancer. Crit Rev Microbiol 2011;37(1):1–14.
[171] Arthur JC, Jobin C. The struggle within: microbial influences on colorectal cancer. Inflamm Bowel Dis 2010;17(1):396–409.
[172] Jala VR, Maturu P, Bodduluri SR, Krishnan E, Mathis S, Subbarao K, et al. Leukotriene B4-receptor-1 mediated host response shapes gut microbiota and controls colon tumor progression. OncoImmunology 2017;6(12):e1361593.
[173] Zhao S, Yao K, Li D, Liu K, Jin G, Yan M, et al. Inhibition of LTA4H by bestatin in human and mouse colorectal cancer. EBioMedicine 2019;44:361–74.
[174] Li W, Liu T, Xiong Y, Lv J, Cui X, He R. Diesel exhaust particle promotes tumor lung metastasis via the induction of BLT1-mediated neutrophilic lung inflammation. Cytokine 2018;111:530–40.
[175] Lv J, Zou L, Zhao L, Yang W, Xiong Y, Li B, et al. Leukotriene B4—leukotriene B4 receptor axis promotes oxazolone-induced contact dermatitis by directing skin homing of neutrophils and CD8+ T cells. Immunology 2015;146(1):50–8.
[176] Hikiji H, Ishii S, Yokomizo T, Takato T, Shimizu T. A distinctive role of the leukotriene B4 receptor BLT1 in osteoclastic activity during bone loss. Proc Natl Acad Sci U S A 2009;106(50):21294–9.
[177] Spite M, Hellmann J, Tang Y, Mathis SP, Kosuri M, Bhatnagar A, et al. Deficiency of the leukotriene B4 receptor, BLT-1, protects against systemic insulin resistance in diet-induced obesity. J Immunol 2011;187(4):1942–9.
[178] Li P, Oh DY, Bandyopadhyay G, Lagakos WS, Talukdar S, Osborn O, et al. LTB4 promotes insulin resistance in obese mice by acting on macrophages, hepatocytes and myocytes. Nat Med 2015;21:239.
[179] Ying W, Wollam J, Ofrecio JM, Bandyopadhyay G, El Ouarrat D, Lee YS, et al. Adipose tissue B2 cells promote insulin resistance through leukotriene LTB4/LTB4R1 signaling. J Clin Invest 2017;127(3):1019–30.
[180] Mothe-Satney I, Filloux C, Amghar H, Pons C, Bourlier V, Galitzky J, et al. Adipocytes secrete leukotrienes: contribution to obesity-associated inflammation and insulin resistance in mice. Diabetes 2012;61(9):2311–9.
[181] Kwon SY, Ro M, Kim JH. Mediatory roles of leukotriene B4 receptors in LPS-induced endotoxic shock. Sci Rep 2019;9(1):5936.
[182] Madeira MFM, Queiroz-Junior CM, Corrêa JD, Werneck SMC, Machado FS, Cunha TM, et al. The role of 5-lipoxygenase in aggregatibacter actinomycetemcomitans-induced alveolar bone loss. J Clin Periodontol 2017;44(8):793–802.
[183] Sasaki F, Koga T, Ohba M, Saeki K, Okuno T, Ishikawa K, et al. Leukotriene B4 promotes neovascularization and macrophage recruitment in murine wet-type AMD models. JCI Insight 2018;3(18):e96902.
[184] Kato K, Yokomizo T, Izumi T, Shimizu T. Cell-specific transcriptional regulation of human leukotriene B4 receptor gene. J Exp Med 2000;192(3):413–20.

[185] Yokomizo T, Kato K, Terawaki K, Izumi T, Shimizu T. A second leukotriene B(4) receptor, BLT2. A new therapeutic target in inflammation and immunological disorders. J Exp Med 2000;192(3):421–32.
[186] Okuno T, Ishitani T, Yokomizo T. Biochemical characterization of three BLT receptors in zebrafish. PLoS ONE 2015;10(3):e0117888.
[187] Wang S, Gustafson E, Pang L, Qiao X, Behan J, Maguire M, et al. A novel hepatointestinal leukotriene B4 receptor: cloning and functional characterization. J Biol Chem 2000;275(52):40686–94.
[188] Tryselius Y, Nilsson NE, Kotarsky K, Olde B, Owman C. Cloning and characterization of cDNA encoding a novel human leukotriene B4 receptor. Biochem Biophys Res Commun 2000;274(2):377–82.
[189] Kamohara M, Takasaki J, Matsumoto M, Saito T, Ohishi T, Ishii H, et al. Molecular cloning and characterization of another leukotriene B4 receptor. J Biol Chem 2000;275(35):27000–4.
[190] John H, Cammann K, Schlegel W. Development and review of radioimmunoassay of 12-S-hydroxyheptadecatrienoic acid. Prostaglandins Other Lipid Mediat 1998;56(2):53–76.
[191] Hamberg MM, Samuelsson B. Prostaglandin endoperoxides. Novel transformations of arachidonic acid in human platelets. Proc Natl Acad Sci U S A 1974;71(9):3400–4.
[192] Hecker M, Haurand M, Ullrich V, Diczfalusy U, Hammarström S. Products, kinetics, and substrate specificity of homogeneous thromboxane synthase from human platelets: development of a novel enzyme assay. Arch Biochem Biophys 1987;254(1):124–35.
[193] Matsunobu T, Okuno T, Yokoyama C, Yokomizo T, Thromboxane A. synthase-independent production of 12-hydroxyheptadecatrienoic acid, a BLT2 ligand. J Lipid Res 2013;54(11):2979–87.
[194] Frömel T, Kohlstedt K, Popp R, Yin X, Awwad K, Barbosa-Sicard E, et al. Cytochrome P4502S1: a novel monocyte/macrophage fatty acid epoxygenase in human atherosclerotic plaques. Basic Res Cardiol 2012;108(1):319.
[195] Bui P, Imaizumi S, Beedanagari SR, Reddy ST, Hankinson O. Human CYP2S1 metabolizes cyclooxygenase- and lipoxygenase-derived eicosanoids. Drug Metab Dispos 2011;39(2):180–90.
[196] Hecker MM, Ullrich V. 12(S)-Hydroxy-5,8,10 (Z,E,E)-heptadecatrienoic acid (HHT) is preferentially metabolized to its 12-keto derivative by human erythrocytes in vitro. Eicosanoids 1988;1(1):19–25.
[197] Jang JH, Wei JD, Kim M, Kim JY, Cho AE, Kim JH. Leukotriene B4 receptor 2 gene polymorphism (rs1950504, Asp196Gly) leads to enhanced cell motility under low-dose ligand stimulation. Exp Mol Med 2017;49:e402.
[198] Iizuka Y, Yokomizo T, Terawaki K, Komine M, Tamaki K, Shimizu T. Characterization of a mouse second leukotriene B4 receptor, mBLT2: BLT2-dependent ERK activation and cell migration of primary mouse keratinocytes. J Biol Chem 2005;280(26):24816–23.
[199] Liu M, Saeki K, Matsunobu T, Okuno T, Koga T, Sugimoto Y, et al. 12-hydroxyheptadecatrienoic acid promotes epidermal wound healing by accelerating keratinocyte migration via the BLT2 receptor. J Exp Med 2014;211(6):1063–78.
[200] Shigematsu M, Koga T, Ishimori A, Saeki K, Ishii Y, Taketomi Y, et al. Leukotriene B4 receptor type 2 protects against pneumolysin-dependent acute lung injury. Sci Rep 2016;6:34560.
[201] Iwamoto S, Koga T, Ohba M, Okuno T, Koike M, Murakami A, et al. Non-steroidal anti-inflammatory drug delays corneal wound healing by reducing production of 12-hydroxyheptadecatrienoic acid, a ligand for leukotriene B4 receptor 2. Sci Rep 2017;7(1):13267.
[202] Ishii Y, Saeki K, Liu M, Sasaki F, Koga T, Kitajima K, et al. Leukotriene B4 receptor type 2 (BLT2) enhances skin barrier function by regulating tight junction proteins. FASEB J 2016;30(2):933–47.
[203] Leguina-Ruzzi A, Diban RO, Velarde V. The 12-HHT/BLT2/NO axis is associated to the wound healing and skin condition in different glycaemic states. Med Sci (Basal) 2019;7(4):65.
[204] Liu M, Shen J, Yuan H, Chen F, Song H, Qin H, et al. Leukotriene B4 receptor 2 regulates the proliferation, migration, and barrier integrity of bronchial epithelial cells. J Cell Physiol 2018;233(8):6117–24.
[205] Chiba T, Nakahara T, Hashimoto-Hachiya A, Yokomizo T, Uchi H, Furue M. The leukotriene B4 receptor BLT2 protects barrier function via actin polymerization with phosphorylation of myosin phosphatase target subunit 1 in human keratinocytes. Exp Dermatol 2016;25(7):532–6.
[206] Endara M, Masden D, Goldstein J, Gondek S, Steinberg J, Attinger C. The role of chronic and perioperative glucose management in high-risk surgical closures: a case for tighter glycemic control. Plast Reconstr Surg 2013;132(4):996–1004.
[207] Luo L, Tanaka R, Kanazawa S, Lu F, Hayashi A, Yokomizo T, et al. A synthetic leukotriene B4 receptor type 2 agonist accelerates the cutaneous wound healing process in diabetic rats by indirect stimulation of fibroblasts and direct stimulation of keratinocytes. J Diabetes Complications 2017;31(1):13–20.

[208] Matsunaga Y, Fukuyama S, Okuno T, Sasaki F, Matsunobu T, Asai Y, et al. Leukotriene B4 receptor BLT2 negatively regulates allergic airway eosinophilia. FASEB J 2013;27(8):3306–14.
[209] Cho KJ, Seo JM, Shin YH, Yoo MH, Park CS, Lee SH, et al. Blockade of airway inflammation and hyperresponsiveness by inhibition of BLT2, a low-affinity leukotriene B4 receptor. Am J Respir Cell Mol Biol 2010;42(3): 294–303.
[210] Tufvesson E, Nihlberg K, Westergren-Thorsson G, Bjermer L. Leukotriene receptors are differently expressed in fibroblast from peripheral versus central airways in asthmatics and healthy controls. Prostaglandins Leukot Essent Fatty Acids 2011;85(2):67–73.
[211] Alberca-Custódio RW, Greiffo FR, MacKenzie B, Oliveira-Junior MC, Andrade-Sousa AS, Graudenz GS, et al. Aerobic exercise reduces asthma phenotype by modulation of the leukotriene pathway. Front Immunol 2016;7:237.
[212] Cho KJ, Seo JM, Lee MG, Kim JH. BLT2 is upregulated in allergen-stimulated mast cells and mediates the synthesis of Th2 cytokines. J Immunol 2010;185(10):6329–37.
[213] Ro M, Lee A-J, Kim J-H. 5-/12-Lipoxygenase-linked cascade contributes to the IL-33-induced synthesis of IL-13 in mast cells, thus promoting asthma development. Allergy 2018;73(2):350–60.
[214] Assi M. The differential role of reactive oxygen species in early and late stages of cancer. Am J Physiol Regul Integr Comp Physiol 2017;313(6):R646–53.
[215] Lee JW, Kim JH. Activation of the leukotriene B4 receptor 2-reactive oxygen species (BLT2-ROS) cascade following detachment confers anoikis resistance in prostate cancer cells. J Biol Chem 2013;288(42):30054–63.
[216] Cho NK, Joo YC, Wei JD, Park JI, Kim JH. BLT2 is a pro-tumorigenic mediator during cancer progression and a therapeutic target for anti-cancer drug development. Am J Cancer Res 2013;3(4):347–55.
[217] Lee J-W, Kim G-Y, Kim J-H. Androgen receptor is up-regulated by a BLT2-linked pathway to contribute to prostate cancer progression. Biochem Biophys Res Commun 2012;420(2):428–33.
[218] Seo JM, Cho KJ, Kim EY, Choi MH, Chung BC, Kim JH. Up-regulation of BLT2 is critical for the survival of bladder cancer cells. Exp Mol Med 2011;43(3):129–37.
[219] Choi JA, Lee JW, Kim H, Kim EY, Seo JM, Ko J, et al. Pro-survival of estrogen receptor-negative breast cancer cells is regulated by a BLT2–reactive oxygen species-linked signaling pathway. Carcinogenesis 2009;31(4): 543–51.
[220] Seo J-M, Park S, Kim J-H. Leukotriene B4 receptor-2 promotes invasiveness and metastasis of ovarian cancer cells through signal transducer and activator of transcription 3 (STAT3)-dependent up-regulation of matrix metalloproteinase 2. J Biol Chem 2012;287(17):13840–9.
[221] Kim H, Choi JA, Park GS, Kim JH. BLT2 up-regulates interleukin-8 production and promotes the invasiveness of breast cancer cells. PLoS ONE 2012;7(11). e49186-e.
[222] Kim GY, Lee JW, Cho SH, Seo JM, Kim JH. Role of the low-affinity leukotriene B4 receptor BLT2 in VEGF-induced angiogenesis. Arterioscler Thromb Vac Biol 2009;29(6):915–20.
[223] Park J, Jang JH, Park GS, Chung Y, You HJ, Kim JH. BLT2, a leukotriene B4 receptor 2, as a novel prognostic biomarker of triple-negative breast cancer. BMB Rep 2018;51(8):373–7.
[224] Park J, Jang JH, Kim JH. Mediatory role of BLT2 in the proliferation of KRAS mutant colorectal cancer cells. Biochim Biophys Acta Mol Cell Res 2019;1866(3):329–36.
[225] Houthuijzen JM, Daenen LGM, Roodhart JML, Oosterom I, van Jaarsveld MTM, Govaert KM, et al. Lysophospholipids secreted by splenic macrophages induce chemotherapy resistance via interference with the DNA damage response. Nat Commun 2014;5:5275.
[226] Sakata K, Dahlén S-E, Bäck M. The contractile action of leukotriene B4 in the guinea-pig lung involves a vascular component. Br J Pharmacol 2004;141(3):449–56.
[227] Pace E, Ferraro M, Vincenzo SD, Bruno A, Giarratano A, Scafidi V, et al. Cigarette smoke increases BLT2 receptor functions in bronchial epithelial cells: in vitro and ex vivo evidence. Immunology 2013;139(2):245–55.
[228] Sánchez-Galán E, Gómez-Hernández A, Vidal C, Martín-Ventura JL, Blanco-Colio LM, Muñoz-García B, et al. Leukotriene B4 enhances the activity of nuclear factor-kappaB pathway through BLT1 and BLT2 receptors in atherosclerosis. Cardiovasc Res 2009;81(1):216–25.

CHAPTER

# 13

# The forkhead box O family in insulin action and lipid metabolism

*Sojin Lee*[a], *Cuiling Zhu*[a,b], *Jun Yamauchi*[a], *Ping Zhu*[a,c], *Xiaoyun Feng*[a,d], *Shen Qu*[b], *H. Henry Dong*[a,e]

[a]Division of Endocrinology and Diabetes, Department of Pediatrics, University of Pittsburgh School of Medicine, Pittsburgh, PA, United States; [b]Department of Endocrinology and Metabolism, Shanghai 10th People's Hospital, Tongji University School of Medicine, Shanghai, China; [c]Department of Endocrinology, Guangzhou Red Cross Hospital, Medical College of Jinan University, Guangzhou, China; [d]Department of Endocrinology and Metabolism, Shanghai General Hospital, Shanghai Jiaotong University, Shanghai, China; [e]Rangos Research Center, Children's Hospital of Pittsburgh of UPMC, Pittsburgh, PA, United States

OUTLINE



Lipid Signaling and Metabolism. http://dx.doi.org/10.1016/B978-0-12-819404-1.00013-0

## The forkhead box O family

The forkhead box O (FoxO) family consists of four proteins including FoxO1, FoxO3, FoxO4, and FoxO6 in mammalian cells [1,2]. Due to the conservation of the "forkhead box" (also known as "winged helix") domain in their protein structures, FoxO proteins act as transcription factors in regulating the expression of target genes, whose functions are instrumental for cell metabolism, growth, differentiation, oxidative stress, survival and apoptosis, autophagy, and aging in mammals [1]. These four FoxO isoforms are encoded by four distinct genes that are situated on different chromosomes, with FoxO1 on chromosome 13, FoxO3 on chromosome 6, FoxO4 on chromosome X, and FoxO6 on chromosome 1 in humans. This inherited difference in their chromosomal locations constitutes a difference in the chromatin context, in which the four FoxO members are differentially regulated. Indeed, the relative expression levels of FoxO1 proteins vary from cell type to cell type as well as from organ to organ, although all four FoxO isoforms are ubiquitously expressed in the body. For example, FoxO1, FoxO3, and FoxO4 are expressed across the brain [3,4]. In contrast, FoxO6 expression is confined to the hippocampus, a region that is responsible for learning and memory consolidation [5]. Despite the structural conservation in the forkhead box domain in the FoxO family, the four FoxO members have nonredundant functions, in keeping with their variable carboxyl domains with affinity of interacting with different cofactors [6]. Indeed, genetically modified mice with whole-body FoxO1 knockout die at embryonic day 11.5, due to defective angiogenesis and vasculogenesis in the embryos [7]. In contrast, mice with homozygous FoxO4 or FoxO6 knockout are viable [8,9]. It appears that FoxO1, as opposed to FoxO4 and FoxO6, are physiologically important for prenatal angiogenesis and vasculogenesis in mammals [7]. Although homozygous FoxO3 knockout mice are viable, female FoxO3 knockout mice are associated with infertility, due to abnormal ovarian follicular development [10]. Phylogenetic studies indicate that FoxO6 is evolutionarily divergent from the other three FoxO members [6]. Indeed, FoxO6 mediates insulin action in a mechanism that is distinct from other three FoxO members [11,12]. Unlike FoxO1, FoxO3, and FoxO4 proteins that undergo insulin-dependent nucleocytoplasmic trafficking, FoxO6 is localized permanently in the nucleus, regardless of insulin action [11,12]. Clearly, the four FoxO members have evolved to possess nonoverlapping functions by targeting different genes, as they differ in tissue distribution, despite sharing some structural similarity and functional redundancy [1]. In this chapter, we focus our review on FoxO1 in integrating insulin and glucagon signaling to hepatic glucose and lipid metabolism, and its contribution to hyperglycemia and hyperlipidemia—the dual pathogenesis in obesity and type 2 diabetes. We will also review the role of FoxO1 in regulating macrophage activation and migration, and the contribution of macrophage FoxO1 deregulation to hepatic inflammation and nonalcoholic fatty liver disease (NAFLD). We will provide mechanistic insights into posttranslational modification of FoxO1 activity in response to nutritional and hormonal cues. Furthermore, we will review clinical evidence that genetic mutations in FoxO1 or altered FoxO1 expression are associated with NAFLD and type 2 diabetes in humans. Finally, we will evaluate whether FoxO1 is a therapeutic target for improving glucose and lipid metabolism in insulin-resistant subjects with metabolic diseases.

## FoxO1 mediates the inhibitory action of insulin or IGF-1 in cells

FoxO1 is the most characterized member in the FoxO subfamily. FoxO1 acts as a substrate of protein kinase B (PKB, also known as Akt) and serum/glucocorticoid-induced kinase (SGK) to mediate the inhibitory action of insulin or IGF-1 on target gene expression in cells. Insulin (or IGF-1) exerts its inhibitory effect on gene expression via a highly conserved sequence (TG/ATTTT/G), termed insulin response element (IRE) in the promoters of target genes [1]. Crystallography studies show that FoxO1 binds as a monomer via its forkhead box domain to the IRE DNA with amino acid residues $Asn^{211}$, $Ser^{212}$, $Arg^{214}$, $His^{215}$, $Asn^{216}$, and $Ser^{218}$ in the major groove of the DNA helix [13]. This effect is amplified by PGC-1$\alpha$ that acts as a coactivator to augment FoxO1 activity [14] and is inhibited by insulin (or IGF-1). FoxO1 activity is tightly controlled by insulin (or IGF-1) in cells. In response to reduced insulin (or IGF-1) action in fasting states, FoxO1 is confined to the nucleus via the characteristic nuclear localization signal (NLS) within its forkhead box domain. FoxO1 binds to the IRE DNA motif in target promoters to enhance target gene expression. In response to increased insulin (or IGF-1) action in fed states, FoxO1 undergoes Akt-dependent phosphorylation at amino acid residues Thr24, Ser256, and Ser319, resulting in FoxO1 nuclear exclusion. This effect serves as an acute mechanism by which insulin (or IGF-1) disables FoxO1 activity by precluding its binding to chromatin DNA in the nucleus, resulting in immediate inhibition of target gene expression at the transcription level [1].

To gain insights into the mechanism by which FoxO1 undergoes insulin-stimulated nucleocytoplasmic trafficking, Kim et al. [11] show that FoxO1 via its nuclear exit signal (NES) forms a complex with the chromosomal maintenance 1 (CRM-1), a chaperone known as exportin-1 that is responsible for binding to the NES motif of a cargo protein and transporting the cargo protein from the nucleus to the cytoplasm [15,16]. In response to insulin stimulation, phosphorylated FoxO1 proteins in complex with CRM-1 are translocated from the nucleus to cytoplasm in cells [11]. Therefore, CRM-1 plays a pivotal role in facilitating insulin-dependent FoxO1 nucleocytoplasmic shuttling (Fig. 13.1). It remains to be determined whether CRM-1 also mediates insulin-stimulated subcellular redistribution of FoxO3 and FoxO4, two other FoxO members with the characteristic NES motif in their carboxyl domains.

In contrast, FoxO6 does not interact with CRM-1, due to the lack of a consensus NES motif in FoxO6 protein. As a result, FoxO6 cannot undergo insulin-mediated nucleocytoplasmic shuttling, remaining permanently in the nucleus [5,11]. This raises a fundamental question as to how FoxO6 mediates the inhibitory effect of insulin on target gene expression. To answer this question, Kim and colleagues [11] show that FoxO6 harbors a consensus 14-3-3 binding motif ($^{23}$RSCTWP$^{28}$) within its forkhead box domain. FoxO6, when phosphorylated by Akt in response to insulin, interacts with 14-3-3, a scaffolding protein that sequesters FoxO6. This effect disables the ability of FoxO6 to bind to target promoters. Their studies support the idea that insulin inhibits FoxO6 activity in the nucleus [11]. In keeping with this idea, Tsai et al. [17] used a site-directed mutagenesis approach to alter the NES motif in FoxO1. As a result, the mutant FoxO1 is unable to undergo insulin-stimulated nucleocytoplasmic trafficking, while still retaining the ability to undergo Akt-mediated phosphorylation. Interestingly, the mutant FoxO1 is still sensitive to insulin inhibition, congruent with the idea that insulin inhibition of FoxO1 activity can take place in the nucleus without necessarily altering FoxO1 subcellular redistribution. When

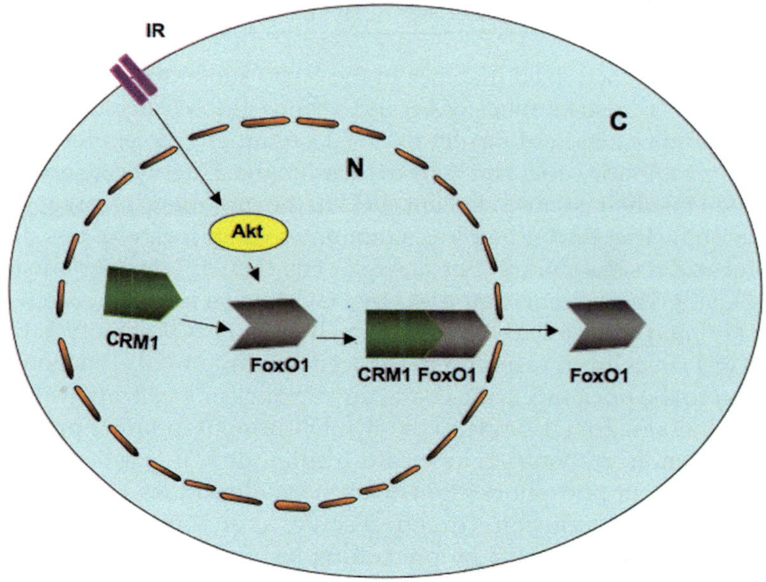


**FIGURE 13.1** **Mechanism of FoxO1 nucleocytoplasmic trafficking.** In response to insulin, FoxO1 is phosphorylated by Akt. Phosphorylated FoxO1 proteins are recognized by CRM1, an exportin that binds to FoxO1 and transports FoxO1 from the nucleus to the cytoplasm. This action results in insulin-dependent inactivation of FoxO1 activity in cells. IR, insulin receptor. *C*, cytoplasm; *N*, nucleus.

phosphorylated in response to insulin, FoxO1 forms a complex with protein 14-3-3 in the nucleus, and this effect prevents FoxO1 binding to target promoters [1,2]. Taken together, these results suggest that the compartmental effect, resulting from insulin-stimulated nuclear exclusion of FoxO1, is not a prerequisite for insulin-mediated inhibition of FoxO1 activity in cells.

What is the fate of FoxO1 proteins after their translocation from the nucleus to the cytoplasm? A prevailing view is that phosphorylated FoxO1 proteins are targeted for ubiquitination, followed by proteosome-mediated degradation in the cytoplasm [18–20]. However, this view has been challenged by Zhao et al. [21], who show that cytosolic FoxO1 proteins are not destined for proteosome-mediated degradation. Instead, cytosolic FoxO1 mediates autophagy in the cytoplasm via its interaction with autophagy-related protein 7 (Atg7) in cells in response to oxidative stress. This effect promotes autophagy-mediated cell death in cancer cells [21]. This observation illustrates a facet of cytosolic FoxO1 function for the induction of autophagy, independently of its transcriptional activity in cancer cells. Nonetheless, it remains to be determined whether this cytosolic FoxO1-mediated autophagy is a general mechanism in noncancer cell types.

## FoxO1 mediates the stimulatory action of glucagon in cells

FoxO1 is regarded as a nutrient sensor. This is exemplified in *Caenorhabditis elegans*, in which the FoxO ortholog Daf16 is markedly upregulated to maintain energy homeostasis for survival in response to food deprivation [22,23]. Likewise, hepatic FoxO1 expression along with its activity is significantly increased in the liver in response to fasting [24]. This effect has been viewed as a mechanism for priming the liver to undergo gluconeogenesis in the fasting state [24,25]. To shed light on the underlying mechanism, Wu et al. [26] show that FoxO1 acts as a substrate of PKA to mediate the stimulatory effect of glucagon on target genes in the liver. In response to glucagon (via cAMP), FoxO1 is phosphorylated by PKA at the amino acid Ser276. This effect promotes FoxO1 nuclear localization and enhances FoxO1 activity in stimulating hepatic expression of PEPCK and G6Pase—two key enzymes in the gluconeogenesis pathway. Genetically modified mice with S276A or S276D allele-specific knock-in are associated with abnormal glucose metabolism, due in part to the reduced responsiveness of the liver with altered FoxO1 activity to glucagon signaling [26]. Diabetic *db/db* mice are associated with abnormally higher basal FoxO1-S276 phosphorylation levels in the liver, coinciding with the prevailing hyperglycemia and hyperglucagonemia [26]. These suggest that glucagon signaling through FoxO1 serves as a counter-regulatory mechanism to promote hepatic gluconeogenesis for maintaining glucose homeostasis under fasting conditions (Fig. 13.2). Consistent with this interpretation, Altomonte and coworkers report that hepatic FoxO1 expression along with its activity is markedly regulated in insulin deficient and insulin-resistant

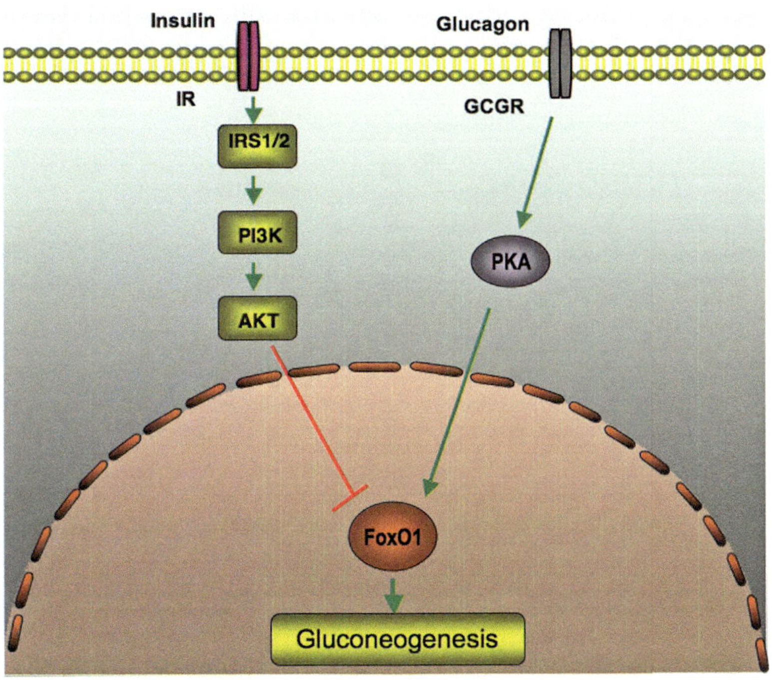


FIGURE 13.2 **Hepatic FoxO1 activity is reciprocally regulated by insulin and glucagon.** Insulin signaling through AKT inhibits FoxO1 activity and this effect acts to suppress gluconeogenesis in the liver in fed states. In contrast, glucagon signaling via PKA augments FoxO1 activity and the resulting action serves to stimulate gluconeogenesis in the liver in fasting states. *GCGR*, glucagon receptor; *IR*, insulin receptor; *IRS1/2*, insulin receptor substrates 1 and 2.

livers of mice [27,28]. Significantly higher FoxO1 expression is also detected in the livers of humans with metabolic disease [29].

## Hepatic FoxO1 expression is regulated by a feedback mechanism

In addition to acute regulation of FoxO1 activity by insulin and glucagon, FoxO1 is subject to feedback regulation at the transcriptional level in the liver (Fig. 13.3). This feedback mechanism is first reported in *Drosophila*, in which dFoxO, the *Drosophila* FoxO1 ortholog, is regulated to promote the expression of its upstream effector gene encoding insulin receptor (dInR) in response to nutrient deprivation [30]. Likewise, Kamagate et al. show that increased FoxO1 activity augments hepatic expression of insulin receptor (Ir) and insulin receptor substrate 2 (Irs2), which in turn inhibit FoxO1 activity in response to attenuated insulin action in the liver [31]. To address the underlying physiology, Kamagate et al. [31] used an adenovirus-mediated gene transfer approach to achieve hepatic production of FoxO1-ADA, a constitutively active FoxO1 mutant that is refractory to insulin inhibition. As expected, hepatic FoxO1-ADA production resulted in marked induction of IR and IRS2. Mice with hepatic FoxO1-ADA production exhibited near complete depletion of hepatic glycogen content, due in part to unrestrained FoxO1 activity in promoting glycogenolysis in the liver. Hepatic FoxO1-ADA production also provoked profound endoplasm reticulum (ER) stress [31]. These results, which are totally unexpected, illustrate the physiological significance of the IR/IRS2-dependent

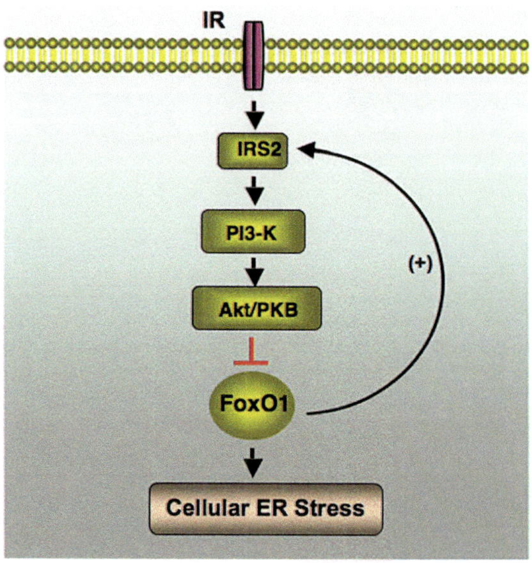


FIGURE 13.3 **FoxO1 is subject to Feedback regulation.** FoxO1 activity is regulated via a feedback loop, in which FoxO1 stimulates hepatic expression of IRS2, whose activity in turn suppresses FoxO1 activity in cells. Such a feedback mechanism is critical for keeping FoxO1 activity in check to prevent the deleterious effect of unbridled FoxO1 activity on cells.

feedback loop for tight regulation of hepatic FoxO1 activity in response to nutrient availability or insulin action. It implies that unchecked FoxO1 activity, resulting from a dislodged FoxO1 feedback loop in insulin-resistant liver, is deleterious to hepatic metabolism, contributing to excessive glycogen breakdown and undue ER stress in the liver.

Such a FoxO1 feedback mechanism is conserved in pancreatic β-cells, in which FoxO1 stimulates IR and IRS2 expression, which in turn suppresses FoxO1 expression [31,32]. As excessive ER stress, caused by FoxO1 overproduction, is associated with β-cell dysfunction and apoptosis [33], it is likely that the IR/IRS2-dependent feedback loop acts to keep FoxO1 activity in check to prevent ER stress for maintaining β-cell function. Indeed, it has been reported that unchecked FoxO1 activity, resulting from the constitutively active FoxO1-ADA allele, is associated with ER stress and metabolic diapause, contributing to abnormal glucose-stimulated insulin secretion profiles in β-cells [31,33,34].

## FoxO1 *trans*-activation versus *trans*-repression mechanism

It is well characterized that FoxO1 functions as a *trans*-activator to enhance its target gene expression via selective binding to its consensus sites, known as the IRE motif, within target promoters. Aside from its *trans*-activation mechanism, FoxO1 acts as a *trans*-repressor to suppress gene expression in a *trans*-repression mechanism. For example, FoxO1 is shown to interact with PPAR-γ, the resulting effect of which contributes to the inhibition of PPAR-γ activity, and this action is counteracted by insulin in adipocytes [35,36]. A similar trans-repression mechanism is seen in HEK293 cells, in which FoxO1 and PPAR-γ physically interact and functionally antagonize each other in a reciprocal manner [37]. Likewise, FoxO1 associates with PPAR-α protein and counteracts PAPR-α activity in the liver [38]. FoxO1 is also shown to compete with carbohydrate response element-binding protein (ChREBP), contributing to the inhibitory effect of FoxO1 on the expression of ChREBP-targeted genes in β-cells [39]. More recently, Langlet et al. [40] report that FoxO1 binds to Sin3a protein, a corepressor in the nucleus, and this action results in the induction of glucokinase (GK) expression in hepatocytes. This effect contributes to postprandial glucose handling in the liver. Thus, FoxO1 can act as *trans*-activator or *trans*-repressor in modulating the expression of its target genes depending on gene contexts and physiological cues [25,35,37–40].

## FoxO1 in gluconeogenesis and its contribution to hyperglycemia in diabetes

FoxO1 is a key regulator of hepatic gluconeogenesis by regulating hepatic expression of PEPCK and G6Pase—two molecular targets of FoxO1 in the liver. In the fasting state, FoxO1 targets PEPCK and G6Pase genes for *trans*-activation, and this effect promotes gluconeogenesis in the liver [14]. Consistent with FoxO1 action, both the PEPCK and G6Pase promoters contain consensus IRE motif for FoxO1 binding [1,2,41]. As a result, FoxO1 gain-of-function in the liver is associated with increased hepatic gluconeogenesis in mice [24,42]. In contrast, FoxO1 loss-of-function in the liver is associated with decreased hepatic gluconeogenesis in mice [27]. Further physiological underpinning for the role of FoxO1 in gluconeogenesis is the observation that genetically modified mice with liver-conditional FoxO1 depletion develop hypoglycemia in response to a prolonged starvation [43]. In fasted liver, FoxO1 is shown to form a complex with PGC-1α, a coactivator that functions to amplify the stimulatory effect of FoxO1 on hepatic PEPCK and G6Pase expression [14].

Interestingly, hepatic FoxO1 activity is differentially regulated by insulin and glucagon—two counter-regulatory hormones that are instrumental for controlling gluconeogenesis in the liver [44–48]. Insulin acts via Akt to inhibit FoxO1 activity, whereas glucagon acts via PKA to enhance FoxO1 activity [1,2,26]. These two opposing signaling pathways converge at FoxO1 to adjust the rate of hepatic glucose production in response to metabolic shift from fasting to refeeding states (Fig. 13.2). It is noteworthy that apart from FoxO1, both FoxO3 and FoxO4 play redundant roles in regulating hepatic gluconeogenesis in mice [49]. There is also evidence that hepatic FoxO6 activity is upregulated in fasted liver to compensate for the loss of FoxO1 function in mediating insulin action on hepatic gluconeogenesis [11,12]. These data illustrate the functional redundancy of the four FoxO isoforms in integrating nutrient and hormonal signaling to gluconeogenesis in the liver.

What is the clinical significance of these findings? While excessive glucose production in the liver is a major contributing factor for the development of fasting hyperglycemia in both type 1 and type 2 diabetes, the underlying mechanism remains incompletely characterized. It is clear that the former is secondary to insulin deficiency and the latter is due to insulin resistance. Of striking significance, abnormally higher FoxO1 activity is detectable in insulin-deficient as well as insulin-resistant liver, coinciding with the development of fasting hyperglycemia in type 1 and type 2 diabetes, respectively [24,27–29,38,50]. Thus, both type 1 and type 2 diabetes share a common pathological denominator of FoxO1 deregulation in the liver. An important implication of these findings is that unchecked FoxO1 activity, resulting from either insulin deficiency or insulin resistance, contributes to hepatic over-expression of PEPCK and G6Pase—two FoxO1 targets whose functions are responsible for catalyzing the rate-limiting steps of hepatic gluconeogenesis [1,2]. This effect serves as a driving force for hepatic glucose overproduction, contributing to fasting hyperglycemia in diabetes. These data have inspired the idea of targeting FoxO1 for ameliorating fasting hyperglycemia in diabetes [51].

## FoxO1 in insulin regulation of hepatic MTP expression and VLDL production

The microsomal triglyceride transfer protein (MTP) is a molecular chaperone that is produced mainly in the liver and to lesser extent in the intestine [52–55]. MTP, when heterodimerized with its small subunit protein disulfide isomerase (PDI) in the endoplasmic reticulum, catalyzes the transfer of triglycerides and phospholipids to the nascent apolipoprotein B (ApoB) in a rate-limiting step for the assembly and maturation of VLDL-TG in the liver [52,56–60]. Similarly, MTP is essential for intestinal assembly and secretion of chylomicrons, TG-rich particles that are critical for the postprandial absorption of lipids and essential vitamins [61,62]. Human subjects with genetic MTP deficiency are associated with abetalipoproteinemia, a rare autosomal recessive disease, due to the defects in the assembly and secretion of TG-rich particles, and severe abnormalities in TG and vitamin metabolism [54,62–65]. Genetic studies indicate that human MTP polymorphism within its promoter at the site −493G/T is linked with altered MTP expression and abnormal lipoprotein metabolism, accounting in part for an increased risk of developing coronary heart disease [66,67]. There is clinical evidence that the deleterious effect of the MTP promoter −493G/T polymorphism on lipoprotein metabolism is strongly compounded by visceral adiposity in human subjects [68,69]. Likewise, preclinical studies show that mice with MTP haploinsufficiency are associated with reduced hepatic VLDL-TG production in $MTP^{+/-}$ heterozygous mice [65]. Conversely, hepatic MTP overproduction contributes to increased VLDL-TG secretion, raising fasting plasma TG levels in mice [70]. Given its critical importance in regulating VLDL-TG secretion, pharmacological inhibition of MTP is being investigated for improving lipoprotein metabolism in humans. Recent clinical trials show that the MTP inhibitor lomitapide is highly efficient in improving blood lipid profiles in patients with familiar hypercholesterolemia [71–73]. However, its long-term therapeutic benefit is associated with the accumulation of hepatic fat and impairment in fat-soluble vitamins and fatty acids absorption, raising the caution for continuous monitoring of hepatic fat accumulation in patients with long-term lomitapide treatment [74].

Interestingly, hepatic MTP expression is negatively regulated by insulin [75,76]. In insulin-resistant states, hepatic MTP is overproduced. This effect, along with increased influx of free fatty acids into the liver, promotes hepatic assembly and secretion of VLDL-TG in obesity and type 2 diabetes. Indeed, both hepatic and intestinal MTP mRNA levels are significantly elevated in insulin-resistant nondiabetic obese Zucker rats with impaired TG metabolism [77], consistent with the notion that insulin exerts an inhibitory action on hepatic MTP expression and VLDL-TG production [75,76]. To address the underlying mechanism, Kamagate and colleagues determined FoxO1-dependent regulation of MTP expression in hepatocytes in response to insulin [43]. They demonstrate that FoxO1 stimulates hepatic MTP expression via its selective binding to the IRE DNA motif within the MTP promoter, and this effect is inhibited by insulin, correlating with the ability of insulin to stimulate FoxO1 phosphorylation and promote its translocation from nucleus to the cytoplasm [1,2]. Site-directed mutagenesis of the IRE DNA motif results in the abrogation of both FoxO1-mediated stimulation and insulin-mediated inhibition of MTP promoter activity in hepatocytes. FoxO1 transgenic overexpression in the liver is associated with increased MTP expression and elevated hepatic VLDL production, rendering FoxO1-transgenic mice to developing fasting hypertriglyceridemia and lipid intolerance. Diabetic *db/db* mice exhibit increased MTP expression, along with increased FoxO1 activity in the liver. Furthermore, RNAi-mediated FoxO1 knockdown in the liver is capable of inhibiting hepatic MTP expression and suppressing VLDL overproduction, contributing to improved triglyceride metabolism in *db/db* mice as well as FoxO1-transgenic mice [43]. Together, these results underscore the importance of FoxO1 in integrating insulin signaling to hepatic regulation of MTP expression and VLDL secretion (Fig. 13.4). These results advance our understanding of the mechanism by which insulin suppresses VLDL-TG production in the liver. In response to postprandial insulin secretion, hepatic VLDL-TG production is suppressed to limit plasma triglyceride excursion in the fed state [78–81]. In contrast, hepatic VLDL-TG production along with increased FoxO1 activity is induced, resulting in increased secretion of VLDL-TG from the liver to other peripheral tissues for providing TG as a fuel source for energy metabolism in the fasting state. These findings also shed light on the close association of insulin resistance to hepatic VLDL-TG overproduction—a major contributing factor for the pathogenesis of fasting hyperlipidemia in obesity and type 2 diabetes [25,82–87].

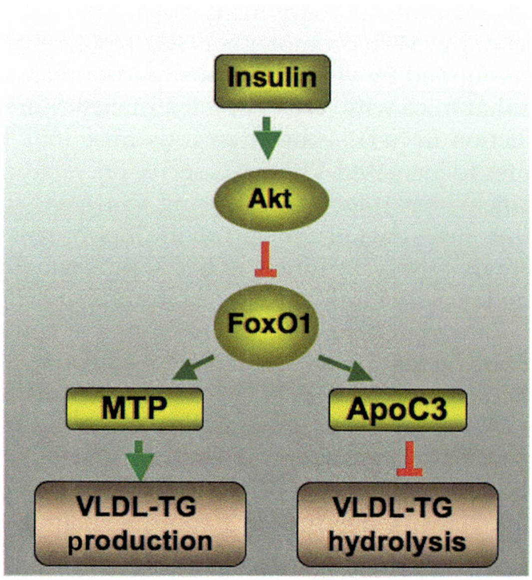


FIGURE 13.4 **FoxO1 regulation of hepatic MTP and ApoC3 expression.** FoxO1 mediates the inhibitory effect of insulin on hepatic expression of MTP and ApoC3, two key factors involved in triglyceride metabolism. Increased FoxO1 activity, resulting from insulin resistance, promotes hepatic MTP and ApoC3 overproduction. These effects contribute to increased production and decreased hydrolysis of triglyceride-rich lipoprotein particles, a major contributing factor for fasting hyperlipidemia in obesity and type 2 diabetes. *ApoC3*, apolipoprotein C3; *MTP*, microsomal triglyceride transfer protein.

## FoxO1 in hepatic ApoC3 production and its contribution to hyperlipidemia

In addition to its role in adjusting the rate of hepatic VLDL-TG production in response to insulin, FoxO1 plays a significant role in regulating hepatic expression of apolipoprotein C3 (ApoC3). Produced mainly in the liver and to a lesser extent in the intestine, ApoC3 is one of the most abundant apolipoproteins in the blood, where it is present as an exchangeable moiety between high-density lipoproteins (HDLs) and TG-rich particles such as VLDL and chylomicrons [88]. ApoC3 acts as an inhibitor of lipoprotein lipase (LPL) and hepatic lipase (HL) to retard the hydrolysis of TG in VLDL and chylomicrons in the post-absorption phase [89–91]. As a result, elevated ApoC3 levels tend to delay the systemic clearance of TG-rich lipoprotein remnants [92,93]. There is emerging evidence that ApoC3 seems to play a role in promoting VLDL-TG assembly and secretion from the liver [94–97]. Furthermore, ApoC3 negatively impacts apolipoprotein E (ApoE)-mediated uptake of TG-rich remnants in the liver. This effect, which is independent of hepatic low-density lipoprotein receptor (LDLR), contributes to the accumulation of TG-rich particles in the blood [98]. Both clinical and preclinical studies indicate that elevated plasma APOC3 levels are inversely correlated with blood TG metabolism. For example, ApoC3 transgenic overproduction impairs TG catabolism, resulting in the premature development of hypertriglyceridemia in ApoC3-transgenic mice [99]. ApoC3 gene knockout improves systemic TG hydrolysis and clearance, lowering plasma TG levels in ApoC3-knockout mice [100–102]. Human subjects harboring genetic APOC3 mutations with reduced APOC3 production are associated with decreased plasma TG levels and reduced risk of coronary artery disease [103–105]. Antisense oligonucleotide-mediated reduction of plasma APOC3 levels results in significantly lower plasma TG levels in both nonhuman primates and patients with familial chylomicronemia [106,107]. Likewise, monoclonal anti-ApoC3 antibody-mediated ApoC3 clearance contributes to significantly reduced plasma levels of TG-rich lipoproteins in mice with hypertriglyceridemia [108]. These clinical and preclinical data lend solid support for the notion that abnormally high APOC3 production is a significant contributing factor for hypertriglyceridemia—an independent risk factor for cardiovascular diseases. This has fueled the development of APOC3 antagonism for ameliorating hypertriglyceridemia to reduce the cardiovascular risk in patients [109,110].

There is anecdotal evidence that human subjects harboring genetic variants (C-482T or T-455C) in the APOC3 promoter with increased APOC3 production are at increased risk of developing NAFLD [111]. However, this finding was contradicted by three independent clinical studies showing that the two common APOC3 mutant alleles (C-482T and T-455C) do not confer a significant risk for NAFLD in humans [112–114]. A recent preclinical study indicates that ApoC3 overproduction neither affects the onset nor the progression of NAFLD in ApoC3-transgenic mice, despite prolonged feeding on a high fat diet [115]. These data suggest that APOC3, whose overproduction is responsible for the pathogenesis of hypertriglyceridemia, is not a causative factor for NAFLD.

Hepatic ApoC3 expression is negatively regulated by insulin [28]. This effect correlates with the presence of the IRE in the promoter of the ApoC3 gene. Indeed, human subjects with homozygous IRE mutant alleles (−455T>C and −482C>T) within the APOC3 promoter have impaired glucose homeostasis and abnormal lipoprotein profiles, independently of gender [112,116,117]. Both insulin deficiency and insulin resistance contribute to ApoC3 overproduction, accounting in part for the development of diabetic dyslipidemia in type 1 and type 2 diabetes [118,119]. These data have led to the hypothesis that the ApoC3 gene is a FoxO1 target. To address this hypothesis, Altomonte and coworkers show that FoxO1 binds to the IRE in the ApoC3 promoter and this effect translates into a stimulatory effect on ApoC3 promoter activity in hepatocytes [28]. FoxO1 also mediates the inhibitory effect of insulin on intestinal ApoC3 expression. As a result, FoxO1-transgenic mice with a constitutively active FoxO1 allele have increased ApoC3 production, contributing to impaired postprandial clearance of TG-rich particles [28]. Hepatic ApoC3 expression along with increased FoxO1 activity becomes upregulated in the liver, contributing to the induction of hypertriglyceridemia in high fructose-fed hamsters [38,120]. Fenofibrate treatment, which corrects hepatic FoxO1 dysregulation, lowers plasma ApoC3 levels, and ameliorates hypertriglyceridemia in high fructose-fed hamsters [38]. Together, these data suggest that the insulin-FoxO1-ApoC3 signaling cascade plays a pivotal role in regulating TG-rich lipoprotein catabolism (Fig. 13.4).

## FoxO1 in hepatic lipogenesis and steatosis

Although FoxO1 contributes to hepatic regulation of lipid metabolism, there is a lack of consensus regarding the role of FoxO1 in hepatic lipogenesis and steatosis. Both gain and loss of FoxO1 function are associated with excessive fat accumulation in the liver. Matsumoto and colleagues [121] report that hepatic lipogenic genes including sterol regulatory element binding protein (SREBP-1c), fatty acid synthase (FAS), and acetyl-CoA carboxylase (ACC) are upregulated in FoxO1-transgenic mice with steatosis [121]. Qu and co-workers show that adenovirus-mediated FoxO1 production results in augmented lipogenesis with concomitant fat accumulation in the liver [24,121]. Furthermore, unchecked FoxO1 activity is detectable in insulin-resistant liver, correlating with the development of steatosis in dietary obese mice and diabetic *db/db* mice [27,28]. Similar results were reproducible in high fructose-fed hamsters [38]. There is clinical evidence that patients with NAFLD are associated with enhanced FoxO1 activity in the liver [29]. These results argue for the idea that FoxO1 gain-of-function is a predisposing factor for NAFLD.

However, this idea is contradicted by the finding that hepatic depletion of FoxO1, FoxO3, and FoxO4 proteins in combination results in severe steatosis in mice on a high fat and high cholesterol diet [122,123]. Although the underlying mechanism remains to be determined, one plausible interpretation is that FoxO1 loss-of-function impairs hepatic VLDL-TG secretion, contributing to the accumulation of fat in the liver in mice with genetic depletion of FoxO1/3/4 proteins. Indeed, FoxO1 signaling through MTP plays an important role for regulating hepatic VLDL-TG production [124]. More recently, Langlet et al. [40] report that FoxO1 inhibits glucokinase (GK) mRNA expression in the liver. This finding leads to the postulation that FoxO1 deficiency would enhance GK activity and increase glucose influx into the liver, contributing to the development of steatosis [125]. The physiological significance of this finding is unclear, as GK is activated at the post-translational level in the postprandial phase [126,127], in which FoxO1 is inactivated by insulin-stimulated phosphorylation and nuclear exclusion in hepatocytes [1]. Furthermore, this finding is at variance with previous observations that FoxO1 deficiency, resulting from hepatic expression of its dominant-negative allele or antisense oligonucleotide-mediated FoxO1 knockdown in liver, is associated with improved insulin action and hepatic lipid metabolism in insulin-resistant obese and diabetic mice [27,128]. FoxO1 haploinsufficiency protects against fat diet-induced insulin resistance and steatosis in FoxO1 heterozygous knockout mice [129]. FoxO1 deficiency also rescues the diabetic phenotype in insulin receptor substrate 2 (Irs2)-deficient diabetic mice [130]. Furthermore, liver-specific FoxO1 knockout results in near normalization of the metabolic disorders in insulin-resistant mice with liver-conditional ablation of Irs1 and Irs2 [131]. Given the fact that FoxO1 does not directly target the lipogenesis pathway for *trans*-activation [132–134], the impact of hepatocyte-expressed FoxO1 on lipogenesis and its contribution to steatosis is likely mediated by other mechanisms, which warrants further investigation.

## FoxO1 in fatty acid oxidation and its contribution to steatosis

Impaired hepatic lipid oxidation is a significant contributing factor for the pathogenesis of NAFLD. Anti-hypertriglyceridemic agents such as fibrates, which enhance fatty acid oxidation, are shown to improve lipid metabolism and ameliorate NAFLD [38,135,136]. To date, the role of FoxO1 in hepatic fatty acid oxidation remains poorly characterized. Qu et al. [38] show that FoxO1 physically binds to PPAR-α and functionally inhibits hepatic PPAR-α activity via a *trans*-repression mechanism. This effect contributes to the suppressive effect of FoxO1 on hepatic fatty acid oxidation [38]. As PPAR-α is the master regulator of fatty acid oxidation, these data suggest that FoxO1 functions as an inhibitor of fatty acid oxidation. Consistent with these interpretation, increased FoxO1 activity is detected along with impaired fatty acid oxidation in insulin-resistant liver of obesity and type 2 diabetes [24,28,31,38,43].

However, Zhang et al. report [137] that FoxO1 stimulates hepatic expression of adipose triacylglycerol lipase (ATGL) and suppresses hepatic expression of the G0/G1 switch gene 2 (G0G2)—the inhibitor of ATGL, a cytoplasmic enzyme that catalyzes the first step of lipolysis. This effect contributes to the induction of hepatic fatty acid oxidation [137]. FoxO1 is also shown to stimulate the expression of ATGL in adipose tissue and promote lipolysis [138]. These findings have led to the idea that ATGL-dependent lipolysis contributes to the stimulatory effect of FoxO1 on fatty acid oxidation in the liver [137]. Nonetheless, the underlying physiological significance of these findings is unclear. Despite FoxO1-mediated induction of ATGL activity, hepatic expression of the key enzymes including PPAR-α and its downstream targets such as CPT1, LCAD, ACOX1, and ACOT1 involved in fatty acid oxidation remain unchanged in response to FoxO1 overproduction [137]. Furthermore, FoxO1 gain-of-function is associated with impaired fatty acid oxidation and increased fat accumulation in the liver [24,29,38,121]. Further research is needed to clarify the role of FoxO1 in fatty acid oxidation and determine its ensuing impact on hepatic lipid metabolism.

## FoxO1 in macrophage activation and its contribution to hepatic inflammation and NAFLD

NAFLD is characterized by steatosis, affecting about 30% of the worldwide population. Its prevalence is increasing along with the rising epidemic of obesity in both adults and children [139–145]. About 25% of patients with steatosis evolve to nonalcoholic steatohepatitis (NASH), a major risk factor for fibrosis, cirrhosis and hepatocellular cancer [141,146–148]. Nevertheless, the precise cause and mechanism of NALFD remains poorly understood [140,149–151]. There is emerging evidence that Kupffer cells play a significant role in the pathogenesis of NAFLD [140,149–152]. Kupffer cells are resident macrophages in liver, constituting about 10% of total liver cell population [153]. In response to inflammatory stimuli such as lipopolysaccharide (LPS) or increased fatty acid influx into the liver, Kupffer cells are activated to secrete inflammatory cytokines [153–155]. This effect dampens insulin sensitivity in hepatocytes, highlighting the importance of Kupffer cells in hepatic inflammation and insulin resistance. Consistent with these findings, several independent studies show that chemical depletion of Kupffer cells reduces hepatic inflammation and improves insulin sensitivity in rodents [156–158]. There are clinical data indicating that patients with NASH are associated with hepatic inflammation and oxidative stress [159–163]. These results have led to the search for genetic factors culpable for linking insulin resistance to inflammatory cytokine production in Kupffer cells. It is anticipated that such factors play a key role in catalyzing the transition from steatosis to NASH by promoting hepatic inflammation and instigating hepatic insulin resistance.

To decipher the molecular linkage between inflammatory cytokine production and insulin resistance, Su et al. [164] determined the effect of FoxO1 on cytokine production profiles in normal and diabetic mice, demonstrating that increased FoxO1 activity, resulting from insulin resistance, contributes to macrophage overproduction of IL-1β, an inflammatory cytokine that is deleterious to hepatic glucose and lipid metabolism. To gain mechanistic insight into

FoxO1-mediated induction of IL-1β overproduction in macrophages, Su et al. [164] show that FoxO1 binds to the IL-1β promoter to stimulate IL-1β promoter activity in cultured macrophages. This effect contributes to increased macrophage IL-1β expression, which is counteracted by insulin. Mutations of the FoxO1 consensus DNA binding motif within IL-1β promoter abrogate FoxO1-mediated induction of IL-1β expression. Under inflammatory conditions, FoxO1 expression is upregulated, contributing to increased production of IL-1β protein in macrophages. Increased FoxO1 activity along with marked upregulation of IL-1β production is seen in peritoneal macrophages of wild-type C57BL/6 mice with prior exposure to LPS or diabetic *db/db* mice. These results characterize FoxO1 as a significant transcription factor that integrates insulin resistance to increased macrophage IL-1β production, contributing to the induction of hepatic inflammation in obesity and type 2 diabetes. In keeping with these findings, Fan et al. [165] show that FoxO1 potentiates inflammation by promoting TLR4 signaling in macrophages. Chung et al. [166] report that FoxO1 regulates allergic asthmatic inflammation by regulating macrophage polarization toward the inflammatory M1 state in alveolar macrophages. Furthermore, Wang and colleagues show that FoxO1 is critical for regulating macrophage polarization to maintain a competent T-cell immune response against Staphylococcus aureus infection [167]. Taken together, these results are consistent with the idea that FoxO1 couples insulin resistance to abnormal macrophage activation, and this effect contributes to the inflammatory cytokine profiles in obesity and type 2 diabetes.

However, there are preclinical observations against this idea. Tsuchiya and colleagues report that myeloid-conditional depletion of FoxO1, FoxO3, and FoxO4 in combination seems to stimulate myeloid cell proliferation and cause oxidative stress in peripheral tissues, aggravating the development of atherosclerosis in LDLR-deficient mice [168]. Kawano et al. [169] show that myeloid depletion of Pdk1-FoxO1 signaling predisposes to adipose tissue inflammation and insulin resistance in mice, indicating that macrophage FoxO1 may play a role in anti-inflammation. Furthermore, FoxO1 is shown to promote anti-inflammatory cytokine IL-10 expression in LPS-stimulated macrophages in culture [170], although this in vitro finding contradicts the literature data that LPS elicits inflammation and perpetuates insulin resistance in vitro and in vivo [164,171–174]. Given the controversy regarding the role of FoxO1 in macrophages, it remains unclear whether FoxO1 in tissue macrophages argues for or protects against tissue inflammation. Further studies are warranted to resolve the controversy and elucidate the physiological role of FoxO1 in macrophages and determine FoxO1 contribution to hepatic inflammation and NAFLD in obesity and type 2 diabetes.

## Association of FoxO polymorphism with metabolic disease and aging

Mussig et al. [175] report that human subjects harboring genetic FOXO1 variants are associated with β-cell dysfunction, impaired glucose tolerance, and type 2 diabetes, independently of the differences in gender. Similar results were obtained by Muller et al., who show that genetic FOXO1 mutations are associated with an increased risk of obesity and type 2 diabetes in Pima Indians [176]. FOXO1 polymorphism is also associated with increased susceptibility to developing diabetic nephropathy [177]. Likewise, FOXO1 is independently associated with the risk of developing type 2 diabetes in a Chinese Han population [178,179]. A caveat of these clinical studies is the lack of determination of FOXO1 expression or activity associated with its polymorphism. Therefore, it remains an open question as to whether FOXO1 gain or loss of function is associated with the risk of type 2 diabetes in humans. On the other hand, there are clinical data showing that the close association of FOXO1 polymorphism with the risk of type 2 diabetes is not apparent in other ethical groups including Caucasian and African-Americans [180].

Clinical studies did not reveal an association between genetic FOXO3 variants and the risk of type 2 diabetes in humans [181]. Instead, FOXO3 polymorphism is consistently correlated with associated with longevity in multiple ethical groups including the ethnic Japanese from Hawaii [182], Italian [183], German [184], Chinese [185,186], American [187], Danish [188], and Jewish [189]. In contrast, there is a lack of association between FOXO1, FOXO4, or FOXO6 and life expectancy in humans [190]. Further physiological underpinning for the role of FOXO3 in longevity is derived from preclinical studies showing that FoxO3 is functionally involved in caloric restriction-mediated lifespan extension in mice [191]. In addition, transgenic FoxO1 overproduction in skeletal muscle has no impact on lifespan extension in mice [192]. Together, these clinical and preclinical studies reveal a distinct role for FOXO3 in the regulation of longevity in both humans and rodents.

## Targeted FoxO1 inhibition for treating metabolic diseases

Hepatic insulin signaling bifurcates at FoxO1 to govern two key metabolic pathways in the liver, glucose production and VLDL-TG secretion [25,193]. This effect helps synchronize hepatic insulin signaling to simultaneously adjust the rates of both hepatic glucose and VLDL production in response to nutrient availability. This FoxO1-dependent mechanism is pivotal for the liver to rapidly adapt to metabolic shift between fasting to feeding states for maintaining normal glucose and lipid homeostasis. However, FoxO1 activity becomes unchecked in obesity and type 2 diabetes, due to the inability of FoxO1 to undergo insulin-stimulated phosphorylation and nucleocytoplasmic trafficking. As a result, FoxO1 is permanently localized in the nucleus, accounting for its constitutive activity in promoting hepatic glucose and VLDL-TG overproduction, contributing to the simultaneous induction of hyperglycemia and hyperlipidemia in obesity and type 2 diabetes [28,43]. This finding is reproducible in multiple animal models and patients with metabolic disease [24,27–29,38,42,43,50,82,121]. Moreover, genetic FoxO1 variants are associated with glucose and lipid disorders, and increased risk of type 2 diabetes in humans [175,176]. These clinical and preclinical data have inspired the idea of inhibiting FoxO1 activity in insulin-resistant liver for treating metabolic disease. It is anticipated that pharmacological FoxO1 inhibition would achieve dual therapeutic effects for correcting hyperglycemia and hyperlipidemia in obesity and type 2 diabetes. Meanwhile, there is evidence that abnormally higher FoxO1 activity, resulting from insulin resistance, promotes proinflammatory cytokine production and aggravates low-grade inflammation, a key pathological component in obesity and type 2 diabetes [164]. It would be of interest to determine whether pharmacological FoxO1 inhibition is sufficient to suppress low-grade inflammation and improve insulin sensitivity, translating into beneficial effects on glucose and lipid metabolism in obesity and type 2 diabetes (Fig. 13.5).

Is FoxO1 a potential therapeutic target for metabolic disease? To answer this question, Altomonte et al. [28] employed an adenovirus-mediated gene transfer approach to achieve

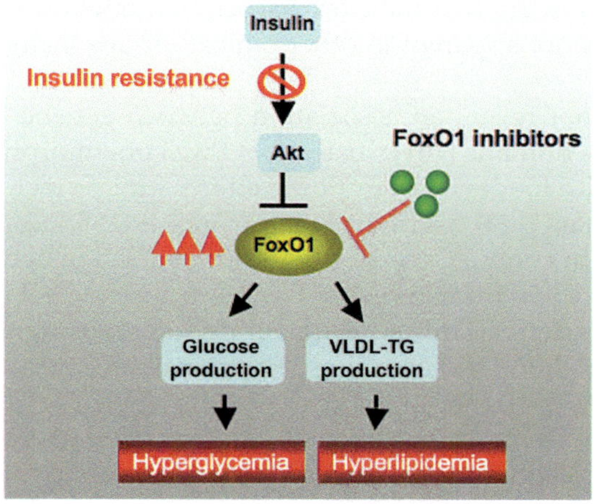


**FIGURE 13.5 Pharmacological FoxO1 inhibition for improving glucose and lipid metabolism.** Insulin signaling bifurcates at FoxO1 to regulate both gluconeogenesis and VLDL-TG production in the liver. This mechanism is crucial for maintaining glucose and lipid homeostasis in a normal physiological state. Under pathological conditions, unchecked FoxO1 activity, resulting from insulin resistance, promotes hepatic overproduction of both glucose and VLDL-TG, contributing to the dual pathogenesis of hyperglycemia and hyperlipidemia in obesity and type 2 diabetes. Pharmacological FoxO1 inhibition with small molecule drugs is expected to ameliorate hyperglycemia and hyperlipidemia in insulin-resistant subjects with metabolic disease.

hepatic production of a FoxO1 dominant-negative allele to inhibit FoxO1 activity in the liver of diabetic *db/db* mice. As a result of insulin resistance, hepatic FoxO1 activity becomes unchecked in the liver and this effect contributes to hepatic glucose overproduction and fasting hyperglycemia in diabetic *db/db* mice. Altomonte et al. [28] demonstrate that FoxO1 inhibition is effective in curbing hepatic glucose overproduction, contributing to the reduction of hyperinsulinemia and hyperglycemia, accompanied by significantly improved glucose tolerance in diabetic *db/db* mice [27]. Samuel et al. [128] used an antisense oligonucleotide-mediated gene silencing approach to suppress FoxO1 activity in dietary obese mice. This approach results in FoxO1 inhibition in the liver and adipose tissue. This study demonstrates that FoxO1 inhibition helps enhance peripheral insulin sensitivity, conferring a significant beneficial effect on glucose and lipid metabolism in high fat-induced obese mice.

To achieve pharmacological inhibition of FoxO1, Nagashima et al. [194] identified a small-molecule compound named AS1842856 from a drug-screening project. They show that AS1842856 has the chemical property of binding to the *trans*-activation domain of FoxO1 protein and inhibiting its transcriptional activity in cultured hepatocytes. When orally administrated, AS1842856 is shown to suppress hepatic gluconeogenesis and reduce fasting hyperglycemia in diabetic *db/db* mice. Likewise, Langlet et al. [40] uncovered an anti-FoxO1 lead compound, named Compound #13 that is capable of suppressing hepatic expression of FoxO1 target genes such as G6Pase in cultured hepatocytes. However, the potential therapeutic effect of Compound #13 remains to be determined in vivo. Together these results provide proof of principle that hepatic FoxO1 activity is amenable to pharmacological inhibition for improving glucose and lipid metabolism in insulin-resistant subjects with metabolic disease (Fig. 13.5). These two studies also lend important support for the growing prosperity of targeting nuclear transcription factors with small molecule drugs for the development of therapeutics for a variety of diseases [195–203].

## Conclusions and perspectives

Clinical and preclinical studies have characterized FoxO1 as a key transcription factor in regulating glucose and VLDL-TG production in the liver in response to physiological cues. Hepatic FoxO1 activity is differentially regulated by insulin and glucagon. Insulin inhibits FoxO1 activity via AKT-mediated FoxO1 phosphorylation and nuclear exclusion, whereas glucagon enhances FoxO1 activity through PKA-dependent phosphorylation and nuclear localization. These two opposing actions are critical for keeping hepatic FoxO1 activity in check to maintain normal glucose and lipid metabolism in healthy individuals. In obesity and type 2 diabetes, FoxO1 becomes deregulated, accounting for its enhanced transcriptional activity for promoting hepatic overproduction of glucose and VLDL-TG production. It follows that hepatic FoxO1 deregulation may be one of the root causes of fasting hyperglycemia and hyperlipidemia in insulin-resistant subjects with obesity and type 2 diabetes. Therefore, it is imperative to test the hypothesis that pharmacological FoxO1 inhibition would ameliorate the dual pathogenesis of fasting hyperglycemia and hyperlipidemia in animal models with obesity and type 2 diabetes.

FoxO1 deregulation is also implicated as a contributing factor for the development of NAFLD in obesity and type 2 diabetes. However, this notion is contradicted by the finding that FoxO1 loss-of-function is also associated with the pathogenesis of NAFLD in mice in response to high fat feeding. Further investigation is needed to clarify the direct contribution of hepatic FoxO1 activity to NAFLD. Likewise, FoxO1 deregulation in macrophages is shown to promote abnormal macrophage activation, contributing to low-grade inflammation in obesity and type 2 diabetes. However, increased FoxO1 activity is also detected along with the induction of anti-inflammatory cytokine production in macrophages in culture. Further research is warranted to elucidate the role of FoxO1 in macrophage activation, polarization, and migration in the liver and extrahepatic tissues in obesity and type 2 diabetes.

Although FoxO1 antagonism is being exploited as a potential therapeutic avenue, caution should be exercised in developing FoxO1-targeted therapeutics. Although the four FoxO members share in common the same forkhead box domain and display some degrees of functional redundancy, they also possess distinct functions, due in part to their variable carboxyl domains. It is important to determine the drug specificity of anti-FoxO1 lead compounds and their potential off-targets in cells, while testing their therapeutic efficacy for treating metabolic disease in animal models. In addition, FoxO1 is ubiquitously expressed in central and peripheral tissues. Apart from its role in integrating insulin signaling to hepatic glucose and lipid metabolism, FoxO1 plays important roles in diverse pathways including cell growth, proliferation, and differentiation, autophagy and oxidative stress in different cell types [1,2]. The roles of FoxO1 in central and peripheral tissues are incompletely characterized. Further research is needed to determine the impact of chronic pharmacological FoxO1 inhibition on central and peripheral tissues in rodent models with obesity and type 2 diabetes.

## Acknowledgments

We thank Rita Miller for critical proofreading of this chapter. This work is supported by National Health Institute grant 1R01DK120310-01A1. C.Z. is a visiting scholar who is sponsored by scholarship #201806260074 from China Scholarship Council.

## References

[1] Accili D, Arden KC. FoxOs at the crossroads of cellular metabolism, differentiation, and transformation. Cell 2004;117:421–6.

[2] Barthel A, Schmoll D, Unterman TG. FoxO proteins in insulin action and metabolism. Trends Endocrinol Metab 2005;16:183–9.

[3] Furuyama T, Nakazawa T, Nakano I, Mori N. Identification of the differential distribution patterns of mRNAs and consensus binding sequences for mouse DAF-16 homologues. Biochem J 2000;349:629–34.

[4] Hoekman MF, Jacobs FM, Smidt MP, Burbach JP. Spatial and temporal expression of FoxO transcription factors in the developing and adult murine brain. Gene Expr Patterns 2006;6:134–40.

[5] Jacobs FM, van der Heide LP, Wijchers PJ, Burbach JP, Hoekman MF, Smidt MP. FoxO6, a novel member of the FoxO class of transcription factors with distinct shuttling dynamics. J Biol Chem 2003;278:35959–67.

[6] Wang M, Zhang X, Zhao H, Wang Q, Pan Y. FoxO gene family evolution in vertebrates. BMC Evol Biol 2009;9:222.

[7] Furuyama T, Kitayama K, Shimoda Y, Ogawa M, Sone K, Yoshida-Araki K, et al. Abnormal angiogenesis in Foxo1 (Fkhr)-deficient mice. J Biol Chem 2004;279:34741–9.

[8] Hosaka T, Biggs WH 3rd, Tieu D, Boyer AD, Varki NM, Cavenee WK, et al. Disruption of forkhead transcription factor (FOXO) family members in mice reveals their functional diversification. Proc Natl Acad Sci U S A 2004;101:2975–80.

[9] Calabuig-Navarro V, Yamauchi J, Lee S, Zhang T, Liu YZ, Sadlek K, et al. Forkhead box O6 (FoxO6) depletion attenuates hepatic gluconeogenesis and protects against fat-induced glucose disorder in mice. J Biol Chem 2015;290:15581–94.

[10] Castrillon DH, Miao L, Kollipara R, Horner JW, DePinho RA. Suppression of ovarian follicle activation in mice by the transcription factor Foxo3a. Science 2003;301:215–8.

[11] Kim DH, Perdomo G, Zhang T, Slusher S, Lee S, Phillips BE, et al. FoxO6 integrates insulin signaling with gluconeogenesis in the liver. Diabetes 2011;60:2763–74.

[12] Kim DH, Zhang T, Lee S, Dong HH. FoxO6 in glucose metabolism. J Diabetes 2013;5:233–40.

[13] Brent MM, Anand R, Marmorstein R. Structural basis for DNA recognition by FoxO1 and its regulation by posttranslational modification. Structure 2008;16:1407–16.

[14] Puigserver P, Rhee J, Donovan J, Walkey CJ, Yoon JC, Oriente F, et al. Insulin-regulated hepatic gluconeogenesis through FOXO1-PGC-1alpha interaction. Nature 2003;423:550–5.

[15] Monecke T, Guttler T, Neumann P, Dickmanns A, Gorlich D, Ficner R. Crystal structure of the nuclear export receptor CRM1 in complex with Snurportin1 and RanGTP. Science 2009;324:1087–91.

[16] Dong X, Biswas A, Suel KE, Jackson LK, Martinez R, Gu H, et al. Structural basis for leucine-rich nuclear export signal recognition by CRM1. Nature 2009;458:1136–41.

[17] Tsai WC, Bhattacharyya N, Han LY, Hanover JA, Rechler MM. Insulin inhibition of transcription stimulated by the forkhead protein Foxo1 is not solely due to nuclear exclusion. Endocrinology 2003;144:5615–22.

[18] Matsuzaki H, Daitoku H, Hatta M, Tanaka K, Fukamizu A. Insulin-induced phosphorylation of FKHR (Foxo1) targets to proteasomal degradation. Proc Natl Acad Sci U S A 2003;100:11285–90.

[19] Aoki M, Jiang H, Vogt PK. Proteasomal degradation of the FoxO1 transcriptional regulator in cells transformed by the P3k and Akt oncoproteins. Proc Natl Acad Sci U S A 2004;101:13613–7.

[20] Huang H, Regan KM, Wang F, Wang D, Smith DI, van Deursen JM, et al. Skp2 inhibits FOXO1 in tumor suppression through ubiquitin-mediated degradation. Proc Natl Acad Sci U S A 2005;102:1649–54.

[21] Zhao Y, Yang J, Liao W, Liu X, Zhang H, Wang S, et al. Cytosolic FoxO1 is essential for the induction of autophagy and tumour suppressor activity. Nat Cell Biol 2010;12:665–75.

[22] Hwangbo DS, Gershman B, Tu MP, Palmer M, Tatar M. Drosophila dFOXO controls lifespan and regulates insulin signalling in brain and fat body. Nature 2004;429:562–6.

[23] Lee SS, Kennedy S, Tolonen AC, Ruvkun G. DAF-16 target genes that control C. elegans life-span and metabolism. Science 2003;300:644–7.
[24] Qu S, Altomonte J, Perdomo G, He J, Fan Y, Kamagate A, et al. Aberrant forkhead box O1 function is associated with impaired hepatic metabolism. Endocrinology 2006;147:5641–52.
[25] Sparks DJ, Dong HH. FoxO1 and hepatic lipid metabolism. Curr Opin Lipidol 2009;20:217–26.
[26] Wu Y, Pan Q, Yan H, Zhang K, Guo X, Xu Z, et al. Novel mechanism of Foxo1 phosphorylation in glucagon signaling in control of glucose homeostasis. Diabetes 2018;67:2167–82.
[27] Altomonte J, Richter A, Harbaran S, Suriawinata J, Nakae J, Thung SN, et al. Inhibition of Foxo1 function is associated with improved fasting glycemia in diabetic mice. Am J Physiol 2003;285:E718–28.
[28] Altomonte J, Cong L, Harbaran S, Richter A, Xu J, Meseck M, et al. Foxo1 mediates insulin action on ApoC-III and triglyceride metabolism. J Clin Invest 2004;114:1493–503.
[29] Valenti L, Rametta R, Dongiovanni P, Maggioni M, Ludovica Fracanzani A, Zappa M, et al. Increased expression and activity of the transcription factor Foxo1 in nonalcoholic steatohepatitis. Diabetes 2008;57:1355–62.
[30] Puig O, Tjian R. Transcriptional feedback control of insulin receptor by dFOXO/FOXO1. Genes Dev 2005;19:2435–46.
[31] Kamagate A, Kim DH, Zhang T, Slusher S, Gramignoli R, Strom SC, et al. FoxO1 links hepatic insulin action to endoplasmic reticulum stress. Endocrinology 2010;151:3521–35.
[32] Tsunekawa S, Demozay D, Briaud I, McCuaig J, Accili D, Stein R, et al. FoxO feedback control of basal IRS-2 expression in pancreatic beta-cells is distinct from that in hepatocytes. Diabetes 2011;60:2883–91.
[33] Martinez SC, Tanabe K, Cras-Meneur C, Abumrad NA, Bernal-Mizrachi E, Permutt MA. Inhibition of Foxo1 protects pancreatic islet beta-cells against fatty acid and endoplasmic reticulum stress-induced apoptosis. Diabetes 2008;57:846–59.
[34] Okamoto H, Hribal ML, Lin HV, Bennett WR, Ward A, Accili D. Role of the forkhead protein FoxO1 in beta cell compensation to insulin resistance. J Clin Invest 2006;116:775–82.
[35] Armoni M, Harel C, Karni S, Chen H, Bar-Yoseph F, Ver MR, et al. FOXO1 represses peroxisome proliferator-activated receptor-gamma1 and -gamma2 gene promoters in primary adipocytes. A novel paradigm to increase insulin sensitivity. J Biol Chem 2006;281:19881–91.
[36] Fan W, Imamura T, Sonoda N, Sears DD, Patsouris D, Kim JJ, et al. FOXO1 transrepresses peroxisome proliferator-activated receptor gamma transactivation, coordinating an insulin-induced feed-forward response in adipocytes. J Biol Chem 2009;284:12188–97.
[37] Dowell P, Otto TC, Adi S, Lane MD. Convergence of peroxisome proliferator-activated receptor gamma and Foxo1 signaling pathways. J Biol Chem 2003;278:45485–91.
[38] Qu S, Su D, Altomonte J, Kamagate A, He J, Perdomo G, et al. PPARα mediates the hypolipidemic action of fibrates by antagonizing FoxO1. Am J Physiol Endocrinol Metab 2007;292:E421–34.
[39] Kibbe C, Chen J, Xu G, Jing G, Shalev A. FOXO1 competes with carbohydrate response element-binding protein (ChREBP) and inhibits thioredoxin-interacting protein (TXNIP) transcription in pancreatic beta cells. J Biol Chem 2013;288:23194–202.
[40] Langlet F, Haeusler RA, Linden D, Ericson E, Norris T, Johansson A, et al. Selective inhibition of FOXO1 activator/repressor balance modulates hepatic glucose handling. Cell 2017;171. 824-835.e818.
[41] Nakae J, Kitamura T, Silver DL, Accili D. The forkhead transcription factor Foxo1 (Fkhr) confers insulin sensitivity onto glucose-6-phosphatase expression. J Clin Invest 2001;108:1359–67.
[42] Matsumoto M, Pocai A, Rossetti L, Depinho RA, Accili D. Impaired regulation of hepatic glucose production in mice lacking the forkhead transcription factor foxo1 in liver. Cell Metab 2007;6:208–16.
[43] Kamagate A, Qu S, Perdomo G, Su D, Kim DH, Slusher S, et al. FoxO1 mediates insulin-dependent regulation of hepatic VLDL production in mice. J Clin Invest 2008;118:2347–64.
[44] O'Brien RM, Lucas PC, Forest CD, Magnuson MA, Granner DK. Identification of a sequence in the PEPCK gene that mediates a negative effect of insulin on transcription. Science 1990;249:533–7.
[45] O'Brien RM, Granner DK. Regulation of gene expression by insulin. Physiol Rev 1996;76:1109–61.
[46] Pilkis SJ, Granner DK. Molecular physiology of the regulation of hepatic gluconeogenesis and glycogenolysis. Annu Rev Physiol 1992;54:885–909.
[47] Unger RH, Orci L. Glucagon and the a cell. New Engl J Med 1981;304:1575–80.
[48] Koo SH, Flechner L, Qi L, Zhang X, Screaton RA, Jeffries S, et al. The CREB coactivator TORC2 is a key regulator of fasting glucose metabolism. Nature 2005;437:1109–11.

[49] Haeusler RA, Kaestner KH, Accili D. FoxOs function synergistically to promote glucose production. J Biol Chem 2010;285:35245–8.
[50] Imae M, Fu Z, Yoshida A, Noguchi T, Kato H. Nutritional and hormonal factors control the gene expression of FoxOs, the mammalian homologues of DAF-16. J Mol Endocrinol 2003;30:253–62.
[51] Kim DH, Zhang T, Ringquist S, Dong HH. Targeting FoxO1 for hypertriglyceridemia. Curr Drug Targets 2011;12:1245–55.
[52] Hussain MM, Shi J, Dreizen P. Microsomal triglyceride transfer protein and its role in apoB-lipoprotein assembly. J Lipid Res 2003;44:22–32.
[53] Hussain MM, Nijstad N, Franceschini L. Regulation of microsomal triglyceride transfer protein. Clin Lipidol 2011;6:293–303.
[54] Raabe M, Veniant MM, Sullivan MA, Zlot CH, Bjorkegren J, Nielsen LB, et al. Analysis of the role of microsomal triglyceride transfer protein in the liver of tissue-specific knockout mice. J Clin Invest 1999;103:1287–98.
[55] Luchoomun J, Zhou Z, Bakillah A, Jamil H, Hussain MM. Assembly and secretion of VLDL in nondifferentiated Caco-2 cells stably transfected with human recombinant ApoB48 cDNA. Arterioscler Thromb Vasc Biol 1997;17:2955–63.
[56] Swift LL, Zhu MY, Kakkad B, Jovanovska A, Neely MD, Valyi-Nagy K, et al. Subcellular localization of microsomal triglyceride transfer protein. J Lipid Res 2003;44:1841–9.
[57] Gordon DA, Wetterau JR, Gregg RE. Microsomal triglyceride transfer protein: a protein complex required for the assembly of lipoprotein particles. Trends Cell Biol 1995;5:317–21.
[58] Hui TY, Olivier LM, Kang S, Davis RA. Microsomal triglyceride transfer protein is essential for hepatic secretion of apoB-100 and apoB-48 but not triglyceride. J Lipid Res 2002;43:785–93.
[59] Manchekar M, Richardson PE, Forte TM, Datta G, Segrest JP, Dashti N. Apolipoprotein B-containing lipoprotein particle assembly: lipid capacity of the nascent lipoprotein particle. J Biol Chem 2004;279:39757–66.
[60] Swift LL, Jovanovska A, Kakkad B, Ong DE. Microsomal triglyceride transfer protein expression in mouse intestine. Histochem Cell Biol 2005;123:475–82.
[61] Harrison EH, Hussain MM. Mechanisms involved in the intestinal digestion and absorption of dietary vitamin A. J Nutr 2001;131:1405–8.
[62] Berriot-Varoqueaux N, Aggerbeck LP, Samson-Bouma M, Wetterau JR. The role of the microsomal triglyceride transfer protein in abetalipoproteinemia. Annu Rev Nutr 2000;20:663–97.
[63] Liao W, Hui TY, Young SG, Davis RA. Blocking microsomal triglyceride transfer protein interferes with apoB secretion without causing retention or stress in the ER. J Lipid Res 2003;44:978–85.
[64] Bjorkegren J, Beigneux A, Bergo MO, Maher JJ, Young SG. Blocking the secretion of hepatic very low density lipoproteins renders the liver more susceptible to toxin-induced injury. J Biol Chem 2002;277:5476–83.
[65] Leung GK, Veniant MM, Kim SK, Zlot CH, Raabe M, Bjorkegren J, et al. A deficiency of microsomal triglyceride transfer protein reduces apolipoprotein B secretion. J Biol Chem 2000;275:7515–20.
[66] Ledmyr H, Karpe F, Lundahl B, McKinnon M, Skoglund-Andersson C, Ehrenborg E. Variants of the microsomal triglyceride transfer protein gene are associated with plasma cholesterol levels and body mass index. J Lipid Res 2002;43:51–8.
[67] Ledmyr H, McMahon AD, Ehrenborg E, Nielsen LB, Neville M, Lithell H, et al. The microsomal triglyceride transfer protein gene-493T variant lowers cholesterol but increases the risk of coronary heart disease. Circulation 2004;109:2279–84.
[68] Juo SH, Han Z, Smith JD, Colangelo L, Liu K. Common polymorphism in promoter of microsomal triglyceride transfer protein gene influences cholesterol, ApoB, and triglyceride levels in young African American men: results from the coronary artery risk development in young adults (CARDIA) study. Arterioscler Thromb Vasc Biol 2000;20:1316–22.
[69] Garcia-Garcia AB, Gonzalez C, Real JT, Martin de Llano JJ, Gonzalez-Albert V, Civera M, et al. Influence of microsomal triglyceride transfer protein promoter polymorphism -493 GT on fasting plasma triglyceride values and interaction with treatment response to atorvastatin in subjects with heterozygous familial hypercholesterolaemia. Pharmacogenet Genomics 2005;15:211–8.
[70] Tietge UJ, Bakillah A, Maugeais C, Tsukamoto K, Hussain M, Rader DJ. Hepatic overexpression of microsomal triglyceride transfer protein (MTP) results in increased in vivo secretion of VLDL triglycerides and apolipoprotein B. J Lipid Res 1999;40:2134–9.
[71] Kolovou GD, Kolovou V, Papadopoulou A, Watts GF. MTP gene variants and response to lomitapide in patients with homozygous familial hypercholesterolemia. J Atheroscler Thromb 2016;23:878–83.

[72] Rader DJ, Kastelein JJ. Lomitapide and mipomersen: two first-in-class drugs for reducing low-density lipoprotein cholesterol in patients with homozygous familial hypercholesterolemia. Circulation 2014;129:1022–32.

[73] Cuchel M, Bloedon LT, Szapary PO, Kolansky DM, Wolfe ML, Sarkis A, et al. Inhibition of microsomal triglyceride transfer protein in familial hypercholesterolemia. N Engl J Med 2007;356:148–56.

[74] Khoury E, Brisson D, Roy N, Tremblay G, Gaudet D. Review of the long-term safety of lomitapide: a microsomal triglycerides transfer protein inhibitor for treating homozygous familial hypercholesterolemia. Expert Opin Drug Saf 2019;18:403–14.

[75] Hagan DL, Kienzle B, Jamil H, Hariharan N. Transcriptional regulation of human and hamster microsomal triglyceride transfer protein genes. Cell type-specific expression and response to metabolic regulators. J Biol Chem 1994;269:28737–44.

[76] Lin MC, Gordon D, Wetterau JR. Microsomal triglyceride transfer protein (MTP) regulation in HepG2 cells: insulin negatively regulates MTP gene expression. J Lipid Res 1995;36:1073–81.

[77] Phillips C, Owens D, Collins P, Tomkin GH. Microsomal triglyceride transfer protein: does insulin resistance play a role in the regulation of chylomicron assembly? Atherosclerosis 2002;160:355–60.

[78] Malmstrom R, Packard CJ, Caslake M, Bedford D, Stewart P, Yki-Jarvinen H, et al. Effects of insulin and acipimox on VLDL1 and VLDL2 apolipoprotein B production in normal subjects. Diabetes 1998;47:779–87.

[79] Malmstrom R, Packard CJ, Watson TD, Rannikko S, Caslake M, Bedford D, et al. Metabolic basis of hypotriglyceridemic effects of insulin in normal men. Arterioscler Thromb Vasc Biol 1997;17:1454–64.

[80] Lewis GF, Steiner G. Acute effects of insulin in the control of VLDL production in humans. Implications for the insulin-resistant state. Diabetes Care 1996;19:390–3.

[81] Sparks JD, Sparks CE. Insulin regulation of triacylglycerol-rich lipoprotein synthesis and secretion. Biochim Biophys Acta 1994;1215:9–32.

[82] Sparks JD, Sparks CE. Overindulgence and metabolic syndrome: is FoxO1 a missing link? J Clin Invest 2008;118:2012–5.

[83] Ota T, Gayet C, Ginsberg HN. Inhibition of apolipoprotein B100 secretion by lipid-induced hepatic endoplasmic reticulum stress in rodents. J Clin Invest 2008;118:316–32.

[84] Taghibiglou C, Carpentier A, Van Iderstine SC, Chen B, Rudy D, Aiton A, Lewis GF, Adeli K. Mechanisms of hepatic very low density lipoprotein overproduction in insulin resistance. Evidence for enhanced lipoprotein assembly, reduced intracellular ApoB degradation, and increased microsomal triglyceride transfer protein in a fructose-fed hamster model. J Biol Chem 2000;275:8416–25.

[85] Hsieh J, Hayashi AA, Webb J, Adeli K. Postprandial dyslipidemia in insulin resistance: mechanisms and role of intestinal insulin sensitivity. Atheroscler Suppl 2008;9:7–13.

[86] Sparks JD, Sparks CE, Adeli K. Selective hepatic insulin resistance, VLDL overproduction, and hypertriglyceridemia. Arterioscler Thromb Vasc Biol 2012;32:2104–12.

[87] Choi SH, Ginsberg HN. Increased very low density lipoprotein (VLDL) secretion, hepatic steatosis, and insulin resistance. Trends Endocrinol Metab 2011;22:353–63.

[88] Jong MC, Hofker MH, Havekes LM. Role of apoCs in lipoprotein metabolism: functional differences between apoC1, apoC2 and apoC3. Arterioscher Thromb Vasc Biol. 1999;19:472–84.

[89] Wang C, McConathy WJ, Kloer HJ, Alaupovic P. Modulation of lipoprotein lipase activity by apolipoproteins: effect of apolipoprotein C-III. J Clin Invest 1985;75:384.

[90] McConathy WJ, Gesquiere JC, Bass H, Tartar A, Fruchart JC. Inhibition of lipoprotein lipase activity by synthetic peptides of apolipoprotein C-III. J Lipid Res 1992;33:995–1003.

[91] Kinnunen PKJ, Ehnholm C. Effect of serum and C apolipoproteins from very low density lipoproteins on human post-heparin plasma hepatic lipase. FEBS Lett 1976;65:354–7.

[92] Wuarfordt SH, Michalopoulos G, Schirmer B. The effect of human C apolipoproteins on the in vitro hepatic metabolism of triglyceride emulsions in the rat. J Biol Chem 1982;257:14642–7.

[93] Mann CJ, Troussard AA, Yen FT, Hannouche N, Najib J, Fruchart JC, Lotteau V, Andre P, Bihain BE. Inhibitory effects of specific apolipoprotein C-III isoforms on the binding of triglyceride-rich lipoproteins to the lipolysis-stimulated receptor. J Biol Chem 1997;272:31348–54.

[94] Qin W, Sundaram M, Wang Y, Zhou H, Zhong S, Chang CC, et al. Missense mutation in APOC3 within the C-terminal lipid binding domain of human ApoC-III results in impaired assembly and secretion of triacylglycerol-rich very low density lipoproteins: evidence that ApoC-III plays a major role in the formation of lipid precursors within the microsomal lumen. J Biol Chem 2011;286:27769–80.

[95] Chan DC, Watts GF, Nguyen MN, Barrett PH. Apolipoproteins C-III and A-V as predictors of very-low-density lipoprotein triglyceride and apolipoprotein B-100 kinetics. Arterioscler Thromb Vasc Biol 2006;26:590–6.
[96] Taskinen MR, Adiels M, Westerbacka J, Soderlund S, Kahri J, Lundbom N, et al. Dual metabolic defects are required to produce hypertriglyceridemia in obese subjects. Arterioscler Thromb Vasc Biol 2011;31:2144–50.
[97] Cohn JS, Patterson BW, Uffelman KD, Davignon J, Steiner G. Rate of production of plasma and very-low-density lipoprotein (VLDL) apolipoprotein C-III is strongly related to the concentration and level of production of VLDL triglyceride in male subjects with different body weights and levels of insulin sensitivity. J Clin Endocrinol Metab 2004;89:3949–55.
[98] Shachter NS. Apolipoproteins C-1 and C-III as important modulators of lipoprotein metabolism. Curr Opin Lipidol 2001;12:297–304.
[99] Ito YN, Azrolan A, O'Connell A, Walsh A, Breslow JL. Hypertriglyceridemia as a result of human apo CIII gene expression in transgenic mice. Science 1990;249:790–3.
[100] Maeda N, Li H, Lee D, Oliver P, Quarfordt SH, Osada J. Targeted disruption of the apolipoprotein C-III gene in mice results in hypotriglyceridemia and protection from postprandial hypertriglyceridemia. J Biol Chem 1994;269:23610–6.
[101] Gerritsen G, Rensen PC, Kypreos KE, Zannis VI, Havekes LM, Willems van Dijk K. ApoC-III deficiency prevents hyperlipidemia induced by apoE overexpression. J Lipid Res 2005;46:1466–73.
[102] Jong MC, Rensen PCN, Dahlmans VE, van der Boom H, van Berkel TJC, Havekes LM. Apolipoprotein C-III deficiency accelerates triglyceride hydrolysis by lipoprotein lipase in wild-type and apoE knockout mice. J Lipid Res 2001;42:1578–85.
[103] Pollin TI, Damcott CM, Shen H, Ott SH, Shelton J, Horenstein RB, et al. A null mutation in human APOC3 confers a favorable plasma lipid profile and apparent cardioprotection. Science 2008;322:1702–5.
[104] Jorgensen AB, Frikke-Schmidt R, Nordestgaard BG, Tybjaerg-Hansen A. Loss-of-function mutations in APOC3 and risk of ischemic vascular disease. N Engl J Med 2014;371:32–41.
[105] Crosby J, Peloso GM, Auer PL, Crosslin DR, Stitziel NO, Lange LA, et al. Loss-of-function mutations in APOC3, triglycerides, and coronary disease. N Engl J Med 2014;371:22–31.
[106] Graham MJ, Lee RG, Bell TA 3rd, Fu W, Mullick AE, Alexander VJ, et al. Antisense oligonucleotide inhibition of apolipoprotein C-III reduces plasma triglycerides in rodents, nonhuman primates, and humans. Circ Res 2013;112:1479–90.
[107] Gaudet D, Brisson D, Tremblay K, Alexander VJ, Singleton W, Hughes SG, et al. Targeting APOC3 in the familial chylomicronemia syndrome. N Engl J Med 2014;371:2200–6.
[108] Khetarpal SA, Zeng X, Millar JS, Vitali C, Somasundara AVH, Zanoni P, et al. A human APOC3 missense variant and monoclonal antibody accelerate apoC-III clearance and lower triglyceride-rich lipoprotein levels. Nat Med 2017;23:1086–94.
[109] Norata GD, Tsimikas S, Pirillo A, Catapano AL. Apolipoprotein C-III: from pathophysiology to pharmacology. Trends Pharmacol Sci 2015;36:675–87.
[110] Khetarpal SA, Qamar A, Millar JS, Rader DJ. Targeting ApoC-III to reduce coronary disease risk. Curr Atheroscler Rep 2016;18:54.
[111] Petersen KF, Dufour S, Hariri A, Nelson-Williams C, Foo JN, Zhang XM, et al. Apolipoprotein C3 gene variants in nonalcoholic fatty liver disease. N Engl J Med 2010;362:1082–9.
[112] Valenti L, Nobili V, Al-Serri A, Rametta R, Leathart JB, Zappa MA, et al. The APOC3 T-455C and C-482T promoter region polymorphisms are not associated with the severity of liver damage independently of PNPLA3 I148M genotype in patients with nonalcoholic fatty liver. J Hepatol 2011;55:1409–14.
[113] Verrijken A, Beckers S, Francque S, Hilden H, Caron S, Zegers D, et al. A gene variant of PNPLA3, but not of APOC3, is associated with histological parameters of NAFLD in an obese population. Obesity (Silver Spring) 2013;21:2138–45.
[114] Sentinelli F, Romeo S, Maglio C, Incani M, Burza MA, Scano F, et al. Lack of effect of apolipoprotein C3 polymorphisms on indices of liver steatosis, lipid profile and insulin resistance in obese Southern Europeans. Lipids Health Dis 2011;10:93.
[115] Cheng X, Yamauchi J, Lee S, Zhang T, Gong Z, Muzumdar R, et al. APOC3 protein is not a predisposing factor for fat-induced nonalcoholic fatty liver disease in mice. J Biol Chem 2017;292:3692–705.
[116] Jansen H, Waterworth DM, Nicaud V, Ehnholm C, Talmud PJ, EARS-II Study Group. Interaction of the common apolipoprotein C-III (APOC3, -482C>T) and hepatic (LIPC, -514C>T) promoter variants affects glucose tolerance in young adults. European Atherosclerosis Research Study II (EARS-II). Ann Hum Genet 2001;65:237–43.

[117] Hegele RA, Connelly PW, Hanley AJG, Sun F, Harris SB, Zinman B. Common genetic variation in the APOC3 promoter associated with variation in plasma lipoproteins. Arterioscher Thromb Vasc Biol 1997;17:2753–8.
[118] Kanter JE, Shao B, Kramer F, Barnhart S, Shimizu-Albergine M, Vaisar T, et al. Increased apolipoprotein C3 drives cardiovascular risk in type 1 diabetes. J Clin Invest 2019;130.
[119] van Hoek M, van Herpt TW, Dehghan A, Hofman A, Lieverse AG, van Duijn CM, et al. Association of an APOC3 promoter variant with type 2 diabetes risk and need for insulin treatment in lean persons. Diabetologia 2011;54:1360–7.
[120] Zhang L, Perdomo G, Kim DH, Qu S, Ringquist S, Trucco M, et al. Proteomic analysis of fructose-induced fatty liver in hamsters. Metabolism 2008;57:1115–24.
[121] Matsumoto M, Han S, Kitamura T, Accili D. Dual role of transcription factor FoxO1 in controlling hepatic insulin sensitivity and lipid metabolism. J Clin Invest 2006;116:2464–72.
[122] Pan X, Zhang Y, Kim HG, Liangpunsakul S, Dong XC. FOXO transcription factors protect against the diet-induced fatty liver disease. Sci Rep 2017;7:44597.
[123] Dong XC. FOXO transcription factors in non-alcoholic fatty liver disease. Liver Res 2017;1:168–73.
[124] Kamagate A, Dong HH. FoxO1 integrates insulin signaling to VLDL production. Cell Cycle 2008;7:3162–70.
[125] Pajvani UB, Accili D. Erratum to: the new biology of diabetes. Diabetologia 2015;58:2683.
[126] Beck T, Miller BG. Structural basis for regulation of human glucokinase by glucokinase regulatory protein. Biochemistry 2013;52:6232–9.
[127] Choi JM, Seo MH, Kyeong HH, Kim E, Kim HS. Molecular basis for the role of glucokinase regulatory protein as the allosteric switch for glucokinase. Proc Natl Acad Sci U S A 2013;110:10171–6.
[128] Samuel VT, Choi CS, Phillips TG, Romanelli AJ, Geisler JG, Bhanot S, et al. Targeting foxo1 in mice using antisense oligonucleotide improves hepatic and peripheral insulin action. Diabetes 2006;55:2042–50.
[129] Nakae J, Kitamura T, Kitamura Y, Biggs WH 3rd, Arden KC, Accili D. The forkhead transcription factor Foxo1 regulates adipocyte differentiation. Dev Cell 2003;4:119–29.
[130] Okamoto H, Nakae J, Kitamura T, Park BC, Dragatsis I, Accili D. Transgenic rescue of insulin receptor-deficient mice. J Clin Invest 2004;114:214–23.
[131] Dong XC, Copps KD, Guo S, Li Y, Kollipara R, DePinho RA, et al. Inactivation of hepatic Foxo1 by insulin signaling is required for adaptive nutrient homeostasis and endocrine growth regulation. Cell Metab 2008;8:65–76.
[132] Zhang W, Patil S, Chauhan B, Guo S, Powell DR, Le J, et al. FoxO1 regulates multiple metabolic pathways in the liver: effects on gluconeogenic, glycolytic, and lipogenic gene expression. J Biol Chem 2006;281:10105–17.
[133] Ling AV, Gearing ME, Semova I, Shin DJ, Clements R, Lai ZW, et al. FoxO1 is required for most of the metabolic and hormonal perturbations produced by hepatic insulin receptor deletion in male mice. Endocrinology 2018;159:1253–63.
[134] Deng X, Zhang W, O-Sullivan I, Williams JB, Dong Q, Park EA, et al. FoxO1 inhibits sterol regulatory element-binding protein-1c (SREBP-1c) gene expression via transcription factors Sp1 and SREBP-1c. J Biol Chem 2012;287:20132–43.
[135] Fernandez-Miranda C, Perez-Carreras M, Colina F, Lopez-Alonso G, Vargas C, Solis-Herruzo JA. A pilot trial of fenofibrate for the treatment of non-alcoholic fatty liver disease. Dig Liver Dis 2008;40:200–5.
[136] Minnich A, Tian N, Byan L, Bilder G. A potent PPARalpha agonist stimulates mitochondrial fatty acid beta-oxidation in liver and skeletal muscle. Am J Physiol Endocrinol Metab 2001;280:E270–279.
[137] Zhang W, Bu SY, Mashek MT, O-Sullivan I, Sibai Z, Khan SA, et al. Integrated regulation of hepatic lipid and glucose metabolism by adipose triacylglycerol lipase and FoxO proteins. Cell Rep 2016;15:349–59.
[138] Chakrabarti P, Kandror KV. FoxO1 controls insulin-dependent adipose triglyceride lipase (ATGL) expression and lipolysis in adipocytes. J Biol Chem 2009;284:13296–300.
[139] Sanyal AJ, Chalasani N, Kowdley KV, McCullough A, Diehl AM, Bass NM, et al. Pioglitazone, vitamin E, or placebo for nonalcoholic steatohepatitis. N Engl J Med 2010;362:1675–85.
[140] Lim JS, Mietus-Snyder M, Valente A, Schwarz JM, Lustig RH. The role of fructose in the pathogenesis of NAFLD and the metabolic syndrome. Nat Rev Gastroenterol Hepatol 2010;7:251–64.
[141] Vuppalanchi R, Chalasani N. Nonalcoholic fatty liver disease and nonalcoholic steatohepatitis: selected practical issues in their evaluation and management. Hepatology 2009;49:306–17.
[142] Burgert TS, Taksali SE, Dziura J, Goodman TR, Yeckel CW, Papademetris X, et al. Alanine aminotransferase levels and fatty liver in childhood obesity: associations with insulin resistance, adiponectin, and visceral fat. J Clin Endocrinol Metab 2006;91:4287–94.
[143] Lerret SM, Skelton JA. Pediatric nonalcoholic fatty liver disease. Gastroenterol Nurs 2008;31:115–9.

[144] Dunn W, Schwimmer JB. The obesity epidemic and nonalcoholic fatty liver disease in children. Curr Gastroenterol Rep 2008;10:67–72.
[145] Nobili V, Alisi A, Raponi M. Pediatric non-alcoholic fatty liver disease: preventive and therapeutic value of lifestyle intervention. World J Gastroenterol 2009;15:6017–22.
[146] Brunt EM. Pathology of nonalcoholic fatty liver disease. Nat Rev Gastroenterol Hepatol 2010;7:195–203.
[147] Nugent C, Younossi ZM. Evaluation and management of obesity-related nonalcoholic fatty liver disease. Nat Clin Pract Gastroenterol Hepatol 2007;4:432–41.
[148] Torres DM, Harrison SA. Diagnosis and therapy of nonalcoholic steatohepatitis. Gastroenterology 2008;134:1682–98.
[149] Olefsky JM, Glass CK. Macrophages, inflammation, and insulin resistance. Annu Rev Physiol 2010;72:219–46.
[150] Speliotes EK, Massaro JM, Hoffmann U, Vasan RS, Meigs JB, Sahani DV, et al. Fatty liver is associated with dyslipidemia and dysglycemia independent of visceral fat: the Framingham Heart Study. Hepatology 2010;51:1979–87.
[151] Fabbrini E, Sullivan S, Klein S. Obesity and nonalcoholic fatty liver disease: biochemical, metabolic, and clinical implications. Hepatology 2010;51:679–89.
[152] Schenk S, Saberi M, Olefsky JM. Insulin sensitivity: modulation by nutrients and inflammation. J Clin Invest 2008;118:2992–3002.
[153] Baffy G. Kupffer cells in non-alcoholic fatty liver disease: the emerging view. J Hepatol 2009;51:212–23.
[154] Tan HH, Fiel MI, Sun Q, Guo J, Gordon RE, Chen LC, Friedman SL, Odin JA, Allina J. Kupffer cell activation by ambient air particulate matter exposure may exacerbate non-alcoholic fatty liver disease. J Immunotoxicol 2009;6:266–75.
[155] Obstfeld AE, Sugaru E, Thearle M, Francisco AM, Gayet C, Ginsberg HN, et al. C-C chemokine receptor 2 (CCR2) regulates the hepatic recruitment of myeloid cells that promote obesity-induced hepatic steatosis. Diabetes 2010;59:916–25.
[156] Huang W, Metlakunta A, Dedousis N, Zhang P, Sipula I, Dube JJ, et al. Depletion of liver Kupffer cells prevents the development of diet-induced hepatic steatosis and insulin resistance. Diabetes 2010;59:347–57.
[157] Lanthier N, Molendi-Coste O, Horsmans Y, van Rooijen N, Cani PD, Leclercq IA. Kupffer cell activation is a causal factor for hepatic insulin resistance. Am J Physiol Gastrointest Liver Physiol 2010;298:G107–16.
[158] Neyrinck AM, Cani PD, Dewulf EM, De Backer F, Bindels LB, Delzenne NM. Critical role of Kupffer cells in the management of diet-induced diabetes and obesity. Biochem Biophys Res Commun 2009;385:351–6.
[159] Gonzalez-Reimers E, Castellano-Higuera A, Aleman-Valls R, Alvarez-Arguelles H, de la Vega-Prieto MJ, Abreu-Gonzalez P, et al. Relation between body fat and liver fat accumulation and cytokine pattern in non-alcoholic patients with chronic HCV infection. Ann Nutr Metab 2009;55:351–7.
[160] Rensen SS, Slaats Y, Nijhuis J, Jans A, Bieghs V, Driessen A, et al. Increased hepatic myeloperoxidase activity in obese subjects with nonalcoholic steatohepatitis. Am J Pathol 2009;175:1473–82.
[161] Wouters K, van Gorp PJ, Bieghs V, Gijbels MJ, Duimel H, Lutjohann D, et al. Dietary cholesterol, rather than liver steatosis, leads to hepatic inflammation in hyperlipidemic mouse models of nonalcoholic steatohepatitis. Hepatology 2008;48:474–86.
[162] Schattenberg JM, Singh R, Wang Y, Lefkowitch JH, Rigoli RM, Scherer PE, et al. JNK1 but not JNK2 promotes the development of steatohepatitis in mice. Hepatology 2006;43:163–72.
[163] Lalor PF, Faint J, Aarbodem Y, Hubscher SG, Adams DH. The role of cytokines and chemokines in the development of steatohepatitis. Semin Liver Dis 2007;27:173–93.
[164] Su D, Coudriet GM, Kim DH, Lu Y, Perdomo G, Qu S, et al. FoxO1 links insulin resistance to proinflammatory cytokine IL-1beta production in macrophages. Diabetes 2009;58:2624–33.
[165] Fan W, Morinaga H, Kim JJ, Bae E, Spann NJ, Heinz S, et al. FoxO1 regulates Tlr4 inflammatory pathway signalling in macrophages. EMBO J 2010;29:4223–36.
[166] Chung S, Lee TJ, Reader BF, Kim JY, Lee YG, Park GY, et al. FoxO1 regulates allergic asthmatic inflammation through regulating polarization of the macrophage inflammatory phenotype. Oncotarget 2016;7:17532–46.
[167] Wang YC, Ma HD, Yin XY, Wang YH, Liu QZ, Yang JB, et al. Forkhead box O1 regulates macrophage polarization following Staphylococcus aureus infection: experimental murine data and review of the literature. Clin Rev Allergy Immunol 2016;51:353–69.
[168] Tsuchiya K, Westerterp M, Murphy AJ, Subramanian V, Ferrante AW Jr, Tall AR, et al. Expanded granulocyte/monocyte compartment in myeloid-specific triple FoxO knockout increases oxidative stress and accelerates atherosclerosis in mice. Circ Res 2013;112:992–1003.

[169] Kawano Y, Nakae J, Watanabe N, Fujisaka S, Iskandar K, Sekioka R, et al. Loss of Pdk1-Foxo1 signaling in myeloid cells predisposes to adipose tissue inflammation and insulin resistance. Diabetes 2012;61:1935–48.

[170] Chung S, Ranjan R, Lee YG, Park GY, Karpurapu M, Deng J, et al. Distinct role of FoxO1 in M-CSF- and GM-CSF-differentiated macrophages contributes LPS-mediated IL-10: implication in hyperglycemia. J Leukoc Biol 2015;97:327–39.

[171] Cani PD, Amar J, Iglesias MA, Poggi M, Knauf C, Bastelica D, et al. Metabolic endotoxemia initiates obesity and insulin resistance. Diabetes 2007;56:1761–72.

[172] Rorato R, Borges BC, Uchoa ET, Antunes-Rodrigues J, Elias CF, Elias LLK. LPS-induced low-grade inflammation increases hypothalamic JNK expression and causes central insulin resistance irrespective of body weight changes. Int J Mol Sci 2017;18:E1431.

[173] Liang H, Hussey SE, Sanchez-Avila A, Tantiwong P, Musi N. Effect of lipopolysaccharide on inflammation and insulin action in human muscle. PLoS ONE 2013;8:e63983.

[174] Ghosh S, Lertwattanarak R, Garduño-García JdeJ, Galeana JJ, Li J, Zamarripa F, et al. Elevated muscle TLR4 expression and metabolic endotoxemia in human aging. J Gerontol A Biol Sci Med Sci 2015;70:232–46.

[175] Mussig K, Staiger H, Machicao F, Stancakova A, Kuusisto J, Laakso M, et al. Association of common genetic variation in the FOXO1 gene with beta-cell dysfunction, impaired glucose tolerance, and type 2 diabetes. J Clin Endocrinol Metab 2009;94:1353–60.

[176] Muller YL, Hanson RL, Wiessner G, Nieboer L, Kobes S, Piaggi P, et al. Assessing FOXO1A as a potential susceptibility locus for type 2 diabetes and obesity in American Indians. Obesity (Silver Spring) 2015;23:1960–5.

[177] Zhao Y, Wei J, Hou X, Liu H, Guo F, Zhou Y, et al. SIRT1 rs10823108 and FOXO1 rs17446614 responsible for genetic susceptibility to diabetic nephropathy. Sci Rep 2017;7:10285.

[178] Li T, Wu X, Zhu X, Li J, Pan L, Li P, et al. Association analyses between the genetic polymorphisms of HNF4A and FOXO1 genes and Chinese Han patients with type 2 diabetes. Mol Cell Biochem 2011;353:259–65.

[179] Gong L, Li R, Ren W, Wang Z, Yang M, Zhang S. The FOXO1 gene-obesity interaction increases the risk of type 2 diabetes mellitus in a chinese han population. J Korean Med Sci 2017;32:264–71.

[180] Karim MA, Craig RL, Wang X, Hale TC, Elbein SC. Analysis of FOXO1A as a candidate gene for type 2 diabetes. Mol Genet Metab 2006;88:171–7.

[181] Nair AK, Sugunan D, Kumar H, Anilkumar G. Association analysis of common variants in FOXO3 with type 2 diabetes in a South Indian Dravidian population. Gene 2012;491:182–6.

[182] Willcox BJ, Donlon TA, He Q, Chen R, Grove JS, Yano K, et al. FOXO3A genotype is strongly associated with human longevity. Proc Natl Acad Sci U S A 2008;105:13987–92.

[183] Anselmi CV, Malovini A, Roncarati R, Novelli V, Villa F, Condorelli G, et al. Association of the FOXO3A locus with extreme longevity in a southern Italian centenarian study. Rejuvenation Res 2009;12:95–104.

[184] Flachsbart F, Caliebe A, Kleindorp R, Blanche H, von Eller-Eberstein H, Nikolaus S, et al. Association of FOXO3A variation with human longevity confirmed in German centenarians. Proc Natl Acad Sci U S A 2009;106:2700–5.

[185] Li Y, Wang WJ, Cao H, Lu J, Wu C, Hu FY, et al. Genetic association of FOXO1A and FOXO3A with longevity trait in Han Chinese populations. Hum Mol Genet 2009;18:4897–904.

[186] Sun L, Hu C, Zheng C, Qian Y, Liang Q, Lv Z, et al. FOXO3 variants are beneficial for longevity in Southern Chinese living in the Red River Basin: a case-control study and meta-analysis. Sci Rep 2015;5:9852.

[187] Pawlikowska L, Hu D, Huntsman S, Sung A, Chu C, Chen J, et al. Association of common genetic variation in the insulin/IGF1 signaling pathway with human longevity. Aging Cell 2009;8:460–72.

[188] Tan Q, Soerensen M, Kruse TA, Christensen K, Christiansen L. A novel permutation test for case-only analysis identifies epistatic effects on human longevity in the FOXO gene family. Aging Cell 2013;12:690–4.

[189] Soerensen M, Dato S, Christensen K, McGue M, Stevnsner T, Bohr VA, et al. Replication of an association of variation in the FOXO3A gene with human longevity using both case-control and longitudinal data. Aging Cell 2010;9:1010–7.

[190] Kleindorp R, Flachsbart F, Puca AA, Malovini A, Schreiber S, Nebel A. Candidate gene study of FOXO1, FOXO4, and FOXO6 reveals no association with human longevity in Germans. Aging Cell 2011;10:622–8.

[191] Shimokawa I, Komatsu T, Hayashi N, Kim SE, Kawata T, Park S, et al. The life-extending effect of dietary restriction requires Foxo3 in mice. Aging Cell 2015;14:707–9.

[192] Chiba T, Kamei Y, Shimizu T, Shirasawa T, Katsumata A, Shiraishi L, et al. Overexpression of FOXO1 in skeletal muscle does not alter longevity in mice. Mech Ageing Dev 2009;130:420–8.

[193] Lee S, Dong HH. FoxO integration of insulin signaling with glucose and lipid metabolism. J Endocrinol 2017;233:R67–79.

[194] Nagashima T, Shigematsu N, Maruki R, Urano Y, Tanaka H, Shimaya A, et al. Discovery of novel forkhead box O1 inhibitors for treating type 2 diabetes: improvement of fasting glycemia in diabetic db/db mice. Mol Pharmacol 2010;78:961–70.

[195] Crunkhorn S. Drug discovery: targeting transcription factors. Nat Rev Drug Discov 2018;18:18.

[196] Laudisi F, Cherubini F, Monteleone G, Stolfi C. STAT3 interactors as potential therapeutic targets for cancer treatment. Int J Mol Sci 2018;19:E1787.

[197] Chong PSY, Chng WJ, de Mel S. STAT3: a promising therapeutic target in multiple myeloma. Cancers (Basel) 2019;11:E731.

[198] Jiang X, Wu M, Xu Z, Wang H, Yu X, Li Z, et al. HJC0152, a novel STAT3 inhibitor with promising anti-tumor effect in gastric cancer. Cancer Manag Res 2018;10:6857–67.

[199] Haftchenary S, Luchman HA, Jouk AO, Veloso AJ, Page BD, Cheng XR, et al. Potent targeting of the STAT3 protein in brain cancer stem cells: a promising route for treating glioblastoma. ACS Med Chem Lett 2013;4:1102–7.

[200] Kitamura H, Ohno Y, Toyoshima Y, Ohtake J, Homma S, Kawamura H, et al. Interleukin-6/STAT3 signaling as a promising target to improve the efficacy of cancer immunotherapy. Cancer Sci 2017;108:1947–52.

[201] Gusarova GA, Wang IC, Major ML, Kalinichenko VV, Ackerson T, Petrovic V, et al. A cell-penetrating ARF peptide inhibitor of FoxM1 in mouse hepatocellular carcinoma treatment. J Clin Invest 2007;117:99–111.

[202] Shukla S, Milewski D, Pradhan A, Rama N, Rice K, Le T, et al. The FOXM1 inhibitor RCM-1 decreases carcinogenesis and nuclear beta-catenin. Mol Cancer Ther 2019;18:1217–29.

[203] Sun L, Ren X, Wang IC, Pradhan A, Zhang Y, Flood HM, et al. The FOXM1 inhibitor RCM-1 suppresses goblet cell metaplasia and prevents IL-13 and STAT6 signaling in allergen-exposed mice. Sci Signal 2017;10:475.

CHAPTER

# 14

# Interplays between nutritional and inflammatory signaling and fat metabolism in pathophysiology of NAFLD

*Juan Zheng*[a,b], *Ting Chen*[a], *Xin Guo*[c], *James M. Ntambi*[d], *Chaodong Wu*[b]

[a]Department of Endocrinology, Union Hospital, Tongji Medical College, Huazhong University of Science and Technology, Wuhan, China; [b]Department of Nutrition, Texas A&M University, College Station, TX, United States; [c]Departments of Nutrition and Food Hygiene, School of Public Health, Shandong University, Jinan, China; [d]Departments of Biochemistry and Nutritional Sciences, University of Wisconsin-Madison, Madison, WI, United States

OUTLINE



Lipid Signaling and Metabolism. http://dx.doi.org/10.1016/B978-0-12-819404-1.00014-2

## Nutritional signaling

The liver is the central organ for handling nutrients including storage of glycerol and fats in its hepatocytes, manufacturing triglycerides and cholesterol, synthesizing glycogen, and producing bile from cholesterol. This property enables the liver to also play a central role in the pathogenesis of various metabolic diseases. As a common health problem associated metabolic dysregulation of nutrients, nonalcoholic fatty liver disease (NAFLD) is characterized by excess accumulation of triglycerides in the hepatocytes due to both increased inflow of free fatty acids (FFAs) and de novo lipogenesis (DNL). Although studying hepatic metabolism in human subjects is challenging, it has been clearly demonstrated that it is total energy intake, but not fat intake per se, that is a key mediator of liver fat content [1]. Glucose is the key nutrient to initiate lipogenesis and promote fat deposition in hepatocytes, which in turn interacts with inflammatory signaling. When glucose levels in the hepatic cells were elevated, certain key enzymes of lipogenesis are activated, including fructose-6-phosphate 2-kinase/fructose-2, 6-bisphosphatase, fatty acid synthase, acetyl-CoA carboxylase, and L-type pyruvate kinase (LPK). The transcription of genes for these enzymes is also induced by a high-carbohydrate diet. This in turn acts to promote the conversion of glucose into triglycerides for long-term storage [2]. Much evidence suggests not only dietary simple sugar such as sucrose and fructose but also some complex carbohydrates from diet may promote the progress of NAFLD. It has been confirmed that the dietary sucrose contributes to the accumulation of lipid in hepatocytes by stimulating DNL [3], leading to the generation of long-chain saturated fatty acids in rats with methionine- and choline-deficient (MCD)-mediated NAFLD. Previous studies have shown that up to 26% hepatic lipid originates from DNL [4], and reducing dietary simple sugar can decrease DNL and inhibit its toxic effect on the liver [5]. Dietary sucrose is capable of inducing steatohepatitis in a manner dependent on fructokinase as this is suggested by the finding that fructokinase deficiency protects mice from steatohepatitis induced by sucrose [6]. Recent studies also have demonstrated that high fructose diets increase the risk factors for cardiovascular disease and metabolic syndrome. The possible underlying mechanisms are attributable to the metabolic effects of dietary fructose resulted from rapid fructose metabolism in the liver. The latter is catalyzed by fructokinase C, which produces the substrates for DNL and leads to elevated uric acid levels [7]. Consistently, a study by Cox et al. has shown that 10-week fructose-beverage-consumption caused significant increases in the levels of inflammatory factors including plasminogen activator inhibitor-1 (PAI-1), monocyte chemoattractant protein-1, and E-selectin; as well as retinol binding protein-4 and the liver enzyme, gamma glutamyl transferase [8,9]. In addition to the findings that simple sugar in diets increases lipogenesis, several complex carbohydrates including aspartame sweetener and caramel colorant, which are rich in advanced glycation end products, likely act to aggravate inflammation and IR [10]. To be noted, studies indicated that it is elevated lipid deposition including saturated fatty acids (SFAs) and monounsaturated fatty acids (MUFAs), more D6D and SCD-1 activities, higher ACC1 expression and DNL in the rats fed a high-carbohydrate diet, but not high-fat diet, that are intensively associated with liver inflammation state [11]. Also, a clinical study comprising of obese patients underwent bariatric surgery suggested that higher carbohydrate intake was significantly associated with a higher risk of inflammation compared with higher fat diet [12]. In terms of the pathophysiology of metabolic inflammation, macrophages accumulated in human adipose tissue have been identified as key factors in inflammation and obesity-induced insulin resistance [13–15]. However, the factors that act to activate ATM have not been fully studied. Recent studies suggest that treating macrophages with mixtures of glucose, insulin, and palmitate produces a "metabolically activated" complex macrophage phenotype distinct from classical activation, contributing to development of liver inflammation as the driver of nonalcoholic steatohepatitis (NASH) [16].

A significant number of studies also have indicated that the composition of dietary fats plays a role in liver fat accumulation. In particular, diets enriched in saturated fat appear to increase liver fat content to a greater extent compared with diets enriched in unsaturated fats [17]. In the context of NAFLD pathogenesis, it is proposed that hepatic steatosis is caused by the accumulation of lipid in hepatocytes secondary to multiple factors, including environmental, metabolic, and genetic background, which may all occur simultaneously. In most cases, insulin resistance results in an increase in lipolysis in the adipose tissue, leading to increased FFA mobilization to the bloodstream and increased influx of FFAs to hepatocytes. In the circulation, the predominant forms of fatty acids are long-chain fatty acids oleate and palmitate [18]. Fatty acids accrue in liver by hepatocellular uptake from the plasma and by de novo biosynthesis [19]. FFAs enter the cells by both passive diffusion and facilitated transport by fatty acid transport protein (FATP) and fatty acid translocase (FAT/CD36) [20]. Increased uptake of plasma FFAs derived from lipolysis significantly contributes to NAFLD development, as this is supported by the finding that liver-specific knockout of fatty acid transporters, FATP-2 [21] and FATP-5, protected mice from development of NAFLD [22]. Excessive FFAs are rapidly esterified within hepatocytes to form diacylglycerides (DAGs) and triacylglycerides (TAGs) [20]. Thus, significant increases in hepatic DAG, TAG, and total lipid content are shown in patients with NASH and, to a greater extent, in patients with NAFLD; although no major increase in hepatic FFAs is noted in NAFLD or NASH patients when compared to normal patients [23]. Cells unable to sequester chemically reactive lipid molecules undergo mitochondrial injury, endoplasmic reticulum (ER) stress [24] and autophagy, which all are critical processes of NASH pathogenesis. In fact, a growing body of evidence suggests that esterification of FFAs into TGs is not a toxic event but rather is part of a detoxification process for the liver [25]. The specific lipotoxic lipids that promote cell injury and NASH phenotype are not yet identified. However, in vitro and in vivo studies have demonstrated that excessive accumulation of nonesterified saturated FFAs, such as palmitate (C16:0) or stearate (C18:0), is hepatotoxic because saturated FFAs are capable of inducing ER stress and hepatocyte lipoapoptosis [26]. In contrast, unsaturated FFAs, such as oleate (C18:1) or palmitoleate (C16:1), are relatively nontoxic largely because unsaturated FFAs commonly are rapidly esterified and incorporated into triglycerides, and act to inhibit lipoapoptosis through blocking ER stress [27].

Cholesterol accumulation in the liver also promotes NASH [28]. Of note, including cholesterol in diets is shown to increase the severity of diet-induced NAFLD phenotype in mice [29]. As amphipathic molecules, bile acids (BAs) are generated from cholesterol oxidation in the liver and also play an essential role in the progression of NAFLD. At physiological concentrations, BAs function as a master regulator in digestion and absorption of lipids. However, excessive amount of BAs exerts detrimental effects on hepatocytes [30]. BAs also help maintain TG homeostasis and act as a metabolic regulator of glucose uptake and lipid metabolism [31]. For instance, BAs are capable of binding and activating farnesoid X receptor

(FXR) in the liver and inhibiting sterol regulatory element-binding protein 1c (SREBP1c)-mediated lipogenesis [32].

## Nutrient sensing and fat accumulation

Excessive accumulation of fatty acids causes liver steatosis, ER stress, and increased production of reactive oxygen species (ROS), thereby bringing about NAFLD phenotype and promoting development of NASH. During hepatic TAG formation, fatty acids are derived from diets, DNL, and lipolysis of fats in adipose tissue. It is now widely accepted that elevated DNL initiates simple steatosis and drives the progression of simple steatosis to NASH when liver lipotoxicity is overt. Of note, the driver of NAFLD/NASH is the type(s) of lipid molecules that are accumulated, but not the quantity of liver fat. Also, how lipid molecules are "packaged" also critically causes the initiation of subcellular injury. In NAFLD patients, the liver reveals increases in hepatic levels of TG, DG, cholesteryl ester (CE), and unesterified cholesterol and lysophosphatidyl-choline (LPC), along with changes in the percentages of polyunsaturated fatty acid (PUFA) in these lipids. Lipotoxicity-related mechanisms of NAFLD can be better explained by the "double-hit" hypothesis. As the "first hit," lipid accumulation increases the vulnerability of the liver. As the "second hit," the reduced glutathione levels due to oxidative stress lead to overactivation of c-Jun N-terminal kinase (JNK)/c-Jun signaling that induces hepatic cell death, pro-inflammatory cytokine expression [33]. Accumulation of toxic levels of ROS is caused by the ineffectual cycling of the ER oxidoreductin (Ero1)-protein disulfide isomerase oxidation cycle through the downstream of the inner membrane mitochondrial oxidative metabolism and Kelch like-ECH-associated protein 1 (Keap1)-Nuclear factor (erythroid-derived 2)-like 2 (Nrf2) pathway [34].

However, multiple findings suggest that FFA-mediated cytotoxicity to hepatocytes is attributable to the generation of the toxic metabolites, such as LPC [35], but not esterified products (TGs). Also, the disposal of fatty acids through formation of triglycerides is generally thought as an adaptive or protective response to a supply of fatty acids that exceeds the capacity to metabolize them [36]. This is in accordance with the finding that FFA levels are increased in the liver of mice fed an MCD diet [37]. Also, the circulating FFAs increase in NAFLD in parallel with the increase in the severity of the disease [38]. Similarly, other lipid metabolites such as ceramides or free cholesterol (FC) are accumulated in NAFLD and contribute to hepatocyte injury and the pathogenesis of the disease. Increased hepatocyte apoptosis and elevated serum caspase-cleaved cytokeratin 18 fragments have also been validated in patients with NAFLD. Moreover, the amount of hepatocyte apoptosis correlates with the severity of NAFLD and contributes to disease pathogenesis [39].

## AMPK

Hepatic fuel selection can change considerably under different nutritional situations. Over the last few years, a growing body of evidence indicates that AMP-activated protein kinase (AMPK), a serine threonine kinase comprising a catalytic $\alpha$ subunit and regulatory $\beta/\gamma$ subunits, is a cellular energy sensor, and represents a point of convergence of regulatory signals monitoring systemic and cellular energy status [40]. AMPK is a key controller of energy balance. Activation of hepatic AMPK leads to a decrease in ATP-consuming pathways and an increase in ATP-producing pathways. AMPK activation may reduce NAFLD by (1) suppressed hepatic lipogenesis, (2) increased fatty acid oxidation (FAO) by increasing mitochondrial function, biogenesis and integrity, (3) suppressed proinflammatory response signaling pathways [41], and (4) induced autophagy [42].

AMPK increases FAO and simultaneously inhibits hepatic lipogenesis due to suppression of fatty acid synthase [43] and SREBP1/2 [44]. Moreover, studies indicated that the activators of AMPK could reduce liver lipids, inhibit hepatic lipogenesis and increase insulin sensitivity [45]. Liver-specific activation of AMPK decreased hepatic lipogenesis in vivo and completely protected against hepatic steatosis [46]. Surprisingly, recent studies indicated that liver-specific AMPKa1a2 KO mice did not result in hyperglycemia or reduction in HGP, and AICAR still suppressed HGP in mice without detectable liver AMPK activity. In addition, the activation of liver AMPK via the AMPK β1-specific activator A769662 did not inhibit HGP in hepatocytes [47], which suggests that AMPK is not necessary for inhibiting hepatic gluconeogenesis and AICAR reduces glucose via an AMPK-independent manner [47–49]. Despite reductions in liver ATP content in humans with NAFLD [50], liver AMPK activity may actually be reduced in rodents [51], suggesting that additional factors are important for controlling AMPK activity in the liver during NAFLD. One possibility is that inflammatory factors known to be elevated with NAFLD, such as lipopolysaccharide (LPS) and tumor necrosis factor-α (TNFα), reduce AMPK activity. Studies found that the reduction of activation of AMPK is in response to some chronic low-grade inflammation induced by TNFα, LPS, or by high-fat diet (HFD)-induced obesity in multiple tissues including skeletal muscle [52], liver [41,53], WAT [54], BAT [55], and macrophages [56], which commonly exists in both mice and humans. These effects of AMPK may be mediated through phosphorylation of ACC. Because the hepatocytes isolated from the ACC knock-in mice exhibit an elevated fatty acid synthesis and a decrease in FAO. It has also been confirmed in young ACC knock-in mice fed a control chow diet, the rapidly developed NAFLD and liver insulin resistance lead to glucose intolerance and early fibrosis [57]. In addition to inhibiting ACC, AMPK may influence the liver fatty acid metabolism by controlling mitochondrial content, quality, and function, which are mediated through the regulation of mitochondrial biogenesis, autophagy (mitophagy), and fission. The downstream mediator on these pathways may be PGC1a, HDAC [58], ULK1 [59], and MFF [58,60]. Recent studies highlight the deregulation of these pathways may lead to an elevation of NAFLD in both mice and humans [41,61,62]. In addition to the increase of mitochondrial content, mitophagy and mitochondrial fission may contribute to this impairment. However, this view has yet to be proven by more detailed studies [63]. A number of direct AMPK activators have been proposed to be therapeutically beneficial for the treatment of NAFLD, including the widely used antidiabetic drug, metformin.

Autophagy, known as a degrading process in lysosomes, can regulate cellular lipid metabolism, ameliorate the state of IR, and mediate overactive innate immunoreaction. It is widely acknowledged that immunological diseases, diabetes, and NAFLD have an intertwined relationship with dysfunctional autophagy [64,65]. Studies identified the relationship between autophagy and lipolysis that mice fed with a high-fat diet or MCD diet had markedly decreased levels of autophagy [66]. Silencing the ATG5 gene with siRNA or treating the mice with 3-aleimidopropionic acids significantly elevated the accumulation of lipid droplets in liver cells. Moreover, recent studies show that the promotion of autophagy by rapamycin was able to negatively regulate lipid deposition and ER stress [67]. It is well documented that autophagy is up regulated during the early stage of NAFLD to prevent lipid accumulation [68] However, as NAFLD progresses, the autophagy process is blocked [69]. Studies have highlighted that AMPK is an autophagy inducer by Atg1/Unc-51-like-kinase1 (ULK1) phosphorylation (Ser 555) [42,70]. Under nutrient sufficiency, high mTOR activity prevents Ulk1 activation by phosphorylating Ulk1 Ser757 and disrupting the interaction between Ulk1 and AMPK. This coordinated phosphorylation is important for Ulk1 in autophagy induction [70].

### AKT-mTOR-SREBP signaling

Sterol regulatory-element binding proteins (SREBPs) are transcription factors that regulate the expression of genes involved in lipid synthesis and function as key nodes of convergence and divergence within global biological signaling networks involved in various physiological and pathophysiological processes. Increased evidence indicates that the AKT/mTOR/SREBP-1 signaling pathway is a key pathway to regulate hepatic cellular lipid metabolism. Distinctive physiological roles of SREBPs have been established: SREBP1a is involved in global lipid synthesis and growth; SREBP1c is involved in fatty acid synthesis and energy storage [71]; and SREBP2 is involved in cholesterol regulation. SREBP1 activation causes lipid-mediated cellular stress (lipotoxicity) that contributes to hepatosteatosis and atherosclerosis, thereby further extending SREBP-related pathology to include inflammation and fibrosis in various organs. Akt activates mTOR in hepatocytes [72]. Study shows that [73]: Notch signaling increases mTOR complex-1 stability, augmenting mTorc1 function and sterol regulatory SREBP1c-mediated lipogenesis while liver-specific ablation of Notch signaling prevents hepatosteatosis by blocking mTOR complex-1 activity.

Insulin physiologically activates Akt through the insulin receptor/PI3K pathway to inactivate FOXO1 by phosphorylating and promoting its nuclear exclusion. FOXO1 induces the expression in the liver of lipogenic genes such as FAS and PANDER, and increased lipid deposition [74]. Under severe insulin resistance, inactivating hepatic FOXO1 via insulin- independent mechanism(s) may be a potential therapeutic strategy for the treatment of NAFLD [75].

## Fat metabolism: FFAs as signaling molecules

Hepatic steatosis is the net outcomes of increased de novo hepatic DNL, increased hepatic uptake of fatty acids, decreased hepatic FAO, and/or decreased export of very low-density lipoproteins (VLDL)-triglycerides from the liver [4,76–82]. Much evidence further supports more important roles for increased DNL and decreased FAO in contributing to hepatic steatosis [83–85]. To be noted, malonyl-CoA, the product of the first step of DNL, is also a powerful inhibitor of carnitine palmitoyltransferase 1a (CPT1a), whose activation stimulates FAO through transferring long-chain FFAs to mitochondria for oxidation. Because of this, nutritional control of DNL has profound effects on regulating development and progression of hepatic steatosis. Under conditions of nutrition stress or excessive intake of nutrients, glucose is a strong signal to stimulate DNL, whose action is mediated largely by activating transcription activity of carbohydrate-responsive element-binding protein (ChREBP) [86]. In response to glucose stimulation, insulin, at the elevated levels, functions as the most powerful lipogenic signal to increase DNL, largely through increasing transcription activity of SREBP1c [87–90]. As the products of DNL, FFAs are substrates for the synthesis of TGs, which are then packaged into VLDL in hepatocytes. After its release, VLDL is hydrolyzed to generate FFAs, whose uptake by fat tissues leads fat storage. Of importance, FFAs also function as signaling molecules to regulate fat metabolic pathways and inflammatory responses in hepatocytes.

As mentioned above, FFAs critically regulate various events related to hepatic steatosis and inflammation. These roles of FFAs have been further manifested by the differences in actions of saturated FFAs vs. unsaturated FFAs. For instance, saturated FFAs are more pro-inflammatory [91], whereas unsaturated FFAs are more associated with metabolic status of obesity. Studies have suggested that there is an uneven accumulation of saturated FAs in animals with steatohepatitis, and this change is related to inhibition of fatty acid desaturase stearoyl Co-A desaturase 1 (SCD1) [92]. SCD is a central lipogenic enzyme that represents a potential target for the control of lipogenesis. Hepatic SCD1 plays a key role in prevention of steatohepatitis by partitioning excess lipid into MUFA that can be safely stored [93]. Hepatic SCD1 deficiency caused a significant reduction in hepatic lipogenic gene expression and reduced DNL associated with reduced hepatic TG secretion [94]. Inhibition of SCD1 by antisense oligonucleotides protected against diet-induced obesity, insulin resistance, and hepatic steatosis also elevated the LPS-induced inflammatory response in peritoneal macrophages through increased activation of TLR4 [95]. However, Liu identified higher SCD2 expression in macrophages than in SCD1; SCD2 may compensate for the loss of SCD1 in macrophages via certain mechanisms [96]. Study also showed that although both oleate and palmitoleate are enzymatic products of SCD1, only oleic, but not palmitoleic or linoleic, acid contributes to the inflammation in RAW macrophages [97]. How this subtle structural difference in the FAs that leads to differential outcomes of inflammation signaling needs further investigation.

FFAs also serve as critical signals to regulate DNL. This role of FFAs is further manifested by the findings that saturated FFAs and unsaturated FFAs reveal distinct effects on DNL. For instance, linoleate [C18:2 (n-6)], EPA [C20:5 (n-3)], and DHA [C22:6 (n-3)], but not stearate and oleate, are shown to decrease the effect of glucose on stimulating FAS expression via mechanisms involving ChREBP [98]. Similarly, at the whole animal level, substitution of linoleic acid with α-linolenic acid or long-chain n-3 PUFA prevents the effect of a high-fat and high fructose diet (HFHF) on inducing hepatic steatosis [99], which is attributable to, in large extent, the effect of PUFAs on suppressing lipogenic gene expression. Unlike PUFAs, monounsaturated FFAs appear to be more lipogenic. This is particularly true for palmitoleate (PO, C16:1 n7). At the whole animal level, supplementation of PO brings about hepatic steatosis, which is accompanied with increased expression of hepatic genes for lipogenesis such as FAS [100]. At the cellular level, PO acts to increase the binding of SREBP1c to FAS promoter in hepatocytes, which accounts for the lipogenic effect of PO. To date, however, it remains to be elucidated how FFAs function as signaling molecules, for example, whether and how different FFAs act though certain receptors and/or mediators to display distinct effects on fat metabolism.

## Glucose and fat metabolism in acetylation

Liver mitochondrial acetyl-CoA and nonenzymatic acetylation may play a pivotal role in regulating metabolic responses of inflammatory stress. For example, a recent study revealed an increase of hepatic acetyl-CoA content resulted from inflammation-induced lipolysis [101]. This accumulation of acetyl-CoA appeared to be due to FAO rather than pyruvate oxidative decarboxylation during the inflammatory stress response. Consistent with this, another study in which $^{13}$C-labeled palmitate and $^{13}$C-labeled glucose were used to test, which is the primary carbon source of acetyl-CoA during the inflammatory stress response, showed that the percentage of $^{13}$C-labeled palmitate-derived acetyl-CoA was definitely more than that from glucose, while the lactate levels were derived almost entirely from the $^{13}$C-labeled glucose. In addition, this study validated that the contribution of glucose to the TCA cycle was reduced by FAO upon examining the isotope-labeling pattern of citrate. Moreover, acetyl-CoA may induce acetylation of malate dehydrogenase 2 (MDH2) and glutamic-oxaloacetic transaminase 2 (GOT2). The latter brings about subsequent improvement of the enzymatic activity of MDH2. As supporting evidence, treatment with acetyl-CoA directly increased not only the MDH2 activity but also the acetylation level of MDH2 and GOT2 [102]. MDH2 and GOT2 are two major mitochondrial enzymes in the MAS pathway that is known to be closely associated with glycolysis [103]. Given this, the acetyl-CoA induced by hepatic FAO is essential to the regulation of hepatic mitochondrial nonenzymatic acetylation, which modulates inflammatory-metabolic reprogramming of glucose and may provide a novel therapeutic direction.

## Lipotoxicity: Oxidative stress, apoptosis and inflammation

Although excessive accumulation of TG in hepatocytes is a hallmark of NAFLD, simple steatosis is not necessarily pathogenic since NAFLD, at the early stage, is reversible with weight loss and exercise. Oxidative stress and release of proinflammatory cytokines, such as TNFα, are major consequences of hepatic lipid overload, and serve as key factors contributing to progression of NAFLD to NASH. Oxidative stress and activation of the unfolded protein response are two well-characterized cell stress pathways that promote cell death in NASH [104]. FFAs and cholesterol, especially when accumulated in the mitochondria, are "aggressive" lipids leading to TNFα and ROS production and acting as early "inflammatory" hits, which act to promote NASH [105]. The concept of lipotoxicity and the involved lipid species have been mentioned before. Initially, excessive amount of FFAs causes lipotoxicity via induction of ROS release, which in turn causes inflammation, apoptosis, and consequently, the progression to NASH and fibrogenesis [106,107].

While the mechanisms underlying NASH are complicated, it is clear that two processes are involved in the pathogenesis of disease: inflammation and fibrogenesis. Hepatic inflammation is a key factor that leads to histological damage and the progression of NASH, resulting in terminal liver diseases, such as cirrhosis, hepatocellular carcinoma, and liver failure. Inflammation is a critical response to tissue damage or infection, in which secreted mediators, such as cytokines, chemokines, and eicosanoids coordinate cellular defenses and tissue repair. Nutrition stress is shown to overly activate immunity, resulting in inflammation and related metabolic diseases such as NAFLD. Compelling evidence shows that the initiation and perpetuation of liver inflammation are crucial to the pathogenesis of NAFLD and NASH [13,53,108–110].

Both liver resident cells (e.g., Kupffer cells, hepatic stellate cells, sinusoidal endothelial cells) and cells which are recruited in response to injury (e.g., monocytes, macrophages, dendritic cells, natural killer cells) release proinflammatory signals and contribute to the apoptotic or necrotic demise of hepatocytes in the pathogenesis of NASH. Two notable features of inflammation in NASH are an increase in infiltrating immune cells (macrophages, lymphocytes, eosinophils, neutrophils, and leukocyte subsets) into hepatic cords [111] and an increase in proinflammatory activation of immune cells. Many patterns of macrophage accumulation are recognized in NASH including clusters termed microgranulomas, fat droplet containing lipogranulomas, and surrounding ballooned hepatocytes termed "satellitosis" [112]. Various studies showed that the severity and composition of the inflammatory infiltrates correlate with the degrees of steatosis and subsequent increases in the risk of progression of NAFLD [113,114]. In fact, the recruitment of extra-hepatic inflammatory cells to the site of inflammation is mainly mediated by the interactions between the chemokines and cytokines that were secreted by activated stellate cells and Kupffer cells, and their ligands [115]. Mouse studies have demonstrated that hepatic inflammasome activation in the liver leads to the expression of proinflammatory cytokines, such as interleukin-1β (IL-1β) and IL-18, and promotes apoptosis through caspase-1 activation [116,117]. Key inflammatory signals, including IL-1β, TLR4, TNFα, and $CD^{11b+}$ and $CD^{11c+}$ macrophages, were particularly associated with liver inflammation. Lanthier et al. have likewise shown that macrophage inflammation of adipose tissue is responsible for the early stages of both hepatic and peripheral insulin resistance in HFD-fed mice, but deletion of adipose macrophages cannot reverse the later phase of NAFLD once liver inflammation is established [118].

Inflammation affecting or infiltrating the liver in NASH may also originate outside the liver. One site of interest is the adipose tissue, particularly visceral adipose tissue that is expanded in NAFLD [110,119]. Macrophage polarization is often associated with the proinflammatory state in adipose tissue. M1 macrophages are considered proinflammatory or "classically activated" because they produce proinflammatory cytokines such as IL-1β, IL-6, IL-8, IL-12, IL-17, and TNFα play a pivotal role in the triggering of tissue injury. In contrast, M2 macrophages differentiate in response to IL-4, IL-13, and IL-10 and are involved in tissue repair and efficient phagocytosis of cellular debris (efferocytosis) [108]. In particular, an increase in M1 macrophages number or M1/M2 ratio triggers the production and secretion of various proinflammatory signals (i.e., adipocytokines). As a result, these inflammatory factors reach the liver, leading to local M1/M2 macrophage polarization and consequent onset of the histological damage characteristic of NAFLD [108]. A study showed that in mice fed an HFD, inflammatory signals shift from adipose tissue to the liver and influence the development of steatohepatitis [120]. When mice were fed with high fat and cholesterol diet, macrophage recruitment and cytokine transcripts were up regulated in adipose tissue at 6–16 weeks, before their appearance in the liver from 16 to 26 weeks [120]. B cell-activating factor (BAFF) is an adipokine belonging to the TNF-ligand family related to impaired insulin sensitivity, and the serum BAFF concentrations are associated with NAFLD severity [121]. Moreover, BAFF deficiency ameliorated obesity-associated insulin resistance and inflammation in VAT and prevented fat accumulation in the liver [121].

Distinct immune signatures and signaling pathways have been found not only in immune cells but also in hepatocytes. In support of this, hepatocyte inflammasome activation appears to be an important link between the initial metabolic stress and subsequent hepatocyte death and stimulation of fibrogenesis in NASH [122]. Stressed hepatocytes release increased number of extracellular vesicles (EVs), which are known to participate in intercellular signaling and coordination of the behavior of immune cell populations via their cargo. Circulating EVs are increased in NASH and correlate with macrophage accumulation in the liver. As such, complex crosstalk between hepatocytes and other liver cells (e.g., macrophages) plays a decisive role in the pathogenesis of NAFLD [123], which is also summarized in Fig. 14.1.

## Inflammatory signaling pathways

Inflammation drives the progression of simple steatosis to NASH. Because of this, numerous studies have focused on elucidating how inflammatory pathways are activated to promote liver inflammatory damage during NAFLD with hope to identify new and effective targets for managing NASH. To date, a number of inflammatory pathways have been functionally validated in the context of NAFLD/NASH pathophysiology (Fig. 14.2).

### *TLR4*

In NAFLD, alterations of gut microbiota and increased intestinal permeability increase exposure of the liver to gut-derived bacterial products, such as LPSs and unmethylated CpG DNA [124]. These products stimulate innate immune receptors, namely Toll-like receptors (TLRs), which activate signaling pathways involved in liver inflammation and fibrogenesis. TLR4 is expressed in all parenchymal and nonparenchymal cell types, and contributes to

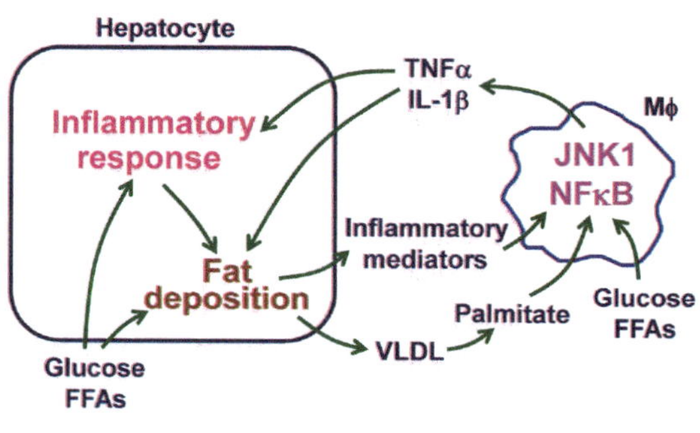


**FIGURE 14.1 Dysregulation of hepatocyte-macrophage crosstalk in pathogenesis of NAFLD.** Under nutrition stress, for example, excessive intake of simple sugar and fats (in particular saturated fats), glucose and saturated free fatty acids (FFAs) act to promote hepatocyte fat deposition (steatosis), and to increase hepatocyte proinflammatory responses. As a result of increased fat deposition, hepatocytes release inflammatory mediator and secrete more very low-density lipoproteins (VLDLs), whose hydrolysis generates palmitate. Through paracrine manners, hepatocyte-derived inflammatory mediators and palmitate act on liver macrophages/Kupffer cells to enhance their inflammatory responses, serving as a key mechanism underlying development and progression of liver inflammation. Meanwhile, glucose and FFAs act on liver macrophages/Kupffer cells to trigger or exacerbate their inflammatory responses. The latter not only contribute to liver inflammation, but also act on hepatocytes to increase hepatocyte inflammatory responses and to enhance hepatocyte fat deposition. Source: *Modified based on Zheng et al. Front Med 2015;9:173–86.*

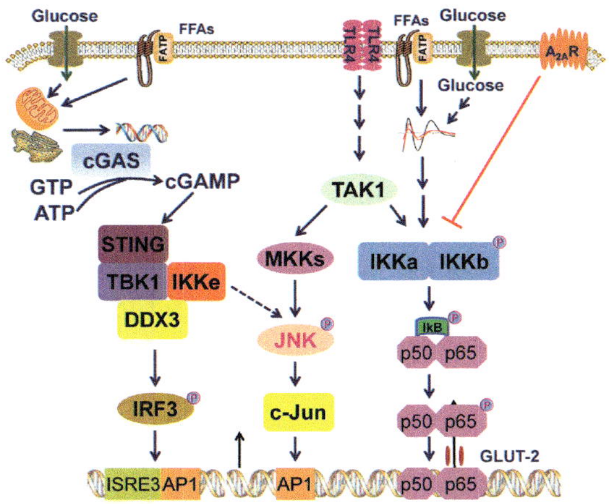


**FIGURE 14.2** **Interplays between nutritional and inflammatory signaling.** When excess in amount, nutrients, for example, glucose and palmitate, act directly and/or indirectly to activate several inflammatory signaling pathways in both hepatocytes and macrophages. Under the same conditions, palmitate, with or without glucose, also causes ER and/or mitochondrial stress, which in turn activates STING to enhance macrophage proinflammatory activation. Similarly, nutrition stress causes circadian clock dysregulation, which functions to enhance macrophage proinflammatory activation. In contrast, $A_{2A}R$, when increased by glucose and/or palmitate, acts in a defensive manner to suppress proinflammatory responses. Overall, the interplays between nutritional and inflammatory signaling underlie development and progression of liver inflammation during NAFLD. See text for details.

tissue damage, liver fibrosis, NASH, and HCC progression. There is accumulating evidence demonstrating that altered TLR4 and Sphingosine kinase1 (SphK1)/sphingosine 1 phosphate (S1P) signaling pathways are key players in the pathogenesis of NAFLD [124,125]. In particular, FFAs are shown to activate proinflammatory signals through TLR4, and act in a synergistic manner with LPS in an animal model of NASH [124–126]. Current evidence suggests that SFAs act as nonmicrobial TLR4 agonists, and trigger inflammatory responses downstream of TLR4 [126]. Additionally, SFAs activate TLR4-dependent signaling in both macrophages and adipocytes. At the whole animal level, TLR4 deficiency protects mice from insulin resistance driven by intravenous lipid infusion [126]. In the liver, the pathological effect (mediating the progression of simple steatosis to NASH) of TLR4 in Kupffer cells is achieved by inducing ROS-dependent activation of X-box binding protein-1 (XBP-1) [127].

### *$A_{2A}R$ signaling pathway*

Adenosine is an endogenous regulator of inflammation and tissue repair. Among the four adenosine receptors, adenosine 2A receptors ($A_{2A}R$) display powerful antiinflammatory effects in macrophages and hepatocytes. This effect is achieved through elevation of Pro-IL1β and NLRP3 and/or increased aspase-1 activation of macrophages [128]. $A_{2A}R$ is a critical part of the physiological negative feedback mechanism for limiting or terminating both tissue-specific and systemic inflammatory responses [129]. $A_{2A}R$-deficient mice have enhanced proinflammatory cytokines accumulation [129]. Moreover, $A_{2A}R$-deficient macrophages exhibited increased proinflammatory responses in vitro, and enhanced fat deposition of wild-type primary hepatocytes in macrophage and hepatocyte cocultures [130]. In primary hepatocytes,

$A_{2A}R$ deficiency increased the proinflammatory responses and enhanced the effect of palmitate on stimulating fat deposition [130]. MCD-fed $A_{2A}R$-disrupted mice revealed increased severity of hepatic steatosis compared with MCD-fed control mice [131]. Moreover, $A_{2A}R$ deficiency significantly increased the abundance of SREBP1c in livers of fasted mice and in hepatocytes upon nutrient deprivation [130]. In the absence of $A_{2A}R$, SREBP1c transcription activity was significantly increased in mouse hepatocytes [130]. Another study showed that $A_{2A}R$ stimulation inhibited NASH development by modulating the responses of CD4+T-helper (Th) cells and inhibiting the exacerbation of the IL-17-induced JNK-dependent lipotoxicity [132]. In vitro, incubation with MCD-mimicking media increased LPS-induced phosphorylation states of JNK p46 and/or NFκB p65 and cytokine mRNAs in control macrophages and RAW264.7 cells, but not primary hepatocytes [131]. Taken together, $A_{2A}R$ acts to suppress inflammation and lipogenesis, thereby playing a protective role in NASH development.

### NFκβ and Jnk signaling pathway

Nuclear factor kappa-light-chain-enhancer of activated B cells (NFκβ), a nuclear transcription factor, is a proinflammatory master switch that controls the production of a host of inflammatory markers and mediators, including C reactive protein (CRP), PAI-1, interleukin-6 (IL-6), TNFα, and IL-1β. NFκβ signaling pathway is prominent in regulating insulin resistance and metabolic homeostasis, along with high levels of systematic and/or local cytokine releases. NFκβ in hepatocytes leads to impaired hepatocytes homeostasis and enhanced liver inflammation, as well as cell death during hepatic injury. The activation of NFκβ induces the production of cytokines and contributes to the recruitment and activation of Kupffer cells to mediate inflammation in NASH [133]. NFκβ upregulates the expression of TNFα, Fas ligand (FasL), and TGFβ, which are considered to be major factors in the response to apoptosis and fibrosis that drive NASH progression [134]. Upon treatment with NFκβ p65 siRNA, HFD-fed (HFPS) mice were protected from hepatic steatosis and insulin resistance [135].

JNK is a mitogen-activated protein kinase (MAPK) family member that is activated by diverse stimuli, including cytokines (such as TNFα and IL-1), ROS, pathogens, toxins, drugs, ER stress, FFAs, and metabolic changes. JNK activation has been shown to contribute to development of NAFLD [136]. JNK mediates cellular responses to a variety of intra- and extracellular stresses. JNK and MAPKs are encoded by three genes, two of which are jnk1 and jnk2 that are expressed in hepatocytes [137]. The role of JNK activation in hepatocytes has been linked to the nuclear hormone receptor peroxisome proliferator-activated receptor-α (PPARα) and fibroblast growth factor 21 (FGF21). JNK1 is also expressed in macrophages. It has been reported that LPS stimulates the phosphorylation of JNK1, which induces macrophage proinflammatory activation [53]. FGF21 induces hepatic expression of peroxisome proliferator-activated receptor gamma coactivator protein-1alpha (PGC-1alpha), a key transcriptional regulator of energy homeostasis, and causes corresponding increases in FAO and tricarboxylic acid cycle flux [138]. Sustained hepatic JNK activity in HFD-fed mice potently repressed PPARα responsive genes through increased expression of a PPARα-corepressor, nuclear receptor corepressor 1 (NCoR1). NCoR1, which interacts with PPARα to suppress its transactivation, binds to the autophagosomal GABARAP family proteins and is degraded by autophagy. Therefore, hepatic JNK activation decreased the expression of PPARα target genes that increase fatty acid β-oxidation, included cytokines (such as TNFα), downregulated FGF21 expression and promoted the development of insulin resistance in the liver [138,139].

TNFα is another inflammatory molecule responsible for inflammation, apoptosis, and fibrosis of hepatocytes [140]. TNFα has been reported to play a role in hepatic insulin resistance through the upregulation of suppressor of cytokine signaling 3 (SOCS 3) [141]. By decreasing TNFα gene transcription with pentoxifylline (PTX, a phosphodiesterase inhibitor), lobular inflammation and fibrosis were significantly improved in patients with NASH [142].

### *cGAS-cGAMP-STING pathway*

The recognition of microbial nucleic acids is a major mechanism by which the immune system detects pathogens. Cyclic GMP-AMP (cGAMP) synthase (cGAS) is a cytosolic DNA sensor that activates innate immune responses through production of the second messenger cGAMP, which activates the adaptor stimulator of interferon genes (STING) [143]. The cGAS-cGAMP-STING pathway mediates type I interferon inflammatory responses in immune cells to defend against viral and bacterial infections. cGAMP is capable of switching antiinflammatory macrophages (M2) back to proinflammatory activation (M1) [144]. Recent studies have shown that this pathway also contributes to increased sterile inflammation, insulin resistance, and the development of NAFLD [145–147]. The levels of STING were increased in liver tissues from patients with NAFLD and mice with HFD-induced steatosis [146]. Loss of STING from macrophages decreased the severity of liver fibrosis and the inflammatory responses in mice [146]. Deficiency of STING prevented lipid accumulation in hepatocytes and protected mice from alcohol-induced liver injury [148]. Chronic exposure to STING agonist led to hepatic steatosis and inflammation in WT mice, but not in STING-deficient mice [147]. To be noted, hepatocyte fat deposition is shown to enhance hepatocyte production of mtDNA, which in turn acts on Kupffer cells to promote TNFα transcripts [149]. STING functions as an mtDNA sensor in the Kupffer cells of liver under lipid overload and induces NFκB-dependent inflammation in NASH [147]. Potential interactions of the cGAS-cGAMP-STING pathway with mTORC1 signaling [150], autophagy [150], mitochondrial damage [151], ER stress [148], and apoptosis [148] have been reported, suggesting an important role of the cGAS-cGAMP-STING pathway in the networking and coordination of these important biological processes. Development of STING inhibitors or manipulation of the cGAS-cGAMP-STING pathway may represent a novel approach to managing NASH in patients.

### *Timed nutrition and inflammatory signaling*

Circadian clocks are affected by many factors, including nutrition and metabolism. Circadian clocks located in cells and tissues throughout the body regulate daily rhythms, including inflammation and metabolism. In mammals, circadian rhythms are also observed in various metabolic processes. For example, lipid levels in the circulation fluctuate with a rhythm that peaks around mid-day [152], and AMPK is similarly regulated by the circadian rhythms of mRNA expression and phosphorylation in mouse liver that peak during the subjective day [153]. This rhythmic regulation of tissue- and cell-specific processes is widely accepted to play a pivotal role in the homeostatic regulation of inflammatory responses, metabolism, and other important physiologic processes. The relationship between metabolism and circadian clock is complicate and intertwined. For example, circadian clock regulates the metabolic state of body and metabolic status feeds back to the clock in turn so that feeding behavior directly confers molecular clock function [154]. Much evidence suggests nutrition stress, in particular nutrition overload, is closely related to circadian clock dysfunction. The specific mechanisms by which circadian clock disruption leads to metabolic disorders remains unclear. However, it has been shown that activation of proinflammatory macrophages and inflammatory signaling plays critical roles in this process. For instance, HFD feeding or palmitate is shown to cause circadian clock dysregulation, thereby leading to activation of proinflammatory signaling through JNK1 and NFκB [155]. The dysregulation of circadian clock aggravates macrophage proinflammatory activation, leading to adipose tissue inflammation and further potentiation of the effect of HFD feeding on inducing increases in body weight, hyperglycemia and IR [155], which all contribute to the pathogenesis of NAFLD.

## Perspective on management of NAFLD and NASH

The number of people who suffer from NAFLD is increasing rapidly due to, in large part, an ongoing pandemic of obesity. To date, however, there is no effective treatment for NAFLD. At this point, lifestyle intervention still remains the cornerstone of NASH treatment. In general, the current approaches for treating NAFLD include dietary intervention, exercise, weight loss agents, insulin sensitizers, and bariatric surgery, which all aiming to reduce hepatic fat accumulation. Current pharmacotherapeutic options for NAFLD included TZD, vitamin E, metformin, and GLP-1 analogue. TZDs play a key role in restoring insulin sensitivity and decreasing adipose tissue inflammation, generating histologic improvements in steatohepatitis. Pioglitazone can be used to treat NASH in patients with type 2 diabetes with biopsy proven NAFLD. Meanwhile, nondiabetic patients can be treated with vitamin E [156]. It is accepted that the current TZDs are first-generation, nonspecific activators of PPARγ, which may account for why utilization of TZDs results in a wide array of deleterious side effects and why currently the use of TZDs for treatment of NAFLD is limited. The development of highly targeted selective PPARδ modulators and dual PPARα/δ agonists might be new cues for their present dilemma [41,157]. Metformin is the first line drug for treatment of type 2 diabetes. Numerous clinical studies have investigated the effectiveness of metformin on patients with NAFLD and NASH [41]. However, further studies are needed to prove histologic improvements in patients with NAFLD. Additional drug candidates include GLP-1/GLP-2/GIP analog, Cilostazol [158], sodium-glucose cotransporter 2 (SGLT2) [159], obeticholic acid (OCA) [160], DPP4 inhibitor, and lipid-lowering agents. Although all of these medicines revealed some improvements in biochemical or histological parameters in NAFLD, their efficacy needs to be proven in large RCTs. OCA is a synthetic BA derivative and FXR agonist that was recently studied in NASH patients without cirrhosis in the FLINT trial (a multicenter, randomized, double-blind, placebo-controlled study). In this study, 283 patients were randomly assigned to receive either a daily dose of 25 mg OCA ($n = 141$) or placebo ($n = 142$) on histologic response (defined as decrease in nonalcoholic fatty liver disease activity score [NAS] by ≥2, with no worsening of fibrosis) for 72 weeks [160,161]. The findings indicated that treatment with OCA improved the histological features of NASH. However, the long-term benefits and safety of OCA need to be further validated. Indeed, there also were cases in which OCA treatment worsened lipid profile, which may warrant a closer monitor of possible cardiovascular consequences [160]. Additionally, in the appendix of FLINT trial participants with or without diabetes appeared to have different histological responses to OCA. In particular, in participants with diabetes, liver histology was improved in 53% of patients who received OCA vs. 19% for placebo whereas in patients without diabetes, liver histology was improved in 37% patients with OCA vs. 23% with placebo [160]. Moreover, OCA appears to be highly promising for the treatment of NASH and its upcoming phase III trial (NCT02548351), with more ambitious end points (including mortality and liver-related morbidity), will hopefully confirm the efficacy of this treatment in NASH patients. It is worth noting that several pharmacological agents were effective only in the animal models of NASH, but failed in human trials. These agents include PDE4 inhibitors [162], caspase inhibitors [163], resveratrol [164], ethyl-eicosapentaenoic acid (synthetic PUFA) [165], anti-TNFα, and probiotics [166].

It was recently suggested that an ideal drug choice for NASH should reduce hepatic inflammation and liver cell injury, and be able to correct the underlying insulin resistance, and deliver antifibrotic effects [167]. Because the immune system plays a critical role in the pathogenesis of NASH, various treatments are being developed to directly or indirectly target the relevant inflammatory pathways. Preclinical studies in animal models or human cells have indicated that reagents that inhibit inflammation pathways might be viable for treating patients with NAFLD and NASH. For example, MK615 suppresses the production of inflammatory cytokines such as TNFα and IL-6 by inactivating NFκB in RAW264.7 cells [168]. MK615 is regarded as a hepato-protective agent, as it has been shown that MK615 treatment decreased liver enzyme levels in patients with NAFLD and improved liver histology in rat models [169]. Pentoxifylline is known to inhibit lipid peroxidation. It also has antiinflammatory properties and acts to prevent TNFα synthesis, which plays a detrimental role in hepatocellular damage, inflammation, and fibrogenesis [170]. One clinical study indicated significant improvement in steatosis, lobular inflammation, and hepatocyte ballooning in NAFLD patients treated with pentoxifylline [170]. The recruitment of inflammatory monocytes and macrophages via chemokine receptor CCR2 as well as of lymphocytes and hepatic stellate cells via CCR5 promote the progression of NASH to fibrosis. Cenicriviroc (CVC), a CCR2/CCR5 dual antagonist, showed potent antiinflammatory and antifibrotic activity in animal models. In a phase 2b clinical trial (CENTAUR) [171] comprising of 289 patients with NASH and fibrosis, CVC consistently brought about liver fibrosis improvement after 1 year of therapy and had an excellent safety profile, leading to the implementation of a phase 3 trial (AURORA) [172]. Considering a key role of cytokines in NAFLD, monoclonal antibodies of cytokines may have a protective effect on the progression of NAFLD. It is worth mentioning that there are already monoclonal antibodies against IL-17 (secukinumab and ixekizumab) that have been released for the treatment of rheumatic diseases resulted from activation of IL-17-axis. In experimental studies, anti-IL-17 antibodies improved hepatic steatosis through suppressing IL-17-related fatty acid metabolism [173]. However, there are no clinical studies that have tested the use of anti-IL-17 antibodies in NAFLD. Although absence of clinical trials and controversial results may be present in some studies, there is no doubt that development of compounds that spare the liver from multiple levels of immune damage is highly promising. Therefore, searching the target of NAFLD immunotherapy could be a new direction in the research for the treatment of NAFLD.

## Conclusion and future directions

Mounting data generated from both clinical and basic research have significantly advanced our understanding of how the interplays between nutritional and inflammatory signaling contribute to development and progression of NAFLD. The pertinent findings also have validated the viability of various new targets for the management of NAFLD. However, it remains a big challenge that the multifactorial mechanisms for the pathogenesis of NAFLD in particular NASH make it difficult develop a single therapy to effectively treat NASH. While considering the beneficial effect of weight loss on improving NAFLD/NASH aspects, it may a time for us to be cautiously optimistic on nutrition approaches, that is, precision nutrition and timed nutrition, as opportunities to bring or open new directions for managingNAFLD and NASH. This, certainly, will not be possible without extensive trails in human populations.

## Conflict of interest

The authors declare that there is no conflict of interest.

## Acknowledgments

This work was supported in whole or in part by grants from the American Diabetes Association (1-17-IBS-145 to C.W.) and the National Institutes of Health (DK095862 to C.W.). C.W. is also supported by the Hatch Program of the National Institutes of Food and Agriculture (NIFA). J.Z. is supported by National Natural Science Foundation (81770772) and Hubei Province Natural Science Foundation (2019CFB701). X.G. is supported by National Natural Science Foundation Youth Project (81803224), Shandong University Basic Research Project (2017TB0028), and Shandong University Young Scholars Future Plan Project (2018WLJH33).

## References

[1] Parry SA, Hodson L. Influence of dietary macronutrients on liver fat accumulation and metabolism. J Investig Med 2017;65:1102–15.

[2] Goodridge AG. Dietary regulation of gene expression: enzymes involved in carbohydrate and lipid metabolism. Annu Rev Nutr 1987;7:157–85.

[3] Parks EJ, Krauss RM, Christiansen MP, Neese RA, Hellerstein MK. Effects of a low-fat, high-carbohydrate diet on VLDL-triglyceride assembly, production, and clearance. J Clin Invest 1999;104:1087–96.

[4] Donnelly KL, Smith CI, Schwarzenberg SJ, Jessurun J, Boldt MD, Parks EJ. Sources of fatty acids stored in liver and secreted via lipoproteins in patients with nonalcoholic fatty liver disease. J Clin Invest 2005;115:1343–51.

[5] Pickens MK, Yan JS, Ng RK, Ogata H, Grenert JP, Beysen C, et al. Dietary sucrose is essential to the development of liver injury in the methionine-choline-deficient model of steatohepatitis. J Lipid Res 2009;50:2072–82.

[6] Ishimoto T, Lanaspa MA, Rivard CJ, Roncal-Jimenez CA, Orlicky DJ, Cicerchi C, et al. High-fat and high-sucrose (western) diet induces steatohepatitis that is dependent on fructokinase. Hepatology 2013;58:1632–43.

[7] Stanhope KL, Schwarz JM, Havel PJ. Adverse metabolic effects of dietary fructose: results from the recent epidemiological, clinical, and mechanistic studies. Curr Opin Lipidol 2013;24:198–206.

[8] Cox CL, Stanhope KL, Schwarz JM, Graham JL, Hatcher B, Griffen SC, et al. Circulating concentrations of monocyte chemoattractant protein-1, plasminogen activator inhibitor-1, and soluble leukocyte adhesion molecule-1 in overweight/obese men and women consuming fructose- or glucose-sweetened beverages for 10 weeks. J Clin Endocrinol Metab 2011;96:E2034–8.

[9] Cox CL, Stanhope KL, Schwarz JM, Graham JL, Hatcher B, Griffen SC, et al. Consumption of fructose- but not glucose-sweetened beverages for 10 weeks increases circulating concentrations of uric acid, retinol binding protein-4, and gamma-glutamyl transferase activity in overweight/obese humans. Nutr Metab (Lond) 2012;9:68.

[10] Nseir W, Nassar F, Assy N. Soft drinks consumption and nonalcoholic fatty liver disease. World J Gastroenterol 2010;16:2579–88.

[11] da Silva-Santi LG, Antunes MM, Caparroz-Assef SM, Carbonera F, Masi LN, Curi R, et al. Liver fatty acid composition and inflammation in mice fed with high-carbohydrate diet or high-fat diet. Nutrients 2016;8:682.

[12] Solga S, Alkhuraishe AR, Clark JM, Torbenson M, Greenwald A, Diehl AM, et al. Dietary composition and nonalcoholic fatty liver disease. Dig Dis Sci 2004;49:1578–83.
[13] Chawla A, Nguyen KD, Goh YP. Macrophage-mediated inflammation in metabolic disease. Nat Rev Immunol 2011;11:738–49.
[14] Lumeng CN, Saltiel AR. Inflammatory links between obesity and metabolic disease. J Clin Invest 2011;121:2111–7.
[15] Wellen KE, Hotamisligil GS. Inflammation, stress, and diabetes. J Clin Invest 2005;115:1111–9.
[16] Kratz M, Coats BR, Hisert KB, Hagman D, Mutskov V, Peris E, et al. Metabolic dysfunction drives a mechanistically distinct proinflammatory phenotype in adipose tissue macrophages. Cell Metab 2014;20:614–25.
[17] Cheng Y, Zhang K, Chen Y, Li Y, Li Y, Fu K, et al. Associations between dietary nutrient intakes and hepatic lipid contents in NAFLD patients quantified by (1)H-MRS and dual-echo MRI. Nutrients 2016;8:527.
[18] Taegtmeyer H, Lam T, Davogustto G. Cardiac metabolism in perspective. Compr Physiol 2016;6:1675–99.
[19] Alves-Bezerra M, Cohen DE. Triglyceride metabolism in the liver. Compr Physiol 2017;8:1–8.
[20] Cazanave SC, Gores GJ. Mechanisms and clinical implications of hepatocyte lipoapoptosis. Clin Lipidol 2010;5:71–85.
[21] Falcon A, Doege H, Fluitt A, Tsang B, Watson N, Kay MA, et al. FATP2 is a hepatic fatty acid transporter and peroxisomal very long-chain acyl-CoA synthetase. Am J Physiol Endocrinol Metab 2010;299:E384–93.
[22] Doege H, Grimm D, Falcon A, Tsang B, Storm TA, Xu H, et al. Silencing of hepatic fatty acid transporter protein 5 in vivo reverses diet-induced non-alcoholic fatty liver disease and improves hyperglycemia. J Biol Chem 2008;283:22186–92.
[23] Puri P, Baillie RA, Wiest MM, Mirshahi F, Choudhury J, Cheung O, et al. A lipidomic analysis of nonalcoholic fatty liver disease. Hepatology 2007;46:1081–90.
[24] Kim JY, Garcia-Carbonell R, Yamachika S, Zhao P, Dhar D, Loomba R, et al. ER stress drives lipogenesis and steatohepatitis via caspase-2 activation of S1P. Cell 2018;175:133–145.e115.
[25] Rinella ME, Sanyal AJ. Management of NAFLD: a stage-based approach. Nat Rev Gastroenterol Hepatol 2016;13:196–205.
[26] Cao J, Dai DL, Yao L, Yu HH, Ning B, Zhang Q, et al. Saturated fatty acid induction of endoplasmic reticulum stress and apoptosis in human liver cells via the PERK/ATF4/CHOP signaling pathway. Mol Cell Biochem 2012;364:115–29.
[27] Akazawa Y, Cazanave S, Mott JL, Elmi N, Bronk SF, Kohno S, et al. Palmitoleate attenuates palmitate-induced Bim and PUMA up-regulation and hepatocyte lipoapoptosis. J Hepatol 2010;52:586–93.
[28] Ioannou GN. The role of cholesterol in the pathogenesis of NASH. Trends Endocrinol Metab 2016;27:84–95.
[29] Asgharpour A, Cazanave SC, Pacana T, Seneshaw M, Vincent R, Banini BA, et al. A diet-induced animal model of non-alcoholic fatty liver disease and hepatocellular cancer. J Hepatol 2016;65:579–88.
[30] Carulli L, Gabbi C, Bertolotti M. Bile acids and nonalcoholic fatty liver disease: an intriguing relationship. Hepatology 2016;63:1739–40.
[31] Kobayashi Y, Hara N, Sugimoto R, Mifuji-Moroka R, Tanaka H, Eguchi A, et al. The associations between circulating bile acids and the muscle volume in patients with non-alcoholic fatty liver disease (NAFLD). Intern Med 2017;56:755–62.
[32] Zhu Y, Liu H, Zhang M, Guo GL. Fatty liver diseases, bile acids, and FXR. Acta Pharm Sin B 2016;6:409–12.
[33] Joshi-Barve S, Barve SS, Amancherla K, Gobejishvili L, Hill D, Cave M, et al. Palmitic acid induces production of proinflammatory cytokine interleukin-8 from hepatocytes. Hepatology 2007;46:823–30.
[34] Engin A. Non-alcoholic fatty liver disease. Adv Exp Med Biol 2017;960:443–67.
[35] Kakisaka K, Cazanave SC, Fingas CD, Guicciardi ME, Bronk SF, Werneburg NW, et al. Mechanisms of lysophosphatidylcholine-induced hepatocyte lipoapoptosis. Am J Physiol Gastrointest Liver Physiol 2012;302:G77–84.
[36] Bril F, Barb D, Portillo-Sanchez P, Biernacki D, Lomonaco R, Suman A, et al. Metabolic and histological implications of intrahepatic triglyceride content in nonalcoholic fatty liver disease. Hepatology 2017;65:1132–44.
[37] Larter CZ, Yeh MM, Cheng J, Williams J, Brown S, dela Pena A, et al. Activation of peroxisome proliferator-activated receptor alpha by dietary fish oil attenuates steatosis, but does not prevent experimental steatohepatitis because of hepatic lipoperoxide accumulation. J Gastroenterol Hepatol 2008;23:267–75.
[38] Feldstein AE, Werneburg NW, Canbay A, Guicciardi ME, Bronk SF, Rydzewski R, et al. Free fatty acids promote hepatic lipotoxicity by stimulating TNF-alpha expression via a lysosomal pathway. Hepatology 2004;40:185–94.
[39] Feldstein AE, Wieckowska A, Lopez AR, Liu YC, Zein NN, McCullough AJ. Cytokeratin-18 fragment levels as noninvasive biomarkers for nonalcoholic steatohepatitis: a multicenter validation study. Hepatology 2009;50:1072–8.

[40] Yan Y, Zhou XE, Xu HE, Melcher K. Structure and physiological regulation of AMPK. Int J Mol Sci 2018;19:3534.
[41] Zheng J, Woo SL, Hu X, Botchlett R, Chen L, Huo Y, et al. Metformin and metabolic diseases: a focus on hepatic aspects. Front Med 2015;9:173–86.
[42] Porcu C, Sideri S, Martini M, Cocomazzi A, Galli A, Tarantino G, et al. Oleuropein induces AMPK-dependent autophagy in NAFLD mice, regardless of the gender. Int J Mol Sci 2018;19:3948.
[43] Cool B, Zinker B, Chiou W, Kifle L, Cao N, Perham M, et al. Identification and characterization of a small molecule AMPK activator that treats key components of type 2 diabetes and the metabolic syndrome. Cell Metab 2006;3:403–16.
[44] Li Y, Xu S, Mihaylova MM, Zheng B, Hou X, Jiang B, et al. AMPK phosphorylates and inhibits SREBP activity to attenuate hepatic steatosis and atherosclerosis in diet-induced insulin-resistant mice. Cell Metab 2011;13:376–88.
[45] Smith BK, Marcinko K, Desjardins EM, Lally JS, Ford RJ, Steinberg GR. Treatment of nonalcoholic fatty liver disease: role of AMPK. Am J Physiol Endocrinol Metab 2016;311:E730–40.
[46] Woods A, Williams JR, Muckett PJ, Mayer FV, Liljevald M, Bohlooly YM, et al. Liver-specific activation of AMPK prevents steatosis on a high-fructose diet. Cell Rep 2017;18:3043–51.
[47] Foretz M, Hebrard S, Leclerc J, Zarrinpashneh E, Soty M, Mithieux G, et al. Metformin inhibits hepatic gluconeogenesis in mice independently of the LKB1/AMPK pathway via a decrease in hepatic energy state. J Clin Invest 2010;120:2355–69.
[48] Hasenour CM, Ridley DE, Hughey CC, James FD, Donahue EP, Shearer J, et al. 5-Aminoimidazole-4-carboxamide-1-beta-D-ribofuranoside (AICAR) effect on glucose production, but not energy metabolism, is independent of hepatic AMPK in vivo. J Biol Chem 2014;289:5950–9.
[49] Miller RA, Chu Q, Xie J, Foretz M, Viollet B, Birnbaum MJ. Biguanides suppress hepatic glucagon signalling by decreasing production of cyclic AMP. Nature 2013;494:256–60.
[50] Schmid AI, Szendroedi J, Chmelik M, Krssak M, Moser E, Roden M. Liver ATP synthesis is lower and relates to insulin sensitivity in patients with type 2 diabetes. Diabetes Care 2011;34:448–53.
[51] Qiang X, Xu L, Zhang M, Zhang P, Wang Y, Wang Y, et al. Demethyleneberberine attenuates non-alcoholic fatty liver disease with activation of AMPK and inhibition of oxidative stress. Biochem Biophys Res Commun 2016;472:603–9.
[52] Steinberg GR, Michell BJ, van Denderen BJ, Watt MJ, Carey AL, Fam BC, et al. Tumor necrosis factor alpha-induced skeletal muscle insulin resistance involves suppression of AMP-kinase signaling. Cell Metab 2006;4:465–74.
[53] Woo SL, Xu H, Li H, Zhao Y, Hu X, Zhao J, et al. Metformin ameliorates hepatic steatosis and inflammation without altering adipose phenotype in diet-induced obesity. PLoS ONE 2014;9:e91111.
[54] Lindholm CR, Ertel RL, Bauwens JD, Schmuck EG, Mulligan JD, Saupe KW. A high-fat diet decreases AMPK activity in multiple tissues in the absence of hyperglycemia or systemic inflammation in rats. J Physiol Biochem 2013;69:165–75.
[55] Mottillo EP, Desjardins EM, Crane JD, Smith BK, Green AE, Ducommun S, et al. Lack of adipocyte AMPK exacerbates insulin resistance and hepatic steatosis through brown and beige adipose tissue function. Cell Metab 2016;24:118–29.
[56] Galic S, Fullerton MD, Schertzer JD, Sikkema S, Marcinko K, Walkley CR, et al. Hematopoietic AMPK beta1 reduces mouse adipose tissue macrophage inflammation and insulin resistance in obesity. J Clin Invest 2011;121:4903–15.
[57] Fullerton MD, Galic S, Marcinko K, Sikkema S, Pulinilkunnil T, Chen ZP, et al. Single phosphorylation sites in Acc1 and Acc2 regulate lipid homeostasis and the insulin-sensitizing effects of metformin. Nat Med 2013;19:1649–54.
[58] Ducommun S, Deak M, Sumpton D, Ford RJ, Nunez Galindo A, Kussmann M, et al. Motif affinity and mass spectrometry proteomic approach for the discovery of cellular AMPK targets: identification of mitochondrial fission factor as a new AMPK substrate. Cell Signal 2015;27:978–88.
[59] Egan DF, Shackelford DB, Mihaylova MM, Gelino S, Kohnz RA, Mair W, et al. Phosphorylation of ULK1 (hATG1) by AMP-activated protein kinase connects energy sensing to mitophagy. Science 2011;331:456–61.
[60] Toyama EQ, Herzig S, Courchet J, Lewis TL Jr, Loson OC, Hellberg K, et al. Metabolism. AMP-activated protein kinase mediates mitochondrial fission in response to energy stress. Science 2016;351:275–81.
[61] Galloway CA, Lee H, Brookes PS, Yoon Y. Decreasing mitochondrial fission alleviates hepatic steatosis in a murine model of nonalcoholic fatty liver disease. Am J Physiol Gastrointest Liver Physiol 2014;307:G632–641.

[62] Gariani K, Philippe J, Jornayvaz FR. Non-alcoholic fatty liver disease and insulin resistance: from bench to bedside. Diabetes Metab 2013;39:16–26.

[63] Day EA, Ford RJ, Steinberg GR. AMPK as a therapeutic target for treating metabolic diseases. Trends Endocrinol Metab 2017;28:545–60.

[64] Gracia-Sancho J, Guixe-Muntet S, Hide D, Bosch J. Modulation of autophagy for the treatment of liver diseases. Expert Opin Investig Drugs 2014;23:965–77.

[65] Sinha RA, Farah BL, Singh BK, Siddique MM, Li Y, Wu Y, et al. Caffeine stimulates hepatic lipid metabolism by the autophagy-lysosomal pathway in mice. Hepatology 2014;59:1366–80.

[66] Singh R, Kaushik S, Wang Y, Xiang Y, Novak I, Komatsu M, et al. Autophagy regulates lipid metabolism. Nature 2009;458:1131–5.

[67] Wang Y, Shi M, Fu H, Xu H, Wei J, Wang T, et al. Mammalian target of the rapamycin pathway is involved in non-alcoholic fatty liver disease. Mol Med Rep 2010;3:909–15.

[68] Cai N, Zhao X, Jing Y, Sun K, Jiao S, Chen X, et al. Autophagy protects against palmitate-induced apoptosis in hepatocytes. Cell Biosci 2014;4:28.

[69] Jiang P, Huang Z, Zhao H, Wei T. Hydrogen peroxide impairs autophagic flux in a cell model of nonalcoholic fatty liver disease. Biochem Biophys Res Commun 2013;433:408–14.

[70] Kim J, Kundu M, Viollet B, Guan KL. AMPK and mTOR regulate autophagy through direct phosphorylation of Ulk1. Nat Cell Biol 2011;13:132–41.

[71] Takashima M, Ogawa W, Emi A, Kasuga M. Regulation of SREBP1c expression by mTOR signaling in hepatocytes. Kobe J Med Sci 2009;55:E45–52.

[72] Rodriguez A, Catalan V, Gomez-Ambrosi J, Garcia-Navarro S, Rotellar F, Valenti V, et al. Insulin- and leptin-mediated control of aquaglyceroporins in human adipocytes and hepatocytes is mediated via the PI3K/Akt/mTOR signaling cascade. J Clin Endocrinol Metab 2011;96:E586–97.

[73] Pajvani UB, Qiang L, Kangsamaksin T, Kitajewski J, Ginsberg HN, Accili D. Inhibition of Notch uncouples Akt activation from hepatic lipid accumulation by decreasing mTorc1 stability. Nat Med 2013;19:1054–60.

[74] Qu S, Altomonte J, Perdomo G, He J, Fan Y, Kamagate A, et al. Aberrant Forkhead box O1 function is associated with impaired hepatic metabolism. Endocrinology 2006;147:5641–52.

[75] Wang J, Yang W, Chen Z, Chen J, Meng Y, Feng B, et al. Long noncoding RNA lncSHGL recruits hnRNPA1 to suppress hepatic gluconeogenesis and lipogenesis. Diabetes 2018;67:581–93.

[76] Sanyal AJ. Mechanisms of disease: pathogenesis of nonalcoholic fatty liver disease. Nat Clin Pract Gastroenterol Hepatol 2005;2:46–53.

[77] Schaffler A, Scholmerich J, Buchler C. Mechanisms of disease: adipocytokines and visceral adipose tissue – emerging role in nonalcoholic fatty liver disease. Nat Clin Pract Gastroenterol Hepatol 2005;2:273–80.

[78] Roden M. Mechanisms of disease: hepatic steatosis in type 2 diabetes – pathogenesis and clinical relevance. Nat Clin Pract Endocrinol Metab 2006;2:335–48.

[79] Tilg H, Moschen AR. Insulin resistance, inflammation, and non-alcoholic fatty liver disease. Trends Endocrinol Metab 2008;19:371–9.

[80] Kawano Y, Cohen DE. Mechanisms of hepatic triglyceride accumulation in non-alcoholic fatty liver disease. J Gastroenterol 2013;48:434–41.

[81] Lambert JE, Ramos-Roman MA, Browning JD, Parks EJ. Increased de novo lipogenesis is a distinct characteristic of individuals with nonalcoholic fatty liver disease. Gastroenterology 2014;146:726–35.

[82] Aroor AR, Habibi J, Ford DA, Nistala R, Lastra G, Manrique C, et al. Dipeptidyl peptidase-4 inhibition ameliorates Western diet-induced hepatic steatosis and insulin resistance through hepatic lipid remodeling and modulation of hepatic mitochondrial function. Diabetes 2015;64:1988–2001.

[83] Horton JD, Shah NA, Warrington JA, Anderson NN, Park SW, Brown MS, et al. Combined analysis of oligonucleotide microarray data from transgenic and knockout mice identifies direct SREBP target genes. Proc Natl Acad Sci U S A 2003;100:12027–32.

[84] Moon YA, Liang G, Xie X, Frank-Kamenetsky M, Fitzgerald K, Koteliansky V, et al. The Scap/SREBP pathway is essential for developing diabetic fatty liver and carbohydrate-induced hypertriglyceridemia in animals. Cell Metab 2012;15:240–6.

[85] Shimomura I, Bashmakov Y, Horton JD. Increased levels of nuclear SREBP-1c associated with fatty livers in two mouse models of diabetes mellitus. J Biol Chem 1999;274:30028–32.

[86] Davies MN, O'Callaghan BL, Towle HC. Glucose activates ChREBP by increasing its rate of nuclear entry and relieving repression of its transcriptional activity. J Biol Chem 2008;283:24029–38.

[87] Koo SH, Dutcher AK, Towle HC. Glucose and insulin function through two distinct transcription factors to stimulate expression of lipogenic enzyme genes in liver. J Biol Chem 2001;276:9437–45.
[88] Stoeckman AK, Towle HC. The role of SREBP-1c in nutritional regulation of lipogenic enzymes gene expression. J Biol Chem 2002;277:27029–35.
[89] Browning JD, Horton JD. Molecular mediators of hepatic steatosis and liver injury. J Clin Invest 2004;114:147–52.
[90] Zhao X, Feng D, Wang Q, Abdulla A, Xie X-J, Zhou J, et al. Regulation of lipogenesis by cyclin-dependent kinase 8-mediated control of SREBP-1. J Clin Invest 2012;122:2417–27.
[91] van Dijk SJ, Feskens EJ, Bos MB, Hoelen DW, Heijligenberg R, Bromhaar MG, et al. A saturated fatty acid-rich diet induces an obesity-linked proinflammatory gene expression profile in adipose tissue of subjects at risk of metabolic syndrome. Am J Clin Nutr 2009;90:1656–64.
[92] Rizki G, Arnaboldi L, Gabrielli B, Yan J, Lee GS, Ng RK, et al. Mice fed a lipogenic methionine-choline-deficient diet develop hypermetabolism coincident with hepatic suppression of SCD-1. J Lipid Res 2006;47:2280–90.
[93] Li ZZ, Berk M, McIntyre TM, Feldstein AE. Hepatic lipid partitioning and liver damage in nonalcoholic fatty liver disease: role of stearoyl-CoA desaturase. J Biol Chem 2009;284:5637–44.
[94] Miyazaki M, Flowers MT, Sampath H, Chu K, Otzelberger C, Liu X, et al. Hepatic stearoyl-CoA desaturase-1 deficiency protects mice from carbohydrate-induced adiposity and hepatic steatosis. Cell Metab 2007;6:484–96.
[95] Brown JM, Chung S, Sawyer JK, Degirolamo C, Alger HM, Nguyen T, et al. Inhibition of stearoyl-coenzyme A desaturase 1 dissociates insulin resistance and obesity from atherosclerosis. Circulation 2008;118:1467–75.
[96] Liu X, Strable MS, Ntambi JM. Stearoyl CoA desaturase 1: role in cellular inflammation and stress. Adv Nutr 2011;2:15–22.
[97] Liu X, Miyazaki M, Flowers MT, Sampath H, Zhao M, Chu K, et al. Loss of stearoyl-CoA desaturase-1 attenuates adipocyte inflammation: effects of adipocyte-derived oleate. Arterioscler Thromb Vasc Biol 2010;30:31–8.
[98] Dentin R, Benhamed F, Pégorier J-P, Foufelle F, Viollet B, Vaulont S, et al. Polyunsaturated fatty acids suppress glycolytic and lipogenic genes through the inhibition of ChREBP nuclear protein translocation. J Clin Invest 2005;115:2843–54.
[99] Jeyapal S, Kona SR, Mullapudi SV, Putcha UK, Gurumurthy P, Ibrahim A, et al. Substitution of linoleic acid with alpha-linolenic acid or long chain n-3 polyunsaturated fatty acid prevents Western diet induced nonalcoholic steatohepatitis. Sci Rep 2018;8:10953.
[100] Guo X, Li H, Xu H, Halim V, Zhang W, Wang H, et al. Palmitoleate induces hepatic steatosis but suppresses liver inflammatory response in mice. PLoS ONE 2012;7:e39286.
[101] Perry RJ, Camporez JG, Kursawe R, Titchenell PM, Zhang D, Perry CJ, et al. Hepatic acetyl CoA links adipose tissue inflammation to hepatic insulin resistance and type 2 diabetes. Cell 2015;160:745–58.
[102] Wang T, Yao W, Li J, He Q, Shao Y, Huang F. Acetyl-CoA from inflammation-induced fatty acids oxidation promotes hepatic malate-aspartate shuttle activity and glycolysis. Am J Physiol Endocrinol Metab 2018;315:E496–510.
[103] Wang C, Chen H, Zhang M, Zhang J, Wei X, Ying W. Malate-aspartate shuttle inhibitor aminooxyacetic acid leads to decreased intracellular ATP levels and altered cell cycle of C6 glioma cells by inhibiting glycolysis. Cancer Lett 2016;378:1–7.
[104] Friedman SL, Neuschwander-Tetri BA, Rinella M, Sanyal AJ. Mechanisms of NAFLD development and therapeutic strategies. Nat Med 2018;24:908–22.
[105] Berlanga A, Guiu-Jurado E, Porras JA, Auguet T. Molecular pathways in non-alcoholic fatty liver disease. Clin Exp Gastroenterol 2014;7:221–39.
[106] Mari M, Caballero F, Colell A, Morales A, Caballeria J, Fernandez A, et al. Mitochondrial free cholesterol loading sensitizes to TNF- and Fas-mediated steatohepatitis. Cell Metab 2006;4:185–98.
[107] Mota M, Banini BA, Cazanave SC, Sanyal AJ. Molecular mechanisms of lipotoxicity and glucotoxicity in nonalcoholic fatty liver disease. Metabolism 2016;65:1049–61.
[108] Alisi A, Carpino G, Oliveira FL, Panera N, Nobili V, Gaudio E. The role of tissue macrophage-mediated inflammation on NAFLD pathogenesis and its clinical implications. Mediators Inflamm 2017;2017:8162421.
[109] Celebi G, Genc H, Gurel H, Sertoglu E, Kara M, Tapan S, et al. The relationship of circulating Fetuin-A with liver histology and biomarkers of systemic inflammation in nondiabetic subjects with nonalcoholic fatty liver disease. Saudi J Gastroenterol 2015;21:139–45.
[110] Farrell GC, van Rooyen D, Gan L, Chitturi S. NASH is an inflammatory disorder: pathogenic, prognostic and therapeutic implications. Gut Liver 2012;6:149–71.

[111] Gadd VL, Skoien R, Powell EE, Fagan KJ, Winterford C, Horsfall L, et al. The portal inflammatory infiltrate and ductular reaction in human nonalcoholic fatty liver disease. Hepatology 2014;59:1393–405.

[112] Liao CY, Song MJ, Gao Y, Mauer AS, Revzin A, Malhi H. Hepatocyte-derived lipotoxic extracellular vesicle sphingosine 1-phosphate induces macrophage chemotaxis. Front Immunol 2018;9:2980.

[113] De Vito R, Alisi A, Masotti A, Ceccarelli S, Panera N, Citti A, et al. Markers of activated inflammatory cells correlate with severity of liver damage in children with nonalcoholic fatty liver disease. Int J Mol Med 2012;30:49–56.

[114] Pais R, Charlotte F, Fedchuk L, Bedossa P, Lebray P, Poynard T, et al. A systematic review of follow-up biopsies reveals disease progression in patients with non-alcoholic fatty liver. J Hepatol 2013;59:550–6.

[115] Saiman Y, Friedman SL. The role of chemokines in acute liver injury. Front Physiol 2012;3:213.

[116] Farrell GC, Haczeyni F, Chitturi S. Pathogenesis of NASH: how metabolic complications of overnutrition favour lipotoxicity and pro-inflammatory fatty liver disease. Adv Exp Med Biol 2018;1061:19–44.

[117] Wree A, McGeough MD, Pena CA, Schlattjan M, Li H, Inzaugarat ME, et al. NLRP3 inflammasome activation is required for fibrosis development in NAFLD. J Mol Med (Berl) 2014;92:1069–82.

[118] Lanthier N. Targeting Kupffer cells in non-alcoholic fatty liver disease/non-alcoholic steatohepatitis: why and how? World J Hepatol 2015;7:2184–8.

[119] Fain JN, Madan AK, Hiler ML, Cheema P, Bahouth SW. Comparison of the release of adipokines by adipose tissue, adipose tissue matrix, and adipocytes from visceral and subcutaneous abdominal adipose tissues of obese humans. Endocrinology 2004;145:2273–82.

[120] Stanton MC, Chen SC, Jackson JV, Rojas-Triana A, Kinsley D, Cui L, et al. Inflammatory signals shift from adipose to liver during high fat feeding and influence the development of steatohepatitis in mice. J Inflamm (Lond) 2011;8:8.

[121] Nakamura Y, Abe M, Kawasaki K, Miyake T, Watanabe T, Yoshida O, et al. Depletion of B cell-activating factor attenuates hepatic fat accumulation in a murine model of nonalcoholic fatty liver disease. Sci Rep 2019;9:977.

[122] Csak T, Ganz M, Pespisa J, Kodys K, Dolganiuc A, Szabo G. Fatty acid and endotoxin activate inflammasomes in mouse hepatocytes that release danger signals to stimulate immune cells. Hepatology 2011;54:133–44.

[123] Ma M, Duan R, Zhong H, Liang T, Guo L. The crosstalk between fat homeostasis and liver regional immunity in NAFLD. J Immunol Res 2019;2019:3954890.

[124] Frasinariu OE, Ceccarelli S, Alisi A, Moraru E, Nobili V. Gut-liver axis and fibrosis in nonalcoholic fatty liver disease: an input for novel therapies. Dig Liver Dis 2013;45:543–51.

[125] Rivera CA, Adegboyega P, van Rooijen N, Tagalicud A, Allman M, Wallace M. Toll-like receptor-4 signaling and Kupffer cells play pivotal roles in the pathogenesis of non-alcoholic steatohepatitis. J Hepatol 2007;47:571–9.

[126] Rocha DM, Caldas AP, Oliveira LL, Bressan J, Hermsdorff HH. Saturated fatty acids trigger TLR4-mediated inflammatory response. Atherosclerosis 2016;244:211–5.

[127] Ye D, Li FY, Lam KS, Li H, Jia W, Wang Y, et al. Toll-like receptor-4 mediates obesity-induced non-alcoholic steatohepatitis through activation of X-box binding protein-1 in mice. Gut 2012;61:1058–67.

[128] Ouyang X, Ghani A, Malik A, Wilder T, Colegio OR, Flavell RA, et al. Adenosine is required for sustained inflammasome activation via the A(2)A receptor and the HIF-1alpha pathway. Nat Commun 2013;4:2909.

[129] Ohta A, Sitkovsky M. Role of G-protein-coupled adenosine receptors in downregulation of inflammation and protection from tissue damage. Nature 2001;414:916–20.

[130] Cai Y, Li H, Liu M, Pei Y, Zheng J, Zhou J, et al. Disruption of adenosine 2A receptor exacerbates NAFLD through increasing inflammatory responses and SREBP1c activity. Hepatology 2018;68:48–61.

[131] Zhou J, Li H, Cai Y, Ma L, Mathews D, Lu B, et al. Mice lacking adenosine 2A receptor reveal increased severity of MCD-induced NASH. J Endocrinol 2019;243:199–209. doi: 10.1530/JOE-19-0198.

[132] Alchera E, Rolla S, Imarisio C, Bardina V, Valente G, Novelli F, et al. Adenosine A2a receptor stimulation blocks development of nonalcoholic steatohepatitis in mice by multilevel inhibition of signals that cause immunolipotoxicity. Transl Res 2017;182:75–87.

[133] Cobbina E, Akhlaghi F. Non-alcoholic fatty liver disease (NAFLD) – pathogenesis, classification, and effect on drug metabolizing enzymes and transporters. Drug Metab Rev 2017;49:197–211.

[134] Hui JM, Hodge A, Farrell GC, Kench JG, Kriketos A, George J. Beyond insulin resistance in NASH: TNF-alpha or adiponectin? Hepatology 2004;40:46–54.

[135] Zeng T, Zhou J, He L, Zheng J, Chen L, Wu C, et al. Blocking nuclear factor-kappa B protects against diet-induced hepatic steatosis and insulin resistance in mice. PLoS ONE 2016;11:e0149677.

[136] Czaja MJ. JNK regulation of hepatic manifestations of the metabolic syndrome. Trends Endocrinol Metab 2010;21:707–13.

[137] Weston CR, Davis RJ. The JNK signal transduction pathway. Curr Opin Genet Dev 2002;12:14–21.
[138] Potthoff MJ, Inagaki T, Satapati S, Ding X, He T, Goetz R, et al. FGF21 induces PGC-1alpha and regulates carbohydrate and fatty acid metabolism during the adaptive starvation response. Proc Natl Acad Sci U S A 2009;106:10853–8.
[139] Bookout AL, de Groot MH, Owen BM, Lee S, Gautron L, Lawrence HL, et al. FGF21 regulates metabolism and circadian behavior by acting on the nervous system. Nat Med 2013;19:1147–52.
[140] Yang YM, Seki E. TNFalpha in liver fibrosis. Curr Pathobiol Rep 2015;3:253–61.
[141] Persico M, Capasso M, Persico E, Svelto M, Russo R, Spano D, et al. Suppressor of cytokine signaling 3 (SOCS3) expression and hepatitis C virus-related chronic hepatitis: insulin resistance and response to antiviral therapy. Hepatology 2007;46:1009–15.
[142] Zein CO, Yerian LM, Gogate P, Lopez R, Kirwan JP, Feldstein AE, et al. Pentoxifylline improves nonalcoholic steatohepatitis: a randomized placebo-controlled trial. Hepatology 2011;54:1610–9.
[143] Chen Q, Sun L, Chen ZJ. Regulation and function of the cGAS-STING pathway of cytosolic DNA sensing. Nat Immunol 2016;17:1142–9.
[144] Downey CM, Aghaei M, Schwendener RA, Jirik FR. DMXAA causes tumor site-specific vascular disruption in murine non-small cell lung cancer, and like the endogenous non-canonical cyclic dinucleotide STING agonist, 2′3′-cGAMP, induces M2 macrophage repolarization. PLoS ONE 2014;9:e99988.
[145] Guo X, Shu C, Li H, Pei Y, Woo SL, Zheng J, et al. Cyclic GMP-AMP ameliorates diet-induced metabolic dysregulation and regulates proinflammatory responses distinctly from STING activation. Sci Rep 2017;7:6355.
[146] Luo X, Li H, Ma L, Zhou J, Guo X, Woo SL, et al. Expression of STING is increased in liver tissues from patients with NAFLD and promotes macrophage-mediated hepatic inflammation and fibrosis in mice. Gastroenterology 2018;155:1971–1984.e4.
[147] Yu Y, Liu Y, An W, Song J, Zhang Y, Zhao X. STING-mediated inflammation in Kupffer cells contributes to progression of nonalcoholic steatohepatitis. J Clin Invest 2019;129:546–55.
[148] Petrasek J, Iracheta-Vellve A, Csak T, Satishchandran A, Kodys K, Kurt-Jones EA, et al. STING-IRF3 pathway links endoplasmic reticulum stress with hepatocyte apoptosis in early alcoholic liver disease. Proc Natl Acad Sci U S A 2013;110:16544–9.
[149] Garcia-Martinez I, Santoro N, Chen Y, Hoque R, Ouyang X, Caprio S, et al. Hepatocyte mitochondrial DNA drives nonalcoholic steatohepatitis by activation of TLR9. J Clin Invest 2016;126:859–64.
[150] Lindqvist LM, Frank D, McArthur K, Dite TA, Lazarou M, Oakhill JS, et al. Autophagy induced during apoptosis degrades mitochondria and inhibits type I interferon secretion. Cell Death Differ 2018;25:782–94.
[151] Mao Y, Luo W, Zhang L, Wu W, Yuan L, Xu H, et al. STING-IRF3 triggers endothelial inflammation in response to free fatty acid-induced mitochondrial damage in diet-induced obesity. Arterioscler Thromb Vasc Biol 2017;37:920–9.
[152] Dallmann R, Viola AU, Tarokh L, Cajochen C, Brown SA. The human circadian metabolome. Proc Natl Acad Sci U S A 2012;109:2625–9.
[153] Lamia KA, Sachdeva UM, DiTacchio L, Williams EC, Alvarez JG, Egan DF, et al. AMPK regulates the circadian clock by cryptochrome phosphorylation and degradation. Science 2009;326:437–40.
[154] Schibler U, Ripperger J, Brown SA. Peripheral circadian oscillators in mammals: time and food. J Biol Rhythms 2003;18:250–60.
[155] Xu H, Li H, Woo SL, Kim SM, Shende VR, Neuendorff N, et al. Myeloid cell-specific disruption of Period1 and Period2 exacerbates diet-induced inflammation and insulin resistance. J Biol Chem 2014;289:16374–88.
[156] Perumpail BJ, Li AA, John N, Sallam S, Shah ND, Kwong W, et al. The role of vitamin E in the treatment of NAFLD. Diseases 2018;6:86. doi: 10.3390/diseases6040086.
[157] Kung J, Henry RR. Thiazolidinedione safety. Expert Opin Drug Saf 2012;11:565–79.
[158] Jung YA, Kim HK, Bae KH, Seo HY, Kim HS, Jang BK, et al. Cilostazol inhibits insulin-stimulated expression of sterol regulatory binding protein-1c via inhibition of LXR and Sp1. Exp Mol Med 2014;46:e73.
[159] Jung CH, Mok JO. The effects of hypoglycemic agents on non-alcoholic fatty liver disease: focused on sodium-glucose cotransporter 2 inhibitors and glucagon-like peptide-1 receptor agonists. J Obes Metab Syndr 2019;28:18–29.
[160] Neuschwander-Tetri BA, Loomba R, Sanyal AJ, Lavine JE, Van Natta ML, Abdelmalek MF, et al. Farnesoid X nuclear receptor ligand obeticholic acid for non-cirrhotic, non-alcoholic steatohepatitis (FLINT): a multicentre, randomised, placebo-controlled trial. Lancet 2015;385:956–65.

[161] Loomba R, Sanyal AJ, Kowdley KV, Terrault N, Chalasani NP, Abdelmalek MF, et al. Factors associated with histologic response in adult patients with nonalcoholic steatohepatitis. Gastroenterology 2019;156:e85.

[162] Ratziu V, Bedossa P, Francque SM, Larrey D, Aithal GP, Serfaty L, et al. Lack of efficacy of an inhibitor of PDE4 in phase 1 and 2 trials of patients with nonalcoholic steatohepatitis. Clin Gastroenterol Hepatol 2014;12:1724–1730.e5.

[163] Ratziu V, Sheikh MY, Sanyal AJ, Lim JK, Conjeevaram H, Chalasani N, et al. A phase 2, randomized, double-blind, placebo-controlled study of GS-9450 in subjects with nonalcoholic steatohepatitis. Hepatology 2012;55:419–28.

[164] Chachay VS, Macdonald GA, Martin JH, Whitehead JP, O'Moore-Sullivan TM, Lee P, et al. Resveratrol does not benefit patients with nonalcoholic fatty liver disease. Clin Gastroenterol Hepatol 2014;12:2092–2103.e1-6.

[165] Sanyal AJ, Abdelmalek MF, Suzuki A, Cummings OW, Chojkier M, Group E-AS. No significant effects of ethyl-eicosapentanoic acid on histologic features of nonalcoholic steatohepatitis in a phase 2 trial. Gastroenterology 2014;147:377–384.e1.

[166] Ilan Y. Review article: novel methods for the treatment of non-alcoholic steatohepatitis – targeting the gut immune system to decrease the systemic inflammatory response without immune suppression. Aliment Pharmacol Ther 2016;44:1168–82.

[167] Pearlman M, Loomba R. State of the art: treatment of nonalcoholic steatohepatitis. Curr Opin Gastroenterol 2014;30:223–37.

[168] Morimoto Y, Kikuchi K, Ito T, Tokuda M, Matsuyama T, Noma S, et al. MK615 attenuates *Porphyromonas gingivalis* lipopolysaccharide-induced pro-inflammatory cytokine release via MAPK inactivation in murine macrophage-like RAW264.7 cells. Biochem Biophys Res Commun 2009;389:90–4.

[169] Hokari A, Ishikawa T, Tajiri H, Matsuda T, Ishii O, Matsumoto N, et al. Efficacy of MK615 for the treatment of patients with liver disorders. World J Gastroenterol 2012;18:4118–26.

[170] Alam S, Nazmul Hasan S, Mustafa G, Alam M, Kamal M, Ahmad N. Effect of pentoxifylline on histological activity and fibrosis of nonalcoholic steatohepatitis patients: a one year randomized control trial. J Transl Int Med 2017;5:155–63.

[171] Friedman S, Sanyal A, Goodman Z, Lefebvre E, Gottwald M, Fischer L, et al. Efficacy and safety study of cenicriviroc for the treatment of non-alcoholic steatohepatitis in adult subjects with liver fibrosis: CENTAUR Phase 2b study design. Contemp Clin Trials 2016;47:356–65.

[172] Tacke F. Cenicriviroc for the treatment of non-alcoholic steatohepatitis and liver fibrosis. Expert Opin Investig Drugs 2018;27:301–11.

[173] Shi W, Zhu Q, Gu J, Liu X, Lu L, Qian X, et al. Anti-IL-17 antibody improves hepatic steatosis by suppressing interleukin-17-related fatty acid synthesis and metabolism. Clin Dev Immunol 2013;2013:253046.

CHAPTER

# 15

# Endocannabinoids: the lipid effectors of metabolic regulation in health and disease

*Tony Jourdan*[a], *Pascal Degrace*[a], *Isabel González-Mariscal*[b], *Gergo Szanda*[c], *Joseph Tam*[d]

[a]UMR 1231 INSERM-UB-Agrosup, Team Pathophysiology of Dyslipidemia, Faculty of Sciences, Dijon, France; [b]Biomedical Research Institute of Malaga-IBIMA, Endocrinology and Nutrition UGC, Regional University Hospital of Malaga, Malaga, Spain; [c]MTA-SE Laboratory of Molecular Physiology, Department of Physiology, Semmelweis University, Budapest, Hungary; [d]Obesity and Metabolism Laboratory, Institute for Drug Research, School of Pharmacy, Faculty of Medicine, The Hebrew University of Jerusalem, Jerusalem, Israel

OUTLINE



Lipid Signaling and Metabolism. http://dx.doi.org/10.1016/B978-0-12-819404-1.00015-4

## Introduction to the endocannabinoid system

*Cannabis sativa* (marijuana) and its psychoactive and medicinal uses have been well known for millennia [1]. It is considered as one of the most ingenious plants; its investigation during the past half-century has led to the discovery of an important homeostatic system, the endocannabinoid system (ECS), which plays a key role in human physiology. Currently, 545 natural compounds have been identified from this plant [2]. Of these, 144 have been isolated and identified as phytocannabinoids [3]. However, our understanding of the mechanisms of action of cannabinoids emerged only during the 1960s, after different phytocannabinoids were characterized by Raphael Mechoulam and his colleagues. Back then, the Mechoulam group isolated and described the correct structure and stereochemistry of cannabidiol (CBD) [4], $\Delta^9$-tetrahydrocannabinol ($\Delta^9$-THC, the main psychoactive component of marijuana) [5,6], $\Delta^8$-tetrahydrocannabinol ($\Delta^8$-THC) [7], cannabigerol (CBG) [8], cannabichromene (CBC) [9], and cannabicyclol (CBL) [10].

After these initial discoveries, almost three decades passed until the binding sites of $\Delta^9$-THC in the brain and peripheral organs were identified, which were then termed as the cannabinoid-1 and -2 receptors ($CB_1R$ and $CB_2R$, respectively) [11–13]. As of recently, their structures have been identified and described by a few groups [14–17]. Despite the fact that both receptors signal mainly via $G_i/G_o$ proteins, they can also activate $G_s$, $G_{q/11}$, as well as G protein-independent signaling pathways [18]. $CB_1R$, the most widely expressed G-protein-coupled receptor (GPCR) in the human brain [19], is primarily localized in the plasma membrane. It is also present in many peripheral organs, such as adipose tissue, liver, muscle, kidney, pancreas, bone and the GI tract [20]. $CB_2R$, on the other hand, is predominantly localized in cells that are associated with the immune system; it is also expressed in many peripheral tissues (as mentioned above), with conflicting evidence regarding its expression in the central nervous system (CNS) [21]. Of the 144 phytocannabinoids present in *Cannabis*, only $\Delta^9$-THC and its less abundant propyl analog, $\Delta^9$-tetrahydrocannabivarin (THCV), bind to $CB_1R$ and $CB_2R$ with high affinity (with agonistic and antagonistic activity for THC and THCV, respectively). In contrast, many studies have shown that the other phytocannabinoids have the ability to bind to several different receptors, ranging from other GPCRs (GPR18, GPR55, and GPR119) to ion channel (thermosensitive transient receptor potential (TRP) channels) and nuclear receptors (peroxisome proliferator-activated receptors, PPARs) (reviewed in Ref. [22]); however, their physiological functions are still largely unknown.

Characterization of $CB_1R$ and $CB_2R$ in mammalian cells initiated the discovery of their endogenous ligands, arachidonoyl ethanolamide (AEA, anandamide) [23] and 2-arachidonoyl glycerol (2-AG) [24,25], both of which are considered as endocannabinoids (eCBs). Whereas AEA is a high-affinity, partial agonist of $CB_1R$ and is barely active at $CB_2R$, 2-AG activates both receptors with moderate-to-low affinity [26,27]. Once eCBs are generated and released, they usually remain attached to the cell membrane due to their lipophilic profile and therefore, they can be taken back up by the cells via a high-affinity transport mechanism [28]. eCBs are synthesized, transported, and inactivated in their respective target tissues differently. AEA is generated from *N*-acyl-phosphatidylethanolamine (NAPE) by NAPE-specific phospholipase D (NAPE-PLD) or via other means [29]. On the other hand, 2-AG is most likely catalyzed from diacylglycerol (DAG) by either DAG lipase (DAGL) α or β [30]. Their clearance depends on cellular uptake and specific enzymatic degradation. Whereas AEA is degraded mainly by membrane-associated fatty-acid amide hydrolase (FAAH) into free arachidonic acid and ethanolamine [31], 2-AG is primarily degraded by monoglyceride lipase (MAGL) into arachidonic acid and glycerol [32]. The eCBs, their receptors and the enzymes/proteins involved in their biosynthesis, transport, and degradation jointly make up the ECS. This endogenous system, acting both in the CNS and peripheral organs, is considered as one of the main physiological systems in humans and animals. Here, we will review its role in modulating metabolic homeostasis in various tissues in health and disease.

## Regulating energy balance by endocannabinoids

The ECS is present and functional in virtually all tissues involved in the regulation of energy metabolism. Net energy flux depends on the balance of food intake and total energy expenditure and both processes are influenced by cannabinoids [33]. $CB_1R$ and eCBs exert pleiotropic actions, which typically favor positive energy balance [34]. Here we discuss the most relevant metabolic effects of eCBs that, together with the ever more dominant sedentary lifestyle, often lead to the manifestation of the so-called thrifty phenotype, as illustrated by the global obesity pandemic [35].

Food intake can be regulated by both the homeostatic (caloric) requirements of the organism and by the drive to "overeat" if palatable food is available. The former is usually (and somewhat mechanistically) categorized as hunger, the latter as appetite and both are typically strengthened by eCBs. Hunger and satiety signals are integrated primarily in the hypothalamus where $CB_1Rs$ modulate both excitatory and inhibitory neurotransmission [36] and thereby regulate the production and/or release of several regulators of energy metabolism. In the hypothalamus, $CB_1Rs$ disinhibit orexigenic melanin-concentrating hormone neurons [37] and directly activate the production of orexigenic endorphins [38]. Although some $CB_1Rs$ in the striatum [39] and in the orexin system [40] exert anorexigenic actions *per se*, such anorexigenic $CB_1Rs$ may be lost in obesity [40].

Activation of brain $CB_1Rs$ characteristically has a net orexigenic/anabolic outcome [21]. For instance, fasting-induced food intake is reduced by brain penetrant $CB_1R$ antagonists (but not by a peripherally restricted antagonist) [41] and $CB_1R$ antagonists effectively reduce body weight in obese patients [42]. Importantly, the level of eCBs in the hypothalamus (sometimes referred to as "eCB tone") readily reacts to fasting [43–45] and it is also increased in genetically induced hyperphagia [46]. Taken together, eCBs in the hypothalamus may be regarded as *bona fide* physiological regulators of food intake [21].

Since food oversupply was typically followed by long periods of scarcity during mammalian evolution, the capacity to consume more calories than what would normally replenish energy stores must have been a highly adaptive trait. eCBs support such an "overeating" ability by augmenting the reward effect of palatable [47,58] (and possibly also of nonpalatable [48]) food via modulating dopamine signaling within the mesolimbic system [49,50] and through $CB_1Rs$ in the pons [51]. (Nota bene: the blockade of $CB_1Rs$ within the mesolimbic pathway has contributed to the alarming psychiatric side-effects of brain-penetrant $CB_1R$ antagonists [52,83]). Interestingly, eCBs seem to modulate food intake via actions on sensory organs as well. Olfactory bulb $CB_1Rs$ increase hunger-driven food intake after fasting [45], whereas eCBs in taste buds selectively sensitize sugar-detecting neurons [53] and may thereby enhance appetite-induced overconsumption [49,54]. In addition, it has been reported that $CB_1R$ blockade or deletion in mice alters oro-sensory detection of fatty acids, suggesting that activation of the ECS could increase sensory stimuli relaying palatability of foods [55]. In this respect, it is particularly noteworthy that in obese individuals salivary eCB levels are higher than in normal-weight subjects and eCB concentration decreases during body-weight reduction [56]. Taken together, these observations suggest that in obesity, eCBs, along with other factors, reinforce hedonistic stimuli and enhance "pleasure" associated with palatable food [49].

Complex interplay exists between the ECS and peripheral hormones conveying information about energy depots and feeding status to the central nervous system. The main anorexigenic (satiety) hormone, leptin, reduces eCB levels in the hypothalamus [46]; this action of the hormone probably contributes to its anorexigenic effects. On the other hand, $CB_1Rs$ are crucial for the development and sustenance of leptin resistance [41], a hallmark of obesity and metabolic syndrome [57,67]. In line with that, $CB_1R$ knock-out animals display augmented sensitivity to leptin and are resistant to diet-induced obesity [41,58]. Interestingly, selective blockade of peripheral $CB_1Rs$ alone, without affecting central receptors, is sufficient to restore hypothalamic leptin sensitivity, reduce food intake and normalize body weight in animal models of obesity [41,59]. Analogous to leptin resistance, eCBs also contribute to insulin resistance via both central [60] and peripheral [61] $CB_1Rs$. It can be argued that eCB-driven insulin resistance is most likely associated with reduced central insulin sensitivity [62] and with a decreased anorexic potential of insulin [63].

Accumulating data indicate that eCBs produced in the gut, in response to meal composition and size, influence additional food intake via the gut-brain axis [64]. More specifically, dietary lipid-induced eCB production in the jejunum selectively enhances further consumption of a fat-rich meal [65]. Cannabinoids also increase the gastric release of the "hunger hormone" ghrelin [66], which in turn, promotes both hunger-induced and motivational food intake and decreases the basal metabolic rate [67]. Finally, since $CB_1Rs$ on vagal terminals increase gastrointestinal transit time [68], they may also enhance digestive and adsorptive efficiency [69].

The ECS also favors positive energy balance by reducing the basal metabolic rate. Neuronal $CB_1Rs$ in the hypothalamus [70,71], in the forebrain [72,73] and on sympathetic neurons [73] decrease energy expenditure via the reduction of sympathetic tone, lipid oxidation, and thermogenesis. $CB_1Rs$ on sympathetic nerve terminals innervating fat depots may be particularly important in determining the basal metabolic rate, since a $CB_1R$-mediated decrease in fat sympathetic tone impairs white-to-beige adipocyte trans-differentiation ("browning") and brown adipose tissue thermogenesis [74,75,78]. (Consequently, $CB_1R$ activation is expected to reduce thermogenesis by UCP1-positive beige adipocytes [76–78].) Moreover, activation of *adipocyte* $CB_1Rs$ may have a similar dampening effect on browning [79,80]. Peripheral $CB_1Rs$ also impair mitochondrial biogenesis in target organs of insulin [81], notably in skeletal muscle, which would naturally further decrease the whole body's thermogenic capacity.

Overall, the ECS may be exploited in more than one way when approaching metabolic disorders. Inhibition of the ECS represents a highly effective treatment in models of obesity and metabolic syndrome [59,82], whereas the activation of $CB_1Rs$ may be useful if increased appetite and the restoration of body weight are the therapeutic goals (albeit careful consideration is imperative in both potential clinical situations [83,84]).

## Endocannabinoids and adipose tissue metabolism

White adipose tissue (WAT) is largely composed of mature adipocytes separated by the stromal vascular fraction, which contains pluripotent mesenchymal stem cells, hematopoietic cells such as monocytes/macrophages and T lymphocytes. In addition, fibroblast, smooth muscle, and endothelial cells as well as collagen fibers can also be present [85,86]. Several studies have highlighted the presence of the ECS within visceral and subcutaneous WAT by detecting AEA and 2-AG [87–89], FAAH and MAGL as well as $CB_1Rs$ and $CB_2Rs$ [80,88–93]. Furthermore, both human and murine adipocytes can produce a significant amount of eCBs [87,88,94]; this production is exacerbated in visceral WAT during obesity [88,90,95,96]. Indeed, WAT appears to be an important source of circulating eCBs. Notably, circulating 2-AG is markedly increased in obese or type-2 diabetic patients, compared with healthy subjects. Specifically, a significant correlation was found between circulating 2-AG and visceral fat mass. However, whether circulating eCBs exert metabolic regulatory functions or only serve as indirect markers of tissue "eCB tone" remains to be determined (for a review, see Ref. [97]). Recently, the presence of an active ECS was also demonstrated in brown adipose tissue (BAT) [98–101]. Ever since the discovery of a functional ECS in adipose tissue, its metabolic influence has been widely investigated; ECS appears to be involved in the control of fat storage, tissue expansion, and thermogenesis. Also, according to the literature, ECS overactivation may be involved in the development of pathological changes in adipose tissue during obesity, consequently inducing remodeling, insulin resistance set up and alteration of secretory functions.

The ECS is involved in glucose uptake by the adipocytes. Indeed, $CB_1R$ stimulation with AEA and 2-AG increases insulin-dependent glucose uptake in 3T3L1 adipocytes and in diet-induced obese mice by facilitating glucose transporter 4 (GLUT4) translocation to the plasma membrane [74,92,102,103]. In vitro experiments showed that these effects could be reversed by $CB_1R$ blockade [102]. However, in vivo data obtained from obese mice showed a contradictory picture, since blocking $CB_1Rs$ led to an induction of GLUT4 expression and increased glycolysis in adipocytes from obese mice [104].

The ECS is also tightly involved in regulating lipogenesis, since stimulating $CB_1R$ in primary mouse adipocytes leads to an increase in lipoprotein lipase (LPL) activity by reducing cyclic adenosine monophosphate (cAMP) availability [88,105] or by increasing the expression of the lipogenic transcription factor SREBP-1c and its targets, such as acetyl-CoA carboxylase 1 (ACC1) and fatty acid synthase (FAS) [106,107].

Some studies also reported that ECS activation in adipose tissue is associated with a decrease in lipolysis, thus limiting fat mobilization under fasting conditions and thereby promoting fat storage. Presynaptic inhibition of noradrenaline release from sympathetic neurons by cannabinoids has been demonstrated in various peripheral tissues [108–111]. Therefore, it is very likely that eCBs also target sympathetic innervation of adipose tissue, thus limiting free fatty acid (FFA) mobilization under fasting conditions. Indeed, it has been reported that $CB_1R$ blockade induces the opposite effect by enhancing sympathetic tone [112]. However, some studies also reported a direct inhibitory effect of eCBs on adipocyte lipolysis as reported in 3T3-L1 [113] and in rat adipose tissue explants [114].

During obesity, however, the picture is quite different, since this pathological context is associated with an increase in basal lipolysis, mainly due to the insulin-resistant state of the hypertrophied WAT. In a study combining in vivo and ex vivo approaches, Muller and colleagues suggested that overactivation of the ECS in WAT might represent one of the mechanisms by which FFA release is increased in obesity, thus demonstrating that $CB_1R$ activation impairs the antilipolytic action of insulin through inhibition of the PI3K/Akt pathway [114]. This result is of particular importance since excessive FFA delivery from WAT is considered an important primary event contributing to the onset of insulin resistance, promoting ectopic fat storage and ultimately lipotoxicity in extra-adipose tissues [114–116]. In addition, it is noteworthy to mention that $CB_2R$ could also be involved in this process, since its activation, which typically reduces cAMP levels, leads to increased lipolysis through protein kinase A (PKA) stimulation and upregulation of the adipose triglyceride lipase (ATGL) protein expression in obese mice [119].

Activation of the ECS in adipocytes is involved in the adipocyte proliferation and differentiation process termed adipogenesis. $CB_1R$ activation in vitro by AEA or HU-210 in primary adipocytes (3T3-F442A and 3T3L1) led to increased proliferation through a peroxisome proliferator-activated receptor gamma-dependent (PPARγ) mechanism [88,117,118]. These effects were specifically mediated by $CB_1R$, since its blockade was able to counteract the proliferative state [88,105]. Interestingly, the regulation of adipocyte differentiation might change, depending on the fat depot. Certainly, adipocytes from visceral fat of $CB_1R$ knockout mice displayed a lower differentiation and a higher apoptotic rate than wild type did. However, it was the opposite in subcutaneous adipocytes [125]. $CB_2R$ signaling might also be involved in the regulation of WAT expansion, since its stimulation in obese mice resulted in reduced body weight associated with a reduced adipocyte cell size [119].

WAT expansion is associated with significant macrophage infiltration and overproduction of inflammatory cytokines, which constitute an important event initiating obesity-associated insulin resistance and metabolic syndrome. Several studies indicate that local inflammation is reversed by $CB_1R$ blockade [120–123]. This effect could result from the inactivation of macrophage $CB_1R$, since its stimulation with AEA markedly induces the expression of *Ccl2*, *Nlrp3*, *Il1b*, and *Tnf* (involved in M1 macrophage polarization) in RAW264.7 cells while its blockade fully reverses these changes [121,122,124]. Similar findings were obtained in obese mice and diabetic rats [120–123]. In addition, a recent study indicated that $CB_1Rs$ are key regulators of a crosstalk between adipocytes, immune cells, and the sympathetic nervous system [80]. The authors suggested that specific inactivation of $CB_1R$ in adipocytes contributes to the adaptive thermogenesis and browning of WAT inducing macrophage polarization into the M2 phenotype, which acts as a local source of catecholamine [80]. In addition, activating the ECS is associated with an alteration of mitochondrial function because it lowers mitochondrial biogenesis [81,125], as well as reduces adipocyte's fatty acid β-oxidation and oxygen consumption [81]. In contrast, $CB_1R$ blockade leads to an increase in AMPK phosphorylation and oxygen consumption by adipocytes [79].

In addition to its role in energy storage, WAT displays some endocrine functions and can secrete various bioactive substances/cytokines, referred to as adipokines. Several studies reported that the production of these factors is affected by the ECS (for a review, see Ref. [126]). The expression of the insulin-sensitizing adipokine, adiponectin, is inhibited following $CB_1R$ activation and is increased by its blockade in 3T3L1 adipocytes [88,127] and in human adipose tissue explants [128]. Consistent with these findings, adiponectin production is restored in obese animals and humans treated with $CB_1R$ antagonists [41,127,129,130]. Similarly, $CB_1R$

appears to control the production of the anorexigenic adipokine leptin, which reduces food intake by acting on hypothalamic receptors. Indeed, incubation of 3T3L1 adipocytes with different $CB_1R$ agonists increases leptin secretion, whereas its blockade produces the opposite effect [41]. A recent study also reported that blocking $CB_1Rs$ significantly decreases the concentration of less favorable adipokines such as resistin, apelin, visfatin, interleukin 6 (IL-6), and interferon-γ, both in subcutaneous and visceral adipose tissue of obese mice [131]. Interestingly, $CB_2R$ agonism appears to lower the protein expression of PPARγ, leptin, IL-6, and tumor necrosis factor α (TNFα) while elevating the protein levels of adiponectin and IL-4 in adipocytes derived from obese patients [132].

To date, we have been addressing WAT functions. However, as mentioned before, the ECS is present and functional within the BAT, where it has an impact on thermogenesis. $CB_1R$ blockade has been shown to enhance energy expenditure and to activate BAT, as indicated by the increased expression of genes involved in thermogenesis [uncoupling protein 1 (UCP1) and proliferator-activated receptor gamma coactivator 1 alpha (PGC-1α)] and the concomitant decrease in lipid droplet size in BAT [75]. Similarly, a recent study using PET imaging reported a positive correlation between $CB_1R$ density and glucose uptake in human BAT under cold exposure [100]. Thermogenesis appears to be activated in part through the action of the CNS and peripheral sympathetic neuron innervation [73–75,100,133,134]. Indeed, activating $\beta_3$-adrenergic receptors results in higher eCB content and increased $CB_1R$ expression in primary brown adipocytes, further suggesting the existence of an intercommunication between the ECS and the sympathetic signals [99]. In addition, several studies also found ECS to be involved in fat browning, the biological process by which white adipocytes develop some brown adipocyte-like features such as thermogenic ability. Krott and colleagues reported an association between ECS upregulation, decreased thermogenesis, lipid droplet formation, and WAT browning [98]. Furthermore, several studies showed that peripheral $CB_1R$ blockade could enhance the formation of UCP1-expressing adipocytes within the WAT, elevate body temperature and decrease adipocyte size [79,125,133]. In addition, the involvement of $CB_2R$-expressed immune cells within the WAT in adipocyte browning seems plausible because $CB_2R$ agonism promotes T lymphocytes, Th2 polarization as well as IL-4 secretion, which, in turn, results in browning of WAT [132].

Finally, the current literature strongly suggests that $CB_1R$ activation in WAT promotes fat expansion and dysfunction of adipocyte metabolism, leading to an alteration of the secretory function of adipose tissue associated with an increased cardiovascular risk (Fig. 15.1). On the other hand, a few data related to $CB_2R$ activation indicate an opposite effect on WAT metabolism.

## Regulation of insulin homeostasis by endocannabinoids

Insulin secretion is a timely and finely regulated process that maintains blood glucose continually within a narrow range (normoglycemia) in healthy nondiabetic individuals [135]. Insulin secretion is glucose dependent. Upon eating, blood glucose rises and glucose is taken up by β-cell through the transporter GLUT2 and is phosphorylated by glucokinase (GCK) to enter glycolysis. Metabolism of glucose in β-cells results in increased ATP content, which is required to initiate the exocytosis of insulin-containing granules [135]. Once secreted,

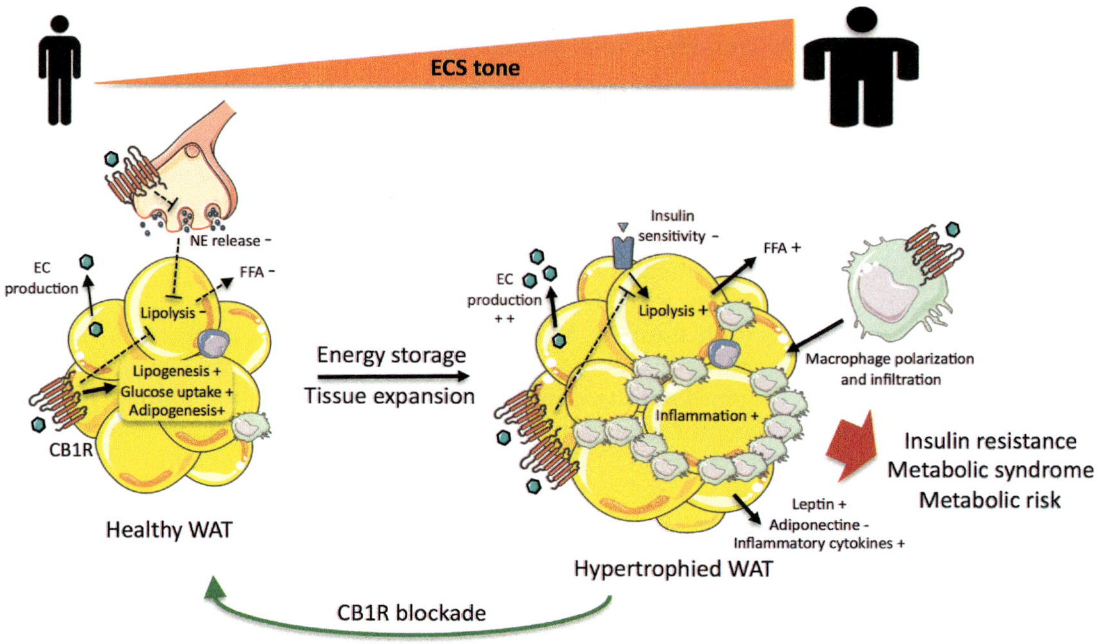


**FIGURE 15.1 Schematic illustrating the contribution of $CB_1R$ in the progressive expansion of white adipose tissue, ultimately leading to increased metabolic risk.** *$CB_1R$*, cannabinoid-1 receptors; *EC*, endocannabinoids; *ECS*, endocannabinoid system; *FFA*, free fatty acids; *NE*, norepinephrine; *WAT*, white adipose tissue. *Source:This figure was created using Servier Medical Art templates, which are licensed under a Creative Commons Attribution 3.0 Unported License; https://smart.servier.com.*

circulating insulin binds to its transmembrane tyrosine kinase receptor, the insulin receptor (IR), which upon insulin binding, is autophosphorylated and initiates intracellular signaling. This leads to the translocation of GLUT4 to the plasma membrane and to blood glucose uptake into the cells of insulin-sensitive tissues (i.e., muscle and adipose tissue). Lowering blood glucose ends insulin secretion and circulating insulin is promptly cleared by the liver. Dysregulation of insulin secretion and/or action leads to diabetes, which is associated with micro- and macrovascular complications and increases the risk of renal failure, blindness, peripheral neuropathy, and premature death, among others. The ECS regulates insulin homeostasis by acting at different ends of this axis.

Human and rodent pancreatic islets contain an autonomous ECS. It is currently accepted that pancreatic islet β-cells express $CB_1R$ and the eCB synthesizing and degradative enzymes [136–138], although some studies showed otherwise [89,136]. eCBs are synthesized on demand and secreted from β-cells in a glucose-dependent manner; when glucose levels rise, β-cells secrete eCBs (mainly 2-AG) [136,138]. Unlike the kidney [139], β-cell $CB_1R$ activation reduces GLUT2 and GCK levels [140,141]. GLUT2 and GCK are the glucose sensors of β-cells; namely, they are necessary for glucose uptake and metabolism in β-cells at a $K_m$ of ~15 mM and ~7 mM, respectively. The ATP obtained from glucose metabolism is utilized for closure of the $K_{ATP}$ channels and for plasma membrane depolarization. Consequently, the voltage-dependent $Ca^{2+}$ channels open, resulting in an influx of $Ca^{2+}$ that triggers exocytosis of insulin-containing granules by $Ca^{2+}$-binding proteins. Activating $CB_1Rs$ indeed reduces $Ca^{2+}$ influx [142,143], subsequently inhibiting insulin secretion [138,140,143,144]. $CB_1R$-dependent reduction of intracellular cAMP is also involved in alteration of insulin secretion [137,138,140,143,144]. Closure of $K_{ATP}$ channels and exocytosis of insulin is amplified by increased intracellular cAMP, which, in turn, activates the cAMP sensor proteins PKA and cAMP-GEF/Epac [135]. Incretins (GIP and GLP-1) are hormones secreted postprandially from L and K cells in the gut. Binding of incretins to their $G_s$ protein-coupled receptors (GIPR and GLP-1R, respectively) in the β-cell activate adenylate cyclase (AC) in a glucose-dependent manner to increase intracellular cAMP and enhance insulin secretion [145,146]. $CB_1R$ activation further inhibits AC activity, reducing intracellular cAMP, thus hindering insulin secretion [137,138,140,143,144]. In agreement with these findings, enhanced insulin secretion has been observed when deleting $CB_1R$ or blocking it pharmacologically [137,138,144,147], although some studies have found that $CB_1R$ can couple to $G_s$ in β-cells [136,148,149]. $CB_1R$ also blocks incretin-mediated glucose-dependent insulin secretion [140,147]. Both eCBs and $CB_1R$-specific synthetic agonists lower GLP-1R-stimulated glucose-dependent intracellular cAMP levels and insulin secretion in insulinoma cell lines and islets, whereas $CB_1R$ inverse agonism partially prevents this inhibition [140,147]. Activating $CB_1Rs$, which are expressed in K cells in the intestine [150], inhibits GIP secretion [150,151] in rodents and humans, further impacting insulin secretion.

β-Cell mass is critical for maintaining insulin homeostasis. When functional β-cell mass drops below a critical level, insulin secretion fails to compensate for elevated blood glucose levels and consequently, diabetes occurs. β-Cell mass is regulated by a balance between cell viability and proliferation, although another less understood mechanism termed transdifferentiation (loss of β-cell identity) exists in diabetes [152]. The IR in β-cells is not required for glucose uptake, but it signals through Akt to regulate these processes. In fact, $CB_1R$ is involved in β-cell viability and proliferation interacting with IR. The direct negative regulation of the IR by $CB_1R$ has been described in β-cells [153]. The $G_i$ subunit of $CB_1R$ directly binds, when activated by eCBs, to the tyrosine kinase domain of the IR, thus blocking its activation by phosphorylation [153]. As a result, the downstream pathway of the IR is inhibited, including blockade of the phosphorylation of Akt, leading to β-cell apoptosis. Persistent hyperglycemia and obesity are associated with stress-induced β-cell apoptosis [155]. Importantly, blockade of $CB_1R$ allows β-cell proliferation and preserves β-cell viability under stress [138,144]. Despite the ECS being involved in the embryonic development of pancreatic isles [154], the role of the ECS in the transdifferentiation of mature β-cells has not been studied.

Inflammation, which occurs in obesity as a chronic event, is associated with damage in the islets, which ultimately causes loss of functional β-cell mass and/or β-cell death [155]. Activating $CB_1R$ in both immune cells and β-cells leads to initiation of inflammation that results in immune cell infiltration in and around islets in obese rodents [121,123,144]. Pharmacological blockade of $CB_1R$ or genetic ablation of $CB_1R$ specifically in macrophages or in β-cells prevents obesity-induced activation of the NLRP3 inflammasome by modulating the mitogen-activated protein kinase (MAPK) signaling pathway. In islet-resident macrophages, $CB_1R$ stimulation induces their polarization to pro-inflammatory M1 macrophages by activating both the NLRP3 inflammasome, which leads to IL-1β secretion and p38 MAPK, which leads to secretion of CCL2, which consequently enhances further macrophage recruitment

[121,123]. Concordantly, genetic ablation of $CB_1R$ in β-cells reduces MAPK activation in diet-induced obese mice and reduces the secretion of pro-inflammatory cytokines from pancreatic islets [144]. Overall, overactivation of $CB_1R$ negatively regulates functional β-cell mass, which is the origin of the circulating insulin.

In a similar fashion as in β-cells, $CB_1R$ also negatively regulates IR activity in insulin-sensitive tissues. As described above, activating $CB_1R$ induces insulin resistance in targeted tissues, such as muscle [156,157] and adipose tissue [114,158], altering glucose uptake and lipid metabolism. Reduction of insulin sensitivity has a negative feedback loop on β-cells by altering glucose homeostasis, thus insulin secretion. In fact, global blockade of $CB_1R$ has been shown to improve insulin sensitivity and glucose uptake [159].

Finally, but not less importantly, a dysregulated ECS alters insulin clearance by the liver. In an elegant genetic approach, hepatocyte-specific overexpression of $CB_1R$ in global $CB_1R$ knockout mice led to downregulation of insulin degradative enzymes (IDE) in liver [160]. The latter has been associated with insulin resistance in liver and with lower IR signaling in hepatocytes, which mimics what occurs in obese mice compared to lean mice. Thus, in obesity, overactivation of $CB_1R$ would further increase hyperinsulinemia and exacerbate insulin resistance (Fig. 15.2).

$CB_2R$ is largely known for its function in the immune system and not much is known about its role in insulin homeostasis. Unlike $CB_1R$, $CB_2R$ is not expressed in β-cells [137,143]. The role of $CB_2R$ in insulin secretion is largely unexplored and remains conflicting [136,142,161].

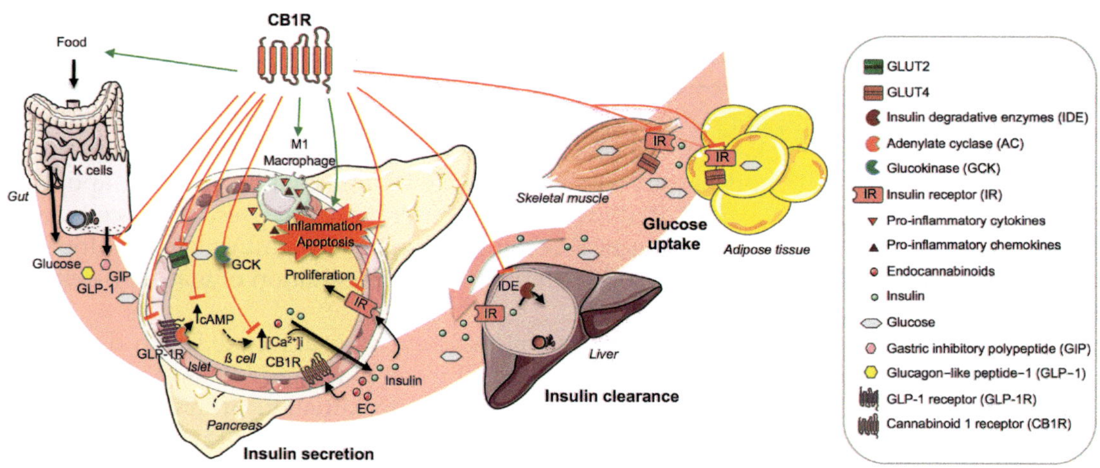


**FIGURE 15.2** **Schematic illustrating the contribution of $CB_1R$ in the regulation of insulin homeostasis** . $CB_1R$ negatively regulates insulin secretion, clearance and action by acting in a variety of peripheral tissues. An overactivation of the $CB_1R$/ECS, as it occurs in obesity, leads to (1) increased appetite and food intake, (2) reduction in the secretion of gastric inhibitory polypeptide (GIP), (3) pancreatic islet inflammation and β-cell dysfunction, and (4) downregulation of insulin clearance by the liver and insulin resistance. *Source:This figure was created using Servier Medical Art templates, which are licensed under a Creative Commons Attribution 3.0 Unported License; https://smart.servier.com.*

## Endocannabinoids and hepatic lipogenesis

Ample evidence indicates that the ECS plays a significant role in the pathogenesis of different liver diseases. However, the initial report on the importance of this system in the liver [107] was completely unexpected, since the liver has been used as a negative control to study the function of neuronal $CB_1Rs$ for many years [162]. Normally, both $CB_1Rs$ and $CB_2Rs$ are expressed at fairly low levels in various cells within the liver (hepatocytes, stellate cells, hepatic myofibroblasts, Kupffer cells, and hepatic vascular endothelial cells) [73,107,129,162]. Both $CB_1R$ and $CB_2R$ expression is upregulated during liver pathologies [163–168]. In addition, eCBs are present in the liver at levels that are comparable to those found in the brain [107,169].

The possibility that the ECS is involved in modulating hepatic *de novo* lipogenesis has been the focus of many recent studies. A role for $CB_1R$ in the hepatic *de novo* lipogenic response to high-fat diet (HFD) was initially reported by Osei-Hyiaman and colleagues, who demonstrated the complete resistance of $CB_1R$ knockout mice to HFD-induced hepatic steatosis [107]. Complementary in vivo, in vitro, and ex vivo observations were reported while trying to activate $CB_1R$, which, in turn, resulted in increased expression of several lipogenic genes in mice and zebrafish larvae livers [107,170,171]. Similarly, a few studies have described the positive correlations between circulating eCBs and the development of fatty liver disease [172–174]. These findings are consistent with studies in humans that suggest that circulating eCBs play a role as biomarkers for visceral obesity and its related metabolic comorbidities [90,96,175–178].

Recent studies also suggest that increased intake of HFD activates the ECS, specifically by downregulating the activity of FAAH, which elevates AEA levels [107]. In fact, an earlier report that found that lipids present in tissue extracts inhibit FAAH activity [179] suggests that consumption of HFD may upregulate the generation of endogenous FAAH inhibitors, which would consequently contribute to activating the ECS by reducing the levels of the degrading enzyme of AEA. Indeed, activating hepatic $CB_1Rs$ increases the synthesis of monounsaturated fatty acids (MUFAs), which potently inhibit FAAH [180], suggesting a positive feedback loop between the hepatic ECS and MUFAs, which contributes to increased *de novo* hepatic lipogenesis during obesity. Similarly to HFD consumption, it has been shown that xenoestrogen bisphenol A (BPA) induces hepatic steatosis via upregulating eCB levels and $CB_1R$ expression as well as reducing FAAH expression [181]. In parallel to AEA/FAAH, inhibiting MAGL results in hepatic elevation of 2-AG, which, in turn, contributes to increased expression of lipogenic genes and proteins and the accumulation of triglycerides in the liver, an effect that was completely dependent on $CB_1R$ [182]. Acutely blocking the two eCB degrading enzymes also increases hepatic triglyceride content [183]. Interestingly, MAGL nullification in mice provided protection from HFD-induced hepatic steatosis, suggesting a $CB_1R$ independent aspect of MAGL in this process [184].

In agreement with the above findings, it was reported that pharmacological blockade of $CB_1Rs$ enhances fatty acid oxidation and reduces hepatic inflammation and *de novo* lipogenesis [41,107,130,170,185–190]. These observations were further confirmed preclinically and clinically, demonstrating the ability of globally acting and peripherally restricted $CB_1R$ antagonists to downregulate the expression levels of hepatic lipogenic genes and to reverse fatty liver disease [41,107,129,130,185,189,191–196]. $CB_1R$ blockade reverses fatty liver disease and hepatic injury, most likely via regulating a cellular defense mechanism against oxidative stress, upregulating AMPK activity, modulating the hepatic expression of activin A and follistatin, as well as inhibiting the recruitment of bone marrow-derived monocytes/macrophages expressing $CB_1Rs$ [197–200]. Interestingly, $CB_1R$ in Kupffer cells does not play a role in fatty liver disease, since its specific genetic deletion in these cells did not protect diet-induced obese mice from developing hepatic steatosis [122]. Enhanced hepatic *de novo* lipogenesis and hepatic steatosis was also reported with chronic alcohol use [201–203], which also increases the activity of the ECS. In fact, mice exposed to an ethanol diet exhibit upregulated $CB_1R$ and 2-AG levels, contributing to their increase in lipogenic genes in the liver, an effect that was absent in mice lacking $CB_1R$ globally or in hepatocytes only [204–206]. In parallel to these results, it has recently been shown that peripherally restricted $CB_1R$ blockade reduces ethanol-induced increases in lipid accumulation in the liver [207].

Although all of the above-mentioned findings strongly imply an important role played by the hepatic ECS in *de novo* lipogenesis, which contributes to fatty liver disease, they do not negate the idea that activating the ECS in extrahepatic organs, such as adipose tissue (as discussed earlier) may influence hepatic lipogenesis indirectly. For instance, the major source of liver fat may come from WAT, in which stimulating $CB_1Rs$ alters the antilipolytic action of insulin and thereby increases the FFA flux from adipose tissue to the liver [114]. Blocking $CB_1Rs$, on the other hand, has been shown to reverse the HFD-induced upregulation of fatty acid translocase/CD36, which mediates the uptake of FFAs from the circulation to the liver [41,185,189]. These findings are also in agreement with the observation that mice lacking $CB_1Rs$ specifically in adipocytes are protected from HFD-induced hepatic steatosis [80].

In contrast to $CB_1R$, the involvement of $CB_2R$ in modulating hepatic *de novo* lipogenesis is controversial and is not fully delineated. It was initially suggested because of the observation that $CB_2R$ has increased expression in hepatocytes from both leptin-deficient *ob/ob* and HFD-induced obese mice [166]. Interestingly, activation of $CB_2R$ enhances liver triglyceride accumulation in diet-induced obese mice [208] and $CB_2R$ knockout mice are known to be resistant to HFD-induced hepatic steatosis [208,209], but are sensitive to alcoholic hepatic injury [206]. In vitro activation of $CB_2R$ dose dependently induces lipid accumulation in hepatocytes in the presence of oleic acid, most likely due to increased FAS expression [210]. On the other hand, treatment of mice with alcohol induces liver injury and steatosis; together with different $CB_2R$ agonists, it reduces hepatic inflammation and steatosis via attenuating the pro-inflammatory phenotype of Kupffer cells as well as by reducing neutrophil infiltration, effects that were shown to induce autophagy [211–213]. Moreover, a study of children with biopsy-proven fatty liver disease revealed that whereas the $CB_2R$ variant carrying an arginine residue at codon 63 is not associated with steatosis, it is associated with the severity of the liver inflammatory component, which could increase the risk of nonalcoholic steatohepatitis [214]. Based on the above, further studies are needed to delineate the specific steatogenic role of $CB_2Rs$.

## Conclusions

The ECS, present in various tissues and organs, is involved in controlling energy and metabolic homeostasis, adiposity, insulin and glucose balance as well as hepatic lipogenesis and fatty acid oxidation (Table 15.1). It has important therapeutic implications specifically in relation to targeting $CB_1Rs$ and $CB_2Rs$. Increased activity of the ECS contributes to the development of the metabolic syndrome, whereas $CB_1R$ antagonism attenuates or prevents the development

TABLE 15.1 List of the most relevant cannabinoid actions on metabolism. For details, please see main text.

| SYSTEM | STRUCTURE/CELL | RECEPTOR | EFFECT | CONSEQUENCE | REMARK / SUSPECTED EFFECT |
|---|---|---|---|---|---|
| **CNS** | *hypothalamus* | $CB_1R$ | MCH ↑ | orexigenic | increases metabolic rate |
| | | | β-endorphin ↑ | orexigenic | may increase metabolic rate |
| | | | orexin ↓ | anorexigenic | subject to remodelling in obesity |
| | *mesolimbic pathway* | | dopamine release ↑ | self-reward ↑ | psychiatric side effects of $CB_1R$ blockade |
| | *ventral striatum* | | GABA release ↓ | anorexigenic | |
| | *forebrain neurons* | | metabolic rate and lipolysis ↓ | susceptibility to DIO ↑ | |
| **sensory organs** | *olfactory bulb* | | odor detection ↑ | orexigenic | |
| | *taste buds* | | sugar sensitivity ↑ | supports reward | |
| **sympathetic NS** | *ganglion neurons* | | metabolic rate and lipolysis ↓ | susceptibility to DIO ↑ | |
| | *postsynaptic axon terminals* | | lipolysis and BAT thermogenesis ↓ | total energy expenditure ↓ | |
| **vagal nerve** | *intestinal axon terminals* | | gastrointestinal motility ↓ | digestive, adsorptive efficiency ↑ | |
| **GI tract** | *parietal cell of stomach* | $CB_1R$ | ghrelin release ↑ | orexigenic, metabolic rate ↑ | |
| | *jejunum* | | fat consumption ↑ | increase of meal size | |
| | *K cells* | | GIP secretion ↓ | insulin secretion ↓ | |
| | *unknown* | $CB_2R$ | intestinal motility ↓ | digestive, adsorptive efficiency ↑ | |
| **WAT** | *adipocyte* | $CB_1R$ | browning potential ↓ | metabolic rate ↓ | $CB_1R$ blockade reverses these effects |
| | | | glucose uptake ↑ | lipogenesis ↑ | |
| | | | cAMP availability ↓ SREBP-1c activity ↑ | lipogenesis ↑ | |
| | | | insulin signaling ↓ | lypolysis ↑ | |
| | | | PPAR-g activity ↑ | adipogenesis ↑ | |
| | | | mitochondrial activity ↓ | fatty acid oxidation ↓ | |
| | | | adiponectin secretion ↓ | insulin resistance ↑ | |
| | | | leptin secretion ↑ | leptin resistance ↑ | |
| | *adipocytes/immune cells* | | pro-inflammatory cytokines ↑ | local inflammation ↑ | |
| | *adipocyte* | $CB_2R$ | adipose triglyceride lipase ↑ | lipolysis ↑ | |
| | *T lymphocytes* | | $Th_2$ polarization | fat browning ↑ | |
| **BAT** | *adipocyte* | $CB_1R$ | UCP1 ↓ and PGC-1a ↓ | thermogenesis ↓ | $CB_1R$ blockade reverses this effect |
| **skeletal muscle** | *muscle cell* | $CB_1R$ | insulin signaling ↓ | glucose uptake ↓ | hyperactivation leads to insulin resistance |
| | | | generation of mitochondria ↓ | heat production | total energy expenditure decreases |
| **pancreas** | *β-cells* | $CB_1R$ | GLP-1R sensitivity ↓ $[cAMP]_i$ ↓ | insulin secretion ↓ | $CB_1R$ blockade reverses |
| | | | GLUT2 expression ↓ | glucose uptake ↓ | hyperactivation leads to β-cell dysfuntion |
| | | | GCK expression ↓ | glucose metabolism ↓ | |
| | | | $[Ca^{2+}]_i$ ↓ | insulin secretion ↓ | |
| | | | apoptosis ↑ | β-cell mass ↓ | hyperactivation leads to β-cell loss |
| | | | pro-inflammatory cytokines ↑ | local inflammation ↑ | hyperactivation leads to islet inflammation |
| | *macrophages* | | pro-inflammatory cytokines ↑ | local inflammation ↑ | hyperactivation leads to islet inflammation |
| **liver** | *hepatocytes* | $CB_1R$ | insulin degradative enzymes ↓ | insulin clearance ↓ | hyperactivation leads to hyperinsulinemia |
| | | | de novo lipogenesis ↑ | hepatic steatosis/alcohol-induced fatty liver ↑ | $CB_1R$ blockade reverses these effects |
| | | | fatty acid beta oxidation ↓ | hepatic steatosis ↑ | |
| | | | fatty acid uptake ↑ | hepatic steatosis ↑ | |
| | *hepatic myofibroblasts* | | inflammation ↑ | NASH ↑ | |
| | | | fibrogenesis ↑ | liver cirrhosis ↑ | |
| | | $CB_2R$ | fibrogenesis ↓ | liver cirrhosis ↓ | |
| | *myofibroblasts/Kupffer cells* | | inflammation ↓ | NASH ↓ | |

of the cardio-metabolic comorbidities associated with it. Since the withdrawal of the globally acting $CB_1R$ inverse agonist, rimonabant, from the market, our understanding of the role of the ECS in metabolic regulation in peripheral tissues has expanded tremendously. To date, a wide variety of drugs that target the $CB_1R$ in periphery are being developed and preclinically tested. Based on these studies and the lack of CNS-induced pyschoactivity, one may postulate that in the next 5 years peripherally restricted $CB_1R$ inhibitors will be translated to clinical evaluation and testing as useful therapeutic agents. As per $CB_2R$, which was originally considered to be devoid of a metabolic role, it is now emerging as a potential player in metabolic regulation due to its (1) antiinflammatory role, (2) effect on hepatic de novo lipogenesis, and (3) involvement in glucose and insulin homeostasis. These findings will eventually promote the preclinical evaluation and clinical testing of $CB_2R$ agonists against the metabolic sequelae. Moreover, the fact that the ECS is dysregulated during metabolic diseases, in which circulating eCBs are most likely reflection of imbalanced "eCB tone" in peripheral tissues, might be not only serve as biomarkers, but also a significant causative role of metabolic abnormalities.

## Acknowledgments

This work was supported in part by the Israel Science Foundation (grant # 158/18) to J.T.; the INSERM (Institut National de la Santé et de la Recherche Médicale), the Regional Council of Bourgogne, the University of Bourgogne-Franche Comté and by a French Government grant managed by the French National Research Agency (ANR) under the program "Investissements d'Avenir" (grant # ANR-11-LABX-0021-01-LipSTIC LabEx) to T.J. and P.D.; the Hungarian National Research, Development and Innovation Office (grant # NKFI-6/FK_124038) to G.S.; the Horizon 2020 program from the European Union (H2020-MSCA-IF-2016, Grant # 748749) to I.G.M.

## References

[1] Abel EL. Cannabis: effects on hunger and thirst. Behav Biol 1975;15(3):255–81.

[2] Hanus LO, Meyer SM, Munoz E, Taglialatela-Scafati O, Appendino G. Phytocannabinoids: a unified critical inventory. Nat Prod Rep 2016;33(12):1357–92. doi: 10.1039/c6np00074f.

[3] Berman P, Futoran K, Lewitus GM, Mukha D, Benami M, Shlomi T, et al. A new ESI-LC/MS approach for comprehensive metabolic profiling of phytocannabinoids in Cannabis. Sci Rep 2018;8(1):14280. doi: 10.1038/s41598-018-32651-4.

[4] Mechoulam R, Shvo Y, Hashish I. The structure of cannabidiol. Tetrahedron 1963;19(12):2073–8.

[5] Gaoni Y, Mechoulam R. Isolation, structure, partial, synthesis of an active constituent of hashish. J Am Chem Soc 1964;86:1646–7.

[6] Mechoulam R, Gaoni Y. The absolute configuration of delta-1-tetrahydrocannabinol, the major active constituent of hashish. Tetrahedron Lett 1967;12:1109–11.

[7] Hively RL, Mosher WA, Hoffmann FW. Isolation of trans-delta-tetrahydrocannabinol from marijuana. J Am Chem Soc 1966;88(8):1832–3.

[8] Gaoni Y, Mechoulam R. The structure and synthesis of cannabigerol, a new hashish constituent. Proc Chem Soc 1964;82.

[9] Gaoni Y, Mechoulam R. Cannabichromene, a new active principle in hashish. Chem Commun 1966;1:20–1.

[10] Crombie L, Ponsford R, Shani A, Yagnitinsky B, Mechoulam R. Hashish components. Photochemical production of cannabicyclol from cannabichromene. Tetrahedron Lett 1968;(55):5771–2.

[11] Devane WA, Dysarz FA III, Johnson MR, Melvin LS, Howlett AC. Determination and characterization of a cannabinoid receptor in rat brain. Mol Pharmacol 1988;34(5):605–13.

[12] Matsuda LA, Lolait SJ, Brownstein MJ, Young AC, Bonner TI. Structure of a cannabinoid receptor and functional expression of the cloned cDNA. Nature 1990;346(6284):561–4. doi: 10.1038/346561a0.

[13] Munro S, Thomas KL, Abu-Shaar M. Molecular characterization of a peripheral receptor for cannabinoids. Nature 1993;365(6441):61–5. doi: 10.1038/365061a0.

[14] Hua T, Vemuri K, Nikas SP, Laprairie RB, Wu Y, Qu L, et al. Crystal structures of agonist-bound human cannabinoid receptor CB1. Nature 2017;547(7664):468–71. doi: 10.1038/nature23272.

[15] Hua T, Vemuri K, Pu M, Qu L, Han GW, Wu Y, et al. Crystal structure of the human cannabinoid receptor CB1. Cell 2016;167(3):750–62. doi: 10.1016/j.cell.2016.10.004. e714.

[16] Li X, Hua T, Vemuri K, Ho JH, Wu Y, Wu L, et al. Crystal structure of the human cannabinoid receptor CB2. Cell 2019;176(3):459–67. doi: 10.1016/j.cell.2018.12.011. e413.

[17] Shao Z, Yin J, Chapman K, Grzemska M, Clark L, Wang J, et al. High-resolution crystal structure of the human CB1 cannabinoid receptor. Nature 2016;doi: 10.1038/nature20613.

[18] Howlett AC. Cannabinoid receptor signaling. Handb Exp Pharmacol 2005;168:53–79.

[19] Mackie K. Distribution of cannabinoid receptors in the central and peripheral nervous system. Handb Exp Pharmacol 2005;(168):299–325.

[20] Tam J, Hinden L, Drori A, Udi S, Azar S, Baraghithy S. The therapeutic potential of targeting the peripheral endocannabinoid/CB1 receptor system. Eur J Intern Med 2018;49:23–9. doi: 10.1016/j.ejim.2018.01.009.

[21] Zou S, Kumar U. Cannabinoid receptors and the endocannabinoid system: signaling and function in the central nervous system. Int J Mol Sci 2018;19(3.)doi: 10.3390/ijms19030833.

[22] Di Marzo V, Piscitelli F. The endocannabinoid system and its modulation by phytocannabinoids. Neurotherapeutics 2015;12(4):692–8. doi: 10.1007/s13311-015-0374-6.

[23] Devane WA, Hanus L, Breuer A, Pertwee RG, Stevenson LA, Griffin G, et al. Isolation and structure of a brain constituent that binds to the cannabinoid receptor. Science 1992;258(5090):1946–9.

[24] Mechoulam R, Ben-Shabat S, Hanus L, Ligumsky M, Kaminski NE, Schatz AR, et al. Identification of an endogenous 2-monoglyceride, present in canine gut, that binds to cannabinoid receptors. Biochem Pharmacol 1995;50(1):83–90.

[25] Sugiura T, Kondo S, Sukagawa A, Nakane S, Shinoda A, Itoh K, et al. 2-Arachidonoylglycerol: a possible endogenous cannabinoid receptor ligand in brain. Biochem Biophys Res Commun 1995;215(1):89–97.

[26] Di Marzo V, De Petrocellis L. Why do cannabinoid receptors have more than one endogenous ligand? Philos Trans R Soc Lond B Biol Sci 2012;367(1607):3216–28. doi: 10.1098/rstb.2011.0382.

[27] Pertwee RG, Howlett AC, Abood ME, Alexander SP, Di Marzo V, Elphick MR, et al. International Union of Basic and Clinical Pharmacology. LXXIX. Cannabinoid receptors and their ligands: beyond CB(1) and CB(2). Pharmacol Rev 2010;62(4):588–631. doi: 10.1124/pr.110.003004.

[28] Fowler CJ. The pharmacology of the cannabinoid system—a question of efficacy and selectivity. Mol Neurobiol 2007;36(1):15–25. doi: 10.1007/s12035-007-0001-6.

[29] Pacher P, Batkai S, Kunos G. The endocannabinoid system as an emerging target of pharmacotherapy. Pharmacol Rev 2006;58(3):389–462. doi: 10.1124/pr.58.3.2.

[30] Murataeva N, Straiker A, Mackie K. Parsing the players: 2-arachidonoylglycerol synthesis and degradation in the CNS. Br J Pharmacol 2014;171(6):1379–91. doi: 10.1111/bph.12411.

[31] McKinney MK, Cravatt BF. Structure and function of fatty acid amide hydrolase. Annu Rev Biochem 2005;74:411–32. doi: 10.1146/annurev.biochem.74.082803.133450.

[32] Dinh TP, Carpenter D, Leslie FM, Freund TF, Katona I, Sensi SL, et al. Brain monoglyceride lipase participating in endocannabinoid inactivation. Proc Natl Acad Sci U S A 2002;99(16):10819–24. doi: 10.1073/pnas.152334899.

[33] Mazier W, Saucisse N, Gatta-Cherifi B, Cota D. The endocannabinoid system: pivotal orchestrator of obesity and metabolic disease. Trends Endocrinol Metab 2015;26(10):524–37. doi: 10.1016/j.tem.2015.07.007.

[34] Simon V, Cota D. Mechanisms in endocrinology: endocannabinoids and metabolism: past, present and future. Eur J Endocrinol 2017;176(6):R309–24. doi: 10.1530/eje-16-1044.

[35] Ng M, Fleming T, Robinson M, Thomson B, Graetz N, Margono C, et al. Global, regional, and national prevalence of overweight and obesity in children and adults during 1980-2013: a systematic analysis for the Global Burden of Disease Study 2013. Lancet 2014;384(9945):766–81. doi: 10.1016/s0140-6736(14)60460-8.

[36] Kano M, Ohno-Shosaku T, Hashimotodani Y, Uchigashima M, Watanabe M. Endocannabinoid-mediated control of synaptic transmission. Physiol Rev 2009;89(1):309–80. doi: 10.1152/physrev.00019.2008.

[37] Jo YH, Chen YJ, Chua SC Jr, Talmage DA, Role LW. Integration of endocannabinoid and leptin signaling in an appetite-related neural circuit. Neuron 2005;48(6):1055–66. doi: 10.1016/j.neuron.2005.10.021.

[38] Koch M, Varela L, Kim JG, Kim JD, Hernandez-Nuno F, Simonds SE, et al. Hypothalamic POMC neurons promote cannabinoid-induced feeding. Nature 2015;519(7541):45–50. doi: 10.1038/nature14260.

[39] Bellocchio L, Lafenetre P, Cannich A, Cota D, Puente N, Grandes P, et al. Bimodal control of stimulated food intake by the endocannabinoid system. Nat Neurosci 2010;13(3):281–3. doi: 10.1038/nn.2494.

[40] Cristino L, Busetto G, Imperatore R, Ferrandino I, Palomba L, Silvestri C, et al. Obesity-driven synaptic remodeling affects endocannabinoid control of orexinergic neurons. Proc Natl Acad Sci U S A 2013;110(24):E2229–2238. doi: 10.1073/pnas.1219485110.

[41] Tam J, Cinar R, Liu J, Godlewski G, Wesley D, Jourdan T, et al. Peripheral cannabinoid-1 receptor inverse agonism reduces obesity by reversing leptin resistance. Cell Metab 2012;16(2):167–79. doi: 10.1016/j.cmet.2012.07.002.

[42] Pi-Sunyer FX, Aronne LJ, Heshmati HM, Devin J, Rosenstock J. RIO-North America Study Group. Effect of rimonabant, a cannabinoid-1 receptor blocker, on weight and cardiometabolic risk factors in overweight or obese patients: RIO-North America: a randomized controlled trial. JAMA 2006;295(7):761–75. doi: 10.1001/jama.295.7.761.

[43] Kirkham TC, Williams CM, Fezza F, Di Marzo V. Endocannabinoid levels in rat limbic forebrain and hypothalamus in relation to fasting, feeding and satiation: stimulation of eating by 2-arachidonoyl glycerol. Br J Pharmacol 2002;136(4):550–7. doi: 10.1038/sj.bjp.0704767.

[44] Matias I, Di Marzo V. Endocannabinoids and the control of energy balance. Trends Endocrinol Metab 2007;18(1):27–37. doi: 10.1016/j.tem.2006.11.006.

[45] Soria-Gomez E, Bellocchio L, Reguero L, Lepousez G, Martin C, Bendahmane M, et al. The endocannabinoid system controls food intake via olfactory processes. Nat Neurosci 2014;17(3):407–15. doi: 10.1038/nn.3647.

[46] Di Marzo V, Goparaju SK, Wang L, Liu J, Batkai S, Jarai Z, et al. Leptin-regulated endocannabinoids are involved in maintaining food intake. Nature 2001;410(6830):822–5. doi: 10.1038/35071088.

[47] Maccioni P, Pes D, Carai MA, Gessa GL, Colombo G. Suppression by the cannabinoid CB1 receptor antagonist, rimonabant, of the reinforcing and motivational properties of a chocolate-flavoured beverage in rats. Behav Pharmacol 2008;19(3):197–209. doi: 10.1097/FBP.0b013e3282fe8888.

[48] Perio A, Barnouin MC, Poncelet M, Soubrie P. Activity of SR 141716 on post-reinforcement pauses in operant responding for sucrose reward in rats. Behav Pharmacol 2001;12(8):641–5.

[49] Egecioglu E, Skibicka KP, Hansson C, Alvarez-Crespo M, Friberg PA, Jerlhag E, et al. Hedonic and incentive signals for body weight control. Rev Endocr Metab Disord 2011;12(3):141–51. doi: 10.1007/s11154-011-9166-4.

[50] Gardner EL. Addictive potential of cannabinoids: the underlying neurobiology. Chem Phys Lipids 2002;121(1–2):267–90.

[51] DiPatrizio NV, Simansky KJ. Activating parabrachial cannabinoid CB1 receptors selectively stimulates feeding of palatable foods in rats. J Neurosci 2008;28(39):9702–9. doi: 10.1523/jneurosci.1171-08.2008.

[52] Richey JM, Woolcott O. Re-visiting the endocannabinoid system and its therapeutic potential in obesity and associated diseases. Curr Diab Rep 2017;17(10):99. doi: 10.1007/s11892-017-0924-x.

[53] Yoshida R, Ohkuri T, Jyotaki M, Yasuo T, Horio N, Yasumatsu K, et al. Endocannabinoids selectively enhance sweet taste. Proc Natl Acad Sci U S A 2010;107(2):935–9. doi: 10.1073/pnas.0912048107.

[54] Sclafani A. Oral and postoral determinants of food reward. Physiol Behav 2004;81(5):773–9. doi: 10.1016/jphysbeh.2004.04.031.

[55] Brissard L, Leemput J, Hichami A, Passilly-Degrace P, Maquart G, Demizieux L, et al. Orosensory detection of dietary fatty acids is altered in CB(1)R(-/-) mice. Nutrients 2018;10(10)doi: 10.3390/nu10101347.

[56] Matias I, Gatta-Cherifi B, Tabarin A, Clark S, Leste-Lasserre T, Marsicano G, et al. Endocannabinoids measurement in human saliva as potential biomarker of obesity. PLoS One 2012;7(7):e42399. doi: 10.1371/journal.pone.0042399.

[57] Myers MG, Cowley MA, Munzberg H. Mechanisms of leptin action and leptin resistance. Annu Rev Physiol 2008;70:537–56. doi: 10.1146/annurev.physiol.70.113006.100707.

[58] Ravinet Trillou C, Delgorge C, Menet C, Arnone M, Soubrie P. CB1 cannabinoid receptor knockout in mice leads to leanness, resistance to diet-induced obesity and enhanced leptin sensitivity. Int J Obes Relat Metab Disord 2004;28(4):640–8. doi: 10.1038/sj.ijo.0802583.

[59] Tam J, Szanda G, Drori A, Liu Z, Cinar R, Kashiwaya Y, et al. Peripheral cannabinoid-1 receptor blockade restores hypothalamic leptin signaling. Mol Metab 2017;6(10):1113–25. doi: 10.1016/j.molmet.2017.06.010.

[60] O'Hare JD, Zielinski E, Cheng B, Scherer T, Buettner C. Central endocannabinoid signaling regulates hepatic glucose production and systemic lipolysis. Diabetes 2011;60(4):1055–62. doi: 10.2337/db10-0962.

[61] Jourdan T, Godlewski G, Kunos G. Endocannabinoid regulation of beta-cell functions: implications for glycaemic control and diabetes. Diabetes Obes Metab 2016;18(6):549–57. doi: 10.1111/dom.12646.

[62] Arnold SE, Arvanitakis Z, Macauley-Rambach SL, Koenig AM, Wang HY, Ahima RS, et al. Brain insulin resistance in type 2 diabetes and Alzheimer disease: concepts and conundrums. Nat Rev Neurol 2018;14(3):168–81. doi: 10.1038/nrneurol.2017.185.

[63] Loh K, Zhang L, Brandon A, Wang Q, Begg D, Qi Y, et al. Insulin controls food intake and energy balance via NPY neurons. Mol Metab 2017;6(6):574–84. doi: 10.1016/j.molmet.2017.03.013.

[64] DiPatrizio NV, Piomelli D. Intestinal lipid-derived signals that sense dietary fat. J Clin Invest 2015;125(3):891–8. doi: 10.1172/jci76302.

[65] DiPatrizio NV, Astarita G, Schwartz G, Li X, Piomelli D. Endocannabinoid signal in the gut controls dietary fat intake. Proc Natl Acad Sci U S A 2011;108(31):12904–8. doi: 10.1073/pnas.1104675108.

[66] Senin LL, Al-Massadi O, Folgueira C, Castelao C, Pardo M, Barja-Fernandez S, et al. The gastric CB1 receptor modulates ghrelin production through the mTOR pathway to regulate food intake. PLoS One 2013;8(11):e80339. doi: 10.1371/journal.pone.0080339.

[67] Cui H, Lopez M, Rahmouni K. The cellular and molecular bases of leptin and ghrelin resistance in obesity. Nat Rev Endocrinol 2017;13(6):338–51. doi: 10.1038/nrendo.2016.222.

[68] Vianna CR, Donato J Jr, Rossi J, Scott M, Economides K, Gautron L, et al. Cannabinoid receptor 1 in the vagus nerve is dispensable for body weight homeostasis but required for normal gastrointestinal motility. J Neurosci 2012;32(30):10331–7. doi: 10.1523/jneurosci.4507-11.2012.

[69] Troy-Fioramonti S, Demizieux L, Gresti J, Muller T, Verges B, Degrace P. Acute activation of cannabinoid receptors by anandamide reduces gastrointestinal motility and improves postprandial glycemia in mice. Diabetes 2015;64(3):808–18. doi: 10.2337/db14-0721.

[70] Cardinal P, Andre C, Quarta C, Bellocchio L, Clark S, Elie M, et al. CB1 cannabinoid receptor in SF1-expressing neurons of the ventromedial hypothalamus determines metabolic responses to diet and leptin. Mol Metab 2014;3(7):705–16. doi: 10.1016/j.molmet.2014.07.004.

[71] Cardinal P, Bellocchio L, Clark S, Cannich A, Klugmann M, Lutz B, et al. Hypothalamic CB1 cannabinoid receptors regulate energy balance in mice. Endocrinology 2012;153(9):4136–43. doi: 10.1210/en.2012-1405.

[72] Jung KM, Clapper JR, Fu J, D'Agostino G, Guijarro A, Thongkham D, et al. 2-Arachidonoylglycerol signaling in forebrain regulates systemic energy metabolism. Cell Metab 2012;15(3):299–310. doi: 10.1016/j.cmet.2012.01.021.

[73] Quarta C, Bellocchio L, Mancini G, Mazza R, Cervino C, Braulke LJ, et al. CB(1) signaling in forebrain and sympathetic neurons is a key determinant of endocannabinoid actions on energy balance. Cell Metab 2010;11(4):273–85. doi: 10.1016/j.cmet.2010.02.015.

[74] Bajzer M, Olivieri M, Haas MK, Pfluger PT, Magrisso IJ, Foster MT, et al. Cannabinoid receptor 1 (CB1) antagonism enhances glucose utilisation and activates brown adipose tissue in diet-induced obese mice. Diabetologia 2011;54(12):3121–31. doi: 10.1007/s00125-011-2302-6.

[75] Boon MR, Kooijman S, van Dam AD, Pelgrom LR, Berbee JF, Visseren CA, et al. Peripheral cannabinoid 1 receptor blockade activates brown adipose tissue and diminishes dyslipidemia and obesity. FASEB J 2014;28(12):5361–75. doi: 10.1096/fj.13-247643.

[76] Cypess AM, Lehman S, Williams G, Tal I, Rodman D, Goldfine AB, et al. Identification and importance of brown adipose tissue in adult humans. N Engl J Med 2009;360(15):1509–17. doi: 10.1056/NEJMoa0810780.

[77] Kalinovich AV, de Jong JM, Cannon B, Nedergaard J. UCP1 in adipose tissues: two steps to full browning. Biochimie 2017;134:127–37. doi: 10.1016/j.biochi.2017.01.007.

[78] Rossi F, Punzo F, Umano GR, Argenziano M, Miraglia Del Giudice E. Role of cannabinoids in obesity. Int J Mol Sci 2018;19(9)doi: 10.3390/ijms19092690.

[79] Perwitz N, Wenzel J, Wagner I, Buning J, Drenckhan M, Zarse K, et al. Cannabinoid type 1 receptor blockade induces transdifferentiation towards a brown fat phenotype in white adipocytes. Diabetes Obes Metab 2010;12(2):158–66. doi: 10.1111/j.1463-1326.2009.01133.x.

[80] Ruiz de Azua I, Mancini G, Srivastava RK, Rey AA, Cardinal P, Tedesco L, et al. Adipocyte cannabinoid receptor CB1 regulates energy homeostasis and alternatively activated macrophages. J Clin Invest 2017;127(11):4148–62. doi: 10.1172/JCI83626.

[81] Tedesco L, Valerio A, Dossena M, Cardile A, Ragni M, Pagano C, et al. Cannabinoid receptor stimulation impairs mitochondrial biogenesis in mouse white adipose tissue, muscle, and liver: the role of eNOS, p38 MAPK, and AMPK pathways. Diabetes 2010;59(11):2826–36. doi: 10.2337/db09-1881.

[82] Horn H, Bohme B, Dietrich L, Koch M. Endocannabinoids in body weight control. Pharmaceuticals (Basel) 2018;11(2)doi: 10.3390/ph11020055.

[83] Christensen R, Kristensen PK, Bartels EM, Bliddal H, Astrup A. Efficacy and safety of the weight-loss drug rimonabant: a meta-analysis of randomised trials. Lancet 2007;370(9600):1706–13. doi: 10.1016/S0140-6736(07)61721-8.

[84] Clark TM, Jones JM, Hall AG, Tabner SA, Kmiec RL. Theoretical explanation for reduced body mass index and obesity rates in cannabis users. Cannabis Cannabinoid Res 2018;3(1):259–71. doi: 10.1089/can.2018.0045.

[85] Sethi JK, Vidal-Puig AJ. Thematic review series: adipocyte biology. Adipose tissue function and plasticity orchestrate nutritional adaptation. J Lipid Res 2007;48:1253–62.

[86] Zwick RK, Guerrero-Juarez CF, Horsley V, Plikus MV. Anatomical, physiological, and functional diversity of adipose tissue. Cell Metab 2018;27(1):68–83. doi: 10.1016/j.cmet.2017.12.002.

[87] Gonthier MP, Hoareau L, Festy F, Matias I, Valenti M, Bes-Houtmann S, et al. Identification of endocannabinoids and related compounds in human fat cells. Obesity (Silver Spring) 2007;15(4):837–45. doi: 10.1038/oby.2007.581.

[88] Matias I, Gonthier MP, Orlando P, Martiadis V, De Petrocellis L, Cervino C, et al. Regulation, function, and dysregulation of endocannabinoids in models of adipose and beta-pancreatic cells and in obesity and hyperglycemia. J Clin Endocrinol Metab 2006;91(8):3171–80. doi: 10.1210/jc.2005-2679.

[89] Starowicz KM, Cristino L, Matias I, Capasso R, Racioppi A, Izzo AA, et al. Endocannabinoid dysregulation in the pancreas and adipose tissue of mice fed with a high-fat diet. Obesity (Silver Spring) 2008;16(3):553–65. doi: 10.1038/oby.2007.106.

[90] Bluher M, Engeli S, Kloting N, Berndt J, Fasshauer M, Batkai S, et al. Dysregulation of the peripheral and adipose tissue endocannabinoid system in human abdominal obesity. Diabetes 2006;55(11):3053–60. doi: 10.2337/db06-0812.

[91] Engeli S, Heusser K, Janke J, Gorzelniak K, Batkai S, Pacher P, et al. Peripheral endocannabinoid system activity in patients treated with sibutramine. Obesity (Silver Spring) 2008;16(5):1135–7. doi: 10.1038/oby.2008.47.

[92] Gasperi V, Fezza F, Pasquariello N, Bari M, Oddi S, Agro AF, et al. Endocannabinoids in adipocytes during differentiation and their role in glucose uptake. Cell Mol Life Sci 2007;64(2):219–29. doi: 10.1007/s00018-006-6445-4.

[93] Sarzani R, Bordicchia M, Marcucci P, Bedetta S, Santini S, Giovagnoli A, et al. Altered pattern of cannabinoid type 1 receptor expression in adipose tissue of dysmetabolic and overweight patients. Metabolism 2009;58(3):361–7. doi: 10.1016/jmetabol.2008.10.009.

[94] Roche R, Hoareau L, Bes-Houtmann S, Gonthier MP, Laborde C, Baron JF, et al. Presence of the cannabinoid receptors, CB1 and CB2, in human omental and subcutaneous adipocytes. Histochem Cell Biol 2006;126(2):177–87. doi: 10.1007/s00418-005-0127-4.

[95] D'Eon TM, Pierce KA, Roix JJ, Tyler A, Chen H, Teixeira SR. The role of adipocyte insulin resistance in the pathogenesis of obesity-related elevations in endocannabinoids. Diabetes 2008;57(5):1262–8. doi: 10.2337/db07-1186.

[96] Engeli S, Bohnke J, Feldpausch M, Gorzelniak K, Janke J, Batkai S, et al. Activation of the peripheral endocannabinoid system in human obesity. Diabetes 2005;54(10):2838–43.

[97] Hillard CJ. Circulating endocannabinoids: from whence do they come and where are they going? Neuropsychopharmacology 2018;43(1):155–72. doi: 10.1038/npp.2017.130.

[98] Krott LM, Piscitelli F, Heine M, Borrino S, Scheja L, Silvestri C, et al. Endocannabinoid regulation in white and brown adipose tissue following thermogenic activation. J Lipid Res 2016;57(3):464–73. doi: 10.1194/jlr.M065227.

[99] Krott LM, Piscitelli F, Heine M, Borrino S, Scheja L, Silvestri C, et al. Endocannabinoid regulation in white and brown adipose tissue following thermogenic activation. J Lipid Res 2016;57(3):464–73. doi: 10.1194/jlr.M065227.

[100] Lahesmaa M, Eriksson O, Gnad T, Oikonen V, Bucci M, Hirvonen J, et al. Cannabinoid type 1 receptors are upregulated during acute activation of brown adipose tissue. Diabetes 2018;67(7):1226. doi: 10.2337/db17-1366.

[101] Perwitz N, Fasshauer M, Klein J. Cannabinoid receptor signaling directly inhibits thermogenesis and alters expression of adiponectin and visfatin. Horm Metab Res 2006;38(5):356–8. doi: 10.1055/s-2006-925401.

[102] Motaghedi R, McGraw TE. The CB1 endocannabinoid system modulates adipocyte insulin sensitivity. Obesity 2008;16(8):1727–34. doi: 10.1038/oby.2008.309.

[103] Pagano C, Pilon C, Calcagno A, Urbanet R, Rossato M, Milan G, et al. The endogenous cannabinoid system stimulates glucose uptake in human fat cells via phosphatidylinositol 3-kinase and calcium-dependent mechanisms. J Clin Endocrinol Metab 2007;92(12):4810–9. doi: 10.1210/jc.2007-0768.

[104] Jbilo O, Ravinet-Trillou C, Arnone M, Buisson I, Bribes E, Peleraux A, et al. The CB1 receptor antagonist rimonabant reverses the diet-induced obesity phenotype through the regulation of lipolysis and energy balance. FASEB J 2005;19(11):1567–9. doi: 10.1096/fj.04-3177fje.

[105] Cota D, Marsicano G, Tschop M, Grubler Y, Flachskamm C, Schubert M, et al. The endogenous cannabinoid system affects energy balance via central orexigenic drive and peripheral lipogenesis. J Clin Invest 2003;112(3):423–31. doi: 10.1172/JCI17725.

[106] Bellocchio L, Cervino C, Vicennati V, Pasquali R, Pagotto U. Cannabinoid type 1 receptor: another arrow in the adipocytes' bow. J Neuroendocrinol 2008;20(Suppl. 1):130–8. doi: 10.1111/j.1365-2826.2008.01682.x.

[107] Osei-Hyiaman D, DePetrillo M, Pacher P, Liu J, Radaeva S, Batkai S, et al. Endocannabinoid activation at hepatic CB1 receptors stimulates fatty acid synthesis and contributes to diet-induced obesity. J Clin Invest 2005;115(5):1298–305. doi: 10.1172/JCI23057.

[108] Ishac EJ, Jiang L, Lake KD, Varga K, Abood ME, Kunos G. Inhibition of exocytotic noradrenaline release by presynaptic cannabinoid CB1 receptors on peripheral sympathetic nerves. Br J Pharmacol 1996;118(8):2023–8. doi: 10.1111/j.1476-5381.1996.tb15639.x.

[109] Ralevic V, Kendall DA. Cannabinoid modulation of perivascular sympathetic and sensory neurotransmission. Curr Vasc Pharmacol 2009;7(1):15–25.

[110] Szabo B, Nordheim U, Niederhoffer N. Effects of cannabinoids on sympathetic and parasympathetic neuroeffector transmission in the rabbit heart. J Pharmacol Exp Ther 2001;297(2):819–26.

[111] Vizi ES, Katona I, Freund TF. Evidence for presynaptic cannabinoid CB(1) receptor-mediated inhibition of noradrenaline release in the guinea pig lung. Eur J Pharmacol 2001;431(2):237–44. doi: 10.1016/s0014-2999(01)01413-3.

[112] Molhoj S, Hansen HS, Schweiger M, Zimmermann R, Johansen T, Malmlof K. Effect of the cannabinoid receptor-1 antagonist rimonabant on lipolysis in rats. Eur J Pharmacol 2010;646(1–3):38–45. doi: 10.1016/j.ejphar.2010.08.006.

[113] Teixeira D, Pestana D, Faria A, Calhau C, Azevedo I, Monteiro R. Modulation of adipocyte biology by delta(9)-tetrahydrocannabinol. Obesity (Silver Spring) 2010;18(11):2077–85. doi: 10.1038/oby.2010.100.

[114] Muller T, Demizieux L, Troy-Fioramonti S, Gresti J, Pais de Barros JP, Berger H, et al. Overactivation of the endocannabinoid system alters the antilipolytic action of insulin in mouse adipose tissue. Am J Physiol Endocrinol Metab 2017;313(1):E26–36. doi: 10.1152/ajpendo.00374.2016.

[115] Girousse A, Tavernier G, Valle C, Moro C, Mejhert N, Dinel AL, et al. Partial inhibition of adipose tissue lipolysis improves glucose metabolism and insulin sensitivity without alteration of fat mass. PLoS Biol 2013;11(2):e1001485. doi: 10.1371/journal.pbio.1001485.

[116] Morigny P, Houssier M, Mouisel E, Langin D. Adipocyte lipolysis and insulin resistance. Biochimie 2016;125:259–66. doi: 10.1016/j.biochi.2015.10.024.

[117] Bouaboula M, Hilairet S, Marchand J, Fajas L, Le Fur G, Casellas P. Anandamide induced PPARgamma transcriptional activation and 3T3-L1 preadipocyte differentiation. Eur J Pharmacol 2005;517(3):174–81. doi: 10.1016/j.ejphar.2005.05.032.

[118] Karaliota S, Siafaka-Kapadai A, Gontinou C, Psarra K, Mavri-Vavayanni M. Anandamide increases the differentiation of rat adipocytes and causes PPARgamma and CB1 receptor upregulation. Obesity (Silver Spring) 2009;17(10):1830–8. doi: 10.1038/oby.2009.177.

[119] Verty ANA, Stefanidis A, McAinch AJ, Hryciw DH, Oldfield B. Anti-obesity effect of the CB2 receptor agonist JWH-015 in diet-induced obese mice. PLoS One 2015;10(11):e0140592. doi: 10.1371/journal.pone.0140592.

[120] Han JH, Shin H, Rho JG, Kim J-E, Son DH, Yoon J, et al. Peripheral cannabinoid 1 receptor blockade mitigates adipose tissue inflammation via NLRP3 inflammasome in mouse models of obesity. Diabetes Obes Metab 2018;20(9):2179–89. doi: 10.1111/dom.13350.

[121] Jourdan T, Godlewski G, Cinar R, Bertola A, Szanda G, Liu J, et al. Activation of the Nlrp3 inflammasome in infiltrating macrophages by endocannabinoids mediates beta cell loss in type 2 diabetes. Nat Med 2013;19(9):1132–40. doi: 10.1038/nm.3265.

[122] Jourdan T, Nicoloro SM, Zhou Z, Shen Y, Liu J, Coffey NJ, et al. Decreasing CB1 receptor signaling in Kupffer cells improves insulin sensitivity in obese mice. Mol Metab 2017;6(11):1517–28. doi: 10.1016/j.molmet.2017.08.011.

[123] Jourdan T, Szanda G, Cinar R, Godlewski G, Holovac DJ, Park JK, et al. Developmental role of macrophage cannabinoid-1 receptor signaling in type-2 diabetes. Diabetes 2017;doi: 10.2337/db16-1199.

[124] Han KH, Lim S, Ryu J, Lee CW, Kim Y, Kang JH, et al. CB1 and CB2 cannabinoid receptors differentially regulate the production of reactive oxygen species by macrophages. Cardiovasc Res 2009;84(3):378–86. doi: 10.1093/cvr/cvp240.

[125] Wagner IV, Perwitz N, Drenckhan M, Lehnert H, Klein J. Cannabinoid type 1 receptor mediates depot-specific effects on differentiation, inflammation and oxidative metabolism in inguinal and epididymal white adipocytes. Nutr Diabetes 2011;1:e16. doi: 10.1038/nutd.2011.12.

[126] van Eenige R, van der Stelt M, Rensen PCN, Kooijman S. Regulation of adipose tissue metabolism by the endocannabinoid system. Trends Endocrinol Metab 2018;29(5):326–37. doi: 10.1016/j.tem.2018.03.001.

[127] Bensaid M, Gary-Bobo M, Esclangon A, Maffrand JP, Le Fur G, Oury-Donat F, et al. The cannabinoid CB1 receptor antagonist SR 141716 increases Acrp30 mRNA expression in adipose tissue of obese fa/fa rats and in cultured adipocyte cells. Mol Pharmacol 2003;63(4):908–14. doi: 10.1124/mol.63.4.908.

[128] Ge Q, Maury E, Rycken L, Gerard J, Noel L, Detry R, et al. Endocannabinoids regulate adipokine production and the immune balance of omental adipose tissue in human obesity. Int J Obes (Lond) 2013;37(6):874–80. doi: 10.1038/ijo.2012.123.

[129] Jourdan T, Djaouti L, Demizieux L, Gresti J, Verges B, Degrace P. CB1 antagonism exerts specific molecular effects on visceral and subcutaneous fat and reverses liver steatosis in diet-induced obese mice. Diabetes 2010;59(4):926–34. doi: 10.2337/db09-1482.

[130] Tam J, Vemuri VK, Liu J, Batkai S, Mukhopadhyay B, Godlewski G, et al. Peripheral CB1 cannabinoid receptor blockade improves cardiometabolic risk in mouse models of obesity. J Clin Invest 2010;120(8):2953–66. doi: 10.1172/jci42551.

[131] Jorgacevic B, Vucevic D, Veskovic M, Mladenovic D, Vukicevic D, Vukicevic RJ, et al. The effect of cannabinoid receptor 1 blockade on adipokine and proinflammatory cytokine concentration in adipose and hepatic tissue in mice with nonalcoholic fatty liver disease. Can J Physiol Pharmacol 2019;97(2):120–9. doi: 10.1139/cjpp-2018-0607.

[132] Rossi F, Bellini G, Luongo L, Manzo I, Tolone S, Tortora C, et al. Cannabinoid receptor 2 as antiobesity target: inflammation, fat storage, and browning modulation. J Clin Endocrinol Metab 2016;101(9):3469–78. doi: 10.1210/jc.2015-4381.

[133] Hsiao WC, Shia KS, Wang YT, Yeh YN, Chang CP, Lin Y, et al. A novel peripheral cannabinoid receptor 1 antagonist, BPR0912, reduces weight independently of food intake and modulates thermogenesis. Diabetes Obes Metab 2015;17(5):495–504. doi: 10.1111/dom.12447.

[134] Verty AN, Allen AM, Oldfield BJ. The effects of rimonabant on brown adipose tissue in rat: implications for energy expenditure. Obesity (Silver Spring) 2009;17(2):254–61. doi: 10.1038/oby.2008.509.

[135] Fu Z, Gilbert ER, Liu D. Regulation of insulin synthesis and secretion and pancreatic beta-cell dysfunction in diabetes. Curr Diabetes Rev 2013;9(1):25–53.

[136] Bermudez-Silva FJ, Suarez J, Baixeras E, Cobo N, Bautista D, Cuesta-Munoz AL, et al. Presence of functional cannabinoid receptors in human endocrine pancreas. Diabetologia 2008;51(3):476–87. doi: 10.1007/s00125-007-0890-y.

[137] Gonzalez-Mariscal I, Krzysik-Walker SM, Doyle ME, Liu QR, Cimbro R, Santa-Cruz Calvo S, et al. Human CB1 receptor isoforms, present in hepatocytes and beta-cells, are involved in regulating metabolism. Sci Rep 2016;6:33302. doi: 10.1038/srep33302.

[138] Kim W, Doyle ME, Liu Z, Lao Q, Shin YK, Carlson OD, et al. Cannabinoids inhibit insulin receptor signaling in pancreatic beta-cells. Diabetes 2011;60(4):1198–209. doi: 10.2337/db10-1550.

[139] Hinden L, Udi S, Drori A, Gammal A, Nemirovski A, Hadar R, et al. Modulation of renal GLUT2 by the cannabinoid-1 receptor: implications for the treatment of diabetic nephropathy. J Am Soc Nephrol 2018;29(2):434–48. doi: 10.1681/ASN.2017040371.

[140] Shin H, Han JH, Yoon J, Sim HJ, Park TJ, Yang S, et al. Blockade of cannabinoid 1 receptor improves glucose responsiveness in pancreatic beta cells. J Cell Mol Med 2018;22(4):2337–45. doi: 10.1111/jcmm.13523.

[141] Vilches-Flores A, Delgado-Buenrostro NL, Navarrete-Vazquez G, Villalobos-Molina R. CB1 cannabinoid receptor expression is regulated by glucose and feeding in rat pancreatic islets. Regul Pept 2010;163(1–3):81–7. doi: 10.1016/j.regpep.2010.04.013.

[142] Juan-Pico P, Fuentes E, Bermudez-Silva FJ, Javier Diaz-Molina F, Ripoll C, Rodriguez de Fonseca F, et al. Cannabinoid receptors regulate $Ca^{2+}$ signals and insulin secretion in pancreatic beta-cell. Cell Calcium 2006;39(2):155–62. doi: 10.1016/j.ceca.2005.10.005.

[143] Nakata M, Yada T. Cannabinoids inhibit insulin secretion and cytosolic $Ca^{2+}$ oscillation in islet beta-cells via CB1 receptors. Regul Pept 2008;145(1–3):49–53. doi: 10.1016/j.regpep.2007.08.009.

[144] Gonzalez-Mariscal I, Montoro RA, Doyle ME, Liu QR, Rouse M, O'Connell JF, et al. Absence of cannabinoid 1 receptor in beta cells protects against high-fat/high-sugar diet-induced beta cell dysfunction and inflammation in murine islets. Diabetologia 2018;61(6):1470–83. doi: 10.1007/s00125-018-4576-4.

[145] Drucker DJ, Philippe J, Mojsov S, Chick WL, Habener JF. Glucagon-like peptide I stimulates insulin gene expression and increases cyclic AMP levels in a rat islet cell line. Proc Natl Acad Sci U S A 1987;84(10):3434–8. doi: 10.1073/pnas.84.10.3434.

[146] Montrose-Rafizadeh C, Egan JM, Roth J. Incretin hormones regulate glucose-dependent insulin secretion in RIN 1046-38 cells: mechanisms of action. Endocrinology 1994;135(2):589–94. doi: 10.1210/endo.135.2.8033807.

[147] Gonzalez-Mariscal I, Krzysik-Walker SM, Kim W, Rouse M, Egan JM. Blockade of cannabinoid 1 receptor improves GLP-1R mediated insulin secretion in mice. Mol Cell Endocrinol 2016;423:1–10. doi: 10.1016/j.mce.2015.12.015.

[148] De Petrocellis L, Marini P, Matias I, Moriello AS, Starowicz K, Cristino L, et al. Mechanisms for the coupling of cannabinoid receptors to intracellular calcium mobilization in rat insulinoma beta-cells. Exp Cell Res 2007;313(14):2993–3004. doi: 10.1016/j.yexcr.2007.05.012.

[149] Li C, Jones PM, Persaud SJ. Cannabinoid receptors are coupled to stimulation of insulin secretion from mouse MIN6 beta-cells. Cell Physiol Biochem 2010;26(2):187–96. doi: 10.1159/000320527.

[150] Moss CE, Marsh WJ, Parker HE, Ogunnowo-Bada E, Riches CH, Habib AM, et al. Somatostatin receptor 5 and cannabinoid receptor 1 activation inhibit secretion of glucose-dependent insulinotropic polypeptide from intestinal K cells in rodents. Diabetologia 2012;55(11):3094–103. doi: 10.1007/s00125-012-2663-5.

[151] Chia CW, Carlson OD, Liu DD, Gonzalez-Mariscal I, Santa-Cruz Calvo S, Egan JM. Incretin secretion in humans is under the influence of cannabinoid receptors. Am J Physiol Endocrinol Metab 2017;313(3):E359–66. doi: 10.1152/ajpendo.00080.2017.

[152] Basile G, Kulkarni RN, Morgan NG. How, when, and where do human beta-cells regenerate? Curr Diab Rep 2019;19(8):48. doi: 10.1007/s11892-019-1176-8.

[153] Kim W, Lao Q, Shin YK, Carlson OD, Lee EK, Gorospe M, et al. Cannabinoids induce pancreatic beta-cell death by directly inhibiting insulin receptor activation. Sci Signal 2012;5(216):ra23. doi: 10.1126/scisignal.2002519.

[154] Malenczyk K, Keimpema E, Piscitelli F, Calvigioni D, Bjorklund P, Mackie K, et al. Fetal endocannabinoids orchestrate the organization of pancreatic islet microarchitecture. Proc Natl Acad Sci U S A 2015;112(45):E6185–6194. doi: 10.1073/pnas.1519040112.

[155] Montane J, Cadavez L, Novials A. Stress and the inflammatory process: a major cause of pancreatic cell death in type 2 diabetes. Diabetes Metab Syndr Obes 2014;7:25–34. doi: 10.2147/DMSO.S37649.

[156] Esposito I, Proto MC, Gazzerro P, Laezza C, Miele C, Alberobello AT, et al. The cannabinoid CB1 receptor antagonist rimonabant stimulates 2-deoxyglucose uptake in skeletal muscle cells by regulating the expression of phosphatidylinositol-3-kinase. Mol Pharmacol 2008;74(6):1678–86. doi: 10.1124/mol.108.049205.

[157] Gonzalez-Mariscal I, Montoro RA, O'Connell JF, Kim Y, Gonzalez-Freire M, Liu QR, et al. Muscle cannabinoid 1 receptor regulates Il-6 and myostatin expression, governing physical performance and whole-body metabolism. FASEB J 2019;33(5):5850–63. doi: 10.1096/fj.201801145R.

[158] Nogueiras R, Veyrat-Durebex C, Suchanek PM, Klein M, Tschop J, Caldwell C, et al. Peripheral, but not central, CB1 antagonism provides food intake-independent metabolic benefits in diet-induced obese rats. Diabetes 2008;57(11):2977–91. doi: 10.2337/db08-0161.

[159] Lindborg KA, Jacob S, Henriksen EJ. Effects of chronic antagonism of endocannabinoid-1 receptors on glucose tolerance and insulin action in skeletal muscles of lean and obese Zucker rats. Cardiorenal Med 2011;1(1):31–44. doi: 10.1159/000322826.

[160] Liu J, Zhou L, Xiong K, Godlewski G, Mukhopadhyay B, Tam J, et al. Hepatic cannabinoid receptor-1 mediates diet-induced insulin resistance via inhibition of insulin signaling and clearance in mice. Gastroenterology 2012;142(5):1218–28. doi: 10.1053/j.gastro.2012.01.032. e1211.

[161] Zhang X, Gao S, Niu J, Li P, Deng J, Xu S, et al. Cannabinoid 2 receptor agonist improves systemic sensitivity to insulin in high-fat diet/streptozotocin-induced diabetic mice. Cell Physiol Biochem 2016;40(5):1175–85. doi: 10.1159/000453171.

[162] Galiegue S, Mary S, Marchand J, Dussossoy D, Carriere D, Carayon P, et al. Expression of central and peripheral cannabinoid receptors in human immune tissues and leukocyte subpopulations. Eur J Biochem 1995;232(1):54–61.

[163] Buckley NE, Hansson S, Harta G, Mezey E. Expression of the CB1 and CB2 receptor messenger RNAs during embryonic development in the rat. Neuroscience 1998;82(4):1131–49.

[164] Floreani A, Lazzari R, Macchi V, Porzionato A, Variola A, Colavito D, et al. Hepatic expression of endocannabinoid receptors and their novel polymorphisms in primary biliary cirrhosis. J Gastroenterol 2010;45(1):68–76. doi: 10.1007/s00535-009-0122-y.

[165] Julien B, Grenard P, Teixeira-Clerc F, Van Nhieu JT, Li L, Karsak M, et al. Antifibrogenic role of the cannabinoid receptor CB2 in the liver. Gastroenterology 2005;128(3):742–55.

[166] Mendez-Sanchez N, Zamora-Valdes D, Pichardo-Bahena R, Barredo-Prieto B, Ponciano-Rodriguez G, Bermejo-Martinez L, et al. Endocannabinoid receptor CB2 in nonalcoholic fatty liver disease. Liver Int 2007;27(2):215–9. doi: 10.1111/j.1478-3231.2006.01401.x.

[167] Teixeira-Clerc F, Belot MP, Manin S, Deveaux V, Cadoudal T, Chobert MN, et al. Beneficial paracrine effects of cannabinoid receptor 2 on liver injury and regeneration. Hepatology 2010;52(3):1046–59. doi: 10.1002/hep.23779.

[168] Xu X, Liu Y, Huang S, Liu G, Xie C, Zhou J, et al. Overexpression of cannabinoid receptors CB1 and CB2 correlates with improved prognosis of patients with hepatocellular carcinoma. Cancer Genet Cytogenet 2006;171(1):31–8. doi: 10.1016/j.cancergencyto.2006.06.014.

[169] Siegmund SV, Qian T, de Minicis S, Harvey-White J, Kunos G, Vinod KY, et al. The endocannabinoid 2-arachidonoyl glycerol induces death of hepatic stellate cells via mitochondrial reactive oxygen species. FASEB J 2007;21(11):2798–806. doi: 10.1096/fj.06-7717com.

[170] Jourdan T, Demizieux L, Gresti J, Djaouti L, Gaba L, Verges B, et al. Antagonism of peripheral hepatic cannabinoid receptor-1 improves liver lipid metabolism in mice: evidence from cultured explants. Hepatology 2012;55(3):790–9. doi: 10.1002/hep.24733.

[171] Migliarini B, Carnevali O. Anandamide modulates growth and lipid metabolism in the zebrafish Danio rerio. Mol Cell Endocrinol 2008;286(1–2 Suppl. 1):S12–16. doi: 10.1016/j.mce.2008.01.021.

[172] Azar S, Sherf-Dagan S, Nemirovski A, Webb M, Raziel A, Keidar A, et al. Circulating endocannabinoids are reduced following bariatric surgery and associated with improved metabolic homeostasis in humans. Obes Surg 2019;29(1):268–76. doi: 10.1007/s11695-018-3517-0.

[173] Westerbacka J, Kotronen A, Fielding BA, Wahren J, Hodson L, Perttila J, et al. Splanchnic balance of free fatty acids, endocannabinoids, and lipids in subjects with nonalcoholic fatty liver disease. Gastroenterology 2010;139(6):1961–71. doi: 10.1053/j.gastro.2010.06.064. e1961.

[174] Zelber-Sagi S, Azar S, Nemirovski A, Webb M, Halpern Z, Shibolet O, et al. Serum levels of endocannabinoids are independently associated with nonalcoholic fatty liver disease. Obesity (Silver Spring) 2017;25(1):94–101. doi: 10.1002/oby.21687.

[175] Abdulnour J, Yasari S, Rabasa-Lhoret R, Faraj M, Petrosino S, Piscitelli F, et al. Circulating endocannabinoids in insulin sensitive vs. insulin resistant obese postmenopausal women. A MONET group study. Obesity (Silver Spring) 2014;22(1):211–6. doi: 10.1002/oby.20498.

[176] Cote M, Matias I, Lemieux I, Petrosino S, Almeras N, Despres JP, et al. Circulating endocannabinoid levels, abdominal adiposity and related cardiometabolic risk factors in obese men. Int J Obes (Lond) 2007;31(4):692–9. doi: 10.1038/sj.ijo.0803539.

[177] Di Marzo V, Cote M, Matias I, Lemieux I, Arsenault BJ, Cartier A, et al. Changes in plasma endocannabinoid levels in viscerally obese men following a 1 year lifestyle modification programme and waist circumference reduction: associations with changes in metabolic risk factors. Diabetologia 2009;52(2):213–7. doi: 10.1007/s00125-008-1178-6.

[178] Knani I, Earley BJ, Udi S, Nemirovski A, Hadar R, Gammal A, et al. Targeting the endocannabinoid/CB1 receptor system for treating obesity in Prader-Willi syndrome. Mol Metab 2016;5(12):1187–99. doi: 10.1016/j.molmet.2016.10.004.

[179] Katayama K, Ueda N, Kurahashi Y, Suzuki H, Yamamoto S, Kato I. Distribution of anandamide amidohydrolase in rat tissues with special reference to small intestine. Biochim Biophys Acta 1997;1347(2–3):212–8.

[180] Liu J, Cinar R, Xiong K, Godlewski G, Jourdan T, Lin Y, et al. Monounsaturated fatty acids generated via stearoyl CoA desaturase-1 are endogenous inhibitors of fatty acid amide hydrolase. Proc Natl Acad Sci U S A 2013;110(47):18832–7. doi: 10.1073/pnas.1309469110.

[181] Martella A, Silvestri C, Maradonna F, Gioacchini G, Allara M, Radaelli G, et al. Bisphenol A induces fatty liver by an endocannabinoid-mediated positive feedback loop. Endocrinology 2016;157(5):1751–63. doi: 10.1210/en.2015-1384.

[182] Ruby MA, Nomura DK, Hudak CS, Mangravite LM, Chiu S, Casida JE, et al. Overactive endocannabinoid signaling impairs apolipoprotein E-mediated clearance of triglyceride-rich lipoproteins. Proc Natl Acad Sci U S A 2008;105(38):14561–6. doi: 10.1073/pnas.0807232105.

[183] Ruby MA, Nomura DK, Hudak CS, Barber A, Casida JE, Krauss RM. Acute overactive endocannabinoid signaling induces glucose intolerance, hepatic steatosis, and novel cannabinoid receptor 1 responsive genes. PLoS One 2011;6(11):e26415. doi: 10.1371/journal.pone.0026415.

[184] Yoshida K, Kita Y, Tokuoka SM, Hamano F, Yamazaki M, Sakimura K, et al. Monoacylglycerol lipase deficiency affects diet-induced obesity, fat absorption, and feeding behavior in CB1 cannabinoid receptor-deficient mice. FASEB J 2019;33(2):2484–97. doi: 10.1096/fj.201801203R.

[185] Cinar R, Godlewski G, Liu J, Tam J, Jourdan T, Mukhopadhyay B, et al. Hepatic cannabinoid-1 receptors mediate diet-induced insulin resistance by increasing de novo synthesis of long-chain ceramides. Hepatology 2014;59(1):143–53. doi: 10.1002/hep.26606.

[186] Jorgacevic B, Vucevic D, Duricic I, Sobajic S, Mladenovic D, Veskovic M, et al. The effect of cannabinoid receptor 1 blockade on hepatic free fatty acid profile in mice with nonalcoholic fatty liver disease. Chem Phys Lipids 2017;204:85–93. doi: 10.1016/j.chemphyslip.2017.03.009.

[187] Osei-Hyiaman D, Liu J, Zhou L, Godlewski G, Harvey-White J, Jeong WI, et al. Hepatic CB1 receptor is required for development of diet-induced steatosis, dyslipidemia, and insulin and leptin resistance in mice. J Clin Invest 2008;118(9):3160–9. doi: 10.1172/JCI34827.

[188] Shi D, Zhan X, Yu X, Jia M, Zhang Y, Yao J, et al. Inhibiting CB1 receptors improves lipogenesis in an in vitro non-alcoholic fatty liver disease model. Lipids Health Dis 2014;13(1):173. doi: 10.1186/1476-511X-13-173.

[189] Tam J, Godlewski G, Earley BJ, Zhou L, Jourdan T, Szanda G, et al. Role of adiponectin in the metabolic effects of cannabinoid type 1 receptor blockade in mice with diet-induced obesity. Am J Physiol Endocrinol Metab 2014;306(4):E457–468. doi: 10.1152/ajpendo.00489.2013.

[190] Wu HM, Yang YM, Kim SG. Rimonabant, a cannabinoid receptor type 1 inverse agonist, inhibits hepatocyte lipogenesis by activating liver kinase B1 and AMP-activated protein kinase axis downstream of Galpha i/o inhibition. Mol Pharmacol 2011;80(5):859–69. doi: 10.1124/mol.111.072769.

[191] Bowles NP, Karatsoreos IN, Li X, Vemuri VK, Wood JA, Li Z, et al. A peripheral endocannabinoid mechanism contributes to glucocorticoid-mediated metabolic syndrome. Proc Natl Acad Sci U S A 2015;112(1):285–90. doi: 10.1073/pnas.1421420112.

[192] Chorvat RJ, Berbaum J, Seriacki K, McElroy JF. JD-5006 and JD-5037: peripherally restricted (PR) cannabinoid-1 receptor blockers related to SLV-319 (Ibipinabant) as metabolic disorder therapeutics devoid of CNS liabilities. Bioorg Med Chem Lett 2012;22(19):6173–80. doi: 10.1016/j.bmcl.2012.08.004.

[193] Despres JP, Golay A, Sjostrom L. Rimonabant in Obesity-Lipids Study Group. Effects of rimonabant on metabolic risk factors in overweight patients with dyslipidemia. N Engl J Med 2005;353(20):2121–34. doi: 10.1056/NEJMoa044537.

[194] Despres JP, Ross R, Boka G, Almeras N, Lemieux I. ADAGIO-Lipids Investigators. Effect of rimonabant on the high-triglyceride/low-HDL-cholesterol dyslipidemia, intraabdominal adiposity, and liver fat: the ADAGIO-Lipids trial. Arterioscler Thromb Vasc Biol 2009;29(3):416–23. doi: 10.1161/ATVBAHA.108.176362.

[195] Gary-Bobo M, Elachouri G, Gallas JF, Janiak P, Marini P, Ravinet-Trillou C, et al. Rimonabant reduces obesity-associated hepatic steatosis and features of metabolic syndrome in obese Zucker fa/fa rats. Hepatology 2007;46(1):122–9. doi: 10.1002/hep.21641.

[196] Wierzbicki AS, Pendleton S, McMahon Z, Dar A, Oben J, Crook MA, et al. Rimonabant improves cholesterol, insulin resistance and markers of non-alcoholic fatty liver in morbidly obese patients: a retrospective cohort study. Int J Clin Pract 2011;65(6):713–5. doi: 10.1111/j.1742-1241.2011.02683.x.

[197] Chang E, Kim DH, Yang H, Lee DH, Bae SH, Park CY. CB1 receptor blockade ameliorates hepatic fat infiltration and inflammation and increases Nrf2-AMPK pathway in a rat model of severely uncontrolled diabetes. PLoS One 2018;13(10):e0206152. doi: 10.1371/journal.pone.0206152.

[198] Hussien NI, El-Kerdasy HI, Ibrahim ME. Protective effect of rimonabant, a canabinoid receptor 1 antagonist, on nonalcoholic fatty liver disease in a rat model through modulation of the hepatic expression of activin A and follistatin. Can J Physiol Pharmacol 2017;95(12):1433–41. doi: 10.1139/cjpp-2017-0070.

[199] Liu J, Godlewski G, Jourdan T, Liu Z, Cinar R, Xiong K, et al. Cannabinoid-1 receptor antagonism improves glycemic control and increases energy expenditure through sirtuin-1/mechanistic target of rapamycin complex 2 and 5′adenosine monophosphate-activated protein kinase signaling. Hepatology 2019;69(4):1535–48. doi: 10.1002/hep.30364.

[200] Mai P, Yang L, Tian L, Wang L, Jia S, Zhang Y, et al. Endocannabinoid system contributes to liver injury and inflammation by activation of bone marrow-derived monocytes/macrophages in a CB1-dependent manner. J Immunol 2015;195(7):3390–401. doi: 10.4049/jimmunol.1403205.

[201] Lieber CS, Schmid R. The effect of ethanol on fatty acid metabolism; stimulation of hepatic fatty acid synthesis in vitro. J Clin Invest 1961;40:394–9. doi: 10.1172/JCI104266.

[202] Muramatsu M, Kuriyama K, Yuki T, Ohkuma S. Hepatic lipogenesis and mobilization of peripheral fats in the formation of alcoholic fatty liver. Jpn J Pharmacol 1981;31(6):931–40.

[203] You M, Fischer M, Deeg MA, Crabb DW. Ethanol induces fatty acid synthesis pathways by activation of sterol regulatory element-binding protein (SREBP). J Biol Chem 2002;277(32):29342–7. doi: 10.1074/jbc.M202411200.

[204] Chanda D, Kim YH, Li T, Misra J, Kim DK, Kim JR, et al. Hepatic cannabinoid receptor type 1 mediates alcohol-induced regulation of bile acid enzyme genes expression via CREBH. PLoS One 2013;8(7):e68845. doi: 10.1371/journal.pone.0068845.

[205] Jeong WI, Osei-Hyiaman D, Park O, Liu J, Batkai S, Mukhopadhyay P, et al. Paracrine activation of hepatic CB1 receptors by stellate cell-derived endocannabinoids mediates alcoholic fatty liver. Cell Metab 2008;7(3):227–35. doi: 10.1016/j.cmet.2007.12.007.

[206] Trebicka J, Racz I, Siegmund SV, Cara E, Granzow M, Schierwagen R, et al. Role of cannabinoid receptors in alcoholic hepatic injury: steatosis and fibrogenesis are increased in CB2 receptor-deficient mice and decreased in CB1 receptor knockouts. Liver Int 2011;31(6):860–70. doi: 10.1111/j.1478-3231.2011.02496.x.

[207] Amato GS, Manke A, Harris DL, Wiethe RW, Vasukuttan V, Snyder RW, et al. Blocking alcoholic steatosis in mice with a peripherally restricted purine antagonist of the type 1 cannabinoid receptor. J Med Chem 2018;61(10):4370–85. doi: 10.1021/acs.jmedchem.7b01820.

[208] Deveaux V, Cadoudal T, Ichigotani Y, Teixeira-Clerc F, Louvet A, Manin S, et al. Cannabinoid CB2 receptor potentiates obesity-associated inflammation, insulin resistance and hepatic steatosis. PLoS One 2009;4(6):e5844. doi: 10.1371/journal.pone.0005844.

[209] Agudo J, Martin M, Roca C, Molas M, Bura AS, Zimmer A, et al. Deficiency of CB2 cannabinoid receptor in mice improves insulin sensitivity but increases food intake and obesity with age. Diabetologia 2010;53(12):2629–40. doi: 10.1007/s00125-010-1894-6.

[210] De Gottardi A, Spahr L, Ravier-Dall'Antonia F, Hadengue A. Cannabinoid receptor 1 and 2 agonists increase lipid accumulation in hepatocytes. Liver Int 2010;30(10):1482–9. doi: 10.1111/j.1478-3231.2010.02298.x.

[211] Denaes T, Lodder J, Chobert MN, Ruiz I, Pawlotsky JM, Lotersztajn S, et al. The cannabinoid receptor 2 protects against alcoholic liver disease via a macrophage autophagy-dependent pathway. Sci Rep 2016;6:28806. doi: 10.1038/srep28806.

[212] Louvet A, Teixeira-Clerc F, Chobert MN, Deveaux V, Pavoine C, Zimmer A, et al. Cannabinoid CB2 receptors protect against alcoholic liver disease by regulating Kupffer cell polarization in mice. Hepatology 2011;54(4):1217–26. doi: 10.1002/hep.24524.

[213] Varga ZV, Matyas C, Erdelyi K, Cinar R, Nieri D, Chicca A, et al. beta-Caryophyllene protects against alcoholic steatohepatitis by attenuating inflammation and metabolic dysregulation in mice. Br J Pharmacol 2018;175(2):320–34. doi: 10.1111/bph.13722.

[214] Rossi F, Bellini G, Alisi A, Alterio A, Maione S, Perrone L, et al. Cannabinoid receptor type 2 functional variant influences liver damage in children with non-alcoholic fatty liver disease. PLoS One 2012;7(8):e42259. doi: 10.1371/journal.pone.0042259.

CHAPTER

# 16

# Gut microbiota interaction in host lipid metabolism

Rachel M. Golonka[a], Ahmed A. Abokor[a], James M. Ntambi[b], Matam Vijay-Kumar[a]

[a]Department of Physiology and Pharmacology, University of Toledo College of Medicine and Life Sciences, Toledo, OH, United States; [b]Departments of Biochemistry and Nutritional Sciences, University of Wisconsin-Madison, Madison, WI, United States

OUTLINE



Lipid Signaling and Metabolism. http://dx.doi.org/10.1016/B978-0-12-819404-1.00016-6

# Introduction

## The liver comprises of several moonlighting duties

The central function of the liver is to process, partition, and metabolize macronutrients delivered from the digestive tract as this serves as a critical energy source for the cells within our body [1]. Over the course of evolution, the liver has further consolidated into becoming a "firewall" against pathogenic microbes and toxins by promoting their clearance [2–5]. Furthermore, the liver demonstrates a substantial amount of moonlighting properties through its extensive involvement in maintaining lipid, cholesterol, protein and glucose homeostasis, along with regulating blood volume, lipogenesis, bile pigment and bile salt production, and the metabolism of xenobiotic compounds [1,6,7]. In addition to metabolic and pathogen clearance functions, the liver also comprises of immunoregulatory operations to prevent unintended organ damage [8]; specifically, there is a relatively large amount of negative regulatory immune cells to balance inflammatory signals, making the liver an immunotolerant organ [9]. This reciprocates with its intricate system of regeneration [10], which is intriguingly a sole property of the liver that is absent from all other organs. It is not surprising then that the term liver means "life" in old Germanic and Romance languages [11]. Likewise, the Greek word "hèpar" for hepatic was associated with "pleasure", and the liver was regarded as the seat of soul and intelligence [11,12].

In theory, the labeling of the liver as the seat of life and soul was due to its richness in blood [13]. Particularly, the liver is strategically placed to serve as the endpoint from dual blood sources: the hepatic artery and the hepatic portal vein [14]. Absorbed intestinal contents released into the hepatic portal vein must first pass through the liver before having permission to enter systemic circulation. The liver then is considered the "first-hit organ" to encounter noxious diet-derived metabolites, gut microbial products, and viable bacteria in cases of intestinal barrier dysfunction, referred to as a "leaky gut" (Fig. 16.1) [14]. In response, the liver utilizes its firearm, phagocytic Kupffer cells (resident hepatic macrophages) to clear gut commensals [4,15]. However, damage to the hepatic firewall impairs clearance of commensals from the blood, which can make the liver susceptible to various gut bacterial-associated conditions, including nonalcoholic fatty liver disease (NAFLD) [16]. Considering that the liver is a primary source for modulating lipid metabolism, understanding the gut–liver axis can help us further understand as to how gut commensals can promote disease pathogenesis, and also allow us to elucidate which bacteria can be utilized to counteract such detrimental outcomes.

## Gut microbiota: an organ within an organ

As mentioned prior, the liver is considered as a firewall against pathogenic microorganisms. More particularly, the liver is the "first hit organ" against microbes that stem from the intestine [4]. These organisms originate from the gut microbiota, a collection of bacteria, archaea, and eukarya colonizing the gastrointestinal tract (GI), which have coevolved with the host over thousands of years to form a mutualistic relationship [17]. The gut microbiota can be thought of as a microbial organ found within a host organ, where its different lineages can communicate with one another and the host [17]. One form of communication is through short-chain fatty acids (SCFAs), including acetate, butyrate, and propionate. These fermented

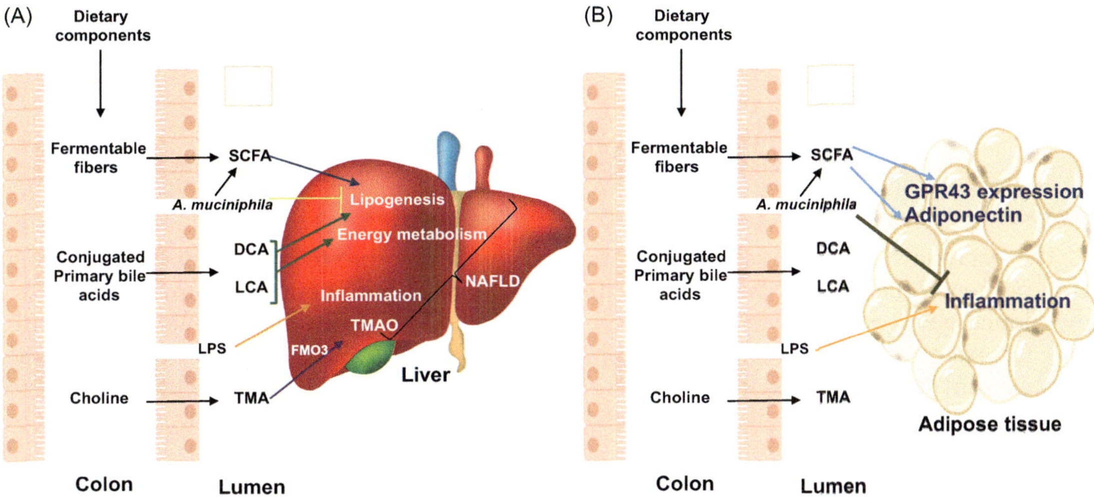


FIGURE 16.1 **Gut bacteria and their metabolites strongly impact hepatic de novo lipogenesis and adipose tissue homeostasis.** Gut dysbiosis can result in increased intestinal permeability, which allows for the leakage of various gut microbes and their metabolites that reach the liver and adipose tissue via enterohepatic circulation. The bacteria and their products can influence the rate of lipogenesis (A) and degree of adipose inflammation (B). Cases of elevated hepatic lipogenesis and inflammation increases the risk for the development of metabolic diseases, including nonalcoholic fatty liver disease (NAFLD). *A. muciniphila* is able to diminish the accumulation of triglycerides and cholesterol and thus might be a possible therapeutic. *DCA*, deoxycholate; *GPR43*, G protein-coupled receptor 43; *LCA*, lithocholate; *LPS*, lipopolysaccharide; *SCFA*, short-chain fatty acids; *TMA*, trimethylamine. Source: *Credit: Rachel M. Golonka.*

products from plant-derived complex carbohydrates can activate intestinal G protein-coupled receptors (GPCRs, i.e., free fatty acids receptors 2 and 3, FFAR2 and FFAR3; *alias* GPR43 and GPR41) and act as histone deacetylase (HDAC) inhibitors [18]. Both GPR41 and GPR43 are expressed by enteroendocrine and pancreatic β-cells, along with GPR43 having further expression on immune cells and adipocytes, while GPR41 is also found on some peripheral neurons [19]. This emphasizes that gut metabolites, and therefore, the gut microbiota, can have a substantial influence on different physiological processes within the host (as summarized in Fig. 16.1).

As said before, we have a symbiotic relationship with our gut microbiota; however, an imbalance in the cross talk and cross-regulation between the host and microbiota can result in gut "dysbiosis". This phenomenon can be classified into three categories: (1) loss of beneficial organisms, (2) excessive growth of pathogenic bacteria, and (3) loss of overall microbial diversity [20]. Despite the adult human gut containing about $10^{14}$ bacterial cells with more than 1000 different bacterial species, an altered ratio between the two most dominant phyla groups (i.e., Bacteroidetes and Firmicutes) is one of the strongest associations with dysbiosis [20–22]. In particular, an increase in the Firmicutes to Bacteroidetes (F/B) ratio has been highly linked to various intra- and extraintestinal disorders, including obesity [21], autism [23], inflammatory bowel disease (IBD) [24], and hypertension [25] [in depth review [26]]. Whether alterations in the gut microbiota are a cause or correlation with these various disease states is still unclear; therefore, distinguishing how the gut microbiota specifically influences different physiological processes is critical. In this chapter, we will focus on how the gut microbiota influences host lipid metabolism, and how this could be associated with different disease states.

# Gut metabolites regulate hepatic lipid metabolism

## Background

The bacteria that reside within the gut microbiota are able to transform both host and environmental (i.e., dietary fermentable fiber) components into new compounds. Earlier, we mentioned about SCFAs, which are fermented products of dietary fibers. In addition to SCFAs, the gut microbiota can generate various other molecules, including secondary bile acids and trimethylamine (TMA), where recent studies have highlighted that these gut metabolites have a strong influence on host lipid metabolism (Fig. 16.1). Sections 2.2–2.4 provide further details.

## Short-chain fatty acids: gut-derived lipogenic and gluconeogenic molecules

Multiple nondigestible carbohydrates, including dietary fibers, are unable to be broken down by the host due to lack of enzyme(s); therefore, these nutrients escape to the colon and are fermented by the gut microbiota into SCFAs. SCFAs are saturated, aliphatic organic acids that contain less than six carbons. These include formate (C1), acetate (C2), propionate (C3), butyrate (C4), isobutryate (C4), valeric acid (C5), and isovaleric acid (C5). Comparatively, the most abundant (>95) are acetate, propionate, and butyrate, which may be due to the metabolic cross-feeding of other gut metabolites; for example, lactate production can serve as a substrate for other bacteria to generate butyrate [27]. Along with modulating the gut microbiota, SCFAs can regulate host physiology, including lipid metabolism. Lu et al. demonstrates that dietary supplementation of either metabolite, or as a combined cocktail, significantly reduces triglyceride and cholesterol levels in diet-induced obesity (DIO) mice [28]. Their data further suggests that this protective effect of SCFAs could be due to SCFAs modulating GPR43 tissue expression and regulating the expressions of distinct adipokines. Specifically, administration of SCFAs to DIO mice increases GPR43 expression in adipose tissue, while reducing its expression in the colon. These results correlate with previous results demonstrating that GPR43-deficient mice become naturally obese, whereas overexpression of GPR43 in adipose tissue protects mice from high-fat DIO [29]. Additionally, Lu et al. [28] showed that SCFA supplementation significantly mitigates leptin expression and augments mRNA levels of adiponectin and resistin. These results are analogous with previous studies demonstrating that leptin levels are decreased while adiponectin is increased during weight loss [30,31].

Considering that adiponectin can activate AMP-activated map kinase (AMPK)-mediated fatty acid oxidation [32,33], this presents one plausible mechanism to the beneficial effects of SCFAs in modulating lipid metabolism to prevent obesity. Alongside, SCFAs have been associated to inducing the expression of the lipoprotein lipase (LPL) inhibitor, fasting-induced adipose factor/angiopoietin-like protein (FIAF/ANGPLT4), via the nuclear receptor peroxisome proliferator-activated receptor $\gamma$ (PPAR$\gamma$) [34], which would generally result in a reduction of triglyceride accumulation. However, there has also been opposing reports that suppression of FIAF and, therefore, the upregulation of LPL is microbiota-dependent [35]. Likewise, the inhibition of FIAF expression is reversed in germ-free (microbiota depleted) mice that are resistant to high-fat DIO, and this is further associated with activation of AMPK in the liver and muscle tissue [36]. Regardless of the contribution of FIAF to hepatic lipid storage, one common denominator in the prevention of lipid accumulation and thus, obesity, is the promotion of AMPK; therefore, further research is necessary to determine other microbiotal-associated factors that influence AMPK.

Despite SCFAs providing a potential advantage toward adipose tissue, these metabolites have questionable interests when looking globally at other organs. Along with modulating adipose tissue metabolism, SCFAs are considered as a major fuel source for colonocytes. Utilizing stable isotopes, den Besten et al. [37] demonstrated that the remainder of SCFAs that are not absorbed by the colonocytes can travel to the liver through the hepatic portal vein and participate in de novo lipogenesis. In particular, acetate and butyrate were primarily considered lipogenic as they promoted hepatic levels of palmitate (C16:0), stearate (C18:0), and neutral lipids, along with elevating triglyceride and cholesterol abundance [37]. Comparatively, their study suggests that propionate is gluconeogenic, which is due to the conversion of propionate into succinate, which subsequently supports gluconeogenesis via oxaloacetate [37]. Likewise, Tirosh et al. recently showed that propionate can also promote glycogenolysis and hyperglycemia by increasing plasma concentrations of glucagon and fatty acid-binding protein 4 (FABP4) [38]. While acetate and butyrate can enter the tricarboxylic acid (TCA) cycle and be converted to oxaloacetate after a few steps, in the end, there is no net carbon transfer for the formation of glucose from acetate and butyrate [37]. Recently, it has been observed that acetate uptake into hypoxic cells induces hyperacetylation, which allows for the activation of lipogenic genes, including acetyl-CoA carboxylase α (*Acaca*) and fatty acid synthase (*Fasn*), and the enhancement of hepatic de novo lipid biosynthesis [39]. More intriguingly, the same study demonstrated that this is a protective mechanism for cancer cells against hypoxia, where acetate can serve as a metabolic precursor for fatty acid synthesis and can function as an epigenetic metabolite for tumor cell survival [39]. In line, considering that cancer cells exhibit the Warburg effect, which is further characterized by hyperactive glucose metabolism [40], the recent study by Liu et al. demonstrating that de novo acetate production is critical for sustaining acetyl-coenzyme A levels and thus, lipogenesis in limited metabolic environments (i.e. mitochondrial dysfunction) [41] could provide a novel pathophysiological mechanism that cancer cells hijack in order to survive. This is further supported in the recent study by Schug et al., where they utilized metabolomics and lipidomics platforms to demonstrate acetate as a nutritional source for cancer cells during hypoxic and lipid-depleted conditions [42]. With this information, future studies can begin to target de novo lipogenesis [43], and in particular, acetate, for anticancer therapy. Not to mention, this could provide a novel link between dietary fibers, SCFAs and cancer cell proliferation, which would be analogous to our groups recent finding on dietary fermentable fiber-induced cholestatic hepatocellular carcinoma (HCC; most common liver malignancy) in gut microbiota dysbiotic mice [44].

Our group has also demonstrated that in a separate subset of gut dysbiotic, Toll-like receptor 5 deficient (*Tlr5*KO; lack innate immune receptor for bacterial flagellin) mice that SCFAs are associated with elevated hepatic lipogenesis mediated by stearoyl-CoA desaturase-1 (SCD1), which is further associated with aggravated metabolic syndrome [45]. Comparatively, $^{1}$H NMR-based metabolic profiling revealed that, in particular, butyrate and propionate were elevated, whereas there was no change in acetate levels in the *Tlr5*KO mice. This correlates with the study by Nishitsuji et al. who observed that Tsumura Suzuki obese diabetes (TSOD) mice have an increase in butyrate and propionate, but a decrease in acetate plasma levels [46]. The TSOD mice also suffer from gut dysbiosis, as indicated with an increase in the Firmicutes to Bacteroidetes (F/B) ratio [46], analogous to the *Tlr5*KO mice. While this could infer butyrate, and maybe propionate, as culprits in metabolic syndrome, there are studies stating that butyrate administration is protective against metabolic syndrome and its indices [47,48]. Likewise, it has been reported that butyrate activates intestinal gluconeogenesis (IGN) gene expression via cAMP-dependent mechanism, and propionate can modulate IGN through the gut-brain axis involving GPR41, which was necessary for the beneficial effects of dietary fibers in alleviating metabolic syndrome [49]. In line, SCFA activation of GPR41 and GPR43 has been shown to stimulate the secretion of glucagon-like peptide-1 (GLP-1), peptide YY (PYY), and other peptide hormones that can travel through the endocrine route and provide beneficial effects on appetite and energy homeostasis [as reviewed in [50]. Comparatively, it has been recently observed that increased production of acetate during gut dysbiosis leads to activation of the parasympathetic nervous system, which in turn promotes indices of metabolic syndrome [51]. In summary, current research highlights acetate and butyrate as lipogenic, while propionate is gluconeogenic; however, acetate seems to have a stronger effect at initiating de novo lipogenesis, which can be detrimental in times of cancer cell proliferation and metabolic syndrome.

## Bile acids and FXR modulation of lipid homeostasis

Conjugated primary bile acids (i.e., glycocholate, taurocholate) are amphiphilic metabolites of hepatic cholesterol catabolism. One of their important functions is facilitating the intestinal assimilation of dietary triglycerides, cholesterol, lipids, and fat-soluble vitamins [52,53]. Approximately 5% of the primary bile acids that are not reabsorbed in the ileal part of the small intestine are further metabolized into more hydrophobic, toxic secondary bile acids by gut commensals that reside in the gut microbiota, such as *Clostridium cluster* XI and XIVa and *Bacteroides* [54]. Specifically, primary bile acids are susceptible to the removal of their 7α-hydroxy group, via the gut microbiota-derived 7α-dehydroxylase enzyme, which allows for the formation of deoxycholate from cholate and lithocholate from chenodeoxycholate [55]. These secondary bile acids are generally excreted through the feces, which represents the major route for cholesterol elimination. Additionally, analogous to primary bile acids, secondary bile acids can get reabsorbed and enter enterohepatic circulation via the hepatic portal vein, allowing for bile acids to revert back to their origin of production. After reaching the liver, these secondary bile acids can activate specific nuclear and G protein-coupled receptors that regulate different physiological processes. The main bile acid sensor and regulator is the farnesoid X receptor (FXR; *alias* NR1H4) as it adjusts the bile acid pool through hepatic and intestinal tissue-specific signaling to suppress the genes (i.e., *Cyp7a1* and *Cyp8b1*) that encode rate-limiting enzymes for hepatic bile acid biosynthesis [56]. Comparatively, bile acid activation of hepatic FXR initiates small heterodimer partner (SHP) mediated inhibition of *Cyp7a1* and *Cyp8b1*, whereas intestinal FXR activation promotes fibroblast growth factor 15 (FGF15) signaling to implement similar suppression of those key bile acid biosynthetic genes [56]. This emphasizes the critical negative feedback loop on bile acid homeostasis, as overproduction of bile acids has been associated with multiple pathological diseases, especially hepatocarcinogenesis [57–62].

Along with modulating its own metabolism, bile acids have been recognized to regulate other metabolic functions, including lipid metabolism. One of the first clues that bile acids may influence lipogenesis was by Watanabe et al. who delineated that FXR activation of SHP can inhibit SREBP1-c, which resulted in a decrease of triglyceride levels [63]. Sterol regulatory element-binding proteins (SREBPs), and in particular SREBP-1, control the expression of genes involved in lipogenesis, such as acetyl-CoA carboxylase (ACC) and fatty acid synthase (FASN) [64]. Similar to the previous study, Yang et al. demonstrated that nonalcoholic fatty liver disease (NAFLD) patient's exhibit diminished hepatic FXR, which was associated with an increased expression of SREBP-1c and FASN and hepatic triglyceride synthesis [65]. Therefore, it could be inferred that the benefits of FXR-SHP-mediated inhibition of SREBP-1c would be due to the diminished expression of further downstream lipogenic genes. Opposing, however, Matsukuma et al. later found direct evidence that chenodeoxycholate activation of the FXR/retinoid X receptor α (RXRα) heterodimer binds to the FASN inverted repeat spaced by one nucleotide (IR-1) on the FASN promoter, leading to an increase in FASN mRNA expression and thus, lipogenesis [66]. This emphasized that SHP modulation of SREBP-1c is not the only mechanism regulating the lipogenic response to bile acids. Regardless, there is a degree in FXR significantly improving hyperlipidemia, which makes FXR agonists (i.e., GW4064) potential therapeutic agents [67]. Another potential therapeutic could be the use of antibiotics as Kuno et al. recently demonstrated that short-term antibiotic-induced gut dysbiosis, resulting in diminished hepatic secondary bile acids, reduced serum glucose and triglyceride levels [68]. Considering the potential role of secondary bile acids, this has opened research to studying the influence of gut microbiota-derived enzymes in lipid metabolism. Bile salt hydrolase (BSH), for example, cleaves the amino acid side chain of glyco- or tauro-conjugated bile acids, which makes primary bile acids susceptible to further bacterial modification to yield secondary bile acids [69]. Joyce et al. showed that elevated BSH activity was associated with weight loss, reduced serum cholesterol and liver triglyceride levels, and modulation of major host adipo-signaling regulators (i.e., *Dgat1*, *Abca1*, *Acox1*, and *ApoA1*) [69]. Overall, bile acids have strong effects on lipid homeostasis, where its metabolism and signaling could be potential therapeutic targets.

## Choline metabolism and trimethylamine impacts fatty acid biosynthesis

The gut microbial *cut* cluster genes encode the glycyl radical enzyme (CutC), the glycyl radical activating protein (CutD), the two-component Rieske-type oxygenase/reductase (CntA/B), and the dioxygenase/oxidoreductase complex (YeaW/X) in order to metabolize dietary precursors, such as choline, choline-containing compounds, betaine and L-carnitine into trimethylamine (TMA) [70,71]. In particular, choline as a source of $-CH_3$ groups is an essential nutrient as it is required for various biological processes including hepatic lipogenesis, as well as the synthesis of very low-density lipoprotein (VLDL), acetylcholine and cell membrane phospholipids [72]. Therefore, bacterial metabolism of choline to TMA can reduce choline bioavailability and impair phosphatidylcholine biosynthesis in the liver. This was shown in a study by Dumas et al., where obesity-predisposed 129S6 mice fed a high-fat diet (HFD) exhibited reduced choline bioavailability, impaired VLDL secretion and increased triglyceride accumulation in the liver; these alterations were correlated with increased urinary TMA levels [73]. These mice also displayed classic symptoms (e.g., hepatic steatosis) of NALFD, which are often seen in mice fed a choline-deficient diet [74].

In addition to TMA, trimethylamine oxide (TMAO) has been identified as an independent risk marker and factor for NAFLD [75]. TMAO production from TMA is due to the hepatic enzyme, flavin-containing monooxygenase 3 (FMO3) [76]. Intriguingly, TMAO can get converted back to TMA in the gut microbiota, predominantly by *Enterobacteriaceae*, which leads to a continuous cycle known as retroconversion [77]. Recently, it has been described that TMAO aggravates liver steatosis by shifting the hepatic bile acid profile toward FXR-antagonistic activity, which would explain, at least in part, the elevation of hepatic triglycerides and lipogenesis in mice fed a HFD [78]. As potentially expected, an immediate therapeutic target would be to diminish the activity of microbiota-regulated FMO3 and thus, the production of TMAO; however, there have been considerable positive and negative side effects reported. On the positive side, Warrier et al. demonstrated that FMO3 deficiency aids in rebalancing cholesterol metabolism through blunting intestinal cholesterol absorption, limiting the production of hepatic oxysterols and cholesteryl esters, and diminishing the bile acid pool [79]. Additionally, the same study showed that FMO3 deficiency suppressed genes involved in hepatic lipogenesis (i.e., FASN); yet, they found that this was independent of the TMA/TMAO axis [79]. A recent study suggests the effects of FMO3 on lipogenesis and gluconeogenesis may be mediated through PPARα and Kruppel-like 15 pathways [80]. Along with lipid synthesis, FMO3 has been associated as a promoter of obesity through its negative regulation on beiging programs in white adipose tissue (WAT), whereas knockdown of FMO3 confers protection against lipid accumulation [81]. Despite these benefits, FMO3 knockdown has been reported to exacerbate hepatic ER stress and inflammation, in part, due to the diminished production of hepatic oxysterols and cholesteryl esters and subsequent liver X receptor (LXR) activation [79]. Alongside, a genetic defect in FMO3 can result in trimethylaminuria (TMAuria), or fish-odor syndrome, as this results in an overabundance of odorous TMA compared to nonodorous TMAO [82]. Therefore, a more attractive therapeutic target would be to inhibit the gut microbial enzymes that generate TMA instead of eliminating FMO3 [82]. Overall, future studies are still warranted to determine the independent and collaborative roles of TMA, TMAO, and FMO3 in influencing host cholesterol metabolism and lipogenesis.

## The gut microbiota-beiging axis

Comparatively, WAT stores energy in the form of triglycerides, whereas brown adipose tissue (BAT) metabolizes fatty acids to produce heat. One of latest promising therapies against obesity and metabolic syndrome is the "browning" of WAT, more popularly known as beiging (Fig. 6.2A). This thermogenesis process is primarily mediated through the action of uncoupling protein-1 (UCP1) as it promotes substrate oxidation, uncoupling of oxidative phosphorylation and the subsequent production of heat [83]. More recently, estrogens have been suggested to induce BAT through various mitochondrial and neurological pathways [as reviewed in [84]]. Additionally, initiation of β-adrenergic signaling [81,85,86] and activation of PPARγ via fibroblast growth factor 21 (FGF21) [87] have been attributed to promoting the conversion of white to BAT (Fig. 16.2A). In particular, FGF21 hepatic expression is highly

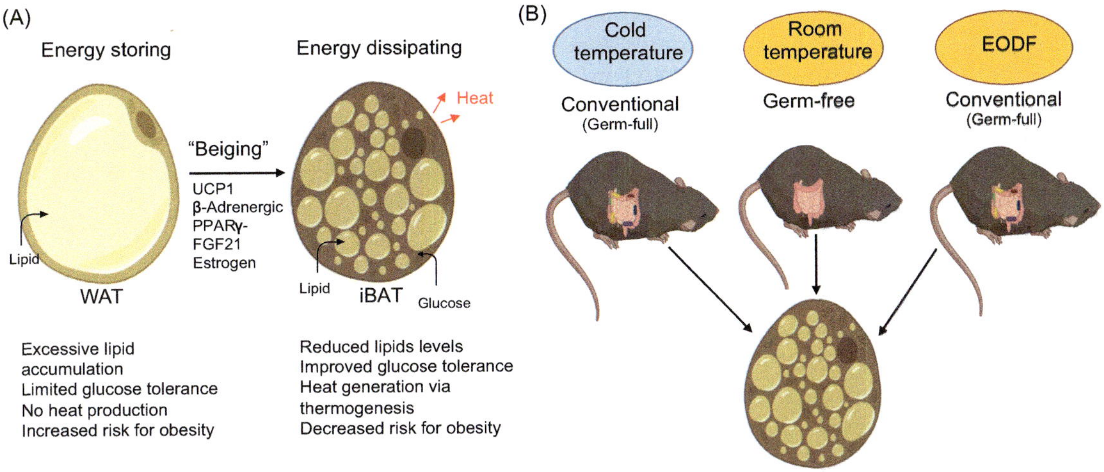


**FIGURE 16.2 Gut microbiota-dependent beiging is triggered by either cold exposure or periodic fasting.** (A) Beiging is a process that involves the formation of inducible brown adipose tissue (iBAT) from white adipose tissue (WAT). This is mediated by various molecules and pathways, including uncoupling protein 1 (UCP1), β-adrenergic signaling, PPARγ-FGF21 signaling, and more recently estrogen-mediated neurological and mitochondrial pathways. iBAT is associated with decreased risk factors for obesity. (B) Conventional mice that are exposed to a cold environment triggers beiging, which is absent in germ-free mice under cold exposure. At room temperature, beiging is inhibited in germ-full but initiated in germ-free mice. When challenged with every other day fasting (EODF) at room temperature, beiging is promoted in conventional mice, but is absent in germ-free mice. This emphasizes that gut microbiota-dependent beiging must be stimulated by an external factor. Source: *Credit: Rachel M. Golonka.*

increased during fasting conditions, which is mediated by PPARγ [88], indicating a potential positive feedback loop among FGF21, PPARγ and beiging. Li et al. intriguingly showed that every other day fasting (EODF) can induce beiging through independent mechanisms of FGF21 and β-adrenergic signaling [89]. They determined that the thermogenesis was gut microbiota-dependent as EODF-treated germ-free mice lacked beiging, whereas germ-free mice supplemented with microbiota from EODF-treated mice recapitulated the browning of adipose tissue [89]. Mechanistically, they suggest that the elevated production of the gut metabolites, acetate and lactate, in EODF-treated conventional mice could be influencing the beiging process, as this was further associated with increased monocarboxylate transporter-1 (*Mct1*; SCFA transporter) expression and alleviated metabolic indices, along with diminished hepatic steatosis and liver injury markers [89]. In line, HFD fed mice administered a nanoparticle containing acetate have improved hepatic function and increased mitochondrial efficiency, which was further associated with browning of WAT that lead to a reduction in body adiposity [90].

In contrast to Li et al., Suárez-Zamorano et al. demonstrated that the commensal microbiota actually impedes beiging, as germ-free mice and mice treated with antibiotics have augmented browning of WAT compared to conventional mice, which was abated in germ-free mice transplanted with a conventional microbiota [91]. They propose that depleting the microbiota enhances thermogenic capacity via increasing eosinophil infiltration and inducing type 2 cytokine signaling and M2 macrophage polarization in WAT [91]. Similarly, Caesar et al. reported that saturated dietary lipids can modulate the gut microbiota to support an elevation in toll-like receptor 4 (TLR4) signaling, which subsequently results in increased WAT inflammation due to accumulated macrophage infiltration [92]. This may explain why germ-free mice have alleviation of WAT inflammation and insulin resistance, which protects them from metabolic syndrome [36]. Yet, other studies suggest that type 2 immune signaling and alternative macrophage activation are unlikely to affect adaptive thermogenesis [93,94]. As another possibility, Virtue et al. have recently shown that tryptophan-derived metabolites produced by the gut microbiota control specific micro-RNAs in white adipocytes, where dysregulation of this connection was essential for the development of obesity, insulin resistance and WAT inflammation [95]. Future studies are required to determine whether the gut microbiota-WAT inflammation axis could be a potent regulator of beiging.

Another method to determine the influence of the gut microbiota on beiging has been by introducing mice to a cold environment. Chevalier et al. showed that cold exposure significantly alters the gut microbiota composition, which was further associated with improved insulin sensitivity and WAT browning [96]. Intriguingly, transferring a "cold microbiota" to germ-free mice helped in prompting fat loss due to increased beiging and energy expenditure [96]. It is further interesting that mice lacking a gut microbiota have impaired UCP1-mediated thermogenesis when exposed to a cold environment [93]. Similar to Suárez-Zamorano et al., though, this study demonstrated that microbiota depletion does not promote beiging at room temperature, whereas thermogenesis capacity of germ-free mice is eliminated when they are exposed to cold challenge [93]. Analogous to EODF, a "cold microbiota" is composed of an elevated F/B ratio but there was also an increase in other secondary bile acid and butyrate-producing bacteria, including various species of Clostridia (i.e., *Lachnospiraceae*) [93,96,97]. The elevation of bile acids is further associated with an increase expression of G protein-coupled bile acid receptor (Gpbar1; *alias* TGR5), AMPK-mediated lipid β-oxidation and BAT, which is suggested to aid in the complete alleviation of DIO [97]. Mechanistically, bile acids can activate the TGR5-cAMP-dependent thyroid hormone activating enzyme type 2 iodothyronine deiodinase (D2) pathway, which is known to promote energy expenditure in BAT [98].

Along with bile acids, conventional mice exposed to a cold challenge have elevated butyrate levels, indicating an increase in fermentation [93,96]. Intriguingly, butyrate supplementation to antibiotic treated mice recovers the thermogenesis that would have been normally loss under the cold environment, indicating that this SCFA plays an influence on thermogenesis and thus beiging [93]. One theory of its benefits on beiging is due to the fact that butyrate is an inhibitor of histone deacetylases (HDACs). Comparatively, HDAC1 and HDAC3 have opposing effects on beiging. HDAC1 blocks the activation of BAT-specific genes (i.e., UCP1, PGC1α), where deleting HDAC1 enhances β-adrenergic signaling for BAT gene expression in brown adipocytes [85]. On the other hand, HDAC3 is required to ensure thermogenic aptitude in BAT, where mice deficient of HDAC3 become severely hypothermic due to the elimination of UCP1 [99]. While butyrate was able to rescue beiging in antibiotic treated mice, future studies are warranted to determine its differential effects on HDAC1/3 in this model. Another plausible means for the benefits of butyrate is its ability to enter the TCA cycle and promote succinate generation, as oxidation by succinate dehydrogenase is required for activation of thermogenesis [100]. Overall, the beneficial effects of the gut microbiota on beiging seems to need a "trigger", whether it be fasting or introduced to a cold environment (Fig. 16.2B), but future research is needed to fill in the missing pieces of this intricate puzzle.

## *Akkermansia muciniphila*, intestinal integrity and adipose tissue metabolism

We have been mentioning about SCFAs, including propionate, which are fermented metabolites from the dietary fibers that we ingest. In addition to this method, de novo propionate production is also possible through a few gut bacterial genera. Despite only representing 1%–3% of the total microbiota, *Akkermansia muciniphila*, under the Verrucomicrobia phylum, is one of the dominant bacterial genera in generating propionate through its mucin degradation activity [101–103]. Intriguingly, mucin degradation by *A. muciniphila* can support the growth of butyrate producing bacteria (i.e. *Anaerostipes caccae*) and subsequently the production of butyrate via the acetyl-CoA pathway [104]. Additionally, there is a positive feedback loop where elevation in the *A. caccae* community promotes the expression of mucolytic enzymes by *A. muciniphila* [104]. Currently, *A. muciniphila* is being introduced as one of the "next-generation beneficial" microbes due to its positive effects on glucose metabolism, lipid metabolism, and intestinal immunity [105]. In line, there is an inverse correlation between *A. muciniphila* abundance and obesity [106–109]. Therefore, it is critical to understand how this bacterial strain is implicated in lipid metabolism and its therapeutic potential.

Schneeberger et al. [109] demonstrated that there is a linear decrease (~100 fold) in *A. muciniphila* abundance when mice are fed an HFD. This type of gut dysbiosis was inversely associated with alterations in lipogenesis, metabolic inflammation, adipogenesis, and fat browning. Specifically, HFD-fed mice had elevated gene expressions for lipogenesis (i.e., *Dgat 2*, *Fasn*) and metabolic inflammation (i.e., *Il6*), but decreased mRNA levels toward adipogenesis (*Pparg*, *Cepba*) and fat browning [109]. Comparatively, *A. mucinphila* supplementation is able to alleviate body weight gain, reduce fat mass, improve glucose and insulin sensitivity, and attenuate fatty acid biosynthesis [110]. Likewise, administration of *A. muciniphila* to HFD-fed mice significantly diminishes adipose tissue inflammation by inducing Foxp3 regulatory T cells (Treg) [111] and stimulates genes involved in adipocyte differentiation, which improves overall adipose tissue metabolism [108]. Another plausible explanation for the benefits of *A. muciniphila* is due to its ability to stabilize gut permeability, which is generally compromised during gut dysbiosis [112,113]. Specifically, *A. muciniphila* treatment restores the thickness of the intestinal barrier, including the mucus layer, which is heavily reduced during HFD feeding [108]. Moreover, supplementation of this bacteria promotes the endocannabinoid system, which includes the upregulation of lipid mediators such as 2-arachidonoylglycerol, 2-palmitoylglycerol, and 2-oleoylglycerol that control barrier function [108]. This is further associated with low levels of serum lipopolysaccharide (LPS) levels and inactivated TLR4 signaling, which is indicative of alleviated metabolic endotoxemia and low-grade chronic inflammation [108,110]. These positive results have lead must research to explore the therapeutic potential of *A. muciniphila*. It is noteworthy that administration of the prebiotic, oligofructose, to HFD-fed mice increases intestinal *A. muciniphila* abundance and this results in the same beneficial effects as direct administration of the bacterium [108]. Additionally, *A. muciniphila*-derived extracellular vesicles (AmEVs) are being utilized to regulate gut permeability. Excitingly, AmEV administration enhances tight junction function in both in vitro [114] and in vivo studies [115]. Furthermore, this therapeutic is associated with reduced body weight gain and improved glucose tolerance in HFD-induced diabetic mice [115]. Intriguingly, the improved gut barrier integrity can be recapitulated when administering a specific outer membrane protein of *A. muciniphila*, termed Amuc100 [116]. This research is very impactful when considering that *A. muciniphila* gradually decreases as we age [117]. Overall, *A. muciniphila* is strongly associated with alleviation of obesity, metabolic endotoxemia, and adipose tissue inflammation.

## Metabolic endotoxemia, inflammation, and hepatic lipogenesis

Metabolic endotoxemia is a chronic, low-grade inflammatory response due to augmented endotoxins that travel into systemic circulation via translocation, which triggers inflammatory tone and initiates metabolic disease symptoms, such as weight gain and insulin resistance [118]. In 2007, Cani et al. first coined this term as they observed HFD fed mice exhibited an elevated abundance of LPS-producing gut bacteria, which further correlated with a two- to threefold increase in plasma levels of this endotoxin, and thus, resulted in insulin resistance and obesity [118]. Recall that LPS is a major outer surface membrane component present in almost all Gram-negative bacteria; therefore, augmented LPS indicates a case of intestinal barrier dysfunction, also known as a "leaky gut". Comparatively, mice treated with antibiotics have an altered gut microbiota profile, which is associated with reduced cecal LPS content and alleviated metabolic endotoxemia [119]. Likewise, in vitro and in vivo studies have shown that CD14 (LPS coreceptor found on monocytes, macrophages, and adipocytes) deficient mice have alleviated metabolic syndrome [120], which supports Cani et al.'s original theory that LPS must activate CD14 for the recruitment of macrophages to induce adipose tissue inflammation during obesity [118]. More particularly, it has been determined that LPS is recognized by a cascade of receptors and accessory proteins, including CD14, LPS binding protein (LBP), and the TLR4–MD-2 complex [121]. This correlates with TLR4-deficient mice being hyporesponsive to LPS, indicating that this endotoxin is required for TLR4 signaling [122]. Hence, much research has explored into the relationship among LPS, TLR4, metabolic endotoxemia and lipid metabolism.

Earlier, we mentioned a study by Caesar et al. who observed that a diet rich in saturated dietary fat can modulate the gut microbiota to support an elevation in TLR4 signaling, which subsequently resulted in increased WAT inflammation during obesity progression [92]. Considering that previous studies have extensively shown that TLR4 signaling is necessary for fatty acid-induced inflammation and insulin resistance [123,124], it was believed that long chain saturated fatty acids (i.e., palmitate) were also agonists of TLR4 [125]. While saturated fatty acids, like palmitate, from a HFD have been shown to promote TLR4-JNK signaling in macrophages [126], this was thought to be independent of TLR4 dimerization and activation [127]. Recently, Lancaster et al. has confirmed that palmitate is not a ligand for TLR4 signaling; yet, they do emphasize that LPS-activation of TLR4 signaling is still required for fatty acid-mediated inflammation [128]. In concordance, Caesar et al. showed that LPS induction of TLR4 promotes downstream TRIF/MYD88 signaling, which subsequently promotes the chemokine, CCL2, and thus, macrophage infiltration [92]. Likewise, since LPS is able to translocate to the liver via the hepatic portal vein, this endotoxin is also able to either directly damage hepatocytes or activate Kupffer cells by interacting with TLR4 complexes, which initiates proinflammatory cytokine production (i.e., TNF-$\alpha$) [129]. In particular, increased quantities of tumor necrosis factor-$\alpha$ (TNF-$\alpha$) has been widely understood to stimulate de novo lipogenesis [130], along with dysregulating insulin receptor (IR) signaling by suppressing phosphorylation of insulin receptor substrate 1 (IRS-1) [131], which may explain the elevated TNF-α levels that are exhibited in patients suffering from obesity and/or diabetes [132].

Considering the detrimental effects of LPS and TLR4 signaling, much research has delved into looking for therapeutics to either mitigate LPS levels or to impede TLR4-mediated signaling pathways. As mentioned prior, administration of antibiotics or deletion of CD14 has been shown to be preventive against metabolic syndrome. Likewise, deficiency of MyD88 has been shown to be protective against metabolic disease [133]; yet, except for TLR3, MyD88 is a common downstream component of toll-like receptors, so it can be difficult to determine whether ablation of TLR4-MyD88 signaling is the true savor against metabolic syndrome. In fact, TLR4 has been actually found to be required for the inhibition of lipogenic genes, where deficiency of TLR4 results in elevated expression of *Fasn*, *Pparγ*, *Dgat1*, and *Scd1*; yet, except for *Pparγ*, the elevation of these lipogenic genes happened only when mice were exposed to fasting conditions [134]. Therefore, instead of targeting TLR4 or its downstream components, it might be more potent to target the trigger of the inflammatory pathways. Therefore, to regulate gut bacterial-derived LPS levels, much research has involved targeting one of the strongest environmental factors on the gut microbiota: our diet. As a preventative measure against obesity and other indices of metabolic syndrome, it is highly recommended to sway away from the "Western-style" diet, which has been associated with increased LPS levels via chylomicrons [135]. In addition, the "Western-style" diet reduces the expression of tight junction-associated proteins (i.e., claudin-1 and occludin) in the colon [136], which would allow for the leak of LPS and other endotoxins. Likewise, this type of diet is capable of causing gut dysbiosis, as indicated by an increase in the Firmicutes to Bacteroidetes ratio [136]. As another preventative option, it has been suggested to ingest probiotics that are able to inhibit the proliferation of Gram-negative bacteria, while also improving the gut mucosa barrier and alleviating intestinal endotoxemia [129]. Intriguingly, it has been observed that normal endotoxemia is modulated during the fast/fed cycle [118]; therefore, understanding the circadian rhythm on the gut microbiota and lipid status may also provide better therapeutic angles, as further explained in the next section.

## Circadian disruption on gut microbiota alters lipid metabolism

We have established that the gut microbiota strongly influences lipid metabolism, especially during times of dysbiosis and impairment of intestinal barrier integrity, which subsequently alters the microbiota composition and levels of gut metabolites that further influence lipid homeostasis. One of strongest factors on the bacterial composition is dietary habits, where ingestion of a HFD can perturb the microbiota into dysbiosis, which increases susceptibility to obesity due to alterations in lipid metabolism. Moreover, the host circadian clock is essential for overall health and physiological homeostasis, where disruption is associated with intestinal dysbiosis [137]. As a quick review, the central clock is located within the suprachiasmatic nucleus (SCN) of the brain, where it is critical for coordinating light-dark and sleep-wake cycles [138]. Projections from the SCN can reach to tissue-specific peripheral clocks, including the liver, which further allows coordination of neural and metabolic programs [139]. Specifically, the transcription factors, brain and muscle ARNT-like 1 (BMAL1) and circadian locomotor output protein kaput (CLOCK), heterodimerize and translocate into the nucleus to activate the core clock genes, period (PER) and cryptochrome (CRY) [139–141]. At the end of the circadian day, PER and CRY heterodimerize and translocate into the nucleus to repress BMAL1/CLOCK and thus, suppress their own expression [139–141]. This constitutes an integral negative feedback loop of the core molecular oscillation system within the host, which is imperative in regulating certain circadian dependent enzymes (i.e., HMGCoA reductase, active nocturnal) and substrates (i.e., bile acids, fast/fed cycle).

As stated, the circadian clock is required in the regulation of various metabolic events, including lipid metabolism. For example, the muscle clock has been shown to promote diurnal cycles of neutral lipid storage while also inhibiting lipid catabolism before the wake cycle through BMAL1-mediated activation of *Dgat2* (catalyzes the final step in triglyceride synthesis) [142]. More recently, it has been realized that the gut microbiota exhibits diurnal oscillations that lead to time-specific compositional and functional profiles over the course of the day [143]. In particular, the circadian clock of the microbiota is influenced by feeding rhythms, where a HFD impairs the diurnal oscillations of the gut microbiota and also the liver [138,143]. Likewise, disruption of the molecular clock, whether by ablation of clock genes or induction of jet lag, leads to aberrant oscillations within the microbiota and promotes dysbiosis, which further results in glucose intolerance and obesity [143]. Intriguingly, inducing the jet lag phenomenon does not drive metabolic syndrome in germ-free mice that lack microbial oscillations, but they can recapitulate the disease upon fecal transplantation [143]. Likewise, HFD-fed mice subjected to time-restricted feeding (TRF; feeding is consolidated to the nocturnal phase) are protected against obesity and metabolic diseases, due to the partial restoration of the cyclical fluctuations [144]. Additionally, TRF reverts from an obesogenic microbiota as indicated by a decrease in the Firmicutes to Bacteroidetes ratio [144]. This emphasizes that fluctuations in the circadian clock of the gut microbiota can greatly affect host physiology.

As predicted, if the gut microbiota itself exhibits diurnal oscillations, this indicates that various gut metabolites could also present level fluctuations throughout the circadian day. Tahara et al. recently demonstrated that SCFAs have significant rhythmicity, where acetate and butyrate levels are higher at the beginning of the nocturnal cycle [145]. Upon administration of a HFD, there is an impairment in the butyrate production oscillation, which was further found to have an impact on the liver circadian clock [138]. This can be reverted through butyrate supplementation, as it was found to increase the *per2:bmal1* mRNA ratio in the liver [138]. In addition to SCFAs, bile acid synthesis and secretion is heavily regulated by the fasting/feeding cycle. Upon eating, the pancreatic hormone, cholecystokinin (CCK), stimulates the release of bile acids into the duodenum to aid in food digestion, where then a small proportion is subjected to biotransformation by the gut microbiota. Intriguingly, through the utilization of *Bsh*-transgenic *E. coli* colonization in germ-free and conventional mice, it has been shown that microbial deconjugation, via BSH, regulates hepatic and ileal clock genes and clock controlled genes that influence lipid (*Ppary*, *Angtpl4*) and cholesterol (*Abcg5/8*) metabolism [69]. As a last note, the production and translocation of gut-derived LPS has been shown to be significantly increased when mice were subjected to circadian disruption via constant light exposure for 4 weeks [146]. This was further associated with an impairment of intestinal integrity, which could result in increased inflammatory tone due to increased TLR4 signaling [121]. Overall, considering that the human lifestyle can consist of chronic work shifts, frequently flying across time zones and overconsumption of fast food, induction of circadian disruption represents a common hallmark of gut dysbiosis-mediated inflammation and metabolic diseases (Fig. 16.3).

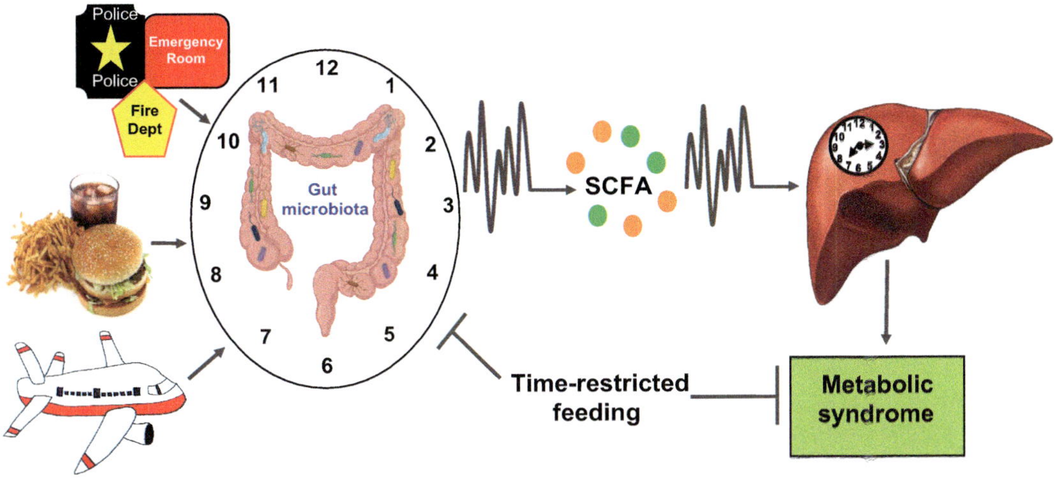


FIGURE 16.3 **Disruption of circadian oscillation within the gut microbiota promotes metabolic syndrome.** Various lifestyle factors, including nocturnal occupations, enrichment of processed foods in the diet, and jet lag from travelling through different time zones, can disturb gut microbiota and short-chain fatty acid (SCFA) diurnal oscillations, which can further impact the hepatic peripheral clock, resulting in metabolic syndrome. Time-restricted feeding can revert the gut microbiota back to homeostasis, which is a potential therapeutic against metabolic syndrome. Source: *Credit: Rachel M. Golonka.*

## Lipidomics: a new tool to study gut microbiota management of lipid profiles

Cellular lipids are highly dynamic macromolecules as they are constantly changing with physiological, pathological, and environmental conditions. Considering that lipids are integrated into multiple, distinct signaling mechanisms, having a tool that can measure alterations in those pathways during disease states is critical. In 2003, the field of lipidomics emerged, which works to study the structure and function of the complete set of lipids (i.e., lipidome) in a given cell or organism as well as to provide a large-scale data set of pathways and networks of cellular lipids in biological systems [147]. Lipidomic analysis of biological samples involves sample preparation, mass spectrometry-based analysis (i.e., MS) and data processing. During MS analysis, the lipid solutions can be analyzed either by shotgun lipidomics [148] or by chromatography-based lipidomics (i.e., liquid chromatography/LC) [149]. Additionally, depending on the method chosen, either global or targeted lipid analysis can be applied to provide the entire or partial cellular lipidome, respectively [147]. Currently, there are many databases that support lipid identification and quantification, data management and lipidome visualization, including LipidHome [150], LIPID MAPS Proteome Databased (LMPD) [151], and Analysis of Lipid Experiments (ALEX) [152]. The application of these analytical chemistry techniques and bioinformatics has further advanced in understanding the roles of lipids in various disease states, including metabolic syndrome, neurological disorders, and cancer, which has further provided a new direction for personalized medicine (as reviewed in [147]).

Recently, lipidomics has been utilized to better elucidate the role of the gut microbiota in host lipid metabolism. For example, Kindt et al. performed a comprehensive, quantitative lipidomic analysis to determine the differential hepatic and plasma lipid profiles between specific-pathogen free (SPF) and germ-free mice [153]. In particular, they observed a systematic shift from monounsaturated to polyunsaturated lipid species between SPF and germ-free mice; specifically, SPF mice exhibited a 1.5-fold increase of palmitoleic acid (C16:1) and glycerophospholipid levels, whereas germ-free mice had elevated arachidonic acid (C20:4), docosahexaenoic acid (C22:6), and phosphatidylinositol lipid levels [153]. They further found that SPF, compared to GF, mice exhibited an increase in fatty acid de novo synthesis, elevated delta-9 desaturation of palmitate to palmitoleate by stearoyl-CoA desaturase-1 (SCD1), and a surge of omega-6 fatty acid elongation [153]. Hence, this study describes that gut microbes induce monounsaturated fatty acid generation and polyunsaturated fatty acid elongation, leading to alterations in glycerophospholipid acyl-chain profiles. Another study by Chakrabarti et al. developed transcriptomics-driven lipidomics (TDL) to predict the impact of *E. coli* colonization from a mouse model of obesity and metabolic disease on lipid metabolism in gnotobiotic mice [154]. They determined that *E. coli* colonization promotes arachidonic acid metabolites but decreases components of glycerophospholipid metabolism, indicating an alteration in the bioavailability and digestion of dietary lipids, which could be further associated with an induction of inflammation [154]. These are promising starts into utilizing lipidomics to understand the influence of the gut microbiota on lipid metabolism.

## Summary and future perspectives

The gut microbiota is uncovering to be a hidden metabolic organ, where alterations to its population can have major beneficial and harmful after effects toward the host. Recent captivating findings demonstrate that the microbiota and its gut metabolites have a strong impact on global lipid metabolism, including lipogenesis and fat storage, which is further associated with various metabolic disorders. Accordingly, much research has delved into exploiting the microbial community in several ways, including (1) utilizing specific probiotics, prebiotics or other natural complex carbohydrates to promote the gut bacteria that are considered "anti-obesity" and/or antiinflammatory and (2) isolating and transferring the microbiome consortium (fecal therapy) from lean individuals to obese individuals with dysbiosis as a means to protect against weight gain. While these are great broad-spectrum approaches, personalized medicine through lifestyle intervention and various platform of omics technologies may be the new direction in alleviating these liver-associated pathologies that have become a major public health issue. To summarize, by clearly understanding the interaction between the gut microbiota and host lipid metabolism, this will allow us to move forward in rebuilding our alliance with the bacteria and thus, improve overall human health.

## Acknowledgments

### *Statement of ethics*

The authors have no ethical conflicts to disclose.

### *Disclosure statement*

The authors have no conflicts of interest to declare.

# References


[1] Trefts E, Gannon M, Wasserman DH. The liver. Curr Biol 2017;27(21):R1147–51.

[2] Broadley SP, Plaumann A, Coletti R, Lehmann C, Wanisch A, Seidlmeier A, et al. Dual-track clearance of circulating bacteria balances rapid restoration of blood sterility with induction of adaptive immunity. Cell Host Microbe 2016;20(1):36–48.

[3] Zeng Z, Surewaard BG, Wong CH, Geoghegan JA, Jenne CN, Kubes P. CRIg functions as a macrophage pattern recognition receptor to directly bind and capture blood-borne gram-positive bacteria. Cell Host Microbe 2016;20(1):99–106.

[4] Balmer ML, Slack E, de Gottardi A, Lawson MA, Hapfelmeier S, Miele L, et al. The liver may act as a firewall mediating mutualism between the host and its gut commensal microbiota. Sci Transl Med 2014;6(237):237ra66.

[5] Helmy KY, Katschke KJ Jr, Gorgani NN, Kljavin NM, Elliott JM, Diehl L, et al. CRIg: a macrophage complement receptor required for phagocytosis of circulating pathogens. Cell 2006;124(5):915–27.

[6] Boyer JL. Bile formation and secretion. Compr Physiol 2013;3(3):1035–78.

[7] Sanders FW, Griffin JL. De novo lipogenesis in the liver in health and disease: more than just a shunting yard for glucose. Biol Rev Camb Philos Soc 2016;91(2):452–68.

[8] Protzer U, Maini MK, Knolle PA. Living in the liver: hepatic infections. Nat Rev Immunol 2012;12:201.

[9] Sun H, Sun C, Tian Z, Xiao W. NK cells in immunotolerant organs. Cell Mol Immunol 2013;10(3):202–12.

[10] Forbes SJ, Newsome PN. Liver regeneration—mechanisms and models to clinical application. Nat Rev Gastroenterol Hepatol 2016;13:473.

[11] Riva MA, Riva E, Spicci M, Strazzabosco M, Giovannini M, Cesana G. "The city of Hepar": rituals, gastronomy, and politics at the origins of the modern names for the liver. J Hepatol 2011;55(5):1132–6.

[12] Tiniakos DG, Kandilis A, Geller SA. Tityus: a forgotten myth of liver regeneration. J Hepatol 2010;53(2):357–61.

[13] Abbort ME. The evolution of modern medicine. Can Med Assoc J 1922;12(3):182–3.

[14] Llorente C, Schnabl B. Fast-track clearance of bacteria from the liver. Cell Host Microbe 2016;20(1):1–2.

[15] Macpherson AJ, Heikenwalder M, Ganal-Vonarburg SC. The liver at the nexus of host-microbial interactions. Cell Host Microbe 2016;20(5):561–71.

[16] Campo L, Eiseler S, Apfel T, Pyrsopoulos N. Fatty liver disease and gut microbiota: a comprehensive update. J Clin Transl Hepatol 2019;7(1):56–60.

[17] Backhed F, Ley RE, Sonnenburg JL, Peterson DA, Gordon JI. Host-bacterial mutualism in the human intestine. Science 2005;307(5717):1915–20.

[18] Thursby E, Juge N. Introduction to the human gut microbiota. Biochem J 2017;474(11):1823–36.

[19] Tang C, Offermanns S. FFA2 and FFA3 in Metabolic Regulation. Handb Exp Pharmacol. 2017;236:205–20.

[20] DeGruttola AK, Low D, Mizoguchi A, Mizoguchi E. Current understanding of dysbiosis in disease in human and animal models. Inflamm Bowel Dis 2016;22(5):1137–50.

[21] Riva A, Borgo F, Lassandro C, Verduci E, Morace G, Borghi E, et al. Pediatric obesity is associated with an altered gut microbiota and discordant shifts in Firmicutes populations. Environ Microbiol 2017;19(1):95–105.

[22] Arumugam M, Raes J, Pelletier E, Le Paslier D, Yamada T, Mende DR, et al. Enterotypes of the human gut microbiome. Nature 2011;473:174.

[23] Williams BL, Hornig M, Parekh T, Lipkin WI. Application of novel PCR-based methods for detection, quantitation, and phylogenetic characterization of Sutterella species in intestinal biopsy samples from children with autism and gastrointestinal disturbances. MBio 2012;3(1).

[24] Sun Y, Li L, Xia Y, Li W, Wang K, Wang L, et al. The gut microbiota heterogeneity and assembly changes associated with the IBD. Sci Rep 2019;9(1):440.

[25] Mell B, Jala VR, Mathew AV, Byun J, Waghulde H, Zhang Y, et al. Evidence for a link between gut microbiota and hypertension in the Dahl rat. Physiol Genomics 2015;47(6):187–97.

[26] Rinninella E, Raoul P, Cintoni M, Franceschi F, Miggiano GAD, Gasbarrini A, et al. What is the healthy gut microbiota composition? A changing ecosystem across age, environment, diet, and diseases. Microorganisms 2019;7(1).

[27] Belenguer A, Duncan SH, Calder AG, Holtrop G, Louis P, Lobley GE, et al. Two routes of metabolic cross-feeding between *Bifidobacterium adolescentis* and butyrate-producing anaerobes from the human gut. Appl Environ Microbiol 2006;72(5):3593–9.

[28] Lu Y, Fan C, Li P, Lu Y, Chang X, Qi K. Short chain fatty acids prevent high-fat-diet-induced obesity in mice by regulating G Protein-coupled receptors and gut microbiota. Sci Rep 2016;6:37589.

[29] Kimura I, Ozawa K, Inoue D, Imamura T, Kimura K, Maeda T, et al. The gut microbiota suppresses insulin-mediated fat accumulation via the short-chain fatty acid receptor GPR43. Nat Commun 2013;4:1829.
[30] Rosenbaum M, Sy M, Pavlovich K, Leibel RL, Hirsch J. Leptin reverses weight loss-induced changes in regional neural activity responses to visual food stimuli. J Clin Invest 2008;118(7):2583–91.
[31] Ahl S, Guenther M, Zhao S, James R, Marks J, Szabo A, et al. Adiponectin levels differentiate metabolically healthy vs unhealthy among obese and nonobese white individuals. J Clin Endocrinol Metab 2015;100(11):4172–80.
[32] Zhao L, Fu Z, Wu J, Aylor KW, Barrett EJ, Cao W, et al. Globular adiponectin ameliorates metabolic insulin resistance via AMPK-mediated restoration of microvascular insulin responses. J Physiol 2015;593(17):4067–79.
[33] Hardie DG, Pan DA. Regulation of fatty acid synthesis and oxidation by the AMP-activated protein kinase. Biochem Soc Trans 2002;30(Pt. 6):1064–70.
[34] Alex S, Lange K, Amolo T, Grinstead JS, Haakonsson AK, Szalowska E, et al. Short-chain fatty acids stimulate angiopoietin-like 4 synthesis in human colon adenocarcinoma cells by activating peroxisome proliferator-activated receptor gamma. Mol Cell Biol 2013;33(7):1303–16.
[35] Backhed F, Ding H, Wang T, Hooper LV, Koh GY, Nagy A, et al. The gut microbiota as an environmental factor that regulates fat storage. Proc Natl Acad Sci U S A 2004;101(44):15718–23.
[36] Backhed F, Manchester JK, Semenkovich CF, Gordon JI. Mechanisms underlying the resistance to diet-induced obesity in germ-free mice. Proc Natl Acad Sci USA 2007;104(3):979–84.
[37] den Besten G, Lange K, Havinga R, van Dijk TH, Gerding A, van Eunen K, et al. Gut-derived short-chain fatty acids are vividly assimilated into host carbohydrates and lipids. Am J Physiol Gastrointest Liver Physiol 2013;.
[38] Tirosh A, Calay ES, Tuncman G, Claiborn KC, Inouye KE, Eguchi K, et al. The short-chain fatty acid propionate increases glucagon and FABP4 production, impairing insulin action in mice and humans. Sci Transl Med 2019;11(489):eaav0120.
[39] Gao X, Lin S-H, Ren F, Li J-T, Chen J-J, Yao C-B, et al. Acetate functions as an epigenetic metabolite to promote lipid synthesis under hypoxia. Nat Commun 2016;7:11960.
[40] Allen AE, Locasale JW. Glucose metabolism in cancer: The saga of pyruvate kinase continues. Cancer Cell 2018;33(3):337–9.
[41] Liu X, Cooper DE, Cluntun AA, Warmoes MO, Zhao S, Reid MA, et al. Acetate production from glucose and coupling to mitochondrial metabolism in mammals. Cell 2018;175(2):502–513.e13.
[42] Schug ZT, Peck B, Jones DT, Zhang Q, Grosskurth S, Alam IS, et al. Acetyl-CoA synthetase 2 promotes acetate utilization and maintains cancer cell growth under metabolic stress. Cancer Cell 2015;27(1):57–71.
[43] Stoiber K, Nagło O, Pernpeintner C, Zhang S, Koeberle A, Ulrich M, et al. Targeting de novo lipogenesis as a novel approach in anti-cancer therapy. Br J Cancer 2017;118:43.
[44] Singh V, Yeoh BS, Chassaing B, Xiao X, Saha P, Aguilera Olvera R, et al. Dysregulated microbial fermentation of soluble fiber induces cholestatic liver cancer. Cell 2018;175(3):679–694.e22.
[45] Singh V, Chassaing B, Zhang L, San Yeoh B, Xiao X, Kumar M, et al. Microbiota-dependent hepatic lipogenesis mediated by stearoyl CoA desaturase 1 (SCD1) promotes metabolic syndrome in TLR5-deficient mice. Cell Metab 2015;22(6):983–96.
[46] Nishitsuji K, Xiao J, Nagatomo R, Umemoto H, Morimoto Y, Akatsu H, et al. Analysis of the gut microbiome and plasma short-chain fatty acid profiles in a spontaneous mouse model of metabolic syndrome. Sci Rep 2017;7(1):15876.
[47] Li HP, Chen X, Li MQ. Butyrate alleviates metabolic impairments and protects pancreatic beta cell function in pregnant mice with obesity. Int J Clin Exp Pathol 2013;6(8):1574–84.
[48] Gao Z, Yin J, Zhang J, Ward RE, Martin RJ, Lefevre M, et al. Butyrate improves insulin sensitivity and increases energy expenditure in mice. Diabetes 2009;58(7):1509–17.
[49] De Vadder F, Kovatcheva-Datchary P, Goncalves D, Vinera J, Zitoun C, Duchampt A, et al. Microbiota-generated metabolites promote metabolic benefits via gut-brain neural circuits. Cell 2014;156(1–2):84–96.
[50] Byrne CS, Chambers ES, Morrison DJ, Frost G. The role of short chain fatty acids in appetite regulation and energy homeostasis. Int J Obes (Lond) 2015;39(9):1331–8.
[51] Perry RJ, Peng L, Barry NA, Cline GW, Zhang D, Cardone RL, et al. Acetate mediates a microbiome-brain-beta-cell axis to promote metabolic syndrome. Nature 2016;534(7606):213–7.
[52] de Aguiar Vallim TQ, Tarling EJ, Edwards PA. Pleiotropic roles of bile acids in metabolism. Cell Metab 2013;17(5):657–69.

[53] Hylemon PB, Zhou H, Pandak WM, Ren S, Gil G, Dent P. Bile acids as regulatory molecules. J Lipid Res 2009;50(8):1509–20.
[54] Ridlon JM, Kang DJ, Hylemon PB. Bile salt biotransformations by human intestinal bacteria. J Lipid Res 2006;47(2):241–59.
[55] Hirano S, Masuda N. Enhancement of the 7 alpha-dehydroxylase activity of a gram-positive intestinal anaerobe by Bacteroides and its significance in the 7-dehydroxylation of ursodeoxycholic acid. J Lipid Res 1982;23(8):1152–8.
[56] Kong B, Wang L, Chiang JY, Zhang Y, Klaassen CD, Guo GL. Mechanism of tissue-specific farnesoid X receptor in suppressing the expression of genes in bile-acid synthesis in mice. Hepatology 2012;56(3):1034–43.
[57] Kim I, Morimura K, Shah Y, Yang Q, Ward JM, Gonzalez FJ. Spontaneous hepatocarcinogenesis in farnesoid X receptor-null mice. Carcinogenesis 2007;28(5):940–6.
[58] Knisely AS, Strautnieks SS, Meier Y, Stieger B, Byrne JA, Portmann BC, et al. Hepatocellular carcinoma in ten children under five years of age with bile salt export pump deficiency. Hepatology 2006;44(2):478–86.
[59] Li G, Kong B, Zhu Y, Zhan L, Williams JA, Tawfik O, et al. Small heterodimer partner overexpression partially protects against liver tumor development in farnesoid X receptor knockout mice. Toxicol Appl Pharmacol 2013;272(2):299–305.
[60] Sun L, Beggs K, Borude P, Edwards G, Bhushan B, Walesky C, et al. Bile acids promote diethylnitrosamine-induced hepatocellular carcinoma via increased inflammatory signaling. Am J Physiol Gastrointest Liver Physiol 2016;311(1):G91–G104.
[61] Wolfe A, Thomas A, Edwards G, Jaseja R, Guo GL, Apte U. Increased activation of the Wnt/beta-catenin pathway in spontaneous hepatocellular carcinoma observed in farnesoid X receptor knockout mice. J Pharmacol Exp Ther 2011;338(1):12–21.
[62] Yang F, Huang X, Yi T, Yen Y, Moore DD, Huang W. Spontaneous development of liver tumors in the absence of the bile acid receptor farnesoid X receptor. Cancer Res 2007;67(3):863–7.
[63] Watanabe M, Houten SM, Wang L, Moschetta A, Mangelsdorf DJ, Heyman RA, et al. Bile acids lower triglyceride levels via a pathway involving FXR, SHP, and SREBP-1c. J Clin Invest 2004;113(10):1408–18.
[64] Brown MS, Goldstein JL. The SREBP pathway: regulation of cholesterol metabolism by proteolysis of a membrane-bound transcription factor. Cell 1997;89(3):331–40.
[65] Yang ZX, Shen W, Sun H. Effects of nuclear receptor FXR on the regulation of liver lipid metabolism in patients with non-alcoholic fatty liver disease. Hepatol Int 2010;4(4):741–8.
[66] Matsukuma KE, Bennett MK, Huang J, Wang L, Gil G, Osborne TF. Coordinated control of bile acids and lipogenesis through FXR-dependent regulation of fatty acid synthase. J Lipid Res 2006;47(12):2754–61.
[67] Zhang Y, Lee FY, Barrera G, Lee H, Vales C, Gonzalez FJ, et al. Activation of the nuclear receptor FXR improves hyperglycemia and hyperlipidemia in diabetic mice. Proc Natl Acad Sci USA 2006;103(4):1006–11.
[68] Kuno T, Hirayama-Kurogi M, Ito S, Ohtsuki S. Reduction in hepatic secondary bile acids caused by short-term antibiotic-induced dysbiosis decreases mouse serum glucose and triglyceride levels. Sci Rep. 2018;8(1):1253.
[69] Joyce SA, MacSharry J, Casey PG, Kinsella M, Murphy EF, Shanahan F, et al. Regulation of host weight gain and lipid metabolism by bacterial bile acid modification in the gut. Proc Natl Acad Sci USA 2014;111(20):7421–6.
[70] Canyelles M, Tondo M, Cedo L, Farras M, Escola-Gil JC, Blanco-Vaca F. Trimethylamine N-oxide: a link among diet, gut microbiota, gene regulation of liver and intestine cholesterol homeostasis and HDL function. Int J Mol Sci. 2018;19(10.).
[71] Schugar RC, Willard B, Wang Z, Brown JM. Postprandial gut microbiota-driven choline metabolism links dietary cues to adipose tissue dysfunction. Adipocyte 2018;7(1):49–56.
[72] Vance DE. Role of phosphatidylcholine biosynthesis in the regulation of lipoprotein homeostasis. Curr Opin Lipidol 2008;19(3):229–34.
[73] Dumas ME, Barton RH, Toye A, Cloarec O, Blancher C, Rothwell A, et al. Metabolic profiling reveals a contribution of gut microbiota to fatty liver phenotype in insulin-resistant mice. Proc Natl Acad Sci USA 2006;103(33):12511–6.
[74] Kim SH, Lim Y, Park JB, Kwak JH, Kim KJ, Kim JH, et al. Comparative study of fatty liver induced by methionine and choline-deficiency in C57BL/6N mice originating from three different sources. Lab Anim Res 2017;33(2):157–64.
[75] Chen YM, Liu Y, Zhou RF, Chen XL, Wang C, Tan XY, et al. Associations of gut-flora-dependent metabolite trimethylamine-N-oxide, betaine and choline with non-alcoholic fatty liver disease in adults. Sci Rep 2016;6:19076.

[76] Janeiro MH, Ramirez MJ, Milagro FI, Martinez JA, Solas M. Implication of trimethylamine N-oxide (TMAO) in disease: potential biomarker or new therapeutic target. Nutrients 2018;10(10).

[77] Hoyles L, Jimenez-Pranteda ML, Chilloux J, Brial F, Myridakis A, Aranias T, et al. Metabolic retroconversion of trimethylamine N-oxide and the gut microbiota. Microbiome 2018;6(1):73.

[78] Tan X, Liu Y, Long J, Chen S, Liao G, Wu S, et al. Trimethylamine N-oxide aggravates liver steatosis through modulation of bile acid metabolism and inhibition of farnesoid X receptor signaling in nonalcoholic fatty liver disease. Mol Nutr Food Res 2019;63:e1900257.

[79] Warrier M, Shih DM, Burrows AC, Ferguson D, Gromovsky AD, Brown AL, et al. The TMAO-generating enzyme flavin monooxygenase 3 is a central regulator of cholesterol balance. Cell Rep 2015;10(3):326–38.

[80] Shih DM, Wang Z, Lee R, Meng Y, Che N, Charugundla S, et al. Flavin containing monooxygenase 3 exerts broad effects on glucose and lipid metabolism and atherosclerosis. J Lipid Res 2015;56(1):22–37.

[81] Schugar RC, Shih DM, Warrier M, Helsley RN, Burrows A, Ferguson D, et al. The TMAO-producing enzyme flavin-containing monooxygenase 3 regulates obesity and the beiging of white adipose tissue. Cell Rep 2017;19(12):2451–61.

[82] Hernandez D, Addou S, Lee D, Orengo C, Shephard EA, Phillips IR. Trimethylaminuria and a human FMO3 mutation database. Hum Mutat 2003;22(3):209–13.

[83] Kajimura S, Spiegelman BM, Seale P. Brown and beige fat: physiological roles beyond heat generation. Cell Metab 2015;22(4):546–59.

[84] Frank AP, Palmer BF, Clegg DJ. Do estrogens enhance activation of brown and beiging of adipose tissues? Physiol Behav 2018;187:24–31.

[85] Li F, Wu R, Cui X, Zha L, Yu L, Shi H, et al. Histone deacetylase 1 (HDAC1) negatively regulates thermogenic program in brown adipocytes via coordinated regulation of histone H3 lysine 27 (H3K27) deacetylation and methylation. J Biol Chem 2016;291(9):4523–36.

[86] Cypess AM, Weiner LS, Roberts-Toler C, Franquet Elia E, Kessler SH, Kahn PA, et al. Activation of human brown adipose tissue by a beta3-adrenergic receptor agonist. Cell Metab 2015;21(1):33–8.

[87] Dutchak PA, Katafuchi T, Bookout AL, Choi JH, Yu RT, Mangelsdorf DJ, et al. Fibroblast growth factor-21 regulates PPARgamma activity and the antidiabetic actions of thiazolidinediones. Cell. 2012;148(3):556–67.

[88] Badman MK, Pissios P, Kennedy AR, Koukos G, Flier JS, Maratos-Flier E. Hepatic fibroblast growth factor 21 is regulated by PPARalpha and is a key mediator of hepatic lipid metabolism in ketotic states. Cell Metab 2007;5(6):426–37.

[89] Li G, Xie C, Lu S, Nichols RG, Tian Y, Li L, et al. Intermittent Fasting Promotes White Adipose Browning and Decreases Obesity by Shaping the Gut Microbiota. Cell Metab 2017;26(4):672–685.e4.

[90] Sahuri-Arisoylu M, Brody LP, Parkinson JR, Parkes H, Navaratnam N, Miller AD, et al. Reprogramming of hepatic fat accumulation and 'browning' of adipose tissue by the short-chain fatty acid acetate. Int J Obes 2016;40:955.

[91] Suárez-Zamorano N, Fabbiano S, Chevalier C, Stojanovic´ O, Colin DJ, Stevanovic´ A, et al. Microbiota depletion promotes browning of white adipose tissue and reduces obesity. Nat Med 2015;21:1497.

[92] Caesar R, Tremaroli V, Kovatcheva-Datchary P, Cani PD, Backhed F. Crosstalk between gut microbiota and dietary lipids aggravates WAT inflammation through TLR signaling. Cell Metab 2015;22(4):658–68.

[93] Li B, Li L, Li M, Lam SM, Wang G, Wu Y, et al. Microbiota depletion impairs thermogenesis of brown adipose tissue and browning of white adipose tissue. Cell Rep 2019;26(10):2720–2737.e5.

[94] Fischer K, Ruiz HH, Jhun K, Finan B, Oberlin DJ, van der Heide V, et al. Alternatively activated macrophages do not synthesize catecholamines or contribute to adipose tissue adaptive thermogenesis. Nat Med 2017;23(5):623–30.

[95] Virtue AT, McCright SJ, Wright JM, Jimenez MT, Mowel WK, Kotzin JJ, et al. The gut microbiota regulates white adipose tissue inflammation and obesity via a family of microRNAs. Sci Transl Med 2019;11(496):eaav1892.

[96] Chevalier C, Stojanovic O, Colin DJ, Suarez-Zamorano N, Tarallo V, Veyrat-Durebex C, et al. Gut microbiota orchestrates energy homeostasis during cold. Cell 2015;163(6):1360–74.

[97] Zietak M, Kovatcheva-Datchary P, Markiewicz LH, Stahlman M, Kozak LP, Backhed F. Altered microbiota contributes to reduced diet-induced obesity upon cold exposure. Cell Metab 2016;23(6):1216–23.

[98] Watanabe M, Houten SM, Mataki C, Christoffolete MA, Kim BW, Sato H, et al. Bile acids induce energy expenditure by promoting intracellular thyroid hormone activation. Nature 2006;439(7075):484–9.

[99] Emmett MJ, Lim HW, Jager J, Richter HJ, Adlanmerini M, Peed LC, et al. Histone deacetylase 3 prepares brown adipose tissue for acute thermogenic challenge. Nature 2017;546(7659):544–8.

[100] Mills EL, Pierce KA, Jedrychowski MP, Garrity R, Winther S, Vidoni S, et al. Accumulation of succinate controls activation of adipose tissue thermogenesis. Nature 2018;560(7716):102–6.
[101] Reichardt N, Duncan SH, Young P, Belenguer A, McWilliam Leitch C, Scott KP, et al. Phylogenetic distribution of three pathways for propionate production within the human gut microbiota. ISME J 2014;8(6):1323–35.
[102] Derrien M, Collado MC, Ben-Amor K, Salminen S, de Vos WM. The Mucin degrader Akkermansia muciniphila is an abundant resident of the human intestinal tract. Appl Environ Microbiol 2008;74(5):1646–8.
[103] Derrien M, Vaughan EE, Plugge CM, de Vos WM. Akkermansia muciniphila gen. nov., sp. nov., a human intestinal mucin-degrading bacterium. Int J Syst Evol Microbiol 2004;54(Pt. 5):1469–76.
[104] Chia LW, Hornung BVH, Aalvink S, Schaap PJ, de Vos WM, Knol J, et al. Deciphering the trophic interaction between Akkermansia muciniphila and the butyrogenic gut commensal Anaerostipes caccae using a metatranscriptomic approach. Antonie Van Leeuwenhoek 2018;111(6):859–73.
[105] Naito Y, Uchiyama K, Takagi T. A next-generation beneficial microbe: Akkermansia muciniphila. J Clin Biochem Nutr 2018;63(1):33–5.
[106] Karlsson CL, Onnerfalt J, Xu J, Molin G, Ahrne S, Thorngren-Jerneck K. The microbiota of the gut in preschool children with normal and excessive body weight. Obesity (Silver Spring) 2012;20(11):2257–61.
[107] Dao MC, Everard A, Aron-Wisnewsky J, Sokolovska N, Prifti E, Verger EO, et al. Akkermansia muciniphila and improved metabolic health during a dietary intervention in obesity: relationship with gut microbiome richness and ecology. Gut 2016;65(3):426–36.
[108] Everard A, Belzer C, Geurts L, Ouwerkerk JP, Druart C, Bindels LB, et al. Cross-talk between Akkermansia muciniphila and intestinal epithelium controls diet-induced obesity. Proc Natl Acad Sci USA 2013;110(22):9066–71.
[109] Schneeberger M, Everard A, Gomez-Valades AG, Matamoros S, Ramirez S, Delzenne NM, et al. Akkermansia muciniphila inversely correlates with the onset of inflammation, altered adipose tissue metabolism and metabolic disorders during obesity in mice. Sci Rep 2015;5:16643.
[110] Zhao S, Liu W, Wang J, Shi J, Sun Y, Wang W, et al. Akkermansia muciniphila improves metabolic profiles by reducing inflammation in chow diet-fed mice. J Mol Endocrinol 2017;58(1):1–14.
[111] Shin NR, Lee JC, Lee HY, Kim MS, Whon TW, Lee MS, et al. An increase in the Akkermansia spp. population induced by metformin treatment improves glucose homeostasis in diet-induced obese mice. Gut 2014;63(5):727–35.
[112] Jiang W, Wu N, Wang X, Chi Y, Zhang Y, Qiu X, et al. Dysbiosis gut microbiota associated with inflammation and impaired mucosal immune function in intestine of humans with non-alcoholic fatty liver disease. Sci Rep 2015;5:8096.
[113] Chelakkot C, Ghim J, Ryu SH. Mechanisms regulating intestinal barrier integrity and its pathological implications. Exp Mol Med 2018;50(8):103.
[114] Ashrafian F, Behrouzi A, Shahriary A, Ahmadi Badi S, Davari M, Khatami S, et al. Comparative study of effect of Akkermansia muciniphila and its extracellular vesicles on toll-like receptors and tight junction. Gastroenterol Hepatol Bed Bench 2019;12(2):163–8.
[115] Chelakkot C, Choi Y, Kim D-K, Park HT, Ghim J, Kwon Y, et al. Akkermansia muciniphila-derived extracellular vesicles influence gut permeability through the regulation of tight junctions. Exp Mol Med 2018;50:e450.
[116] Plovier H, Everard A, Druart C, Depommier C, Van Hul M, Geurts L, et al. A purified membrane protein from Akkermansia muciniphila or the pasteurized bacterium improves metabolism in obese and diabetic mice. Nat Med 2017;23(1):107–13.
[117] Collado MC, Derrien M, Isolauri E, de Vos WM, Salminen S. Intestinal integrity and Akkermansia muciniphila, a mucin-degrading member of the intestinal microbiota present in infants, adults, and the elderly. Appl Environ Microbiol 2007;73(23):7767–70.
[118] Cani PD, Amar J, Iglesias MA, Poggi M, Knauf C, Bastelica D, et al. Metabolic endotoxemia initiates obesity and insulin resistance. Diabetes 2007;56(7):1761–72.
[119] Cani PD, Bibiloni R, Knauf C, Waget A, Neyrinck AM, Delzenne NM, et al. Changes in gut microbiota control metabolic endotoxemia-induced inflammation in high-fat diet-induced obesity and diabetes in mice. Diabetes 2008;57(6):1470–81.
[120] Luche E, Cousin B, Garidou L, Serino M, Waget A, Barreau C, et al. Metabolic endotoxemia directly increases the proliferation of adipocyte precursors at the onset of metabolic diseases through a CD14-dependent mechanism. Mol Metab 2013;2(3):281–91.
[121] Park BS, Lee JO. Recognition of lipopolysaccharide pattern by TLR4 complexes. Exp Mol Med 2013;45:e66.

[122] Hoshino K, Takeuchi O, Kawai T, Sanjo H, Ogawa T, Takeda Y, et al. Cutting edge: Toll-like receptor 4 (TLR4)-deficient mice are hyporesponsive to lipopolysaccharide: evidence for TLR4 as the Lps gene product. J Immunol 1999;162(7):3749–52.

[123] Holland WL, Bikman BT, Wang LP, Yuguang G, Sargent KM, Bulchand S, et al. Lipid-induced insulin resistance mediated by the proinflammatory receptor TLR4 requires saturated fatty acid-induced ceramide biosynthesis in mice. J Clin Invest 2011;121(5):1858–70.

[124] Shi H, Kokoeva MV, Inouye K, Tzameli I, Yin H, Flier JS. TLR4 links innate immunity and fatty acid-induced insulin resistance. J Clin Invest 2006;116(11):3015–25.

[125] Osborn O, Olefsky JM. The cellular and signaling networks linking the immune system and metabolism in disease. Nat Med 2012;18:363.

[126] Huang S, Rutkowsky JM, Snodgrass RG, Ono-Moore KD, Schneider DA, Newman JW, et al. Saturated fatty acids activate TLR-mediated proinflammatory signaling pathways. J Lipid Res 2012;53(9):2002–13.

[127] Lancaster GI, Langley KG, Berglund NA, Kammoun HL, Reibe S, Estevez E, et al. Evidence that TLR4 Is Not a Receptor for Saturated Fatty Acids but Mediates Lipid-Induced Inflammation by Reprogramming Macrophage Metabolism. Cell Metab 2018;27(5):1096–1110.e5.

[128] Lancaster GI, Langley KG, Berglund NA, Kammoun HL, Reibe S, Estevez E, et al. Evidence that TLR4 is not a receptor for saturated fatty acids but mediates lipid-induced inflammation by reprogramming macrophage metabolism. Cell Metab 2018;27(5):1096–1110.e5.

[129] Xue L, He J, Gao N, Lu X, Li M, Wu X, et al. Probiotics may delay the progression of nonalcoholic fatty liver disease by restoring the gut microbiota structure and improving intestinal endotoxemia. Sci Rep 2017;7:45176.

[130] Feingold KR, Grunfeld C. Tumor necrosis factor-alpha stimulates hepatic lipogenesis in the rat in vivo. J Clin Invest 1987;80(1):184–90.

[131] Hotamisligil GS, Peraldi P, Budavari A, Ellis R, White MF, Spiegelman BM. IRS-1-mediated inhibition of insulin receptor tyrosine kinase activity in TNF-alpha- and obesity-induced insulin resistance. Science 1996;271(5249):665–8.

[132] Miyazaki Y, Pipek R, Mandarino LJ, DeFronzo RA. Tumor necrosis factor α and insulin resistance in obese type 2 diabetic patients. Int J Obes 2003;27(1):88–94.

[133] Everard A, Geurts L, Caesar R, Van Hul M, Matamoros S, Duparc T, et al. Intestinal epithelial MyD88 is a sensor switching host metabolism towards obesity according to nutritional status. Nat Commun 2014;5:5648.

[134] Pang S, Tang H, Zhuo S, Zang YQ, Le Y. Regulation of fasting fuel metabolism by toll-like receptor 4. Diabetes 2010;59(12):3041–8.

[135] Pendyala S, Walker JM, Holt PR. A high-fat diet is associated with endotoxemia that originates from the gut. Gastroenterology 2012;142(5):1100–1101.e2.

[136] Kim KA, Gu W, Lee IA, Joh EH, Kim DH. High fat diet-induced gut microbiota exacerbates inflammation and obesity in mice via the TLR4 signaling pathway. PLoS ONE 2012;7(10):e47713.

[137] Voigt RM, Summa KC, Forsyth CB, Green SJ, Engen P, Naqib A, et al. The circadian clock mutation promotes intestinal dysbiosis. Alcohol Clin Exp Res 2016;40(2):335–47.

[138] Leone V, Gibbons SM, Martinez K, Hutchison AL, Huang EY, Cham CM, et al. Effects of diurnal variation of gut microbes and high-fat feeding on host circadian clock function and metabolism. Cell Host Microbe 2015;17(5):681–9.

[139] Hastings MH, Maywood ES, Brancaccio M. Generation of circadian rhythms in the suprachiasmatic nucleus. Nat Rev Neurosci 2018;19(8):453–69.

[140] Huang N, Chelliah Y, Shan Y, Taylor CA, Yoo SH, Partch C, et al. Crystal structure of the heterodimeric CLOCK:BMAL1 transcriptional activator complex. Science 2012;337(6091):189–94.

[141] Kume K, Zylka MJ, Sriram S, Shearman LP, Weaver DR, Jin X, et al. mCRY1 and mCRY2 are essential components of the negative limb of the circadian clock feedback loop. Cell 1999;98(2):193–205.

[142] Dyar KA, Hubert MJ, Mir AA, Ciciliot S, Lutter D, Greulich F, et al. Transcriptional programming of lipid and amino acid metabolism by the skeletal muscle circadian clock. PLoS Biol 2018;16(8):e2005886.

[143] Thaiss CA, Zeevi D, Levy M, Zilberman-Schapira G, Suez J, Tengeler AC, et al. Transkingdom control of microbiota diurnal oscillations promotes metabolic homeostasis. Cell 2014;159(3):514–29.

[144] Zarrinpar A, Chaix A, Yooseph S, Panda S. Diet and feeding pattern affect the diurnal dynamics of the gut microbiome. Cell Metab 2014;20(6):1006–17.

[145] Tahara Y, Yamazaki M, Sukigara H, Motohashi H, Sasaki H, Miyakawa H, et al. Gut microbiota-derived short chain fatty acids induce circadian clock entrainment in mouse peripheral tissue. Sci Rep 2018;8(1):1395.

[146] Deaver JA, Eum SY, Toborek M. Circadian disruption changes gut microbiome taxa and functional gene composition. Front Microbiol 2018;9:737.

[147] Yang K, Han X. Lipidomics: techniques, applications, and outcomes related to biomedical sciences. Trends Biochem Sci 2016;41(11):954–69.

[148] Han X, Yang K, Gross RW. Multi-dimensional mass spectrometry-based shotgun lipidomics and novel strategies for lipidomic analyses. Mass Spectrom Rev 2012;31(1):134–78.

[149] Cajka T, Fiehn O. Comprehensive analysis of lipids in biological systems by liquid chromatography-mass spectrometry. Trends Analyt Chem 2014;61:192–206.

[150] Foster JM, Moreno P, Fabregat A, Hermjakob H, Steinbeck C, Apweiler R, et al. LipidHome: a database of theoretical lipids optimized for high throughput mass spectrometry lipidomics. PLoS ONE 2013;8(5):e61951.

[151] Cotter D, Maer A, Guda C, Saunders B, Subramaniam S. LMPD: LIPID MAPS proteome database. Nucleic Acids Res 2006;34(Database issue):D507–10.

[152] Husen P, Tarasov K, Katafiasz M, Sokol E, Vogt J, Baumgart J, et al. Analysis of lipid experiments (ALEX): a software framework for analysis of high-resolution shotgun lipidomics data. PLoS ONE 2013;8(11):e79736.

[153] Kindt A, Liebisch G, Clavel T, Haller D, Hörmannsperger G, Yoon H, et al. The gut microbiota promotes hepatic fatty acid desaturation and elongation in mice. Nat Commun 2018;9(1):3760.

[154] Chakrabarti A, Membrez M, Morin-Rivron D, Siddharth J, Chou CJ, Henry H, et al. Transcriptomics-driven lipidomics (TDL) identifies the microbiome-regulated targets of ileal lipid metabolism. NPJ Syst Biol Appl 2017;3(1):33.

CHAPTER

# 17

# Insights into the metabolism of lipids in obesity and diabetes

*Ademola O. Ayeleso*[b], *Mashudu G. Matumba*[a], *James M. Ntambi*[c], *Emmanuel Mukwevho*[a]

[a]Department of Biochemistry, Faculty of Natural and Agricultural Science, North West University, Mmabatho, South Africa; [b]Department of Biochemistry, Faculty of Science, Adeleke University, Ede, Osun State, Nigeria; [c]Departments of Biochemistry and Nutritional Sciences, University of Wisconsin-Madison, Madison, WI, United States

OUTLINE



## Introduction

Obesity is a metabolic disorder involving excessive amount of body fat. It is as a result of irregular energy intake and energy balance, changes in gut microbiota and improper diet with the influence of genetic makeup and environmental factors [1]. Obesity is becoming a global epidemic in both children and adults throughout the world. It was initially considered to be problem with a high-income countries but obesity is now on the increase in low- and middle-income countries. Obesity is associated with different kinds of complications,

Lipid Signaling and Metabolism. http://dx.doi.org/10.1016/B978-0-12-819404-1.00017-8

affecting people's quality of life with increasing death rates. It is linked with numerous ailments which include type 2 diabetes, certain cancers, and cardiovascular diseases (CVDs).

Diabetes mellitus is described as metabolic disease that is characterized by chronic hyperglycemia due to defects in insulin secretion (pancreas cannot produce sufficient insulin), insulin action (the body may not effectively use the insulin it produces), or both [2]. It is among the most chronic metabolic disorders, affecting adult population in Africa and the world at large. Continuous increase in blood glucose in the body may result in much serious damage to most of the body's systems, especially the nerves and blood vessels. In diabetes, chronic hyperglycemia is linked with long-term damage, dysfunction, and organs failure, especially the eyes, kidneys, nerves, heart, and blood vessels [3]. The increased risk of developing a number of serious life-threatening health problems in diabetic patients results into higher cost of medical care, reduced quality of life and high mortality [4]. Diabetes mellitus is majorly classified as Type 1 or insulin-dependent diabetes mellitus or Type 2 or noninsulin-dependent diabetes mellitus. Type 1 diabetes is caused by autoimmune β-cell destruction that leads to absolute insulin deficiency and type 2 diabetes ranges from predominantly insulin resistance with relative insulin deficiency to predominantly an insulin secretory defect with insulin resistance [3].

Lipids are the fourth major group of molecules found in all cells with a common feature of low solubility in water. In general, lipids are crucial molecules of the cell and perform three major biological functions in the cell. The first function is that lipid molecules which are in the form of lipid bilayers are essential components of biological membranes. Second, lipids containing hydrocarbon chains serves as energy stores. Thirdly, many intra- and intercellular signaling events involve lipid molecules [5]. Furthermore, this class of biomolecules includes the main constituent of a large tissue (adipose tissue), membrane components, hormones, vitamins, and a plethora of other substances possessing important biological properties [6]. However, lipids that serve as energy storage are associated with pathogenesis of obesity and type 2 diabetes. Furthermore, interferences in the signaling pathways that lead to the intracellular communication of the cell can also lead to accumulation of unwanted molecules in the cell. For example, in diabetes, the intracellular communication in the insulin signaling pathway such as TNF-alpha malfunctions leading to the accumulation of sugar in the bloodstream. Furthermore, accumulation of lipid species such as ceramides, long chain fatty acids, and diacylglycerol are also known to interfere with the same insulin signaling pathways. Consistently, research has demonstrated that excess and accumulation of various lipids in the cell contributes to the pathogenesis of metabolic disorders such as obesity and type 2 diabetes [7]. The increase in the scourge of obesity has led to extensive pleas for consistent monitoring of changes in overweight and obesity prevalence in all populations [8].

Diet and habitual activities are methods that are mostly recommended to manage both diabetes and obesity with positive outcomes at clinical level. Conventional drugs are available for the management of obesity and diabetes widely used in first world countries with often reported side effects and expensive to purchase. On the contrary, in Africa where people are in low income ranges, depend heavily on herbal medicines such as *Moringa* and others as a first line of treatment. The mechanisms by which the plants and their bioactive compounds exert their potential therapeutic effects is poorly understood necessitating further studies to provide more detailed mechanisms of action especially on the rise of obesity and diabetes in Africa which might provide a much better and safe option.

This review provides an insight on the roles of lipid metabolism in obesity and diabetes especially in Africa. The potential therapeutic effects of plant-derived compounds in the treatment of these metabolic diseases are discussed. Potential mechanisms of action of the plant bioactive compounds such as targeting $Ca^{2+}$/calmodulin-dependent protein kinase II (CaMK II) and stearoyl-CoA desaturase (SCD) as avenues for therapeutic approach in the management of obesity and diabetes are also enumerated in the chapter. This chapter therefore gives some insights on the potential therapeutic nature of plant-derived compounds which may be better treatment than conventional drugs as line of treatment on the face of the scourge of obesity and diabetes worldwide if much attention is given to them.

# Obesity and diabetes

## Global status of obesity and diabetes

Globally, diabetes has risen from 4.7 % in 1980 to about 8.5% in 2014 of adults aged 18 years and older suffered from diabetes. In 2016, diabetes was the direct cause of 1.6 million deaths worldwide and in 2012, high blood glucose was the cause of another 2.2 million deaths [9]. According to IDF report in 2013, the global prevalence of diabetes in adults (20–79 years old) was 8.3% (382 million people), 14 million more men than women (198 million men vs 184 million women) and the majority between the ages 40 and 59 years with expected increase in number beyond 592 million by 2035 [10]. In 2010, overweight and obesity were estimated to cause about up to 3.4 million deaths, 3.9% of years of life lost, and 3.8% of disability-adjusted life-years globally. According to World Health Organization, more than 1.9 billion adults aged 18 years and older were overweight in 2016 and of these, over 650 million adults were obese in 2016. Also, 9% of adults aged 18 years and over (39% of men and 40% of women) were overweight which indicates that about 13% of the world's adult population (11% of men and 15% of women) were obese in 2016 [11].

## Obesity and diabetes on the rise in sub-Saharan Africa

Sub-Sahara African countries were initially plagued with diseases associated with malnutrition and infectious disease such as malaria, tuberculosis, and many others. However, presently, these countries have to deal with the additional challenges posed by the high and rising burden of noncommunicable diseases such as diabetes and its major risk factor which is obesity [12,13]. This new trend has been linked to the changing demographic profile together with rapid urbanization and changing lifestyles in both rural and urban settings in Africa [12,13]. The lifestyle changes in the recent century have contributed significantly to the rise in noncommunicable diseases such as diabetes and obesity. The increase in the prevalence of these diseases are due in part, to modern high calories diet and reduced physical inactivity [13].

In South Africa, rapid urbanization is leading to the consumption of more westernized diets comprising mainly energy-dense processed foods containing high amounts of fats and carbohydrates [14]. Furthermore, out-of-home foods referring to takeaway and fast foods have turn out to be popular in recent decades and research has demonstrated that they are a key driver in the increasing levels of overweight and obesity due to their unfavorable nutritional content [15]. Studies have shown that most of the processed foods contain high calories which play a role in stimulating the obesity epidemic.

In today's fast-paced lifestyle, technology has advanced to a point where it reduces levels of physical activity in daily routines, especially for people with sedentary occupations [16]. In the past, African populations performed high levels of physical activity through "hunter gather" activities in the form of occupational and transportation-related activities, with few opportunities for leisure time activity or sedentary behavior [17]. Urbanization and its effect on lifestyle behaviors has reduced physical activity amongst African population, about 20.9% of the African population in 2010 was classified as physical inactive, with women (24.4%) having higher levels of physical inactivity than men (17.3%) [17]. In South Africa, about one in three adolescents watch television for more than 3 h daily and more than one in two adult women and one in three adult men are either overweight or obese due to reduced physical activity [18]. In sub-Saharan Africa, physical activity in urban areas is commonly poor compared to rural areas. This is due to the activities performed by rural dwellers, for example, they are mostly involved in high intensity activity including farming and long distance walking [17]. These changes in life styles in Africa from a "hunter-gather" activities to modern sedentary lifestyle have led to many reported cases of diabetes and obesity incidences not only in South Africa but the whole of sub-Saharan Africa and the rest of the world.

## Obesity and lipid metabolism

Obesity as previously described is characterized by increased body weight and abnormal development of adipose tissue with excessive lipid storage, accompanied by a dramatic dysregulation of the endocrine function of adipose tissue [19,20]. There are multiple factors that play a crucial role in the pathogenesis of obesity, namely genetics, socioeconomic status, environment and individual life style [21]. Over 50% of South African adult women and 30% of adult men are either overweight or obese. More than one in two adult women and one in three adult man are either overweight or obese [18]. Furthermore, recent estimates demonstrated that obesity currently affects more than 600 million people worldwide [22]. Obesity is a major risk factor for the development of type 2 diabetes, CVDs, cancer, disability, hypertension, and stroke [23,21].

Obesity results from an energy imbalance between caloric intake and caloric expenditure or insufficient physical activity [20]. Dietary approaches to limit fat intake are commonly prescribed to achieve the hypo caloric conditions necessary for weight loss. But dietary fat restriction is often accompanied by increased carbohydrate intake, which can dramatically increase endogenous fatty acid synthesis depending upon carbohydrate composition [24]. Since both dietary and endogenously synthesized fatty acids contribute to the whole-body fatty acid pool, obesity can therefore result from increased hepatic de novo lipogenesis [24]. Elucidating the mechanisms of lipid-dependent coordination of metabolism promises invaluable insights into the understanding of metabolic diseases and may contribute to the development of a new generation of preventative and therapeutic approaches.

## Diabetes and lipid metabolism

Diabetes mellitus is a chronic metabolic disease characterized by blood glucose level higher than normal resulting in hyperglycemia. This disease is associated with impaired insulin signaling pathway either through insulin resistance or autoimmune distraction of insulin production by the β-cells in the pancreas [25]. Furthermore, there are two types of diabetes mellitus namely type-1 diabetes (T1D) also known as insulin-dependent and type 2 diabetes (T2D) also known as insulin-dependent [26]. Type 2 diabetes mellitus is a major cause of morbidity and mortality worldwide, and the prevalence is set to increase dramatically over the coming decades. It constitutes about 90% of all diabetes cases. Therefore, understanding the metabolic pathways that lead to type 2 diabetes is therefore an important healthcare objective [27].

Research has reported that, the link between obesity and type 2 diabetes is through insulin resistance [7,28]. Insulin resistance is a condition by which the liver, skeletal muscle, and adipose tissue do not properly respond to the signal sent out by the hormone insulin. These tissues are metabolically active and their impaired insulin action is called systemic insulin resistance. Systemic insulin resistance differs amongst the tissues [29]. In skeletal muscle, insulin resistance is characterized by a decrease in glucose transport and a reduction in muscle glycogen synthesis in response to circulating insulin [29,30]. Insulin sensitivity has been reported to decrease in myocytes obtained from obese individuals or cultured myocytes in the presence of adipocytes-derived lipids [29]. Most studies use saturated fatty acid called palmitic acid to induce insulin resistance in myocytes [31]. The use of saturated fatty acids to induce insulin resistance in myocytes further confirms the concept that accumulation of excess lipids or their metabolic derivatives lead to decrease in insulin signaling in skeletal muscle [29].

In the liver, insulin resistance is selective in that insulin fails to inhibit gluconeogenesis, but continues to encourage fatty acid synthesis. Thus, the point at which insulin signaling is disrupted in obesity is downstream of insulin receptor activation [29]. It has also been reported that insulin resistance is a result of accumulation of intracellular lipids metabolites such as fatty acyl CoAs and diacylglycerol in liver tissues [27]. Moreover, accumulation of intracellular lipid metabolites activate a serine kinase cascade involving activated protein kinase C (PKC-ε), leading to decreased tyrosine phosphorylation of insulin receptor substrate (IRS-2), a key mediator of insulin action in the liver [27]. From these findings, accumulation of intracellular lipid metabolites shows to be the main driver to stimulate insulin resistance not obesity per se.

In adipose tissue, insulin resistance is demonstrated as impaired insulin-stimulated glucose transport, as well as impaired inhibition of lipolysis [29]. In a normal condition, the action of insulin to lower blood glucose levels results from suppression of hepatic glucose production and increased glucose uptake into muscle and adipose tissue [32]. Furthermore, insulin promotes adipocytes triglyceride stores by a number of mechanisms, including stimulating triglyceride synthesis (lipogenesis) and inhibiting lipolysis. Insulin also increases the uptake of fatty acids derived from circulating lipoproteins by stimulating lipoprotein lipase activity in adipose tissue [32]. In a condition of insulin resistance, such mechanisms are known to be impaired due to obesity (Fig. 17.1).

## Role of $Ca^{2+}$/calmodulin-dependent protein kinase II (CaMKII) in obesity and diabetes

CaMKII is a serine/threonine specific protein kinase which is a multicomplex enzyme that regulates many cellular functions beneficial to individuals including alleviation of symptoms

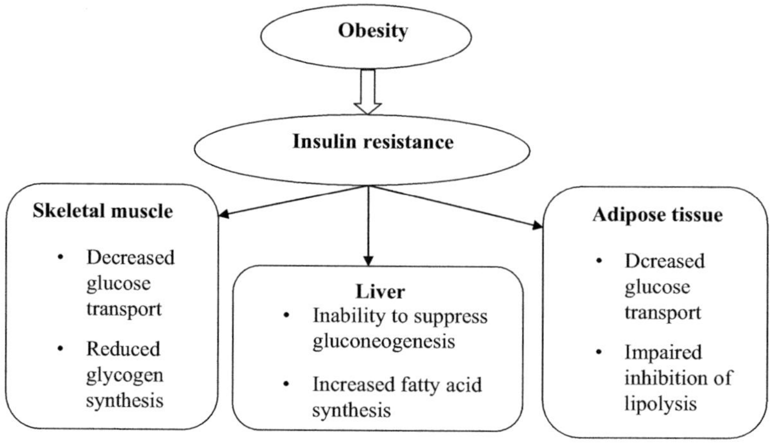


**FIGURE 17.1** **Diagram showing the relation between obesity and insulin resistance.**

associated with diabetes and obesity. CaMKII is a widely studied multicomplex enzyme in the brain, adipose, liver, and skeletal muscle mostly in its association with memory and metabolic diseases such as obesity and diabetes. In obesity and diabetes, CaMKII regulate the cell by controlling the mitochondria, a cell powerhouse, resulting in altering the ATP levels from organic substrates such as lipids and carbohydrates. Furthermore, CaMKII also regulates glucose transport through the increase in GLUT4, the major glucose transporter in the cell, which is seen as a therapeutic target in the design of pharmacological intervention in the management of diabetes [33].

In our laboratory, we have shown that CaMKII activation through exercise could help to regulate lipids and genes associated with obesity and type 2 diabetes. Our study showed that CaMKII activation by exercise increased the expression of carnitine palmitoyltransferase (CPT-1) gene in rat skeletal muscle [7]. Lipid metabolism is regulated by nuclear respiratory factor (NRF-1), a transcriptional factor that regulates genes that are responsible for lipid oxidation and this is controlled by a set of mitochondrial enzymes which include CPT-1 [7]. CPT-1 is an enzyme that controls the transport of long chain fatty acids across the mitochondrial membrane and results in the production of ATP. We further investigated the modulatory roles of exercise-induced CaMKII activation on the levels of adiposomes and ATP concentration [34].

We have also shown that oleic acid and palmitoleic acid levels were increased following exercise-induced CaMKII activity [34]. It has been reported that palmitoleic acid has beneficial effect on hyperglycemia, hypertriglyceridemia and body weight, which may help to improve glucose and lipid metabolism [35]. Studies have also shown that palmitoleic acid increased glucose transport into skeletal muscle cells which is beneficial in reducing diabetes through a process mediated by up-regulation of glucose transporter 4 (GLUT4) [36,37]. In our research group, it has also been shown that CaMKII activation by exercise increased the expression of GLUT4 gene in rat skeletal muscle [34]. GLUT4 is a glucose transporter responsible for insulin-regulated glucose transport into the cells. Upregulation of GLUT4 expression alleviates hyperglycemia and diabetes. It can be deduced that CaMKII activation could provide an avenue for new therapeutic approach in the management of metabolic diseases which include type 2 diabetes and obesity.

### *Stearoyl-CoA desaturase in obesity and diabetes*

There are many enzymes involved in the pathway of lipid synthesis and of interest is the SCD. SCD is a central enzyme in lipid metabolism that catalyzes the delta 9-desaturation of the saturated fatty acids (SAFs) palmitate (16:0) and stearate (18:0) to the monounsaturated fatty acids (MUFAs) palmitoleate (16:1n7) and oleate (18:1n9), respectively [37a–40]. These MUFAs, mainly oleate, are predominant components of cellular and circulating free fatty acids, triglycerides, cholesterol esters, sphingolipids, and phospholipids. At first glance, SCD would be considered a housekeeping enzyme because it synthesizes oleate, a well-known MUFA that is abundant in many dietary sources. However, numerous studies have shown that SCD is a very highly regulated enzyme that features in many physiological processes suggesting that endogenously synthesized MUFA may have signaling properties that regulate metabolism. A proper ratio of SFA to MUFA contributes to membrane fluidity, and oleate has also been implicated as a mediator of signal transduction, cellular differentiation, and metabolic homeostasis [38,41].

SCD expression is very sensitive to multiple dietary (i.e., glucose, fructose, cholesterol, saturated fat) and hormonal factors (i.e., insulin, leptin) [37a,42]. Such changes in phenotypes have never been achieved by knocking out other SCD isoforms (SCD2, SCD3, and SCD4). SCD activity due to SCD1 expression is elevated in obese and insulin resistant states, suggesting that excess MUFA synthesis may contribute to the development of these disease states [37a,38]. In the absence of SCD activity, as it occurred in SCD1$^{-/-}$ mice, endogenous MUFA production is decreased, resulting in reduced lipid accumulation due to decreased fatty acid synthesis and increased fatty acid oxidation. The metabolic alterations in SCD1$^{-/-}$ mice resulted in a lean phenotype, providing protection from diet-induced obesity and insulin resistance [40,42,43]. This indicates that proper endogenous MUFA synthesis via SCD1 is an important metabolic control point influencing the cellular decision between energy storage and catabolism. Monounsaturated fatty acids generated via stearoyl CoA desaturase-1 regulate endocannabinoid signaling system by acting as endogenous inhibitors of fatty acid amide hydrolase [44].

Although numerous studies have shown SCD1 deficiency to elicit favorable metabolic changes, other studies have also shown that SCD1 deficiency is not protective against the pathophysiological consequences of certain lipotoxic insults [43,45]. Thus, both excessive and insufficient MUFA synthesis may be causative for disease depending upon the environmental or genetic stimulus. Since SCD11$^{-/-}$ mice have reduced capacity for MUFA synthesis, these mice are presumed to have an increased dietary requirement for unsaturated fat. We reported that mice with a hepatic-specific inactivation of SCD1 (LKO) maintained on a high carbohydrate diet for 10 days had decreased sterol regulatory element binding protein-1c (SREBP-1c) and carbohydrate response element binding protein (ChREBP)-mediated de novo fatty acid synthesis [46]. Cholesterol synthesis was not affected. In addition, the LKO mice developed a complex phenotype involving hypoglycemia and decreased hepatic glycogen accumulation compared with control mice [46]. These phenotypes were largely prevented, independent of food intake, by supplementing the diet with dietary oleate but not stearate, indicating that an increase in the stearate:oleate ratio induces these metabolic responses. However, the molecular mechanisms by which the change in the stearate:oleate ratio reduces hepatic de novo lipogenesis and adiposity are not fully understood.

### *Plant-derived compounds in alleviation of obesity and diabetes*

Many patients in Africa use medicinal plants as the first-line treatment for many chronic medical conditions including metabolic syndrome related disorders which include obesity and diabetes. This is due to lack of access to conventional health care which is expensive for a larger part of the African population and thus, contributes to the persistent use of herbal medicines [47]. Africa is rich in the diversity of the normal flora as such rich in herbal medicine. For example, one of such herbal medicines widely used is *Moringa oleifera* Lam. *M. oleifera* Lam is a multipurpose medicinal plant that is well known for its numerous pharmacological activities in Africa and Asia folkloric medicine [48]. The leaves of *M. oleifera* are eaten in both raw and cooked form in African countries, such as Ghana, Ethiopia, Nigeria, and Malawi and it believed to have antidiabetic and antioxidant potentials [48]. However, the mechanisms of action of herbal medicines in Africa are not well known, and hence, more mechanistic studies to understand how herbal medicines exert their antidiabetic effects are necessary. The naturally occurring bioactive compounds that are present in plants (often referred to as phytochemicals) are known to be responsible for the health promoting effects of medicinal plants. Examples of such plant-derived compounds that are involved in the alleviation of obesity and diabetes are discussed in the following section.

## Curcumin in obesity and diabetes

The management of obesity has focused on dietary and lifestyle modifications such as restricting caloric intake and increasing physical activity [49]. An area that has recently aroused considerable research interest is investigating the potential role of spices, particularly the Asian spice turmeric for combating obesity [49]. Therefore, one such bioactive compound that our research group is presently working on is curcumin. Curcumin is an active ingredient of the dietary spice turmeric and is extracted from the rhizomes of *Curcuma longa*, a plant in the ginger family [50–52].

Evidence suggests that curcumin regulates lipid metabolism pathways which play a central role in the development of obesity and its complications [49]. Curcumin increases the adiponectin expression in inflammation-related obesity. It was pointed out that the adiponectin production, which increases due to effect of curcumin, may have a positive effect against obesity by decreasing NF-kB activity [52]. A study by Kuo and coworkers showed that obese mice fed with curcumin had significant weight loss and significantly reduced serum triglycerides (TG) levels.

Diabetic mice models were also employed to investigate curcumin's ability as an antidiabetic agent. In a streptozotocin-induced diabetic mouse model, administration of curcumin markedly reversed glucose intolerance, hyperglycemia and hypoinsulinemia [50,53]. Furthermore, in high fat diet induced obesity and insulin resistance in mice, its oral administration proved to be effective in reducing glucose intolerance [50]. Curcumin also modulated several important molecular targets [53]. In some of the molecular targets, curcumin has been found to be involved in the activation of liver enzymes associated with gluconeogenic, glycolysis, and lipid metabolic processes [50]. In addition, it was found to induce peroxisome

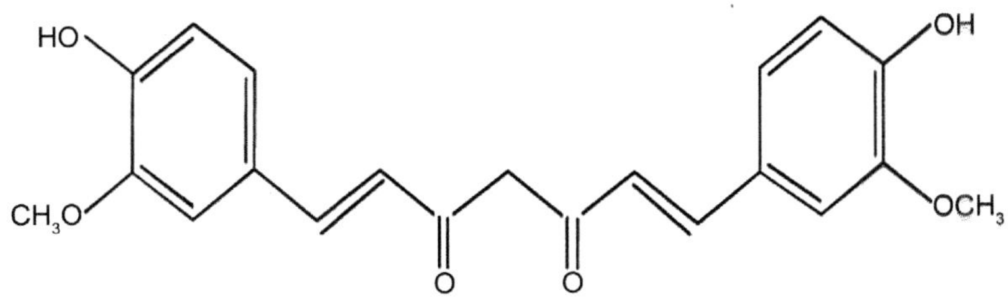

FIGURE 17.2 **The chemical structure of curcumin [55].**

proliferator-activated receptor gamma (PPAR-γ) activation and elevate the levels of plasma insulin activity [53]. A study conducted on type 2 diabetic KK-Ay mice found that curcumin suppressed increased blood glucose level via PPAR-γ activation [52]. In our research group, we have recently shown that long-term neonatal administration of curcumin protected against nonalcoholic steatohepatitis in high-fructose-fed adolescent rats [54]. In all these studies curcumin alleviates some of the symptoms of both diabetes and obesity. One of the means by which curcumin alleviates these diseases is through influencing the lipid profiles of cells especially in reducing the levels of those lipids that are known to increase the symptoms of both diabetes and obesity (Fig. 17.2).

## Oleanolic acid in fructose-induced neonatal metabolic derangements

Fructose involvement in diabetes and obesity has been described well with conflicting results on the subject. Worldwide, a large proportion of processed foods, canned foods, and carbonated beverages contain high fructose corn syrup or other fructose-derived sweeteners [56]. As such, high fructose corn syrup today is believed to be the most commonly consumed sugar [57]. In animal, when there is an oversupply of dietary carbohydrate, the excess carbohydrate is converted into triglycerides. This involves the synthesis of fatty acids from acetyl-CoA and esterification of fatty acids in the production of triglycerides, a process called lipogenesis [5]. This suggests that high consumption of dietary carbohydrate, that is, fructose may play a significant role in the development of obesity.

Oleanolic acid is a pentacyclic triterpenoid that is found in plants (abundant in oleaceae family such as the olive plant) and exists in nature as a free acid or as an aglycone of triterpenoid saponins [58]. The presence of oleanolic acid in many fruits and vegetables such as olive leaves, mistletoe sprouts, grapes, cloves, and pomegranate flowers has provided a wide range of pharmacological and biochemical effects including antihypolipidemic and hypoglycemic effects [59]. Oleanolic acid has been shown to ameliorate obesity-associated insulin resistance and hyperlipidemia [59]. The effect of neonatal oral administration of oleanolic acid on the levels of some biochemical parameters in the plasma such as cholesterol, insulin, glucose, triglycerides as well as insulin resistance (HOMA-IR) and glucose tolerance in high fructose-fed rats has been shown in our laboratory [60]. The results revealed that treatment with oleanolic acid during a critical window of developmental plasticity in rats protected against the development of fructose diet-induced health outcomes associated with metabolic dysfunction in male and female Sprague Dawley rats. We have also shown that high fructose-feeding suppressed the expression of some genes that are associated with diabetes and

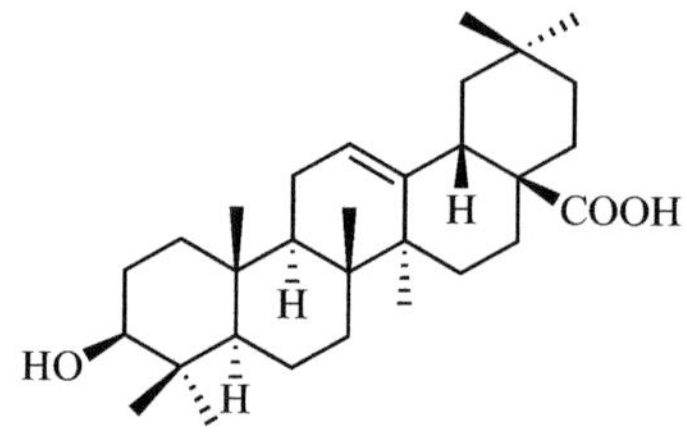

FIGURE 17.3 **The chemical structure of oleanolic acid [58].**

obesity such as *glut-4, nrf-1,* and *cpt-1* genes, and increased the expression of *acc-1* gene in rats. However, treatment with oleanolic acid was able to improve the expression of these genes [61]. Acetyl CoA carboxylase, which has two isoforms ACC-1 and ACC-2, is a major enzyme in energy balance that controls the synthesis of malonyl-CoA, an allosteric inhibitor of CPT-1 [62]. The down regulation of ACC-1 and increased expression of CPT-1 leads to decrease in lipid accumulation and obesity [63]. In another study conducted by our research group, the levels of inflammatory cytokines such as TNF-α and IL-6 were increased due to high fructose consumption as inflammation plays a critical role in the pathogenesis of T2D and obesity [64]. The levels of the inflammatory cytokines were reduced following treatment of fructose-induced metabolic derangements with oleanolic acid [64]. It has been reported that oleanolic acid has the ability to improve insulin action, inhibit gluconeogenesis and promote glucose utilization [58]. Oleanolic acid has strongly shown to be a promising therapeutic agent for the treatment of diabetes and obesity by affecting lipids profiles in the cell by reducing lipids that are associated to cause these conditions in the body (Fig. 17.3).

## Conclusion and future perspectives

Obesity and diabetes are critical metabolic conditions that have attracted much research interest as they place heavy burdens on health delivery systems worldwide and now at epidemic levels. Medicinal plants have been used from ages as a remedy for many pathologies for generations to generations over the centuries without much of scientific evidence on their mechanisms of action. Several studies on the modulatory role of medicinal plants and their bioactive compounds on lipid levels in the blood and tissues as well as genes/enzymes associated with lipid metabolism have shown promising findings in the management of obesity and diabetes. The degree in which medicinal plants exert its properties in alleviating the symptoms of obesity and diabetes has been very comparable to pharmacological drugs. Although pharmacological drugs are widely used in Europe and other first world countries, several side effects have been reported in their use compared to the natural remedies broadly used in Africa, with less reported side effects. This promising finding in the use of medicinal plant extracts or other natural components may be more beneficial than medicines currently and widely used pharmacological drugs as they have less side effects. Further research needs to be conducted on these herbal medicines to give more detailed mechanisms through which they exert their modulatory effects in alleviating these metabolic diseases. Future studies may focus more on SCD and CaMKII as therapeutic targets in the management of obesity and diabetes.

# References

[1] Sivamaruthi BS, Kesika P, Suganthy N, Chaiyasut C. A review on role of microbiome in obesity and antiobesity properties of probiotic supplements. BioMed Res Int 2019;2019:3291367. 20 pp.

[2] Kharroubi AT, Darwish HM. Diabetes mellitus: the epidemic of the century. World J Diabetes 2015;6(6):850–67.

[3] American Diabetes Association. Diagnosis and classification of diabetes mellitus. Diabetes Care 2014;37(Suppl. 1):S81–90.

[4] IDF DIABETES ALAS. 2017. Epidemiology & research. (Diabetic Complication-Chapter 5) pg 82-95.

[5] Voet D, Voet JG, Pratt CW. Principles of biochemistry. Wiley; 2008.

[6] Mougios V. Exercise biochemistry. Human Kinetics Publishers; 2019.

[7] Joseph JS, Ayeleso AO, Mukwevho E. Role of exercise-induced calmodulin protein kinase (CAMK) II activation in the regulation of omega-6 fatty acids and lipid metabolism genes in rat skeletal muscle. Physiol Res 2017;66.

[8] Ng M, Fleming T, Robinson M, Thomson B, Graetz N, Margono C, et al. Global, regional, and national prevalence of overweight and obesity in children and adults during 1980-2013: a systematic analysis for the Global Burden of Disease Study 2013. Lancet 2014;384:766–81.

[9] Ogurtsova K, da Rocha Fernandes J, Huang Y, Linnenkamp U, Guariguata L, Cho NH, et al. IDF Diabetes Atlas: global estimates for the prevalence of diabetes for 2015 and 2040. Diabetes Res Clin Pract 2017;128:40–50.

[10] Poirier P, Giles TD, Bray GA, Hong Y, Stern JS, Pi-Sunyer FX, et al. Obesity and cardiovascular disease: pathophysiology, evaluation, and effect of weight loss. Circulation 2006;113:898–918.

[11] World Health Organization. Obesity and overweight. . https://www.who.int/news-room/fact-sheets/detail/obesity-and-overweight.

[12] Mbanya JC, Assah FK, Saji J, Atanga EN. Obesity and type 2 diabetes in sub-Sahara Africa. Curr Diabetes Rep 2014;14:501.

[13] Price AJ, Crampin AC, Amberbir A, Kayuni-Chihana N, Musicha C, Tafatatha T, et al. Prevalence of obesity, hypertension, and diabetes, and cascade of care in sub-Saharan Africa: a cross-sectional, population-based study in rural and urban Malawi. Lancet Diabetes Endocrinol 2018;6:208–22.

[14] Nienaber-Rousseau C, Sotunde O, Ukegbu P, Myburgh P, Wright H, Havemann-Nel L, et al. Socio-demographic and lifestyle factors predict 5-year changes in adiposity among a group of black South African adults. Int J Environ Res Public Health 2017;14:1089.

[15] Janssen HG, Davies IG, Richardson LD, Stevenson L. Determinants of takeaway and fast food consumption: a narrative review. Nutr Res Rev 2018;31:16–34.

[16] Junco JJ, Slaga TJ. Prevention of obesity, diabetes, and cancer with lifestyle intervention strategies. In: Carkeet C, Grann K, Randolph RK, editors. Phytochemicals: Health Promotion and Therapeutic Potential. CRC Press; 2012. 113, [chapter 7].

[17] Goedecke JH, Mtintsilana A, Dlamini SN, Kengne AP. Type 2 diabetes mellitus in African women. Diabetes Res Clin Pract 2017;123:87–96.

[18] Lambert E, Kolbe-Alexander T. Innovative strategies targeting obesity and non-communicable diseases in South Africa: what can we learn from the private healthcare sector? Obes Rev 2013;14:141–9.

[19] Gary-Bobo M, Elachouri G, Gallas JF, Janiak P, Marini P, Ravinet-Trillou C, et al. Rimonabant reduces obesity-associated hepatic steatosis and features of metabolic syndrome in obese Zucker fa/fa rats. Hepatology 2007;46:122–9.

[20] Ranjani H, Mehreen T, Pradeepa R, Anjana RM, Garg R, Anand K, et al. Epidemiology of childhood overweight & obesity in India: a systematic review. Indian J Med Res 2016;143:160.

[21] Smith KB, Smith MS. Obesity statistics. Primary Care 2016;43:121–35.

[22] Verma S, Hussain ME. Obesity and diabetes: an update. Diabetes Metab Syndr 2017;11:73–9.

[23] Weisberg SP, Leibel R, Tortoriello DV. Dietary curcumin significantly improves obesity-associated inflammation and diabetes in mouse models of diabesity. Endocrinology 2008;149:3549–58.

[24] Flowers MT, Ntambi JM. Stearoyl-CoA desaturase and its relation to high-carbohydrate diets and obesity. Biochim Biophys Acta 2009;1791:85–91.

[25] Alismail H, Jin S. Microenvironmental stimuli for proliferation of functional islet β-cells. Cell Biosci 2014;4:12.

[26] Kim MS, Hur HJ, Kwon DY, Hwang J-T. Tangeretin stimulates glucose uptake via regulation of AMPK signaling pathways in C2C12 myotubes and improves glucose tolerance in high-fat diet-induced obese mice. Mol Cell Endocrinol 2012;358:127–34.

[27] Petersen KF, Shulman GI. Etiology of insulin resistance. Am J Med 2006;119:S10–6.

[28] Saltiel AR, Kahn CR. Insulin signalling and the regulation of glucose and lipid metabolism. Nature 2001;414:799.

[29] Hardy OT, Czech MP, Corvera S. What causes the insulin resistance underlying obesity? Curr Opin Endocrinol Diabetes Obes 2012;19:81.

[30] Ter Horst KW, Schene MR, Holman R, Romijn JA, Serlie MJ. Effect of fructose consumption on insulin sensitivity in nondiabetic subjects: a systematic review and meta-analysis of diet-intervention trials 2. Am J Clin Nutr 2016;104:1562–76.

[31] Zhou QG, Hou FF, Guo ZJ, Liang M, Wang GB, Zhang X. 1,25-Dihydroxyvitamin D improved the free fatty-acid-induced insulin resistance in cultured C2C12 cells. Diabetes Metab Res Rev 2008;24:459–64.

[32] Kahn BB, Flier JS. Obesity and insulin resistance. J Clin Invest 2000;106:473–81.

[33] Mukwevho E, Kohn TA, Lang D, Nyatia E, Smith J, Ojuka EO. Caffeine induces hyperacetylation of histones at the MEF2 site on the Glut4 promoter and increases MEF2A binding to the site via a CaMK-dependent mechanism. Am J Physiol Endocrinol Metab 2008;294:E582–8.

[34] Mukwevho E, joseph JS. Calmodulin dependent protein kinase II activation by exercise regulates saturated & unsaturated fatty acids and improves some metabolic syndrome markers. Life Sci 2014;111:53–61.

[35] Yang Z-H, Miyahara H, Hatanaka A. Chronic administration of palmitoleic acid reduces insulin resistance and hepatic lipid accumulation in KK-A y Mice with genetic type 2 diabetes. Lipids Health Dis 2011;10:120.

[36] Ojuka EO, Jones TE, Nolte LA, Chen M, Wamhoff BR, Sturek M, et al. Regulation of GLUT4 biogenesis in muscle: evidence for involvement of AMPK and $Ca^{2+}$. Am J Physiol Endocrinol Metab 2002;282:E1008–13.

[37] Ojuka EO, Jones TE, Han D-H, Chen M, Holloszy JO. Raising $Ca^{2+}$ in L6 myotubes mimics effects of exercise on mitochondrial biogenesis in muscle. FASEB J 2003;17:675–81.

[37a] Hulver MW, JR Berggren, Carper MJ, Miyazaki M, JM Ntambi, Hoffman EP, Thyfault JP, Stevens R, Dohm GL, Houmard JA, Muoio DM. Elevated stearoyl-CoA desaturase-1 expression in skeletal muscle contributes to abnormal fatty acid partitioning in obese humans. *Cell Metab* 2005;**4**:251–61.

[38] Miyazaki M, Kim Y-C, Gray-Keller MP, Attie AD, Ntambi JM. The biosynthesis of hepatic cholesterol esters and triglycerides is impaired in mice with a disruption of the gene for stearoyl-CoA desaturase 1. J Biol Chem 2000;275:30132–8.

[39] Jiang G, Li Z, Liu F, Ellsworth K, Dallas-Yang Q, Wu M, et al. Prevention of obesity in mice by antisense oligonucleotide inhibitors of stearoyl-CoA desaturase-1. J Clin Invest 2005;115:1030–8.

[40] Sampath H, Flowers MT, Liu X, Paton CM, Sullivan R, Chu K, et al. Skin-specific deletion of stearoyl-CoA desaturase-1 alters skin lipid composition and protects mice from high fat diet-induced obesity. J Biol Chem 2009;284:19961–73.

[41] Cohen P, Ntambi JM, Friedman JM. Stearoyl-CoA desaturase-1 and the metabolic syndrome. Curr Drug Targets Immune Endocr Metabol Disord 2003;3:271–80.

[42] Ntambi JM, Miyazaki M, Stoehr JP, Lan H, Kendziorski CM, Yandell BS, et al. Loss of stearoyl-CoA desaturase-1 function protects mice against adiposity. Proc Natl Acad Sci 2002;99:11482–6.

[43] Flowers JB, Rabaglia ME, Schueler KL, Flowers MT, Lan H, Keller MP, et al. Loss of stearoyl-CoA desaturase-1 improves insulin sensitivity in lean mice but worsens diabetes in leptin-deficient obese mice. Diabetes 2007;56:1228–39.

[44] Liu J, Cinar R, Xiong K, Godlewski G, Jourdan T, Lin Y, et al. Monounsaturated fatty acids generated via stearoyl CoA desaturase-1 are endogenous inhibitors of fatty acid amide hydrolase. Proc Natl Acad Sci 2013;110:18832–7.

[45] Macdonald ML, van Eck M, Hildebrand RB, Wong BW, Bissada N, Ruddle P, et al. Despite antiatherogenic metabolic characteristics, SCD1-deficient mice have increased inflammation and atherosclerosis. Arterioscler Thromb Vas Biol 2009;29:341–7.

[46] Miyazaki M, Flowers MT, Sampath H, Chu K, Otzelberger C, Liu X, et al. Hepatic stearoyl-CoA desaturase-1 deficiency protects mice from carbohydrate-induced adiposity and hepatic steatosis. Cell Metab 2007;6:484–96.

[47] Baldé N, Youla A, Baldé M, Kaké A, Diallo M, Baldé M, et al. Herbal medicine and treatment of diabetes in Africa: an example from Guinea. Diabetes Metab 2006;32:171–5.

[48] Atolani O, Olorundare OE, Anoka AN, Osin AO, Biliaminu S. Antioxidant, proteinase inhibitory and membrane stabilization potentials of *Moringa oleifera* seed oil. FABAD J Pharm Sci 2018;43:1–13.

[49] Alappat L, Awad AB. Curcumin and obesity: evidence and mechanisms. Nutr Rev 2010;68:729–38.

[50] Ghosh S, Banerjee S, Sil PC. The beneficial role of curcumin on inflammation, diabetes and neurodegenerative disease: a recent update. Food Chem Toxicol 2015;83:111–24.

[51] Gokce EC, Kahveci R, Gokce A, Sargon MF, Kisa U, Aksoy N, et al. Curcumin attenuates inflammation, oxidative stress, and ultrastructural damage induced by spinal cord ischemia–reperfusion injury in rats. J Stroke Cerebrovasc Dis 2016;25:1196–207.

[52] Kocaadam B, Şanlier N. Curcumin, an active component of turmeric (*Curcuma longa*), and its effects on health. Crit Rev Food Sci Nutr 2017;57:2889–95.

[53] Shishodia S, Sethi G, Aggarwal BB. Curcumin: getting back to the roots. Ann N Y Acad Sci 2005;1056:206–17.

[54] Ibrahim KG, Chivandi E, Nkomozepi P, Matumba MG, Mukwevho E, Erlwanger KH. The long-term protective effects of neonatal administration of curcumin against nonalcoholic steatohepatitis in high-fructose-fed adolescent rats. Physiol Rep 2019;7:e14032.

[55] Kawamori T, Lubet R, Steele VE, Kelloff GJ, Kaskey RB, Rao CV, et al. Chemopreventive effect of curcumin, a naturally occurring anti-inflammatory agent, during the promotion/progression stages of colon cancer. Cancer research 1999;59:597–601.

[56] Saad AF, Dickerson J, Kechichian TB, Yin H, Gamble P, Salazar A, et al. High-fructose diet in pregnancy leads to fetal programming of hypertension, insulin resistance, and obesity in adult offspring. Am J Obstet Gynecol 2016;215:378e1–6.

[57] Softic S, Cohen DE, Kahn CR. Role of dietary fructose and hepatic de novo lipogenesis in fatty liver disease. Dig Dis Sci 2016;61:1282–93.

[58] Ayeleso T, Matumba M, Mukwevho E. Oleanolic acid and its derivatives: biological activities and therapeutic potential in chronic diseases. Molecules 2017;22:1915.

[59] de Melo CL, Queiroz MGR, Fonseca SG, Bizerra AM, Lemos TL, Melo TS, et al. Oleanolic acid, a natural triterpenoid improves blood glucose tolerance in normal mice and ameliorates visceral obesity in mice fed a high-fat diet. Chem Biol Interact 2010;185:59–65.

[60] Nyakudya TT, Mukwevho E, Erlwanger KH. The protective effect of neonatal oral administration of oleanolic acid against the subsequent development of fructose-induced metabolic dysfunction in male and female rats. Nutr Metab 2018;15:82.

[61] Molepo M, Ayeleso A, Nyakudya T, Erlwanger K, Mukwevho E. A study on neonatal intake of oleanolic acid and metformin in rats (Rattus norvegicus) with metabolic dysfunction: implications on lipid metabolism and glucose transport. Molecules 2018;23:2528.

[62] Zang Y, Wang T, Xie W, Wang-Fischer YL, Getty L, Han J, et al. Regulation of acetyl CoA carboxylase and carnitine palmitoyl transferase-1 in rat adipocytes. Obes Res 2005;13:1530–9.

[63] Kinfe HH, Belay YH, Joseph JS, Mukwevho E. Evaluation of the Influence of thiosemicarbazone–triazole hybrids on genes implicated in lipid oxidation and accumulation as potential anti-obesity agents. Bioorg Med Chem Lett 2013;23:5275–8.

[64] Matumba MG, Ayeleso AO, Nyakudya T, Erlwanger K, Chegou NN, Mukwevho E. Long-term impact of neonatal intake of oleanolic acid on the expression of AMP-activated protein kinase, adiponectin and inflammatory cytokines in rats fed with a high fructose diet. Nutrients 2019;11:226.

CHAPTER

# 18

# Lipid metabolic features of skeletal muscle in pathological and physiological conditions

Yura Son[a], Chad M. Paton[a,b]

[a]Department of Foods & Nutrition, University of Georgia, Athens, GA, United States;
[b]Department of Food Science & Technology, University of Georgia, Athens, GA, United States

OUTLINE



Lipid Signaling and Metabolism. http://dx.doi.org/10.1016/B978-0-12-819404-1.00018-X

## Lipid metabolic pathway in skeletal muscle

Triglycerides (TGs) are carried in blood by chylomicrons and very low-density lipoproteins (VLDLs), coming from diet or liver, respectively. Plasma TGs are hydrolyzed into free fatty acids (FFAs) and monoacylglycerol by lipoprotein lipase (LPL), which is located on the capillary endothelial membrane. In addition, FFAs are also derived from adipose tissue through the hydrolysis of stored TGs during lipolysis. These FFAs form a complex with albumin in plasma and they are transported into tissues where the rate of FFA transport is dependent on several factors, including the extracellular FA concentration, FFA type, and tissue metabolic demand [1,2]. The transport of FFAs can occur by simple diffusion; however, it is primarily mediated by membrane-associated fatty acid-binding proteins, also called FFA transporters [3].

## Transport of FAs

Three proteins have been identified that participate in FFA transport across the muscle cell membrane (Fig. 18.1). They are (1) fatty acid translocase/cluster of differentiation 36 (FAT/CD36); (2) fatty acid transport proteins (FATPs); and (3) plasma membrane-associated fatty acid-binding protein (FABPpm). Each transporter has different characteristics and effectiveness, and it is believed that all three participate in a coordinated fashion to regulate FFA entry into skeletal muscle. While it has been shown that genetic deletion of one or more of these three proteins is insufficient to prevent FFA uptake, the rate of transport is clearly and consistently diminished by their (individual) absence [4–6].

### FAT/CD36

FAT/CD36 is located within an intracellular pool in skeletal muscle and is translocated from vesicles to the cell membrane upon stimulation, such as with muscle contraction or insulin signaling. In addition to cell membrane associated FAT/CD36, it is also found associated with the mitochondrial membrane and it regulates the transport of FAs into mitochondria. Inhibiting mitochondrial FAT/CD36 almost completely blocks mitochondrial lipid oxidation in skeletal muscle [7]. FAT/CD36 knockout mice showed decreases in FFA uptake of approximately 40%–75% in skeletal muscle [4]. When FAT/CD36 was re-expressed, the rate of FFA utilization significantly increased compared to FAT/CD36 knockout mice, and plasma FFA concentration was similar to the levels in wild-type mice [4]. Conversely, FFA uptake in FAT/CD36 overexpressed mice was increased 3-fold [8–10].

### FATP

The FATP group of transporters includes six proteins (FATP1-6), and each FATP has different expression patterns, tissue distribution, location, and FFA transport capacity. Among the six isoforms, FATP1 and FATP4 are the major isoforms expressed in skeletal muscle [11]. The exact mechanism of FATP activity in skeletal muscle remains unclear, but it is generally accepted that they mediate FFA translocation from the extracellular to intracellular compartments. Thus, it has been suggested that FATP facilitates FFA transport by functioning as

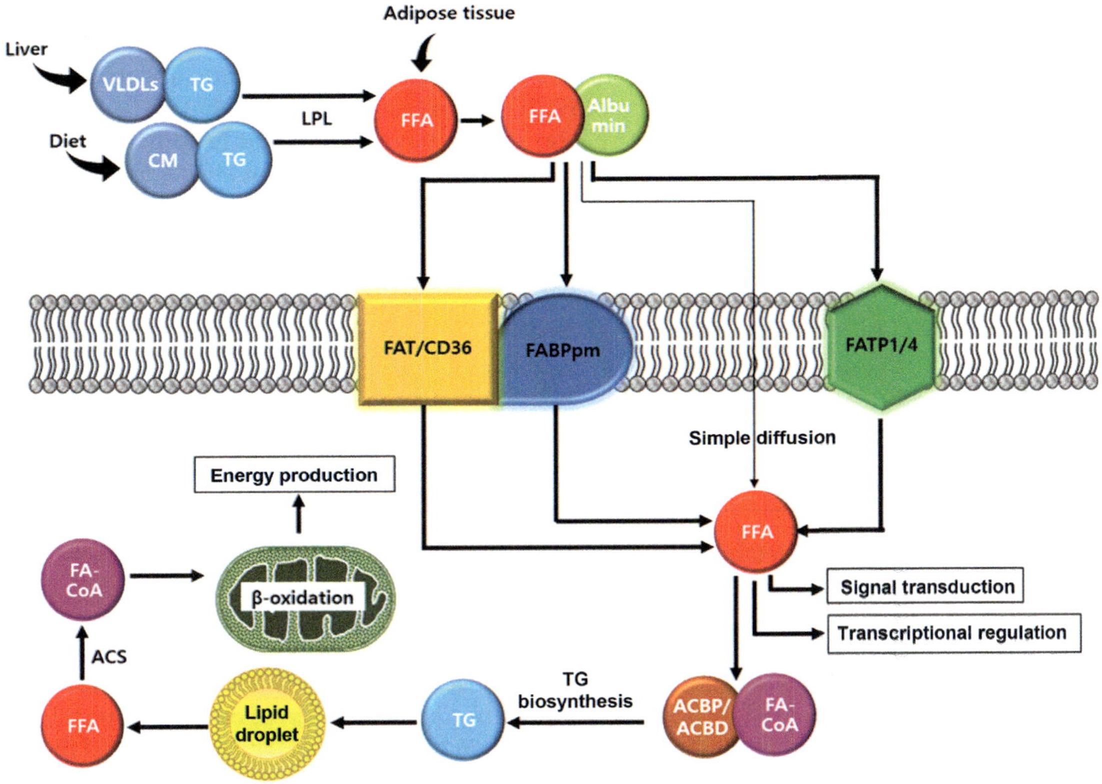


FIGURE 18.1 **Transport of FFAs and their roles in skeletal muscle.** Plasma TGs from liver or diet are hydrolyzed into FFAs by LPL. FFAs are also derived from adipose tissue. These FFAs form a complex with albumin in plasma, and they are transported into skeletal muscle through simple diffusion or FFA transporters including FAT/CD36, FABPpm, and FATP1/4. Transported FFAs mediate signal transduction and regulate gene expression. In addition, FFAs undergo TG biosynthesis, and newly formed TGs are stored in lipid droplets. Stored TGs are hydrolyzed into FFAs, and then FFAs are esterified to form FA-CoA by ACS. Fatty acyl-CoA enters into mitochondria and produce energy through β-oxidation. *VLDLs*, very low-density lipoproteins; *CM*, chylomicrons; *TG*, triglycerides; *LPL*, lipoprotein lipase; *FFA*, free fatty acids; *FA-CoA*, fatty acyl-CoA; *ACS*, ACBP, fatty acyl-CoA binding protein; *ACBD*, fatty acyl-CoA binding domain protein; *ACS*, long chain fatty acyl-CoA synthethase.

a membrane-associated CoA-synthetase [12]. In skeletal muscle isolated from rats, overexpression of FATP1 and FATP4 increased palmitate transport rates by 25% and 38%, respectively [13].

## FABPpm

FABPpm is located on the plasma membrane in skeletal muscle. Antibody-mediated inhibition of FABPpm significantly decreased palmitate uptake in rat skeletal muscles, suggesting that FABPpm, at least in part, contributes to overall FFA uptake [14,15]. However, FFA transport was observed when both FAT/CD36 and FABPpm or only FAT/CD36 are expressed [16]. Overexpression of FABPpm in rats was not sufficient to meet parallel rises in palmitate transport and metabolism [14], which suggests that it may not be the primary means of transport. Another study also showed that overexpression of FABPpm increased the rates of FFA transport by +79%, but the increase in activity was significantly less than would be expected given the increased amount of FABPpm protein, suggesting that FABPpm activity may be regulated [15]. Interestingly, several studies have shown that FABPpm stimulates lipid oxidation, even though FABPpm by itself does not have significant effects on FFA transport. Compared to FAT/CD36 or FATP, FABPpm has been shown to be the least efficient FFA transporter because the rate of palmitate transport in FABPpm overexpressed mice was two times lower than in FAT/CD36 or FATP transfected mice [17]. Even though FABPpm is a less efficient transport of FFAs, the rates of fat oxidation in mice overexpressing FABPpm were more than three times those in FATP transfected mice [17]. Thus, it is suggested that the rates of FFA transport and oxidation are not 1:1, and FABPpm may provide additional effects on FFA transport and oxidation in addition to membrane transport [13].

## Regulation of FFA transport

The levels of muscle metabolic activity play an important role in the regulation of FFA transport. It has been shown that preventing voluntary muscle contraction by the denervation of the sciatic nerve significantly deceases FFA transport (−39%), and that this decrease was associated with reduced levels of plasmalemmal FAT/CD36 (−24%) and FABPpm (−28%) [18]. The study also found that chronic muscle stimulation induces FFA transport through FAT/CD36 and FABPpm. Seven days of chronic stimulation of hind limb muscles in rats increased the plasmalemmal levels of FAT/CD36 by 42% and FABPpm by 13% [18–20]. Furthermore, chronic contraction-induced FFA transport resulted in increased expression of FAT/CD36 and a subcellular redistribution of FAT/CD36 [8].

Exercise training enhances skeletal muscle use of FFAs as energy sources. Aerobic exercise training (AET) significantly improves lipid oxidation in skeletal muscle by increasing expression of genes of FFA transport across both plasma and mitochondrial membranes in skeletal muscle [18]. AET increases the relative proportion of energy from β-oxidation by increasing protein expression of membrane bound lipid binding proteins such as FAT/CD36, FATP4, and FABPpm. Since CD36 is involved in translocation of FFAs into mitochondria as well, AET facilitates the translocation of CD36 from the intracellular compartment to the inner mitochondrial membrane in both human and rodent muscle. Previous studies have demonstrated the importance of CD36 during exercise by utilizing the CD36 knockout/overexpression model [9,21,22]. In exercise trained CD36 knockout mice, FFA transport (−41%), lipid oxidation (−37%), and exercise duration (-44%) decreased, whereas carbohydrate metabolism was increased 2-fold in skeletal muscle [9]. Taken together, it is evident that the activity of CD36 can be elevated by AET, and exercise induced CD36 activity increases lipid oxidation.

In addition to chronic adaptation, there are acute factors that have been shown to induce membrane translocation of FFA transporters. Muscle contraction promotes the movement of FFAs into muscle cells because of the increased demand of lipid oxidation and insulin stimulation [8]. This acutely regulates the rates of FFA transport (i.e., within minutes) by increasing the translocation of CD36. In rats, electrically stimulated muscles showed significantly increased palmitate uptake. The rates of palmitate uptake increased by approximately 50%

within 3 min compared to the resting status, and returned to normal levels within 5 min post-contraction [23]. There is a positive correlation between muscle contraction rate and palmitate uptake where it was estimated that contraction (i.e., energy expenditure) increased 28% and 56% at 20 and 40 tetani/min, respectively. The muscle contraction induced-increased palmitate uptake is associated with elevated FAT/CD36 expression. It is important to note that contraction for as little as 30 min was sufficient to result in increased palmitate uptake by 65% and FAT/CD36 expression by 40% [23].

Within their role of regulating uptake, FFA transporters control the rate of FFA entry into the cell. They are located on the cell membrane as well as in intracellular vesicles, thus, it is reasonable to conclude that shuttling within the cell provides a mechanism to control the amount of membrane-associated transporter in order to control or regulate the rate of FFA uptake (much like GLUT4 translocation). Translocating FFA transporters to the sarcolemma increases FFA uptake, whereas removal from the sarcolemma to endosome deceases FFA uptake [24]. Muscle contraction has been shown to upregulate FFA transport by reversibly translocating FAT/CD36 and FABPpm from an intracellular endosomal compartment to the sarcolemma [20]. Muscle contraction also activates the AMP-activated protein kinase (AMPK) pathway, which induces the translocation of FAT/CD36 and FABPpm. However, AMPK is not only activated by muscle contraction, but also from energy deficit (i.e., fasting or calorie restriction) or pharmacologically (i.e., AICAR or oligomycin). Regardless of the cause for AMPK activation, it is a potent facilitator of the translocation of FAT/CD36 and FABPpm to the membrane.

As with insulin-stimulated glucose disposal, the rate of post-prandial FFA transport can be increased in response to insulin through activation of phosphoinositide-3-kinase (PI3K). When insulin binds to its receptor, it activates PI3K. Activated PI3K phosphorylates protein kinase B (Akt), which induces reversible translocation of CD36 from endosomes to the plasma membrane and GLUT4 translocation. A previous study demonstrated that insulin signaling in rat skeletal muscle increased plasma membrane FAT/CD36 [3]. Insulin signaling induces both CD36 and GLUT4 recruitment to sarcolemma at the same time, resulting in the increases in both glucose and FFA uptake [25]. Despite the fact that FAT/CD36 and GLUT4 translocation occur simultaneously, the intermediate signals between the two pathways are still unclear. Once FFAs are transported into skeletal muscle, the fate of the FFAs depends on metabolic demands and lipid availability within the skeletal muscle. FFAs are used as substrates in energy production, for phospholipid synthesis as membrane constituents, and for signaling mediators [26].

## Signal transduction mediator

FFAs are known to mediate signal transduction through several methods, including protein kinase C (PKC), phospholipase C (PLC) mediated $Ca^{2+}$ release, and as hormones (e.g., prostaglandins, leukotrienes, and thromboxanes). PKC is an enzyme that transduces various signals activating lipid hydrolysis [27]. Several PKC isoforms are expressed in skeletal muscle including conventional ($\alpha$, $\beta$1, $\beta$2, and $\gamma$), novel ($\delta$, $\varepsilon$, $\theta$, and $\eta$), and atypical ($\zeta$ and $\iota/\lambda$) PKCs [10]. Across most isoforms, fatty acyl-CoA has been shown to increase PKC activity in skeletal muscle, which then phosphorylates insulin receptors with a subsequent decrease in their tyrosine kinase activity. This inhibits the activation of PI3K, causing the inhibition of GLUT4 translocation to the cell surface which is linked to the reduction of insulin sensitivity [28]. High-fat diets induce insulin resistance by activating PKCθ in skeletal muscle. In PKCθ knockout mice, high-fat diets did not induce insulin resistance, possibly indicating that PKCθ impairs insulin signaling [29].

PLC hydrolyzes the membrane phospholipid PIP2, thereby forming inositol trisphosphate (IP3) and diacylglycerol (DAG). DAG activates PKC to induce the release of intracellular $Ca^{2+}$ from the sarcoplasmic reticulum. IP3 also binds to the IP3 receptor (IP3R) and $Ca^{2+}$ is released through activated IP3R from the sarcoplasmic reticulum. The released $Ca^{2+}$ induces muscle contraction by participating in excitation-contraction coupling, which regulates muscle contraction by stabilizing the actin–myosin cross-bridge [30–32]. Released $Ca^{2+}$ enters into mitochondria and regulates the tricarboxylic acid (TCA) cycle by activating pyruvate dehydrogenase phosphatase (protein phosphatase 2C) to reverse the effect of pyruvate dehydrogenase kinase, thereby activating pyruvate dehydrogenase complex and glycolysis [33]. 20-Carbon omega-3 or omega-6 polyunsaturated fatty acids (PUFAs), such as arachidonic acid (AA) and eicosapentaenoic acid (EPA) are substrates for cyclooxygenase and lipoxygenase-mediated pathways to synthesize eicosanoids. Eicosanoids are signaling molecules that have various biological functions such as regulating inflammation, immune responses, pain perception, cell growth, and blood pressure. Eicosanoids consist of prostaglandins (PGs), thromboxanes (TXs), and leukotrienes (LTs). In the enzymatic pathway, PUFAs are modified by cyclooxygenases and/or lipoxygenase, forming the initial intermediates in biosynthesis [34]. TXs are also synthesized via the PG pathway by TX synthase [35]. PUFAs such as AA are converted into 5-hydroperoxyeicosatetraenoic acid (5-HPETE) by lipoxygenase, and the 5-HPETE forms leukotriene A4 (LTA4) by the same enzyme, lipoxygenase. LTA4 is unstable, but it forms stable LTs through the activity of leukotriene A4 hydrolase (LTB4) and by reacting with glutathione (LTC4, LTD4, and LTE4). It is generally thought that PUFAs can regulate skeletal muscle DNA transcriptional activity either directly or indirectly via their conversion to PG or LT [36–38].

## Transcriptional regulation of lipid metabolism

Metabolic activity in skeletal muscle is controlled by several factors, one of which is the expression of genes involved in lipid metabolism. While it is well-known that energy expenditure can be a driving force for increasing synthesis of ATP producing enzymes, influx of lipid into muscle can also directly impact the metabolic phenotype of the tissue [39]. In the former, much of the changes are seen in catabolic pathways, whereas with the latter, anabolic changes are observed [40,41]. FFAs regulate gene expression affecting lipogenesis in skeletal muscle, such as stearoyl-CoA desaturase-1 (SCD1) (−9 desaturase), acyl-CoA:diacylglycerol acyltransferase 1 (DGAT1), sterol regulatory element-binding protein-1c (SREBP-1c), acetyl-CoA carboxylase (ACC), fatty acid synthase (FAS), glycerol-3-phosphate acyltransferase (GPAT) and the −6 and −5 desaturases [42–45]. PUFAs, or their CoA esters that are produced from TG metabolism, are known to enter the nucleus in concert with the FABP5 lipid binding protein, to induce ligand-activated transcriptional activation of peroxisome proliferator-activated receptor δ (PPARδ). Additionally, PUFA-derived PGs or LTs are thought to be ligands for transcription factors, which bind to specific regions of DNA in promoter/enhancer regions to either suppress or activate gene expression [46].

## SREBPs

Sterol regulatory element-binding proteins (SREBPs) are transcription factors that regulate FFA and cholesterol biosynthesis in lipogenic tissues, and there is some evidence that they are active in skeletal muscle as well [47,48]. There are three major SREBP isoforms: SREBP-1a, -1c, and -2. SREBP-1a and -1c regulate the genes involved in FA biosynthesis. SREBP-1a is a prenatal isoform whose expression declines with postnatal development and is taken over by the SREBP-1c isoform. It is expressed as an inactive precursor that is post-translationally processed by phosphorylation, acetylation, and ubiquitination. It has been known that these post-translational modifications regulate the stability/transcriptional activity of SREBPs. The transcriptional activity of SREBPs is critically regulated by acetylation via P300/CBP acetyltransferase [44,49–51]. SREBP-2 is primarily involved in cholesterol biosynthesis and increases expression of genes encoding the LDL receptor, 3-hydroxy-3-methylglutaryl-CoA (HMG-CoA) reductase, HMG-CoA synthase, and others involved in cholesterol and lipoprotein generation.

Insulin is thought to be the most significant inducer of SREBP activity, however it is known that liver X receptor and FFAs can modify its expression, processing, and/or its DNA binding activity [52]. PUFAs activate the movement of cholesterol from the plasma membrane to the endoplasmic reticulum (ER) membrane. This increases the level of cholesterol in the ER membrane, decreasing the levels of SREBP in the nucleus. It has been found that PUFAs suppress the transcription of fatty acid synthesis genes by reducing the transcription of the SREBP-1 gene or the maturation of SREBP-1 protein [53,54].

## Nuclear factor κB

Oversupply of FFAs, either chronically or acutely, activate nuclear factor κB (NF-κB) to promote the transcription of proinflammatory genes which are associated with decreased insulin sensitivity [55]. NF-κB also plays a role in the regulation of muscle atrophy through an inflammation-dependent mechanism. NF-κB in skeletal muscle enhances the degradation of muscle protein, inflammation, fibrosis, and decreases the capacity for muscle regeneration after injury or atrophy [56]. Various fFAs activate NF-κB in skeletal muscle, with long-chain saturated fatty acids such as palmitate and stearate being the most potent inducers. One study suggested that there is a causal relationship between the activation of NF- κB and insulin resistance in skeletal muscle, where it was demonstrated that blocking NF-κB activation protected the skeletal muscle cells from palmitate-induced insulin resistance [57]. Interestingly, among PUFAs, AA (omega-6 fatty acid, 20:4n6) is capable of activating NF-κB at higher doses, however EPA (omega-3 fatty acid, 20:5n3) is not [58]. While it is known that AA metabolites are capable of inducing inflammation through the prostaglandin pathway, AA, or its 18-carbon precursor linoleic acid (18:2n6) is necessary but not sufficient to induce inflammation in vivo [59]. Similarly, EPA and its metabolites are generally regarded as anti-inflammatory, however regardless of FA composition (i.e., saturated fatty acid (SFA), monounsaturated fatty acid (MUFA), n3, or n6 PUFA), excess energy delivery to cells or tissues seems to be the factor that drives the pro-inflammatory response in skeletal muscle as well as other tissues.

## Liver X receptors

Liver X receptors (LXRs) are nuclear receptors that heterodimerize with retinoid X receptors (RXRs). LXRs regulate the genes involved in cholesterol and lipid metabolism such as SREBP-1c, FAS, and SCD-1, and they function in concert with SREBP-1. By activating SREBP-1c expression, LXRs increase de novo FA synthesis and serum TG levels [60]. Recently, it was found that LXRs also regulate glucose metabolism, participating in the regulation of GLUT4 gene expression [61]. Two isoforms have been found and both (LXRα and LXRβ) are expressed in skeletal muscle. Previous work has shown that saturated medium-chain FFAs induced the activation of LXRα, but long-chain FFAs had no effect on LXRα [62]. This might be because LXRα acts as a receptor for saturated FFAs or fatty acyl-CoA's of medium-chain length [62]. Activation of LXRs by LXR agonists, such as T0901317, increased the expression of FAT/CD36, GLUT1, GLUT4, SREBP-1c, PPARγ, carnitine palmitoyl transferase-Ib (CPT-Ib), uncoupling protein-2 (UCP-2), and uncoupling protein-3 (UCP-3) in human skeletal muscle cells [61]. As a result, the capacity for de novo lipid synthesis and β-oxidation would be expected to increase. While it may appear that these two pathways are in opposition to one another (i.e., catabolic and anabolic), they can actually be seen to compliment on another by virtue of their ability to store energy (anabolic) for future use when necessary in the muscle (catabolic). Under conditions of continual or regular turnover of stored lipid, such as with AET, this dual nature would prove beneficial. However, in the absence of increased energy expenditure, this process would be expected to induce muscle lipid accumulation and insulin resistance. Another study found that T090317 failed to induce LXR target genes in mice skeletal muscle, including ATP binding cassette transporter ABCA1, apolipoprotein E (apoE), SREBP-1c, and FAS in LXRα/β knockout mice indicating the specificity and importance of LXR in muscle lipid metabolism [63]. Finally, T0901317 increased the gene expression of ABCA1, apoE, SREBP-1c, SCD-1, and FAS in C2C12 murine myoblast cells [63], suggesting that LXRs play a critical role in regulating cholesterol and lipid metabolism in skeletal muscle.

## Retinoid X receptors

RXRs belong to the family of nuclear receptors that bind to retinoids, such as 9-*cis*-retinoic acid (9-*cis*-RA). However, it is unknown whether 9-*cis*-RA is the innate ligand for RXRs. Studies have shown that unsaturated fatty acids can activate RXRs [64,65]. Specifically, these studies detected unsaturated fatty acids such as linoleic, linolenic, and docosahexaenoic acids in the ligand-binding domain of RXRs [64,65]. Typically, RXR binds to its DNA response elements as half of a heterodimer with retinoic acid receptor (RAR) or peroxisome proliferator activated receptors (PPAR). Furthermore, RXR can cooperate with other hormone receptors such as thyroid hormone receptor, LXR, and the vitamin D3 receptor. Additionally, RXR can regulate transcriptional activity as a homodimer (RXR–RXR). There are several RXR isoforms, but RXRγ is the major isoform found in skeletal muscle [66,67]. In fact, RXRγ transcript expression has been shown to regulate myogenic differentiation and TG metabolism in skeletal muscle [68,69]. Activated RXRγ has been shown to induce myogenesis associated genes, such as paired box gene 3 (Pax3), paired box gene 7 (Pax7), myogenic factor 5 (Myf5), mesenchyme homeobox 2 (Meox2), and myoblast determination protein 1 (MyoD) [68,70,71]. RXRγ knockout mice showed decreased plasma TG levels during fasting (1.6-fold) and increased LPL activity (1.4-fold) compared to wild-type mice in skeletal muscle [69]. Furthermore, RXRγ deficient mice had lower weight gain and leptin levels than normal mice after 14 weeks on a high fat diet. This study suggested that RXRγ deficient mice were protected from weight gain due to increased LPL activity in skeletal muscle.

## Peroxisome proliferator activated receptors

PPARα and PPARδ are highly expressed in oxidative tissues such as brown adipose tissue, skeletal muscle, and cardiac muscle where they regulate lipid oxidation. They possess a ligand binding domain and a DNA binding domain to regulate transcriptional activity in a gene and fatty acid-dependent manner. PPARs form a heterodimer with RXR and bind to PPAR response elements (PPRE) in the promoter and/or gene body of their target genes. Historically, it was generally regarded that PPAR binding to the PPRE primarily resulted in transcriptional activation. However, it is now known that ligand binding to PPARs, most notably PPARδ, can function as an agonist, antagonist, or inverse agonist [72–74]. To add additional complexity, these functions can occur with or without ligand binding and appear to be isoform, gene, and tissue dependent [75]. It is also important to note that while there is overlap in PPARα and PPARδ target genes and pathways, the two are indispensable. In the absence of either isoform, there appears to be some compensatory activity and it was shown that PPARα knockout mice are still viable and fertile, albeit with reduced metabolic function [76]. However, PPARδ global deletion results in >50% embryonic lethality and diminished capacity in the viable offspring [77]. PPARγ is highly expressed in adipose tissue and liver where it regulates adipocyte differentiation and genes related to lipid storage [78]. PPARγ expression is increased in the skeletal muscle of obese, insulin resistant humans as the tissue shifts away from oxidation and more toward lipid deposition [79].

PPARα is a major regulator of fatty acid oxidative genes, as PPARα knockout mice showed reduced lipid oxidation rates during fasting. In addition, PPARα increased the expression of CPT-Ib and malonyl-CoA decarboxylase, which increased the rates of lipid oxidation in human skeletal muscle cells [80]. Another supporting study showed that PPARα mRNA levels were positively correlated with the mRNA levels of fatty acid metabolic genes, including CD36, CPT-Ib, uncoupling protein-2 (UCP-2), uncoupling protein-3 (UCP-3), and LPL in human skeletal muscle. However, the study showed that increased PPARα mRNA levels did not correlate with insulin sensitivity or BMI [81]. Therefore, this study suggested that PPARα regulates lipid metabolism in skeletal muscle, but not glucose metabolism. However, this concept may not be absolute because other studies have shown that a PPARα agonist (Tesaglitazar and WY14,643) decreased blood glucose and insulin levels in obese or T2D rodents [82,83]. Additionally, others have shown that PPARα knockout mice were not affected by high-fat diet induced-insulin resistance with respect to glucose and insulin tolerance [84]. Although it is clear that PPARα regulates lipid metabolism in skeletal muscle, its role in glucose metabolism is less clear.

For the last decade, it has been generally accepted that PPARδ regulates genes that control Lipid oxidation, mitochondrial biogenesis, and muscle fiber type switching. PPARδ transgenic mice showed enhanced FA transport and oxidation, and changes in muscle fiber types with a switch from glycolytic (type 2) to oxidative (type 1) muscle fibers [85–87]. Conversely, muscle specific PPARδ knockout mice showed impaired FA metabolism, and muscle fiber type switching from type 1 to type 2 [86,88]. While this fiber type switching phenomenon has not been observed in humans, there evidence to suggest that PPARδ agonists induce a more oxidative phenotype of skeletal muscle [89].

PPARδ activation enhances the expression of forkhead box O1 (Foxo1), a transcription factor involved in energy metabolism and metabolic adaptation in liver, adipose tissue, and muscle [90]. Foxo1 is a fasting-induced transcription factor that activates several genes, including pyruvate dehydrogenase kinase 4 (PDK4), CD36, and LPL, resulting in increases in lipid oxidation and glucose sparing. Recent advancements in the understanding of PPARδ's role in skeletal muscle suggest that is plays a primary role in suppressing glucose utilization and enhancing lipid oxidation [91]. With prolonged exercise, the result is a significant enhancement in aerobic exercise time to exhaustion. It should also be noted that at face value, it would seem that glucose sparing would be counterproductive when seeking to reduce hyperglycemia. However, skeletal muscle insulin resistance and the resulting hyperglycemia is more likely to be a result of increased lipid deposition in skeletal muscle; therefore, methods to increase lipid turnover and lipid oxidation should lead to a reduction in blood glucose levels. In light of these results, there is increasing interest in the role of PPARδ in enhancing muscle metabolic activity both as a therapeutic target and for exercise/physical performance outcomes.

PPARδ also increases peroxisome proliferator-activated receptor gamma co-activator 1-alpha (PGC-1α) expression which enhances mitochondrial biogenesis in skeletal muscle and increases in the proportion of type 1 muscle fibers. Importantly, PGC-1α also acts as a transcriptional co-activator of PPARδ and the RXR complex at the PPRE. In general, PGC-1α upregulates genes involved in FFA transport, such as FAT/CD36, FATP1 and 4, and FABPpm. In addition, PGC-1α co-activates nuclear respiratory factor (NRF) 1 and 2 as well as mitochondrial transcription factor A (Tfam), which increases mitochondrial DNA replication and transcription [92]. Taken together, the PGC-1α-PPARδ signaling pathway appears to be a master regulator of lipid metabolism and mitochondrial biogenesis in skeletal muscle.

## Intracellular fatty acyl-CoA synthesis

Fatty acyl-CoAs are oxidized in the mitochondria through the β-oxidation pathway. Therefore, FFAs in the cytosol of skeletal muscle must be activated by esterification to coenzyme A and then transported into the mitochondrial membrane for ATP production. The first step in the process involves the FFAs from plasma that are converted into activated fatty acyl-CoA by long chain fatty acyl-CoA synthetase (ACSL). Inhibition of ACSL causes a reduction of mitochondrial fatty acyl-CoA content and reduces the rate of lipid oxidation in skeletal muscle. Deficiency of ACSL1, one of the ACSL isoforms, results in a 60–85% reduction in β-oxidation in mice skeletal muscle, which results in increased dependence on glucose oxidation with subsequent lipid droplet accumulation. In addition, ACSL1 knockout mice showed decreased exercise capacity, measured as voluntary wheel running and grip strength [93]. Most studies investigated ACSL function in rodent liver with only a handful of studies that have assessed ACSL in skeletal muscle [94,95]. ACSL1, -3 and -6 are predominantly expressed in muscle [93].

Fasting for 48 h upregulated the mRNA expression of ACSL1 and downregulated the mRNA expression of ACSL6 in rats and re-feeding showed the opposite pattern. ACSL1 knockout mice showed impaired lipid oxidation and increased glucose metabolism [96] and ACSL6 mRNA expression was changed by nutritional status in skeletal muscle [97]. ACSL6 knock-down in rat skeletal muscle decreased lipid accumulation in lipid droplets, and increased mitochondrial oxidative capacity and lipid oxidation by activating the AMPK/PGC-1α pathway [97]. Interestingly, one study has found opposite results to those outlined above. The study found that fasting and chronic aerobic training results in decreases in ACSL6 mRNA expression, but acute lipid intake increases the expression [97]. Therefore, more studies are needed to understand the functional mechanism of ACSL6 in skeletal muscle and its role in physiological and pathological lipid metabolism. Unlike ACSL6, ACSL3 was not affected by nutritional changes [97] and the mechanism(s) controlling its regulation have not been fully elucidated. Based on the limited information to date, it is becoming clear that ACSL isoforms play multiple roles in regulating FA metabolism in skeletal muscle, however more studies are required, especially in humans.

Unbound fatty acyl-CoA rarely exists in the cytosol because most is bound to fatty acyl-CoA binding protein (ACBP). ACBP is the only bona fide fatty acyl-CoA binding protein known to exist, however, it has recently been shown that there are several other proteins that contain acyl-CoA binding sites as an internal domain (Table 18.1). These proteins are called acyl-CoA binding domain proteins (ACBDs), and they belong to the family of intracellular lipid binding proteins wherein ACBP is a member [98]. ACBP is expressed in a broad range of species, including plants, yeast, and humans, and it has a high sequence similarity, suggesting that ACBP's function is species-independent [99]. ACBP delivers fatty acyl-CoA for synthesis of phospholipids, glycerolipids, cholesterol esters, and TGs. Interestingly, the expression of ACBP differs according to skeletal muscle fiber types [100] where it showed the highest expression in soleus muscle (type 1) and the lowest expression in white gastrocnemius (type 2) in rats (soleus > red gastrocnemius > extensor digitorum longus > white gastrocnemius). This is not surprising as type 1 fibers are more dependent on β-oxidation and type 2 fibers are more dependent on glycolysis for energy production. In addition, obese rats showed higher fatty acyl-CoA/ACBP levels in all muscles compared to lean rats [100], likely due to increased lipid deposition in all muscle tissues. Thus, the fatty acyl-CoA/ACBP complex appears to be an important indicator of intracellular lipid metabolism and oxidative capacity in skeletal muscle of metabolically healthy organisms.

## Triglyceride synthesis

FFAs enter the cell and are immediately esterified with CoA. Next, the fatty acyl-CoAs are transported to the endoplasmic reticulum to undergo TG or phospholipid biosynthesis, after which they are stored in intracellular lipid droplets or membranes [105–107]. It is important to note that only after entering into the lipid droplet pool can TG be used for energy production in the mitochondria. After initial esterification, fatty acyl-CoA is converted into lysophosphatidic acid (LPA) by glycerol-3-phosphate acyltransferases (GPATs); this process is the rate-limiting step of TG synthesis. Then, acylglcerol-3-phosphate acyltransferases (AGPATs) convert LPA to phosphatidic acid (PA). Next, PA is converted into DAG by phosphatidic acid

**TABLE 18.1** Acyl-CoA binding domain proteins (ACBDs).

| | Location | Function |
|---|---|---|
| ACBD1 [101] | Cytosolic, and the nucleus of rat liver Cells | Pancreatic secretion, postprandial Release of cholecystokinin, and Mediator in corticotropin-dependent Adrenal steroidogenesis |
| ACBD2 [101] | Peroxisomes | Metabolism of unsaturated fatty acids |
| ACBD3 [101,102] | Golgi, cytoplasm, endoplasmic Reticulum, and mitochondria | Steroidogenesis, iron homeostasis, cell apoptosis, and neuronal Differentiation |
| ACBD4 [101] | Unknown | Acyl-CoA dependent lipid Metabolism |
| ACBD5 [101] | Neurospheres | Brain growth and development/Peroxisome-dependent lipid metabolic pathway |
| ACBD6 [101,103] | Bone marrow, spleen, placenta, cord Blood, circulating CD34(+), progenitors, and embryonic-like stem cells derived from placenta | Hematopoiesis and blood vessel development |
| ACBD7 [101,104] | Spleen, thymus, and brain | Participating in the hypothalamic leptin signaling pathway |

phosphatase (PAP). PA also participates in the synthesis of acidic phospholipids, phosphatidylglycerol, or phosphatidylinositol and cardiolipin. Finally, DGAT1 catalyzes DAG to TG [108]. DGAT1 knockout mice fed high fat diet showed significantly decreased expression of PPARα, δ, and γ genes and lipid oxidation associated genes including CD36, LPL, adipose triglyceride lipase (ATGL), PDK4, acyl-CoA oxidase (AOX), CPT-Ib, GLUT1, GLUT4, and PGC-1α versus wild-type mice fed a high-fat diet [109]. Interestingly, the same research team also found that overexpression of DGAT1 increased the amount of TG in skeletal muscle and decreased DAG and ceramide (i.e., reaction intermediates). At the same time, enhanced insulin sensitivity was observed in mice and differentiated C2C12 myocytes that overexpressed DGAT1 [108]. Therefore, it may be likely that DGAT1 plays a key role in the 'athlete's paradox' wherein endurance-trained athletes who have a greater oxidative capacity and insulin sensitivity, also have high intramyocellular lipid content. In endurance trained athletes, synthesized TGs are rapidly and routinely turned over due to habitual exercise training. This is not the same pattern as what is seen with high-fat diet induced muscle TG accumulation where muscle TG accumulates and is rarely turned over due to low energy expenditure and elevated energy consumption.

With the channeling of FFAs into TG synthesis prior to entry into β-oxidation, this process may enable some measure of control over entry into the mitochondria from an extracellular lipid supply. While this theory has been posited for several years, there is still insufficient data to conclusively determine the stepwise movement of FFAs/fatty acyl CoAs between intracellular pools. However, it is clear that FFA utilization and storage in skeletal muscle are tightly regulated processes that are designed to prevent excessive lipid accumulation into the mitochondria [110] wherein lipid droplets play a significant role in the regulation of lipid homeostasis in skeletal muscle.

## Fatty acid β-oxidation

The first step in the process of lipid oxidation starts with hydrolysis of stored TG in lipid droplets to DAG and FFA by ATGL. Next, DAG is further hydrolyzed by hormone sensitive lipase (HSL), forming monoacylglycerol (MAG) and FFA, and then MAG is hydrolyzed to glycerol and FFA by monoacylglycerol lipase (MAGL). These processes produce three FFAs that are oxidized in the mitochondria. However, they must first be esterified with CoA to form fatty acyl-CoA by long chain fatty acyl-CoA synthetase in the cytosol. Next, the fatty acyl-CoA is shuttled to the outer membrane of the mitochondria where CoA is exchanged with free carnitine via the enzyme CPT-Ib. This is the rate limiting step in β-oxidation and is regulated by CPT-Ib expression level, malonyl-CoA (from lipid biosynthesis), and carnitine availability. The activity of CPT-Ib transfers the fatty acyl from CoA to carnitine, thereby generating a fatty acyl carnitine and free CoA in the cytosol. Then, the acyl-carnitine is able to diffuse through the outer membrane and is then transported through the inner mitochondrial membrane to the matrix via carnitine acyl translocase in a concentration-dependent manner. Finally, the acyl carnitine is converted into a fatty acyl-CoA by carnitine acyl transferase-II (CPT-II) within the matrix, with free carnitine as a reaction product (the free carnitine is pumped out of the matrix in antiporter fashion coupled with acyl carnitine movement into the matrix). Once in the matrix, the acyl-CoA is free to undergo β-oxidation.

In previous in vivo studies, muscle CPT-Ib knockout mice displayed decreased lipid oxidation with concomitant increased lipid accumulation in skeletal muscle [111,112]. With high intensity exercise (~90% VO2 max), the majority of carnitine is found as fatty acyl-carnitine thereby creating a potentially rate-limiting factor in β-oxidation during high intensity exercise. However, caution concerning this effect should be used as this had not been demonstrated in vivo and is possible that even at low concentrations of free carnitine, there is sufficient transport of fatty acids across the mitochondrial membrane to maintain energy production during high intensity exercise. It is more likely that O2 delivery is a limiting factor in energy production at maximal exercise which confounds the question of whether or not the rate of carnitine recycling is limiting, versus the dependence on β-oxidation when O2 efficiency is limiting.

## Skeletal muscle fiber type-dependent lipid metabolism

Skeletal muscle fiber types can be broadly classified into three groups: type 1 (slow/oxidative, red color), type 2A (fast/oxidative and glycolytic, red color), and type 2B (fast/glycolytic, white color), although there appears to be a continuum of at least eight types between type 1 and type 2B in mice [113–116] and possibly seven in humans [117–119]. Each muscle fiber

TABLE 18.2 Histological and biochemical characteristics of skeletal muscle fibers [67,121].

| | Type 1 | Type 2A | Type 2B |
|---|---|---|---|
| Color | Red | Red | White |
| Contraction time | Slow | Fast | Very fast |
| Resistance to fatigue | High | Medium | Low |
| Energy generation | Oxidative | Oxidative and glycolytic | Glycolytic |
| Myoglobin content | High | Medium | Low |
| Mitochondria content | High | Medium | Low |
| Capillary density | High | Intermediate | Low |
| Force production | Low | High | Very high |
| Activity | Long-term aerobic | Long-term anaerobic | Short-term anaerobic |
| Exercise example | 10,000 m running | 400 m running | 100 m running |

type has different characteristics when it comes to metabolism (oxidative or glycolytic), contraction (myosin heavy chain (MHC) or ATPase isoforms), and fatigue resistance (Table 18.2). Type 1 muscle fibers show higher oxidative capacity and higher fatigue resistance compared to type 2A or type 2B muscle fibers. In addition, type 1 muscle fibers mostly use FFAs for energy production whereas type 2 muscle fibers rely on glycolytic metabolism using glucose as a substrate for energy production [120]. Therefore, various biological and biochemical characteristics of different muscle fiber types can contribute to their metabolic function.

It is known that FFA transport capacity is considerably different between type 1 and type 2 fibers [122]. In rodents, oxidative muscle fibers such as soleus and red gastrocnemius have significantly higher cytosolic FABP concentration compared to glycolytic white gastrocnemius (about 7-fold and 3-fold, respectively) [123]. Furthermore, mRNA and protein expression of FATP and FAT/CD36 in soleus were shown to be significantly higher than flexor digitorum brevis (FDB) [124]. In addition, it has been shown that type 1 fibers have greater rates of palmitate transport than type 2 fibers due to the higher availability of FFA transporters [116,124]. Another study found similar results in human skeletal muscle [121], although it is important to note that humans possess a far greater content of type 1 fibers than rodents [125]. In fact, human skeletal muscle is further distinguished histologically by a more uniform distribution of type 1 and -2A/B fibers within a given muscle compared to rodents [116,117]. However, given these species-dependent differences, the metabolic function within a given type of fiber remains consistent between species. Vistisen et al. observed the expression of FAT/CD36 in human vastus lateralis muscle via immunofluorescence staining method. Subjects had markedly higher FAT/CD36 expression in type 1 compared with type 2 fibers [126]. Another study reported that FABPpm protein content in red muscle fiber was greater than in white muscle, and the amount of FABPpm in red muscle was positively correlated with succinate dehydrogenase content but not in white muscle from rats [127]. Succinate dehydrogenase plays a significant role in energy production by participating in both the TCA cycle and the electron transport chain and is an indicator of oxidative capacity. Based on these facts, the degree of dependence on lipid metabolism in skeletal muscle depends on muscle fiber type, with type 1 muscle fiber possessing a higher lipid oxidation capacity and FFA transport potential [120].

Another factor that limits the rate of lipid oxidation is muscle mitochondrial density. Type 1 and type 2 muscle fibers have distinctly different amounts of mitochondria and subsequent capacity to oxidize FFAs. Mitochondrial density is vastly higher in type 1 fibers than type 2 fibers in all species assessed due in large part to the inherent oxidative properties of type 1 fibers [128,129]. It was observed that porcine red skeletal muscle had higher mitochondrial content (2-fold) compared to white muscle, though both types of muscle showed similar mitochondrial protein composition. In other words, mitochondria of type 1 and type 2 fibers are essentially the same, there is just more mitochondria in the former. Despite the similarity in protein content, the red muscle appeared to exhibit higher protein expression of the entire β-oxidation pathway and showed significantly higher maximal respiration at state 3 (ADP-stimulated respiration) of palmitoyl-carnitine + malate [130]. On the other hand, in the presence of glycerol-3-phosphate, maximal respiration at state 3 was 4–10 times higher in the mitochondria from type 2 fibers in rats and rabbits than type 1 fibers [131–134]. Glycerol-3-phosphate is a glycolytic intermediate and continues through the TCA cycle, which is highly active in both type 1 and type 2 fibers. Therefore, it is likely that type 2 fibers can produce more energy through glycolysis whereas type 2 fibers may be limited in their oxidative capacity. These data accurately illustrate the metabolic characteristics of type 1 and type 2 muscle fibers and in the light of these differences, caution should be used when employing rodent models to study human skeletal muscle lipid metabolism.

Similar to lipid oxidation, muscle fiber type impacts TG synthesis capacity as well. It has been shown that the rate of TG synthesis is higher in type 1 fibers than type 2 fibers [135], and when viewed in light of their substrate preference, it is not surprising that TG synthesis is positively correlated with β-oxidation capacity [136]. Mouse soleus muscle displayed higher mRNA expression of lipid synthesis genes compared to extensor digitorum longus muscle (EDL) [137], which is mostly composed of type 2 fibers. Additionally, soleus muscle intramyocellular lipid droplets were higher than in EDL muscle when the mice were fed a high-fat diet [137]. Type 1 fibers have greater capacity to utilize TG for energy and therefore, they possess a higher endogenous capacity for TG synthesis than type 2 fibers.

While the molecular mechanisms of skeletal muscle myogenesis and differentiation are well characterized, is not fully understood how muscle satellite (stem) cells are specified into fiber type. What is known is that various transcriptional co-activators and repressors are recruited during myogenesis to induce contractile properties and metabolic function specific to either type 1 or type 2 fibers. Myosin-2 (Myh2) activation by four and a half LIM domain protein 3 (FHL3) and cyclic AMP-response element binding protein-1 (CREB1) is known to induce type 2 fiber differentiation whereas MyoD and nuclear factor of activated T-cells, cytoplasmic 1 (NFATC1) activity on the myosin heavy chain β (Myh7) promoter induces type 1 differentiation [118]. In addition, the activation of PGC-1α induces mitochondrial biogenesis and respiration, and appears to be necessary and sufficient to induce fiber type switching in mice [138].

Mice over-expressing PGC-1α showed increased expression of oxidative fibers including myoglobin, MHC1, and MHC2x and decreased expression of glycolytic fiber genes (MHC2a, MHC2b, and CASQ-1) [139]. The expression of the genes involved in lipid oxidation such as cytochrome c oxidase 2 and 4, and citrate synthase were significantly higher in PGC-1α transgenic animals than wild-type animals [139]. PGC-1α is a transcriptional co-activator in skeletal muscle where it forms a complex along with PPARδ. As with PGC-1α transgenic mice, similar effects on fiber type and oxidative capacity were seen when PPARδ was constitutively in skeletal muscle [89]. Conversion of differentiated type 2 to type 1 fibers has not been observed in humans and is questionable in rodents, however it is generally accepted that myogenic differentiation of satellite cells and their subsequent fiber type selection can be impacted by environmental factors, such as diet, in both humans and rodents. While the biochemical mechanisms of skeletal muscle lipid metabolism and signaling has been well-studied in fully differentiated muscle cells, very little is known about how and when dietary fat impacts myogenesis.

## Angiopoietin-like proteins as mediators of integrative metabolism of lipids

Angiopoietin-like proteins (ANGPTLs) are a family of secreted proteins with diverse functions. Among the ANGPTLs, it is well established that ANGPTL3, ANGPTL4, and ANGPTL8 regulate lipid metabolism in various tissues [140]. ANGPTL3, ANGPTL4, and ANGPTL8 inhibit the activity of LPL which hydrolyzes TG into glycerol and FFAs in the capillary endothelium. ANGPTL3 and ANGPTL4 are composed of an N-terminal coiled-coil domain and a C-terminal fibrinogen-like domain, while ANGPTL8 consists of an N-terminal domain, but lacks the C-terminal domain. To inhibit LPL activity, ANGPTL3 and ANGPTL4 need to be cleaved to generate the active N-terminal domain, which reduces the activity of LPL by converting active LPL dimers into inactive monomers. ANGPTL8 does not need to be cleaved because it lacks a C-terminal domain. However, ANGPTL8 promotes ANGPTL3 cleavage, after which the activated ANGPTL3 binds to ANGPTL8, forming a complex to inhibit LPL activity [141–143].

Many studies have shown that ANGPTL3, ANGPTL4, and ANGPTL8 are important regulators of LPL. Mice overexpressing ANGPTL3 displayed hypertriglyceridemia, whereas ANGPTL3 knockout resulted in lower plasma TGs with increased LPL activity in mice [144,145]. Likewise, ANGPTL4 overexpression or injection of ANGPTL4 recombinant protein significantly elevated plasma TG levels and reduced LPL activity, while deletion of the gene lowered plasma TGs in mice by increasing LPL activity [145,146]. Overexpression of ANGPTL8 significantly elevated plasma TG levels [143,147]. However, unlike ANGPTL3 and 4, the absence of ANGPTL8 had no effect on LPL activity, despite its effects on plasma TG levels. It has been found that the LPL inhibitory effect of ANGPTL8 occurs only when ANGPTL3 is present [143].

ANGPTL3, -4, and -8 are differentially regulated by nutritional status in specific tissues (Fig. 18.2). Fasting induces the elevation of LPL activity in skeletal muscle to generate energy through lipolysis of very low-density lipoprotein (VLDL)-derived TG. Simultaneously, TG hydrolysis is decreased in WAT by reduced LPL activity with the tissue-specific regulation attributed to ANGPTL3–4–8 coordination [141]. In the fed (post-absorptive) state, LPL activity increases in white adipose tissue (WAT) to uptake chylomicrons and VLDL TGs from de novo lipogenesis (i.e., carbohydrate derived), while it decreases in skeletal muscle [148]. An ANGPTL3–4–8 model has been suggested to explain these tissue-specific patterns of regulation of LPL activity [149]. Based on the proposed model, ANGPTL3 is stable regardless of the nutritional states in tissues. During fasting, ANGPTL4 is selectively upregulated in WAT and inhibits LPL activity. This results in the localization of TGs to muscles with decreased

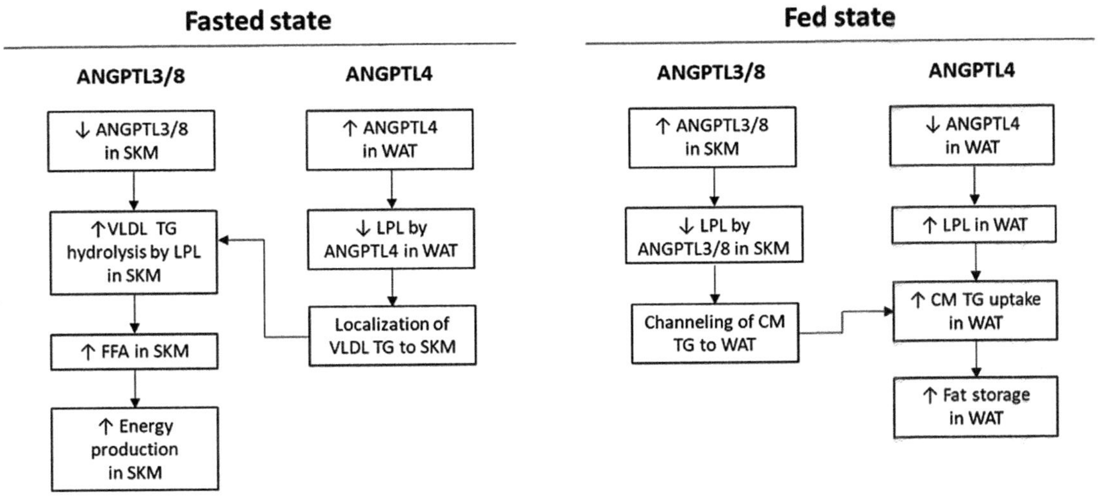


FIGURE 18.2 **The regulation of ANGPTL3-4-8 in response to nutritional status in a tissue-specific manner.** During the fasted state, ANGPTL4 is upregulated in WAT, resulting in decreased LPL activity and the localization of VLDL TGs to muscle. LPL hydrolyzes TGs into FFAs in the skeletal muscle endothelium, and hydrolyzed FFAs enter skeletal muscle. In the fed state, ANGPTL8 is increased in skeletal muscle, and the ANGPTL3-8 complex is activated to inhibit muscle LPL activity. Thus, TGs are redirected from skeletal muscle to WAT, and TG uptake increases in WAT. *CM*, chylomicron; *VLDL*, very low density lipoprotein; *WAT*, white adipose tissue; *TG*, triglycerides; *SKM*, skeletal muscle; *LPL*, lipoprotein lipase.

ANGPTL8 levels. Accordingly, TGs are hydrolyzed by LPL in the skeletal muscle endothelium, and hydrolyzed FFAs enter skeletal muscle for energy production. Conversely, during feeding, ANGPTL8 is selectively increased in skeletal muscle, and this increase activates the ANGPTL3–8 complex to inhibit LPL activity in skeletal muscle. As a result, TG uptake in WAT increases as a result of redirection of TG from skeletal muscles to WAT. Therefore, at present, it is believed that ANGPTL4 is a fasting-induced LPL inhibitor expressed largely in WAT and liver, whereas ANGPTL8 is a feeding-induced LPL inhibitor. However, despite the growing understanding of the ANGPTL3–4–8 model, there are still several limitations to fully describing their function.

## Significance of ANGPTL3/4/8 in skeletal muscle

ANGPTL3, ANGPTL4, and ANGPTL8 play important roles in the regulation of lipid metabolism by inhibiting LPL activity, which results in decreased plasma TG clearance. It is observed that ANGPTL3, ANGPTL4, and ANGPTL8 are all expressed in skeletal muscle, which is one of the largest metabolic tissues. Nevertheless, ANGPTL3, ANGPTL4, and ANGPTL8 activities in skeletal muscle have not been extensively investigated. It has been reported that dietary fat affects ANGPTL3, ANGPTL4, and ANGPTL8 responses in various tissues as well as their appearance in plasma [150,151]. Many studies have shown that FFAs activate the transcription and secretion of ANGPTL4 through PPARs. One study reported that short-chain fatty acids induce the activation of PPARγ but not PPARα or δ, resulting in the activation of ANGPTL4 in human colon cells [152]. Additionally, it was found that long-chain fatty acids activate PPARδ, but not PPARα or PPARγ, and the activated PPARδ strongly induces ANGPTL4 secretion in human myotubes [153]. One study investigated the effect of different FFAs on the expression of ANGPTL4 in rat hepatoma cells and found that docosahexaenoic acid (22:6n3) induced the highest ANGPTL4 expression (70-fold), followed by linoleic acid (18:2n6) (27-fold), and oleic acid (18:1n9) (15-fold) [153]. However, the patterns of plasma ANGPTL3, ANGPTL4, and ANGPTL8 levels did not match their associated tissue expression levels. A clinical study found that a 7-day, n3 PUFA-enriched diet, reduced plasma ANGPTL3, and -8 responses following high saturated fat meals in adult females that corresponded to a significant decrease in postprandial TGs after the diet [154]. However, there was no effect on plasma ANGPTL4, nor was there a postprandial relationship in males. Taken together, different FFAs seem to differentially affect ANGPTL3, ANGPTL4, and ANGPTL8 in a tissue-specific manner. However, the mechanisms of action underlying these responses are not yet clear and more studies in skeletal muscle are warranted.

## Summary and future directions

The increased prevalence of obesity related disorders has been attributed to an increase in fat content of skeletal muscle with a concomitant decrease in energy expenditure [155]. The deposition in lipids in skeletal muscle is not intrinsically detrimental to muscle metabolic function, as numerous studies have established elevated lipogenic activity and TG content of endurance exercisers [156–158]. Instead, it appears that high-fat diet induced changes in biological function occur in concert with decreased lipid turnover (i.e., TG hydrolysis and subsequent β-oxidation) which is observed as increased oxidative stress [159], muscle wasting, insulin resistance, and decreased muscle mass [160]. Furthermore, high-fat diet impairs lipid metabolism and mitochondrial function in skeletal muscle [161,162], suggesting that dietary fat could significantly affect skeletal muscle physiology.

Since overweight/obesity affects more than 30% of the world population, methods to reduce obesity and its related complications are needed to ensure a healthier population. We have found that skeletal muscle development is a key process that provides protection from obesity-related disorders, and the sooner in life muscle is created, the better the protection. Furthermore, the amount of muscle developed in the first 5 years of life plays a major role in the quality of life for the next 5–7 decades [163,164]. Skeletal muscle comprises approximately 40% of body mass and it is the largest metabolic organ in the human body [165]. It is the primary site for glucose clearance and is responsible for reducing blood glucose levels following high carbohydrate meals [166]. It is known that free fatty acids (FFAs) play a significant role in biological and physiological function in skeletal muscle [167]. However, their impact on skeletal muscle-specific metabolic function is far less studied relative to adipose tissue and liver. Emerging evidence suggests that high-fat diet induced increases in circulating FFAs may impair differentiation of muscle stem cells at various stages of life, ranging from gestation through extended aging. Given the fact that the Western Diet and lifestyle is unlikely to change, understanding how lipid metabolism is linked to skeletal muscle differentiation is crucial in order to promote healthy aging across the lifespan [168]. Therefore, future studies aimed at determining the impact of diet/lifestyle on muscle differentiation and the associated factors affecting metabolic function in muscle are needed. Caution should be used though when focusing on rodent models due to the impact of fiber-type differences compared to humans on metabolic function.

## References


[1] Spector AA. Plasma free fatty acid and lipoproteins as sources of polyunsaturated fatty acid for the brain. J Mol Neurosci 2001;16(2–3):159–65.

[2] Hamilton JA. Fatty acid transport: difficult or easy? J Lipid Res 1998;39(3):467–81.

[3] Watt MJ, Hoy AJ. Lipid metabolism in skeletal muscle: generation of adaptive and maladaptive intracellular signals for cellular function. Am J Physiol Endocrinol Metab 2011;302(11):E1315–28.

[4] Bonen A, et al. A null mutation in skeletal muscle FAT/CD36 reveals its essential role in insulin-and AICAR-stimulated fatty acid metabolism. Am J Physiol Endocrinol Metab 2007;292(6):E1740–9.

[5] Black PN, et al. Targeting the fatty acid transport proteins (FATP) to understand the mechanisms linking fatty acid transport to metabolism. Immunol Endocr Metab Agents Med Chem 2009;9(1):11–7.

[6] Storch J, Thumser AE. The fatty acid transport function of fatty acid-binding proteins. Biochim Biophys Acta 2000;1486(1):28–44.

[7] Holloway GP, et al. FAT/CD36-null mice reveal that mitochondrial FAT/CD36 is required to upregulate mitochondrial fatty acid oxidation in contracting muscle. Am J Physiol Regul Integr Comp Physiol 2009;297(4):R960–7.

[8] Holloway GP, et al. Fatty acid transport in skeletal muscle: role in energy provision and insulin resistance. Physiol Rev 2010;5(5):731–45.

[9] McFarlan JT, et al. In vivo, fatty acid translocase (CD36) critically regulates skeletal muscle fuel selection, exercise performance, and training-induced adaptation of fatty acid oxidation. J Biol Chem 2012;287(28): 23502–16.

[10] Perrini S, et al. Exercise-induced protein kinase C isoform-specific activation in human skeletal muscle. Diabetes 2004;53(1):21–4.

[11] Kiens B, Alsted TJ, Jeppesen J. Factors regulating fat oxidation in human skeletal muscle. Obes Rev 2011;12(10):852–8.

[12] Abumrad N, Coburn C, Ibrahimi A. Membrane proteins implicated in long-chain fatty acid uptake by mammalian cells: CD36, FATP and FABPm. Biochim Biophys Acta 1999;1441(1):4–13.

[13] Nickerson JG, et al. Greater transport efficiencies of the membrane fatty acid transporters FAT/CD36 and FATP4 compared with FABPpm and FATP1 and differential effects on fatty acid esterification and oxidation in rat skeletal muscle. J Biol Chem 2009;284(24):16522–30.

[14] Turcotte LP, et al. Muscle palmitate uptake and binding are saturable and inhibited by antibodies to FABPPM. Mol Cell Biochem 2000;210(1–2):53–63.

[15] Clarke DC, et al. Overexpression of membrane-associated fatty acid binding protein (FABPpm) in vivo increases fatty acid sarcolemmal transport and metabolism. Physiol Genomics 2004;17(1):31–7.

[16] Chabowski A, et al. Evidence for concerted action of FAT/CD36 and FABPpm to increase fatty acid transport across the plasma membrane. Prostaglandins Leukot Essent Fatty Acids 2007;77(5–6):345–53.

[17] Berk P, et al. Plasma membrane fatty acid-binding protein and mitochondrial glutamic-oxaloacetic transaminase of rat liver are related. Proc Natl Acad Sci USA 1990;87(9):3484–8.

[18] Koonen DP, et al. Different mechanisms can alter fatty acid transport when muscle contractile activity is chronically altered. Am J Physiol Endocrinol Metab 2004;286(6):E1042–9.

[19] Bonen A, et al. Muscle contractile activity increases fatty acid metabolism and transport and FAT/CD36. Am J Physiol 1999;276(4):E642–9.

[20] Glatz JF, Luiken JJ. From fat to FAT (CD36/SR-B2): understanding the regulation of cellular fatty acid uptake. Biochimie 2017;136:21–6.

[21] Yanai H, et al. Attenuated aerobic exercise capacity in CD36 deficiency. J Med Genet 2007;44(7):445–7.

[22] Fujitani M, et al. CD36, but not GPR120, is required for efficient fatty acid utilization during endurance exercise. Biosci Biotechnol Biochem 2014;78(11):1871–8.

[23] Bonen A, et al. Acute regulation of fatty acid uptake involves the cellular redistribution of fatty acid translocase. J Biol Chem 2000;275(19):14501–8.

[24] Heather LC, et al. Differential translocation of the fatty acid transporter, FAT/CD36, and the glucose transporter, GLUT4, coordinates changes in cardiac substrate metabolism during ischemia and reperfusion. Circ Heart Fail 2013;6(5):1058–66.
[25] Bonen A, et al. Regulation of fatty acid transport by fatty acid translocase/CD36. Proc Nutr Soc 2004;63(2): 245–9.
[26] Baskin KK, Winders BR, Olson EN. Muscle as a "mediator" of systemic metabolism. Cell Metab 2015;21(2): 237–48.
[27] Newton AC. Protein kinase C: structure, function, and regulation. J Biol Chem 1995;270(48):28495–8.
[28] C, S.-P., Protein kinase C and lipid-induced insulin resistance in skeletal muscle. Ann N Y Acad Sci, 2002; 967: 146–57.
[29] Bakker W, et al. Protein kinase C activation induces insulin-mediated constriction of muscle resistance arteries. Diabetes 2008;57(3):706–13.
[30] Kuo IY, Ehrlich BE. Signaling in muscle contraction. Cold Spring Harb Perspect Biol 2015;7(2). p. a006023-a006023.
[31] Fletcher JE, et al. Fatty acids modulate calcium-induced calcium release from skeletal muscle heavy sarcoplasmic reticulum fractions: implications for malignant hyperthermia. Biochem Cell Biol 1990;68(10):1195–201.
[32] Katz AM, et al. Fatty acid effects on calcium influx and efflux in sarcoplasmic reticulum vesicles from rabbit skeletal muscle. Biochim Biophys Acta 1982;687(1):17–26.
[33] Wan B, et al. Regulation of citric acid cycle by calcium. J Biol Chem 1989;264(23):13430–9.
[34] Li Y, Kang JX, Leaf A. Differential effects of various eicosanoids on the production or prevention of arrhythmias in cultured neonatal rat cardiac myocytes. Prostaglandins 1997;54(2):511–30.
[35] Tapiero H, et al. Polyunsaturated fatty acids (PUFA) and eicosanoids in human health and pathologies. Biomed Pharmacother 2002;56(5):215–22.
[36] Bagga D, et al. Differential effects of prostaglandin derived from w-6 and w-3 polyunsaturated fatty acids on COX-2 expression and IL-6 secretion. Proc Natl Acad Sci USA 2003;100(4):1751–6.
[37] Moonen H, et al. Effects of polyunsaturated fatty acids on prostaglandin synthesis and cyclooxygenase-mediated DNA adduct formation by heterocyclic aromatic amines in human adenocarcinoma colon cells. Mol Carcinog 2004;40(3):180–8.
[38] Maaløe T, et al. The effect of n-3 polyunsaturated fatty acids on leukotriene B4 and leukotriene B5 production from stimulated neutrophil granulocytes in patients with chronic kidney disease. Prostaglandins Leukot Essent Fatty Acids 2011;85(1):37–41.
[39] Duplus E, Glorian M, Forest C. Fatty acid regulation of gene transcription. J Biol Chem 2000;275(40):30749–52.
[40] Jump DB. Fatty acid regulation of gene transcription. Crit Rev Clin Lab Sci 2004;41(1):41–78.
[41] Ryan RO, van der Horst DJ. Lipid transport biochemistry and its role in energy production. Annu Rev Entomol 2000;45(1):233–60.
[42] Rogowski MP, et al. SCD1 activity in muscle increases triglyceride PUFA content, exercise capacity, and PPAR expression in mice. J Lipid Res 2013;54(10):2636–46.
[43] Currie E, et al. Cellular fatty acid metabolism and cancer. Cell Metab 2013;18(2):153–61.
[44] Damiano F, et al. Translational control of the sterol-regulatory transcription factor SREBP-1 mRNA in response to serum starvation or ER stress is mediated by an internal ribosome entry site. Biochem J 2010;429(3):603–12.
[45] Bandyopadhyay GK, et al. Increased malonyl-CoA levels in muscle from obese and type 2 diabetic subjects lead to decreased fatty acid oxidation and increased lipogenesis; thiazolidinedione treatment reverses these defects. Diabetes 2006;55(8):2277–85.
[46] Wei H, et al. Transcriptional response of porcine skeletal muscle to feeding a linseed-enriched diet to growing pigs. J Anim Sci Biotechnol 2016;7:6–16.
[47] Dessalle K, et al. SREBP-1 transcription factors regulate skeletal muscle cell size by controlling protein synthesis through myogenic regulatory factors. PLoS One 2012;7(11):e50878.
[48] Commerford SR, et al. In vivo regulation of SREBP-1c in skeletal muscle: effects of nutritional status, glucose, insulin, and leptin. Am J Physiol Regul Integr Comp Physiol 2004;287(1):R218–27.
[49] Giandomenico V, et al. Coactivator-dependent acetylation stabilizes members of the SREBP family of transcription factors. Mol Cell Biol 2003;23(7):2587–99.
[50] Kotzka J, et al. Preventing phosphorylation of sterol regulatory element-binding protein 1a by MAP-kinases protects mice from fatty liver and visceral obesity. PLoS One 2012;7(2):e32609.
[51] Shao W, Espenshade PJ. Expanding roles for SREBP in metabolism. Cell Metab 2012;16(4):414–9.

[52] Ulven SM, et al. LXR is crucial in lipid metabolism. Prostaglandins Leukot Essent Fatty Acids 2005;73(1): 59–63.
[53] Kim H-J, Miyazaki M, Ntambi JM. Dietary cholesterol opposes PUFA-mediated repression of the stearoyl-CoA desaturase-1 gene by SREBP-1 independent mechanism. J Lipid Res 2002;43(10):1750–7.
[54] Pégorier J-P, May CdL, Girard J. Control of gene expression by fatty acids. J Nutr 2004;134(9):2444S–9S.
[55] Silveira LR, et al. Updating the effects of fatty acids on skeletal muscle. J Cell Physiol 2008;217(1):1–12.
[56] Li H, Malhotra S, Kumar A. Nuclear factor-kappa B signaling in skeletal muscle atrophy. J Mol Med (Berl) 2008;86(10):1113–26.
[57] Hommelberg PP, et al. Fatty acid-induced NF-kB activation and insulin resistance in skeletal muscle are chain length dependent. Am J Physiol Endocrinol Metab 2009;296(1):E114–20.
[58] Camandola S, et al. Nuclear factor kB is activated by arachidonic acid but not by eicosapentaenoic acid. Biochem Biophys Res Commun 1996;229(2):643–7.
[59] Vaughan RA, et al. A high linoleic acid diet does not induce inflammation in mouse liver or adipose tissue. Lipids 2015;50(11):1115–22.
[60] Krook A. Can the liver X receptor work its magic in skeletal muscle too? J Diabetologia 2006;49(5):819–21.
[61] Kase ET, et al. Skeletal muscle lipid accumulation in type 2 diabetes may involve the liver X receptor pathway. Diabetes 2005;54(4):1108–15.
[62] Bedi S, et al. Fatty acid binding profile of the liver X receptor . J Lipid Res 2017;58(2):393–402.
[63] Muscat GE, et al. Regulation of cholesterol homeostasis and lipid metabolism in skeletal muscle by liver X receptors. J Biol Chem 2002;277(43):40722–8.
[64] Keller HR, et al. Fatty acids and retinoids control lipid metabolism through activation of peroxisome proliferator-activated receptor-retinoid X receptor heterodimers. Proc Natl Acad Sci USA 1993;90(6):2160–4.
[65] Goldstein JT, et al. Isolation and characterization of unsaturated fatty acids as natural ligands for the retinoid-X receptor. Arch Biochem Biophys 2003;420(1):185–93.
[66] Amengual J, et al. Retinoic acid treatment increases lipid oxidation capacity in skeletal muscle of mice. Obesity 2008;16(3):585–91.
[67] Rhee E-J, Plutzky J. Retinoid metabolism and diabetes mellitus. Diabetes Metab J 2012;36(3):167–80.
[68] Georgiades P, Brickell PM. Differential expression of the rat retinoid X receptor gene during skeletal muscle differentiation suggests a role in myogenesis. Dev Dyn 1997;210(3):227–35.
[69] Haugen BR, et al. Retinoid X receptor -deficient mice have increased skeletal muscle lipoprotein lipase activity and less weight gain when fed a high-fat diet. Endocrinology 2004;145(8):3679–85.
[70] Le May M, et al. Contribution of retinoid X receptor signaling to the specification of skeletal muscle lineage. J Biol Chem 2011;286(30):26806–12.
[71] Reijntjes S, Francis-West P, Mankoo BS. Retinoic acid is both necessary for and inhibits myogenic commitment and differentiation in the chick limb. Int J Dev Biol 2010;54(1):125.
[72] Gilde A, Van Bilsen M. Peroxisome proliferator-activated receptors (PPARS): regulators of gene expression in heart and skeletal muscle. Acta Physiol Scand 2003;178(4):425–34.
[73] Kersten S, Desvergne B, Wahli W. Roles of PPARs in health and disease. Nature 2000;405(6785):421–4.
[74] Ehrenborg E, Krook A. Regulation of skeletal muscle physiology and metabolism by peroxisome proliferator-activated receptor . Pharmacol Rev 2009;61(3):373–93.
[75] Kota BP, Huang TH-W, Roufogalis BD. An overview on biological mechanisms of PPARs. Pharmacol Res 2005;51(2):85–94.
[76] Muoio DM, et al. Fatty acid homeostasis and induction of lipid regulatory genes in skeletal muscles of peroxisome proliferator-activated receptor (PPAR) knock-out mice evidence for compensatory regulation by PPAR . J Biol Chem 2002;277(29):26089–97.
[77] Liu Y, et al. The role of ppar- in metabolism, inflammation, and cancer: many characters of a critical transcription factor. Int J Mol Sci 2018;19(11):3339.
[78] Koves TR, et al. Peroxisome proliferator-activated receptor- co-activator 1 -mediated metabolic remodeling of skeletal myocytes mimics exercise training and reverses lipid-induced mitochondrial inefficiency. J Biol Chem 2005;280(39):33588–98.
[79] Park KS, et al. PPAR-γ gene expression is elevated in skeletal muscle of obese and type II diabetic subjects. Diabetes 1997;46(7):1230–4.
[80] Yu G-S, Lu Y-C, Gulick T. Co-regulation of tissue-specific alternative human carnitine palmitoyltransferase Iβ gene promoters by fatty acid enzyme substrate. J Biol Chem 1998;273(49):32901–9.

[81] Zhang J, et al. Human skeletal muscle PPAR expression correlates with fat metabolism gene expression but not BMI or insulin sensitivity. am J Physiol Endocrinol Metab 2004;286(2):E168–75.
[82] Wallenius K, et al. The PPAR / agonist, tesaglitazar, improves insulin mediated switching of tissue glucose and free fatty acid utilization in vivo in the obese Zucker rat. PPAR Res 2013;2013.
[83] Kim H, et al. Peroxisome proliferator–activated receptor- agonist treatment in a transgenic model of type 2 diabetes reverses the lipotoxic state and improves glucose homeostasis. Diabetes 2003;52(7):1770–8.
[84] Guerre-Millo M, et al. PPAR- –null mice are protected from high-fat diet–induced insulin resistance. Diabetes 2001;50(12):2809–14.
[85] Barish GD, Narkar VA, Evans RM. PPAR : a dagger in the heart of the metabolic syndrome. J Clin Invest 2006;116(3):590–7.
[86] Wang Y-X, et al. Regulation of muscle fiber type and running endurance by PPAR . PLoS Biol 2004;2(10): e294.
[87] Kleiner S, et al. PPAR agonism activates fatty acid oxidation via PGC-1 but does not increase mitochondrial gene expression and function. J Biol Chem 2009;284(28):18624–33.
[88] Wang Y-X, et al. Peroxisome-proliferator-activated receptor activates fat metabolism to prevent obesity. Cell 2003;113(2):159–70.
[89] Narkar VA, et al. AMPK and PPAR agonists are exercise mimetics. Cell 2008;134(3):405–15.
[90] Bastie CC, et al. FoxO1 stimulates fatty acid uptake and oxidation in muscle cells through CD36-dependent and-independent mechanisms. J Biol Chem 2005;280(14):14222–9.
[91] Fan W, et al. PPAR promotes running endurance by preserving glucose. Cell Metab 2017;25(5):1186–93. e4.
[92] Jornayvaz FR, Shulman GI. Regulation of mitochondrial biogenesis. Essays Biochem 2010;47:69–84.
[93] Zhao L, et al. Defective fatty acid oxidation in mice with muscle-specific acyl-CoA synthetase 1 deficiency increases amino acid use and impairs muscle function. J Biol Chem 2019;294(22):8819–33.
[94] Li LO, et al. Compartmentalized acyl-CoA metabolism in skeletal muscle regulates systemic glucose homeostasis. Diabetes 2015;64(1):23–35.
[95] Yan S, et al. Long-chain acyl-CoA synthetase in fatty acid metabolism involved in liver and other diseases: an update. World J Gastroenterol 2015;21(12):3492.
[96] Mashek DG, Li LO, Coleman RA. Rat long-chain acyl-CoA synthetase mRNA, protein, and activity vary in tissue distribution and in response to diet. J Lipid Res 2006;47(9):2004–10.
[97] Teodoro BG, et al. Long-chain acyl-CoA synthetase 6 regulates lipid synthesis and mitochondrial oxidative capacity in human and rat skeletal muscle. J Physiol 2017;595(3):677–93.
[98] Neess D, et al. Long-chain acyl-CoA esters in metabolism and signaling: role of acyl-CoA binding proteins. Prog Lipid Res 2015;59:1–25.
[99] Knudsen J, et al. Role of acyl-CoA binding protein in acyl-CoA metabolism and acyl-CoA–mediated cell signaling. J Nutr 2000;130(2):294S–8S.
[100] Franch J, et al. Acyl-CoA binding protein expression is fiber type-specific and elevated in muscles from the obese insulin-resistant Zucker rat. Diabetes 2002;51(2):449–54.
[101] Fan J, et al. Acyl-coenzyme A binding domain containing 3 (ACBD3; PAP7; GCP60): an emerging signaling molecule. Prog Lipid Res 2010;49(3):218–34.
[102] Chen Y, et al. Maturation and activity of sterol regulatory element binding protein 1 is inhibited by acyl-CoA binding domain containing 3. PLoS One 2012;7(11):e49906.
[103] Soupene E, Kuypers FA. Ligand binding to the ACBD6 protein regulates the acyl-CoA transferase reactions in membranes. J Lipid Res 2015;56(10):1961–71.
[104] Lanfray D, et al. Involvement of the Acyl-CoA binding domain containing 7 in the control of food intake and energy expenditure in mice. Elife 2016;5:e11742.
[105] Basseri S, Austin RC. Endoplasmic reticulum stress and lipid metabolism: mechanisms and therapeutic potential. Biochem Res Int 2012;2012.
[106] Guo Y, et al. Lipid droplets at a glance. J Cell Sci 2009;122(6):749–52.
[107] Welte MA, Gould AP. Lipid droplet functions beyond energy storage. Biochim Biophys Acta Mol Cell Biol Lipids 2017;1862(10):1260–72.
[108] Liu L, et al. Upregulation of myocellular DGAT1 augments triglyceride synthesis in skeletal muscle and protects against fat-induced insulin resistance. J Clin Invest 2007;117(6):1679–89.
[109] Liu L, et al. DGAT1 deficiency decreases PPAR expression and does not lead to lipotoxicity in cardiac and skeletal muscle. J Lipid Res 2011;52(4):732–44.

[110] Morales PE, Bucarey JL, Espinosa A. Muscle lipid metabolism: role of lipid droplets and perilipins. J Diabetes Res 2017;2017.
[111] Altamimi TR, Lopaschuk GD. Role of carnitine in modulation of muscle energy metabolism and insulin resistance. Carnitine metabolism and human nutritionCRC Press; 2016. p. 30–55.
[112] Warfel JD, et al. A low fat diet ameliorates pathology but retains beneficial effects associated with CPT1b knockout in skeletal muscle. PLoS One 2017;12(12):e0188850.
[113] Augusto V, Padovani CR, Campos GR. Skeletal muscle fiber types in C57BL6J mice. J Braz J Morphol Sci 2004;21(2):89–94.
[114] Sher J, Cardasis C. Skeletal muscle fiber types in the adult mouse. Acta Neurol Scand 1976;54(1):45–56.
[115] Kammoun M, et al. A simplified immunohistochemical classification of skeletal muscle fibres in mouse. Eur J Histochem 2014;58.(2).
[116] Schiaffino S, Reggiani C. Fiber types in mammalian skeletal muscles. Physiol Rev 2011;91(4):1447–531.
[117] Scott W, Stevens J, Binder–Macleod SA. Human skeletal muscle fiber type classifications. Phys Ther 2001;81(11):1810–6.
[118] Talbot J, Maves L. Skeletal muscle fiber type: using insights from muscle developmental biology to dissect targets for susceptibility and resistance to muscle disease. Wiley Interdiscip Rev Dev Biol 2016;5(4):518–34.
[119] Galpin AJ, et al. Human skeletal muscle fiber type specific protein content. Anal Biochem 2012;425(2):175–82.
[120] Schiaffino S, Hanzlíková V, Pierobon S. Relations between structure and function in rat skeletal muscle fibers. J Cell Biol 1970;47(1):107–19.
[121] Juel C. Membrane Transport in Human Skeletal Muscle; 2012.
[122] Dyck D, et al. Functional differences in lipid metabolism in resting skeletal muscle of various fiber types. Am J Physiol 1997;272(3):E340–51.
[123] Kiens B. Skeletal muscle lipid metabolism in exercise and insulin resistance. Physiol Rev 2006;86(1):205–43.
[124] Bonen A, et al. Palmitate transport and fatty acid transporters in red and white muscles. Am J Physiol 1998;275(3):E471–8.
[125] Holloszy JO, Coyle EF. Adaptations of skeletal muscle to endurance exercise and their metabolic consequences. J Appl Physiol Respir Environ Exerc Physiol 1984;56(4):831–8.
[126] Vistisen B, et al. Sarcolemmal FAT/CD36 in human skeletal muscle colocalizes with caveolin-3 and is more abundant in type 1 than in type 2 fibers. J Lipid Res 2004;45(4):603–9.
[127] Turcotte LP, Srivastava AK, Chiasson J-L. Fasting increases plasma membrane fatty acid-binding protein (FABP(PM)) in red skeletal muscle. Mol Cell Biochem 1997;166(1–2):153–8.
[128] Philippi M, Sillau A. Oxidative capacity distribution in skeletal muscle fibers of the rat. J Exp Biol 1994;189(1):1–11.
[129] Capó LA, Sillau A. The effect of hyperthyroidism on capillarity and oxidative capacity in rat soleus and gastrocnemius muscles. J Physiol 1983;342(1):1–14.
[130] Glancy B, Balaban RS. Protein composition and function of red and white skeletal muscle mitochondria. Am J Physiol Cell Physiol 2011;300(6):C1280–90.
[131] Jackman MR, Willis WT. Characteristics of mitochondria isolated from type I and type IIb skeletal muscle. Am J Physiol 1996;270(2):C673–8.
[132] Ponsot E, et al. Mitochondrial tissue specificity of substrates utilization in rat cardiac and skeletal muscles. J Cell Physiol 2005;203(3):479–86.
[133] Bahi L, et al. Differential effects of thyroid hormones on energy metabolism of rat slow-and fast-twitch muscles. J Cell Physiol 2005;203(3):589–98.
[134] Picard M, Hepple RT, Burelle Y. Mitochondrial functional specialization in glycolytic and oxidative muscle fibers: tailoring the organelle for optimal function. am J Physiol Cell Physiol 2011;302(4):C629–41.
[135] Li X, et al. Skeletal muscle lipid droplets and the athlete's paradox. Cells 2019;8(3):249.
[136] McGarry JD, Mannaerts G, Foster DW. A possible role for malonyl-CoA in the regulation of hepatic fatty acid oxidation and ketogenesis. J Clin Invest 1977;60(1):265–70.
[137] Komiya Y, et al. Mouse soleus (slow) muscle shows greater intramyocellular lipid droplet accumulation than EDL (fast) muscle: fiber type-specific analysis. J Muscle Res Cell Motil 2017;38(2):163–73.
[138] Lin J, et al. Transcriptional co-activator PGC-1 drives the formation of slow-twitch muscle fibres. Nature 2002;418(6899):797.
[139] Zhang L, et al. Skeletal muscle-specific overexpression of PGC-1α induces fiber-type conversion through enhanced mitochondrial respiration and fatty acid oxidation in mice and pigs. Int J Biol Sci 2017;13(9):1152.

[140] Oike Y, et al. Angiopoietin-like proteins: potential new targets for metabolic syndrome therapy. Trends Mol Med 2005;11(10):473–9.
[141] Li Y, Teng C. Angiopoietin-like proteins 3. 4 and 8: regulating lipid metabolism and providing new hope for metabolic syndrome. J Drug Target 2014;22(8):679–87.
[142] Chi X, et al. ANGPTL8 promotes the ability of ANGPTL3 to bind and inhibit lipoprotein lipase. Mol Metab 2017;6(10):1137–49.
[143] Haller JF, et al. ANGPTL8 requires ANGPTL3 to inhibit lipoprotein lipase and plasma triglyceride clearance. J Lipid Res 2017;58(6):1166–73.
[144] Su X, Peng D-q. New insights into ANGPLT3 in controlling lipoprotein metabolism and risk of cardiovascular diseases. Lipids Health Dis 2018;17(1):12.
[145] Köster A, et al. Transgenic angiopoietin-like (angptl) 4 overexpression and targeted disruption of angptl4 and angptl3: regulation of triglyceride metabolism. Endocrinology 2005;146(11):4943–50.
[146] Lichtenstein L, et al. Angptl4 protects against severe pro-inflammatory effects of dietary saturated fat by inhibiting lipoprotein lipase-dependent uptake of fatty acids in mesenteric lymph node macrophages. Cell Metab 2010;12(6):580.
[147] Wang Y, et al. Mice lacking ANGPTL8 (Betatrophin) manifest disrupted triglyceride metabolism without impaired glucose homeostasis. Proc Natl Acad Sci USA 2013;110(40):16109–14.
[148] Sugden MC, Holness MJ, Howard RM. Changes in lipoprotein lipase activities in adipose tissue, heart and skeletal muscle during continuous or interrupted feeding. Biochem J 1993;292(1):113–9.
[149] Zhang R. The ANGPTL3-4-8 model, a molecular mechanism for triglyceride trafficking. Open Biol 2016;6(4):150272.
[150] Mehta N, et al. Differential association of plasma angiopoietin-like proteins 3 and 4 with lipid and metabolic traits. Arterioscler Thromb Vasc Biol 2014;34(5):1057–63.
[151] Barja-Fernandez S, et al. Plasma ANGPTL-4 is associated with obesity and glucose tolerance: cross-sectional and longitudinal findings. Mol Nutr Food Res 2018;62(10):1800060.
[152] Alex S, et al. Short-chain fatty acids stimulate angiopoietin-like 4 synthesis in human colon adenocarcinoma cells by activating peroxisome proliferator-activated receptor . Mol Cell Biol 2013;33(7):1303–16.
[153] Brands M, et al. Omega-3 long-chain fatty acids strongly induce angiopoietin-like 4 in humans. J Lipid Res 2013;54(3):615–21.
[154] Kaviani S, et al. A 7-day high-PUFA diet reduces angiopoietin-like protein 3 and 8 responses and postprandial triglyceride levels in healthy females but not males: a randomized control trial. BMC Nutr 2019;5(1):1.
[155] Choi M-S, et al. , High-fat diet decreases energy expenditure and expression of genes controlling lipid metabolism, mitochondrial function and skeletal system development in the adipose tissue, along with increased expression of extracellular matrix remodelling-and inflammation-related genes. Br J Nutr 2015;113(6):867–77.
[156] Goodpaster BH, et al. Skeletal muscle lipid content and insulin resistance: evidence for a paradox in endurance-trained athletes. J Clin Endocrinol Metab 2001;86(12):5755–61.
[157] Tarnopolsky MA, et al. Influence of endurance exercise training and sex on intramyocellular lipid and mitochondrial ultrastructure, substrate use, and mitochondrial enzyme activity. Am J Physiol Regul Integr Comp Physiol 2007;292(3):R1271–8.
[158] Li X, et al. Skeletal Muscle Lipid Droplets and the Athlete's Paradox. Cells 2019;8(3):249.
[159] Pinho RA, et al. High-fat diet induces skeletal muscle oxidative stress in a fiber type-dependent manner in rats. Free radical biology medicine 2017;110:381–9.
[160] Capllonch G, et al. Long-term high-fat-diet feeding induces skeletal muscle mitochondrial biogenesis in rats in a sex-dependent and muscle-type specific manner. Nutr Metab (Lond) 2012;9:15.
[161] Miotto PM, LeBlanc PJ, Holloway GP. High-fat diet causes mitochondrial dysfunction as a result of impaired ADP sensitivity. Diabetes 2018;67(11):2199–205.
[162] Sparks LM, et al. A high-fat diet coordinately downregulates genes required for mitochondrial oxidative phosphorylation in skeletal muscle. Diabetes 2005;54(7):1926–33.
[163] Tieland M, Trouwborst I, Clark BC. Skeletal muscle performance and ageing. J Cachexia Sarcopenia Muscle 2018;9(1):3–19.
[164] McGregor RA, Cameron-Smith D, Poppitt SD. It is not just muscle mass: a review of muscle quality, composition and metabolism during ageing as determinants of muscle function and mobility in later life. Longev Healthspan 2014;3(1):9–19.

[165] Zurlo F, et al. Skeletal muscle metabolism is a major determinant of resting energy expenditure. J Clin Invest 1990;86(5):1423–7.

[166] Didyk RB, et al. Effect of immobilization on glucose transporter expression in rat hindlimb muscles. Metabolism 1994;43(11):1389–94.

[167] Swaminath G. Fatty acid binding receptors and their physiological role in type 2 diabetes. Archiv Pharmazie 2008;341(12):753–61.

[168] Yun K, Wold B. Skeletal muscle determination and differentiation: story of a core regulatory network and its context. Curr Opin Cell Biol 1996;8(6):877–89.

CHAPTER

# 19

# Sphingolipid mediators of cell signaling and metabolism

*Agnieszka Dobrzyn*[a], *Justyna Janikiewicz*[a], *Zuzanna Tracz-Gaszewska*[b], *Anna Filip*[b], *Aneta M. Dobosz*[a], *Ewelina Krogulec*[a], *Pawel Dobrzyn*[b]

[a]Laboratory of Cell Signaling and Metabolic Disorders, Nencki Institute of Experimental Biology, Polish Academy of Sciences, Warsaw, Poland; [b]Laboratory of Medical Molecular Biochemistry, Nencki Institute of Experimental Biology, Polish Academy of Sciences, Warsaw, Poland

OUTLINE



Lipid Signaling and Metabolism. http://dx.doi.org/10.1016/B978-0-12-819404-1.00019-1

## Introduction

High circulating levels of free fatty acids (FAs) and abnormalities of the regulation of lipid signaling and deposition have been shown to contribute to the onset of metabolic dysfunction. Several observations suggest that the prolonged channeling of FAs for storage in non-adipose tissues, such as the muscle, heart, pancreas, and vasculature, overwhelms the oxidative capacity of these organs, drives aberrant bioactive lipid production, and causes diabetes-related complications, a phenomenon that is known as lipotoxicity [1].

Information about the biological mechanisms by which the oversupply and spillover of lipids impair the function of peripheral tissues remains obscure. Several mechanisms, including endoplasmic reticulum (ER) stress, inflammation, maladaptive autophagy, mitochondrial dysfunction, and aberrant lipid signaling, have been proposed to initiate the development of metabolic diseases [2]. Sphingolipids remain a relatively minor class of lipid species in mammalian cells, but they have gained considerable attention recently because of emerging evidence that links them with lipotoxicity and a reduction of insulin sensitivity. Vast changes in sphingolipid pools and distribution occur in obese humans and rodents [3]. Moreover, the sphingoid backbone relies on the availability of saturated FAs, which are widely recognized to be more toxic than their unsaturated counterparts [4].

The present chapter summarizes our recent understanding of different sphingolipid derivatives in cell-signaling responses of various tissues and their role in the pathogenesis of metabolic diseases, including type 2 diabetes (T2D) and metabolic syndrome. We also provide an overview of the role of sphingolipids in the control of lipotoxicity at the molecular and cellular levels and relevant animal models and human studies. Additionally, we discuss mutual relationships between the actions of sphingolipid and obesity, insulin resistance, muscle function, β-cell exhaustion, the inflammatory response, vascular complications, and cardiac failure. A better awareness of sphingolipid metabolism may reveal potential pharmacological targets for the treatment of obesity-derived T2D and other metabolic disorders.

## Sphingolipid metabolism and turnover

Sphingolipids are a diverse group of lipid derivatives that are typically constructed on the 18-carbon aliphatic amino-alcohol sphingosine [5]. The sphingoid backbone can be either generated from de novo synthesis that is initiated on the cytoplasmic face of the ER or salvaged via the breakdown of more complex sphingolipids in the lysosomal and cytoplasmic membranes [6]. The first step of the de novo pathway is catalyzed by serine palmitoyltransferase (SPT), which requires the condensation of L-serine with palmitoyl-coenzyme A (CoA) to form 3-ketosphinganine (Fig. 19.1). 3-Ketosphinganine is rapidly reduced to dihydrosphingosine,

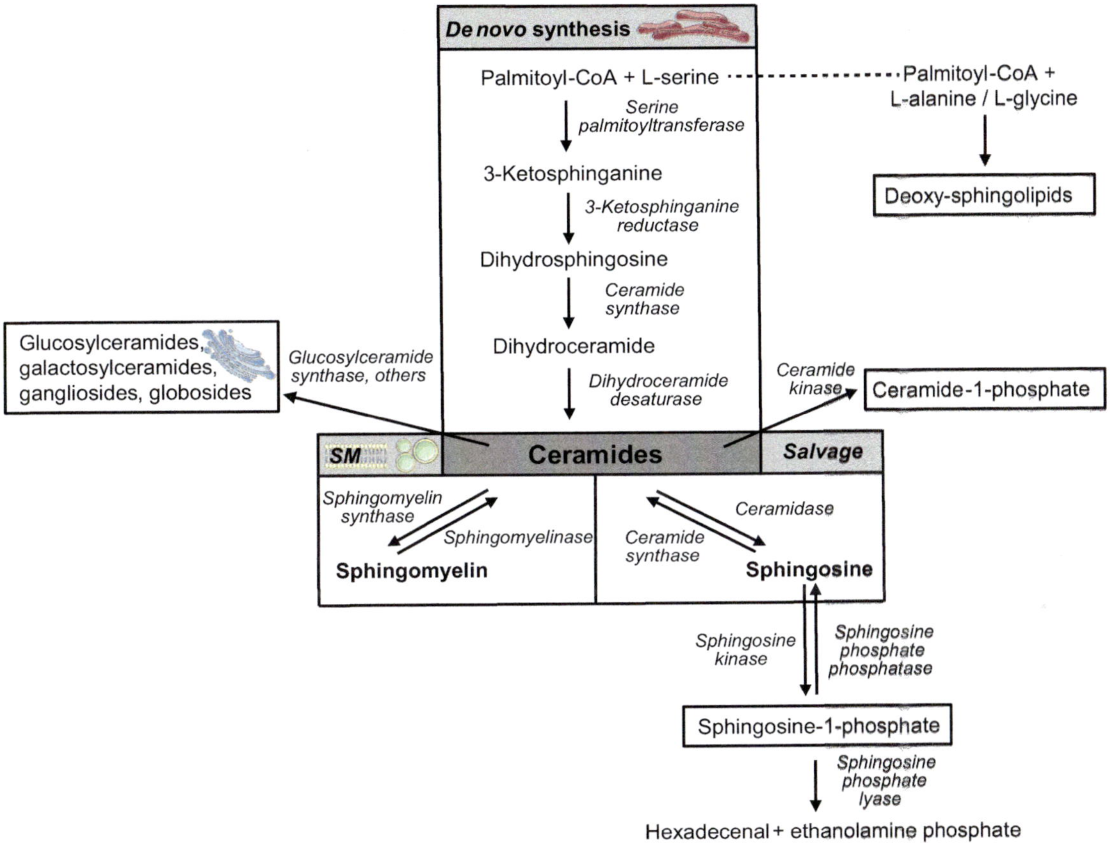


FIGURE 19.1 **Metabolism of sphingolipids.** During the first step of de novo synthesis, palmitoyl-CoA is condensed with L-serine to form 3-ketosphinganine. The product of this reaction is enzymatically reduced to dihydrosphingosine. In turn, a family of six ceramide synthase isoforms gives rise to dihydroceramides, which are finally desaturated into ceramides by dihydroceramide desaturase. Ceramides are transported to the Golgi apparatus where they are metabolized into glucosylceramides and higher glycosphingolipids (e.g., gangliosides and globosides) or turned into sphingomyelin via activation of the sphingomyelinase (SMase) pathway in lysosomal and plasma membranes. Glycosphingolipids, sphingomyelin, and ceramides are converted to sphingosine in late endosomes and lysosomes. Sphingosine exits the lysosome and undergoes re-acylation back into ceramide through the salvage pathway or forms sphingosine-1-phosphate, which is terminally degraded by sphingosine phosphate lyase.

which then undergoes *N*-acylation by ceramide synthase (CerS) isoforms to form dihydroceramide. Ceramide synthase isoforms are responsible for the FA composition of ceramides based on their affinity for various FA moieties with chain lengths that range from 14 to 28 carbons and varying degrees of saturation [7,8]. Finally, dihydroceramide desaturase 1 (DES1) produces ceramides by introducing a *trans* double bond at the 4–5 position of the sphingoid base backbone. Once synthesized, ceramides are transported with assistance from ceramide transport protein (CERT) to the Golgi apparatus where they become precursors of several bioactive sphingolipid metabolites [9]. For example, sphingomyelin synthases (SMSs) produce sphingomyelin from phosphatidylcholine and ceramides. Glucosyl-CerS converts ceramides into glycosphingolipids, including gangliosides and globosides. All of these metabolites can be further catabolized by multiple ceramidases that hydrolyze ceramides to the free sphingoid base [6]. In turn, sphingosine kinases (SphKs) are responsible for the phosphorylation of sphingosine to form sphingosine-1-phosphate (S1P; [1]). The salvage pathway also involves the degradation of sphingomyelin into ceramide through actions of sphingomyelinases (SMases; [3,8]).

## Divergence of bioactive sphingolipid molecules in islets of Langerhans

### Ceramide as a principal contributor to lipotoxicity in pancreatic β-cells: evidence and mechanisms

Ceramides can be generated either in response to inflammatory cytokines or by the excessive deposition of saturated fats. Among the lipid metabolites that are produced during lipotoxicity, ceramides are involved in deleterious effects of glucolipotoxicity on pancreatic β-cells and ultimately the pathogenesis of T2D and its complications [3,6]. Accumulating evidence indicates that lipotoxicity triggers both de novo ceramide synthesis and activation of the SMase pathway in pancreatic β-cells, resulting in ceramide accumulation and a reduction of ceramide translocation toward the Golgi apparatus [9]. Furthermore, lipotoxicity impairs either vesicular- or CERT-mediated ceramide flow from the ER to the Golgi apparatus through a decrease in Akt levels, which in turn can inhibit vesicular traffic and downregulate the amount of active CERT [10]. Studies that have been performed with animal models and cultured cells have shown that ceramides play an important role in β-cell physiology by controlling proliferation, apoptosis, the inflammatory response, adipokine release, and insulin synthesis [11]. Intracellular ceramide accumulation induces various independent effects that possibly contribute to a reduction of insulin secretory capacity and the programmed death of pancreatic β-cells [12,13]. Excessive exposure to long-chain saturated FAs (e.g., palmitate, stearate, arachidate, and linocerate) but not unsaturated FAs induces the de novo synthesis of ceramides in β-cells to activate SPT and CerS [13]. Although the specific role of each ceramide derivative in β-cell dysfunction remains to be clarified, β-cell lipotoxicity was shown to be associated with changes in C18:0, C22:0, and C24:1 ceramide levels and the augmentation of CerS4 expression [7]. In contrast, CerS2-deficient mice were unable to produce C22–C24 ceramide moieties and did not develop defects of insulin secretion [14].

## Role of ceramide in the control of insulin biosynthesis and secretion

Ceramide synthesis is known to promote pancreatic β-cell lipoapoptosis. However, several studies have reported that ceramides can also act on the insulin synthesis apparatus by decreasing insulin gene expression and causing insulin secretion defects [15–17]. Ceramide that was generated from palmitate-induced lipotoxicity decreased the gene expression of insulin, and this effect was prevented by inhibitors of de novo ceramide synthesis (Fig. 19.2; [16]). The inhibition of ceramidase, which consequently led to ceramide accumulation, augmented the lowering effects of exogenous palmitate on insulin gene expression [18]. Interestingly, ceramide-mediated impairments in insulin gene expression were unrelated to mRNA instability but rather related to the strong inhibition of glucose-stimulated insulin promoter activity [16]. This inhibition is mediated by the lower binding activity of key regulatory transcription factors, including pancreas-duodenum homeobox 1 (PDX-1) and MafA. PDX-1 is

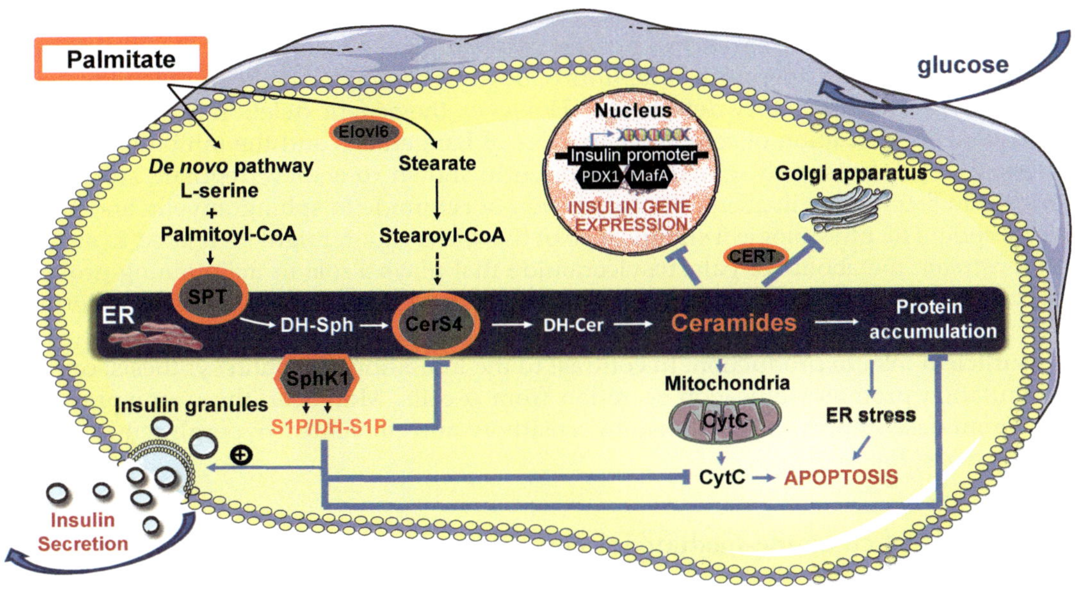


**FIGURE 19.2 The role of sphingolipid biostat in pancreatic β-cell function and failure during palmitate-induced lipotoxicity.** Palmitate has been shown to induce ceramide accumulation through the stimulation of enzymes from the de novo synthesis pathway, such as SPT and CerS4, and inhibit ceramide translocation from the ER to Golgi apparatus by blocking CERT. Ceramide accumulation reduces insulin gene expression and leads to β-cell apoptosis by inducing ER stress and cytochrome C (CytC) release from mitochondria. Palmitate can also induce the production of S1P and DH-S1P through the phosphorylation of DH-Sph by SphK1. The accumulation of these sphingoid base phosphates in the ER can protect β-cells against ceramide-dependent β-cell apoptosis that is induced by palmitic acid. Sphingoid base phosphates affect ceramide synthesis by downregulating CerS4 activity, restoring protein trafficking in the ER, alleviating ER stress, and inhibiting CytC release from mitochondria. *ER, endoplasmic reticulum; SPT, serine palmitoyltransferase; CerS4, ceramide synthase 4; Elovl6, fatty acid elongase 6; CERT, ceramide transporter; DH-Sph, dihydrosphingosine-1-phosphate; DH-Cer, dihydro-ceramide; PDX-1, pancreas-duodenum homeobox 1; S1P, sphingosine-1-phosphate; SphK1, sphingosine kinase 1.*

affected by its ability to translocate to the nucleus, whereas MafA becomes regulated at the overall level of mRNA and protein expression [19]. The precise mechanisms by which ceramide impairs the binding of PDX-1 and MafA to the insulin promoter remain unclear, but some hypotheses that are based on known ceramide targets have been proposed. One of the suggested targets is the extracellular signal-regulated kinase (ERK) pathway, which plays a role in the glucose-dependent regulation of insulin gene transcription. The protein phosphatase 2 (PP2A)-driven dephosphorylation of protein kinases ERK1/2 was shown to mediate the lowering effect of palmitate-induced ceramide accumulation on insulin gene expression (Fig. 2; [20]). Another study found that c-jun N-terminal kinase (JNK), which is known to be affected by ceramides [21], can also repress insulin gene expression by inhibiting PDX-1 binding to the insulin promoter [22]. Moreover, ceramide transport from the ER to Golgi apparatus was proposed to be involved in the regulation of insulin gene expression. The pharmacological inhibition of SMS or the downregulation of CERT enhanced the palmitate-induced downregulation of insulin gene expression and β-cell dysfunction by affecting the binding of transcription factors to the insulin promoter [15]. In contrast to strong evidence that ceramides are modulators of signaling that is involved in the transcriptional regulation of *Ins*, ceramide analogs were shown to exert only marginal effects on glucose-stimulated insulin secretion [23]. Furthermore, inhibitors of de novo ceramide synthesis did not overcome lipotoxicity-derived secretory defects [24]. However, the genetic deletion of SMS1 led to the subsequent accumulation of ceramides in islets of Langerhans and the inhibition of insulin secretion [17]. Interestingly, glucose was recently shown to acutely regulate sphingolipid turnover in β-cells by enhancing the conversion of ceramide to sphingomyelin and galactosylceramide. The physiological significance of this observation might be related to sulfatide, a downstream metabolite of galactosylceramide that plays a role in maintaining proinsulin reserves and secretory granule content [25]. Diabetes has been described as a "bihormonal disease," in which glucagon secretion from pancreatic α-cells remains unsuppressed because of insufficient insulin production. In contrast to the inhibition of insulin synthesis, ceramide accumulation promotes glucagon secretion from α-cells. Moreover, the overexpression of acid ceramidase improved α-cell insulin sensitivity and prevented hyperglycemia in obese diabetic mice [26].

## Mechanisms of ceramide-mediated β-cell apoptosis

The first studies that evaluated the consequences of ceramide accumulation on metabolic disease *in vivo* reported that β-cell lipoapoptosis that was observed in islets in obese Zucker Diabetic Fatty (ZDF) rats resulted from the overproduction of ceramide that was derived especially from long-chain FAs. Furthermore, ZDF rats were treated with the SPT inhibitor cycloserine, followed by ceramide depletion, which was associated with a reduction of the apoptosis of pancreatic islets [27]. Since then, various clinical and experimental studies have demonstrated that ceramides induce β-cell apoptosis through multiple mechanisms that involve the activation of extrinsic and intrinsic branches [28]. This can be achieved by inducing ER stress, increasing lipotoxicity, or disrupting mitochondrial function with the subsequent development of oxidative damage [11,29]. One of the major mechanisms by which ceramide causes β-cell apoptosis was proposed to involve the activation of PP2A and dephosphorylation of Akt [30], which has been widely acknowledged in muscles. Furthermore, ceramides caused ER stress in pancreatic β-cells by inefficient or slowed protein misfolding through downregulation of the sarcoplasmic/endoplasmic reticulum calcium pump (SERCA2) and the depletion of lumenal ER $Ca^{2+}$ [31,32]. The activation of $Ca^{2+}$-independent phospholipase A2, followed by an increase in sphingomyelin hydrolysis and ceramide generation, has been reported to cause ER stress-induced β-cell apoptosis. This effect was subsequently prevented by the inhibition of neutral SMase [33]. Conversely, enhancing the conversion of ceramides into glucosylceramide by the overexpression of glucosylceramide synthase protected against apoptosis by reducing ER stress and defective protein trafficking [34].

Ceramides are synthesized in the ER, but they can be transported to other cellular organelles, including mitochondria, which also contain enzymatic machinery to produce sphingolipid metabolites [35]. In fact, mitochondrial dysfunction and the incomplete oxidation of FAs were triggered by high-fat diet-induced ceramide accumulation. The inhibition of de novo sphingolipid synthesis decreased ceramide content and reversed faulty mitochondria overload in high-fat diet-fed mice [36]. Ultimately, the genetic deletion of SMS1 resulted in the intra-islet accumulation of ceramides and vast abnormalities in mitochondria activity, including a decrease in adenosine triphosphate production, hyperpolarized membrane potential, and an excessive abundance of reactive oxygen species (ROS). The disruption of mitochondrial function was followed by deficiencies in insulin secretion in pancreatic β-cells [17]. Furthermore, long-term exposure of INS-1 and INS 832/13 cells to either C2-ceramide or water-soluble cationic ceramide affected the condition of mitochondria and β-cell viability through cytochrome leakage and a reduction of membrane integrity [37,38].

## Regulation of islet dysfunction by deoxy-sphingolipids

The flexibility of SPT allows it to condensate palmitoyl-CoA with L-alanine or glycine instead of L-serine, giving rise to an atypical category of neurotoxic deoxy-sphingolipids (1-deoxySLs). These metabolites can be further *N*-acylated to dihydroceramides, but because of the missing hydroxyl group in position C1, they are neither formed into complex sphingolipids nor effectively degraded by S1P-lyase through the canonical catabolic pathway [39]. Plasma levels of 1-deoxySLs were significantly elevated in non-diabetic patients with metabolic syndrome and T2D and streptozotocin-injected rats [40,41] and were shown to be novel and independent predictors of the development of T2D in obese individuals and in an asymptomatic population [5,42]. Moreover, 1-deoxySLs promoted the death of INS-1E cells and primary rodent islets. The preincubation of INS-1E cells with 1-deoxysphinganine reduced their metabolic activity, compromised glucose-stimulated insulin secretion, and caused cytoskeletal aberrations in close proximity to insulin-containing vesicles. These changes were reflected by activation of the Rac1, JNK kinase, and p38 mitogen-activated protein kinase (MAPK) signaling cascades [43].

## Effect of glycosphingolipid-dependent lipotoxicity in pancreatic islets

Metabolomic analyses have shown that glycosphingolipids are mediators of pancreatic islets demise [44]. The expression of glucosylceramide GM3 synthase mRNA was significantly elevated in obese ZDF rat and *ob/ob* mouse models of T2D [45]. The pharmacological lowering of glycosphingolipids by chronic glucosylceramide synthase inhibition in high-fat diet-fed mice and ZDF rats resulted in the partial preservation of islet architecture and secretory function and improved insulin sensitivity [46,47]. Additionally, the C16:0, C22:0, and C24:0 glycosylated derivatives of ceramide were upregulated in palmitate-treated murine islets and MIN6 insulinoma cells, followed by a reduction of ER stress and a slower rate of protein trafficking [12,29]. High-spatial-resolution lipid imaging mass spectrometry analysis showed the even localization of specific C34:1 GM3 throughout human pancreas tissue and the periphery in wildtype and *ob/ob* mouse islets, which coincided with the pattern of distribution of α-cells. This abundance was correlated with a higher level of GM3 synthase [48].

## Regulatory role of sphingomyelin in β-cell failure

The initiation of lipotoxic events in pancreatic β-cells that are caused by ER stress and a lower rate of ER-to-Golgi apparatus protein trafficking is also accompanied by alterations of sphingomyelin content in the ER. Levels of the C24:0 sphingomyelin metabolite and total very-long-chain unsaturated C24:1, C25:1, and C26:1 sphingomyelin species were significantly increased by chronic palmitic acid treatment in MIN6 cells and murine islets [12,29]. The reductions of ER sphingomyelin content within whole islets and in MIN6 cells were partially reversed by glucosylceramide synthase overexpression. Nevertheless, the loss of sphingomyelin in the ER resulted in alterations of membrane dynamics and disruptions of lipid rafts and contributed to the defective packing of secretory cargo and vesicle budding from the ER membrane [29]. Additionally, the genetic knockdown or pharmacological inhibition of SMS1/SMS2 significantly reduced insulin secretion in vitro in INS-1 cells and in SMS1 knockout (KO) mouse islets [17,49]. Insulin secretory granules are particularly enriched in sphingomyelin species, specifically those that contain C16:1 and C18:1 side chains that further favor membrane fusion [50].

## Sphingosine-1-phosphate improves pancreatic islet function and survival

Sphingolipid metabolites have been reported to cause pancreatic islet failure and insulin insensitivity, but S1P has also been reported to modulate lipid-derived β-cell survival, preserve insulin secretion, and maintain mitochondrial homeostasis [51,52]. Palmitate treatment elevated the accumulation of S1P through the upregulation of activity of the SphK1 pathway, which abrogated lipotoxicity in MIN6 and INS-1 β-cell lines and murine islets [51,53]. Moreover, circulating S1P levels increased in obese mouse models and in obese human individuals and correlated with insulin resistance, adiposity, and the occurrence of T2D [28,54]. Although S1P and SphK1 share a pro-survival role in β-cells under lipotoxic conditions, elevated levels of S1P antagonized insulin-mediated cell growth and anti-apoptotic signaling via stimulation of the $S1P_2$ receptor subtype [28]. SphK2 was shown to be pro-apoptotic, and $Sphk2^{-/-}$ mice were protected from obesity, insulin resistance, and a diabetic phenotype that was induced by a high-fat diet [55,56]. Furthermore, SphK2 deficiency prevented the loss of β-cell mass, restored insulin production, and attenuated mitochondrial apoptotic pathway activation during β-cell lipotoxicity [56]. Therefore, a relative intracellular balance between SphK1 and SphK2 in β-cells appears to be necessary to prevent the onset and development of T2D.

## Sphingolipids and skeletal muscle metabolism

### Sphingolipids, oxidative stress, and skeletal muscle contractile function and fatigue

Skeletal muscles are exposed to various stressors, including strenuous exercise, heat stress, oxidative stress, and chronic inflammatory disease, such as T2D, obesity, heart failure, and cancer. The impact of these abnormalities on skeletal muscles is mediated by such cytokines as interleukin-1β (IL-1β) and tumor necrosis factor (TNF; [57]). Disease-associated cytokines were also shown to stimulate oxidative stress in skeletal muscles [58]. The excessive generation of ROS and nitric oxide (NO) derivatives or deficiencies in antioxidant defense systems, in turn, stimulate sphingolipid turnover and the generation of bioactive sphingolipids, such as sphingosine, S1P, and ceramide [57]. The activity of SMase3, a muscle-specific SMase, is stimulated by TNF in mouse myoblasts [59] and human satellite cells (i.e., muscle stem cells; [60]), probably via ROS and NO species [61]. Proinflammatory cytokines are strongly upregulated in obesity [62]. The levels of CerS1 and derived C18:0 ceramide were elevated in skeletal muscles in obese mice [63]. The content of ceramide in skeletal muscles in obese insulin-resistant patients was also elevated [64]. Additionally, the accumulation of ceramide, sphingosine, and S1P was observed during prolonged exercise in human skeletal muscles [65,66]. Strenuous exercise can also cause the accumulation of ROS in fibers of working muscles [67]. Thus, sphingolipid metabolites appear to act as second messengers that further increase oxidative activity in skeletal muscles. Ceramide, for example, impairs the mitochondrial electron transport chain, resulting in the massive production of ROS [68,69]. Nikolova-Karakashian and Reid [58] suggested that sphingosine and S1P alter the proper function of mitochondria via the dysregulation of calcium homeostasis. Junctional membranes of transverse tubules (T-tubules) are structures that regulate $Ca^{2+}$ levels in muscle cells and mediate the rapid transfer of action potentials. T-tubules have a high content of sphingosine. Sphingosine decreases $Ca^{2+}$ release from the sarcoplasmic reticulum through the blockade of sarcoplasmic reticulum ryanodine receptors [70,71]. S1P has the opposite effect, increasing cytosolic $Ca^{2+}$ levels by binding to $S1P_3$ and $S1P_2$ receptors in skeletal muscles [72,73] to modulate muscle contractions and force, along with sphingosine. Thus, sphingolipids likely play a prominent role in regulating mitochondrial $Ca^{2+}$ mobilization.

Ceramide and SMase were also shown to impair contractile function and contribute to the fatigue-associated depression of tetanic force of skeletal muscles through an increase in oxidative activity [74]. Neutralizing muscle-derived oxidants with the antioxidant *N*-acetylcysteine delayed the development of fatigue during exercise [75]. Disturbances of $Ca^{2+}$ release lead to the failure of action potential conductance and consequently the fatigue of skeletal muscles. SMase activity was shown to lessen the force and $Ca^{2+}$ sensitivity of the contractile apparatus of the diaphragm in mice [76]. In contrast, S1P, which mobilizes cellular $Ca^{2+}$, slowed the fatigue-associated loss of force during muscle contraction of the mouse extensor digitorum longus muscle [77]. Sphingolipids modulate the function of skeletal muscles, and sphingolipid signaling appears to be liable for the weakness and fatigue of skeletal muscles in response to chronic inflammatory disease. Empinado et al. [78] reported ceramide-mediated dysfunction of the diaphragm following heart failure, which could result from muscle cell apoptosis and impairments in contractile function [79,80]. TNFα-induced ceramide synthesis was shown to cause rat L6 myotube atrophy in in vitro studies. Mice that were inoculated with colon C26 adenocarcinoma developed pronounced muscle atrophy, marked by high ceramide levels in skeletal muscles [81]. The S1P/S1P$_2$ receptor axis was also shown to promote skeletal muscle atrophy that was associated with C26 adenocarcinoma in mice [82].

## Sphingolipids, skeletal muscle differentiation, and regeneration

Sphingolipids likely regulate skeletal muscle development and regeneration with regard to their role in modulating skeletal muscle cell proliferation, differentiation, and apoptosis. Ceramide 1-phosphate stimulates the proliferation of mouse C2C12 myoblasts via the lysophosphatidic acid signaling axis [83] and through activation of the phosphatidylinositol 3-kinase/Akt (PI3K/Akt), MAPK3/1, ERK1/2, and mechanistic/mammalian target of rapamycin signal transduction pathways [84]. In contrast, exogenous S1P has been shown to inhibit the proliferation and induce the differentiation of C2C12 cells through activation of the ERK1/2 and MAPK signaling pathways via S1P$_2$ receptors [85]. S1P has also been shown to promote cytoskeletal remodeling during myoblast differentiation via upregulation of the gap junction protein connexin-43 [86]. The depletion of SphK, the enzyme that generates S1P, stimulates the proliferation of C2C12 myoblasts and inhibits myogenic differentiation [87]. However, ceramide exerts opposite effects and has been shown to inhibit the differentiation of myoblasts [88] and induce apoptotic signaling in differentiated myotubes [89].

The S1P/S1P$_2$ receptor axis has also been shown to be a major inducer of satellite cell proliferation during skeletal muscle regeneration. This occurs mainly via the S1P$_2$/signal transducer and activator of transcription 3 (STAT3) pathway [90], which was demonstrated in dystrophic muscles [91]. However, S1P/S1P$_3$ receptor-mediated regeneration of the soleus muscle was observed after myotoxic injury in rats [92]. Additionally, [93,94] showed that S1P mediates the epidermal growth factor-stimulated proliferation of satellite cells and confirmed that S1P inhibition impaired muscle regeneration. Moreover, S1P increased satellite cell motility, which is required for the recruitment of these cells at the site of muscle damage, through S1P$_1$ and S1P$_4$ receptors [95]. S1P can also be secreted from muscle cells through adenosine triphosphate-binding cassette transporters and act via S1P receptors as autocrine or paracrine messengers to promote muscle cell proliferation, fiber growth, repair, and resistance to fatigue (for review, see [96]).

## Metabolic substrate uptake

For proper functioning, skeletal muscle cells require an unlimited availability and effective uptake of metabolic substrates. Glucose and amino acid uptake by skeletal muscles is largely controlled by sphingolipid signaling. The S1P/S1P$_2$ receptor axis positively regulates insulin-induced glucose uptake via the inhibition of protein tyrosine phosphatase-1B, which in turn promotes the activating *trans*-phosphorylation of insulin receptors [97]. Conversely, ceramide reduces insulin-dependent glucose uptake through impairments in the translocation of glucose transporter type 4 (GLUT4), the loss of Rho family guanosine triphosphatase Rac activation, actin remodeling [98], and the failure of Akt membrane recruitment [99]. Ceramide suppresses amino acid uptake and consequently protein synthesis in rat myotubes by downregulating sodium-coupled neutral amino acid transporter 2 [100].

## Sphingolipids and insulin resistance

Skeletal muscles are responsible for up to 90% of the systemic response to insulin and thus are a key driver of normoglycemia [101]. The cytokine-mediated increase in ceramide content impairs glucose uptake in this tissue and leads to insulin resistance in obesity and diabetes. In obesity, excessive lipids accumulate in non-adipose peripheral tissues, including skeletal muscles, and fuel the de novo synthesis of sphingolipid derivatives, such as ceramide [101]. An elevated supply of plasma free FAs increased ceramide content in different types of skeletal muscles [102]. Human studies have shown a strong correlation between high ceramide concentrations in skeletal muscles in obese subjects and a loss of insulin sensitivity [64]. C18-ceramide species are abundant in skeletal muscles and potent skeletal muscle-specific mediators of insulin resistance [63,103]. Additionally, the levels of CERT, which directs ceramide from the ER toward its turnover to other sphingolipid species in the Golgi apparatus, decrease under lipotoxic conditions, causing ceramide accumulation and the disruption of insulin signaling [104]. Two significant players of insulin signaling (Akt and insulin receptor substrate 1) have been proposed as targets of ceramide. The short-term actions of ceramide cause Akt inhibition via the rapid activation of PP2A [105] or protein kinase Cζ (PKCζ; [106,107]). In caveolae (i.e., the specific membrane sites for stress-induced ceramide synthesis in muscle cells; [108]), both Akt and PKCζ are recruited and sequestrated by ceramides to block insulin signaling [109]. The long-term actions of ceramide that lead to insulin resistance target insulin receptor substrate 1 through two inhibitory axes: (1) interferon-induced, double-stranded RNA-activated protein kinase/JNK [110] and (2) Pbx transcription factors [111]. Skeletal muscle-associated insulin resistance also appears to be a consequence of the ceramide-mediated apoptosis of L6 myotubes under lipotoxic conditions, which is likely attributable to cellular remodeling that accompanies this type of cell death [89]. Limited by accumulating ceramide amino acid uptake and protein synthesis in skeletal muscle leads to muscle waste and the loss of lean body mass in diabetic patients [112].

Sphingomyelin synthases metabolize ceramide to sphingomyelin, and the inhibition of SMS activity in muscle cells disturbs insulin signaling [113]. Low plasma levels of C16:1 sphingomyelin species are correlated with the risk of T2D [114]. Independent of ceramide, GM3 has been shown to inhibit skeletal muscle-associated insulin signaling in an animal model [115].

In summary, sphingolipids are bioactive compounds that have emerged as major regulators of skeletal muscle contractile function, metabolic substrate uptake, development, and regeneration. However, sphingolipids are also responsible for skeletal muscle dysfunction that is observed in response to stress factors or chronic inflammatory disease (Fig. 19.3). Insulin resistance is a particular manifestation of the pathological action of sphingolipids in this tissue and increases the risk of health problems, such as diabetes, heart failure, stroke, and cancer.

# Role of sphingolipids in adipose tissue metabolism

## Adipogenesis

Specific species of sphingolipids and ceramides mediate lipotoxicity in muscles and the heart [116], but the contribution of specific lipotoxic species to white adipose tissue (WAT)

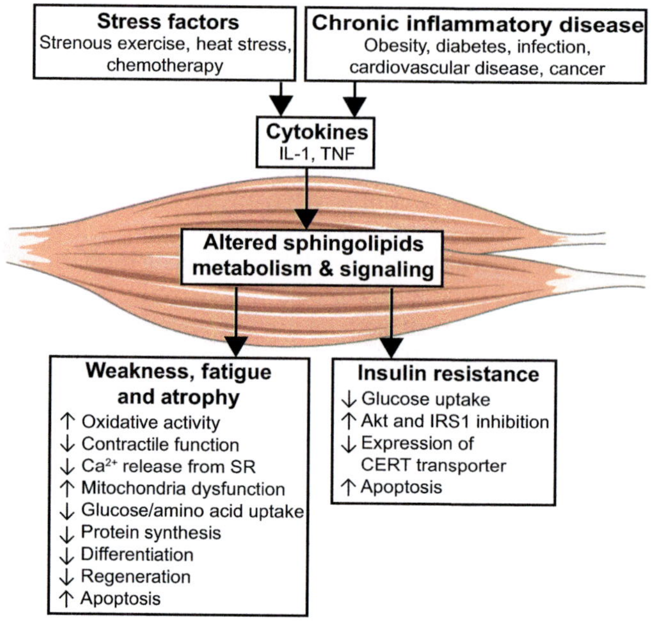


FIGURE 19.3 **Sphingolipid-induced dysfunction of skeletal muscle in response to stress factors and chronic inflammatory disease.** Various environmental stressors and chronic inflammatory disease affect sphingolipid metabolism in skeletal muscles via cytokines. The modulation of different cellular processes by alterations of sphingolipid content leads to pathological changes in skeletal muscles, such as weakness and atrophy, or skeletal muscle-associated insulin resistance. See main text for detailed descriptions of signaling pathways that are modulated by sphingolipids.

differentiation and dysfunction is still not well defined [117]. Numerous studies have reported that the inhibition of SPT activity with myriocin, an SPT-specific inhibitor, significantly attenuates symptoms of diet-induced obesity (DIO), including decreases in ceramide accumulation, decreases in adipose tissue, smaller adipocytes, and improvements in metabolic function [118,119]. More detailed information about the function of SPT in adipose tissue has come from studies of mice with adipocyte-specific deletion of the *Sptlc1* and *Sptlc2* genes [120,121], which encode proteins of long-chain base subunits 1 and 2 of SPT, respectively. Sptlc1-deficient mice initially developed adipose tissue and exhibited a striking age-dependent loss of adipose tissue accompanied by evidence of adipocyte death, an increase in macrophage infiltration, and tissue fibrosis. Adipocyte differentiation was unaffected by Sptlc1 deletion [120]. Adipocyte-specific Sptlc2 KO mice exhibited a marked reduction of adipose tissue mass and an increase in the number of brown/beige adipocytes [121]. Mechanistically, S1P was reduced in adipose tissues in aSptlc2 KO mice, which inhibited adipocyte proliferation and differentiation via the downregulation of $S1P_1$ receptors and lower activity of peroxisome proliferator-activator receptor γ (PPAR-γ). Additionally, the downregulation of sterol regulatory element-binding protein 1c (SREBP-1c) prevented adipogenesis in aSptlc2 KO adipocytes [121]. Collectively, these findings suggest that the tight regulation of de novo sphingolipid biosynthesis and S1P signaling plays an important role in adipogenesis, adipocyte survival, and normal metabolic function [120,121].

The enzyme DES1 catalyzes the last step in the main ceramide biosynthetic pathway. DES1 expression is specifically decreased in adipose tissue in obese patients and murine models of genetic and nutritional obesity. Loss-of-function experiments showed that pharmacologically or genetically ablated DES1 in preadipocytes prevented adipogenesis and decreased lipogenesis [117]. This was associated with elevations of oxidative stress, cellular death, and blockade of the cell cycle. These effects were coupled with higher dihydroceramide and dihydrosphingomyelin content. Moreover, the pharmacological inhibition of DES1 impaired adipocyte differentiation. These findings identified DES1 as a new potential target to restore proper adipose tissue function [117].

The involvement of S1P and S1P receptors in the proliferation and differentiation of white and brown adipocytes has been recently demonstrated [122,123]. A high-fat diet induced the accumulation of large adipocytes, an effect that was blunted by the S1P receptor agonist FTY720, which inhibited adipogenesis and promoted lipolysis [124]. FTY720 also significantly decreased lipid accumulation in maturing preadipocytes and downregulated the transcriptional levels of PPARγ, C/EBPα, and adiponectin, which are markers of adipogenic differentiation. These results indicate that S1P receptor inhibition prevents obesity by modulating adipogenesis and lipolysis [124]. During the differentiation of 3T3-L1 cells and adipose tissue in *ob/ob* mice, SphK1 and SphK2, two isotypes of sphingosine kinase, were transcriptionally upregulated, and cellular S1P levels increased [125]. In DIO mice, SphK1 deficiency increased markers of adipogenesis and adipose gene expression. Aging SphK2$^{-/-}$ mice were protected from metabolic decline and obesity compared with wildtype mice. Fifty-two-week-old male SphK2$^{-/-}$ mice exhibited decreases in body weight and fat mass and increases in glucose tolerance and insulin sensitivity compared with control mice [55]. These findings underscore the important role of S1P metabolism in the regulation of adipocyte function.

SMS1 catalyzes the conversion of ceramide to sphingomyelin. Mice with global SMS1 KO were shown to exhibit lower sphingolipid levels in adipose tissue. SMS1 KO mice exhibited the loss of epididymal WAT mass [17]. No changes were observed in the expression of factors that are required for adipogenesis, suggesting that adipocyte differentiation proceeds normally in WAT cells in SMS1 KO mice. In vivo analyses indicated that FA uptake in WAT but not in the liver decreased in SMS1 KO mice compared with wildtype mice. Altogether, these data suggest that SMS1 is crucial for maintaining WAT function [126].

## Insulin signaling and inflammation

The levels of ceramides increase in rodent WAT in response to a high-fat diet concomitantly with the onset of insulin resistance [127]. Several studies found that the inhibition of myriocin-induced de novo ceramide production attenuated symptoms of DIO, including improvements in insulin signaling through Akt, the abolishment of adipose tissue inflammation, and improvements in metabolic function [36,118,119]. Furthermore, in 3T3-L1 adipocytes and brown adipose tissue, ceramides impair insulin-stimulated GLUT4 expression and glucose uptake [128]. Ceramides also mediate the effect of TNFα on GLUT4 mRNA content in these cells [128].

Whole-body and adipose tissue-specific inhibition or deletion of SPT increased insulin sensitivity by reducing adipose sphingolipids. This manipulation also induced the secretion of insulin-sensitizing adipokines [129]. Similarly, CerS5 KO mice exhibited improvements in glucose tolerance and lower adipose tissue inflammation after high-fat diet feeding. This indicates improvements in overall adipose tissue health after high-fat diet feeding in CerS5 KO mice [130]. These findings corroborate previous observations that the tissue-specific ablation of CerSs in murine brown adipose tissue led to an increase in energy expenditure and insulin sensitivity by enhancing thermogenesis in brown adipose tissue [131]. The pharmacological inhibition of CerS was also shown to stimulate the browning of WAT and increased insulin sensitivity, indicating that high rates of ceramide synthesis may indeed impair the metabolic health of adipose tissue [129].

SphK1/S1P signaling induces adipose inflammation and reduces adipogenesis, leading to adipose tissue dysfunction and subsequent insulin resistance [132]. In DIO mice, SphK1 deficiency increased adipose gene expression of the anti-inflammatory molecules IL-10 and adiponectin and decreased expression of the proinflammatory molecules TNFα and IL-6. These changes were associated with an increase in insulin signaling in adipose and muscle tissues and improved systemic insulin sensitivity and glucose tolerance in $SphK1^{-/-}$ mice. The specific pharmacological inhibition of SK1 in wildtype DIO mice also reduced adipocyte inflammation and improved overall glucose homeostasis [132]. Similar effects were observed in 52-week-old male $Sphk2^{-/-}$ mice [55], thus emphasizing the importance of S1P and SphK in proper adipose insulin signaling.

## Sphingolipids as regulators of adipocyte lipid metabolism

Sphingolipid metabolic pathways affect lipid turnover in adipocytes. The best-known pathway involves the regulation of lipolysis by sphingolipids. Extracellular S1P time- and concentration-dependently stimulated lipolysis in white adipocytes through the adenylyl cyclase-cyclic adenosine monophosphate (cAMP) signaling pathways [133]. The stimulation of $\beta_3$-adrenergic receptors induces the expression of SphK1 and increases S1P production in adipocytes in a manner that depends on hormone-sensitive lipase activity, which ultimately leads to an increase in expression of the proinflammatory cytokine IL-6 [134]. These findings demonstrate that SphK1 is a critical mediator of lipolysis-triggered inflammation in adipocytes. FTY720 treatment promoted lipolysis in adipocytes in high-fat diet-fed mice by inducing the expression of hormone-sensitive lipase, adipose triglyceride lipase, and perilipin, which are regulators of lipolysis [124]. Moreover, analyses of lipid metabolic gene expression showed increases in expression of the adipose triglyceride lipase gene and protein levels in adipocytes in 52-week-old male $Sphk2^{-/-}$ compared with control mice [55]. Overall, these findings clearly indicate that lipolysis is affected by SphK signaling.

Interestingly, the lipotoxic effects of C16 ceramide are partially attributable to the inhibition of fatty acid β-oxidation in adipose tissue [135,131]. Fenretinide (FEN) is a synthetic retinoid that inhibits ceramide biosynthesis. The blockade of ceramide synthesis in adipose tissue in FEN-treated obese mice was associated with the complete normalization of impairments in mitochondrial β-oxidation and tricarboxylic acid cycle flux [136]. This strongly suggests that FEN treatment in vivo can alleviate disturbances in adipose tissue mitochondrial function that result from a high-fat diet.

Excess membrane sphingomyelin in adipocytes can downregulate SREBP-1, PPARγ, cAMP response element binding protein (CREB), and Ras/Raf/MEK/ERK and upregulate SREBP-2 gene expression [137]. In vitro studies that used cultured adipocytes revealed that SMS1 depletion reduced FA uptake [126]. The upregulation of FA synthesis and degradation of triglycerides were observed in WAT in CerS5 KO mice, both of which led to the production of FAs. Phosphoenolpyruvate carboxykinase 1 was induced in WAT in CerS5 KO mice providing glycerol 3-phosphate for TAG synthesis and FAs re-esterification [138]. Collectively, these data indicate that sphingolipids are important contributors to the dysregulation of lipogenic genes in obesity.

## Sphingolipids in the cardiovascular system

### Ceramides in cardiac pathology

In several animal studies, ceramide accumulation in the heart was shown to correlate with pathophysiological events and heart dysfunction [139]. High myocardial ceramide levels were associated with cardiac dysfunction in Akita Ins2 mice, ZDF rats [140], and *ob/ob* mice [141]. The pharmacological or genetic reduction of cardiac ceramide levels in these models improved cardiac function. In myocardial left ventricle biopsies from patients with chronic ischemia, ceramide levels were higher in biopsies from subjects with impairments in heart function [142]. The accumulation of cardiac ceramide in the post-ischemic heart was mediated by acid SMase [142]. The blockade of SPT by myriocin reduced ceramide accumulation in ischemic cardiomyopathy, together with a reduction of ventricular remodeling, fibrosis, and macrophage content following myocardial infraction [143]. Genetic deletion of the *Sptlc2* gene preserved cardiac function after myocardial infarction. Finally, in vitro studies showed that changes in ceramide synthesis are linked to hypoxia and inflammation [143].

The possible direct connection between ceramide and lipotoxic cardiomyopathy is not fully understood. The accumulation of ceramide was reported to be accompanied by cardiomyocyte apoptosis, and the blockade of ceramide biosynthesis reduced cardiomyocyte apoptosis in ZDF rats, MHC-ACS1 mice, and *ob/ob* mice [140,141]. Palmitate-induced cell death in cardiomyocytes is mediated by a reduction of mitochondrial activity. Cyclosporin inhibits the development of mitochondrial transition pores and blocks palmitate-induced alterations of mitochondrial function and palmitate-induced cell death [144]. Alterations of mitochondrial activity may affect energetic substrate utilization in cardiomyocytes. The inhibition of SPT resulted in an increase in FAs and increase in glucose oxidation in isolated perfused hearts in mice, improved systolic function, and improved survival rates. These studies indicate that the lipotoxicity of ceramides may be associated with glucose and FA metabolism variability rather than direct cardiomyocyte apoptosis [145].

### Sphingolipids in cardioprotection

The activation of SphK/S1P-mediated signaling has been shown to exert cardioprotective effects in response to acute ischemia/reperfusion injury [146]. The application of exogenous S1P or GM1, a ganglioside that activates SphK1 and elevates endogenous S1P production, to cultured cardiac myocytes that were subjected to hypoxia or the treatment of isolated hearts either before ischemia or at the onset of reperfusion exerted prosurvival effects in mice. Interestingly, GM1 treatment reduced cardiac injury through PKCε, whereas S1P exerted cardioprotective effects through a PKCε-independent pathway [147]. Endogenous S1P is transported from cardiomyocytes and exerts cardioprotective effects by binding to S1P receptors on the membrane surface [148]. Consistent with these findings, ischemia inhibited SphK activity and reduced S1P content in the heart [149]. SphK1-deficient cardiomyocytes were susceptible to ischemia/reperfusion injury, and overexpression of the SphK1 gene induced cardioprotective effects and prevented ischemic heart failure [150]. A recent study found that S1P activated STAT3 to protect against de novo acute heart failure by improving heart rate, with no changes in left ventricular pressure [151].

Unclear are the specific S1P receptors that play a key role in cardioprotection in the heart. Three of the five known S1P receptors—$S1P_1$, $S1P_2$, and $S1P_3$—are expressed in the adult mammalian heart [139]. $S1P_1$ receptor-specific agonists protected adult mouse cardiomyocytes from hypoxia, whereas the $S1P_1$ receptor antagonist FTY720 suppressed the cardioprotective effect of S1P [152]. The combined deletion of $S1P_2$ and $S1P_3$ receptors augmented infarct size in mice that were subjected to ischemia/reperfusion injury [153]. In these hearts, Akt activation was markedly attenuated compared with wildtype mice, but the absence of either receptor subtype alone did not affect either infarct size or Akt activation after ischemia/reperfusion injury. S1P augmented Akt activity in control murine myocytes but was ineffective in double KO cells. These observations suggest overlapping roles of S1P receptor isoforms, and less abundant cardiac myocyte S1P receptors (i.e., $S1P_2$ and $S1P_3$) may also be necessary for cell survival during ischemia/reperfusion injury [139].

## Ceramides and vascular function

Ceramide and its derivatives are implicated in vascular and cardiac morphogenesis, the maintenance of vascular tone, vascular permeability, inflammation, aging, and atherosclerosis [154,155]. Many studies of the vascular system have focused on the role of ceramides, particularly S1P, which is present at high concentrations in the circulatory system but has low tissue levels. Therefore, the access of plasma S1P to receptors that are expressed on endothelial cells (ECs) and vascular smooth muscle cells (VSMCs) appears to play a more important role in controlling cellular balance than the production of sphingolipids in these cells. Through interactions with G protein-coupled receptors, S1P regulates such cellular processes as cell growth, survival, proliferation, and migration. In the vascular system, five G protein-coupled receptors of S1P have been identified: $S1P_1$–$S1P_5$. Endothelial cells express $S1P_1$ and $S1P_3$ receptors, whereas the expression of $S1P_1$–$S1P_3$ has been found in VSMCs (Fig. 19.4). The impact of S1P on the blood vessel wall depends on the level of expression of particular receptors and receptor-S1P interactions. Interestingly, depending on the localization of the vessel, S1P has differential effects on this tissue [156].

## Vascular reactivity

In the vasculature, S1P has been shown to influence vascular reactivity. In ECs, S1P receptors are involved in mediating vasodilation, whereas S1P receptors that are expressed on VSMCs regulate vasoconstriction [157]. Many studies have indicated a role for SphK in the Akt-mediated activation of endothelial nitric oxide synthase (eNOS) and regulation of NO production in ECs [158,159]. The role of eNOS-mediated NO production in platelet

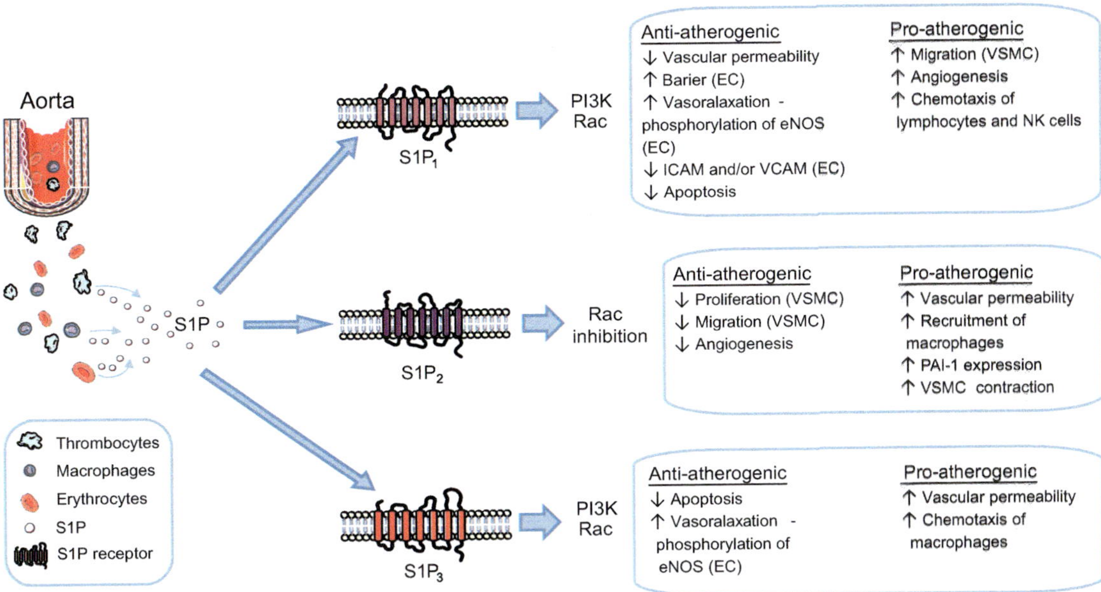


**FIGURE 19.4 S1P signaling through S1P$_1$, S1P$_2$, and S1P$_3$ receptors in the cardiovascular system.** S1P is secreted by different blood cells (erythrocytes, thrombocytes, and macrophages) and in significantly lower amounts by endothelial cells and vascular smooth muscle cells. S1P receptors activate different but partially overlapping signaling pathways to induce complex effects on the vasculature. *S1P, sphingosine-1-phosphate; VSMC, vascular smooth muscle cell; EC, endothelial cell.*

aggregation, vascular remodeling, and angiogenic processes is widely recognized [160]. In the vascular system, the expression of three S1P receptors regulate multiple vascular signaling pathways. The impact of S1P on NO-dependent signaling in arteries appears to influence posttranslational protein modification, extracellular matrix proteins cross-linkage, and arterial wall stiffness [161]. Nevertheless, unknown is whether S1P$_1$ or S1P$_3$ is involved in eNOS-dependent NO production [162,163].

The short-term exposure of ECs to ceramides resulted in a reduction of endothelium-dependent vasorelaxation [164,165]. Moreover, the palmitate-induced dysfunction of ECs was prevented through the inhibition of ceramide production [166]. Similar effects were observed in pulmonary vascular dysfunction. The lipopolysaccharide-induced dysfunction of ECs was prevented by the inhibition of acid SMase activity, thus confirming the aggravating role of ceramides in vascular dysfunction [167]. Many reports have shown that the activity of neutral SMase increases with age, causing an elevation of ceramide production followed by eNOS inactivation and a reduction of NO release. Endothelial dysfunction leads to impairments in vasoconstriction mediator production and bioavailability and thus an imbalance in VSMC signaling [168].

## Vascular remodeling

S1P is abundantly and intracellularly stored in platelets and released into the extracellular environment upon stimulation [169]. The binding of S1P to S1P$_1$ receptors results in Rac activation and VSMC chemotaxis [170]. S1P receptors that are expressed on VSMCs, mainly $S1P_1$ but also $S1P_3$, are indispensable for S1P-dependent VSMC migration through activation of the ERK1/2, MAPK, and JNK pathways. In contrast to $S1P_1$ and $S1P_3$ receptors (i.e., pro-migratory receptors), $S1P_2$ receptors (i.e., anti-migratory receptor) reduce S1P-mediated VSMC migration. Diabetic conditions enhance the expression of $S1P_1$ and $S1P_3$ receptors and thus VSMC migration. Vascular smooth muscle cells that were stimulated by the chemoattractant platelet-derived growth factor exhibited a migratory phenotype that was attributable to SphK activation and S1P synthesis [171]. Angiotensin II is involved in the development of cardiovascular disease and was shown to induce the epigenetic regulation of $S1P_1$ expression through $H_2O_2$-mediated apurinic/apyrimidinic endonuclease/redox factor-1 translocation. This finding may indicate the mechanism by which angiotensin II is involved in neointima formation [172].

## Sphingolipids and vascular disorders

S1P is a constituent of high-density lipoproteins (HDLs) and contributes to their beneficial effects on the cardiovascular system. Such disorders as T2D and coronary artery disease are characterized by a lower blood plasma content of HDL-S1P [173]. The severity of coronary atherosclerosis is negatively correlated with serum levels of HDL-S1P. Moreover, a decrease in S1P content, which is related to alterations of HDL levels, resulted in impairments in NO-dependent vasodilatation in patients with coronary artery disease [174]. S1P interacts with the TNFα-dependent signaling pathway and exerts anti-inflammatory effects on TNFα-induced inflammation in VSMCs. The S1P-dependent pathway through $S1P_2$ receptors activates phosphatase and tensin homolog (PTEN), which negatively regulates the PI3K pathway and influences cell proliferation [175]. Other studies have shown that S1P activates the PI3K/Akt pathway in VSMCs [176]. The HDL-S1P complex inhibits the induction of inducible nitric oxide synthase and matrix metalloproteinase 9 by VSMCs after treatment with the proinflammatory agent TNF-α. This effect was entirely dependent on S1P content in the HDL-S1P complex [175]. Overall, these findings suggest that HDL-S1P is an independent predictor of coronary in-stent restenosis [177].

Ceramides promote lipoprotein infiltration into the vessel wall [178]. The development of atherosclerosis involves ceramide-dependent pathways, such as ceramide-induced foam cell apoptosis, interactions with ROS and NO, and the synthesis of inflammatory cytokines. Sphingolipids are involved in processes that regulate immune and inflammatory responses. The vascular S1P gradient participates in controlling lymphocyte circulation from lymphoid organs into the circulation and the recruitment of lymphocytes to sites of inflammation [179]. The vascular S1P gradient also participates in elevations of TNF-α expression via the stimulation of monocytes/macrophages [180]. Moreover, S1P through $S1P_1$ receptors is required for the ability of natural killer cells to egress from lymphoid tissue and bone marrow [181]. Although S1P is generally atheroprotective, S1P levels and S1P receptor expression under inflammatory conditions undergo changes that lead to modifications of EC and lymphocyte behavior. Under conditions of symptoms of metabolic syndrome, the modified activation of different S1P receptors affects atheroma formation. $S1P_1$ and $S1P_3$ receptors that are expressed on ECs mediate anti-atherosclerotic effects via the stimulation of NO production and inhibition of intracellular adhesion molecule, vascular cell adhesion molecule, and monocyte chemoattractant peptide-1. The attenuation of $S1P_2$ signaling contributes to the enhancement of B1 lymphocytes, which are recognized as atheroprotective cells [182]. In a classic mouse model of atherosclerosis, the deletion of $S1P_2$ in apolipoprotein E KO mice decreased the development of atherosclerosis [183]. Conversely, the expression of $S1P_2$ on macrophages augments pro-atherogenic actions [174]. Depending on its expression site, a dual role for $S1P_3$ receptors has been shown. $S1P_3$ receptors that are expressed on VSMCs contribute to anti-atherogenic effects, whereas $S1P_3$ receptor expression on macrophages contributes to pro-atherogenic effects [174]. Ceramide production increases as a result of the oxidized low-density lipoprotein-dependent activation of SMase. Hydrolysis of the low-density lipoprotein–sphingomyelin complex was proposed to be involved in ceramide accumulation in atheroma. Ceramides may be promising predictive markers of disease development and progression after adjusting for age, sex, body mass index, the occurrence of cardiovascular events, smoking, level of low-density lipoprotein cholesterol, HDL cholesterol, triglycerides, serum glucose, and a family history of coronary artery disease [184].

## Conclusion

Sphingolipid content and composition undergo significant changes in tissues during the course of metabolic diseases. Among lipid metabolites, accumulating evidence indicates that ceramides are the main factors that are involved in lipotoxicity. The mechanisms of action of ceramides include the inhibition of insulin and growth factor signaling, impairments in mitochondrial lipid oxidation, ER stress, and ultimately the induction of apoptosis. The inhibition of ceramide synthesis has been shown to have beneficial effects in rodent models of atherosclerosis, insulin resistance, diabetes, and cardiomyopathy. Notably, however, ceramides are essential membrane components that are necessary for proper cell functioning. Moreover, some classes of sphingolipids might play fundamental constitutive roles or even mediate protective responses in cells. Therefore, lowering the toxic effects of sphingolipids by metabolizing them into less toxic lipid species (e.g., S1P and sphingomyelin) may counteract ceramide lipotoxicity, but one unresolved issue is whether the inhibition of these metabolites can exacerbate pathological conditions. If so, then can the inhibition of specific undesirable sphingolipids be achieved without altering the concentrations of beneficial sphingolipids? How can the tissue-specific effects of intracellular ceramides be dissected from systemic impairments in metabolic homeostasis? As long as these issues are unresolved, the effectiveness of such therapies, especially in humans, remains questionable.

## Acknowledgments

We thank Michael Arends for proofreading of the manuscript. Servier Medical Art artwork licensed under a Creative Common Attribution 3.0 Generic License (http://smart.servier.com) was used in the figures of this chapter. This work was supported by the National Science Centre, Poland (grant numbers UMO-2013/10/E/NZ3/00670 (A.D.), UMO-2011/03/B/NZ4/03055 (A.D.), UMO-2015/19/D/NZ4/03705 (J.J.), UMO-2017/27/N/NZ3/01987 (A.M.D.), UMO-2017/26/D/NZ4/00696 (A.F.), UMO-2014/13/B/NZ4/00199 (P.D.), and UMO-2016/22/E/NZ4/00650 (P.D.)) and the National Centre for Research and Development, Poland (grant number STRATEGMED3/305813/2/NCBR/2017).

## References

[1] Veret J, Bellini L, Giussani P, Ng C, Magnan C, Le Stunff H. Roles of sphingolipid metabolism in pancreatic β cell dysfunction induced by lipotoxicity. J Clin Med 2014;3:646–62.
[2] Janikiewicz J, Hanzelka K, Kozinski K, Kolczynska K, Dobrzyn A. Islet β-cell failure in type 2 diabetes—within the network of toxic lipids. Biochem Biophys Res Commun 2015;460:491–6.
[3] Holland WL, Summers SA. Sphingolipids, insulin resistance, and metabolic disease: new insights from in vivo manipulation of sphingolipid metabolism. Endocrinology 2008;29:381–402.
[4] Gault CR, Obeid LM, Hannun YA. An overview of sphingolipid metabolism: from synthesis to breakdown. Adv Exp Med Biol 2010;688:1–23.
[5] Mwinyi J, Boström A, Fehrer I, Othman A, Waeber G, Marti-Soler H, et al. Plasma 1-deoxysphingolipids are early predictors of incident type 2 diabetes mellitus. PLoS One 2017;12:e0175776.
[6] Galadari S, Rahman A, Pallichankandy S, Galadari A, Thayyullathil F. Role of ceramide in diabetes mellitus: evidence and mechanisms. Lipids Health Dis 2013;12:98.
[7] Veret J, Coant N, Berdyshev EV, Skobeleva A, Therville N, Bailbé D, et al. Ceramide synthase 4 and de novo production of ceramides with specific N-acyl chain lengths are involved in glucolipotoxicity-induced apoptosis of INS-1 beta-cells. Biochem J 2011;438:177–89.
[8] Cowart LA. Sphingolipids: players in the pathology of metabolic disease. Trends Endocrinol Metab 2009;20: 34–42.
[9] Bellini L, Campana M, Mahfouz R, Carlier A, Véret J, Magnan C, et al. Targeting sphingolipid metabolism in the treatment of obesity/type 2 diabetes. Expert Opin Ther Targets 2015;19:1037–50.
[10] Gjoni E, Brioschi L, Cinque A, Coant N, Islam MN, Ng CK, et al. Glucolipotoxicity impairs ceramide flow from the endoplasmic reticulum to the golgi apparatus in INS-1 beta-cells. PLoS One 2014;9:e110875.
[11] Yaribeygi H, Bo S, Ruscica M, Sahebkar A. Ceramides and diabetes mellitus: an update on the potential molecular relationships. Diabet Med 2019;25 doi: 10.1111/dme.13943.
[12] Boslem E, Meikle PJ, Biden TJ. Roles of ceramide and sphingolipids in pancreatic β-cell function and dysfunction. Islets 2012;4:177–87.
[13] Yoon SO, Gong DB, Dong JB, Eun-Young P, Hee-Sook J. Fatty acid-induced lipotoxicity in pancreatic beta-cells during development of type 2 diabetes. Front Endocrinol 2018;9:384.
[14] Park JW, Park WJ, Kuperman Y, Boura-Halfon S, Pewzner-Jung Y, Futerman AH. Ablation of very long acyl chain sphingolipids causes hepatic insulin resistance in mice due to altered detergent-resistant membranes. Hepatology 2013;57:525–32.
[15] Guo J, Qian Y, Xi X, Hu X, Zhu J, Han X. Blockage of ceramide metabolism exacerbates palmitate inhibition of pro-insulin gene expression in pancreatic beta-cells. Mol Cell Biochem 2010;338:283–90.
[16] Kelpe CL, Moore PC, Parazzoli SD, Wicksteed B, Rhodes CJ, Poitout V. Palmitate inhibition of insulin gene expression is mediated at the transcriptional level via ceramide synthesis. J Biol Chem 2003;278:30015–21.
[17] Yano M, Watanabe K, Yamamoto T, Ikeda K, Senokuchi T, Lu M, et al. Mitochondrial dysfunction and increased reactive oxygen species impair insulin secretion in sphingomyelin synthase 1-null mice. J Biol Chem 2011;286:3992–4002.
[18] Guo J, Zhu JX, Deng XH, Hu XH, Zhao J, Sun YJ, et al. Palmitate-induced inhibition of insulin gene expression in rat islet β-cells involves the ceramide transport protein. Cell Physiol Biochem 2010;26:717–28.
[19] Hagman DK, Hays LB, Parazzoli SD, Poitout V. Palmitate inhibits insulin gene expression by altering PDX-1 nuclear localization and reducing MafA expression in isolated rat islets of Langerhans. J Biol Chem 2005;280: 32413–8.
[20] Khoo S, Griffen SC, Xia Y, Baer RJ, German MS, Cobb MH. Regulation of insulin gene transcription by ERK1 and ERK2 in pancreatic β cells. J Biol Chem 2003;278:32969–77.
[21] Mathias S, Peña LA, Kolesnick RN. Signal transduction of stress via ceramide. Biochem J 1998;335:465–80.
[22] Kaneto H, Xu G, Fujii N, Kim S, Bonner-Weir S, Weir GC. Involvement of c-Jun N-terminal kinase in oxidative stress-mediated suppression of insulin gene expression. J Biol Chem 2002;277:30010–8.
[23] Major CD, Gao ZY, Wolf BA. Activation of the sphingomyelinase/ceramide signal transduction pathway in insulin-secreting β-cells: role in cytokine-induced β-cell death. Diabetes 1999;48:1372–80.
[24] Watson ML, Macrae K, Marley AE, Hundal HS. Chronic effects of palmitate overload on nutrient-induced insulin secretion and autocrine signalling in pancreatic MIN6 beta cells. PLoS One 2011;6:e25975.

[25] Pearson GL, Mellett N, Chu KY, Boslem E, Meikle PJ, Biden TJ. A comprehensive lipidomic screen of pancreatic β-cells using mass spectroscopy defines novel features of glucose-stimulated turnover of neutral lipids, sphingolipids and plasmalogens. Mol Metab 2016;5:404–14.
[26] Pearson MJ. Ceramide accumulation in the alpha cell drives glucagon secretion and hyperglycemia. The University of Texas Southwestern Medical Center at Dallas. PhD dissertation (https://utswmed-ir.tdl.org/handle/2152.5/4246).
[27] Shimabukuro M, Higa M, Zhou YT, Wang MY, Newgard CB, Unger RH. Lipoapoptosis in beta-cells of obese prediabetic fa/fa rats. Role of serine palmitoyltransferase overexpression. J Biol Chem 1998;273:32487–90.
[28] Japtok L, Schmitz EI, Fayyaz S, Krämer S, Hsu LJ, Kleuser B. Sphingosine 1-phosphate counteracts insulin signaling in pancreatic β-cells via the sphingosine 1-phosphate receptor subtype 2. FASEB J 2015;29:3357–69.
[29] Boslem E, Weir JM, MacIntosh G, Sue N, Cantley J, Meikle PJ, et al. Alteration of endoplasmic reticulum lipid rafts contributes to lipotoxicity in pancreatic β-cells. J Biol Chem 2013;288:26569–82.
[30] Lang F, Ullrich S, Gulbins E. Ceramide formation as a target in beta-cell survival and function. Expert Opin Ther Targets 2011;15:1061–71.
[31] Preston AM, Gurisik E, Bartley C, Laybutt DR, Biden TJ. Reduced endoplasmic reticulum (ER)-to-Golgi protein trafficking contributes to ER stress in lipotoxic mouse beta cells by promoting protein overload. Diabetologia 2009;52:2369–73.
[32] Hara T, Mahadevan J, Kanekura K, Hara M, Lu S, Urano F. Calcium efflux from the endoplasmic reticulum leads to β-cell death. Endocrinology 2014;155:758–68.
[33] Lei X, Barbour SE, Ramanadham S. Group VIA $Ca^{2+}$-independent phospholipase A2 (iPLA2beta) and its role in beta-cell programmed cell death. Biochimie 2010;92:627–37.
[34] Boslem E, MacIntosh G, Preston AM, Bartley C, Busch AK, Fuller M, et al. A lipidomic screen of palmitate-treated MIN6 β-cells links sphingolipid metabolites with endoplasmic reticulum (ER) stress and impaired protein trafficking. Biochem J 2011;435:267–76.
[35] Janikiewicz J, Szymański J, Malinska D, Patalas-Krawczyk P, Michalska B, Duszy ski J, et al. Mitochondria-associated membranes in aging and senescence: structure, function, and dynamics. Cell Death Dis 2018;9:332.
[36] Ussher JR, Koves TR, Cadete VJ, Zhang L, Jaswal JS, Swyrd SJ, et al. Inhibition of de novo ceramide synthesis reverses dietinduced insulin resistance and enhances whole-body oxygen consumption. Diabetes 2010;59:2453–64.
[37] Veluthakal R, Palanivel R, Zhao Y, McDonald P, Gruber S, Kowluru A. Ceramide induces mitochondrial abnormalities in insulin-secreting INS-1 cells: potential mechanisms underlying ceramide-mediated metabolic dysfunction of the beta cell. Apoptosis 2005;10:841–50.
[38] Syed I, Szulc ZM, Ogretmen B, Kowluru A. L-threo-C6-pyridinium-ceramide bromide, a novel cationic ceramide, induces NADPH oxidase activation, mitochondrial dysfunction and loss in cell viability in INS 832/13 β-cells. Cell Physiol Biochem 2012;30:1051–8.
[39] Lone MA, Santos T, Alecu I, Silva LC, Hornemann T. 1-Deoxysphingolipids. Biochim Biophys Acta Mol Cell Biol Lipids 2019;1864:512–21.
[40] Othman A, Rütti MF, Ernst D, Saely CH, Rein P, Drexel H, et al. Plasma deoxysphingolipids: a novel class of biomarkers for the metabolic syndrome? Diabetologia 2012;55:421–31.
[41] Wei N, Pan J, Pop-Busui R, Othman A, Alecu I, Hornemann T, et al. Altered sphingoid base profiles in type 1 compared to type 2 diabetes. Lipids Health Dis 2014;13:161.
[42] Othman A, Saely CH, Muendlein A, Vonbank A, Drexel H, von Eckardstein A, et al. Plasma 1-deoxysphingolipids are predictive biomarkers for type 2 diabetes mellitus. BMJ Open Diabetes Res Care 2015;3:e000073.
[43] Zuellig RA, Hornemann T, Othman A, Hehl AB, Bode H, Güntert T, et al. Deoxysphingolipids, novel biomarkers for type 2 diabetes, are cytotoxic for insulin-producing cells. Diabetes 2014;63:1326–39.
[44] Ross JS, Russo SB, Chavis GC, Cowart LA. Sphingolipid regulators of cellular dysfunction in Type 2 diabetes mellitus: a systems overview. Clin Lipidol 2014;9:553–69.
[45] Tagami S, Inokuchi JiJ, Kabayama K, Yoshimura H, Kitamura F, Uemura S, et al. Ganglioside GM3 participates in the pathological conditions of insulin resistance. J Biol Chem 2002;277:3085–92.
[46] Aerts JM, Ottenhoff R, Powlson AS, Grefhorst A, van Eijk M, Dubbelhuis PF, et al. Pharmacological inhibition of glucosylceramide synthase enhances insulin sensitivity. Diabetes 2007;56:1341–9.
[47] Zhao H, Przybylska M, Wu IH, Zhang J, Siegel C, Komarnitsky S, et al. Inhibiting glycosphingolipid synthesis improves glycemic control and insulin sensitivity in animal models of type 2 diabetes. Diabetes 2007;56:1210–8.
[48] Prentice BM, Hart NJ, Phillips N, Haliyur R, Judd A, Armandala R, et al. Imaging mass spectrometry enables molecular profiling of mouse and human pancreatic tissue. Diabetologia 2019;62:1036–47.

[49] Subathra M, Qureshi A, Luberto C. Sphingomyelin synthases regulate protein trafficking and secretion. PLoS One 2011;6:e23644.

[50] MacDonald MJ, Ade L, Ntambi JM, Ansari IU, Stoker SW. Characterization of phospholipids in insulin secretory granules and mitochondria in pancreatic beta cells and their changes with glucose stimulation. J Biol Chem 2015;290:11075–92.

[51] Veret J, Coant N, Gorshkova IA, Giussani P, Fradet M, Riccitelli E, et al. Role of palmitate-induced sphingoid base-1-phosphate biosynthesis in INS-1 β-cell survival. Biochim Biophys Acta 2013;1831:251–62.

[52] Hong SW, Lee J, Kwon H, Park SE, Rhee EJ, Park CY, et al. Deficiency of sphingosine-1-phosphate reduces the expression of prohibitin and causes β-cell impairment via mitochondrial dysregulation. Endocrinol Metab 2018;33:403–12.

[53] Qi Y, Chen J, Lay A, Don A, Vadas M, Xia P. Loss of sphingosine kinase 1 predisposes to the onset of diabetes via promoting pancreatic β-cell death in diet-induced obese mice. FASEB J 2013;27:4294–304.

[54] Kowalski GM, Carey AL, Selathurai A, Kingwell BA, Bruce CR. Plasma sphingosine-1-phosphate is elevated in obesity. PLoS One 2013;8:e72449.

[55] Ravichandran S, Finlin BS, Kern PA, Özcan S. Sphk2$^{-/-}$ mice are protected from obesity and insulin resistance. Biochim Biophys Acta Mol Basis Dis 2019;1865:570–6.

[56] Song Z, Wang W, Li N, Yan S, Rong K, Lan T, et al. Sphingosine kinase 2 promotes lipotoxicity in pancreatic β-cells and the progression of diabetes. FASEB J 2019;33:3636–46.

[57] Nikolova-Karakashian MN, Reid MB. Sphingolipid metabolism, oxidant signaling, and contractile function of skeletal muscle. Antioxid Redox Signal 2011;15:2501–17.

[58] Dun YL, Zhou XL, Guan HS, Yu GL, Li CX, Hu T, et al. Low molecular weight guluronate prevents TNF-α-induced oxidative damage and mitochondrial dysfunction in C2C12 skeletal muscle cells. Food Funct 2015;6:3056–64.

[59] Moylan JS, Smith JD, Wolf Horrell EM, McLean JB, Deevska GM, Bonnell MR, et al. Neutral sphingomyelinase-3 mediates TNF-stimulated oxidant activity in skeletal muscle. Redox Biol 2014;2:910–20.

[60] Dargelos E, Renaud V, Decossas M, Bure C, Lambert O, Poussard S. Caveolae-mediated effects of TNF-α on human skeletal muscle cells. Exp Cell Res 2018;370:623–31.

[61] Castillo SS, Levy M, Thaikoottathil JV, Goldkorn T. Reactive nitrogen and oxygen species activate different sphingomyelinases to induce apoptosis in airway epithelial cells. Exp Cell Res 2007;313:2680–6.

[62] Schmidt FM, Weschenfelder J, Sander C, Minkwitz J, Thormann J, Chittka T, et al. Inflammatory cytokines in general and central obesity and modulating effects of physical activity. PLoS One 2015;10:e0121971.

[63] Turpin-Nolan SM, Hammerschmidt P, Chen W, Jais A, Timper K, Awazawa M, et al. CerS1-derived C18:0 ceramide in skeletal muscle promotes obesity-induced insulin resistance. Cell Rep 2019;26:1–10.

[64] Adams JM, Pratipanawatr T, Berria R, Wang E, DeFronzo RA, Sullards MC, Mandarino LJ. Ceramide content is increased in skeletal muscle from obese insulin-resistant humans. Diabetes 2004;53:25–31.

[65] Helge JW, Dobrzyn A, Saltin B, Gorski J. Exercise and training effects on ceramide metabolism in human skeletal muscle. Exp Physiol 2004;89:119–27.

[66] Baranowski M, Błachnio-Zabielska AU, Charmas M, Helge JW, Dela F, Książek M, et al. Exercise increases sphingoid base-1-phosphate levels in human blood and skeletal muscle in a time- and intensity-dependent manner. Eur J Appl Physiol 2015;115:993–1003.

[67] Reid MB. Reactive oxygen species as agents of fatigue. Med Sci Sport Exerc 2016;48:2239–46.

[68] Marí M, Fernández-Checa JC, Morales A, García-Ruiz C, Colell A. Direct effect of ceramide on the mitochondrial electron transport chain leads to generation of reactive oxygen species. J Biol Chem 1997;272:11369–77.

[69] Smith ME, Tippetts TS, Brassfield ES, Tucker BJ, Ockey A, Swensen AC, et al. Mitochondrial fission mediates ceramide-induced metabolic disruption in skeletal muscle. Biochem J 2013;456:427–39.

[70] Sabbadini RA, Betto R, Teresi A, Fachechi-Cassano G, Salviati G. The effects of sphingosine on sarcoplasmic reticulum membrane calcium release. J Biol Chem 1992;267:15475–84.

[71] Sharma C, Smith T, Li S, Schroepfer GJ, Needleman DH. Inhibition of Ca2+ release channel (ryanodine receptor) activity by sphingolipid bases: mechanism of action. Chem Phys Lipids 2000;104:1–11.

[72] Bencini C, Squecco R, Piperio C, Formigli L, Meacci E, Nosi D, et al. Effects of sphingosine 1-phosphate on excitation-contraction coupling in mammalian skeletal muscle. J Muscle Res Cell Motil 2003;24:539–54.

[73] Meacci E, Cencetti F, Formigli L, Squecco R, Donati C, Tiribilli B, et al. Sphingosine 1-phosphate evokes calcium signals in C2C12 myoblasts via Edg3 and Edg5 receptors. Biochem J 2002;362:349–57.

[74] Loehr JA, Abo-Zahrah R, Pal R, Rodney GG. Sphingomyelinase promotes oxidant production and skeletal muscle contractile dysfunction through activation of NADPH oxidase. Front Physiol 2015;5:530.

[75] McKenna MJ, Medved I, Goodman CA, Brown MJ, Bjorksten AR, Murphy KT, et al. N-acetylcysteine attenuates the decline in muscle $Na^+$ $K^+$-pump activity and delays fatigue during prolonged exercise in humans. J Physiol 2006;576:279–88.

[76] Ferreira LF, Moylan JS, Stasko S, Smith JD, Campbell KS, Reid MB. Sphingomyelinase depresses force and calcium sensitivity of the contractile apparatus in mouse diaphragm muscle fibers. J Appl Physiol 2012;112:1538–45.

[77] Danieli-Betto D, Germinario E, Esposito A, Megighian A, Midrio M, Ravara B, et al. Sphingosine 1-phosphate protects mouse extensor digitorum longus skeletal muscle during fatigue. Am J Physiol Cell Physiol 2005;288:C1367–73.

[78] Empinado HM, Deevska GM, Nikolova-Karakashian M, Yoo JK, Christou DD, Ferreira LF. Diaphragm dysfunction in heart failure is accompanied by increases in neutral sphingomyelinase activity and ceramide content. Eur J Heart Fail 2014;16:519–25.

[79] Dalla Libera L, Vescovo G, Volterrani M. Physiological basis for contractile dysfunction in heart failure. Curr Pharm Des 2008;14:2572–81.

[80] Vescovo G, Ambrosio GB, Dalla Libera L. Apoptosis and changes in contractile protein pattern in the skeletal muscle in heart failure. Acta Physiol Scand 2001;171:305–10.

[81] De Larichaudy J, Zufferli A, Serra F, Isidori AM, Naro F, Dessalle K, et al. TNF-α- and tumor-induced skeletal muscle atrophy involves sphingolipid metabolism. Skelet Muscle 2012;2:2.

[82] Pierucci F, Frati A, Battistini C, Matteini F, Iachini MC, Vestri A, et al. Involvement of released sphingosine 1-phosphate/sphingosine 1-phosphate receptor axis in skeletal muscle atrophy. Biochim Biophys Acta Mol Basis Dis 2018;1864:3598–614.

[83] Bernacchioni C, Cencetti F, Ouro A, Bruno M, Gomez-Muñoz A, Donati C, Bruni P. Lysophosphatidic acid signaling axis mediates ceramide 1-phosphate-induced proliferation of C2C12 myoblasts. Int J Mol Sci 2018;19:E139.

[84] Gangoiti P, Bernacchioni C, Donati C, Cencetti F, Ouro A, Gómez-Muñoz A, Bruni P. Ceramide 1-phosphate stimulates proliferation of C2C12 myoblasts. Biochimie 2012;94:597–607.

[85] Donati C, Meacci E, Nuti F, Becciolini L, Farnararo M, Bruni P. Sphingosine 1-phosphate regulates myogenic differentiation: a major role for S1P 2 receptor. FASEB J 2005;19:449–51.

[86] Squecco R, Sassoli C, Nuti F, Martinesi M, Chellini F, Nosi D, et al. Sphingosine 1-phosphate induces myoblast differentiation through Cx43 protein expression: a role for a gap junction-dependent and -independent function. Mol Biol Cell 2006;17:4896–910.

[87] Meacci E, Nuti F, Donati C, Cencetti F, Farnararo M, Bruni P. Sphingosine kinase activity is required for myogenic differentiation of C2C12 myoblasts. J Cell Physiol 2008;214:210–20.

[88] Mebarek S, Komati H, Naro F, Zeiller C, Alvisi M, Lagarde M, et al. Inhibition of de novo ceramide synthesis upregulates phospholipase D and enhances myogenic differentiation. J Cell Sci 2007;120:407–16.

[89] Turpin SM, Lancaster GI, Darby I, Febbraio MA, Watt MJ. Apoptosis in skeletal muscle myotubes is induced by ceramides and is positively related to insulin resistance. Am J Physiol Endocrinol Metab 2006;291:E1341–50.

[90] Saba JD, de la Garza-Rodea AS. S1P lyase in skeletal muscle regeneration and satellite cell activation: exposing the hidden lyase. Biochim Biophys Acta 2013;1831:167–75.

[91] Loh KC, Leong WI, Carlson ME, Oskouian B, Kumar A, Fyrst H, et al. Sphingosine-1-phosphate enhances satellite cell activation in dystrophic muscles through a S1PR2/STAT3 signaling pathway. PLoS One 2012;7:e37218.

[92] Danieli-Betto D, Peron S, Germinario E, Zanin M, Sorci G, Franzoso S, et al. Sphingosine 1-phosphate signaling is involved in skeletal muscle regeneration. Am J Physiol Cell Physiol 2010;298:C550–8.

[93] Nagata Y, Partridge TA, Matsuda R, Zammit PS. Entry of muscle satellite cells into the cell cycle requires sphingolipid signaling. J Cell Biol 2006;174:245–53.

[94] Nagata Y, Ohashi K, Wada E, Yuasa Y, Shiozuka M, Nonomura Y, et al. Sphingosine-1-phosphate mediates epidermal growth factor-induced muscle satellite cell activation. Exp Cell Res 2014;326:112–24.

[95] Calise S, Blescia S, Cencetti F, Bernacchioni C, Donati C, Bruni P. Sphingosine 1-phosphate stimulates proliferation and migration of satellite cells. Role of S1P receptors. Biochim Biophys Acta Mol Cell Res 2012;1823:439–50.

[96] Cordeiro AV, Silva VRR, Pauli JR, da Silva ASR, Cintra DE, Moura LP, et al. The role of sphingosine-1-phosphate in skeletal muscle: physiology, mechanisms, and clinical perspectives. J Cell Physiol 2019;234:10047–59.

[97] Rapizzi E, Taddei ML, Fiaschi T, Donati C, Bruni P, Chiarugi P. Sphingosine 1-phosphate increases glucose uptake through trans-activation of insulin receptor. Cell Mol Life Sci 2009;66:3207–18.

[98] JeBailey L, Wanono O, Niu W, Roessler J, Rudich A, Klip A. Ceramide- and oxidant-induced insulin resistance involve loss of insulin-dependent Rac-activation and actin remodeling in muscle cells. Diabetes 2007;56:394–403.
[99] Hajduch E, Balendran A, Batty IH, Litherland GJ, Blair AS, Downes CP, et al. Ceramide impairs the insulin-dependent membrane recruitment of protein kinase B leading to a loss in downstream signalling in L6 skeletal muscle cells. Diabetologia 2001;44:173–83.
[100] Hyde R, Hajduch E, Powell DJ, Taylor PM, Hundal HS. Ceramide down-regulates System A amino acid transport and protein synthesis in rat skeletal muscle cells. FASEB J 2005;19:461–3.
[101] Bandet CL, Tan-Chen S, Bourron O, Le Stunff H, Hajduch E. Sphingolipid metabolism: new insight into ceramide-induced lipotoxicity in muscle cells. Int J Mol Sci 2019;20:479.
[102] Zabielski P, Błachnio-Zabielska AU, Wójcik B, Chabowski A, Górski J. Effect of plasma free fatty acid supply on the rate of ceramide synthesis in different muscle types in the rat. PLoS One 2017;12:e0187136.
[103] Perreault L, Newsom SA, Strauss A, Kerege A, Kahn DE, Harrison KA, et al. Intracellular localization of diacylglycerols and sphingolipids influences insulin sensitivity and mitochondrial function in human skeletal muscle. JCI Insight 2018;3:96805.
[104] Bandet CL, Mahfouz R, Véret J, Sotiropoulos A, Poirier M, Giussani P, et al. Ceramide transporter CERT is involved in muscle insulin signaling defects under lipotoxic conditions. Diabetes 2018;67:1258–71.
[105] Stratford S, Hoehn KL, Liu F, Summers SA. Regulation of insulin action by ceramide: dual mechanisms linking ceramide accumulation to the inhibition of Akt/protein kinase B. J Biol Chem 2004;279:36608–15.
[106] Powell DJ, Hajduch E, Kular G, Hundal HS. Ceramide disables 3-phosphoinositide binding to the pleckstrin homology domain of protein kinase B (PKB)/Akt by a PKCzeta-dependent mechanism. Mol Cell Biol 2003;23:7794–808.
[107] Mahfouz R, Khoury R, Blachnio-Zabielska A, Turban S, Loiseau N, Lipina C, et al. Characterising the inhibitory actions of ceramide upon insulin signaling in different skeletal muscle cell models: a mechanistic insight. PLoS One 2014;9:e101865.
[108] Razani B, Woodman SE, Lisanti MP. Caveolae: from cell biology to animal physiology. Pharmacol Rev 2002;54:431–67.
[109] Hajduch E, Turban S, Le Liepvre X, Le Lay S, Lipina C, Dimopoulos N, et al. Targeting of PKCζ and PKB to caveolin-enriched microdomains represents a crucial step underpinning the disruption in PKB-directed signalling by ceramide. Biochem J 2008;410:369–79.
[110] Hage Hassan R, Pacheco De Sousa AC, Mahfouz R, Hainault I, Blachnio-Zabielska A, Bourron O, et al. Sustained action of ceramide on the insulin signaling pathway in muscle cells: implication of the double-stranded RNA-activated protein kinase. J Biol Chem 2016;291:3019–29.
[111] Cimmino I, Lorenzo V, Fiory F, Doti N, Ricci S, Cabaro S, et al. A peptide antagonist of Prep1-p160 interaction improves ceramide-induced insulin resistance in skeletal muscle cells. Oncotarget 2017;8:71845–58.
[112] Kim KS, Park KS, Kim MJ, Kim SK, Cho YW, Park SW. Type 2 diabetes is associated with low muscle mass in older adults. Geriatr Gerontol Int 2014;14:115–21.
[113] Park M, Kaddai V, Ching J, Fridianto KT, Sieli RJ, Sugii S, et al. A role for ceramides, but not sphingomyelins, as antagonists of insulin signaling and mitochondrial metabolism in C2C12 myotubes. J Biol Chem 2016;291:23978–88.
[114] Floegel A, Stefan N, Yu Z, Mühlenbruch K, Drogan D, Joost HG, et al. Identification of serum metabolites associated with risk of type 2 diabetes using a targeted metabolomic approach. Diabetes 2013;62:639–48.
[115] Yamashita T, Hashiramoto A, Haluzik M, Mizukami H, Beck S, Norton A, et al. Enhanced insulin sensitivity in mice lacking ganglioside GM3. Proc Natl Acad Sci USA 2003;100:3445–9.
[116] Chavez JA, Summers SA. A ceramide-centric view of insulin resistance. Cell Metab 2012;15:585–94.
[117] Barbarroja N, Rodriguez-Cuenca S, Nygren H, Camargo A, Pirraco A, Relat J, et al. Increased dihydroceramide/ceramide ratio mediated by defective expression of degs1 impairs adipocyte differentiation and function. Diabetes 2015;64:1180–92.
[118] Yang G, Badeanlou L, Bielawski J, Roberts AJ, Hannun YA, Samad F. Central role of ceramide biosynthesis in body weight regulation, energy metabolism, and the metabolic syndrome. Am J Physiol Endocrinol Metab 2009;297:E211–24.
[119] Camell CD, Nguyen KY, Jurczak MJ, Christian BE, Shulman GI, Shadel GS, et al. Macrophage-specific de novo synthesis of ceramide is dispensable for inflammasome-driven inflammation and insulin resistance in obesity. J Biol Chem 2015;290:29402–13.
[120] Alexaki A, Clarke BA, Gavrilova O, Ma Y, Zhu H, Ma X, et al. De novo sphingolipid biosynthesis is required for adipocyte survival and metabolic homeostasis. J Biol Chem 2017;292:3929–39.

[121] Lee SY, Lee HY, Song JH, Kim GT, Jeon S, Song YJ, et al. Adipocyte-specific deficiency of de novo sphingolipid biosynthesis leads to lipodystrophy and insulin resistance. Diabetes 2017;66:2596–609.

[122] Kitada Y, Kajita K, Taguchi K, Mori I, Yamauchi M, Ikeda T, et al. Blockade of sphingosine 1-phosphate receptor 2 signaling attenuates high-fat diet-induced adipocyte hypertrophy and systemic glucose intolerance in mice. Endocrinology 2016;157:1839–51.

[123] Gohlke S, Zagoriy V, Cuadros Inostroza A, Méret M, Mancini C, Japtok L, et al. Identification of functional lipid metabolism biomarkers of brown adipose tissue aging. Mol Metab 2019;24:1–17.

[124] Moon MH, Jeong JK, Lee JH, Park YG, Lee YJ, Seol JW, et al. Antiobesity activity of a sphingosine 1-phosphate analogue FTY720 observed in adipocytes and obese mouse model. Exp Mol Med 2012;44:603–14.

[125] Hashimoto T, Igarashi J, Kosaka H. Sphingosine kinase is induced in mouse 3T3-L1 cells and promotes adipogenesis. J Lipid Res 2009;50:602–10.

[126] Yano M, Yamamoto T, Nishimura N, Gotoh T, Watanabe K, Ikeda K, et al. Increased oxidative stress impairs adipose tissue function in sphingomyelin synthase 1 null mice. PLoS One 2013;8:e61380.

[127] Turner N, Kowalski GM, Leslie SJ, Risis S, Yang C, Lee-Young RS, et al. Distinct patterns of tissue-specific lipid accumulation during the induction of insulin resistance in mice by high-fat feeding. Diabetologia 2013;56:1638–48.

[128] Long SD, Pekala PH. Regulation of GLUT4 mRNA stability by tumor necrosis factor-alpha: alterations in both protein binding to the 3′ untranslated region and initiation of translation. Biochem Biophys Res Commun 1996;220:949–53.

[129] Chaurasia B, Kaddai VA, Lancaster GI, Henstridge DC, Sriram S, Galam DL, et al. Adipocyte ceramides regulate subcutaneous adipose browning, inflammation, and metabolism. Cell Metab 2016;24:820–34.

[130] Gosejacob D, Jäger PS, Vom Dorp K, Frejno M, Carstensen AC, Köhnke M, et al. Ceramide synthase 5 is essential to maintain C16:0-ceramide pools and contributes to the development of diet-induced obesity. J Biol Chem 2016;291:6989–7003.

[131] Turpin SM, Nicholls HT, Willmes DM, Mourier A, Brodesser S, Wunderlich CM, et al. Obesity-induced CerS6-dependent C16:0 ceramide production promotes weight gain and glucose intolerance. Cell Metab 2014;20:678–86.

[132] Wang J, Badeanlou L, Bielawski J, Ciaraldi TP, Samad F. Sphingosine kinase 1 regulates adipose proinflammatory responses and insulin resistance. Am J Physiol Endocrinol Metab 2014;306:E756–68.

[133] Jun DJ, Lee JH, Choi BH, Koh TK, Ha DC, Jeong MW, et al. Sphingosine-1-phosphate modulates both lipolysis and leptin production in differentiated rat white adipocytes. Endocrinology 2006;147:5835–44.

[134] Zhang W, Mottillo EP, Zhao J, Gartung A, VanHecke GC, Lee JF, et al. Adipocyte lipolysis-stimulated interleukin-6 production requires sphingosine kinase 1 activity. J Biol Chem 2014;289:32178–85.

[135] Raichur S, Wang ST, Chan PW, Li Y, Ching J, Chaurasia B, et al. CerS2 haploinsufficiency inhibits beta-oxidation and confers susceptibility to diet-induced steatohepatitis and insulin resistance. Cell Metab 2014;20:687–95.

[136] Mcilroy GD, Tammireddy SR, Maskrey BH, Grant L, Doherty MK, Watson DG, et al. Fenretinide mediated retinoic acid receptor signalling and inhibition of ceramide biosynthesis regulates adipogenesis, lipid accumulation, mitochondrial function and nutrient stress signalling in adipocytes and adipose tissue. Biochem Pharmacol 2016;100:86–97.

[137] Makdissy N, Haddad K, Mouawad C, Popa I, Younsi M, Valet P, et al. Regulation of SREBPs by sphingomyelin in adipocytes via a caveolin and Ras-ERK-MAPK-CREB signaling pathway. PLoS One 2015;10:e0133181.

[138] Tordjman J, Chauvet G, Quette J, Beale EG, Forest C, Antoine B. Thiazolidinediones block fatty acid release by inducing glyceroneogenesis in fat cells. J Biol Chem 2003;278:18785–90.

[139] Park TS, Goldberg IJ. Sphingolipids, lipotoxic cardiomyopathy, and cardiac failure. Heart Fail Clin 2012;8:633–41.

[140] Basu R, Oudit GY, Wang X, Zhang L, Ussher JR, Lopaschuk GD, et al. Type 1 diabetic cardiomyopathy in the Akita (Ins2WT/C96Y) mouse model is characterized by lipotoxicity and diastolic dysfunction with preserved systolic function. Am J Physiol Heart Circ Physiol 2009;297:H2096–108.

[141] Dobrzyn P, Dobrzyn A, Miyazaki M, Ntambi JM. Loss of stearoyl-CoA desaturase 1 rescues cardiac function in obese leptin-deficient mice. J Lipid Res 2010;51:2202–10.

[142] Klevstig M, Stahlman M, Lundqvist A, Scharin-Täng M, Fogelstrand P, Adiels M, et al. Targeting acid sphingomyelinase reduces cardiac ceramide accumulation in the post-ischemic heart. J Mol Cell Cardiol 2016;93:69–72.

[143] Ji R, Akashi H, Drosatos K, Liao X, Jiang H, Kennel PJ, et al. Increased de novo ceramide synthesis and accumulation in failing myocardium. JCI Insight 2017;2:e82922.

[144] Kong JY, Rabkin SW. Palmitate-induced apoptosis in cardiomyocytes is mediated through alterations in mitochondria: prevention by cyclosporin A. Biochim Biophys Acta 2000;1485:45–55.

[145] Park TS, Hu Y, Noh HL, Drosatos K, Okajima K, Buchanan J, et al. Ceramide is a cardiotoxin in lipotoxic cardiomyopathy. J Lipid Res 2008;49:2101–12.
[146] Karliner JS. Sphingosine kinase and sphingosine 1-phosphate in the heart: a decade of progress. Biochim Biophys Acta 2013;183:203–12.
[147] Karliner JS, Honbo N, Summers K, Gray MO, Goetzl EJ. The lysophospholipids sphingosine-1-phosphate and lysophosphatidic acid enhance survival during hypoxia in neonatal rat cardiac myocytes. J Mol Cell Cardiol 2001;33:1713–7.
[148] Tao R, Zhang J, Vessey DA, Honbo N, Karliner JS. Deletion of the sphingosine kinase-1 gene influences cell fate during hypoxia and glucose deprivation in adult mouse cardiomyocytes. Cardiovasc Res 2007;74:56–63.
[149] Vessey DA, Kelley M, Li L, Zhou HZ, Zhu BQ, Karliner JS. Role of sphingosine kinase activity in protection of heart against ischemia reperfusion injury. Med Sci Monit 2006;12:BR318–24.
[150] Duan HF, Wang H, Yi J, Zhang QW, Li LB, Zhang T, et al. Adenoviral gene transfer of sphingosine kinase 1 protects heart against ischemia/reperfusion-induced injury and attenuates its postischemic failure. *Hum Gene Ther* 2007;18:1119–28.
[151] Deshpande GP, Imamdin A, Lecour S, Opie LH. Sphingosine-1-phosphate (S1P) activates STAT3 to protect against de novo acute heart failure (AHF). Life Sci 2018;196:127–32.
[152] Zhang J, Honbo N, Goetzl EJ, Chatterjee K, Karliner JS, Gray MO. Signals from type 1 sphingosine 1-phosphate receptors enhance adult mouse cardiac myocyte survival during hypoxia. Am J Physiol Heart Circ Physiol 2007;293:H3150–8.
[153] Theilmeier G, Schmidt C, Herrmann J, Keul P, Schäfers M, Herrgott I, et al. High-density lipoproteins and their constituent, sphingosine-1-phosphate, directly protect the heart against ischemia/reperfusion injury in vivo via the S1P3 lysophospholipid receptor. Circulation 2006;114:1403–9.
[154] Summers SA. Could ceramides become the new cholesterol? Cell Metab 2018;27:276–80.
[155] Kurz J, Parnham MJ, Geisslinger G, Schiffmann S. Ceramides as novel disease Biomarkers. Trends Mol Med 2019;25:20–32.
[156] Hannun YA, Obeid LM. Sphingolipids and their metabolism in physiology and disease. Nat Rev Mol Cell Biol 2018;19:175–91.
[157] Cantalupo A, Di Lorenzo A. S1P signaling and de novo biosynthesis in blood pressure homeostasis. J Pharmacol Exp Ther 2016;358:359–70.
[158] Igarashi J, Bernier SG, Michel T. Sphingosine 1-phosphate and activation of endothelial nitric-oxide synthase. Differential regulation of Akt and MAP kinase pathways by EDG and bradykinin receptors in vascular endothelial cells. J Biol Chem 2001;276:12420–6.
[159] Tölle M, Klöckl L, Wiedon A, Zidek W, van der Giet M, Schuchardt M. Regulation of endothelial nitric oxide synthase activation in endothelial cells by S1P1 and S1P3. Biochem Biophys Res Commun 2016;476:627–34.
[160] Zhao Y, Vanhoutte PM, Leung SW. Vascular nitric oxide: Beyond eNOS. J Pharmacol Sci 2015;129:83–94.
[161] Santhanam L, Tuday EC, Webb AK, Dowzicky P, Kim JH, Oh YJ, et al. Decreased S-nitrosylation of tissue transglutaminase contributes to age-related increases in vascular stiffness. Circ Res 2010;107:117–25.
[162] Igarashi J, Miyoshi M, Hashimoto T, Kubota Y, Kosaka H. Hydrogen peroxide induces S1P1 receptors and sensitizes vascular endothelial cells to sphingosine 1-phosphate, a platelet-derived lipid mediator. Am J Physiol Cell Physiol 2007;292:C740–8.
[163] Schmitz EI, Potteck H, Schüppel M, Manggau M, Wahydin E, Kleuser B. Sphingosine 1-phosphate protects primary human keratinocytes from apoptosis via nitric oxide formation through the receptor subtype S1P(3). Mol Cell Biochem 2012;371:165–76.
[164] Zhang DX, Zou AP, Li PL. Ceramide-induced activation of NADPH oxidase and endothelial dysfunction in small coronary arteries. Am J Physiol Heart Circ Physiol 2003;284:H605–12.
[165] Didion SP, Faraci FM. Ceramide-induced impairment of endothelial function is prevented by CuZn superoxide dismutase overexpression. Arterioscler Thromb Vasc Biol 2005;25:90–5.
[166] Symons JD, Abel ED. Lipotoxicity contributes to endothelial dysfunction: a focus on the contribution from ceramide. Rev Endocr Metab Disord 2013;14:59–68.
[167] Pandolfi R, Barreira B, Moreno E, Lara-Acedo V, Morales-Cano D, Martinez-Ramas A, et al. Role of acid sphingomyelinase and IL-6 as mediators of endotoxin-induced pulmonary vascular dysfunction. Thorax 2017;72:460–71.
[168] Cogolludo A, Villamor E, Perez-Vizcaino F, Moreno L. Ceramide and regulation of vascular tone. Int J Mol Sci 2019;18:20.

[169] Becker RC, Sexton T, Smyth SS. Translational implications of platelets as vascular first responders. Circ Res 2018;122:506–22.

[170] Fegley AJ, Tanski WJ, Roztocil E, Davies MG. Sphingosine-1-phosphate stimulates smooth muscle cell migration through Gai-and PI3-kinase-dependent p38MAPK activation. J Surg Res 2003;113:32–41.

[171] Duru EA, Fu Y, Davies MG. Role of S-1-P receptors and human vascular smooth muscle cell migration in diabetes and metabolic syndrome. J Surg Res 2012;177:75–82.

[172] Lee DY, Won KJ, Lee KP, Jung SH, Baek S, Chung HW, et al. Angiotensin II facilitates neointimal formation by increasing vascular smooth muscle cell migration: Involvement of APE/Ref-1-mediated overexpression of sphingosine-1-phosphate receptor 1. Toxicol Appl Pharmacol 2018;347:45–53.

[173] Argraves KM, Sethi AA, Gazzolo PJ, Wilkerson BA, Remaley AT, Tybjaerg-Hansen A, et al. S1P, dihydro-S1P and C24:1-ceramide levels in the HDL-containing fraction of serum inversely correlate with occurrence of ischemic heart disease. Lipids Health Dis 2011;9:10–70.

[174] Potì F, Simoni M, Nofer JR. Atheroprotective role of high-density lipoprotein (HDL)-associated sphingosine-1-phosphate (S1P). Cardiovasc Res 2014;103:395–404.

[175] Keul P, Polzin A, Kaiser K, Gräler M, Dannenberg L, Daum G, et al. Potent anti-inflammatory properties of HDL in vascular smooth muscle cells mediated by HDL-S1P and their impairment in coronary artery disease due to lower HDL-S1P: a new aspect of HDL dysfunction and its therapy. FASEB J 2019;33:1482–95.

[176] Tanimoto T, Lungu AO, Berk BC. Sphingosine 1-phosphate transactivates the platelet-derived growth factor β receptor and epidermal growth factor receptor in vascular smooth muscle cells. *Circ Res* 2004;94:1050–8.

[177] Jing XD, Wei XM, Deng SB, Du JL, Liu YJ, She Q. The relationship between the high-density lipoprotein (HDL)-associated sphingosine-1-phosphate (S1P) and coronary in-stent restenosis. Clin Chim Acta 2015;446:248–52.

[178] Li W, Yang X, Xing S, Bian F, Yao W, Bai X, et al. Endogenous ceramide contributes to the transcytosis of oxLDL across endothelial cells and promotes its subendothelial retention in vascular wall. Oxid Med Cell Longev 2014;2014:823071.

[179] Swan DJ, Kirby JA, Ali S. Vascular biology: the role of sphingosine 1-phosphate in both the resting state and inflammation. J Cell Mol Med 2010;14:2211–22.

[180] Keul P, Lucke S, von Wnuck Lipinski K, Bode C, Gräler M, Heusch G, et al. Sphingosine-1-phosphate receptor 3 promotes recruitment of monocyte/macrophages in inflammation and atherosclerosis. Circ Res 2011;108: 314–23.

[181] Jenne CN, Enders A, Rivera R, Watson SR, Bankovich AJ, Pereira JP, et al. T-bet-dependent S1P5 expression in NK cells promotes egress from lymph nodes and bone marrow. J Exp Med 2009;206:2469–81.

[182] Grimm M, Tischner D, Troidl K, Albarrán Juárez J, Sivaraj KK, Ferreirós Bouzas N, et al. S1P2/G12/13 signaling negatively regulates macrophage activation and indirectly shapes the atheroprotective B1-cell population. Arterioscler Thromb Vasc Biol 2016;36:37–48.

[183] Jiang XC, Liu J. Sphingolipid metabolism and atherosclerosis. Handb Exp Pharmacol 2013;216:133–46.

[184] Meeusen JW, Donato LJ, Bryant SC, Baudhuin LM, Berger PB, Jaffe AS. Plasma ceramides. Arterioscler Thromb Vasc Biol 2018;38:1933–9.

CHAPTER

# 20

# Role of bile acid receptors in the regulation of cardiovascular diseases

*Yuji Shiozaki, Audrey L. Keenan, Makoto Miyazaki*

Division of Renal Diseases and Hypertension, University of Colorado-Denver, Aurora, CO, United States

OUTLINE



Lipid Signaling and Metabolism. http://dx.doi.org/10.1016/B978-0-12-819404-1.00020-8

## Bile acid receptors in the regulation of cardiovascular diseases bile acids and bile acid receptors

Bile acids are cholesterol-derived molecules that serve as both amphipathic molecules for lipid and fat-soluble vitamin absorption through micelle formation with lipids as well as ligand molecules of bile acids receptors for signal transduction between organs [1–3]. Primary bile acids are synthesized from cholesterol in hepatocytes, stored in the gall bladder and secreted to the small intestine via bile ducts in response to ingestion of foods. Secondary bile acids are produced from primary bile acids by intestinal bacteria. Primary bile acids are absorbed mainly by active transport in the small intestine and secondary bile acids are absorbed largely by passive absorption in the colon, and afterward are recycled back to the liver [1,2]. More than 95% of bile acids are recycled and less than 10% are excreted as fecal matter (Fig. 20.1) [3]. The major primary bile acids include cholic acid (CA) and chenodeoxycholic acid (CDCA), while lithocholic acid (LCA), deoxycholic acid (DCA), and ursodeoxycholic acid (UDCA) are major secondary bile acids [1,3]. Both primary and secondary bile acids can serve as ligands to bile acid receptors by regulating physiological pathways in lipid and carbohydrate homeostasis [1,2]. Bile acid receptors localize mainly in the nucleus but are also located in cell surface membranes [1,2]. Farnesoid X-activated receptor (FXR/NR1H4), pregnane X receptor (PXR/NR1I2), vitamin D receptor (VDR/NR1I1) and constitutive androstane receptor (CAR/NR1I3) work as nuclear receptors activated by bile acids [1,2,4]. Cell surface bile acid receptors include G-protein-coupled bile acid receptors (TGR5) and muscarinic receptors [1,2]. Activation and inhibition of these bile acid receptors through gene editing and drug treatment have been shown to affect the development of a number of metabolic and cardiovascular diseases such as dyslipidemia, liver disease, diabetes mellitus, and atherosclerosis [2,5].

## Atherosclerosis and vascular calcification

Cardiovascular diseases such as myocardial infarction, angina, heart failure, stroke, and atherosclerosis are associated with a high risk of morbidity and mortality globally [7–11]. Atherosclerosis is a disease in patients with diabetes mellitus, obesity, dyslipidemia, hypertension, and aging that forms plaques including lipids, cells, debris, and scar tissue [7,8,12]. Atherosclerosis develops through lipid deposition, inflammation, vascular smooth muscle cell (VSMC) proliferation, monocyte recruitment, and formation of foam cells, which have abundant lipid droplets and are derived from VSMCs or macrophages [7,8]. Vascular calcification is a common vascular disease that accumulates phosphate-calcium deposits in the medial and intimal layers of arteries, and is one of the most common complications in patients with chronic kidney disease (CKD), hyperphosphatemia, dyslipidemia, diabetes mellitus, and aging [11,13–15]. Vascular calcification is caused by transdifferentiation of VSMCs into osteoblastic-like cells. Transdifferentiated VSMCs express osteoblastic markers such as alkali phosphatase (ALP), osteopontin (OPN), and osteocalcin (OCN), and transcription factors such as Msx2 and osterix [11,13]. Atherosclerosis and vascular calcification can occur individually or simultaneously, but both contribute to arterial stiffness associated

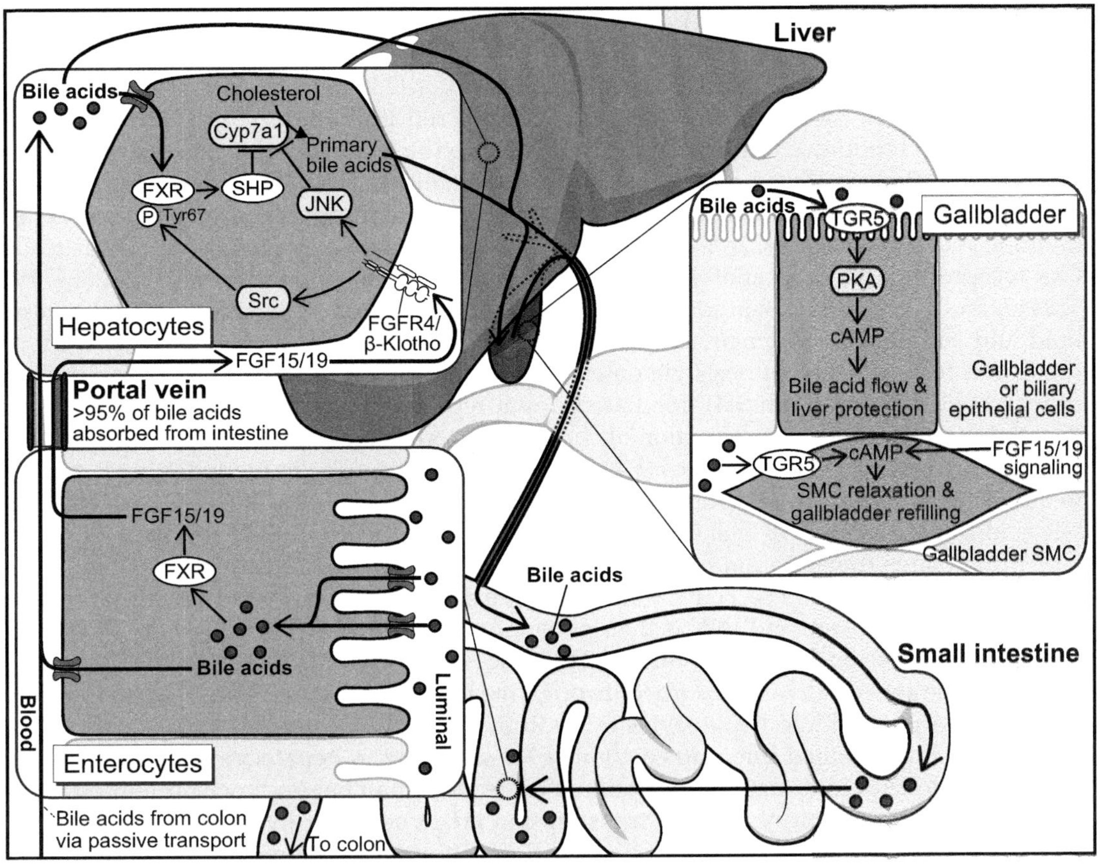


FIGURE 20.1 **FXR and TGR5 signaling through enterohepatic circulation of bile acids.** Bile acids are secreted to the duodenum from the gallbladder through bile ducts in response to food digestion, and form micelles with lipids for lipid absorption. Over 95% of secreted bile acids are reabsorbed by active and passive transport in the intestine. Bile acids that are incorporated to the intracellular ileum through bile acid transporters at the apical membrane activate intestinal FXR signaling, and FGF15/19 are then secreted from ileal cells. Reabsorbed bile acids and secreted FGF15/19 reach the liver through the portal vein and activate FGF15/19-FGFR4/β-Klotho-JNK and -Src-FXR phosphorylation and the bile acid-FXR-SHP pathway to repress Cyp7a1, which produces primary bile acids from cholesterol in hepatocytes. On the other hand, bile acids and FGF15/19 induce gallbladder smooth muscle cell relaxation and gallbladder filling by TGR5 and FGF15/19 signaling [6]. In addition, TGR5 in the gallbladder controls bile acid flow and liver protection [5].

with an increase in cardiovascular mortality [9,16]. Activation or deactivation of bile acid receptors by chemicals or gene modification modulates the development of both atherosclerosis and vascular calcification along with significant alterations in lipid metabolism [17–21]. Therefore, discovering the functions and modulators of bile acid receptors is valuable in finding treatments for preventing atherosclerosis and vascular calcification. We aim to review bile acid receptor signaling in lipid metabolism and the development of cardiovascular diseases.

## Farnesoid X-activated receptor signaling and functions

FXR was the first identified nuclear bile acid receptor [2,22]. Two isoforms of FXR, FXRα, and FXRβ, exist in rodents, but only FXRα is expressed in humans [23]. FXR forms a heterodimer with retinoid X receptor (RXR) and binds to the FXR response element on genomic DNA [22]. FXR is expressed predominantly in the small intestine, liver and kidney, but also in the adrenal glands, cardiovascular tissues, thyroid gland, lungs, skin, adipose tissues, and immune cells [2,23,24]. CDCA, DCA, cholic acid, and LCA are major bile acid ligands for the FXR receptor [4]. Bile acid analogues such as INT-747, GW4064, fexaramine, GSK2324, Way 362450, and PX102 have been identified as synthetic FXR-specific ligands [2]. FXR controls lipid and carbohydrate homeostasis [25]. A major function of FXR is to provide feedback regulation for bile acid synthesis via enterohepatic circulation to repress excess synthesis of primary bile acids through SHP-mediated down-regulation of cholesterol 7-α-hydroxylase (Cyp7a1), which catalyzes production of cholic acid from cholesterol as the rate-limiting step of bile acid synthesis in hepatocytes (Fig. 20.1) [1,2,5]. In response to food, bile acids secreted to the small intestine are incorporated to the intracellular ileum via apical sodium-dependent bile acid transporters (ASBT/SLC10A2), and FXR then binds to bile acid ligands. Activated FXR upregulates transcription of the FGF15 (in rodents) or FGF19 (the human ortholog of the rodent FGF15) gene [1]. FGF15/19 are secreted from ileal cells and reach hepatocytes through the portal vein, and JNK is then phosphorylated and MAPK activation inhibits Cyp7a1 gene expression [1,26,27]. While FXR-FGF15/19-FGFR4/β-Klotho signaling occurs in ileal cells, bile acids that are reabsorbed through the intestine and reach the liver via the portal vein are incorporated into hepatocytes by sodium taurocholate co-transporting polypeptide (NTCP) [1]. Incorporated bile acids activate FXR signaling in hepatocytes [2]. Activation of FXR in hepatocytes induces a transcriptional repressor, small heterodimer protein (SHP), resulting in the repression of Cyp7a1 transcription [1]. A study of liver- or intestine-specific FXR null mice suggests that the intestinal FXR-FGF15 pathway is more important than the hepatic FXR-SHP pathway in repressing Cyp7a1 transcription [2,28]. In contrast to Cyp7a1, regulation of Cyp8a1 is controlled more by liver FXR signaling than intestinal FXR-FGF15 signaling [28]. However, interestingly, recent reports indicate a novel mechanism by which the FXR-FGF15/19-FGFR4/β-Klotho pathway activates nonreceptor tyrosine kinase (Src)-dependent FXR phosphorylation at tyrosine (Y) 67 in FXR signaling of hepatocytes [29]. Src-dependent phosphorylation of FXR Y67 controls the nuclear translocation of FXR, inducing the transcription of SHP to inhibit Cyp7a1 expression [29]. Hepatic FXR-SHP and intestinal FXR-FGF15/19 pathways may be linked to each other in the control of bile acid synthesis. In addition to the regulation of bile acid homeostasis, FXR directly and indirectly regulates the transcription of target genes related to lipid homeostasis such as SR-B1, apolipoprotein (Apo)A1, ApoCII, ApoCIII, phospholipid transfer protein, sterol regulatory element binding protein (SREBP)-1c, and very-low-density lipoprotein (VLDL) [25,30–32]. $FXR^{-/-}$ (FXR KO) disrupts the inhibitory effect on bile acid and lipid homeostasis and elevates serum bile acids, cholesterol, triglycerides, and increases hepatic cholesterol, triglycerides, and serum lipoproteins [33]. Activation of FXR by bile acids and the synthetic activators affects metabolic diseases such as obesity and diabetes [34–37]. Treatment with an FXR agonist, GW4064, improves hyperglycemia and hyperlipidemia in wild type and diabetic db/db mice, but FXR KO exhibits glucose intolerance and insulin resistance [34]. FXR is also associated with gallbladder disorders. GW4064 inhibits cholesterol gallstone disease by restoring cholesterol solubility [38]. In addition to the lipid-lowering effects, FXR agonists elicit antiinflammatory effects by inhibiting NF-κB-mediated cytokine and chemokine production [39].

## FXR functions and the development of cardiovascular diseases

A number of studies have suggested an association between FXR and the development of cardiovascular diseases, especially atherosclerosis. Male $FXR^{-/-}$; $ApoE^{-/-}$ double knockout (DKO) mice exhibit upregulation of inflammatory genes such as TNFα and MAC1 and develop severe atherosclerosis compared with ApoE single KO mice [40], whereas female DKO mice have significantly reduced atherosclerosis. In addition, FXR deficiency attenuates atherosclerotic formations in male LDLR KO mice [41]. In contrast, activation of FXR by selective agonists, INT-747 and WAY362450, inhibits atherosclerosis in ApoE KO and LDLR KO mice [42–44]. Since $SHP^{-/-}$; $ApoE^{-/-}$ (DKO) mice and $SHP^{-/-}$; $LDLR^{-/-}$ (DKO) mice do not exhibit the WAY362450-mediating antiatherosclerotic effect, the FXR-SHP axis serves as a critical antiatherosclerotic signal under activation of FXR [45]. In addition, CDCA improves HIV protease-induced atherosclerosis and dyslipidemia in ApoE mice [46]. INT-747 and PX20606 block atherosclerosis in cholesterol ester transfer protein (CETP) transgenic (tg); LDLR KO mice [47]. Activation of the FGF15-Scr-hepatic FXR Y67 phosphorylation pathway prevents atherosclerosis in ApoE KO mice [48]. Adenoviral overexpression of FXR wild type mice, but not Y67F mutant mice, represses bile acid synthesis and atherosclerosis. In contrast, shRNA-based Src knockdown exacerbated atherosclerosis with an upregulation of bile acid synthesis in ApoE KO mice [48]. This study indicates that signaling of the intestinal FGF15-FGFR/β-Klotho-Src-FXR phosphorylation axis and the hepatic FXR-SHP axis may cooperate to prevent atherosclerosis through FXR activation. Broadly classifying the effects of FXR agonists on atherosclerosis, FXR agonists exert strong lipid (cholesterol, bile acids, LDL, and HDL)-lowering effects but also antiinflammatory (TNF-α, IL-1β, IL-6, and MCP-1) effects through the down-regulation of Nf-κB signaling, which is associated with a reduction in atherosclerosis [39,43–46,48]. Another FXR agonist, oleanolic acid, inhibits ox-LDL-induced cell apoptosis through FXR-dependent upregulation of angiotensin (Ang)-(1–7), a bioactive peptide in the renin–angiotensin system (RAS) [49], in HUVEC and high fat diet-induced atherosclerotic in vivo models [50]. FXR agonists have a protective effect on atherosclerosis and vascular calcification. Activation of FXR by the selective agonist INT-747 protects against aortic vascular calcification in ApoE KO mice with 5/6 nephrectomy, a CKD model [51]. FXR mRNA and protein are upregulated time-dependently during osteoblast generation in bovine calcifying vascular cells (CVCs). ApoE KO mice with CKD exhibit an atherosclerotic phenotype and vascular calcification, but INT-747 reduced calcified lesions without affecting plaque formation. INT-747 inhibits phosphate-induced JNK activation and upregulation of osteogenic markers such as ALP, OPN, and OCL, and osteogenic transcriptional factors including Msx2 and osterix [51]. An in vitro study showed that FXR dominant negative mutant overexpression eliminates inhibition of phosphate-induced calcification under INT-747 treatment [51]. These reports suggest that FXR agonists can exert valuable effects on protecting against cardiovascular diseases. However, some reports showed that attenuation of atherosclerosis by FXR deficiency has opposite results [41,52]. In addition, FXR is associated with production of a proatherosclerotic factor, trimethylamine-N-oxide (TMAO), through upregulation of flavin monooxygenase 3 (FMO3) [53]. Resveratrol reduced TMAO-induced atherosclerosis in ApoE KO mice through inhibition of the FXR-FGF15 axis and suppression of TMAO production [54]. To conclude whether activation or inhibition of FXR is better for prevention of cardiovascular diseases, further studies of FXR agonists and antagonists are needed along with an analysis of detailed metabolic states for each disease.

## G-protein-coupled bile acid receptor (TGR5) signaling and functions

TGR5 is a G protein-coupled cell surface receptor specifically activated by bile acids and highly expressed in the intestine, immune cells and gallbladder, but not in hepatocytes [2]. TGR5 regulates bile acid and glucose homeostasis and energy regulation [1,55]. Both primary and secondary bile acids including LCA, DCA, CDCA, and CA work as endogenous ligands for TGR5 [2]. Several synthetic ligands including INT-767 and INT-777 have also been identified [2]. TGR5 KO mice experience loss of incretion of the bile acid pool in the gallbladder filling and gallbladder relaxation in response to administration of the TGR5 specific agonist INT-777, indicating that TGR5 controls bile acid storage and secretion [56] (Fig. 20.1). This effect is based on TGR5-dependent cyclic AMP production and subsequent relaxation of gallbladder smooth muscle cells. Secretion of glucagon-like peptide-1 (GLP-1) from enteroendocrine L cells is a major TGR5 function [55]. Activation of TGR5 signaling increases the ATP/ADP ratio and calcium levels to stimulate GLP-1. INT-777 increases energy expenditure, reduces hepatic steatosis and adiposity, and improves insulin sensitivity and glucose tolerance in diet-induced obese mice [55]. TGR5 mediates bile acid-induced energy expenditure in brown adipose tissue and skeletal muscles through bile acid-TGR5-cAMP signaling-dependent activation of type2 iodothyronine deiodinase (D2) [57]. TGR5 is also important for the antiinflammation effect of bile acids. TGR5 activation by INT-777 inhibits LPS and lipid loading-induced inflammation through cAMP production-dependent reduction of serum tumor necrosis factor-$\alpha$ (TNF-$\alpha$) and inhibition of NF-$\kappa$B [58]. TGR5 also negatively regulates hepatic and gastric inflammation by antagonizing NF-$\kappa$B in LPS-treated mouse models [59,60]. The epithelial barrier is TGR5-dependent, and the integrity of the intestinal barrier and immune response are lost in TGR5 KO mice with experimental colitis [61]. A TGR5 agonist, RO5527239, provides hepatoprotection through TGR5-dependent regulation of the biliary epithelium barrier function [62]. TGR5 deficiency enhances diethylnitrosamine (DEN)-induced liver carcinoma by loss of TGR5-dependent inhibition of STAT3 signaling [63]. Therefore, TGR5 is a valuable target for treatment against several inflammatory diseases.

## Effects of TGR5-specific activation and dual activation of TGR5 and FXR in the development of atherosclerosis

TGR5 plays an important role against the development of atherosclerosis as well as in FXR signaling. Administration of a specific TGR5 activator, INT-777, inhibits atherosclerosis and plaque macrophage content in LDLR KO mice but not in TGR5$^{-/-}$; LDLR$^{-/-}$ DKO mice [58]. A dual activator of FXR and TGR5, INT-767, strongly inhibited atherosclerotic lesions and inflammatory cytokine production through TGR5-PKA-dependent inhibition of NF-κB signaling in ApoE KO and LDLR KO mice [64]. INT-767 also reduced high fat-induced adiposity and hyperlipidemia in $LDLR^{-/-}$ mice. $FXR^{-/-}$; $LDLR^{-/-}$ (FXR DKO), $TGR5^{-/-}$; $LDLR^{-/-}$ (TGR5 DKO), and $FXR^{-/-}$; $TGR5^{-/-}$; $LDLR^{-/-}$ triple KO (TKO) mice were generated to determine whether INT-767 action occurs through FXR or TGR5 [65]. The study revealed that the lipid-lowering effect of INT-767 is specifically mediated by FXR and not by TGR5. The inhibition of aortic inflammatory cytokine expression, NF-κB activation and atherosclerosis were observed in both FXR DKO and TGR5 DKO mice with INT-767 treatment, but not in TKO mice [65]. These results demonstrate that FXR and TGR5 dual activation is required for the antiinflammatory and the antiatherosclerotic effects of INT-767 to occur. In addition, simultaneous activation of FXR and TGR5 may be more effective for treatment of atherosclerosis than targeting either receptor alone.

## Pregnane X receptor (PXR) signaling, functions and cardiovascular diseases

PXR is identified as an orphan nuclear receptor activated by steroids such as pregnenolone and progesterone [66,67]. PXR also forms a heterodimer complex with RXR [67]. PXR is expressed in the gastrointestinal tract, liver and brain [2]. Bile acids such as LCA, 3-keto-LCA, CDCA, DCA, and CA activate PXR [2,4]. PXR works broadly as a xenobiotics sensor and induces genes related to phase I and phase II drug metabolism, which modify and conjugate xenobiotics and natural compounds including steroids and bile acids [2,68]. LCA induces bile acid toxicities such as inflammation, liver injury, cholestasis, gallstone formation, and carcinogenesis [69–71]. PXR mediates detoxification of LCA through induction of Cyp3A, which catalyzes phase I oxidation [71]. PXR null mice have LCA-induced liver damage [71]. PXR also mediates activation of genes related to cholesterol synthesis and affects the development of atherosclerosis [72]. Administration of a PXR agonist, pregnenolone 16α-carbonitrile (PCN), induces hypercholesterolemia and incretion of VLDL and LDL in wild-type mice and aggravates atherosclerosis in ApoE KO mice [72]. PCN increases CD36 expression and lipid content in macrophages, shown with Oil Red O staining [72]. In contrast, PXR deficiency attenuates atherosclerosis, lipid uptake, foam cell formation and lipid content in peritoneal macrophages in ApoE KO mice [73]. In addition, PXR; ApoE DKO mice have decreased CD36 protein expression in atherosclerotic lesions compared with ApoE single KO mice [73]. Human PXR, but not mouse PXR, is activated by Bisphenol A (BPA) [74], which is a chemical that is broadly used in polycarbonate plastics and associated with elevated CVD risk [75–77]. Administration of BPA to ApoE KO mice expressing humanized PXR increases atherosclerotic formations and induces Cyp3A11, MDR1a, and CD36, lipid content and foam cell formation in macrophages without alteration of plasma lipids and cholesterol [78]. BPA exposure also affects atherosclerosis during the perinatal period in ApoE KO mice expressing humanized PXR [79]. Perinatal BPA exposure modifies epigenomic states of the CD36 promoter with an increase in the transcriptional activator H3K4me3, and a decrease in the transcriptional suppressor H3K27me3 [79]. These results suggest that the activation of PXR aggravates atherosclerosis through the activation of lipid uptake into macrophages. There is little information on the association between PXR and vascular calcification, but a recent report shows that warfarin, which activates PXR [67,80], also activates phosphate-induced calcification via PXR activation [81].

## Constitutive androstane receptor signaling, functions and atherosclerosis

CAR is xenobiotics sensor, similar to PXR, but is also activated by bile acid ligands such as CA, 6-keto-LCA and 12-keto-LCA [2]. CAR binds to a specific element of DNA with RXR as a heterodimer [67,82]. CAR is expressed primarily in the liver and intestinal enterocytes [2,67]. CAR and PXR have overlapping ligands and detoxify the endogenous molecules LCA, bilirubin, and xenobiotics [2,67,83,84]. 1,4-bis[2-(3,5-dichloropyridyloxy)]benzene (TCPOBOP) is commonly used as a CAR agonist [2,4,67]. Administration of TCPOBOP to ob/ob mice, a genetically induced obesity and insulin resistance model, ameliorates insulin resistance, fatty liver and activation of β-oxidation [85]. In addition, CAR deficiency blocks the beneficial metabolic effects of TCPOBOP in ob/ob mice [85]. High fat diet-induced obesity and insulin resistance in C57BL/6J mice are also attenuated by treatment with TCPOBOP [86]. TCPOBOP reduces plasma cholesterol and atherosclerosis in LDLR KO and ApoE KO mice [87,88]. TCPOBOP treatment alters genes related to hepatic lipogenesis and cholesterol synthesis. Baf60a, a Brg/Brm-associated factors (Bafs) family protein, has been identified as a CAR target gene, which is sensitive to diet-induced hypercholesterolemia and forms a feedforward regulatory loop through CAR activation [89]. In ApoE KO mice, hepatic Baf60a deficiency reduces diet-induced atherosclerosis associated with a significant down-regulation of CAR and CAR-target genes involved in bile acid metabolism [89]. Since very few studies regarding the effects of CAR activation or inhibition in the regulation of atherosclerosis have been reported, further mechanistic studies are required to elucidate CAR actions in the development of cardiovascular diseases.

## Vitamin D receptor signaling, functions and cardiovascular diseases

VDR is expressed in the intestine, kidneys, bone, parathyroid gland, adipocytes, VSMCs, and monocytes [2,90]. Vitamin D forms a heterodimer with RXR and binds to the vitamin D response element to activate gene transcription [91]. In addition, calcitriol (1α,25-(OH)2D3), bile acids, LCA and 3-keto-LCA activate VDR [2]. Vitamin D and VDR play critical roles in mineral metabolism and bone mineralization [92,93]. Vitamin D deficiency is associated with rickets, osteomalacia [92], and incident cardiovascular disease as shown by clinical research [94,95]. Administration of a VDR activator in patients with end-stage renal disease (ESRD) improves survival outcomes compared with hemodialysis patients with no vitamin D [96–98]. Supplementation of Vitamin D to hypercholesterolemic swine promotes cholesterol efflux, reduces transcription of TNF-α and Il-1A in macrophages, and reduces atherosclerotic lesions compared to swine with vitamin D deficiency [99,100]. Cotreatment of calcitriol and bexarotene (RXR ligand) inhibits atherosclerosis in ApoE KO mice [101]. VDR is critical for endothelial function. Endothelial cell-specific VDR KO mice have reduced endothelial NO synthesis (eNOS), hypertension, cardiac hypertrophy and upregulated type-A natriuretic peptide (ANP) and type-B natriuretic peptide (BNP) compared to wild-type mice after infusion of angiotensin II (Ang II) [102]. In addition, endothelial-VDR KO exaggerates Ang II-induced cardiac fibrosis along with the transcription of genes Col1a1 and Col3a1. Adenoviral overexpression of VDR in ApoE KO mice reduced total cholesterol, TG and LDL, improved eNOS production and attenuated atherosclerotic plaque formation [103]. VDR KO; ApoE

DKO mice showed accelerated atherosclerosis and upregulated inflammatory genes such as IL-6, VCAM-1, and ICAM-1 [104]. Macrophage VDR is also critical for protecting against atherosclerosis [105]. Macrophage-specific VDR deficiency induces atherosclerotic formation and insulin resistance, cholesterol uptake into macrophages, and the upregulation of the endoplasmic reticulum stress (ER stress) pathway through activation of calmodulin kinase II signaling (CaMKII) [105]. In LDLR KO mice with CKD induced by 5/6 nephrectomy, administration of the VDR agonists calcitriol and paricalcitol attenuated vascular calcification and reduced aortic expression of osteoblastic markers such as osterix, MSX2 and OCN [106]. In contrast, supplementation with cholecalciferol (vitamin D3) increased vascular calcification, fibrosis, oxidative stress and osteoblastic genes such as Bmp-2, Msx2, Runx2, and ALP in ob/ob mice [107].

## Conclusion

In conclusion, activation and inhibition of bile acid receptors by agonists, antagonists, and gene modification affect the progression of cardiovascular diseases including atherosclerosis and vascular calcification by altering lipid profiles, inflammatory responses, oxidative stress, and other mechanisms (Tables 20.1 and 20.2). A number of growing chronic diseases such as obesity, diabetes mellitus, CKD and aging are associated with bile acid levels and cardiovascular complications. Although modulation of bile acid receptors is an attractive drug target for treatment of cardiovascular and metabolic diseases, bile acid receptor signaling and functions are complex and must be elucidated in more detail.

**TABLE 20.1** Bile acid receptors and atherosclerosis.

| Signaling | Effect | | | Model | References |
|---|---|---|---|---|---|
| *FXR* | Inhibition | Ligand | WAY362450 | ApoE KO mice and LDLR KO mice | [43,44] |
| | | | Px20606 | CETPtg; LDLR KO mice | [47] |
| | | | CDCA | ApoE KO mice with Ritonavir | [46] |
| | | | Oleanolic acid | High fat diet-fed rabbits | [50] |
| | | Gene modification | Overexpression (OE) of FXR wild type | ApoE KO mice | [48] |
| | | Other | Resveratrol | ApoE KO mice | [54] |
| | Promotion | Gene modification | c-SRC knockdown | ApoE KO mice | [48] |
| | Inhibition and Promotion | Gene modification | FXR KO | ApoE KO mice and LDLR KO mice | [40,41,52] |
| *TGR5* | Inhibition | Ligand | INT-777 | LDLR KO mice | [58] |

*(Continued)*

TABLE 20.1 Bile acid receptors and atherosclerosis. (*Cont.*)

| Signaling | Effect | | | Model | References |
|---|---|---|---|---|---|
| *FXR* and *TGR* | Inhibition | Ligand | INT-767 | ApoE KO mice and LDLR KO mice | [64,65] |
| *PXR* | Inhibition | Gene modification | PXR KO | ApoE KO mice | [73] |
| | Promotion | Ligand | Pregnenolone 16-carbonitrile | ApoE KO mice | [72] |
| | | | Bisphenol A | Human PXR tg; ApoE KO mice | [78,79] |
| *CAR* | Inhibition | Ligand | TCPOBOP | ApoE KO mice and LDLR KO mice | [87,88] |
| | | Gene modification | Baf60a (CAR cofactor) KO | ApoE KO mice | [89] |
| *VDR* | Inhibition | Ligand | Calcitriol (1α,25-$(OH)_2D_3$) | High cholesterol diet-fed swine | [99,100] |
| | | | Cotreatment of calcitriol and bexarotene (RXR ligand) | ApoE KO mice | [101] |
| | | Gene modification | Endothelial progenitor cell-specific VDR OE | ApoE KO mice | [103] |
| | Promotion | Gene modification | Systemic VDR KO | ApoE KO mice | [104] |
| | | | Macrophage-specific VDR KO | ApoE KO mice and LDLR KO mice | [105] |

TABLE 20.2 Bile acid receptors and cardiovascular calcification.

| Signaling | Effect | | | Model | References |
|---|---|---|---|---|---|
| *FXR* | Inhibition | Ligand | INT-747 | ApoE KO mice with CKD and cultured VCs | [51] |
| *PXR* | Promotion | Activation | R-warfarin | Cultured aortic valve interstitial cells | [81] |
| *VDR* | Inhibition | Ligand | Calcitriol and paricalcitol | LDLR KO mice with CKD | [106] |
| | Promotion | Ligand | Cholecalciferol | ob/ob mice | [107] |

## Acknowledgments

The authors' work was supported by grants from NIH R01DK096030, R01HL117062, R01HL133545, and R01 HL132318 to M.M.

## References


[1] Kuipers F, Bloks VW, Groen AK. Beyond intestinal soap–bile acids in metabolic control. Nat Rev Endocrinol 2014;10(8):488–98.
[2] Schaap FG, Trauner M, Jansen PL. Bile acid receptors as targets for drug development. Nat Rev Gastroenterol Hepatol 2014;11(1):55–67.
[3] Molinaro A, Wahlstrom A, Marschall HU. Role of bile acids in metabolic control. Trends Endocrinol Metab 2018;29(1):31–41.
[4] Geier A, et al. Principles of hepatic organic anion transporter regulation during cholestasis, inflammation and liver regeneration. Biochim Biophys Acta 2007;1773(3):283–308.
[5] Li T, Chiang JY. Bile acid signaling in metabolic disease and drug therapy. Pharmacol Rev 2014;66(4):948–83.
[6] Choi M, et al. Identification of a hormonal basis for gallbladder filling. Nat Med 2006;12(11):1253–5.
[7] Basatemur GL, et al. Vascular smooth muscle cells in atherosclerosis. Nat Rev Cardiol 2019;118(12):692–702.
[8] Libby P, et al. Atherosclerosis. Nat Rev Dis Primers 2019;5(1):56.
[9] Blacher J, et al. Arterial calcifications, arterial stiffness, and cardiovascular risk in end-stage renal disease. Hypertension 2001;38(4):938–42.
[10] Block GA, et al. Mortality effect of coronary calcification and phosphate binder choice in incident hemodialysis patients. Kidney Int 2007;71(5):438–41.
[11] Shanahan CM, et al. Arterial calcification in chronic kidney disease: key roles for calcium and phosphate. Circ Res 2011;109(6):697–711.
[12] Jiang Y, Ji JY. Understanding lamin proteins and their roles in aging and cardiovascular diseases. Life Sci 2018;212:20–9.
[13] Rogers MA, Aikawa E. Cardiovascular calcification: artificial intelligence and big data accelerate mechanistic discovery. Nat Rev Cardiol 2019;16(5):261–74.
[14] Wang Z, et al. Mechanisms of matrix vesicles mediating calcification transition in diabetic plaque. Heart Lung Circ 2019;29(1):112–7.
[15] Liu Y, et al. Prelamin A accelerates vascular calcification via activation of the DNA damage response and senescence-associated secretory phenotype in vascular smooth muscle cells. Circ Res 2013;112(10):e99–e109.
[16] Durham AL, et al. Role of smooth muscle cells in vascular calcification: implications in atherosclerosis and arterial stiffness. Cardiovasc Res 2018;114(4):590–600.
[17] Pols TW. TGR5 in inflammation and cardiovascular disease. Biochem Soc Trans 2014;42(2):244–9.
[18] Hageman J, et al. A role of the bile salt receptor FXR in atherosclerosis. Arterioscler Thromb Vasc Biol 2010;30(8):1519–28.
[19] Porez G, et al. Bile acid receptors as targets for the treatment of dyslipidemia and cardiovascular disease. J Lipid Res 2012;53(9):1723–37.
[20] Ryan PM, Stanton C, Caplice NM. Bile acids at the cross-roads of gut microbiome-host cardiometabolic interactions. Diabetol Metab Syndr 2017;9:102.
[21] Levi M. Role of bile acid-regulated nuclear receptor FXR and G protein-coupled receptor TGR5 in regulation of cardiorenal syndrome (cardiovascular disease and chronic kidney disease). Hypertension 2016;67(6):1080–4.
[22] Forman BM, et al. Identification of a nuclear receptor that is activated by farnesol metabolites. Cell 1995;81(5): 687–93.
[23] Otte K, et al. Identification of farnesoid X receptor beta as a novel mammalian nuclear receptor sensing lanosterol. Mol Cell Biol 2003;23(3):864–72.
[24] Li YT, et al. Farnesoid x receptor ligands inhibit vascular smooth muscle cell inflammation and migration. Arterioscler Thromb Vasc Biol 2007;27(12):2606–11.
[25] Claudel T, Staels B, Kuipers F. The Farnesoid X receptor: a molecular link between bile acid and lipid and glucose metabolism. Arterioscler Thromb Vasc Biol 2005;25(10):2020–30.

[26] Holt JA, et al. Definition of a novel growth factor-dependent signal cascade for the suppression of bile acid biosynthesis. Genes Dev 2003;17(13):1581–91.
[27] Inagaki T, et al. Fibroblast growth factor 15 functions as an enterohepatic signal to regulate bile acid homeostasis. Cell Metab 2005;2(4):217–25.
[28] Kim I, et al. Differential regulation of bile acid homeostasis by the farnesoid X receptor in liver and intestine. J Lipid Res 2007;48(12):2664–72.
[29] Byun S, et al. Postprandial FGF19-induced phosphorylation by Src is critical for FXR function in bile acid homeostasis. Nat Commun 2018;9(1):2590.
[30] Chiang JY. Bile acid regulation of gene expression: roles of nuclear hormone receptors. Endocr Rev 2002;23(4): 443–63.
[31] Lee FY, et al. FXR, a multipurpose nuclear receptor. Trends Biochem Sci 2006;31(10):572–80.
[32] Lefebvre P, et al. Role of bile acids and bile acid receptors in metabolic regulation. Physiol Rev 2009;89(1): 147–91.
[33] Sinal CJ, et al. Targeted disruption of the nuclear receptor FXR/BAR impairs bile acid and lipid homeostasis. Cell 2000;102(6):731–44.
[34] Zhang Y, et al. Activation of the nuclear receptor FXR improves hyperglycemia and hyperlipidemia in diabetic mice. Proc Natl Acad Sci U S A 2006;103(4):1006–11.
[35] Cipriani S, et al. FXR activation reverses insulin resistance and lipid abnormalities and protects against liver steatosis in Zucker (fa/fa) obese rats. J Lipid Res 2010;51(4):771–84.
[36] Watanabe M, et al. Lowering bile acid pool size with a synthetic farnesoid X receptor (FXR) agonist induces obesity and diabetes through reduced energy expenditure. J Biol Chem 2011;286(30):26913–20.
[37] Fang S, et al. Intestinal FXR agonism promotes adipose tissue browning and reduces obesity and insulin resistance. Nat Med 2015;21(2):159–65.
[38] Moschetta A, Bookout AL, Mangelsdorf DJ. Prevention of cholesterol gallstone disease by FXR agonists in a mouse model. Nat Med 2004;10(12):1352–8.
[39] Wang YD, et al. Farnesoid X receptor antagonizes nuclear factor kappaB in hepatic inflammatory response. Hepatology 2008;48(5):1632–43.
[40] Hanniman EA, et al. Loss of functional farnesoid X receptor increases atherosclerotic lesions in apolipoprotein E-deficient mice. J Lipid Res 2005;46(12):2595–604.
[41] Zhang Y, et al. FXR deficiency causes reduced atherosclerosis in Ldlr$^{-/-}$ mice. Arterioscler Thromb Vasc Biol 2006;26(10):2316–21.
[42] Mencarelli A, et al. Antiatherosclerotic effect of farnesoid X receptor. Am J Physiol Heart Circ Physiol 2009;296(2):H272–81.
[43] Flatt B, et al. Discovery of XL335 (WAY-362450), a highly potent, selective, and orally active agonist of the farnesoid X receptor (FXR). J Med Chem 2009;52(4):904–7.
[44] Hartman HB, et al. Activation of farnesoid X receptor prevents atherosclerotic lesion formation in LDLR$^{-/-}$ and apoE$^{-/-}$ mice. J Lipid Res 2009;50(6):1090–100.
[45] Hartman HB, Lai K, Evans MJ. Loss of small heterodimer partner expression in the liver protects against dyslipidemia. J Lipid Res 2009;50(2):193–203.
[46] Mencarelli A, et al. The bile acid sensor FXR protects against dyslipidemia and aortic plaques development induced by the HIV protease inhibitor ritonavir in mice. PLoS ONE 2010;5(10):e13238.
[47] Hambruch E, et al. Synthetic farnesoid X receptor agonists induce high-density lipoprotein-mediated transhepatic cholesterol efflux in mice and monkeys and prevent atherosclerosis in cholesteryl ester transfer protein transgenic low-density lipoprotein receptor (-/-) mice. J Pharmacol Exp Ther 2012;343(3):556–67.
[48] Byun S, et al. Phosphorylation of hepatic farnesoid X receptor by FGF19 signaling-activated Src maintains cholesterol levels and protects from atherosclerosis. J Biol Chem 2019;294(22):8732–44.
[49] Zhang F, et al. Angiotensin-(1-7): new perspectives in atherosclerosis treatment. J Geriatr Cardiol 2015;12(6): 676–82.
[50] Pan Y, et al. Oleanolic acid protects against pathogenesis of atherosclerosis, possibly via FXR-mediated angiotensin (Ang)-(1-7) upregulation. Biomed Pharmacother 2018;97:1694–700.
[51] Miyazaki-Anzai S, et al. Farnesoid X receptor activation prevents the development of vascular calcification in ApoE-/- mice with chronic kidney disease. Circ Res 2010;106(12):1807–17.
[52] Guo GL, et al. Effects of FXR in foam-cell formation and atherosclerosis development. Biochim Biophys Acta 2006;1761(12):1401–9.

[53] Bennett BJ, et al. Trimethylamine-N-oxide, a metabolite associated with atherosclerosis, exhibits complex genetic and dietary regulation. Cell Metab 2013;17(1):49–60.

[54] Chen ML, et al. Resveratrol Attenuates Trimethylamine-N-Oxide (TMAO)-Induced Atherosclerosis by Regulating TMAO Synthesis and Bile Acid Metabolism via Remodeling of the Gut Microbiota. MBio 2016;7(2): e02210–5.

[55] Thomas C, et al. TGR5-mediated bile acid sensing controls glucose homeostasis. Cell Metab 2009;10(3):167–77.

[56] Li T, et al. The G protein-coupled bile acid receptor, TGR5, stimulates gallbladder filling. Mol Endocrinol 2011;25(6):1066–71.

[57] Watanabe M, et al. Bile acids induce energy expenditure by promoting intracellular thyroid hormone activation. Nature 2006;439(7075):484–9.

[58] Pols TW, et al. TGR5 activation inhibits atherosclerosis by reducing macrophage inflammation and lipid loading. Cell Metab 2011;14(6):747–57.

[59] Wang YD, et al. The G-protein-coupled bile acid receptor, Gpbar1 (TGR5), negatively regulates hepatic inflammatory response through antagonizing nuclear factor kappa light-chain enhancer of activated B cells (NF-kappaB) in mice. Hepatology 2011;54(4):1421–32.

[60] Guo C, et al. The G-protein-coupled bile acid receptor Gpbar1 (TGR5) inhibits gastric inflammation through antagonizing NF-kappaB signaling pathway. Front Pharmacol 2015;6:287.

[61] Cipriani S, et al. The bile acid receptor GPBAR-1 (TGR5) modulates integrity of intestinal barrier and immune response to experimental colitis. PLoS ONE 2011;6(10):e25637.

[62] Merlen G, et al. TGR5-dependent hepatoprotection through the regulation of biliary epithelium barrier function. Gut 2019;69(1):146–57.

[63] Chen WD, et al. Deficiency of G-protein-coupled bile acid receptor Gpbar1 (TGR5) enhances chemically induced liver carcinogenesis. Hepatology 2013;57(2):656–66.

[64] Miyazaki-Anzai S, et al. Dual activation of the bile acid nuclear receptor FXR and G-protein-coupled receptor TGR5 protects mice against atherosclerosis. PLoS ONE 2014;9(9):e108270.

[65] Miyazaki-Anzai S, et al. Simultaneous inhibition of FXR and TGR5 exacerbates atherosclerotic formation. J Lipid Res 2018;59(9):1709–13.

[66] Kliewer SA, et al. An orphan nuclear receptor activated by pregnanes defines a novel steroid signaling pathway. Cell 1998;92(1):73–82.

[67] Willson TM, Kliewer SA. PXR. CAR and drug metabolism. Nat Rev Drug Discov 2002;1(4):259–66.

[68] Lehmann JM, et al. The human orphan nuclear receptor PXR is activated by compounds that regulate CYP3A4 gene expression and cause drug interactions. J Clin Invest 1998;102(5):1016–23.

[69] Palmer RH. Gallstones produced experimentally by lithocholic acid in rats. Science 1965;148(3675):1339–40.

[70] Reddy BS, Watanabe K. Effect of cholesterol metabolites and promoting effect of lithocholic acid in colon carcinogenesis in germ-free and conventional F344 rats. Cancer Res 1979;39(5):1521–4.

[71] Xie W, et al. An essential role for nuclear receptors SXR/PXR in detoxification of cholestatic bile acids. Proc Natl Acad Sci U S A 2001;98(6):3375–80.

[72] Zhou C, et al. Activation of PXR induces hypercholesterolemia in wild-type and accelerates atherosclerosis in apoE deficient mice. J Lipid Res 2009;50(10):2004–13.

[73] Sui Y, et al. Deficiency of PXR decreases atherosclerosis in apoE-deficient mice. J Lipid Res 2011;52(9):1652–9.

[74] Takeshita A, et al. Bisphenol-A, an environmental estrogen, activates the human orphan nuclear receptor, steroid and xenobiotic receptor-mediated transcription. Eur J Endocrinol 2001;145(4):513–7.

[75] Han C, Hong YC. Bisphenol A, hypertension, and cardiovascular diseases: epidemiological, laboratory, and clinical trial evidence. Curr Hypertens Rep 2016;18(2):11.

[76] Bae S, et al. Associations of bisphenol A exposure with heart rate variability and blood pressure. Hypertension 2012;60(3):786–93.

[77] Melzer D, et al. Urinary bisphenol A concentration and risk of future coronary artery disease in apparently healthy men and women. Circulation 2012;125(12):1482–90.

[78] Sui Y, et al. Bisphenol A increases atherosclerosis in pregnane X receptor-humanized ApoE deficient mice. J Am Heart Assoc 2014;3(2):e000492.

[79] Sui Y, et al. Perinatal bisphenol A exposure increases atherosclerosis in adult male PXR-humanized mice. Endocrinology 2018;159(4):1595–608.

[80] Rulcova A, et al. Stereoselective interactions of warfarin enantiomers with the pregnane X nuclear receptor in gene regulation of major drug-metabolizing cytochrome P450 enzymes. J Thromb Haemost 2010;8(12):2708–17.

[81] Yu Z, et al. Warfarin calcifies human aortic valve interstitial cells at high-phosphate conditions via pregnane X receptor. J Bone Miner Metab 2019;37(6):944–56.
[82] Baes M, et al. A new orphan member of the nuclear hormone receptor superfamily that interacts with a subset of retinoic acid response elements. Mol Cell Biol 1994;14(3):1544–52.
[83] Huang W, et al. Induction of bilirubin clearance by the constitutive androstane receptor (CAR). Proc Natl Acad Sci USA 2003;100(7):4156–61.
[84] Zhang J, et al. The constitutive androstane receptor and pregnane X receptor function coordinately to prevent bile acid-induced hepatotoxicity. J Biol Chem 2004;279(47):49517–22.
[85] Dong B, et al. Activation of nuclear receptor CAR ameliorates diabetes and fatty liver disease. Proc Natl Acad Sci USA 2009;106(44):18831–6.
[86] Gao J, et al. The constitutive androstane receptor is an anti-obesity nuclear receptor that improves insulin sensitivity. J Biol Chem 2009;284(38):25984–92.
[87] Sberna AL, et al. Constitutive androstane receptor activation stimulates faecal bile acid excretion and reverse cholesterol transport in mice. J Hepatol 2011;55(1):154–61.
[88] Sberna AL, et al. Constitutive androstane receptor activation decreases plasma apolipoprotein B-containing lipoproteins and atherosclerosis in low-density lipoprotein receptor-deficient mice. Arterioscler Thromb Vasc Biol 2011;31(10):2232–9.
[89] Meng ZX, et al. A diet-sensitive BAF60a-mediated pathway links hepatic bile acid metabolism to cholesterol absorption and atherosclerosis. Cell Rep 2015;13(8):1658–69.
[90] Norman AW. Minireview: vitamin D receptor: new assignments for an already busy receptor. Endocrinology 2006;147(12):5542–8.
[91] Haussler MR, et al. The nuclear vitamin D receptor controls the expression of genes encoding factors which feed the "Fountain of Youth" to mediate healthful aging. J Steroid Biochem Mol Biol 2010;121(1–2):88–97.
[92] Verstuyf A, et al. Vitamin D: a pleiotropic hormone. Kidney Int 2010;78(2):140–5.
[93] Yoshizawa T, et al. Mice lacking the vitamin D receptor exhibit impaired bone formation, uterine hypoplasia and growth retardation after weaning. Nat Genet 1997;16(4):391–6.
[94] Cigolini M, et al. Serum 25-hydroxyvitamin D3 concentrations and prevalence of cardiovascular disease among type 2 diabetic patients. Diabetes Care 2006;29(3):722–4.
[95] Wang TJ, et al. Vitamin D deficiency and risk of cardiovascular disease. Circulation 2008;117(4):503–11.
[96] Artaza JN, Mehrotra R, Norris KC. Vitamin D and the cardiovascular system. Clin J Am Soc Nephrol 2009;4(9):1515–22.
[97] Kovesdy CP, Kalantar-Zadeh K. Vitamin D receptor activation and survival in chronic kidney disease. Kidney Int 2008;73(12):1355–63.
[98] Teng M, et al. Activated injectable vitamin D and hemodialysis survival: a historical cohort study. J Am Soc Nephrol 2005;16(4):1115–25.
[99] Gupta GK, et al. Decreased expression of vitamin D receptors in neointimal lesions following coronary artery angioplasty in atherosclerotic swine. PLoS ONE 2012;7(8):e42789.
[100] Yin K, et al. Vitamin D protects against atherosclerosis via regulation of cholesterol efflux and macrophage polarization in hypercholesterolemic swine. Arterioscler Thromb Vasc Biol 2015;35(11):2432–42.
[101] Lin LM, et al. Coadministration of VDR and RXR agonists synergistically alleviates atherosclerosis through inhibition of oxidative stress: An in vivo and in vitro study. Atherosclerosis 2016;251:273–81.
[102] Ni W, et al. Elimination of vitamin D receptor in vascular endothelial cells alters vascular function. Hypertension 2014;64(6):1290–8.
[103] Xiang W, et al. Intravenous transfusion of endothelial progenitor cells that overexpress vitamin D receptor inhibits atherosclerosis in apoE-deficient mice. Biomed Pharmacother 2016;84:1233–42.
[104] Bozic M, et al. Impaired vitamin D signaling in endothelial cell leads to an enhanced leukocyte-endothelium interplay: implications for atherosclerosis development. PLoS ONE 2015;10(8):e0136863.
[105] Oh J, et al. Deletion of macrophage Vitamin D receptor promotes insulin resistance and monocyte cholesterol transport to accelerate atherosclerosis in mice. Cell Rep 2015;10(11):1872–86.
[106] Mathew S, et al. Vitamin D receptor activators can protect against vascular calcification. J Am Soc Nephrol 2008;19(8):1509–19.
[107] Carmo LS, et al. Expansive Vascular Remodeling and Increased Vascular Calcification Response to Cholecalciferol in a Murine Model of Obesity and Insulin Resistance. Arterioscler Thromb Vasc Biol 2019;39(2):200–11.

CHAPTER

# 21

# Molecular mechanisms underlying effects of $n-3$ and $n-6$ fatty acids in cardiovascular diseases

*Denny Joseph Manual Kollareth*[a], *Chuchun L. Chang*[a], *Hylde Zirpoli*[a], *Richard J. Deckelbaum*[a,b]

[a]Institute of Human Nutrition, Columbia University Irving Medical Center, New York, NY, United States; [b]Department of Pediatrics, Vagelos College of Physicians and Surgeons, Columbia University Irving Medical Center, New York, NY, United States

OUTLINE



Lipid Signaling and Metabolism. http://dx.doi.org/10.1016/B978-0-12-819404-1.00021-X

## Abbreviations

**AA** arachidonic acid
**CE** cholesteryl ester
**DHA** docosahexaenoic acid
**EPA** eicosapentaenoic acid
**EC** endothelial cells
**HDL** high density lipoprotein
**LA** linoleic acid
**LDL** low density lipoprotein
**PUFA** polyunsaturated fatty acids
**TG** triglyceride
**VLDL** very low density lipoprotein
**VSMC** vascular smooth muscle cell

## Polyunsaturated fatty acids and cardiovascular diseases: an overview

Fatty acids have traditionally been considered as calorically dense nutrients. Over recent decades, they have been recognized as major biologic mediators; awareness on the role of fatty acids in human health and disease prevention has been markedly increased [1–3]. Omega$-3$ ($n-3$) and omega$-6$ ($n-6$) are the two major families of polyunsaturated fatty acids (PUFA), and are classified based on the position of the last double bond in the terminal methyl group [4–6]. These fatty acids are essential components of membrane phospholipids and are also precursors to a large number of bioactive lipid mediators [7,8].

Alpha-linolenic acid (ALA, 18:3, $n-3$), eicosapentaenoic acid (EPA, 20:5, $n-3$), and docosahexaenoic acid (DHA, 22:6, $n-3$) are the prominent representatives of the $n-3$ family, while linoleic acid (LA, 18:2$n-6$) and arachidonic acid (AA, 20:4$n-6$) are the major $n-6$ PUFA. LA and ALA are considered essential fatty acids because they cannot be synthesized by humans and thus must be obtained through diet. Long chain $n-3$ (DHA, EPA) and $n-6$ (AA) PUFA can be supplied either from the diet or synthesized in the liver from ALA and LA, respectively [9–16]. By the action of a series of desaturases and elongases, LA is converted into AA, and ALA is converted into EPA and further into DHA [17–22]. The efficiency of this conversion is very low in humans and the most effective way of increasing tissue levels of long chain $n-3$ and $n-6$ PUFA are through dietary intake [4,5,23]. Most plants produce much more LA than ALA, and therefore, LA is usually the most prevalent PUFA in human diets [24]. Foods rich in LA include corn and sunflower oils, whereas flaxseed and canola oils are particularly high in ALA. The major sources of DHA and EPA are from fatty fish and fish oil supplements [25–29].

$n-3$ and $n-6$ PUFA, having multiple roles in membrane structure, lipid metabolism, blood clotting, blood pressure, and in particular inflammation, have both been linked to the reduction in cardiovascular diseases (CVD) [25,30–32]. Over five decades ago, Keys and his colleagues reported that blood cholesterol levels were correlated positively with high intake of saturated fatty acids; as well, they observed a relationship between blood cholesterol levels and the types of fatty acids ingested [33]. These findings led to the conclusion that dietary LA would reduce blood cholesterol levels and the development of atherosclerosis, resulting in recommendations for increasing consumption of $n-6$ PUFA, which equated with increases in LA consumption [34]. Others described the rarity of CVD in Greenland Eskimos, a population that consumed a diet high in whale, seal, and fish [35]. Epidemiological studies showed that serum cholesterol and triglycerides (TG) were lower in Greenland Inuit than in age-matched residents of Denmark and the incidence of myocardial infarction (MI) was much lower among Greenland population compared with the Danes [36–38]. Through a series of investigations, the "Eskimo factor" that apparently protected the Eskimos from the damages of CVD was proposed to be $n-3$ fatty acids, in particular, DHA and EPA. Thus, $n-3$ PUFA were linked to the CVD benefits observed with fish and fish oil [39]. ALA intake has been inversely associated with CVD events [40]; however, because of the absence of convincing randomized trials with clinically relevant endpoints, its role in cardiovascular protection is less clear than that of DHA and EPA [41]. Though the effects of $n-6$ fatty acids in CVD protection are controversial, LA and AA have been associated with lower risk of CVD, due to their action on multiple molecular pathways against CVD pathophysiology [23,42].

The present review will focus on describing the molecular mechanisms underlying the protective effects of $n-3$ and $n-6$ fatty acids in CVD. We will also discuss the importance of PUFA derived anti-inflammatory/pro-resolving eicosanoids and docosanoids in inflammation and atherosclerosis (Table 21.1). We will also briefly include the impact of $n-3$ and $n-6$ fatty acids on CVD outcomes based upon selected clinical trials.

## Cardiovascular regulatory mechanisms of $n-3$ and $n-6$ fatty acids

### Reduced lipid levels

Epidemiological and dietary interventions have shown that intake of PUFA significantly alters serum lipid profiles [43]. One of the most consistent and best recognized cardiovascular-related effects of $n-3$ fatty acids is in reduction of fasting and post-prandial serum TG and free fatty acid levels [44,45]. For example, fish oil supplementation significantly reduced blood levels of total cholesterol and low density lipoprotein (LDL)-cholesterol in older adults (above 50 years of age) with hypertension and hypercholesterolemia [46]. Many observational studies suggest advantages of $n-6$ fatty acid consumption on blood lipid profiles [47]. The Keys and Hegsted equations are used to relate differences in serum cholesterol levels to dietary fat intake. Both indicate that saturated fatty acids elevate serum cholesterol, while PUFA lower serum cholesterol [33,48]. These effects are mediated through reduction in the synthesis of TG and hence lower hepatic secretion rates of very low density lipoproteins (VLDL), through the interface with transcription factors that control the expression of enzymes responsible for both TG assembly and fatty acid oxidation [49]. PUFA also increase expression of hepatic LDL receptor (LDLr), a key factor that regulates blood LDL-cholesterol levels [50,51].

#### *Molecular pathways*

##### Transcriptional regulation

Peroxisome proliferator activated receptors (PPARs) are ligand-activated transcription factors, which regulate the expression of genes involved in lipid and glucose metabolism and that are associated with various cardiovascular risk factors [52–55]. $n-3$ (DHA, EPA, ALA) and $n-6$ fatty acids (LA, AA) activate the three PPAR isoforms (PPAR$\alpha$, PPAR$\beta$, PPAR$\gamma$); the

TABLE 21.1 Cardiovascular effects of $n-3$ and $n-6$ polyunsaturated fatty acids.

| | $n-3$ Fatty Acids | $n-6$ Fatty Acids |
|---|---|---|
| Reduced lipid levels | | |
| β-oxidation of fatty acids in mitochondria | ↑ | ↑ |
| Fatty acid transport to various tissues | ↑ | ↑ |
| Lipogenesis and hepatic VLDL secretion | ↓ | ↓ |
| Hepatic LDL receptor expression | ↑ | ↑ |
| ApoC-III levels | ↓ | ↓ |
| **Anti-hypertensive effects** | | |
| Renin–angiotensin–aldosterone system | ↓ | - |
| Nitric oxide synthesis | ↑ | ↑ |
| **Suppression of cardiac arrhythmias** | | |
| Sodium channel function | ↓ | ↓ |
| Cytosolic calcium overload in ischemia | ↓ | ↓ |
| **Inflammation and atherosclerosis** | | |
| Nuclear factor-κB activation | ↓ | ↓ |
| Pro-inflammatory cytokines | ↓ | ↓ |
| Cellular adhesion molecule expression | ↓ | ↓ |
| **Additional contributors of atherosclerosis** | | |
| Lipoprotein lipase activity | ↓ | ↓ |
| Reverse cholesterol transport | ↑ | ↑ |
| Plaque stability | ↑ | - |
| **Lipid mediators** | | |
| Specialized pro-resolving mediators | ↑ | ↑ |
| Anti-inflammatory eicosanoids | ↑ | ↑ |
| Pro-inflammatory eicosanoids | ↓ | ↑ |
| Nitric oxide synthesis | ↑ | ↑ |
| **Reduced platelet aggregation** | | |
| Platelet activation and aggregation | ↓ | ↓ |

*VLDL*, very low density lipoproteins; *LDL*, low density lipoproteins.

specificity of the activation among these fatty acids is different and tissue-specific [56–58]. Activated PPARs increase synthesis of enzymes involved in lipid catabolism and promote β-oxidation of fatty acids in mitochondria and peroxisomes [52,57]. PPARα controls gene expression of rate-limiting enzymes of peroxisomal β-oxidation, including acyl-CoA oxidase 1 (*ACOX1*) and L-bifunctional enzyme (*EHHADH*) [59]. PPARα also controls the expression levels of acyl CoA dehydrogenases, which catalyze the initial step in β-oxidation of fatty acids in mitochondria [59,60]. Mitochondrial carnitine palmitoyltransferase (CPT)-1 and -2 that allow the incorporation of activated long-chain fatty acids into the mitochondria for β-oxidation are targets of PPARα ([61]; [59]). Through these pathways, PUFA promote β-oxidation of fatty acids in peripheral tissues, resulting in increased metabolism of circulating TG in chylomicrons and VLDL. This results in diminished traffic of non-esterified fatty acids to liver, causing reduction in substrate availability for TG synthesis and decreasing hepatic production of VLDL [62]. PPARα also regulates fatty acid transport protein (FATP), fatty acid binding protein-5 (FABP5) and fatty acid translocase (FAT/CD36), which are involved in the transport of fatty acids to various tissues [59]. PPARγ increases adipocyte differentiation and decreases inflammatory response in cardiovascular cells, particularly in endothelial cells (EC) ([63]; [52]). Other cardiovascular protective mechanisms mediated through PPARα and PPARγ are described in later sections of this review. Considerably less is known about the function of PPARβ on lipid homeostasis in humans. PPARβ is ubiquitously expressed and its activation in animal models improves lipid homeostasis and insulin sensitivity [64]. Targeted activation of PPARβ in adipose tissues improved overall lipid profiles and reduced plasma TG levels in mice [65].

Another key family of transcription factors that regulate the expression of genes involved in cholesterol, TG and fatty acid synthesis promoting lipogenesis, are the sterol regulatory element binding proteins (SREBPs). SREBPs, membrane bound transcription factors, have a key role in pathways related to cardiovascular health [66,67]. Before activation, SREBPs are bound to membranes of the endoplasmic reticulum and under conditions of low levels of cell fatty acids (SREBP-1) or low cholesterol supply (SREBP-2), are cleaved to generate the mature active forms that translocate to the nucleus and upregulate expression of lipogenic enzymes, by binding to sterol regulatory element (SRE) or its related sequences [68–70]. SREBP-1c plays a crucial role in the dietary regulation of hepatic lipogenic genes, while SREBP-2 is actively involved in the transcription of cholesterogenic enzymes [67,71–73]. $n-3$ and $n-6$ PUFA inhibit SREBP-1 gene transcription, enhance SREBP-1c mRNA turnover, and inhibit conversion of SREBP-1 to its active form in liver cells, resulting in the inhibition of lipogenesis. Consequently, PUFA decrease fatty acid synthesis and storage in liver, reduce hepatic secretion of VLDL-cholesterol, resulting in a secondary decrease of TG-rich lipoproteins in plasma [67,74–76]. In hepatocytes, $n-3$ PUFA are more potent inhibitors of SREBP-1 expression compared to $n-6$ PUFA. Diets rich in $n-3$ or $n-6$ fatty acids significantly reduced hepatic pre-mature and nuclear SREBP-1, the more prominent reduction being with $n-3$ fatty acids [77,78]. PUFA decreased SREBP-1c promoter activity with the order of potency being AA > EPA > DHA > LA [67]. Liver X receptor(LXR)/retinoid X receptor (RXR) is a dominant activator for the expression of SREBP-1c in liver cells, and PUFA antagonize LXR/RXR leading to suppression of SREBP-1c, minimizing lipogenesis [57,62,67,68].

#### Upregulating LDL receptors

LDL receptors (LDLr) are present on the cell membranes of hepatic and other cells and mediate endocytosis of cholesterol-rich LDL that facilitates LDL oxidation and secretion to bile [79]. $n-3$ and $n-6$ PUFA both upregulate LDLr activity, protein, and mRNA levels in vitro and in vivo [50,79,80]. $n-6$ PUFA (LA) intake significantly reduced LDL-ApoB pool size and increased LDL fractional catabolic rates (FCR) compared to saturated fatty acid intake in guinea pigs [81]. The effects of PUFA on LDLr activity may be mediated through the modification of hepatocyte membrane fluidity [50]. In vitro binding studies demonstrated that alterations in LDL uptake associated with dietary fatty acid composition can also be attributed to an increase in number of LDLr [82].

#### Reducing ApoC-III

ApoC-III inhibits the turnover of plasma TG by mechanisms including inhibition of lipoprotein lipase (LPL) activity, delayed hepatic clearance of TG-rich lipoproteins, and enhanced VLDL secretion [83]. $n-3$ PUFA reduce ApoC-III gene expression by down-regulation of PPARα-mediated transcription [84], resulting in enhanced LPL-mediated lipolysis of TG-rich lipoproteins and removal of remnant lipoproteins from plasma [85,86].

#### Reducing remnant lipoprotein levels

Remnant lipoproteins (RLP) produced from TG-rich chylomicrons and VLDL exert atherogenic effects and are considered an important risk factor for CVD, in particular for sudden cardiac death [87,88]. In hypertriglyceridemic and diabetic patients, DHA and EPA supplementation reduced RLP levels in plasma [89,90] and this was ascribed to TG lowering effects mediated by these fatty acids [89].

## Vascular endothelial function and blood pressure

High blood pressure (BP) is a major but often underappreciated risk factor for CVD. Epidemiological and clinical studies show that consumption of $n-3$ PUFA reduces BP in hypertensive subjects and patients with other CVD risk factors, such as overweight and hyperlipidemia [91–93]. Dietary DHA in spontaneously hypertensive rats lowered BP and reduced angiotensin-II mediated vasoconstriction [94]. By a combined analysis of 36 randomized trials, fish oil intake, especially in older and hypertensive subjects, reduced systolic BP by 2.1 mmHg and diastolic BP by 1.6 mmHg [95]. In another study, an inverse relationship was found between dietary LA uptake and BP [96].

### *Molecular pathways*

PUFA regulate BP through multiple mechanisms. Angiotensin converting enzyme (ACE) converts angiotensin-I into angiotensin-II, which controls BP by modulating renin–angiotensin–aldosterone system. $n-3$ fatty acids inhibit ACE, leading to reduced production of angiotensin-II, vascular relaxation and decreased aldosterone formation [91,97,98]. Both calcium and protein kinase C (PKC) are involved in important signaling pathways for smooth muscle contraction. EPA inhibited intracellular calcium mobilization and PKC activation, reducing vasoconstriction [99]. DHA and EPA prevented excessive growth of smooth muscles by inhibiting transforming growth factor beta (TGF-β) synthesis and promoting vascular smooth muscle cell (VSMC) apoptosis, which led to decreases of vascular wall fibrosis and secondary hypertension development ([91]; [100]).

$n-3$ and $n-6$ fatty acids also regulate BP through increasing endothelial nitric oxide (NO) levels and improving endothelial functions in vitro and in vivo [101–104]. DHA and EPA increased endothelial NO synthase (eNOS) activity and endothelium dependent relaxation of coronary arteries [105,106]. At a resting state, eNOS co-localizes with caveolin-1 in cell membranes. EPA induced eNOS dissociation from caveolin-1 and its calcium-independent activation and translocation to the cytosol [107]. $n-3$ and $n-6$ PUFA also increased NO levels in macrophages by increasing inducible NO synthase (iNOS) expression [101,108]. Further, BP lowering effects of $n-3$ PUFA are also mediated through their effects on membrane ion channel functions. DHA dilated blood vessels by activating large-conductance calcium-dependent potassium (Slo1 BK) channels, providing a critical vasodilatory influence [109].

Dietary LA affects ionic fluxes across cell membranes and is linked to a fall in BP [110]. LA and its metabolites suppress cardiac sodium channel function and alter red and white blood cell sodium transport processes. LA diols and triols inhibited tubular sodium reabsorption, and thereby facilitated sodium excretion in salt-sensitive individuals [111]. BP lowering effects of LA may also be mediated through changes in metabolism of prostaglandins (PGs) [96] that may influence BP through the control of vascular tone, sodium excretion and renin release.

Increasing evidence suggests that PPARγ is also involved in the regulation of vascular function and BP, in addition to its well-recognized role in lipid metabolism [112]. Employing PPARγ agonists, recent studies have demonstrated molecular mechanisms of PPARγ mediated hypotensive effects—increase in endothelial functions and eNOS activity [113,114]; inhibition of angiogenesis [115]; inhibitory effects on angiotensin-II induced aldosterone synthase and aldosterone secretion [116]; and modulatory effects on aorta and mesenteric vessels by the inhibition of PI3K/Akt/4E-BP1 and ERK1/2 signaling pathways [117]. Though $n-3$ and $n-6$ PUFA are ligands for PPARγ [56,58], studies describing PPARγ-mediated hypotensive effects of PUFA are still needed.

## Cardiac arrhythmias

Approximately 80–90% of sudden cardiac deaths are linked to ventricular arrhythmias, a process involving electrophysiological mechanisms controlling muscle contraction [11]. Chronic feeding with $n-3$ fatty acids before experimental MI reduced mortality, infarct size, and ventricular arrhythmias [118]. Epidemiological studies also indicated a possible anti-arrhythmic effect of $n-3$ fatty acids [21]. Replacing saturated fats in the diet with $n-6$ fatty acids also reduced the incidence and severity of arrhythmias occurring in ischemic rats [119].

### *Molecular pathways*

$n-3$ and $n-6$ PUFA affect the conductance of ion channels indirectly by altering the physical state of the membrane phospholipids or by binding directly to ion channel proteins [120]. PUFA bind to a specific location (probably the domain1-segment6 region) on the sodium channel protein that prolongs the duration of its inactivated state resulting in a longer refractory period. Thus, an increased voltage is required for depolarization, which reduces heart rate [121–124]. $n-3$ fatty acids may also exert an anti-arrhythmic effect by inhibiting voltage-dependent L-type calcium channels, leading to a reduction of free cytosolic calcium levels during periods of ischemic stress [125,126]. Anti-arrhythmic effects of $n-3$ fatty acids are also mediated through their action on autonomic tone in humans, especially an increase in vagal tone [127]. EPA reduced ischemic ventricular fibrillation, a major cause of death after acute MI, by inhibiting cardiac $K_{ATP}$ channels and altering monophasic action potentials via down-regulation of Kir6.2 expression, a component of sarcolemmal $K_{ATP}$ channels [128]. Cytochrome P450 monoxygenases converts DHA and EPA to epoxyeicosatetraenoic acids (EpETEs) and epoxydocosapentaenoic acids (EpDPAs), which function as potent cardiovascular lipid mediators [129]. These epoxides are inhibitors of cardiac arrhythmias and protect cardiomyocytes against calcium overload [11,130]. AA prevented ischemia-induced heart arrhythmia by modulating the activity of cardiac sodium channels, causing reduction in the electrical excitability of cardiac myocytes [131].

## Atherosclerosis

Atherosclerosis, a major cause of CVD, is a progressive disease characterized by the accumulation of lipids in the arterial wall leading to development of atherosclerotic plaques. Atherosclerosis is also considered an inflammatory disease [132]. Lipid deposition is an initial step in atherogenesis that begins with the entry of lipoproteins, mainly LDL into the arterial wall. Lipid deposition initiates pro-inflammatory cascades that attract monocytes into subendothelial space. Infiltrated monocytes differentiate into macrophages that take up lipoproteins and become foam cells. These foam cells make up the fatty streak, the precursor of an atherosclerotic plaque. Foam cell formation is a key event in both early and late atherosclerotic lesions [133–135]. With time, the atherosclerotic plaque may become unstable, leading to plaque rupture, platelet aggregation, thrombosis, and coronary artery occlusion causing MI [135a]. $n-3$ and $n-6$ PUFA, especially DHA and EPA, have been extensively reported to play atheroprotective roles [133,135a,135–139].

## Inflammation and atherosclerosis

### *Molecular pathways*

#### Inhibiting production of pro-inflammatory cytokines and adhesion molecules

One of the major mechanisms underlying atheroprotective effects of $n-3$ fatty acids are mediated by repressing the gene transcription and inhibiting the production of pro-inflammatory cytokines. $n-3$ fatty acids down-regulate the activity of nuclear factor-κB (NFκB), which plays a key role in the regulation of gene expression in inflammatory responses and has been implicated in the pathogenesis of atherosclerosis and CVD [140,141]. ALA rich formulae down-regulated NFκB activation and inhibited adhesion molecules expression [142,143]. The transcription factor NFκB, found in a dimeric form in the cytoplasm of non-stimulated cells, is inactive when associated with IκB inhibitors. Phosphorylation of IκB results in its dissociation from the NFκB complex and activation of NFκB. This process allows NFκB to translocate into the nucleus and to activate the transcription of genes for pro-inflammatory cytokines, including tumor necrosis factor-α (TNF-α), interleukin-1β (IL-1β), interleukin−6 (IL−6) and cyclooxygenase-2 (COX-2) [144]. PPARα and PPARγ, activated by PUFA, exhibit robust anti-inflammatory properties and inhibit the expression of several inflammatory genes by inhibiting NFκB activation [145–147]. ALA is far less effective in reducing levels of pro-inflammatory cytokines compared to DHA and EPA [148]. The inhibitory effect of PPARα and PPARγ on NFκB is mainly attributed to increased levels of IκB-α, a negative regulator of NFκB [63,146,149]. PPARγ also inhibits activation of NFκB by decreasing reactive oxygen species generation [150].

The inhibition of NFκB activation by DHA and EPA is also mediated through the inhibition of toll-like receptors (TLRs). TLR4 is a key receptor on which both infectious and noninfectious stimuli converge to induce a pro-inflammatory response [151]. Once activated by extracellular stimuli, TLR4 phosphorylates and dissociates IκB, resulting in NFκB translocation to nuclei and activation of inflammatory genes. Myeloid differentiation primary response gene 88 (MyD88) is a cell membrane associated adapter protein used by TLR4 in the signaling cascade of NFκB activation. TLR4, MyD88 and other signaling proteins associate into lipid rafts to initiate inflammatory signaling cascades. DHA and EPA disrupt this lipid raft formation in inflammatory cells, inhibiting inflammatory signaling mediated through TLR4 and NFκB [24].

DHA is identified as a functional ligand for the G protein-coupled receptor 120 (GPCR120) to directly transduce anti-inflammatory effects and beneficial metabolic profiles in different experimental models (Bäck, 2017; [152]). After ligand stimulation, GPCR120 couples to β-arrestin 2, followed by receptor endocytosis and inhibition of TGF-β activated kinase 1 (TAK1) activation. This inhibits the downstream signaling, resulting in the inhibition of TLR and TNF-α pro-inflammatory signaling pathways, repressing macrophage-induced tissue inflammation [152].

High levels of soluble adhesion molecules are a risk factor for the development of atherosclerosis. These molecules stimulate the adhesive interaction between monocytes and EC, promoting atherosclerosis [24]. DHA and EPA reduce the expression of vascular cell adhesion molecule (VCAM-1) and intercellular adhesion molecule (ICAM-1) in the blood stream and on the surface of EC and monocytes [133,153–154]. Inhibitory mechanisms of $n-3$ fatty acids on TNF-α-induced expression of adhesion molecules are mediated through suppression of NFκB activation [155,156].

Epidemiologic studies do not support the prevailing hypothesis that $n-6$ PUFA promote inflammation [47]. LA significantly reduced levels of C-reactive protein, an inflammatory marker, that are upregulated in CVD in Japanese men [157]. A study in healthy subjects in the United States showed that high intake of $n-6$ fatty acids did not inhibit the anti-inflammatory effects of $n-3$ fatty acids and the combination of both types of fatty acids was associated with significantly reduced levels of inflammation [158]. Intake of a high-fat diet enriched with $n-6$ PUFA prevented the development of atherosclerotic lesions in LDLr knockout mice [137]. LA was found to decrease VCAM-1 and ICAM-1 expression in human coronary artery EC [159]. $n-6$ PUFA reduced the production of inflammatory factors in EC by down-regulating NFκB [136,160]. The protective action of PUFA is directly related to the number of double carbon bonds in the molecule [136]. De Caterina et al. hypothesized that double carbon bonds could scavenge reactive oxygen molecules and decrease generation of hydrogen peroxide that activates NFκB [136,160].

## Effects of PUFA on additional contributors of atherogenesis

### *Molecular pathways*

#### Inhibiting lipoprotein lipase activity

LPL is well known for its lipolytic action in blood lipoprotein TG catabolism. In contrast to its conventional role in catalyzing lipolysis, aortic expression of LPL is a marker of atherosclerosis [161]. LPL promotes adhesion of monocytes to EC. LPL also acts as a molecular bridge for various lipoproteins and promotes their binding and internalization through their receptors or selective uptake of cholesteryl esters (CE), promoting foam cell formation [45,162]. $n-3$ fatty acids decreased mRNA and protein expression of LPL and significantly inhibited LPL activity in macrophages, resulting in reduced arterial LDL whole particle uptake and

LDL-CE selective uptake [163,164]. $n-3$ fatty acids also decreased LPL protein expression within the aortic media and limited LPL expression to the aortic intima directing lipid deposition away from the aortic media [161,165]. Angiopoietin-like proteins (ANGPTL3 and ANGPTL4) are central players in various aspects of lipid metabolism and are known to increase lipid levels in plasma. ANGPTL4, a downstream target of PPARγ, is a regulator of LPL activity [166]. ANGPTL4 irreversibly inhibits LPL activity by converting active LPL dimers into inactive monomers, suppressing foam cell formation and progression of atherosclerosis [167]. Both $n-3$ and $n-6$ PUFA increased ANGPTL4 expression in vitro and in vivo [168].

### Enhancing reverse cholesterol transport

Reverse cholesterol transport (RCT), a mechanism by which excess cholesterol in peripheral tissues is transported to liver for biliary excretion, slows foam cell formation and development of atherosclerosis [169,170]. $n-3$ fatty acids beneficially affect high density lipoproteins (HDL) remodeling through lecithin cholesteryl acyl transferase (LCAT) and cholesteryl ester transfer protein (CETP), facilitating scavenger receptor B1 (SR-B1) and LDLr mediated hepatic uptake of plaque-derived excess cholesterol [170]. Dietary supplementation of fish oil promoted RCT by enhancement of hepatic excretion of macrophage-derived and HDL-derived cholesterol [171]. Fish oil increased the gene expression of *Abcg5/g8*, key proteins regulating hepatic cholesterol secretion into bile, and also downregulated intestinal *Npc1l1*, which reduces intestinal reabsorption of biliary HDL-derived cholesterol [171]. As reviewed previously, pharmacological and genetic modulation of AA metabolome might also affect RCT. $n-6$ PUFA were shown to lower plasma LDL-cholesterol and plasma total cholesterol to HDL-cholesterol ratio [172].

$n-3$ PUFA rich diets increase plasma concentrations of HDL cholesterol that are correlated with decreased risk of CVD [50,173]. The expression of ApoA-I and ApoA-II, the major apolipoproteins of HDL, is controlled by PPAR activated pathways. Fish oil interventions enhanced serum and hepatic ApoA-1 mRNA expression in obese-insulin resistant rats [174,175]. Induction of ApoA-I has been shown to influence the anti-oxidative functions of HDL. ApoA-I stabilizes the enzymatic activity of paraoxonase-1 (PON1), an antioxidant enzyme that associates with HDL particles. PON1 prevents oxidative modification of LDLs, detoxifies oxidized LDLs (oxLDL), inhibits uptake of oxLDLs by macrophages and reduces macrophage oxidative stress [170]. PON1 also increases cholesterol efflux capacity of macrophages [176].

LXRs contribute to regulation of free cholesterol levels in blood and protect cells from cholesterol overload by stimulating RCT and activating cholesterol conversion to bile acids in liver [177]. However, LXR agonists cause hepatic steatosis and $n-3$ rich diets reduced the effects of LXR agonists on increasing liver TG and blunted upregulation of SREBP-1c and fatty acid synthase mRNA expression in mice [178]. In cultured hepatocytes, LA increased LXRα expression mediated by PPARs [179]. LXRα upregulates synthesis of cholesterol 7α-hydroxylase (CYP7), a rate limiting enzyme in the pathway of cholesterol to bile acid conversion [180], resulting in enhanced biliary secretion of cholesterol. This is another example of cross-talk between fatty acid and cholesterol regulation of lipid metabolism.

### Improving plaque stability

Atherosclerotic plaque rupture is an inflammatory event involving the release of matrix metalloproteinases (MMPs) from inflammatory macrophages within the blood vessel wall, resulting in acute coronary syndrome [181]. DHA and EPA stabilize atherosclerotic plaques by decreasing infiltration of inflammatory cells or their activity in plaques and also by thickening the plaque fibrous cap [24]. Vulnerable plaques have high expression of matrix MMPs that weakens the fibrous cap predisposing it to rupture. EPA attenuated TNFα-induced production of MMPs through PPARα-dependent inhibition of NFκB [182].

#### Pro-atherogenic effects of n−6 PUFA

*n*−6 PUFA also exhibit pro-atherogenic effects. Oxidative susceptibility of LDL is influenced by a number of factors, including its antioxidant and fatty acid content [183]. Diets enriched with LA increased the LA content of LDL and its susceptibility to oxidation, which promotes vascular inflammation [42,184]. Oxidatively modified LDL (oxLDL) is capable of stimulating cell expression of a variety of cytokines and adhesion molecules that enhance monocyte chemotaxis, transmigration, and adhesion and formation of foam cells [183,185]. Exposure of EC to oxLDL also reduced production of NO, a key mediator of blood vessel relaxation [186].

## PUFA-derived lipid mediators

Chronic unresolved inflammation contributes to the development of advanced atherosclerosis. Resolution of inflammation is an active and highly coordinated process that involves several chemical mediators and cell types [187]. Specialized pro-resolving mediators (SPMs) represent a family of resolution effectors that include DHA and EPA derived resolvins (RvDs and RvEs, respectively), maresins (MaRs) and protectins (PDs), and AA derived lipoxins (LXs), produced by the action of different enzymes such as COX and lipoxygenases (LOX) [187–192]. SPMs control local inflammatory response, improve atherosclerotic changes and exert cardioprotective actions through their anti-inflammatory and pro-resolving actions [24] (Fig. 21.1).

Circulating levels of inflammation-resolving lipid mediator, RvD1, were decreased in patients with acutely symptomatic carotid disease [193]. In rabbits fed a high-fat high-cholesterol diet, RvE1 treatment reduced atherogenesis [194]. SPMs increased fibrous cap thickness and collagen synthesis in plaques promoting more stable plaque phenotype [189]. The molecular mechanisms through which SPMs exert their atheroprotective responses include activation of GPCRs and down-regulation of pro-inflammatory cytokine production, adhesion molecule expression, and leukocyte-EC adhesion in response to a variety of inflammatory stimuli [195].

### *Molecular pathways*

#### SPMs

*Resolvins*: RvDs and RvEs broadly reduce VSMC responses and modulate vascular injury. They inhibit VSMC proliferation, migration, monocyte adhesion, superoxide production and pro-inflammatory gene expression [194,196]. RvD1 protected EC and their barrier function from disruption by inflammatory mediators via mechanisms, involving the suppression of xanthine oxidase-mediated reactive oxygen species production and blocking protein tyrosine phosphatase SHP2 inactivation [197]. Aspirin triggered RvD1 (AT-RvD1) attenuated platelet-derived growth factor-induced VSMC migration via GPCR ALX/FPR2 and also by increasing the levels of cAMP and protein kinase A [198]. RvD1 protected against IL-17 driven pathological inflammation in human EC by its effects on GSK−3β/C/EBPβ pathways, which control the expression of developmental endothelial locus-1 (Del-1), an anti-inflammatory protein [199]. RvD2 increased endothelial NO production, decreased leukocyte endothelial interactions and modulated leukocyte receptor expression [200]. RvD2 also stimulated prostacyclins (PGI2), a vasodilator and anti-inflammatory mediator, through activating GPCR [200].

*Maresins*: MaRs exert homeostatic actions on vascular cells that counteract pro-inflammatory signals [189]. MaR1 attenuated TNF-α-induced NFκB activation through inhibition of Ik Kinase (IKK) phosphorylation, IκB-α degradation, and nuclear translocation of the NFκB p65 subunit [201]. MaR1 also attenuated TNF-α-induced monocyte adhesion and reactive oxygen species generation in both EC and VSMC associated with down-regulated expression of the adhesion molecule E-selectin and NADPH oxidases [201].

*Lipoxins:* Lipoxin A4 (LXA4) increased adenosine triphosphate-binding cassette transporter A1 expression and promoted cholesterol efflux through an LXRα pathway in macrophage derived foam cells [202]. The let-7 microRNA (miRNA) family plays a key role in modulating inflammatory responses and let-7 levels are decreased in diabetic human carotid plaques. LXA4 restored let-7 miRNA levels, reduced the levels of inflammatory cytokines and also suppressed VSMC proliferation and migration, which are key features of atherosclerosis [203,204]. Aspirin triggered lipoxins (ATL)-induced effective inhibition of VSMC migration mediated through the receptor FPR2/ALX [205].

### Anti-inflammatory eicosanoids from AA

Apart from LXA4, AA is also the precursor of anti-inflammatory eicosanoids [17,153,206]. PGI2 is a major product of COX-2 catalyzed metabolism of AA in the endothelium. It is a potent vasodilator that affects both systemic and pulmonary circulations [207]. The actions of PGI2 are mediated through PGI2 GPCR, known as the IP receptor [208]. The downstream effects of PGI2 from IP receptors are mediated through PPARγ [209]. PGI2 protected against atherogenic processes by limiting platelet adhesion and aggregation, leukocyte adhesion to the endothelium and VSMC proliferation in plaques [207,208]. PGI2 increases NO synthesis by increasing expression of eNOS through cAMP pathway, producing vasorelaxation [208,210]. PGI2 induces CE hydrolase activity, which catalyzes the first step in the removal of cholesterol from foam cells [211]. The metabolism of AA by cytochrome P450 epoxygenases leads to the formation of biologically active epoxyeicosatrienoic acids (EETs). EETs inhibit vascular inflammation, reduce cytokine induced adhesion molecule expression and prevent leukocyte adhesion to the vascular wall by inhibition of NFκB. EETs prevented TNF-α induced degradation of IκB-α, inhibiting NFκB activation and nuclear translocation [212]. EETs are also potential candidates for endothelium-derived hyperpolarizing factor that relax VSMC by activating calcium-sensitive potassium channels [213,214] (Fig. 21.1).

### Pro-inflammatory eicosanoids from AA

AA is also a substrate for the production of a wide variety of pro-inflammatory, vasoconstrictive, and pro-aggregatory eicosanoids, such as various PGs, thromboxanes (TXs) and leukotrienes (LTs), which are implicated in pathological processes such as inflammation [215]. PGE2 is responsible for the induction of its own production as well as induction of pro-inflammatory cytokines and interleukins [216]. TXs mediate several pathophysiological

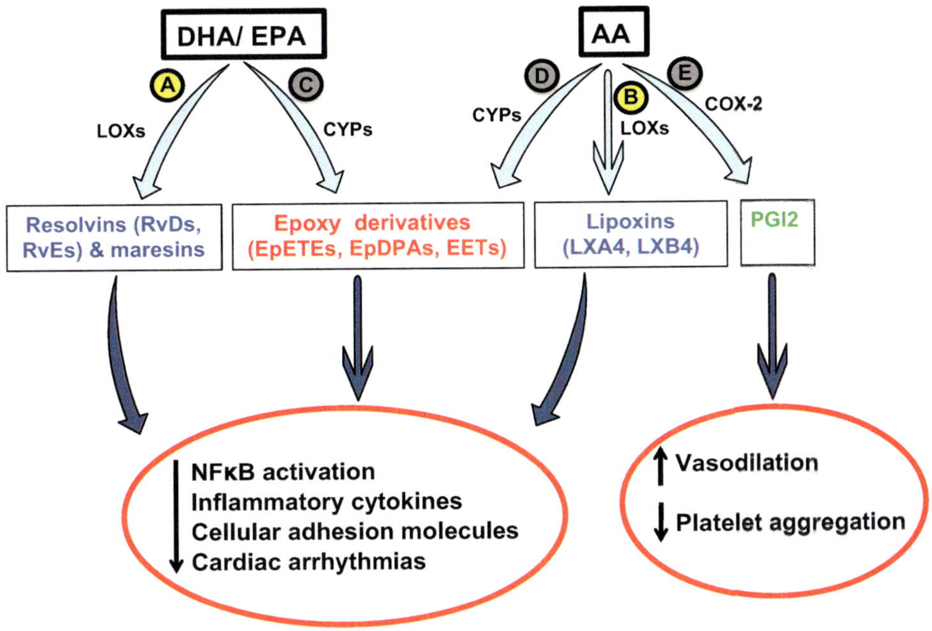


**FIGURE 21.1 Protective effects of lipid mediators from $n-3$ and $n-6$ polyunsaturated fatty acids in cardiovascular diseases.** (A, B) Specialized pro-resolving mediators (resolvins, maresins, lipoxins) produced from DHA, EPA, and AA by the action of LOXs reduce inflammation and slow pro-atherogenic changes; (C, D) epoxyderivatives produced from DHA, EPA, and AA by the action of CYPs exert cardioprotective actions by reducing inflammation and inhibiting cardiac arrhythmias; (E) PGI2 produced from AA by the action of COX-2 increases vasodilation and reduces platelet aggregation. *DHA*, docosahexaenoic acid; *EPA*, eicosapentaenoic acid; *AA*, arachidonic acid; *CYPs*, cytochrome P450 epoxygenases; *LOXs*, lipoxygenases; *NFκB*, nuclear factor-κB; *COX-2*, cyclooxygenase-2; *PGI2*, prostacyclins; *NO*, nitric oxide; *EpDPAs*, epoxydocosapentaenoic acids; *EpETEs*, epoxyeicosatetraenoic acids; *EETs*, epoxyeicosatrienoic acids; *RvDs*, D-series resolvins; *RvEs*, E-series resolvins; *LXA4*, lipoxins A4; *LXB4*, lipoxins B4.

responses, including platelet adhesion and aggregation, smooth muscle contraction and proliferation, and activation of endothelial inflammatory responses. The actions of TXs are mediated through binding to cell surface receptors [217,218]. LTs function in normal host defense and are involved in inflammatory diseases with a major role in development of atherosclerosis. LTs exert their biological effects by binding to specific GPCR [219].

Since AA produces pro-inflammatory lipid mediators and dietary LA can be converted to AA, arguments were raised against high dietary LA consumption, assuming LA and AA will increase the risk for CVD. In contrast, evidence derived from randomized trials, case-control and cohort studies and animal feeding experiments indicates that consumption of $n-6$ PUFA reduces CVD risk [220]. Production of AA from LA is tightly regulated and LA consumption was not shown to alter tissue AA content in humans [221]. In humans, high levels of plasma AA were associated with decreased levels of pro-inflammatory markers and increased levels of anti-inflammatory markers [222]. For these reasons, the American Heart Association recommends that consumption of at least 5–10% of energy from $n-6$ PUFA reduce the risk of CVD [42,220].

## Reduced platelet aggregation

Thrombosis remains an underlying pathology responsible for morbidity and mortality in CVD. Though platelet adhesion, activation, and aggregation play central roles in hemostasis, the same processes may also cause thrombosis and vessel occlusion at the site of ruptured atherosclerotic lesions leading to MI and stroke [223].

### *Molecular pathways*

DHA and EPA competitively inhibit generation of TXA2, a potent platelet aggregator. High $n-3$ PUFA intake favors the replacement of AA in cell membrane phospholipids, which inhibit the COX-1 pathway resulting in reduced TXA2 synthesis [224]. Furthermore, DHA and EPA act as TXA2 antagonists and reduce platelet aggregation by interacting directly with the human platelet TXA2/PGH2 receptor [225]. Decreased platelet aggregation by $n-3$ PUFA has also been attributed to an increase in NO synthesis in EC. PGD2 inhibits platelet activation through its binding to the D-prostanoid receptor 1 (DP1) and subsequent elevation of cAMP [226]. PGI2 activates adenylate cyclase in the platelets via IP receptor and in turn antagonizes platelet aggregation. PGE2 expresses both pro- and anti-aggregatory effects depending on its concentration through the binding of one or more of its PG receptors [226].

Platelet-derived growth factor (PDGF) is a chemoattractant for neutrophils, monocytes, fibroblasts and VSMC, and mediates proliferation of vascular endothelial and intimal smooth muscle cells. Dietary $n-3$ fatty acids down-regulated PDGF expression, suggesting anti-fibrotic and anti-atherogenic actions of $n-3$ fatty acids [227]. Abnormally elevated concentrations of the fibrinolysis inhibitor, plasminogen activator inhibitor-1 (PAI-1), have been associated with thrombotic disease [228] and $n-3$ PUFA reduced PAI-1 levels in human serum [229].

## PUFA and CVD outcomes: updates

Large scale randomized controlled trials – the DART trial [230], the GISSI-Prevenzione trial [231], the JELIS studies [232], and the GISSI-HF trial [233]—have supported beneficial effects of $n-3$ fatty acids in CVD. Systemic review and meta-analysis on the effects of $n-3$ fatty acids on CVD endpoints showed that consumption of $n-3$ fatty acids reduce the risk for CVD [234,235]. However, outcomes of many clinical trials evaluating $n-3$ fatty acid therapy on CVD have been conflicting. Double-blind trials – the OMEGA [236], the Alpha Omega [237], the SU.FOL.OM3 [238], the ORIGIN [239], and the Risk and Prevention study [240] – did not show any effect on major cardiovascular endpoints. Further, meta-analyses and systemic reviews also found no association between $n-3$ PUFA and CVD [241–244]. Two recent randomized trials—the ASCEND trial [245] and the VITAL clinical trial [246]—showed that supplementation with $n-3$ fatty acids did not result in a lower incidence of major cardiovascular events. However, in VITAL trial, analyses of the components of the primary composite cardiovascular endpoint suggested that the risk of MI was lower in the $n-3$ group than in the placebo group. Results from a recent large clinical trial (REDUCE-IT Cardiovascular Outcomes Study) strongly supported high dose EPA-mediated cardioprotective effects in decreasing a substantial number of CVD endpoints in patients on statins with established CVD with or without diabetes [247]. REDUCE-IT results prompt consideration of a potential need for higher doses of $n-3$ fatty acids in future CVD studies.

The relationship between intake of $n-6$ PUFA and risk for CVD have been less extensively studied. Recent meta-analyses have shown positive association between consumption of $n-6$ PUFA and reduction of major cardiovascular events [248,249]. In contrast, Ramsden et al. pointed out that randomized controlled trials, in which dietary conditions specifically increased $n-6$ PUFA without a concurrent increase in $n-3$ PUFA, increased the risk for coronary heart disease, MI, and death from all causes [250–252].

PUFA in general bring about a favorable impact on CVD risk factors such as hypertension, dyslipidemia, and atherosclerosis [220,253,254]. The inconsistent results from clinical trials warrant a better understanding of the effects of PUFA on the subtypes of CVD and their use in primary and secondary prevention.

## Conclusions

Pathways outlined in this review suggest that both $n-3$ and $n-6$ PUFA have the potential to improve cardiovascular health. Anti-inflammatory, hypolipidemic, anti-aggregatory, and anti-arrhythmic effects of these PUFA mediate cardioprotection (Table 21.1). A number of pathways whereby $n-3$ and $n-6$ PUFA may influence plasma lipid levels, platelet aggregation, cardiac arrhythmias, inflammation and atherosclerosis are reviewed. We have described molecular mechanisms whereby $n-3$ and $n-6$ fatty acids increase transcription of genes involved in lipolysis and fatty acid transport by activating PPARs and inhibit expression of genes involved in lipid synthesis through SREBP-mediated pathways. $n-3$ PUFA are more potent inhibitors of SREBP-1 than the $n-6$ PUFA. PUFA also upregulate hepatic LDL receptors and reduce ApoCIII levels, thereby enhancing lipid catabolism and clearance of TG rich lipoproteins from plasma. $n-3$ and $n-6$ PUFA play a role in regulating BP by improving endothelial functions via increasing NO production, and they also influence membrane ion channels, together contributing vasodilatory effect and fall in BP. $n-3$ and $n-6$ fatty acids bind to cardiac sodium channels proteins, reducing electrical excitability of cardiac myocytes, thus mediating anti-arrhythmic effects. A large body of evidence suggests anti-inflammatory effects of $n-3$ PUFA contribute to their protective action in atherogenesis. Specifically, $n-3$ fatty acids down-regulate production of inflammatory cytokines by inhibiting NFκB activation through mechanisms involving PPARs and TLR4. Finally, recent discovered lipid mediators, SPMs, and epoxides produced from DHA, EPA and AA have been shown to facilitate a wide spectrum of anti-inflammatory and pro-resolving pathways, offering cardiovascular protection. Though many of the lipid mediators produced from AA are pro-inflammatory in nature, experimental evidence indicates that consumption of $n-6$ fatty acids reduces CVD. Apart from anti-inflammatory effects, both $n-3$ and $n-6$ PUFA inhibit LPL activity and enhance RCT, further contributing to the anti-atherogenic effects. Together, $n-3$ and $n-6$ fatty acids and their derivatives reduce platelet activation and aggregation, inhibiting thrombosis.

Though the beneficial effects of both $n-3$ and $n-6$ PUFA are well recognized in CVD, conflicting results from clinical trials on $n-3$ and $n-6$ PUFA suggest that further studies are required to optimize the therapeutic strategy for their use in CVD. Randomized controlled trials assessing the impact of $n-3$ and $n-6$ PUFA on CVD morbidity and mortality need to be investigated in different populations at risk for CVD before universal recommendations can be made. Further, studies determining the direct effects of PUFA mediated activation of PPARs on mechanisms associated with CVD merits further investigation. Since chronic inflammation is a hallmark of atherosclerosis and CVD, understanding how resolution is affected in CVD and approaches to promote resolution may offer opportunities to combat inflammatory pathways associated with atherosclerosis and CVD. Further insights into the complexities of docosanoid and eicosanoid metabolism and studies focusing on SPM biosynthesis and pro-resolution mechanisms will help the development of novel pro-resolution therapeutics for CVD.

With the recent promising results of the REDUCE-IT trial showing CVD protective effects of EPA only [247], future studies will need to emphasize more differences between EPA and DHA, not only on clinical outcomes but better comparative delineation of their distinct molecular mechanisms. DHA and EPA have distinct tissue distributions as well as disparate effects on membrane structure and lipid dynamics, rates of lipid oxidation, and signal transduction pathways [102,255]. Studies assessing whether DHA and EPA have differential effects on CVD and how these fatty acids affect processes associated with CVD provide a robust area for future investigations. Linking better understanding of molecular mechanisms of EPA, DHA and $n-6$ PUFA to future clinical trials assessing the impact of specific PUFA on CVD morbidity and mortality will contribute to the formulation of more precise recommendations in different populations.

## References


[1] Kaur N, Chugh V, Gupta AK. Essential fatty acids as functional components of foods—a review. J Food Sci Technol 2012;1–15. doi: 10.1007/s13197-012-0677-0.

[2] Melo HM, Santos LE, Ferreira ST. Diet-derived fatty acids, brain inflammation, and mental health. Front Neurosci 2019;13:265.

[3] Nicolaou A, Mauro C, Urquhart P, Marelli-Berg F. Polyunsaturated fatty acid-derived lipid mediators and T cell function. Front Immunol 2014;5:75.

[4] Ha AW, Kim WK. Intake ratio and major food sources of n-3 and n-6 fatty acids in Korea: a study based on the sixth Korea national health and nutrition examination survey (2013–2014). Asia Pac J Clin Nutr 2018;27:433–40.

[5] Mori TA. Marine OMEGA-3 fatty acids in the prevention of cardiovascular disease. Fitoterapia 2017;123:51–8.

[6] Russo GL. Dietary n-6 and n-3 polyunsaturated fatty acids: from biochemistry to clinical implications in cardiovascular prevention. Biochem Pharmacol 2009;77:937–46.

[7] Dyall SC. Long-chain omega-3 fatty acids and the brain: a review of the independent and shared effects of EPA, DPA and DHA. Front Aging Neurosci 2015;7doi: 10.3389/fnagi.2015.00052.

[8] Setty BNY, Betal SG, Miller RE, Brown DS, Meier M, Cahill M, et al. Relationship of omega-3 fatty acids DHA and EPA with the inflammatory biomarker hs-CRP in children with sickle cell anemia. Prostaglandins Leukot Essent Fatty Acids 2019;146:11–8.

[9] Abedi E, Sahari MA. Long-chain polyunsaturated fatty acid sources and evaluation of their nutritional and functional properties. Food Sci Nutr 2014;2:443–63.

[10] Dyall SC. Interplay between n-3 and n-6 long-chain polyunsaturated fatty acids and the endocannabinoid system in brain protection and repair. Lipids 2017;52:885–900.

[11] Jump DB, Depner CM, Tripathy S. Omega-3 fatty acid supplementation and cardiovascular disease. J Lipid Res 2012;53:2525–45.

[12] Pastor Ó, Guzmán-Lafuente P, Serna J, Muñoz-Hernández M, López Neyra A, García-Rozas P, et al. A comprehensive evaluation of omega-3 fatty acid supplementation in cystic fibrosis patients using lipidomics. J Nutr Biochem 2019;63:197–205.

[13] Saini RK, Keum Y-S. Omega-3 and omega-6 polyunsaturated fatty acids: Dietary sources, metabolism, and significance – a review. Life Sci 2018;203:255–67.

[14] Skiba G, Poławska E, Sobol M, Raj S, Weremko D. Omega-6 and omega-3 fatty acids metabolism pathways in the body of pigs fed diets with different sources of fatty acids. Arch Anim Nutr 2015;69:1–16.

[15] Sokoła-Wysoczańska E, Wysoczański T, Wagner J, Czy K, Bodkowski R, Lochyński S, et al. Polyunsaturated fatty acids and their potential therapeutic role in cardiovascular system disorders—a review. Nutrients 2018;10(pii):E1561.

[16] Tocher DR, Betancor MB, Sprague M, Olsen RE, Napier JA. Omega-3 long-chain polyunsaturated fatty acids, EPA and DHA: bridging the gap between supply and demand. Nutrients 2019;11:E89.

[17] Calder PC. Polyunsaturated fatty acids and inflammation. Biochem Soc Trans 2005;33:423–7.

[18] Deckelbaum RJ, Torrejon C. The omega-3 fatty acid nutritional landscape: health benefits and sources. J Nutr 2012;142:587S–91S.

[19] Joshi K, Gadgil M, Pandit A, Otiv S, Kothapalli KSD, Brenna JT. Dietary pattern regulates fatty acid desaturase 1 gene expression in Indian pregnant women to spare overall long chain polyunsaturated fatty acids levels. Mol Biol Rep 2019;46:687–93.

[20] Li T, Ruan D-G, Lin Z-M, Liu T-Y, Wang K, Xu X-Y, et al. Endurance training counteracts the high-fat diet-induced profiling changes of ω-3 polyunsaturated fatty acids in skeletal muscle of middle-aged rats. Front Physiol 2019;10.

[21] Maki KC, Dicklin MR. Omega-3 fatty acid supplementation and cardiovascular disease risk: glass half full or time to nail the coffin shut? Nutrients 2018;10.

[22] Shahidi F, Ambigaipalan P. Omega-3 polyunsaturated fatty acids and their health benefits. Annu Rev Food Sci Technol 2018;9:345–81.

[23] Czernichow S, Thomas D, Bruckert E. n-6 Fatty acids and cardiovascular health: a review of the evidence for dietary intake recommendations. Br J Nutr 2010;104:788–96.

[24] Calder PC. Omega-3 fatty acids and inflammatory processes: from molecules to man. Biochem Soc Trans 2017;45:1105–15.

[25] Allayee H, Roth N, Hodis HN. Polyunsaturated fatty acids and cardiovascular disease: implications for nutrigenetics. J Nutr Nutr 2009;2:140–8.

[26] Calder PC. Intravenous lipid emulsions to deliver bioactive omega-3 fatty acids for improved patient outcomes. Mar Drugs 2019;17:E274.

[27] Godos J, Castellano S, Galvano F, Grosso G. Chapter 13 - Linking omega-3 fatty acids and depression,. In: Watson RR, Preedy VR, editors. omega fatty acids in brain and neurological health. 2nd ed. Academic Press; 2019. p. 199–212.

[28] Molina-Leyva I, Molina-Leyva A, Bueno-Cavanillas A. Efficacy of nutritional supplementation with omega-3 and omega-6 fatty acids in dry eye syndrome: a systematic review of randomized clinical trials. Acta Ophthalmol 2017;95:e677–85.

[29] Starling P, Charlton K, McMahon AT, Lucas C. Fish intake during pregnancy and foetal neurodevelopment—a systematic review of the evidence. Nutrients 2015;7:2001–14.

[30] Darwesh AM, Sosnowski DK, Lee TYT, Keshavarz-Bahaghighat H, Seubert JM. Insights into the cardioprotective properties of n-3 PUFA against ischemic heart disease via modulation of the innate immune system. Chem Biol Interact 2019;308:20–44.

[31] Sanders TAB. Omega-6 fatty acids and cardiovascular disease. Circulation 2019;139:2437–9.

[32] Zirpoli H, Abdillahi M, Quadri N, Ananthakrishnan R, Wang L, Rosario R, et al. Acute administration of n-3 rich triglyceride emulsions provides cardioprotection in murine models after ischemia-reperfusion. PLoS One 2015;10:e0116274.

[33] Keys A, Anderson JT, Grande F. Prediction of serum-cholesterol responses of man to changes in fats in the diet. Lancet 1957;273:959–66.

[34] Jandacek RJ. Linoleic acid: a nutritional quandary. Healthcare (Basel) 2017;5doi: 10.3390/healthcare5020025.

[35] Sinclair HM. Deficiency of essential fatty acids and atherosclerosis, etcetera. Lancet 1956;270:381–3.

[36] Bang HO, Dyerberg J. Plasma lipids and lipoproteins in Greenlandic west coast Eskimos. Acta Med Scand 1972;192:85–94.

[37] Dyerberg J, Bang HO, Stoffersen E, Moncada S, Vane JR. Eicosapentaenoic acid and prevention of thrombosis and atherosclerosis? Lancet 1978;2:117–9.

[38] Dyerberg J, Bang HO, Hjorne N. Fatty acid composition of the plasma lipids in Greenland Eskimos. Am J Clin Nutr 1975;28:958–66.

[39] Bang HO, Dyerberg J. Lipid metabolism and ischemic heart disease in Greenland Eskimos. Adv Food Nutr Res 1980;1–22.

[40] Fleming JA, Kris-Etherton PM. The evidence for α-linolenic acid and cardiovascular disease benefits: comparisons with eicosapentaenoic acid and docosahexaenoic acid. Adv Nutr 2014;5:863S–76S.

[41] Lee JH, O'Keefe JH, Lavie CJ, Marchioli R, Harris WS. Omega-3 fatty acids for cardioprotection. Mayo Clin Proc 2008;83:324–32.

[42] Harris WS, Mozaffarian D, Rimm E, Kris-Etherton P, Rudel LL, Appel LJ, et al. Omega-6 fatty acids and risk for cardiovascular disease: a science advisory from the American heart association nutrition subcommittee of the council on nutrition, physical activity, and metabolism; council on cardiovascular nursing; and council on epidemiology and prevention. Circulation 2009;119:902–7.

[43] Ander BP, Dupasquier CM, Prociuk MA, Pierce GN. Polyunsaturated fatty acids and their effects on cardiovascular disease. Exp Clin Cardiol 2003;8:164–72.

[44] Bays HE, Tighe AP, Sadovsky R, Davidson MH. Prescription omega-3 fatty acids and their lipid effects: physiologic mechanisms of action and clinical implications. Expert Rev Cardiovasc Ther 2008;6:391–409.

[45] Torrejon C, Jung UJ, Deckelbaum RJ. n-3 Fatty acids and cardiovascular disease: actions and molecular mechanisms. Prostaglandins Leukot Essent Fatty Acids 2007;77:319–26.

[46] Shen T, Xing G, Zhu J, Zhang S, Cai Y, Li D, et al. Effects of 12-week supplementation of marine Omega-3 PUFA-based formulation Omega3Q10 in older adults with prehypertension and/or elevated blood cholesterol. Lipids Health Dis 2017;16:253.

[47] Khandelwal S, Kelly L, Malik R, Prabhakaran D, Reddy S. Impact of omega-6 fatty acids on cardiovascular outcomes: a review. J Preventive Cardiol 2013;2:325–36.

[48] Hegsted DM, McGandy RB, Myers ML, Stare FJ. Quantitative effects of dietary fat on serum cholesterol in man. Am J Clin Nutr 1965;17:281–95.

[49] Massaro M, Scoditti E, Carluccio MA, De Caterina R. Basic mechanisms behind the effects of n-3 fatty acids on cardiovascular disease. Prostaglandins Leukot Essent Fatty Acids 2008;79:109–15.

[50] Fernandez ML, West KL. Mechanisms by which dietary fatty acids modulate plasma lipids. J Nutr 2005;135: 2075–8.

[51] Ishida T, Ohta M, Nakakuki M, Kami H, Uchiyama R, Kawano H, et al. Distinct regulation of plasma LDL cholesterol by eicosapentaenoic acid and docosahexaenoic acid in high fat diet-fed hamsters: participation of cholesterol ester transfer protein and LDL receptor. Prostaglandins Leukot Essent Fatty Acids 2013;88:281–8.

[52] Khan SA, Vanden Heuvel JP. Role of nuclear receptors in the regulation of gene expression by dietary fatty acids (review). J Nutr Biochem 2003;14:554–67.

[53] Lee W-S, Kim J. Peroxisome proliferator-activated receptors and the heart: lessons from the past and future directions. PPAR Res 2015;271983. https://doi.org/10.1155/2015/271983.

[54] Michalik L, Auwerx J, Berger JP, Chatterjee VK, Glass CK, Gonzalez FJ, et al. International Union of Pharmacology. LXI. Peroxisome proliferator-activated receptors. Pharmacol Rev 2006;58:726–41.

[55] Strand E, Lysne V, Grinna ML, Bohov P, Svardal A, Nygård O, et al. Short-term activation of peroxisome proliferator-activated receptors and induces tissue-specific effects on lipid metabolism and fatty acid composition in male wistar rats. PPAR Res 2019;2019:8047627.

[56] Korbecki J, Bobi ski R, Dutka M. Self-regulation of the inflammatory response by peroxisome proliferator-activated receptors. Inflamm Res 2019;68:443–58.

[57] Sampath H, Ntambi JM. Polyunsaturated fatty acid regulation of genes of lipid metabolism. Annu Rev Nutr 2005;25:317–40.

[58] Xu HE, Lambert MH, Montana VG, Parks DJ, Blanchard SG, Brown PJ, et al. Molecular recognition of fatty acids by peroxisome proliferator-activated receptors. Mol Cell 1999;3:397–403.

[59] Pawlak M, Lefebvre P, Staels B. Molecular mechanism of PPARα action and its impact on lipid metabolism, inflammation and fibrosis in non-alcoholic fatty liver disease. J Hepatol 2015;62:720–33.

[60] Thorpe C, Kim JJ. Structure and mechanism of action of the acyl-CoA dehydrogenases. FASEB J 1995;9: 718–25.

[61] Barrero MJ, Camarero N, Marrero PF, Haro D. Control of human carnitine palmitoyltransferase II gene transcription by peroxisome proliferator-activated receptor through a partially conserved peroxisome proliferator-responsive element. Biochem J 2003;369:721–9.

[62] Calvo M, Martinez Cruz M, Torres W, Chávez M, Luzardo E, Villasmil N, et al. Omega-3 polyunsaturated fatty acids and cardiovascular health: a molecular view into structure and function. Vessel Plus 2017;1:116–28.

[63] Hamblin M, Chang L, Fan Y, Zhang J, Chen YE. PPARs and the cardiovascular system. Antioxid Redox Signal 2009;11:1415–52.

[64] Cox RL. Rationally designed PPARδ-specific agonists and their therapeutic potential for metabolic syndrome. Proc Natl Acad Sci USA 2017;114:3284–5.
[65] Wang Y-X, Lee C-H, Tiep S, Yu RT, Ham J, Kang H, et al. Peroxisome-proliferator-activated receptor delta activates fat metabolism to prevent obesity. Cell 2003;113:159–70.
[66] Bertolio R, Napoletano F, Mano M, Maurer-Stroh S, Fantuz M, Zannini A, et al. Sterol regulatory element binding protein 1 couples mechanical cues and lipid metabolism. Nat Commun 2019;10:1–11.
[67] Yoshikawa T, Shimano H, Yahagi N, Ide T, Amemiya-Kudo M, Matsuzaka T, et al. Polyunsaturated fatty acids suppress sterol regulatory element-binding protein 1c promoter activity by inhibition of liver x receptor (LXR) binding to lxr response elements. J Biol Chem 2002;277:1705–11.
[68] Caputo M, Zirpoli H, Torino G, Tecce MF. Selective regulation of UGT1A1 and SREBP-1c mRNA expression by docosahexaenoic, eicosapentaenoic, and arachidonic acids. J Cell Physiol 2011;226:187–93.
[69] Huang L-H, Chung H-Y, Su H-M. Docosahexaenoic acid reduces sterol regulatory element binding protein-1 and fatty acid synthase expression and inhibits cell proliferation by inhibiting pAkt signaling in a human breast cancer MCF-7 cell line. BMC Cancer 2017;17. https://doi.org/10.1186/s12885-017-3936-7.
[70] Worgall TS, Sturley SL, Seo T, Osborne TF, Deckelbaum RJ. Polyunsaturated fatty acids decrease expression of promoters with sterol regulatory elements by decreasing levels of mature sterol regulatory element-binding protein. J Biol Chem 1998;273:25537–40.
[71] Horton JD, Goldstein JL, Brown MS. SREBPs: activators of the complete program of cholesterol and fatty acid synthesis in the liver. J Clin Invest 2002;109:1125–31.
[72] Linden AG, Li S, Choi HY, Fang F, Fukasawa M, Uyeda K, et al. Interplay between ChREBP and SREBP-1c coordinates postprandial glycolysis and lipogenesis in livers of mice. J Lipid Res 2018;59:475–87.
[73] Xu X, So J-S, Park J-G, Lee A-H. Transcriptional control of hepatic lipid metabolism by SREBP and ChREBP. Semin Liver Dis 2013;33:301–11.
[74] Deckelbaum RJ, Johnson RA, Worgall TS. Unsaturated fatty acids inhibit sterol regulatory element-dependent gene expression: a potential mechanism contributing to hypertriglyceridemia in fat-restricted diets. Proc Soc Exp Biol Med 2000;225:184–6.
[75] Dentin R, Benhamed F, Pégorier J-P, Foufelle F, Viollet B, Vaulont S, et al. Polyunsaturated fatty acids suppress glycolytic and lipogenic genes through the inhibition of ChREBP nuclear protein translocation. J Clin Invest 2005;115:2843–54.
[76] Worgall TS, Johnson RA, Seo T, Gierens H, Deckelbaum RJ. Unsaturated fatty acid-mediated decreases in sterol regulatory element-mediated gene transcription are linked to cellular sphingolipid metabolism. J Biol Chem 2002;277:3878–85.
[77] Schmitz G, Ecker J. The opposing effects of n-3 and n-6 fatty acids. Prog Lipid Res 2008;47:147–55.
[78] Xu J, Nakamura MT, Cho HP, Clarke SD. Sterol regulatory element binding protein-1 expression is suppressed by dietary polyunsaturated fatty acids. A mechanism for the coordinate suppression of lipogenic genes by polyunsaturated fats. J Biol Chem 1999;274:23577–83.
[79] Yu-Poth S, Yin D, Kris-Etherton PM, Zhao G, Etherton TD. Long-chain polyunsaturated fatty acids upregulate LDL receptor protein expression in fibroblasts and HepG2 cells. J Nutr 2005;135:2541–5.
[80] Kurushima H, Hayashi K, Toyota Y, Kambe M, Kajiyama G. Comparison of hypocholesterolemic effects induced by dietary linoleic acid and oleic acid in hamsters. Atherosclerosis 1995;114:213–21.
[81] Fernandez ML, Lin EC, McNamara DJ. Differential effects of saturated fatty acids on low density lipoprotein metabolism in the guinea pig. J Lipid Res 1992;33:1833–42.
[82] Fernandez ML, McNamar DJ. Dietary fat-mediated changes in hepatic apoprotein B/E receptor in the guinea pig: effect of polyunsaturated, monounsaturated, and saturated fat. Metabolism 1989;38:1094–102.
[83] Qamar A, Khetarpal SA, Khera AV, Qasim A, Rader DJ, Reilly MP. Plasma apoc-iii levels, triglycerides, and coronary artery calcification in type 2 diabetics. Arterioscler Thromb Vasc Biol 2015;35:1880–8.
[84] Olivieri O, Martinelli N, Sandri M, Bassi A, Guarini P, Trabetti E, et al. Apolipoprotein C-III, n-3 polyunsaturated fatty acids, and "insulin-resistant" T-455C APOC3 gene polymorphism in heart disease patients: example of gene-diet interaction. Clin Chem 2005;51:360–7.
[85] Sahebkar A, Simental-Mendía LE, Mikhailidis DP, Pirro M, Banach M, Sirtori CR, et al. Effect of omega-3 supplements on plasma apolipoprotein C-III concentrations: a systematic review and meta-analysis of randomized controlled trials. Ann Med 2018;50:565–75.
[86] Taskinen M-R, Packard CJ, Borén J. Emerging evidence that apoc-III inhibitors provide novel options to reduce the residual CVD. Curr Atheroscler. Rep 2019;21:27.

[87] Nakamura N, Hamazaki T, Ohta M, Okuda K, Urakaze M, Sawazaki S, et al. Joint effects of HMG-CoA reductase inhibitors and eicosapentaenoic acids on serum lipid profile and plasma fatty acid concentrations in patients with hyperlipidemia. Int J Clin Lab Res 1999;29:22–5.

[88] Oi K, Shimokawa H, Hiroki J, Uwatoku T, Abe K, Matsumoto Y, et al. Remnant lipoproteins from patients with sudden cardiac death enhance coronary vasospastic activity through upregulation of Rho-kinase. Arterioscler Thromb Vasc Biol 2004;24:918–22.

[89] Kelley DS, Siegel D, Vemuri M, Chung GH, Mackey BE. Docosahexaenoic acid supplementation decreases remnant-like particle-cholesterol and increases the (n-3) index in hypertriglyceridemic men. J Nutr 2008;138:30–5.

[90] Satoh N, Shimatsu A, Kotani K, Sakane N, Yamada K, Suganami T, et al. Purified eicosapentaenoic acid reduces small dense LDL, remnant lipoprotein particles, and C-reactive protein in metabolic syndrome. Diabetes Care 2007;30:144–6.

[91] Cabo J, Alonso R, Mata P. Omega-3 fatty acids and blood pressure. Br J Nutr 2012;2(107 Suppl):S195–200.

[92] Mori TA. Omega-3 fatty acids and hypertension in humans. Clin Exp Pharmacol Physiol 2006;33:842–6.

[93] Perez-Martinez P, Katsiki N, Mikhailidis DP. The role of n-3 fatty acids in cardiovascular disease: back to the future. Angiology 2019:3319719842005. https://doi.org/10.1177/0003319719842005.

[94] Lai L, Wang R, Jiang W, Yang X, Song J, Li X, et al. Effects of docosahexaenoic acid on large-conductance $Ca^{2+}$-activated $K^+$ channels and voltage-dependent $K^+$ channels in rat coronary artery smooth muscle cells. Acta Pharmacol Sin 2009;30:314–20.

[95] Geleijnse JM, Giltay EJ, Grobbee DE, Donders ART, Kok FJ. Blood pressure response to fish oil supplementation: metaregression analysis of randomized trials. J Hypertens 2002;20:1493–9.

[96] Miura K, Stamler J, Nakagawa H, Elliott P, Ueshima H, Chan Q, et al. International Study of Macro-Micronutrients and Blood Pressure Research Group. Relationship of dietary linoleic acid to blood pressure. The International Study of Macro-Micronutrients and Blood Pressure Study [corrected]. Hypertension 1979;52:408–14.

[97] Das UN. Essential fatty acids and their metabolites could function as endogenous HMG-CoA reductase and ACE enzyme inhibitors, anti-arrhythmic, anti-hypertensive, anti-atherosclerotic, anti-inflammatory, cytoprotective, and cardioprotective molecules. Lipids Health Dis 2008;7:37.

[98] Jayasooriya AP, Begg DP, Chen N, Mathai ML, Sinclair AJ, Wilkinson-Berka J, et al. Omega-3 polyunsaturated fatty acid supplementation reduces hypertension in TGR(mRen-2)27 rats. Prostaglandins Leukot Essent Fatty Acids 2008;78:67–72.

[99] Nyby MD, Hori MT, Ormsby B, Gabrielian A, Tuck ML. Eicosapentaenoic acid inhibits $Ca^{2+}$ mobilization and PKC activity in vascular smooth muscle cells. Am J Hypertens 2003;16:708–14.

[100] Das UN. Long-chain polyunsaturated fatty acids interact with nitric oxide, superoxide anion, and transforming growth factor-beta to prevent human essential hypertension. Eur J Clin Nutr 2004;58:195–203.

[101] de Lima TM, de Sa Lima L, Scavone C, Curi R. Fatty acid control of nitric oxide production by macrophages. FEBS Lett 2006;580:3287–95.

[102] Preston Mason R. New insights into mechanisms of action for omega-3 fatty acids in atherothrombotic cardiovascular disease. Curr Atheroscler Rep 2019;21:2.

[103] Shimokawa H, Vanhoutte PM. Dietary omega 3 fatty acids and endothelium-dependent relaxations in porcine coronary arteries. Am J Physiol 1989;256:H968–973.

[104] ebrowska A, Mizia-Stec K, Mizia M, G sior Z, Poprz cki S. Omega-3 fatty acids supplementation improves endothelial function and maximal oxygen uptake in endurance-trained athletes. Eur J Sport Sci 2015;15:305–14.

[105] Stebbins CL, Stice JP, Hart CM, Mbai FN, Knowlton AA. Effects of dietary decosahexaenoic acid (DHA) on eNOS in human coronary artery endothelial cells. J Cardiovasc Pharmacol Ther 2008;13:261–8.

[106] Zanetti M, Gortan Cappellari G, Barbetta D, Semolic A, Barazzoni R. Omega 3 polyunsaturated fatty acids improve endothelial dysfunction in chronic renal failure: role of eNOS activation and of oxidative stress. Nutrients 2017;9. E895.

[107] Hirafuji M, Machida T, Hamaue N, Minami M. Cardiovascular protective effects of n-3 polyunsaturated fatty acids with special emphasis on docosahexaenoic acid. J Pharmacol Sci 2003;92:308–16.

[108] Zehr KR, Walker MK. Omega-3 polyunsaturated fatty acids improve endothelial function in humans at risk for atherosclerosis: a review. Prostaglandins Other Lipid Mediat 2018;134:131–40.

[109] Hoshi T, Wissuwa B, Tian Y, Tajima N, Xu R, Bauer M, et al. Omega-3 fatty acids lower blood pressure by directly activating large-conductance Ca2 + -dependent K+ channels. Proc Natl Acad Sci USA 2013;110:4816–21.

[110] el Ashry A, Heagerty AM, Ollerenshaw JD, Thurston H. The effect of dietary linoleic acid on blood pressure and erythrocyte sodium transport. J Hum Hypertens 1989;3:9–15.

[111] Dreisbach AW, Rice JC, Japa S, Newman JW, Sigel A, Gill RS, et al. Salt loading increases urinary excretion of linoleic acid diols and triols in healthy human subjects. Hypertension 2008;1979(51):755–61.

[112] Ketsawatsomkron P, Pelham CJ, Groh S, Keen HL, Faraci FM, Sigmund CD. Does peroxisome proliferator-activated receptor-γ (pparγ) protect from hypertension directly through effects in the vasculature? J Biol Chem 2010;285:9311–6.

[113] Kleinhenz JM, Kleinhenz DJ, You S, Ritzenthaler JD, Hansen JM, Archer DR, et al. Disruption of endothelial peroxisome proliferator-activated receptor-γ reduces vascular nitric oxide production. Am J Physiol Heart Circ Physiol 2009;297:H1647–54.

[114] Nagao S, Yamaguchi T. PPAR-γ agonists in polycystic kidney disease with frequent development of cardiovascular disorders. Curr Mol Pharmacol 2012;5:292–300.

[115] Xin X, Yang S, Kowalski J, Gerritsen ME. Peroxisome proliferator-activated receptor γ ligands are potent inhibitors of angiogenesis in vitro and in vivo. J Biol Chem 1999;274:9116–21.

[116] Ivanova EA, Parolari A, Myasoedova V, Melnichenko AA, Bobryshev YV, Orekhov AN. Peroxisome proliferator-activated receptor (PPAR) gamma in cardiovascular disorders and cardiovascular surgery. J Cardiol 2015;66:271–8.

[117] Benkirane Karim, Viel Émilie C, Amiri Farhad, Schiffrin Ernesto L. Peroxisome proliferator–activated receptor γ regulates angiotensin ii–stimulated phosphatidylinositol 3-kinase and mitogen-activated protein kinase in blood vessels *in vivo*. Hypertension 2006;47:102–8.

[118] von Schacky C. Omega-3 fatty acids: anti-arrhythmic, pro-arrhythmic, or both? Front Physiol 2012;3doi: 10.3389/fphys.2012.00088.

[119] McLennan PL. Relative effects of dietary saturated, monounsaturated, and polyunsaturated fatty acids on cardiac arrhythmias in rats. Am J Clin Nutr 1993;57:207–12.

[120] Kang JX, Leaf A. Antiarrhythmic effects of polyunsaturated fatty acids. Recent studies. Circulation 1996;94:1774–80.

[121] Elinder F, Liin SI. Actions and mechanisms of polyunsaturated fatty acids on voltage-gated ion channels. Front Physiol 2017;8:43.

[122] Kang JX, Leaf A. Evidence that free polyunsaturated fatty acids modify $Na^+$ channels by directly binding to the channel proteins. Proc Natl Acad Sci USA 1996;93:3542–6.

[123] Xiao Y-F, Sigg DC, Leaf A. The Antiarrhythmic effect of n-3 polyunsaturated fatty acids: modulation of cardiac ion channels as a potential mechanism. J Membr Biol 2005;206:141–54.

[124] Xiao YF, Ke Q, Wang SY, Auktor K, Yang Y, Wang GK, et al. Single point mutations affect fatty acid block of human myocardial sodium channel alpha subunit Na+ channels. Proc Natl Acad Sci USA 2001;98:3606–11.

[125] Pound EM, Kang JX, Leaf A. Partitioning of polyunsaturated fatty acids, which prevent cardiac arrhythmias, into phospholipid cell membranes. J Lipid Res 2001;42:346–51.

[126] Xiao YF, Gomez AM, Morgan JP, Lederer WJ, Leaf A. Suppression of voltage-gated L-type $Ca^{2+}$ currents by polyunsaturated fatty acids in adult and neonatal rat ventricular myocytes. Proc Natl Acad Sci USA 1997;94:4182–7.

[127] Christensen JH, Schmidt EB. Autonomic nervous system, heart rate variability and n-3 fatty acids. J Cardiovasc Med (Hagerstown) 2007;8(Suppl 1):S19–22.

[128] Tsuburaya R, Yasuda S, Ito Y, Shiroto T, Gao JY, Ito K, et al. Eicosapentaenoic acid reduces ischemic ventricular fibrillation via altering monophasic action potential in pigs. J Mol Cell Cardiol 2011;51:329–36.

[129] Endo J, Arita M. Cardioprotective mechanism of omega-3 polyunsaturated fatty acids. J Cardiol 2016;67:22–7.

[130] Spector AA, Kim H-Y. Cytochrome P450 epoxygenase pathway of polyunsaturated fatty acid metabolism. Biochim Biophys Acta 2015;1851:356–65.

[131] Tallima H, El Ridi R. Arachidonic acid: physiological roles and potential health benefits – a review. J Adv Res 2017;11:33–41.

[132] Fava C, Montagnana M. atherosclerosis is an inflammatory disease which lacks a common anti-inflammatory therapy: how human genetics can help to this issue. A narrative review. Front Pharmacol 2018;9doi: 10.3389/fphar.2018.00055.

[133] Chang CL, Deckelbaum RJ. Omega-3 fatty acids: mechanisms underlying "protective effects" in atherosclerosis. Curr Opin Lipidol 2013;24:345–50.

[134] Libby P, Buring JE, Badimon L, Hansson GK, Deanfield J, Bittencourt MS, et al. Atherosclerosis. Nat Rev Dis Primer 2019;5:56.

[135] Liu L, Hu Q, Wu H, Xue Y, Cai L, Fang M, et al. Protective role of n6/n3 PUFA supplementation with varying DHA/EPA ratios against atherosclerosis in mice. J Nutr Biochem 2016;32:171–80.

[135a] Bäck M. Omega-3 fatty acids in atherosclerosis and coronary artery disease. Future Sci OA 2017;3:FSO236.

[136] De Caterina R, Liao JK, Libby P. Fatty acid modulation of endothelial activation. Am J Clin Nutr 2000;71. 213S–23S.

[137] Machado RM, Nakandakare ER, Quintao ECR, Cazita PM, Koike MK, Nunes VS, et al. Omega-6 polyunsaturated fatty acids prevent atherosclerosis development in LDLr-KO mice, in spite of displaying a pro-inflammatory profile similar to trans fatty acids. Atherosclerosis 2012;224:66–74.

[138] Saraswathi V, Gao L, Morrow JD, Chait A, Niswender KD, Hasty AH. Fish oil increases cholesterol storage in white adipose tissue with concomitant decreases in inflammation, hepatic steatosis, and atherosclerosis in mice. J Nutr 2007;137:1776–82.

[139] Sato T, Horikawa M, Takei S, Yamazaki F, Ito TK, Kondo T, et al. Preferential incorporation of administered eicosapentaenoic acid into thin-cap atherosclerotic plaques. Arterioscler Thromb Vasc Biol 2019;39:1802–16.

[140] Campos-Staffico AM, Costa APR, Carvalho LSF, Moura FA, Santos SN, Coelho-Filho OR, et al. Omega-3 intake is associated with attenuated inflammatory response and cardiac remodeling after myocardial infarction. Nutr J 2019;18:1–8.

[141] Kromhout D, Yasuda S, Geleijnse JM, Shimokawa H. Fish oil and omega-3 fatty acids in cardiovascular disease: do they really work? Eur Heart J 2012;33:436–43.

[142] Hassan A, Ibrahim A, Mbodji K, Coëffier M, Ziegler F, Bounoure F, et al. An α-linolenic acid-rich formula reduces oxidative stress and inflammation by regulating NF-κB in rats with TNBS-induced colitis. J Nutr 2010;140:1714–21.

[143] Marion-Letellier R, Savoye G, Ghosh S. Polyunsaturated fatty acids and inflammation. IUBMB Life 2015;67:659–67.

[144] Rogero MM, Calder PC. Obesity, inflammation, toll-like receptor 4 and fatty acids. Nutrients 2018;10(pii):E432.

[145] Nakanishi A, Tsukamoto I. n-3 polyunsaturated fatty acids stimulate osteoclastogenesis through PPARγ-mediated enhancement of c-Fos expression, and suppress osteoclastogenesis through PPARγ-dependent inhibition of NFkB activation. J Nutr Biochem 2015;26:1317–27.

[146] Scirpo R, Fiorotto R, Villani A, Amenduni M, Spirli C, Strazzabosco M. Stimulation of nuclear receptor PPAR-γ limits NF-kB-dependent inflammation in mouse cystic fibrosis biliary epithelium. Hepatology 2015;62:1551–62.

[147] Zúñiga J, Cancino M, Medina F, Varela P, Vargas R, Tapia G, et al. n-3 PUFA supplementation triggers ppar-α activation and PPAR-α/NF-κB Interaction: anti-inflammatory implications in liver ischemia-reperfusion injury. PLoS ONE 2011;6.

[148] Wallace FA, Miles EA, Calder PC. Comparison of the effects of linseed oil and different doses of fish oil on mononuclear cell function in healthy human subjects. Br J Nutr 2003;89:679–89.

[149] de Winther MPJ, Kanters E, Kraal G, Hofker MH. Nuclear factor kappaB signaling in atherogenesis. Arterioscler Thromb Vasc Biol 2005;25:904–14.

[150] Lee KS, Kim SR, Park SJ, Park HS, Min KH, Jin SM, et al. Peroxisome proliferator activated receptor-gamma modulates reactive oxygen species generation and activation of nuclear factor-kappaB and hypoxia-inducible factor 1alpha in allergic airway disease of mice. J Allergy Clin Immunol 2006;118:120–7.

[151] Molteni M, Gemma S, Rossetti C. The role of Toll-like receptor 4 in infectious and noninfectious inflammation. Mediators Inflamm 2016;2016:6978936.

[152] Oh DY, Talukdar S, Bae EJ, Imamura T, Morinaga H, Fan W, et al. GPR120 is an omega-3 fatty acid receptor mediating potent anti-inflammatory and insulin-sensitizing effects. Cell 2010;142:687–98.

[153] Calder PC. Marine omega-3 fatty acids and inflammatory processes: Effects, mechanisms and clinical relevance. Biochim Biophys Acta 2015;1851:469–84.

[154] Ramji DP. Polyunsaturated fatty acids and atherosclerosis: insights from pre-clinical studies. Eur J Lipid Sci Technol 2019;121:1800029.

[155] Goua M, Mulgrew S, Frank J, Rees D, Sneddon AA, Wahle KWJ. Regulation of adhesion molecule expression in human endothelial and smooth muscle cells by omega-3 fatty acids and conjugated linoleic acids: Involvement of the transcription factor NF-κB? Prostaglandins Leukot Essent Fatty Acids 2008;78:33–43.

[156] Yang Y-C, Lii C-K, Wei Y-L, Li C-C, Lu C-Y, Liu K-L, et al. Docosahexaenoic acid inhibition of inflammation is partially via cross-talk between Nrf2/heme oxygenase 1 and IKK/NF-κB pathways. J Nutr Biochem 2013;24:204–12.

[157] Poudel-Tandukar K, Nanri A, Matsushita Y, Sasaki S, Ohta M, Sato M, et al. Dietary intakes of α-linolenic and linoleic acids are inversely associated with serum C-reactive protein levels among Japanese men. Nutr Res 2009;29:363–70.

[158] Pischon Tobias, Hankinson Susan E, Hotamisligil Gökhan S, Rifai Nader, Willett Walter C, Rimm Eric B. Habitual dietary intake of n-3 and n-6 fatty acids in relation to inflammatory markers among us men and women. Circulation 2003;108:155–60.

[159] Reissig D, Rassoul F, Salvetter J, Wagner O, Richter V. Effect of fatty acids on expression of endothelial leukocyte adhesion molecules. Eur J Nutr 2003;42:224–7.

[160] Sacks FM, Campos H. Polyunsaturated fatty acids, inflammation, and cardiovascular disease: time to widen our view of the mechanisms. J Clin Endocrinol Metab 2006;91:398–400.

[161] Chang CL. Lipoprotein lipase: new roles for an "old" enzyme. Curr Opin Clin Nutr Metab Care 2019;22:111–5.

[162] Seo T, Qi K, Chang C, Liu Y, Worgall TS, Ramakrishnan R, et al. Saturated fat-rich diet enhances selective uptake of LDL cholesteryl esters in the arterial wall. J Clin Invest 2005;115:2214–22.

[163] Chang CL, Seo T, Du CB, Accili D, Deckelbaum RJ. n-3 Fatty acids decrease arterial low-density lipoprotein cholesterol delivery and lipoprotein lipase levels in insulin-resistant mice. Arterioscler Thromb Vasc Biol 2010;30:2510–7.

[164] Chang CL, Seo T, Matsuzaki M, Worgall TS, Deckelbaum RJ. n-3 fatty acids reduce arterial LDL-cholesterol delivery and arterial lipoprotein lipase levels and lipase distribution. Arterioscler Thromb Vasc Biol 2009;29:555–61.

[165] Sudheendran S, Chang CC, Deckelbaum RJ. N-3 vs saturated fatty acids: effects on the arterial wall. Prostaglandins Leukot Essent Fatty Acids 2010;82:205–9.

[166] Yoshida K, Shimizugawa T, Ono M, Furukawa H. Angiopoietin-like protein 4 is a potent hyperlipidemia-inducing factor in mice and inhibitor of lipoprotein lipase. J Lipid Res 2002;43:1770–2.

[167] Georgiadi A, Wang Y, Stienstra R, Tjeerdema N, Janssen A, Stalenhoef A, et al. Overexpression of angiopoietin-like protein 4 protects against atherosclerosis development. Arterioscler Thromb Vasc Biol 2013;33:1529–37.

[168] Brands M, Sauerwein H, Ackermans M, Kersten S, Serlie M. Omega 3 long chain fatty acids strongly induce ANGPTL4 in humans. J Lipid Res 2013;54doi: 10.1194/jlr.M030049.

[169] Peng Y, Xu J, Zeng Y, Chen L, Xu XL. Polydatin attenuates atherosclerosis in apolipoprotein E-deficient mice: role of reverse cholesterol transport. Phytomedicine 2019;62:152935.

[170] Pizzini A, Lunger L, Demetz E, Hilbe R, Weiss G, Ebenbichler C, et al. The role of omega-3 fatty acids in reverse cholesterol transport: a review. Nutrients 2017;9. E1099.

[171] Nishimoto T, Pellizzon MA, Aihara M, Stylianou IM, Billheimer JT, Rothblat G, et al. Fish oil promotes macrophage reverse cholesterol transport in mice. Arterioscler Thromb Vasc Biol 2009;29:1502–8.

[172] Demetz E, Schroll A, Auer K, Heim C, Patsch JR, Eller P, et al. The arachidonic acid metabolome serves as a conserved regulator of cholesterol metabolism. Cell Metab 2014;20:787–98.

[173] Chan DC, Watts GF, Nguyen MN, Barrett PHR. Factorial study of the effect of n-3 fatty acid supplementation and atorvastatin on the kinetics of HDL apolipoproteins A-I and A-II in men with abdominal obesity. Am J Clin Nutr 2006;84:37–43.

[174] Schoonjans K, Staels B, Auwerx J. Role of the peroxisome proliferator-activated receptor (PPAR) in mediating the effects of fibrates and fatty acids on gene expression. J Lipid Res 1996;37:907–25.

[175] Xie X, Zhang T, Zhao S, Li W, Ma L, Ding M, et al. Effects of n-3 polyunsaturated fatty acids high fat diet intervention on the synthesis of hepatic high-density lipoprotein cholesterol in obesity-insulin resistance rats. Lipids Health Dis 2016;15:81.

[176] Berrougui H, Loued S, Khalil A. Purified human paraoxonase-1 interacts with plasma membrane lipid rafts and mediates cholesterol efflux from macrophages. Free Radic Biol Med 2012;52:1372–81.

[177] Baranowski M. Biological role of liver X receptors. J Physiol Pharmacol Off J Pol Physiol Soc 2008;59(Suppl 7):31–55.

[178] Jung UJ, Millman PN, Tall AR, Deckelbaum RJ. n-3 Fatty acids ameliorate hepatic steatosis and dysfunction after LXR agonist ingestion in mice. Biochim Biophys Acta 2011;1811:491–7.

[179] Tobin KA, Steineger HH, Alberti S, Spydevold O, Auwerx J, Gustafsson JA, et al. Cross-talk between fatty acid and cholesterol metabolism mediated by liver X receptor-alpha. Mol Endocrinol 2000;14:741–52.

[180] Zhu R, Ou Z, Ruan X, Gong J. Role of liver X receptors in cholesterol efflux and inflammatory signaling (review). Mol Med Rep 2012;5:895–900.

[181] Thies F, Pillon C, Moliere P, Lagarde M, Lecerf J. Preferential incorporation of sn-2 lysoPC DHA over unesterified DHA in the young rat brain. Am J Physiol 1994;267:R1273–1279.

[182] Matsumoto M, Sata M, Fukuda D, Tanaka K, Soma M, Hirata Y, et al. Orally administered eicosapentaenoic acid reduces and stabilizes atherosclerotic lesions in ApoE-deficient mice. Atherosclerosis 2008;197:524–33.

[183] Tsimikas S, Philis-Tsimikas A, Alexopoulos S, Sigari F, Lee C, Reaven PD. LDL isolated from Greek subjects on a typical diet or from American subjects on an oleate-supplemented diet induces less monocyte chemotaxis and adhesion when exposed to oxidative stress. Arterioscler Thromb Vasc Biol 1999;19:122–30.

[184] Louheranta AM, Porkkala-Sarataho EK, Nyyssönen MK, Salonen RM, Salonen JT. Linoleic acid intake and susceptibility of very-low-density and low density lipoproteins to oxidation in men. Am J Clin Nutr 1996;63:698–703.

[185] Dufour D, Khalil A, Nuyens V, Rousseau A, Delporte C, Noyon C, et al. Native and myeloperoxidase-oxidized low-density lipoproteins act in synergy to induce release of resolvin-D1 from endothelial cells. Atherosclerosis 2018;272:108–17.

[186] Tsimikas S. Oxidative biomarkers in the diagnosis and prognosis of cardiovascular disease. Am J Cardiol 2006;98:9P–17P.

[187] Serhan CN. Novel pro-resolving lipid mediators in inflammation are leads for resolution physiology. Nature 2014;510:92–101.

[188] Fredman G, Hellmann J, Proto JD, Kuriakose G, Colas RA, Dorweiler B, et al. An imbalance between specialized pro-resolving lipid mediators and pro-inflammatory leukotrienes promotes instability of atherosclerotic plaques. Nat Commun 2016;7:12859.

[189] Fredman G, Spite M. Specialized pro-resolving mediators in cardiovascular diseases. Mol Aspects Med 2017;58:65–71.

[190] Irún P, Lanas A, Piazuelo E. Omega-3 polyunsaturated fatty acids and their bioactive metabolites in gastrointestinal malignancies related to unresolved inflammation. A review. Front Pharmacol 2019;10.

[191] Serhan CN, de la Rosa X, Jouvene C. Novel mediators and mechanisms in the resolution of infectious inflammation: evidence for vagus regulation. J Intern Med 2019;286:240–58.

[192] Serhan CN, Chiang N, Dalli J. New pro-resolving n-3 mediators bridge resolution of infectious inflammation to tissue regeneration. Mol Aspects Med 2017;64:1–17.

[193] Bazan HA, Lu Y, Jun B, Fang Z, Woods TC, Hong S. Circulating inflammation-resolving lipid mediators RvD1 and DHA are decreased in patients with acutely symptomatic carotid disease. Prostaglandins Leukot Essent Fatty Acids 2017;125:43–7.

[194] Hasturk H, Abdallah R, Kantarci A, Nguyen D, Giordano N, Hamilton J, et al. Resolvin E1 (RvE1) attenuates atherosclerotic plaque formation in diet and inflammation-induced atherogenesis. Arterioscler Thromb Vasc Biol 2015;35:1123–33.

[195] de Gaetano M, McEvoy C, Andrews D, Cacace A, Hunter J, Brennan E, et al. Specialized pro-resolving lipid mediators: modulation of diabetes-associated cardio-, reno-, and retino-vascular complications. Front Pharmacol 2018;9:1488.

[196] Miyahara T, Runge S, Chatterjee A, Chen M, Mottola G, Fitzgerald JM, et al. D-series resolvin attenuates vascular smooth muscle cell activation and neointimal hyperplasia following vascular injury. FASEB J 2013;27:2220–32.

[197] Chattopadhyay R, Raghavan S, Rao GN. Resolvin D1 via prevention of ROS-mediated SHP2 inactivation protects endothelial adherens junction integrity and barrier function. Redox Biol 2017;12:438–55.

[198] Mottola G, Chatterjee A, Wu B, Chen M, Conte MS. Aspirin-triggered resolvin D1 attenuates PDGF-induced vascular smooth muscle cell migration via the cyclic adenosine monophosphate/protein kinase A (cAMP/PKA) pathway. PLoS One 2017;12:e0174936.

[199] Maekawa T, Hosur K, Abe T, Kantarci A, Ziogas A, Wang B, et al. Antagonistic effects of IL-17 and D-resolvins on endothelial Del-1 expression through a GSK-3β-C/EBPβ pathway. Nat Commun 2015;6:8272.

[200] Spite M, Norling LV, Summers L, Yang R, Cooper D, Petasis NA, et al. Resolvin D2 is a potent regulator of leukocytes and controls microbial sepsis. Nature 2009;461:1287–91.

[201] Chatterjee A, Sharma A, Chen M, Toy R, Mottola G, Conte MS. The pro-resolving lipid mediator maresin 1 (MaR1) attenuates inflammatory signaling pathways in vascular smooth muscle and endothelial cells. PLoS One 2014;9:e113480.

[202] Sha Y-H, Hu Y-W, Gao J-J, Wang Y-C, Ma X, Qiu Y-R, et al. Lipoxin A4 promotes ABCA1 expression and cholesterol efflux through the LXRα signaling pathway in THP-1 macrophage-derived foam cells. Int J Clin Exp Pathol 2015;8:6708–15.

[203] Brennan E, Wang B, McClelland A, Mohan M, Marai M, Beuscart O, et al. Protective effect of let-7 miRNA family in regulating inflammation in diabetes-associated atherosclerosis. Diabetes 2017;66:2266–77.

[204] Brennan EP, Mohan M, McClelland A, de Gaetano M, Tikellis C, Marai M, et al. Lipoxins protect against inflammation in diabetes-associated atherosclerosis. Diabetes 2018;67:2657–67.
[205] Petri MH, Laguna-Fernandez A, Tseng C-N, Hedin U, Perretti M, Bäck M. Aspirin-triggered 15-epi-lipoxin $A_4$ signals through FPR2/ALX in vascular smooth muscle cells and protects against intimal hyperplasia after carotid ligation. Int J Cardiol 2015;179:370–2.
[206] Calder PC. Long-chain polyunsaturated fatty acids and inflammation. Scand J Food Nutr 2006;50:54–61.
[207] Fink AN, Frishman WH, Azizad M, Agarwal Y. Use of prostacyclin and its analogues in the treatment of cardiovascular disease. Heart Dis Hagerstown Md 1999;1:29–40.
[208] Kawabe J-I, Ushikubi F, Hasebe N. Prostacyclin in vascular diseases. Recent insights and future perspectives. Circ J 2010;74:836–43.
[209] Mohite A, Chillar A, So S-P, Cervantes V, Ruan K-H. Novel mechanism of the vascular protector prostacyclin: regulating microRNA expression. Biochemistry 2011;50:1691–9.
[210] Dorris SL, Peebles RS. PGI2 as a regulator of inflammatory diseases. Mediators Inflamm 2012;926968. https://doi.org/10.1155/2012/926968.
[211] Stitham J, Midgett C, Martin KA, Hwa J. Prostacyclin: an inflammatory paradox. Front Pharmacol 2011;2:24.
[212] Node K, Huo Y, Ruan X, Yang B, Spiecker M, Ley K, et al. Anti-inflammatory properties of cytochrome p450 epoxygenase-derived eicosanoids. Science 1999;285:1276–9.
[213] Oltman CL, Weintraub NL, VanRollins M, Dellsperger KC. Epoxyeicosatrienoic acids and dihydroxyeicosatrienoic acids are potent vasodilators in the canine coronary microcirculation. Circ Res 1998;83:932–9.
[214] Rosolowsky M, Campbell WB. Synthesis of hydroxyeicosatetraenoic (HETEs) and epoxyeicosatrienoic acids (EETs) by cultured bovine coronary artery endothelial cells. Biochim Biophys Acta 1996;1299:267–77.
[215] Patterson, E., Wall, R., Fitzgerald, G.F., Ross, R.P., Stanton, C., 2012. Health implications of high dietary omega-6 polyunsaturated Fatty acids. J. Nutr. Me[256], 539426. https://doi.org/10.1155/2012/539426.
[216] Innes JK, Calder PC. Omega-6 fatty acids and inflammation. Prostaglandins Leukot Essent Fatty Acids 2018;132:41–8.
[217] Ricciotti E, FitzGerald GA. Prostaglandins and Inflammation. Arterioscler Thromb Vasc Biol 2011;31:986–1000.
[218] Thomas DW, Rocha PN, Nataraj C, Robinson LA, Spurney RF, Koller BH, et al. Proinflammatory actions of thromboxane receptors to enhance cellular immune responses. J Immunol 2003;171:6389–95.
[219] Jo-Watanabe A, Okuno T, Yokomizo T. The role of leukotrienes as potential therapeutic targets in allergic disorders. Int J Mol Sci 2019;20. E3580.
[220] Harris WS, Shearer GC. Omega-6 fatty acids and cardiovascular disease: friend, not foe? Circulation 2014;130:1562–4.
[221] Sarkkinen ES, Agren JJ, Ahola I, Ovaskainen ML, Uusitupa MI. Fatty acid composition of serum cholesterol esters, and erythrocyte and platelet membranes as indicators of long-term adherence to fat-modified diets. Am J Clin Nutr 1994;59:364–70.
[222] Ferrucci L, Cherubini A, Bandinelli S, Bartali B, Corsi A, Lauretani F, et al. Relationship of plasma polyunsaturated fatty acids to circulating inflammatory markers. J Clin Endocrinol Metab 2006;91:439–46.
[223] Adili R, Hawley M, Holinstat M. Regulation of platelet function and thrombosis by omega-3 and omega-6 polyunsaturated fatty acids. Prostaglandins Other Lipid Mediat 2018;139:10–8.
[224] Walker CG, West AL, Browning LM, Madden J, Gambell JM, Jebb SA, et al. The pattern of fatty acids displaced by EPA and DHA following 12 months supplementation varies between blood cell and plasma fractions. Nutrients 2015;7:6281–93.
[225] Parent CA, Lagarde M, Venton DL, Le Breton GC. Selective modulation of the human platelet thromboxane A2/prostaglandin H2 receptor by eicosapentaenoic and docosahexaenoic acids in intact platelets and solubilized platelet membranes. J Biol Chem 1992;267:6541–7.
[226] Yeung J, Hawley M, Holinstat M. The expansive role of oxylipins on platelet biology. J Mol Med Berl Ger 2017;95:575–88.
[227] Kaminski WE, Jendraschak E, Kiefl R, von Schacky C. Dietary omega-3 fatty acids lower levels of platelet-derived growth factor mRNA in human mononuclear cells. Blood 1993;81:1871–9.
[228] Kohler HP, Grant PJ. Plasminogen-activator inhibitor type 1 and coronary artery disease. N Engl J Med 2000;342:1792–801.
[229] Mehta J, Lawson D, Saldeen TJ. Reduction in plasminogen activator inhibitor-1 (PAI-1) with omega-3 polyunsaturated fatty acid (PUFA) intake. Am Heart J 1988;116:1201–6.

[230] Burr ML, Fehily AM, Gilbert JF, Rogers S, Holliday RM, Sweetnam PM, et al. Effects of changes in fat, fish, and fibre intakes on death and myocardial reinfarction: diet and reinfarction trial (DART). Lancet 1989;2:757–61.

[231] GISSI-Prevenzione Investigators. Dietary supplementation with n-3 polyunsaturated fatty acids and vitamin E after myocardial infarction: results of the GISSI-Prevenzione trial. Gruppo Italiano per lo Studio della Sopravvivenza nell'Infarto miocardico. Lancet 1999;354:447–55.

[232] Yokoyama M, Origasa H, Matsuzaki M, Matsuzawa Y, Saito Y, Ishikawa Y, et al. Japan EPA lipid intervention study (JELIS) Investigators, Effects of eicosapentaenoic acid on major coronary events in hypercholesterolaemic patients (JELIS): a randomised open-label, blinded endpoint analysis. Lancet 2007;369:1090–8.

[233] Tavazzi L, Maggioni AP, Marchioli R, Barlera S, Franzosi MG, Latini R, et al. Gissi-HF Investigators, Effect of n-3 polyunsaturated fatty acids in patients with chronic heart failure (the GISSI-HF trial): a randomised, double-blind, placebo-controlled trial. Lancet 2008;372:1223–30.

[234] Gobbo LCD, Imamura F, Aslibekyan S, Marklund M, Virtanen JK, Wennberg M, et al. ω-3 Polyunsaturated fatty acid biomarkers and coronary heart disease: pooling project of 19 cohort studies. JAMA Intern Med 2016;176:1155–66.

[235] Wang C, Harris WS, Chung M, Lichtenstein AH, Balk EM, Kupelnick B, et al. n−3 Fatty acids from fish or fish-oil supplements, but not α-linolenic acid, benefit cardiovascular disease outcomes in primary- and secondary-prevention studies: a systematic review. Am J Clin Nutr 2006;84:5–17.

[236] Rauch B, Schiele R, Schneider S, Diller F, Victor N, Gohlke H, et al. OMEGA Study Group, OMEGA, a randomized, placebo-controlled trial to test the effect of highly purified omega-3 fatty acids on top of modern guideline-adjusted therapy after myocardial infarction. Circulation 2010;122:2152–9.

[237] Kromhout D, Giltay EJ, Geleijnse JM. Alpha Omega Trial Group, n-3 fatty acids and cardiovascular events after myocardial infarction. N Engl J Med 2010;363:2015–26.

[238] Galan P, Kesse-Guyot E, Czernichow S, Briancon S, Blacher J, Hercberg S. SU.FOL.OM3 Collaborative Group. Effects of B vitamins and omega 3 fatty acids on cardiovascular diseases: a randomised placebo controlled trial. BMJ 2010;341:c6273.

[239] Gerstein HC, Bosch J, Dagenais GR, Díaz R, Jung H, Maggioni AP, ORIGIN Trial Investigators. et al. Basal insulin and cardiovascular and other outcomes in dysglycemia. N Engl J Med 2012;367:319–28.

[240] Roncaglioni MC, Tombesi M, Avanzini F, Barlera S, Caimi V, Risk and Prevention Study Collaborative Group. et al. n-3 fatty acids in patients with multiple cardiovascular risk factors. N Engl J Med 2013;368:1800–8.

[241] Abdelhamid AS, Brown TJ, Brainard JS, Biswas P, Thorpe GC, Moore HJ, et al. Omega-3 fatty acids for the primary and secondary prevention of cardiovascular disease. Cochrane Database Syst Rev 2018;7:CD003177.

[242] Aung T, Halsey J, Kromhout D, Gerstein HC, Marchioli R, Tavazzi L, et al. Associations of omega-3 fatty acid supplement use with cardiovascular disease risks: meta-analysis of 10 trials involving 77 917 individuals. JAMA Cardiol 2018;3:225–33.

[243] Rizos EC, Ntzani EE, Bika E, Kostapanos MS, Elisaf MS. Association between omega-3 fatty acid supplementation and risk of major cardiovascular disease events: a systematic review and meta-analysis. JAMA 2012;308:1024–33.

[244] Rizos EC, Elisaf MS. Does supplementation with omega-3 PUFA add to the prevention of cardiovascular disease? Curr Cardiol Rep 2017;19:47.

[245] Bowman L, Mafham M, Wallendszus K, Stevens W, Buck G, Barton J, ASCEND Study Collaborative Group. et al. Effects of n-3 fatty acid supplements in diabetes mellitus. N Engl J Med 2018;379:1540–50.

[246] Manson JE, Cook NR, Lee I-M, Christen W, Bassuk SS, Mora S, et al. VITAL Research Group. Marine n-3 fatty acids and prevention of cardiovascular disease and cancer. N Engl J Med 2019;380:23–32.

[247] Bhatt DL, Steg PG, Miller M, Brinton EA, Jacobson TA, Ketchum SB, et al. REDUCE-IT Investigators. Cardiovascular risk reduction with icosapent ethyl for hypertriglyceridemia. N Engl J Med 2019;380:11–22.

[248] Marklund M, Wu JHY, Imamura F, Del Gobbo LC, Fretts A, de Goede J, et al. Biomarkers of dietary omega-6 fatty acids and incident cardiovascular disease and mortality. Circulation 2019;139:2422–36.

[249] Sacks Frank M, Lichtenstein Alice H, Wu Jason HY, Appel Lawrence J, Creager Mark A, Kris-Etherton Penny M, et al. Dietary fats and cardiovascular disease: a presidential advisory from the american heart association. Circulation 2017;136:e1–e23.

[250] Ramsden CE, Zamora D, Majchrzak-Hong S, Faurot KR, Broste SK, Frantz RP, et al. Re-evaluation of the traditional diet-heart hypothesis: analysis of recovered data from Minnesota Coronary Experiment (1968-73). BMJ 2016;353:i1246.

[251] Ramsden CE, Zamora D, Leelarthaepin B, Majchrzak-Hong SF, Faurot KR, Suchindran CM, et al. Use of dietary linoleic acid for secondary prevention of coronary heart disease and death: evaluation of recovered data from the Sydney Diet Heart Study and updated meta-analysis. BMJ 2013;346:e8707.

[252] Ramsden CE, Hibbeln JR, Majchrzak SF, Davis JM. n-6 fatty acid-specific and mixed polyunsaturate dietary interventions have different effects on CHD risk: a meta-analysis of randomised controlled trials. Br J Nutr 2010;104:1586–600.

[253] De Caterina R. n-3 fatty acids in cardiovascular disease. N Engl J Med 2011;364:2439–50.

[254] Deckelbaum RJ, Calder PC. Different outcomes for omega-3 heart trials: why? Curr Opin Clin Nutr Metab Care 2012;15:97.

[255] Mozaffarian D, Wu JHY. (n-3) fatty acids and cardiovascular health: are effects of EPA and DHA shared or complementary? J Nutr 2012;142:614S–25S.

CHAPTER

# 22

# Lipid metabolism and signaling in cancer

*Elena Piccinin*[a], *James M. Ntambi*[b], *Antonio Moschetta*[a,c,d]

[a]Clinica Medica Cesare Frugoni, Department of Interdisciplinary Medicine, University of Bari Aldo Moro, Bari, Italy; [b]Departments of Biochemistry and Nutritional Sciences, University of Wisconsin-Madison, Madison, WI, United States; [c]INBB, National Institute for Biostructures and Biosystems, Rome, Italy; [d]IRCCS Istituto Tumori Giovanni Paolo II, Bari, Italy

OUTLINE



Cancer is a group of diseases characterized by out-of-control cell growth. Tumor cells have the potential to invade other parts of the body and initiate metastasis formation. In contrast with the previous belief, in which tumors were depicted as an insular mass of proliferating cells, nowadays cancer is considered as a complex tissue composed of multiple distinct cell types that participate in heterotypic interactions with one another. Within one organ it is possible to recognize over 100 types of cancers and subtypes, and this reflects upon the different behaviors and the contrasting response to therapies of the same histological tumors [1].

An increasing body of evidence suggests that an "emerging" hallmark of cancer is involved in the pathogenesis of some and possibly all cancers: the capability to reprogram cellular metabolism, thereby replacing the normal one in order to effectively promote neoplastic proliferation. Specifically, the increase in lipid metabolism is a remarkable feature of cancer metabolism. Since lipids are important building blocks of new organelles and cells, highly proliferative cancer cells show a strong lipid and cholesterol avidity, which they satisfy by either increasing the uptake of exogenous (or dietary) lipids and lipoproteins or by overactivating

Lipid Signaling and Metabolism. http://dx.doi.org/10.1016/B978-0-12-819404-1.00022-1

their endogenous synthesis. In normal conditions, lipids are primarily digested and absorbed in the small intestine, before undergoing a series of enzymatic reactions to become biologically active. Then, lipids are delivered to their sites of action in the body, by crossing the liver. In contrast to most normal human cells which prefer exogenous sources, cancer cells synthetize lipids de novo and often exhibit a shift toward fatty acids (FA) or cholesterol synthesis [2,3].

Lipids encompass a vast class of biomolecules, including phospholipids, FA, sphingolipids, triglycerides, cholesterol, and cholesteryl esters, which are broadly distributed in cellular organelles, where they carry out multiple functions. Indeed, lipids can be incorporated into membranes, fitted into signaling pathways or oxidized as an energy source to carbon dioxide and water.

The synthesis of new fatty acids begins with citrate. Both glycolysis and glutaminolysis provides citrate, an intermediate in the Krebs cycle, and NADPH reductive power. The subcellular localization of citrate determines its metabolic fate: mitochondrial citrate is fed into the Krebs cycle, whereas cytoplasmic citrate is used for FA synthesis. Several steps are then necessary to convert citrate to bioactive fatty acids. From citrate, the enzyme ATP-citrate lyase (ACLY) generates acetyl-CoA, that is, subsequently converted into malonyl-CoA by the enzyme acetyl-CoA carboxylase (ACC). Then, fatty acid synthase (FASN) catalyze repeated condensation of acetyl-CoA and malonyl-CoA to generate a palmitic acid (C16:0). Afterward, different elongases and desaturases intervene to transform palmitic acid into a wide spectrum of saturated and unsaturated FA. Stearoyl-CoA desaturase (SCD) is one of the principal mammalian desaturases which, by the introduction of a double bond, is responsible for the rate-limiting step in the synthesis of monounsaturated fatty acids (MUFA). These steps engaging lipogenic enzymes involved in FA synthesis are often upregulated in tumors and can thus be considered as a typical feature of most cancers [3a]. Once in the bioactive pool, FA can be esterified in order to generate triacylglycerols (TG) or sterol esters (SE) which in abundance can be stored in lipid droplets (LD). LD's are nowadays considered as distinctive characteristic of cancer aggressiveness and it has been shown that LD-rich cancer cells are more resistant to chemotherapy [4–6]. However, intermediates of this pathway can be esterified into different phosphoglycerides to phospholipids, whose synthesis is upregulated in high proliferative cancer cells in order to provide structural lipids for membrane biogenesis and to facilitate the formation of detergent resistant microdomains for signal transduction and intracellular trafficking. It is important to note that mammalian cells have a restricted capacity to synthesize polyunsaturated FA, as they lack the Δ12 desaturase. Therefore, enhanced de novo lipogenesis results in an enrichment of cancer cell membranes with saturated and/or MUFA. Since these FA are less prone to lipid peroxidation than polyunsaturated acyl chains, de novo FA synthesis seems to make cancer cells more resistant to oxidative stress-induced cell death [7].

Another class of lipids fundamental for membrane function are sterols, in particular cholesterol and cholesteryl esters. Cholesterol is synthetized from acetyl-CoA via the mevalonate pathway, an enzymatic cascade where the rate-limiting reduction of hydroxylmethyl glutaryl-coenzyme A (HMG-CoA) to mevalonate is catalyzed by HMG-CoA reductase. Then, the conversion of mevalonate to cholesterol involves a coordinate action of different enzymes involved in this pathway, regulated by sterol regulatory element-binding proteins (SREBP) transcription factor family [8]. Beside SREBP, also liver X receptor (LXR) plays a crucial role as a key regulator of cholesterol homeostasis cholesterol is a fundamental component of mammalian cell membrane, as it modulates the fluidity of the lipid bilayer. Moreover, it is an essential precursor of bile acids and steroid hormones, like estrogen and progesterone. Altered cholesterol homeostasis, with deregulated intracellular cholesterol levels, may lead to cancer progression [9–12].

Newly generated lipid molecules can also mediate signal transduction in cancer cells. For instance, lysophosphatidic acid (LPA) is a lipid messenger mediating signaling through a family of G-protein-coupled receptors to promote cancer aggressiveness [13]. Moreover, the upregulation of prostaglandins (mainly PGE2) in cancer cells not only promotes tumor growth in a paracrine manner but also coordinates the complex dialogue between tumor cells and the surrounding stromal cells. This crosstalk results in immunosuppression promotion, thus enabling the evasion from the immune system attack [14]. Furthermore, PGE2 was found to promote the differentiation of monocytes into tumor-associated suppressive macrophages in cervical tumors and a proangiogenic activity of this molecule has also been demonstrated in different cancers [15,16].

Lastly, fatty acids can be used as a major source of energy. Recent studies have demonstrated the important contribution of fatty acid oxidation in tumorigenesis. Fatty acids are a rich energy source that can yield up to two times more ATP than carbohydrates when needed. Fatty acids can undergo oxidation in the mitochondria or by cytoplasmic lipophagy, a new fatty acid catabolic process. Although most tumors rely on glycolysis and glutaminolysis to support the energetic needs as well as biosynthetic requirements, some types of tumors, as prostate cancer, display increased dependence on fatty acid oxidation as their main source of energy [17,18]. Moreover, fatty acid oxidation results in an enriched disposal of NADPH, crucial for the function of many anabolic enzymes to sustain large-scale biosynthetic programs in many cancer cells. NADPH also plays a fundamental role in quenching reactive oxidative stress in cancer cells. Indeed, blocking glioma tumor's fatty acid oxidation leads to rapid depletion of NADPH that finally results in the rise of reactive oxidative species content and consequently an increase in apoptosis rate [19].

All of these data suggest a critical importance of FA and cholesterol for proliferating cells and the metabolic reprogramming of lipid metabolism is now emerging as one of the essential features of cancer (Fig. 22.1). Thus, limiting the availability of FA and cholesterol could provide novel therapeutic strategies in cancer treatments.

## LXR and cholesterol homeostasis

Inside the mammalian cell, lipids can exert their function by binding to specific receptors in the nucleus, the nuclear receptors (NRs). NRs are ligand activated transcription factors that regulate the expression of target genes to affect a plethora of cellular processes [20]. LXRs are members of the NR superfamily that play a pivotal role in lipid metabolism [21]. LXR family consists of two isotypes, LXRα and LXRβ, that are differentially and spatially located despite an extensive sequence homology, with LXRα highly expressed in active metabolic tissue, as liver, intestine and macrophages, while LXRβ has a ubiquitous expression [22]. Both LXRs are activated by cholesterol derivatives, including oxysterols and 24(S),25-epoxycholesterol [23].

Coordinate actions of LXR pathway play a pivotal role in cholesterol homeostasis. High levels of intracellular cholesterol drive cells to actively synthesize oxysterols, thus mediating the activation of LXR transcriptional network to induce cholesterol efflux and reduce cholesterol influx and synthesis as well as conversion to bile acids in mice [24]. Therefore,

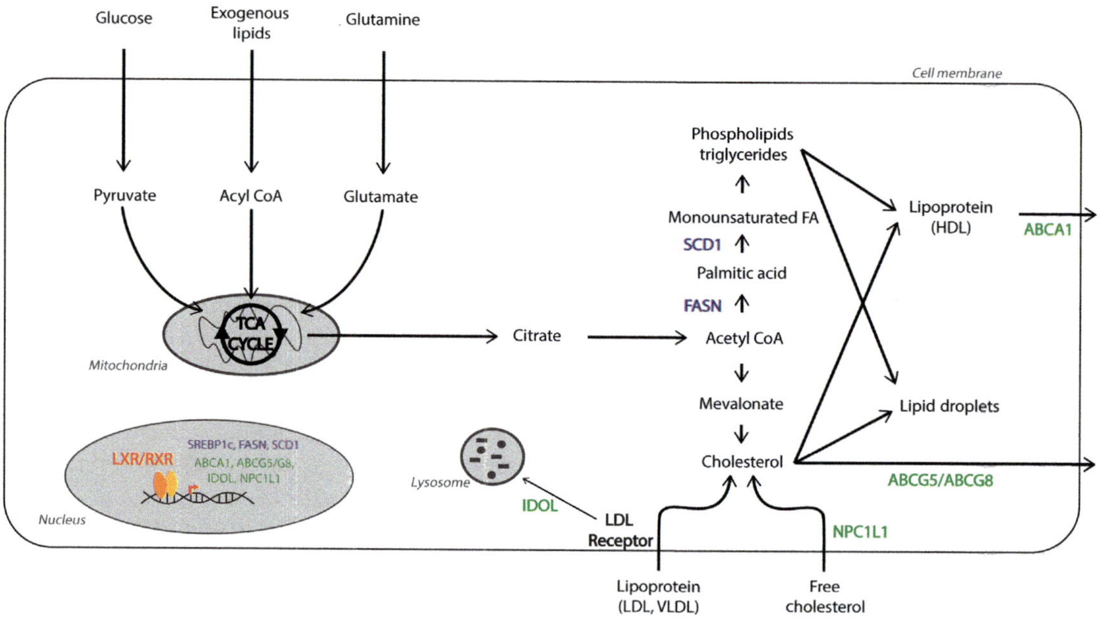


**FIGURE 22.1** **Schematic overview of the pathways involved in lipid metabolism frequently altered in cancer cells.** When activated, LXR induces the transcription of the principal genes involved in de novo lipogenesis (in *blue* [*gray* in print version]) and in cholesterol metabolism (in *green* [*light gray* in print version]). Abbreviations: *ABCA1*, ATP-binding cassette transporter A1; *ABCG5/G8*, ATP-binding cassette transporter G5/G8; *IDOL*, E3 ubiquitin ligase inducible degrader of LDLR; *LXR*, liver X receptor; *NPC1L1*, Niemann-Pick C1-Like 1; *RXR*, retinoid X receptor; *SCD1*, stearoyl-CoA desaturase1; *SREBP1c*, sterol regulatory element-binding transcription factor 1c.

LXRs activation results in a clear elimination of cholesterol from the body coupled with improvement of plasma lipoprotein profile. LXRs also exert important effect on fatty acids and phospholipids metabolism [25].

The first clue of the role of LXR in maintaining cholesterol homeostasis came from mice lacking LXRα: when fed with a cholesterol-enriched diet, they accumulated large amount of cholesterol in the liver. This study led to the identification of the first LXR target gene in the mouse, Cytochrome P450 7A1 (CYP7A1) which catalyzes the rate-limiting step in the production of bile acids and excretion of fecal sterol [26].

Additionally, hepatic LXR activation promotes biliary cholesterol excretion through ABCG5 and ABCG8 (ATP-binding cassette subfamily G members 5 and 8) [26,27]. Subsequent evidences from in vivo studies display a crucial role of LXR in promoting reverse cholesterol transport (RCT), a process by which excess cholesterol in peripheral tissues is transferred to HDL and delivered to the liver for excretion. Specifically, dietary cholesterol is absorbed and efflux from enterocytes to the circulation as HDL via the LXR target ABCA 1 (ATP-binding cassette subfamily A members 1), whereas the efflux from enterocytes into the lumen and transintestinal cholesterol excretion are mediated by ABCG5 and ABCG8 [28,29]. Indeed, mice with intestinal specific LXR activation display upregulated ABCG5/ABCG8 and ABCA1 transporters that brings to diminished cholesterol absorption together with enhanced RCT in absence of overt liver steatosis [28].

Recent evidence indicates that LXR controls intestinal cholesterol absorption by promoting the transcription of the E3 ubiquitin ligase inducible degrader of LDLR (IDOL) that ubiquitinates LDLR degradation, thus diminishing LDL cholesterol uptake [30]. Furthermore, LXR activation downregulates the expression of Niemann–Pick C1-like 1 (NPC1L1), thus limiting intestinal cholesterol absorption. In this view, the intestine emerges as a key player in the control of cholesterol homeostasis. Recent evidence demonstrated that constitutive activation of LXRα in the gut protects against atherosclerosis stimulating RCT without affecting liver lipid accumulation [28].

Beside the major role in modulating cholesterol homeostasis, LXRs regulates hepatic de novo lipogenesis. Precisely, the oral administration of LXR agonist to mice results in the transcription of principal FA biosynthetic genes as well as increased VLDL secretion which results in elevated levels of plasma triglycerides [25]. Active LXRs promote the transcription of SREBP1c gene, the master regulator of FA metabolism. Moreover, LXR induce, directly and via SREBP1c, the expression of several lipogenic genes, including FASN (fatty acids synthase) and SCD1 (stearoyl-coenzyme A desaturase I), implicated in the synthesis of FA required to build complex lipids necessary to sustain cell proliferation [25,31].

Finally, LXR activation increases lysophosphatidylcholine acyltransferase 3 (LPCAT3) expression, a phospholipid-remodeling enzyme which incorporates polyunsaturated fatty acids at the sn-2 site of lysophospholipids. Through the modulation of phospholipid composition in response to fluctuating cellular levels of sterol, LXR is able to provide protection against lipid stress induced by saturated free fatty acids in vitro or by hepatic lipid accumulation in vivo that may eventually damage biological membranes [32].

## LXR in cancer metabolism

During cancer cell proliferation, there is a net uncoupling between intracellular cholesterol increase and LXRs activation, resulting from the reduced intracellular oxysterol concentration [24]. Indeed, many recent studies on pharmacological LXRs agonists in several organs depict an antiproliferative effect of LXR due to alterations of tumor metabolism and microenvironment as well as of growth and death pathways [33–37].

The role of LXR in carcinogenesis has been deeply investigated in several type of tumors, such as glioblastoma, breast cancer, prostate cancer, and colon cancer.

Glioblastoma, the most common primary brain tumor and among lethal cancer in adults, mainly relies on cholesterol uptake for survival that is provided by the increased LDLR expression driven by a mutated and constitutively activated form of epithelial growth factor receptor (EGFR). Indeed, glioblastoma cells showed suppressed cholesterol and LXR ligand synthesis. Therefore, LXR activation induces IDOL-dependent degradation of LDLR, thus causing tumor regression and prolonged survival [38,39].

Also, in endocrine related cancers, such as breast and prostate ones, tumor proliferation is suppressed by LXR activation. Particularly, the insight into the investigation of cholesterol involvement in breast tumorigenesis comes from the observation that women with hypercholesterolemia and/or metabolic syndrome are more prone to develop breast cancer. Using a mouse model of breast cancer and related metastasis, it emerges that oxysterols can selectively modulate estrogen receptor (ER) in order to promote tumor growth, and their catabolism is associated with a better survival outcome [40]. However, in both in vitro and in vivo breast cancer models, LXR activation displays a marked antiproliferative effect in ER-positive breast cancer cells than ER-negative ones, probably due to the LXR ability in regulating the expression of the estrogen sulfotransferase (EST) [41,42]. Notably, the antiproliferative capability of LXR seems to be independent of lipid biosynthesis. Indeed, several studies on breast cancer cell lines reported that treatment with LXR agonist induces the expression of p53 and the concomitant downregulation of genes involved in replication, and other growth-related processes, thereby blocking proliferation [42,43].

Prostate cancer cell proliferation is similarly suppressed by LXR agonists. The expression of S-phase kinase associated protein 2 (SKP2), a protein which targets p27 for degradation, is reduced by ligand treatment. The enhanced p27 expression results in a decreased percentage of S-phase cells, finally diminishing tumor cells proliferation [44]. Moreover, cholesterol depletion induced by LXR alters membrane lipid raft signaling via AKT, thereby blocking cells survival [45]. Particularly, LXR activation inhibits cell proliferation by increasing lipogenic activity and the downregulation of fatty acid synthase disrupted the antiproliferative effects [46]. Besides, the antitumorigenic effects of LXR agonists on prostate cancer cells may also involve effects on the expression of genes involved in androgen metabolism and activity, thereby hindering androgen signaling and cancer progression [47].

The tumor protective action of LXR has been also observed in gastrointestinal cancer. Indeed, LXR agonists block the transactivation activity of β-catenin, a key mediator of Wnt signaling which is frequently mutated pathway in intestinal tumorigenesis [35]. Furthermore, selective intestinal LXR activation exerts a tumor protective role against colon cancer, given to its ability to drive cholesterol depletion [33]. At the same time, LXRαβ-*null* mice display a significant increase of proliferation markers compared to wild type counterpart [36].

Finally, LXR expression has been proposed as a prognostic marker for postoperative outcome in HCC patients: subjects with elevated hepatic expression of LXR exhibited an higher overall survival rate compared to those in which low LXR level are detected [48].

Overall, these observations endorse the antitumoral power of LXR signaling by controlling cholesterol homeostasis and cancer cell proliferation. Due to their action on both cell cycle and tumor environment, basic and translational studies into the roles of LXR and their ligands as novel antitumoral strategies will affect the treatment of a multitude of cancers.

Different studies, including clinical trials, on LXR agonists have been conducted. Most of them pointed to an atheroprotective consequence of LXR activation, although side effects have been described, regarding especially hypertriglyceridemia and hepatic steatosis [49,50]. This accumulation of lipids is explained by the upregulation of SREBP1c and SCD1, direct targets of LXR involved in de novo lipogenesis program [51]. Thereby, the systemic LXR activation has recognized to have both desirable and undesirable effects [52]. The latter could be avoided when SCD1 is ablated [51]. Perhaps, restricting LXR activation to specific tissue may circumvent SREBP1c and SCD1 induction. Alternatively, LXR agonists together with SCD1 inhibitors could represent a promising approach to block malignant proliferation and avoid upregulation of lipid synthesis.

## SCD1 and fatty acids homeostasis

The ratio of saturated fatty acids (SFA) to MUFA is important to modulate phospholipid composition, and an unbalanced ratio toward SFA production is linked to multiple pathological conditions including obesity, diabetes, atherosclerosis, and cancer [53]. The content and distribution of SFA and MUFA is tightly regulated inside the cell. One of the central regulators of the fatty acid composition is SCD1, an enzyme committed to a critical step in the biosynthesis of MUFA.

SCD1 is one of the LXR target genes which plays a pivotal role in the regulation of fuel metabolism, thus representing a potential target for the control of lipogenesis. It is a transmembrane protein, mainly located at ER organelles, that catalyzes conversion of saturated fatty acids to MUFA, that is, palmitic acid (C16:0) to palmitoleic acid (C16:1 n-7) or stearic acid (C18:0 FA) to oleic acid (C18:1 n-9). This is achieved by the introduction of the cis-double bond between carbon 9 and 10 of acyl-CoA [54]. MUFA are then preferentially incorporated into different lipid species, including triglycerides, cholesterol esters, and phospholipids [55].

To date, four mouse isoforms (1–4) and two human isoforms (1 and 5) have been recognized, with SCD1 highly conserved and ubiquitously expressed in both organisms. In adult mice, Scd2 isoform is expressed in most tissues except liver, whereas Scd3 and Scd4 expression is more restricted to some gland and heart, respectively. Differently, SCD5 is unique to primates and highly expressed in brain and pancreas. Despite the distinct distribution pattern, the different isoforms share the same enzymatic function [56].

Due to its implication in physiological and pathological conditions, SCD1 is tightly regulated at either transcriptional and posttranscriptional levels. Usually highly expressed in metabolic tissues, such as liver and adipose tissue, SCD1 expression increases in response to high-carbohydrate diet feeding [57]. Cholesterol promotes SCD1 transcription via either SREBP-1c and LXRα, whereas polyunsaturated fatty acids (PUFA) hinders SCD1 levels and activity, through both a SREBP-dependent and -independent mechanism [26,58,59].

The first evidences into SCD1 function originated from whole-body knock out mouse models. Scd1 global deletion prevents liver steatosis and adiposity in several mice models, including high fat diet and high carbohydrate diet feeding [60,61], thus suggesting that MUFA are fundamental for lipid-promoted weight gain and liver fat accumulation. Notably, the protection against fat accumulation could also be ascribed to the improved insulin sensitivity and the increased energy expenditure of whole-body Scd1 knock out mice.

Intriguingly, despite the reduced hepatic lipogenesis, Scd1 null mice displayed a concomitant increase of lipogenic pathways in white adipose tissue, thus suggesting that SCD1 deficiency may modulate fatty acids synthesis via different pathways in metabolic tissues [62]. Interestingly, when SCD1 is specifically deleted in the liver, mice are no longer protected from high fat diet induced steatosis, thus suggesting that MUFAs derived from endogenous synthesis in extra-hepatic tissues may lead to different compensatory mechanisms [63].

Finally, even in mice gut epithelium, the expression of SCD1 is able to drive de novo synthesis of MUFAs, as indicated by the low levels of palmitoleic and oleic acids observed in mice with intestinal-specific SCD1 ablation [64], which led to the speculation that the loss of protection against steatosis previously described could be attributable to intestinal SCD1 action.

## SCD1 in cancer metabolism

Metabolism in cancer cells is characterized by a shift toward abnormally high rate of aerobic glycolysis and fatty acids biosynthesis. The main products of glucose-derived de novo lipogenesis are SFA, which are abundantly present in cancer cells though, an increased content of MUFA has been observed in tumoral specimens. Interestingly, MUFA represent a more favored substrates for complex lipid biosynthesis than SFA [65].

Consistently, analysis of the fatty acids profile in serum and erythrocytes revealed an increased MUFA to SFA ratio in patients with pancreatic, breast, and prostate cancer. The accumulation of MUFA species overlaps with SCD1 overexpression observed in cancer tissues [66–69].

Particularly, SCD1 may contribute to tumorigenesis not only by providing MUFA for lipid formation, but also by controlling the fatty acids biosynthetic rate. Indeed, SCD1 is directly and indirectly involved in the regulation of ACC, the enzyme that catalyze the first critical step of the lipogenesis pathway. ACC is regulated allosterically by SFA and via phosphorylation mediated by 5′-adenosine monophosphate-activated protein kinase (AMPK). By promoting MUFA synthesis, SCD1 expression avoids the accumulation of SFA inside the cell, thus preventing the inhibition of ACC. Moreover, SCD1 expression leads to a decrease in the activity of AMPK that in turn prevents the downregulation of ACC function [70,71].

In mammary tumors, an altered SFA to MUFA ratio can be used as risk factor and predictor of survival [72]. Correspondingly, SCD1 inhibition in cancer cells is associated with a downregulation of cell proliferation and metastatic potential, probably attributable to a decreased β-catenin translocation to the nucleus [73]. Conversely, oleic acid supplementation promotes apoptosis of breast cancer cells with Her-2/neu oncogene amplification [74].

Additionally, SCD1 overexpression predicts a poor clinical outcome in lung cancer [75]. Specifically, upregulated SCD1 activity in lung adenocarcinoma has been associated with the promotion and stabilization of several oncogenes, including the effectors of the Hippo pathway and EGFR/PI3K/AKT signals [76,77].

However, lack of SCD1 and concomitant SFA accretion cause the activation of a stress response program, culminating with the unfolded protein response (UPR). The UPR is a prosurvival mechanism triggered by accumulation of unfolded or misfolded proteins in the endoplasmic reticulum (ER), a condition referred to as ER stress. Over the past decades, UPR has been largely associated with the acquisition of several malignant characteristics that allow inflammatory conditions and cancer progression [78,79].

Intriguingly, the loss of intestinal SCD1, together with the resulting absence of oleic acid, induces a proinflammatory environment that finally promote cancer development [64,80]. Moreover, a reduced SCD1 stained area has been detected in immunohistology of prostate carcinoma compared to normal prostate epithelium [81].

Anyway, in liver tumor SCD1-mediated ER stress regulates sorafenib sensitivity. Thereby, targeting SCD1 with a specific inhibitor reverse sorafenib resistance in HCC, finally resulting in cell growth suppression [82].

Although several studies have been carried out in the last decades, a clear picture on the relationship between SCD1 activity and cancer has not been depicted yet. Many contradictory results may arise from the different model used for the analysis as well as the timing of evaluations, since the dependence on specific metabolic pathways may change from the early tumorigenic events to established form of cancer.

Moreover, to our knowledge, any study on the induction of SCD1 expression by LXR activation has been conducted yet. If LXR agonists represent a promising therapy against cancer, the implication of this activation on SCD1 induction and consequent contribution to tumorigenesis has not been clarified. Although nowadays targeting SCD1 appears as a promising avenue in cancer treatment, more studies are necessary to fully elucidate the consequence of this inhibition as well as combined therapy designed to promote LXR activation while blocking SCD1 activity.

## Conclusion

Basic researches on the role of lipid metabolism in cancer have led to the identification of two important players, LXR and SCD1. While LXR activation have been established to protect against tumor development and progression by depleting cellular cholesterol reservoirs, the contribution of SCD1 and its product MUFA to cancer is still not well defined (Fig. 22.2).

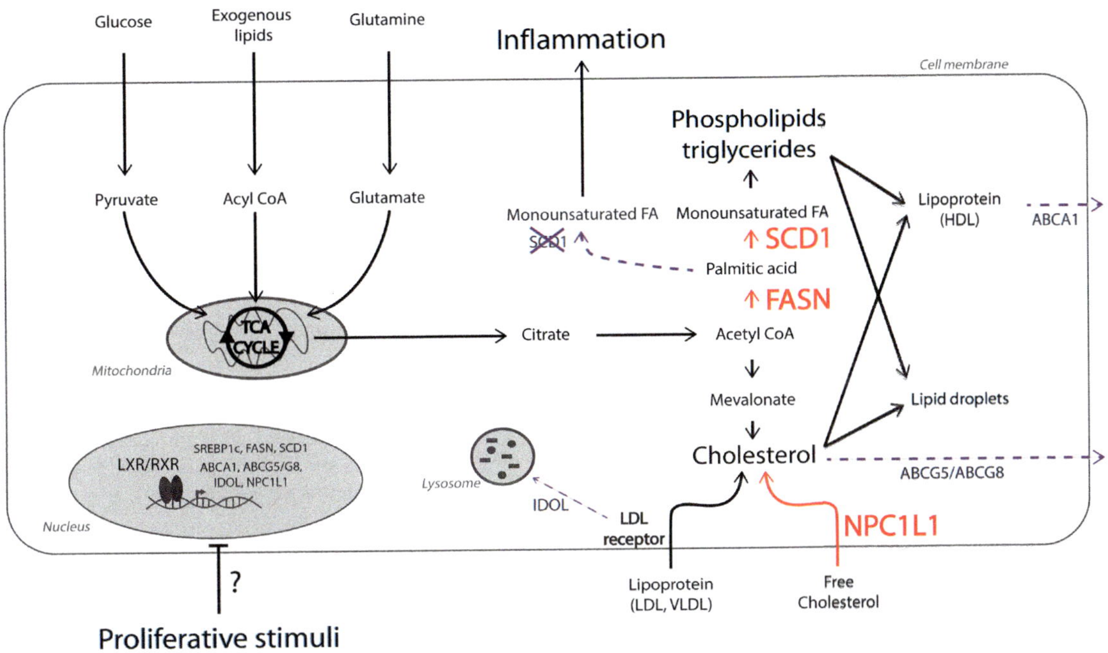


**FIGURE 22.2 Regulation of lipids metabolism in cancer cells.** In cancer cells exposed to proliferative stimuli LXR transcriptional activity is usually repressed, thus resulting in a decreased activation (ABCA1, ABCG5/G8, IDOL) or repression (LDL receptor, NCP1L1) of its target genes involved in cholesterol homeostasis. This results in a net increase of intracellular cholesterol content. No evidence exists regarding the contribution of LXR to de novo lipogenesis pathway in cancer. In overt forms of cancer, genes of de novo lipogenesis pathway (FASN, SCD1) are often upregulated to fulfill new triglycerides and phospholipids to rapidly growing cells. However, when SCD1 is deleted, the monounsaturated to saturated fatty acid ratio decrease, thus promoting inflammation that finally contribute to cancer onset. Despite large plethora of studies on this topic, a clear picture of SCD1 in cancer development and progression is still missing. Genes downregulated in cancer are displayed in *purple* (*gray* in print version), while genes upregulated in *red* (*light gray* in print version). Abbreviations: *ABCA1*, ATP-binding cassette transporter A1; *ABCG5/G8*, ATP-binding cassette transporter G5/G8; *IDOL*, E3 ubiquitin ligase inducible degrader of LDLR; *LXR*, liver X receptor; *NPC1L1*, Niemann-Pick C1-Like 1; *RXR*, retinoid X receptor; *SCD1*, stearoyl-CoA desaturase1; *SREBP1c*, sterol regulatory element-binding transcription factor 1c.

Notably, when active in normal conditions, LXR is not only able to support cholesterol metabolism, but it is also involved in the promotion of de novo lipogenesis pathway, mainly via direct and indirect regulation of SCD1. This "double role" of LXR reflects in the undesired consequences of its agonists, resulting in increased triglyceridemia and hepatic steatosis. However, since today, no study reporting the effect of LXR-mediated SCD1 activation in cancer has been reported. Since drug discovery programs are becoming more adept as our understanding of biological processes growth, new analyses are required to deepen our knowledge on lipid metabolism in cancer. Specifically, some issues are required to be addressed, especially given the contradictory results obtained. For instance, is the tumor microenvironment influencing the cellular metabolism and, therefore, the tumorigenic programs? Is it possible to avoid the negative effects of LXR agonist via a restricted activation in specific tissue? Does the activation of LXR in cancer have an effect also on fatty acids synthesis via SCD1? If so, are combined therapy aimed to activate LXR while blocking SCD1 activity the future perspective in cancer treatment? In vivo studies evaluating both LXR and SCD1 concomitant actions in tumor is the first step toward the development of specific and effective therapies.

## References


[1] Hanahan D, Weinberg RA. The hallmarks of cancer. Cell 2000;100(1):57–70.

[2] Ookhtens M, et al. Liver and adipose tissue contributions to newly formed fatty acids in an ascites tumor. Am J Physiol 1984;247(1 Pt 2):R146–53.

[3] Kuzu OF, Noory MA, Robertson GP. The role of cholesterol in cancer. Cancer Res 2016;76(8):2063–2070.

[3a] Menendez JA, Lupu R. Fatty acid synthase and the lipogenic phenotype in cancer pathogenesis. Nat Rev Cancer 2007;7:763–77.

[4] Qiu B, et al. HIF2alpha-dependent lipid storage promotes endoplasmic reticulum homeostasis in clear-cell renal cell carcinoma. Cancer Discov 2015;5(6):652–67.

[5] Guillaumond F, et al. Cholesterol uptake disruption, in association with chemotherapy, is a promising combined metabolic therapy for pancreatic adenocarcinoma. Proc Natl Acad Sci USA 2015;112(8):2473–8.

[6] de Gonzalo-Calvo D, et al. Intratumor cholesteryl ester accumulation is associated with human breast cancer proliferation and aggressive potential: a molecular and clinicopathological study. BMC Cancer 2015;15:460.

[7] Rysman E, et al. De novo lipogenesis protects cancer cells from free radicals and chemotherapeutics by promoting membrane lipid saturation. Cancer Res 2010;70(20):8117–26.

[8] Sato R. Sterol metabolism and SREBP activation. Arch Biochem Biophys 2010;501(2):177–81.

[9] Smith B, Land H. Anticancer activity of the cholesterol exporter ABCA1 gene. Cell Rep 2012;2(3):580–90.

[10] Montero J, et al. Mitochondrial cholesterol contributes to chemotherapy resistance in hepatocellular carcinoma. Cancer Res 2008;68(13):5246–56.

[11] Freed-Pastor WA, et al. Mutant p53 disrupts mammary tissue architecture via the mevalonate pathway. Cell 2012;148(1–2):244–58.

[12] Sorrentino G, et al. Metabolic control of YAP and TAZ by the mevalonate pathway. Nat Cell Biol 2014;16(4):357–66.

[13] Ren J, et al. Lysophosphatidic acid is constitutively produced by human peritoneal mesothelial cells and enhances adhesion, migration, and invasion of ovarian cancer cells. Cancer Res 2006;66(6):3006–14.

[14] Wang D, Dubois RN. Eicosanoids and cancer. Nat Rev Cancer 2010;10(3):181–93.

[15] Heusinkveld M, et al. M2 macrophages induced by prostaglandin E2 and IL-6 from cervical carcinoma are switched to activated M1 macrophages by CD4+ Th1 cells. J Immunol 2011;187(3):1157–65.

[16] Chang SH, et al. Role of prostaglandin E2-dependent angiogenic switch in cyclooxygenase 2-induced breast cancer progression. Proc Natl Acad Sci USA 2004;101(2):591–6.

[17] Zha S, et al. Peroxisomal branched chain fatty acid beta-oxidation pathway is upregulated in prostate cancer. Prostate 2005;63(4):316–23.

[18] Carracedo A, Cantley LC, Pandolfi PP. Cancer metabolism: fatty acid oxidation in the limelight. Nat Rev Cancer 2013;13(4):227–32.

[19] Pike LJ. Growth factor receptors, lipid rafts and caveolae: an evolving story. Biochim Biophys Acta 2005;1746(3):260–73.
[20] Evans RM, Mangelsdorf DJ. Nuclear receptors, RXR, and the Big Bang. Cell 2014;157(1):255–66.
[21] Wang B, Tontonoz P. Liver X receptors in lipid signalling and membrane homeostasis. Nat Rev Endocrinol 2018;14:452–63.
[22] Repa JJ, et al. Regulation of mouse sterol regulatory element-binding protein-1c gene (SREBP-1c) by oxysterol receptors, LXRalpha and LXRbeta. Genes Dev 2000;14(22):2819–30.
[23] Janowski BA, et al. An oxysterol signalling pathway mediated by the nuclear receptor LXR alpha. Nature 1996;383(6602):728–31.
[24] Bovenga F, Sabba C, Moschetta A. Uncoupling nuclear receptor LXR and cholesterol metabolism in cancer. Cell Metab 2015;21(4):517–26.
[25] Schultz JR, et al. Role of LXRs in control of lipogenesis. Genes Dev 2000;14(22):2831–8.
[26] Peet DJ, et al. Cholesterol and bile acid metabolism are impaired in mice lacking the nuclear oxysterol receptor LXR alpha. Cell 1998;93(5):693–704.
[27] Yu L, et al. Stimulation of cholesterol excretion by the liver X receptor agonist requires ATP-binding cassette transporters G5 and G8. J Biol Chem 2003;278(18):15565–70.
[28] Lo Sasso G, et al. Intestinal specific LXR activation stimulates reverse cholesterol transport and protects from atherosclerosis. Cell Metab 2010;12(2):187–93.
[29] Graf GA, et al. Coexpression of ATP-binding cassette proteins ABCG5 and ABCG8 permits their transport to the apical surface. J Clin Invest 2002;110(5):659–69.
[30] Zelcer N, et al. LXR regulates cholesterol uptake through Idol-dependent ubiquitination of the LDL receptor. Science 2009;325(5936):100–4.
[31] Joseph SB, et al. Direct and indirect mechanisms for regulation of fatty acid synthase gene expression by liver X receptors. J Biol Chem 2002;277(13):11019–25.
[32] Rong X, et al. LXRs regulate ER stress and inflammation through dynamic modulation of membrane phospholipid composition. Cell Metab 2013;18(5):685–97.
[33] Lo Sasso G, et al. Liver X receptors inhibit proliferation of human colorectal cancer cells and growth of intestinal tumors in mice. Gastroenterology 2013;144(7):1497–507. 1497-1507.e1-13.
[34] Pencheva N, et al. Broad-spectrum therapeutic suppression of metastatic melanoma through nuclear hormone receptor activation. Cell 2014;156(5):986–1001.
[35] Uno S, et al. Suppression of beta-catenin signaling by liver X receptor ligands. Biochem Pharmacol 2009;77(2): 186–95.
[36] Vedin LL, Gustafsson JA, Steffensen KR. The oxysterol receptors LXRalpha and LXRbeta suppress proliferation in the colon. Mol Carcinog 2013;52(11):835–44.
[37] Zhang W, et al. Liver X receptor activation induces apoptosis of melanoma cell through caspase pathway. Cancer Cell Int 2014;14(1):16.
[38] Guo D, et al. An LXR agonist promotes glioblastoma cell death through inhibition of an EGFR/AKT/SREBP-1/LDLR-dependent pathway. Cancer Discov 2011;1(5):442–56.
[39] Villa GR, et al. An LXR-cholesterol axis creates a metabolic co-dependency for brain cancers. Cancer Cell 2016;30(5):683–93.
[40] Nelson ER, et al. 27-Hydroxycholesterol links hypercholesterolemia and breast cancer pathophysiology. Science 2013;342(6162):1094–8.
[41] Gong H, et al. Estrogen deprivation and inhibition of breast cancer growth in vivo through activation of the orphan nuclear receptor liver X receptor. Mol Endocrinol 2007;21(8):1781–90.
[42] Vedin LL, et al. The oxysterol receptor LXR inhibits proliferation of human breast cancer cells. Carcinogenesis 2009;30(4):575–9.
[43] Nguyen-Vu T, et al. Liver x receptor ligands disrupt breast cancer cell proliferation through an E2F-mediated mechanism. Breast Cancer Res 2013;15(3):R51.
[44] Fukuchi J, et al. Antiproliferative effect of liver X receptor agonists on LNCaP human prostate cancer cells. Cancer Res 2004;64(21):7686–9.
[45] Pommier AJ, et al. Liver X Receptor activation downregulates AKT survival signaling in lipid rafts and induces apoptosis of prostate cancer cells. Oncogene 2010;29(18):2712–23.
[46] Kim KH, et al. Inhibitory effect of LXR activation on cell proliferation and cell cycle progression through lipogenic activity. J Lipid Res 2010;51(12):3425–33.

[47] Chuu CP, et al. The liver X receptor agonist T0901317 acts as androgen receptor antagonist in human prostate cancer cells. Biochem Biophys Res Commun 2007;357(2):341–6.

[48] Long H, et al. Tumor LXR expression is a prognostic marker for patients with hepatocellular carcinoma. Pathol Oncol Res 2018;24(2):339–44.

[49] Joseph SB, et al. Synthetic LXR ligand inhibits the development of atherosclerosis in mice. Proc Natl Acad Sci USA 2002;99(11):7604–9.

[50] Terasaka N, et al. T-0901317, a synthetic liver X receptor ligand, inhibits development of atherosclerosis in LDL receptor-deficient mice. FEBS Lett 2003;536(1–3):6–11.

[51] Chu K, et al. Stearoyl-coenzyme A desaturase 1 deficiency protects against hypertriglyceridemia and increases plasma high-density lipoprotein cholesterol induced by liver X receptor activation. Mol Cell Biol 2006;26(18):6786–98.

[52] Peter A, et al. Induction of stearoyl-CoA desaturase protects human arterial endothelial cells against lipotoxicity. Am J Physiol Endocrinol Metab 2008;295(2):E339–49.

[53] Ntambi JM, Miyazaki M. Regulation of stearoyl-CoA desaturases and role in metabolism. Prog Lipid Res 2004;43(2):91–104.

[54] Flowers MT, Ntambi JM. Role of stearoyl-coenzyme A desaturase in regulating lipid metabolism. Curr Opin Lipidol 2008;19(3):248–56.

[55] Miyazaki M, Kim YC, Ntambi JM. A lipogenic diet in mice with a disruption of the stearoyl-CoA desaturase 1 gene reveals a stringent requirement of endogenous monounsaturated fatty acids for triglyceride synthesis. J Lipid Res 2001;42(7):1018–24.

[56] Aljohani AM, Syed DN, Ntambi JM. Insights into stearoyl-CoA desaturase-1 regulation of systemic metabolism. Trends Endocrinol Metab 2017;28(12):831–42.

[57] Ntambi JM, et al. Differentiation-induced gene expression in 3T3-L1 preadipocytes. Characterization of a differentially expressed gene encoding stearoyl-CoA desaturase. J Biol Chem 1988;263(33):17291–300.

[58] Ntambi JM, Bene H. Polyunsaturated fatty acid regulation of gene expression. J Mol Neurosci 2001;16(2–3):273–8. [discussion 279–284].

[59] Bene H, Lasky D, Ntambi JM. Cloning and characterization of the human stearoyl-CoA desaturase gene promoter: transcriptional activation by sterol regulatory element binding protein and repression by polyunsaturated fatty acids and cholesterol. Biochem Biophys Res Commun 2001;284(5):1194–8.

[60] Ntambi JM, et al. Loss of stearoyl-CoA desaturase-1 function protects mice against adiposity. Proc Natl Acad Sci USA 2002;99(17):11482–6.

[61] Jiang G, et al. Prevention of obesity in mice by antisense oligonucleotide inhibitors of stearoyl-CoA desaturase-1. J Clin Invest 2005;115(4):1030–8.

[62] Burhans MS, et al. Hepatic oleate regulates adipose tissue lipogenesis and fatty acid oxidation. J Lipid Res 2015;56(2):304–18.

[63] Miyazaki M, et al. Hepatic stearoyl-CoA desaturase-1 deficiency protects mice from carbohydrate-induced adiposity and hepatic steatosis. Cell Metab 2007;6(6):484–96.

[64] Ducheix S, et al. Deletion of stearoyl-CoA desaturase-1 from the intestinal epithelium promotes inflammation and tumorigenesis, reversed by dietary oleate. Gastroenterology 2018;155(5). 1524-1538.e9.

[65] Igal RA. Stearoyl-CoA desaturase-1: a novel key player in the mechanisms of cell proliferation, programmed cell death and transformation to cancer. Carcinogenesis 2010;31(9):1509–15.

[66] Chavarro JE, et al. Blood levels of saturated and monounsaturated fatty acids as markers of de novo lipogenesis and risk of prostate cancer. Am J Epidemiol 2013;178(8):1246–55.

[67] Pala V, et al. Erythrocyte membrane fatty acids and subsequent breast cancer: a prospective Italian study. J Natl Cancer Inst 2001;93(14):1088–95.

[68] Wood CB, et al. Increase of oleic acid in erythrocytes associated with malignancies. Br Med J (Clin Res Ed) 1985;291(6489):163–5.

[69] Guo S, et al. Significantly increased monounsaturated lipids relative to polyunsaturated lipids in six types of cancer microenvironment are observed by mass spectrometry imaging. Sci Rep 2014;4:5959.

[70] Scaglia N, Igal RA. Stearoyl-CoA desaturase is involved in the control of proliferation, anchorage-independent growth, and survival in human transformed cells. J Biol Chem 2005;280(27):25339–49.

[71] Scaglia N, Chisholm JW, Igal RA. Inhibition of stearoylCoA desaturase-1 inactivates acetyl-CoA carboxylase and impairs proliferation in cancer cells: role of AMPK. PLoS ONE 2009;4(8):e6812.

[72] Chajes V, et al. Association between serum trans-monounsaturated fatty acids and breast cancer risk in the E3N-EPIC Study. Am J Epidemiol 2008;167(11):1312–20.
[73] Mauvoisin D, et al. Decreasing stearoyl-CoA desaturase-1 expression inhibits beta-catenin signaling in breast cancer cells. Cancer Sci 2013;104(1):36–42.
[74] Menendez JA, et al. Oleic acid, the main monounsaturated fatty acid of olive oil, suppresses Her-2/neu (erbB-2) expression and synergistically enhances the growth inhibitory effects of trastuzumab (Herceptin) in breast cancer cells with Her-2/neu oncogene amplification. Ann Oncol 2005;16(3):359–71.
[75] Huang J, et al. SCD1 is associated with tumor promotion, late stage and poor survival in lung adenocarcinoma. Oncotarget 2016;7(26):39970–9.
[76] Noto A, et al. Stearoyl-CoA-desaturase 1 regulates lung cancer stemness via stabilization and nuclear localization of YAP/TAZ. Oncogene 2017;36(32):4573–84.
[77] She K, et al. SCD1 is required for EGFR-targeting cancer therapy of lung cancer via re-activation of EGFR/PI3K/AKT signals. Cancer Cell Int 2019;19:103.
[78] Urra H, et al. Endoplasmic reticulum stress and the hallmarks of cancer. Trends Cancer 2016;2(5):252–62.
[79] Liu X, et al. Hepatic oleate regulates liver stress response partially through PGC-1alpha during high-carbohydrate feeding. J Hepatol 2016;65(1):103–12.
[80] Chen C, et al. Metabolomics reveals that hepatic stearoyl-CoA desaturase 1 downregulation exacerbates inflammation and acute colitis. Cell Metab 2008;7(2):135–47.
[81] Moore S, et al. Loss of stearoyl-CoA desaturase expression is a frequent event in prostate carcinoma. Int J Cancer 2005;114(4):563–71.
[82] Ma MKF, et al. Stearoyl-CoA desaturase regulates sorafenib resistance via modulation of ER stress-induced differentiation. J Hepatol 2017;67(5):979–90.

CHAPTER

# 23

# Altered lipid metabolic homeostasis in the pathogenesis of Alzheimer's disease

*Oana C. Marian*[a], *Collin Tran*[a], *Anthony S. Don*[a,b]

[a]Centenary Institute, The University of Sydney, Camperdown, NSW, Australia; [b]NHMRC Clinical Trials Centre, The University of Sydney, Camperdown, NSW, Australia

OUTLINE



Lipid Signaling and Metabolism. http://dx.doi.org/10.1016/B978-0-12-819404-1.00023-3

## Abbreviations

**AA** arachidonic acid
**Aβ** amyloid β
**AD** Alzheimer's disease
**APP** amyloid precursor protein
**β-CTF** β-C-terminal fragment of APP
**CSF** cerebrospinal fluid
**DAG** diacylglycerol
**DHA** docosahexaenoic acid
**DLB** dementia with Lewy bodies
**FTLD** frontotemporal lobar degeneration
**GD** Gaucher's disease
**LSD** lysosomal storage disease
**MAG** monoacylglycerol
**MCI** mild cognitive impairment
**NFT** neurofibrillary tangle
**NPC** Niemann-Pick type C
**PA** phosphatidic acid
**PC** phosphatidylcholine
**PD** Parkinson's disease
**PE** phosphatidylethanolamine
**PS** phosphatidylserine
**S1P** sphingosine 1-phosphate
**SNP** single-nucleotide polymorphism
**Synj1** synaptojanin 1
**TBI** traumatic brain injury

# Introduction

The brain is the second most lipid-rich organ in the body after adipose tissue. This is primarily attributed to myelin, the multilayered membrane structure that surrounds and insulates neuronal axons in vertebrates (Fig. 23.1). Myelin is comprised 70%–80% by dry weight of lipids, and the lipid:protein ratio is higher than in other cell membranes [1,2]. In addition to lipid-rich myelin, a structurally and functionally diverse array of lipids generated by neurons and glial cells is critical for brain development, vesicle formation, neurotransmitter function, synapse formation, cell–cell communication, and regulation of inflammation.

Dysfunctional lipid metabolism triggers neurodegeneration, as is clearly apparent in lysosomal storage diseases (LSDs). This is a group of over 70 disorders, each associated with homozygous loss of function mutations in a particular lysosomal enzyme or transporter [5,6]. There are 10 distinct LSDs caused by defects in sphingolipid catabolic enzymes that result in cellular accumulation of undergraded lipids. These sphingolipidoses are generally associated with neurological deficits, often severe neurodegenerative phenotypes that are fatal early in life. Examples include Gaucher disease (glucocerebrosidase deficiency), Farber disease (acid ceramidase deficiency), metachromatic leukodystrophy (arylsulfatase A deficiency) and Tay-Sachs disease (β-hexosaminidase deficiency). Several leukodystrophies (demyelinating disorders) are attributed to defects in nonlysosomal lipid metabolism, such as X-linked adrenoleukodystrophy [7] (defect in the peroxisomal ATP-binding cassette transporter D1) and hypomyelinating leukodystrophy [8,9] (defect in sphingolipid desaturase DEGS1). The

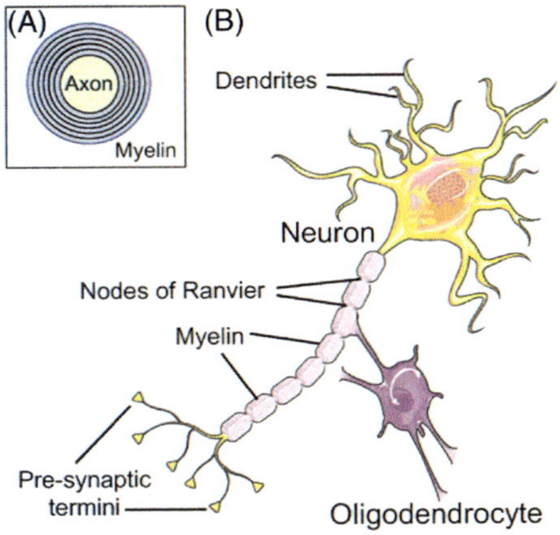


FIGURE 23.1 **Structure of myelin.** Myelin is created by specialized glial cells called oligodendrocytes in the central nervous system, and Schwann cells in the peripheral nervous system, which wrap their cell membrane multiple times around the neuronal axon (A). This limits the passage of ions across the membrane of the axon to gaps in the myelin called Nodes of Ranvier (B), greatly increasing the speed and energy efficiency of electrochemical conduction along the axon. Each neuron is myelinated by multiple oligodendrocytes spaced along its length, and any given oligodendrocyte may myelinate up to 50 axons in its vicinity [3,4].

severe consequences of these lipid metabolic deficiencies on neurological functions highlight the importance of lipid homeostasis for brain development and function.

AD and other neurodegenerative diseases of ageing are often thought of as diseases of defective proteostasis, due to the abnormal accumulation and aggregation of misfolded peptides and proteins, namely amyloid-β (Aβ) and hyperphosphorylated tau in AD. In the 1990s it was discovered that the greatest genetic risk determinant for AD is inheritance of the ε4 allele of the apolipoprotein E (*APOE*) gene [10], elaborated below. While this generated interest in the contribution of altered lipid metabolism to AD, research efforts have focused mostly on reconciling lipid-related risk alleles such as *APOE* with Aβ aggregation, rather than exploring defective lipid metabolism as an independent driver of AD pathogenesis.

It was recently discovered that heterozygous loss of function mutations in the *GBA* gene, encoding the lysosomal lipid hydrolase glucocerebrosidase, are a common genetic cause of Parkinson's disease (PD) [11–13] and dementia with Lewy bodies (DLB) [14]. Homozygous GBA defects cause the LSD Gaucher disease. This observation has precipitated a greater focus on the relationship between lipid metabolic defects, including lysosomal storage defects, and neurodegenerative diseases of ageing. Altered membrane lipid composition and catabolism leads to pathogenic accumulation of Aβ and α-synuclein (α-synuclein protein aggregates are characteristic of PD and DLB), suggesting that early alterations to lipid metabolism could trigger protein aggregation and cellular dysfunction in dementias.

## Alzheimer's disease

AD is the most common form of dementia, accounting for 60%–80% of dementia cases [15]. AD was the sixth leading cause of death in the United States in 2015 [15], while dementia as a whole was the second leading cause of death in Australia in 2017, according to cause of death records held by the Australian Bureau of Statistics. Dementia refers to a group of diseases characterized by pathological abnormalities in the brain and progressive loss of neurological functions [15]. The term encompasses Alzheimer's disease (AD), vascular dementia, PD, Huntington's disease, frontotemporal lobar degeneration (FTLD), DLB, and Creutzfeldt–Jakob disease. Individuals most commonly present with a mixed phenotype involving pathological abnormalities in the brain and neurological changes that are characteristic of more than one type of dementia. AD frequently co-occurs with vascular dementia or DLB [16].

AD is characterized by progressive cognitive decline that typically starts with loss of short-term memory and subtle changes to cognitive ability. As the disease develops, affected individuals have difficulty performing routine tasks, become confused, and exhibit behavioral and personality changes. Patients eventually become bed-ridden, lose their ability to communicate, have difficulty swallowing and as a result, commonly die of infections such as pneumonia [15,17]. Loss of neurological function is caused by brain atrophy attributed to loss of neurons and neuronal synapses. The most heavily affected regions are the hippocampus, entorhinal cortex, and associated temporal cortex [18,19].

## Pathological hallmarks of AD

Two major pathological hallmarks are used to definitively diagnose AD: extracellular plaques comprised of aggregated Aβ peptides, and intracellular neurofibrillary tangles (NFTs), comprised primarily of hyperphosphorylated tau protein [17,20]. Aβ peptides are 38–42 amino acid peptides (most commonly 40 or 42 amino acids) derived by sequential proteolytic cleavage of the amyloid precursor protein (APP) by β-secretase (the enzyme BACE1), then γ-secretase (Fig. 23.2). Alternatively, APP can be cleaved by α-secretase in the nonamyloidogenic pathway [21]. Tau is a cytoskeletal protein involved in microtubule stabilization [22] and its phosphorylation may represent a neuronal response to stress [23]. The presence of Aβ and NFTs may be detected through positron emission topography (PET) imaging [24] or postmortem examination [20]. These, and other biochemical and pathological features of AD, begin to develop many years prior to cognitive symptoms.

Other pathophysiological and biochemical characteristics of AD include loss of neuronal synapses, which is a close correlate of cognitive decline [25,26], loss of cholinergic neurons in the basal forebrain [27], neuroinflammation associated with astrogliosis and microgliosis (the proliferation and accumulation of reactive astrocytes and microglia) [28], cerebral hypometabolism [29], and loss of myelin [30,31].

The most widely accepted hypothesis for AD pathogenesis—the amyloid cascade hypothesis—postulates that the accumulation of cytotoxic oligomers of Aβ trigger NFT formation, synapse loss, neuroinflammation, and cognitive decline [17,34]. This hypothesis is strongly supported by the genetics of inherited AD, described below. However, neuropathological staging suggests that Aβ plaques and NFTs develop independently of one another, and that NFTs may precede significant Aβ deposition [20,35]. NFTs track more closely with cognitive deterioration than Aβ [36].

NFTs are also a defining feature of FTLD and chronic traumatic encephalopathy [37,38], and observed in some LSDs. In addition to AD, Aβ deposits are a characteristic feature of cerebral amyloid angiopathy [39], also observed in many cases of chronic traumatic encephalopathy [37,38], and found in some LSDs.

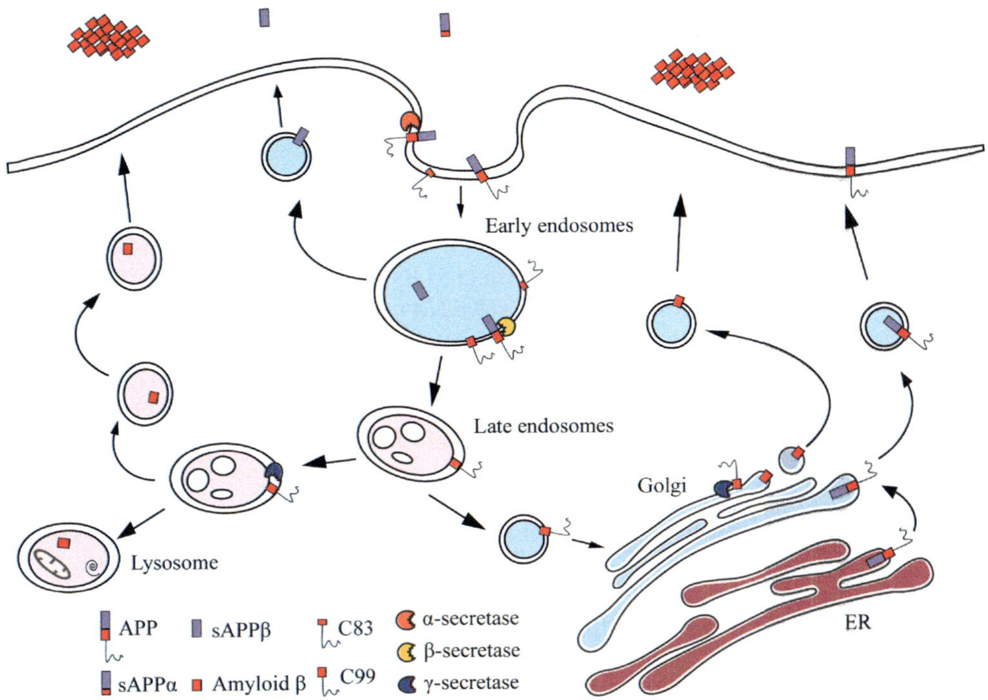


**FIGURE 23.2 APP processing to Aβ in late endosomes.** APP is a membrane protein with a single transmembrane domain whose exact physiological functions are subject to ongoing investigation but are probably at least partially mediated by its proteolytic cleavage products [21]. Nonamyloidogenic APP proteolysis by α-secretase occurs in the plasma membrane, and cleaves APP within the Aβ peptide domain, preventing Aβ formation. Amyloidogenic cleavage of APP by β-secretase (BACE1) takes place in endosomes [32] (BACE1 is most active at acidic pH), forming the sAPPβ and C-terminal fragments (βCTF or C99). The γ-secretase complex, which includes presenilin-1 or -2 together with three other protein subunits, cleaves the C99 C-terminal fragment generated by β-secretase, forming Aβ [33]. Cholesterol and sphingolipid-rich lipid raft domains may drive localization of APP to endosomes, favoring amyloidogenic processing.

## AD therapeutics

There are currently only four drugs approved for the treatment of AD [40]. These drugs provide temporary symptomatic relief but do not significantly slow or reverse the disease process and become ineffective as the disease progresses. Three drugs target cholinesterase to compensate for loss of cholinergic systems. The other is an antagonist of the NMDA receptors, which inhibits excitotoxic neuron death. A significant number of clinical trials aimed at inhibiting the formation of Aβ plaques, or enhancing their clearance with antibodies, have all failed to date [34,40]. This has led a growing body of researchers to question whether Aβ is in fact a driver of AD pathogenesis, or simply a biomarker of neuronal dysfunction. It is also recognized that cognitively normal individuals may present with a high Aβ burden [41]. If we assume that Aβ is indeed a primary driver of synapse loss and cognitive decline in AD, other factors must sensitize the brain to the cytotoxicity of Aβ. Alternatively, factors such as altered lipid metabolism and dysfunction of the endosomal and lysosomal systems with ageing may be interpreted as a root cause of the disease.

## A note on studies of lipid alterations in AD

In considering results for lipid quantification and enzyme expression in neurodegenerative diseases such as AD, it is important to appreciate that the cellular composition and physiological environment changes considerably in heavily affected brain regions. In AD, the hippocampus and temporal cortex are heavily affected by neuron loss and gliosis. This changes the cellular composition, resulting in changes to lipid profiles. This technical problem in analysis of postmortem brain tissue samples can be overcome by analyzing these brain regions in donors who died early in the pathogenesis of the disease, and prior to a clinical diagnosis of AD [19,20]. Alternatively, one might analyze brain regions that are afflicted by AD but not subject to the same extent of atrophy as the hippocampus, for example frontal cortex [42–46].

## Genetics implicates altered lipid metabolism in the etiology of AD

It is important to distinguish the early-onset, inherited form of AD (familial AD) from the later-onset, sporadic form, hereafter referred to as late-onset AD (LOAD). Familial AD is estimated to account for less than 1% of cases [47]. It is caused by autosomal dominant mutations in the *APP*, presenilin-1 (*PSEN1*), or presenilin-2 (*PSEN2*) genes that invariably cause disease in middle age. Most familial AD mutations increase the propensity for generation of Aβ peptide from APP [17,48,49].

LOAD typically occurs after age 65 but the course of the disease, neuropathology, and symptoms are otherwise very similar to familial AD. The greatest risk factor for AD is ageing. In the USA, 3% of 65–74-year olds are living with Alzheimer's; in those aged 85 or older, this number increases 10-fold [50]. A number of single-nucleotide polymorphisms (SNPs) that increase or decrease the odds of developing LOAD have now been identified, mostly through genome-wide association studies [48,51]. A SNP in the *APP* gene that impedes Aβ generation and decreases the risk for LOAD has been identified [52], underscoring the importance of APP metabolism to Aβ in the etiology of AD. Many of the other gene variants that affect risk for LOAD have been shown to impact on Aβ formation, oligomerization, and clearance, as elaborated below. However, gene ontology studies indicate that late-onset risk genes are primarily involved in regulation of lipid metabolism, endocytosis and endosomal trafficking, and inflammation [48,51]. Genetic variants affecting lipid metabolism are described below.

### APOE4

The dominant genetic risk determinant for LOAD is inheritance of the ε4 allele of the *APOE* gene (*APOE4*), encoding a major apolipoprotein of the central nervous system (CNS) and the circulation. There are three common alleles of the *APOE* gene, ε2, ε3, and ε4, which occur with frequencies of approximately 8%, 78%, and 14%, respectively, in the general population [53]. Therefore, approximately one quarter of the population carries at least one ε4 allele, which increases their risk of AD threefold, while inheritance of two ε4 alleles increases the risk 12-fold relative to noncarriers [10]. Age of AD onset is earlier in ε4 carriers [53]. Conversely, carriers of the ε2 allele have decreased risk for AD and increased age at disease onset [54]. The relatively high prevalence and odds ratio of the *APOE4* allele make it the most important genetic risk factor for AD. The *APOE4* allele is also an established genetic risk factor for DLB, whereas *APOE2* is protective [55–58], and while the diseases often co-occur in the same patient [20], the sensitizing effect of the *APOE* allelic variants for the two diseases appears to be independent [55–57].

ApoE is a 34 kDa protein. The functional differences between the three allelic variants are created by the presence of either a cysteine or arginine residue at positions 112 and 158 in the 299 amino acid sequence: Cys112 and Cys158 in ApoE2, Cys112, and Arg158 in ApoE3, Arg112, and Arg158 in ApoE4 [53]. The various functions of ApoE and differences between the three allelic variants, including regulation of brain Aβ burden, is elaborated in the following section. The reader is also referred to a number of reviews on this topic [53,59,60].

## CLU

Genome wide association studies have also demonstrated significant effects of several different SNPs in the *CLU* gene on risk for LOAD [61,62], reviewed in Ref. [63]. Most of these SNPs are intronic. *CLU* encodes another major CNS apolipoprotein, ApoJ (also known as clusterin). The effect size for the *CLU* gene variants is much lower than for *APOE*, however, some of the variants are very common, for example, the rs1136000 SNP was estimated to occur in 88% of Caucasians and increase risk 1.16-fold [63].

## ABCA7

SNPs near the gene encoding the lipid transporter ATP-binding cassette protein A7 (ABCA7) increase risk for AD [48,64,65]. As with the *CLU* variants, *ABCA7* variants are reasonably common but effect size is modest [48]. ABCA7 is highly expressed in microglia [66], the macrophage-like cells of the CNS, and involved in the efflux of cellular lipids into lipoprotein particles [67].

## PLD3

Rare SNPs in the *PLD3* gene (phospholipase D3) that increase AD risk have been identified. The association of these coding variants in *PLD3* with sporadic AD has been controversial, with several studies supporting [68–71] and others failing to demonstrate the association [72–74]. Nonetheless, the *PLD3* V232M variant is also associated with poor cognitive function [75] and cognitive ageing [76], while loss of PLD3 function enhances Aβ formation from APP [68,77,78]. Members of the phospholipase D family hydrolyse the phosphodiester bond on phosphatidylcholine (PC), releasing the choline headgroup and forming phosphatidic acid (PA). However, the specific activity of PLD3 is poorly defined and it may function as a 5′ exonuclease rather than a phospholipase [79]. PLD3 is known to regulate endosomal sorting [77], potentially through its phospholipase activity [78,80].

## TREM2

*TREM2* (triggering receptor expressed on myeloid cells 2) encodes a receptor on microglia [81]. Rare SNPs producing R47H and R62H amino acid changes are associated with significantly increased risk for LOAD [82–84]. SNPs in *TREM2* have also been associated with FTLD [85], while homozygous *TREM2* deficiency produces the early-onset demyelinating dementia Nasu-Hakola disease [86]. While often regarded as an inflammation-associated rather than a lipid biology gene, a number of studies have shown that TREM2 recognizes anionic lipids released by damaged cells or myelin [87,88]. Lipid sensing by TREM2 is required for the microglial response to demyelination and for subsequent remyelination [87], as well as the microglial response to Aβ, promoting Aβ phagocytosis and clearance [88]. The R47H variant that increases AD risk impairs lipid recognition by TREM2, which is speculated to impair the capacity for microglia to sense damaged cells or myelin [88]. Two research groups subsequently reported that ApoE, ApoAI, and ApoJ are ligands for TREM2, proposing that this facilitates Aβ uptake by microglia and demonstrating that the SNPs in TREM2 that are associated with AD compromise ApoE binding [89,90] (Fig. 8.3). ApoE binding by TREM2 was dependent on lipidation in one study [89], but independent of lipidation in another [90]. Given that TREM2 was also shown to bind lipids [87,88], it seems likely that the physiological function of TREM2 is binding of lipidated apolipoproteins.

## PLCG2

Rare coding variants in the *PLCG2* gene that protect against AD were very recently identified and verified in large cohort studies [84,91,92]. This gene encodes phospholipase Cγ2 (PLCγ2), a phosphoinositide-specific PLC isoform that hydrolyses phosphatidylinositol 4,5-bisphosphate to inositol 1,4,5-triphosphate ($IP_3$) and diacylglycerol (DAG) [93]. In the CNS, it is almost exclusively expressed in microglia [94,95]. The protective P522R variant enhances the enzyme's phospholipase activity [95].

# Apolipoproteins and AD

Apolipoproteins are lipid-binding proteins that complex with lipids to form lipoprotein particles, which transport lipids in the extracellular space or circulation. CNS lipoprotein particles are similar to high-density lipoprotein particles in the periphery, and there are no low or very low-density lipoprotein particles in the CNS [96–98]. Nine apolipoproteins have been detected at the protein level in the CNS, of which ApoE, ApoA-I, ApoD, and ApoJ are the most abundant [99,100]. The major cell type expressing both *APOE* and *APOJ* in the CNS is the astrocyte [94,101], however all CNS cell types can express and secrete ApoE. Both ApoE and ApoJ are strongly upregulated and secreted in response to nerve injury [102–104], where they are believed to facilitate repair by clearing lipid debris and redistributing lipids to regenerating neurites [105,106].

Lipoprotein particles, both in the CNS and the circulation, bind to various lipoprotein receptors. ApoE is a ligand for the low-density lipoprotein (LDL) receptor family, of which low-density lipoprotein receptor (LDLR), ApoE receptor 2 (ApoER2), LDL receptor-related protein (LRP1), and very low-density lipoprotein receptor (VLDLR) are expressed in the CNS [97]. Endocytosis of CNS lipoprotein particles facilitates lipid uptake in cells expressing the LDL family receptors, and the LRP1 receptor is much more rapidly endocytosed than the other receptors upon ApoE binding [97]. ApoE receptors also initiate signal transduction processes with effects on synaptic plasticity and neuronal survival [60].

Cells secrete ApoE as a nascent lipoprotein, which is lipidated during and after secretion. In the CNS, ApoE lipidation is mediated by the ATP-binding cassette (ABC) family proteins ABCA1 and ABCG1 [99,107–110], which directly transfer cellular lipids to lipoproteins [111]. Lipidation by ABCA1 is necessary for the stability of ApoE in vivo. ApoE levels are 3-to 5-fold lower in brain parenchyma and 50-fold lower in CSF of mice lacking ABCA1 [107,112]. In a large Danish population study, individuals heterozygous for a loss-of-function *ABCA1* allele were shown to be at significantly increased risk of developing AD, and had significantly decreased levels of ApoE in the plasma [113].

Cholesterol and PC are most efficiently effluxed to ApoE2, and least efficiently effluxed to ApoE4 in vitro [110,114]. Accordingly, ApoE is less lipidated in humans with the ε4 allele [115], and ApoE particles in the cerebrospinal fluid (CSF) of people without dementia are larger in those with an ε2/ε3 genotype, and smaller in those with ε3/ε4 or ε4/ε4 genotypes, compared to the most common ε3/ε3 genotype [116]. The same is observed in mice transgenically expressing the human ApoE isoforms [117]. ApoE content in the CSF and plasma is also higher in ε2 carriers and lower in ε4 carriers compared to the ε3/ε3 genotype [118].

Levels of the abundant lipid PC are lower in the CSF of AD patients compared to cognitively normal individuals, suggesting a lipoprotein deficit [119]. This was confirmed by demonstrating that the capacity for human CSF to mediate cholesterol efflux from ABCA1-expressing cells is reduced in AD patients compared to cognitively normal participants [119]. This was independent of both the levels of ApoE and ApoA-I in the CSF, and *APOE* genotype, indicative of a generalized CSF lipoprotein deficit in AD.

## Effect of APOE genotype and lipidation on Aβ clearance

Aβ burden and clearance through the CSF is directly dependent on ApoE and influenced by *APOE* genotype. Aβ plaque load is highest in the brains of those with an ε4/ε4 genotype, followed by ε3/ε4, then ε3/ε3, and is lowest in ε2 allele carriers [41,120–122]. Low CSF Aβ is associated with the presence of Aβ plaques in the brain, as a consequence of poor Aβ clearance through CSF [120]. Accordingly, ε2 allele carriers have higher CSF Aβ than those with an ε3/ε3 genotype, while levels are lowest in ε4/ε4 homozygotes [120]. In mice expressing human ApoE variants in place of mouse ApoE, expression of human ApoE2 increases Aβ clearance through the CSF and reduces brain Aβ burden, whereas expression of ApoE4 decreases clearance and increases brain Aβ burden, in comparison to ApoE3 [117,120,123]. Paradoxically, the absence of ApoE in mice prevents fibrillar Aβ deposition [124]. However, this phenotype is reversed by deletion of both ApoE and ApoJ, suggesting that these apolipoproteins act together to promote Aβ clearance [125]. It is possible that deletion of either ApoE or ApoJ alone leads to compensatory changes in lipoprotein expression that inhibit Aβ production or enhance its clearance.

Aβ burden is much higher in an amyloidogenic mouse model lacking ABCA1 [126], while heterozygous loss of ABCA1 increases Aβ burden and significantly slows Aβ clearance in

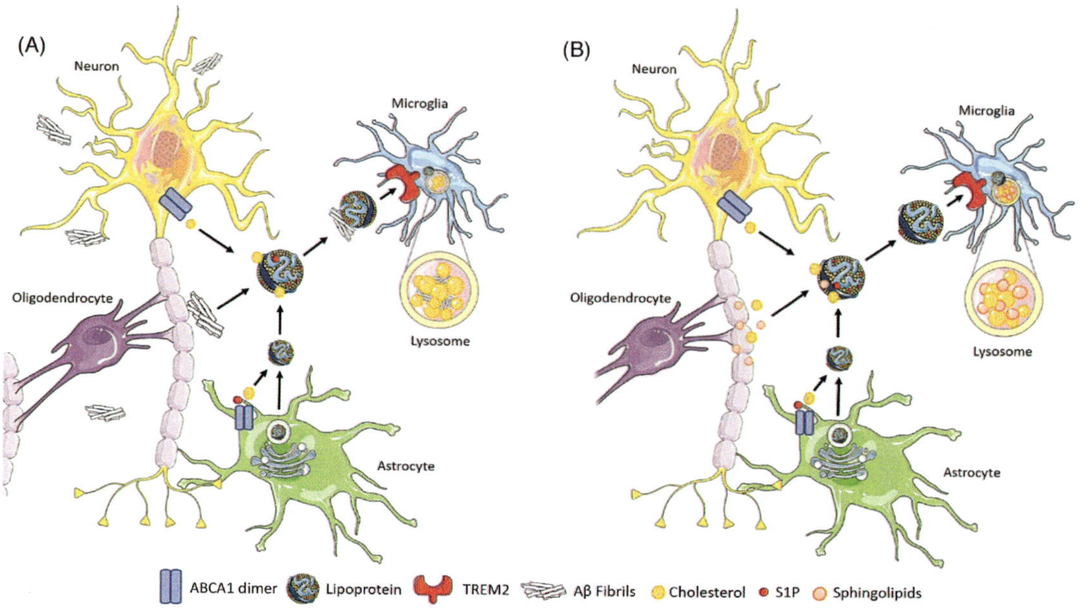


FIGURE 23.3 **Apolipoprotein and TREM2 variants as risk factors for AD.** ApoE and ApoJ are components of lipoproteins that transport lipids between cells. These apolipoproteins are produced by all major CNS cell types but are most abundantly expressed by astrocytes. ABCA1, ABCA7, and other members of the ATP binding cassette superfamily transfer lipids to nascent apolipoproteins, forming lipoprotein particles. In addition to transporting lipids between cells, these lipoproteins carry lipid molecules with important physiological signaling properties, such as S1P. TREM2 is a microglial receptor that participates in both lipid and Aβ turnover and catabolism by binding to apolipoproteins and/or lipids. (A) The proposed role for lipoproteins and TREM2 in the clearance and microglial degradation of Aβ. Poorly lipidated ApoE is less efficient at binding and clearing Aβ. Relatively poor lipidation of ApoE4 may also result in less efficient cholesterol efflux from neurons, thereby promoting amyloidogenic processing of APP. (B) The role of lipoproteins and TREM2 receptor in the clearance and lysosomal degradation of lipids, particularly myelin lipids. In addition to the injury scenario, apolipoproteins and their receptors on phagocytic cells almost certainly mediate constitutive turnover of brain lipids.

targeted replacement mice with human ApoE4, but not those expressing human ApoE3 [127]. These experiments indicate either that ApoE must be lipidated to efficiently clear Aβ (Fig. 23.3) or that poorly lipidated ApoE promotes amyloidogenesis. Higher Aβ burden in humans with ABCA7 variants [122] and Aβ-producing mouse models lacking ABCA7 [128–130] provides further evidence supporting lipoprotein lipidation and Aβ turnover as a potential mechanism through which these gene loci influence risk of developing AD [122,128–130]. Increased Aβ accumulation in mice lacking ABCA7 was attributed to reduced Aβ uptake and clearance by macrophages in one study [128], and in another to increased Aβ production from APP [129]. Interestingly, deletion of the lipoprotein receptor LRP1 in neurons reverses the increased Aβ fibril deposition caused by ApoE4 expression in mice [131]. One explanation may be that LRP1 deletion in neurons increases cerebral ApoE levels, so promoting Aβ clearance. Alternatively, ApoE-mediated delivery of lipids to neurons plays an important role in promoting Aβ aggregation.

## Lipid delivery by ApoE is necessary for neuroprotection, synapse formation, and memory

ApoE3 lipoproteins have been shown to promote neurite outgrowth and sprouting, whereas ApoE4 either had a lesser effect or inhibited neurite outgrowth [132,133]. This may be as simple as an essential requirement for cholesterol, delivered as a complex with ApoE from glial cells to neurons, for formation of new synapses [134]. ApoE2 and ApoE3 also protect neurons and their synapses against Aβ [135,136] or growth factor withdrawal [137], whereas ApoE4 is less effective or ineffective. The protective effect of ApoE3 requires lipids and/or cholesterol, the LRP1 receptor [135,137], and was shown in one study to involve a PKCε signaling pathway [135]. In the other direction (neurons to glial cells), ApoE and ApoD cooperatively facilitate the transport of excess lipids from neurons for storage in glia, which appears important for protecting neurons against oxidative stress [138].

The *APOE4* allele accelerates the rate of cognitive decline in mild-cognitive impairment (MCI) [139,140] and has been associated with accelerated memory decline in older adults without existing impairments [141–146], although not all studies support the latter association [139]. Memory and cognition are worse in targeted Apoe replacement mice expressing human ApoE4, compared to ApoE3 [147,148]. Promoting ApoE lipidation with the retinoid X receptor (RXR) agonist bexarotene, which enhances ABCA1 and ABCG1 expression, reverses the adverse effect of transgenic ApoE4 expression on memory and neuropathology in mice [149], as does the ABCA1 agonist CS-6253 [150]. Likewise, administration of the liver X receptor (LXR) agonist T0901317, which increases both ABCA1 and ApoE levels, to mouse models of AD expressing APP with familial AD mutations improves their performance in memory tests [151,152].

ApoE4 also alters the response to injury independently of dementia. Outcomes for traumatic brain injury (TBI) are worse in *APOE4* carriers [153], and a much higher proportion of *APOE4* carriers with TBI exhibit Aβ pathology than noncarriers [154]. Severe TBI is also a risk factor for AD [155]. Carriers of the *APOE4* allele appear to have altered neural response to injury [156], with a mouse model showing *APOE4* carriers had significantly increased phosphorylated tau levels compared to *APOE3* carriers in response to repetitive TBI [157].

## Phosphoinositide dysregulation by ApoE4 and presenilin-1 mutations

Lipidomic studies have indicated that *APOE* genotype does not have a profound effect on brain lipid composition, either in humans [158–160], or in mice expressing human ApoE variants in place of murine Apoe [161,162]. One study showed that ApoE4 had no significant effect on brain lipid levels in cognitively normal individuals, but was associated with increased cholesterol and ceramide levels in subjects with AD [159]. However, Zhu et al. reported that levels of the signaling lipid phosphatidylinositol 4,5-bisphosphate ($PIP_2$) are reduced in the brains of *APOE4* allele carriers (compared to noncarriers), and in mice in which murine Apoe is replaced with human ApoE4 (compared to ApoE3). This was attributed to higher levels of the phosphatidylinositol phosphatase synaptojanin 1 (Synj1), which dephosphorylates $PIP_2$, in ApoE4 carriers or knock-in mice. Heterozygous loss of Synj1 restored $PIP_2$ levels and rescued cognitive deficits in ApoE4 knock-in mice [163]. Heterozygous deletion of Synj1 also rescued memory and synaptic deficits in an amyloidosis mouse model [164]. Similarly, blast-induced TBI increased $PIP_2$ in ApoE3 but not ApoE4 targeted replacement mice. The lack of $PIP_2$ induction in ApoE4 knock-ins led to higher levels of tau phosphorylation, which was dependent on Synj1 [157].

Significantly reduced $PIP_2$ levels were reported in the frontal and parietal cortex of LOAD cases [163,165], and in cells expressing mutant forms of presenilin-1 that cause familial AD (compared to WT presenilin-1) [166], although not in a presenilin-1 mutant mouse model [165]. The $PIP_2$ deficit, but not Aβ, caused ion channel dysfunction in cells expressing mutant presenilin-1 [166]. Oligomeric Aβ was also shown to decrease $PIP_2$ and suppress long-term potentiation, an important electrophysiological measure of memory, in cultured hippocampal neurons [167]. As with the above models, this was suppressed by heterozygous loss of Synj1. However, these authors also proposed that $PIP_2$ deficiency is mediated through a pathway involving $Ca^{2+}$-activated phospholipase C (PLC) [167].

Specific PLC family members hydrolyse $PIP_2$, yielding the second messengers DAG and $IP_3$ [93]. A number of studies have shown increased DAG levels in the brains of subjects with AD and MCI [168,169]. Increased monoacylglycerol (MAG) was reported in one of these studies [169], indicating that the increased DAG is not derived from MAG acylation. In prior studies looking at protein levels, increased levels of PLCγ (specific isoform not specified) were observed in the hippocampus and temporal cortex of AD patients in one study [170], whereas in another, PLCγ1 activity was decreased and PLCδ1 activity increased in AD temporal cortex [171]. As noted above, rare variants in PLCγ2 that protect against AD have been identified, but their functional significance remains to be determined.

## Myelin lipids and peroxisomal deficits

Extensive myelination is a distinguishing feature of the human brain and almost certainly a key determinant of our advanced cognitive abilities [172]. The major lipid constituents in myelin are cholesterol, the glycosphingolipids galactosylceramide (GalCer; also known as cerebroside) and sulfatide, and ethanolamine plasmalogens [1,2]. The concentration of unesterified cholesterol is higher in the brain than any other organ and this cholesterol is believed to be derived mostly from de novo synthesis, as there is very limited exchange of lipids and lipoproteins between the CNS and the circulation [173,174]. GalCer and sulfatide are derivatives of ceramide synthesized only by oligodendrocytes in the CNS (Fig. 23.4). They are essential for myelin stability [175,176]. Plasmalogens are glycerophospholipids that are defined by a vinyl ether bond at the sn-1 position of glycerol, rather than the more common ester bond [177] (Fig. 23.4). They have either a choline phosphate or ethanolamine phosphate headgroup, but ethanolamine plasmalogens are particularly enriched in myelin and essential for myelin stability and function [178,179]. The acyl chain in the sn-2 position is bent in phosphatidylethanolamine (PE), whereas it is oriented perpendicular to the membrane in plasmalogens, allowing for tighter membrane packing and lower fluidity [2,180]. This may be important for myelin structure and function.

Loss of myelin lipids in AD, including cholesterol, was reported 30 years ago by Svennerholm and colleagues [182,183]. Subsequent studies using mass spectrometry have confirmed pronounced loss of the myelin lipid sulfatide very early in AD pathogenesis [43,44,184]. More

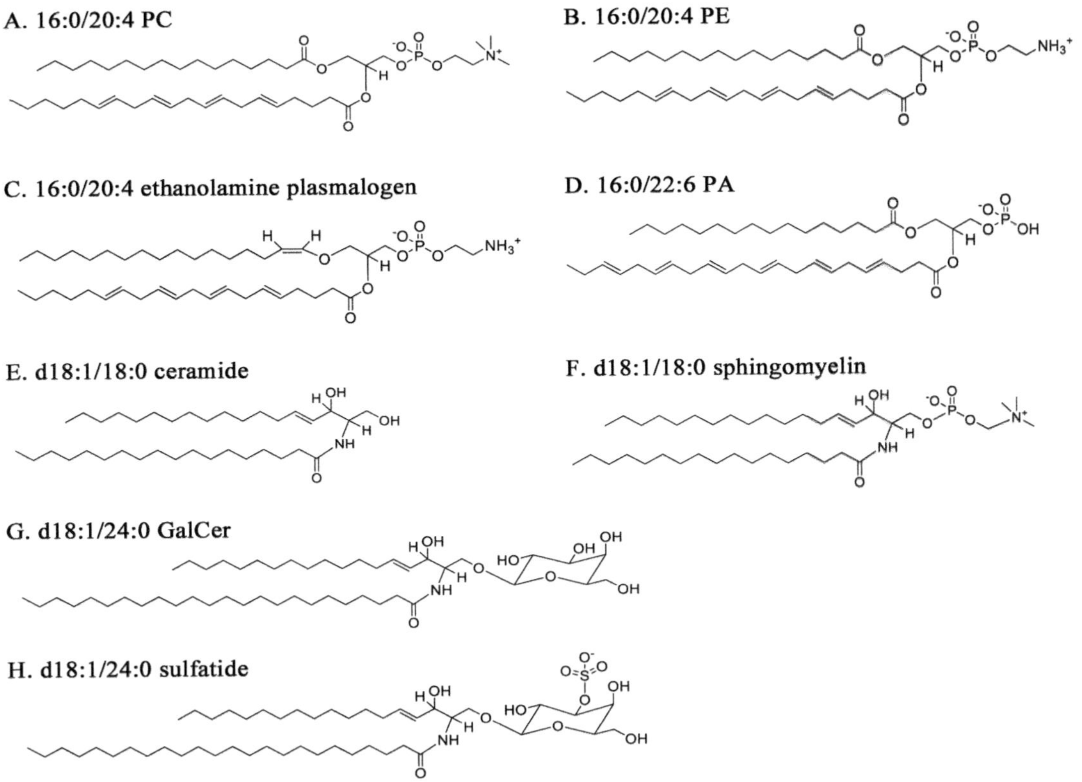


FIGURE 23.4 **Structure of common sphingolipids and phospholipids.** Common membrane phospholipids have two acyl chains linked to the sn-1 and sn-2 positions on glycerol, and the different classes are derived by addition of different headgroups to the sn-3 position. (A) Phosphatidylcholine (PC), (B) phosphatidylethanolamine (PE), (C) ethanolamine plasmalogen with 16:0 and 20:4 (arachidonic acid) fatty acyl chains, (D) phosphatidic acid (PA) with 16:0 and 22:6 (DHA) acyl chains. Sphingolipids are a different family of lipids characterized by a serine (or less commonly an alanine) headgroup [181]. The basic sphingolipid unit is ceramide, which is comprised of a sphingoid base chain (usually sphingosine in mammals) linked through an amide bond to a variable length fatty acid. As with the glycerolipids, addition of different headgroups yields the different classes of sphingolipids. (E) Ceramide and (F) sphingomyelin (SM) with d18:1/18:0 structure; (G) GalCer and (H) sulfatide with d18:1/24:0 structure.

recently, loss of GalCer and very long chain ceramides that are the precursor for GalCer and sulfatide biosynthesis in oligodendrocytes was shown [43,185] (Fig. 23.5). This was associated with reduced activity of the enzyme ceramide synthase 2, which catalyses very long chain ceramide synthesis in oligodendrocytes, in multiple brain regions [43].

Ethanolamine plasmalogens are also heavily depleted in AD [45,168,186,187]. Plasmalogen loss was not observed in PD or Huntington's and is specific to brain regions that are heavily affected in AD [45,186]. In some studies changes to ethanolamine plasmalogens were observed in grey but not white matter [168,187], while in another the changes were more pronounced in white matter [45]. The initial steps of plasmalogen synthesis leading to formation of the characteristic vinyl ether bond take place in peroxisomes [188]. Selective loss of plasmalogens but not PE could be an indirect consequence of myelin degradation, or indicative

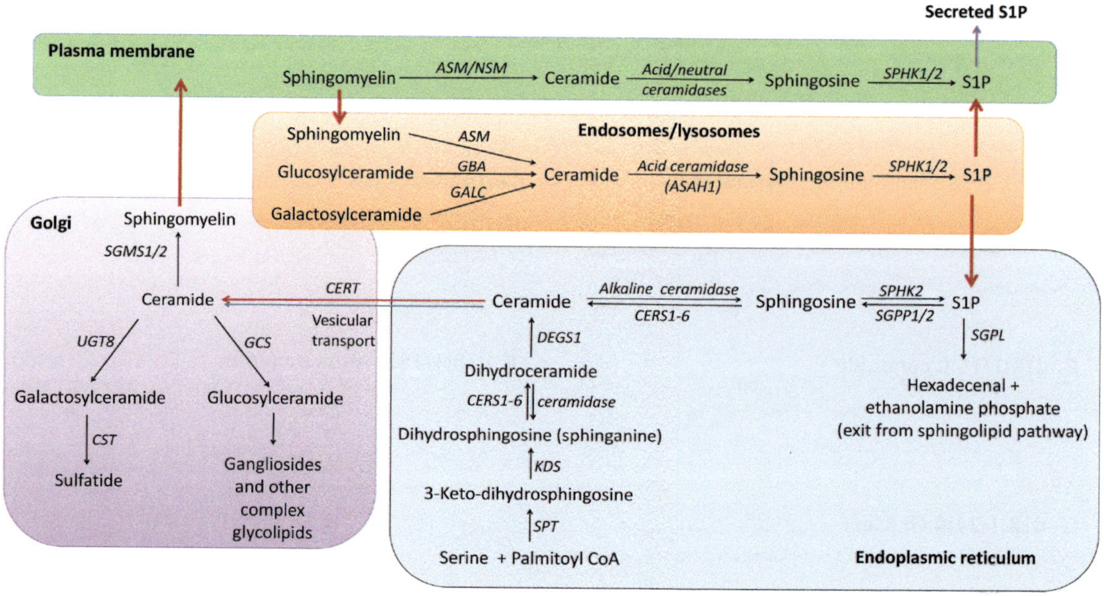


**FIGURE 23.5** **Sphingolipid biosynthesis and catabolism.** Sphingolipid biosynthesis takes place in the endoplasmic reticulum and Golgi apparatus. Sphingomyelin and glycosphingolipids synthesized at the Golgi apparatus are transferred to the plasma membrane to carry out their cellular and physiological functions. Degradation of these lipids takes place in acidic endosomes and lysosomes, and defects in enzymes catalyzing sphingolipid degradation give rise to lysosomal storage diseases. For simplicity, enzymes that degrade complex glycosphingolipids (gangliosides) to glucosylceramide, or sulfatide to galactosylceramide, are not shown. *Red arrows* (*gray* in print version) indicate transport; *black arrows* are biochemical reactions. Enzyme names are shown in italics: ASAH1, acid ceramidase; ASM, acid sphingomyelinase; CERS1-6, ceramide synthases 1-6; CERT, ceramide transfer protein; CST, cerebroside sulfotransferase; DEGS1, dihydroceramide desaturase 1; GALC, galactosylceramidase; GBA, glucocerebrosidase; GCS, glucosylceramide synthase; KDS, 3-ketodihydrosphingosine reductase; NSM, neutral sphingomyelinase; SGMS1/2, sphingomyelin synthases 1 and 2; SGPL, sphingosine 1-phosphate lyase; SGPP1/2, sphingosine 1-phosphate phosphatases 1 and 2; SPHK1/2, sphingosine kinases 1 and 2; SPT, serine palmitoyltransferase; UGT8, UDP-Galactose Ceramide Galactosyltransferase.

of a biosynthetic deficiency caused by a deficit in peroxisomes. Alternatively, oxidative stress may deplete plasmalogens, as the vinyl ether bond in plasmalogens is particularly prone to oxidative damage [177].

Several studies have also reported reduced circulating levels of ethanolamine plasmalogen, but not PE, in subjects with AD [189–192]. In fact, plasmalogen levels were shown to decrease as a function of dementia severity [189]. In a 24-week clinical trial testing dietary plasmalogen supplementation for AD, some improvement in memory tests was observed with plasmalogen supplementation in females with mild AD, however, there was no improvement in the primary mini mental stage examination results [192].

A significant role for ApoE in myelin lipid turnover is suggested by the observation that *ApoE* knockout mice show elevated levels of the myelin lipids GalCer and sulfatide in their dorsal root ganglia [193], and increased sulfatide in multiple brain regions [194]. Sulfatide is transported on ApoE particles in the CSF, and sulfatide content in the CSF is significantly higher in carriers of the *APOE4* allele, suggesting that *APOE* genotype affects myelin lipid turnover [194].

The depletion of both plasmalogens and myelin sphingolipids in AD suggests a generalized myelin deficit. One might assume that this myelin degradation is secondary to neurodegeneration. However, a large and growing body of evidence from magnetic resonance imaging (MRI) indicates that loss of myelin structural integrity is an early feature of AD pathogenesis [195–198]. Similarly, loss of myelin lipids occurs early in AD pathogenesis, including at the mild cognitive impairment (MCI) stage that precedes AD [43–45]. Myelin loss has also been reported as an early pathological feature in several AD mouse models [199,200], preceding Aβ plaque deposition in the aggressive 3 × Tg model (APP, PS1, and tau mutant transgenes) [200].

In addition to plasmalogen synthesis, peroxisomes perform other metabolic functions that are critical for brain health and myelin integrity, including the β-oxidation of very long chain fatty acids (such as 22–26 carbon fatty acids in myelin) and branched chain fatty acids, synthesis of the essential fatty acid docosahexaenoic acid (DHA), and catalase-mediated inactivation of hydrogen peroxide [188,201]. Kou et al. provided convincing evidence for peroxisome deficiency in AD by showing a deficiency of plasmalogens, an increase in very long chain fatty acids (C24:0 and C26:0), and reduced peroxisome content in neuronal processes (but increased peroxisomal density in neuronal cell bodies), in the frontal cortex of AD brains [202]. These deficits were associated with hyperphosphorylated tau pathology but not Aβ plaques. Wood et al. showed similar lipid changes in their study [168].

One research group has proposed that glucose hypometabolism and bioenergetic insufficiency in the ageing brain trigger a switch to utilization of fatty acids derived from myelin as a ketogenic energy source, leading to myelin loss in normal ageing and AD [203–205]. This was proposed to involve age-dependent activation of phospholipase A2, which catalyses hydrolysis of fatty acids from the sn-2 position of glycerophospholipids, and sphingomyelinase, which hydrolyses sphingomyelin to ceramide [203]. In support of this hypothesis, plasma levels of medium-chain acylcarnitines and ketones (β-hydroxybutyrate), which are indicative of fatty acid β-oxidation, are significantly lower in AD patients compared to controls [206]. Importantly, acylcarnitine levels in the plasma were positively correlated with cognitive scores. This indicates reduced availability of peripheral ketones as an energy source for the brain in AD. Although β-hydroxybutyrate levels trended lower in MCI compared to controls, they were not significantly lower, suggesting that this is not an early metabolic change in AD pathogenesis [206]. Loss of myelin does occur as a feature of normal ageing, particularly in the late-myelinating regions that are most heavily affected in AD [207], and this age-dependent myelin loss is accelerated by the presence of the *APOE4* allele [208,209]. Given the importance of myelin and oligodendrocytes for neuronal function, oligodendrocyte pathology and myelin loss with ageing could be an important trigger for neurodegeneration.

## Ceramides and sphingosine 1-phosphate (S1P)

Studies with human brain tissue consistently show elevated ceramides in AD, including at the very early stages of AD pathogenesis [44,46,210,211]. Ceramide content is also higher in the CSF of AD patients [212]. In considering ceramide biology in the CNS, it is important to consider the cellular source of major ceramides. C16 and C18 ceramides are characteristic of neurons and reproducibly shown to increase with AD pathology. The very long chain C24 forms are highly enriched in myelin and have consequently been shown in at least one study to decrease in the AD-affected brain [43]. Higher sphingomyelinase activity, converting sphingomyelin to ceramide, may drive higher ceramide content in AD [46,211,213]. However, higher acid ceramidase activity, which degrades ceramide to sphingosine, has also been demonstrated [46].

Phosphorylation of sphingosine by sphingosine kinases 1 and 2 produces the signaling lipid S1P (Fig. 23.5), a potent agonist for a family of five G protein-coupled receptors that regulate both vascular and neural physiology [214,215]. Sphingosine phosphorylation to S1P is appreciably reduced in AD [42,46] and in the preclinical AD state [42], attributed to reduced activity and protein levels of sphingosine kinases 1 and 2 [42,216]. Thus, sphingolipid metabolism appears to be blocked at the point of sphingosine conversion to S1P, which is necessary for catabolism of all sphingolipids. Overall, this shift in the ceramide:S1P "rheostat" in favor of ceramide would be expected to promote endoplasmic reticulum stress and apoptosis in both neurons and oligodendrocytes [213,217–219]. This has been demonstrated in multiple sclerosis, where higher CSF ceramide levels were causally linked to impaired mitochondrial respiration and apoptosis in neurons [220]. A number of studies have also reported that ceramide and sphingomyelin levels increase with normal ageing in murine and human brains [158,210,221], while S1P levels in the hippocampus decrease as a function of age in cognitively normal females [158]. These age-dependent changes could sensitize the brain to neurodegeneration.

S1P is present at high levels in the circulation, where it is important for vascular barrier function and cardioprotection [222]. These cardiovascular-protective effects of S1P are dependent on its binding to ApoM within high-density lipoprotein particles [223,224]. Similarly, the physiological functions of S1P in the CNS are at-least partially attributed to its association with HDL-like particles [225,226].

Recent research has demonstrated very clear associations between circulating ceramide levels and cardiometabolic conditions such as obesity, diabetes, and nascent cardiovascular disease [227,228]. Some studies have reported that circulating ceramides, and/or the ceramide:sphingomyelin ratio, are higher in AD than cognitively normal control individuals [229–231]. High circulating ceramides were also shown to be strong predictors of risk for memory impairment and AD [232–234]. The results of these cohort studies on ceramides in dementia are subject to some variability. For example, one cohort showed an association between ceramides and incident memory impairment in females [232], while another showed that this relationship only holds for males [234].

## Sphingolipids and cholesterol promote amyloidogenic processing of APP

Together with cholesterol, SM and ceramide are required for formation of detergent-resistant sub-domains of the plasma membrane termed lipid rafts, which concentrate particular signal transduction molecules [235]. Processing of APP in the amyloidogenic pathway by the protease BACE1 and the $\gamma$-secretase complex occurs inside lipid raft domains, while non-amyloidogenic processing of APP by $\alpha$-secretase (ADAM10 and ADAM17 proteases) occurs outside of lipid rafts [236–239] (Fig. 23.2). Much of this evidence is based on experiments showing that high membrane cholesterol content promotes cleavage of APP via the amyloidogenic pathway [239,240], and increases Aβ fibril formation [241]. Polyunsaturated fatty acids promote the processing of APP through the nonamyloidogenic pathway in a cell model [242], whereas dietary hypercholesterolemia promotes amyloidogenic processing of APP and Aβ deposition in transgenic mice expressing human APP with familial AD mutations [243]. Ceramide also enhances Aβ formation by increasing the half-life of BACE1 [244].

The sphingolipid family also encompasses a diverse array of complex glycolipids called gangliosides, which are characterized by a glycan headgroup linked to a membrane-embedded ceramide [245]. The ganglioside glycan structure protrudes from the membrane into the extracellular space and includes one or more sialic acid moieties, conferring a negative charge. Gangliosides are highly enriched in the brain grey matter [246]. Interaction of Aβ peptides with gangliosides, particularly the ganglioside GM1, promotes Aβ oligomerization [247]. GM1 is one of the most abundant gangliosides in the brain [245], and thought to act as an endogenous seed for Aβ oligomerization in lipid rafts at presynaptic neuronal membranes. Aβ has been hypothesized to adopt an altered conformation through tight binding with GM1, resulting in very high aggregation potential [248–250]. Accordingly, inhibition of ganglioside synthesis at the point of glucosylceramide formation inhibits APP processing to Aβ in cultured human neurons [251].

Aβ complexed with GM1 has been found in brains exhibiting early signs of AD [252,253]. Higher levels of GM3, the structural precursor to GM2 and GM1 gangliosides have been demonstrated in the entorhinal cortex of AD patients and familial AD mouse models [254], and both GM3 and GM2 have been observed in increased levels in the serum of human AD subjects compared to normal controls [255]. Accumulation of GM3 and GM2 gangliosides is associated with neurite death and apoptosis [256,257].

## The polyunsaturated fatty acid DHA

The polyunsaturated, ω3 fatty acid DHA (Fig. 23.4) often receives special attention in the literature for its essential role in brain health. DHA is a major constituent of brain phospholipids, and its highly unsaturated acyl chain (22 carbons with 6 double bonds; designated 22:6) may be important for conferring membrane fluidity [258]. It is synthesized in the endoplasmic reticulum and peroxisomes from the essential fatty acid linolenic acid (18:3) [188]. A diverse range of lipid mediators, many of which signal through specific G protein-coupled receptors, are generated from metabolism of the polyunsaturated fatty acids arachidonic acid (AA; 20:4), DHA, and eicosapentaenoic acid (20:5). These polyunsaturated fatty acids are released from membrane phospholipids by phospholipase A2. AA and DHA metabolites are termed eicosanoids and docosanoids, respectively [258–260], and are major physiological regulators of inflammation, as well as being critical for postsynaptic signaling processes [259]. Eicosanoids include the proinflammatory mediators prostaglandins, thromboxanes, and leukotrienes, whereas docosanoids are involved in the resolution of inflammation, and hence encompass the resolvin family of lipid mediators [260]. DHA deficiency is believed to produce a proinflammatory milieu, resulting from increased AA metabolism to eicosanoids [261,262]. Eicosanoids and docosanoids are particularly short-lived molecules whose levels in the brain were shown to change immediately upon death and dissection, and indeed during storage at −80°C [263]. Thus, reliably quantifying these lipid mediators in postmortem brain samples is very difficult.

Epidemiological research has suggested that high plasma DHA levels, associated with consumption of fatty fish or other sources such as walnuts, reduces the risk of developing AD [264–268] or all-cause dementia [268–270]. It is also well established that DHA consumption reduces risk for cardiovascular disease [271,272] and a range of inflammatory and autoimmune conditions [260]. However, not all studies support the association between high intake of DHA and other polyunsaturated fatty acids, and protection against AD [273,274]. It is also important to note that many of the positive associations are related to consumption of fatty fish rather than polyunsaturated fats or DHA specifically [275]. A number of clinical trials have investigated dietary DHA supplementation as a treatment for people with MCI, early AD, or even early memory impairment that may progress to AD. These clinical trials have so far failed to produce positive outcomes [276–278], even using doses that saturate plasma DHA levels [277,278].

Despite the strong interest in DHA and its importance for brain physiology, there is little evidence for a specific DHA deficiency in AD brains [275]. Several studies have reported reduced DHA content in the hippocampus, CSF, and other affected brain regions of AD patients [160,168,279,280], however, this is not unique to DHA, as significant loss of AA, [94,280,281] has also been demonstrated. One study reported an increase in adrenic acid (22:4) as the only significant change to fatty acid composition in AD brains [282], while another showed loss of adrenic acid [280]. Other studies have demonstrated that reduced DHA content is most likely due to a generalized deficiency of PE plasmalogens [85,187,280]. Some of these differences may be attributed to differences in specific brain regions studied, although all studies include regions affected by AD. A number of groups have proposed that loss of both plasmalogens and polyunsaturated fatty acids in AD is attributed to oxidative stress [187,280]. Like the plasmalogen vinyl ether bond, the sites of unsaturation in polyunsaturated fatty acids are particularly prone to oxidative modification.

## AD is associated with cerebrovascular disease

AD commonly occurs with cerebrovascular disease, and a body of evidence supports the possibility that vascular disease is a significant driving influence for AD. This is particularly true of lipid biology links, since *APOE4* is not only the major risk determinant for AD, but also a major risk gene for cardiovascular disease [283]. Atherosclerosis, and the combination of hypercholesterolemia and hypertension, are strong risk factors for AD [284,285]. Further increased risk of AD has been observed among atherosclerotic ε4 allele carriers when compared with individuals with ε3/ε3 genotype [285]. Circulating cholesterol and d18:1/16:0 ceramide are lower in carriers of the ε2 allele, and higher in ε4 carriers, compared to the most common ε3/ε3 genotype [286]. High serum cholesterol in midlife has been identified as an independent risk factor for AD [284,287]; hence research has explored the possibility of lowering circulating cholesterol to decrease risk of AD. Cholesterol-lowering drugs have been associated with reduced risk of AD in cross-sectional epidemiological studies [288,289]; however, subsequent randomized control studies failed to show a link between circulating cholesterol and AD risk, and no improvement in cognition with statin administration [290–293].

Prospective studies demonstrating that dietary factors such as consumption of fatty fish and polyunsaturated fats protect against, while saturated fats or cholesterol increase risk for AD and dementia, support the vascular link; as do studies demonstrating that higher serum ceramides predict increased risk of conversion from normal ageing to cognitive impairment. It is evident that impaired lipid metabolism is a key driving factor for both cardiometabolic disease and AD, and these disease states have common genetic and dietary bases. However, vascular dementia is a separate entity to AD, with a distinct etiology in vascular disease derived from cerebral infarcts and occlusions that can be present anywhere in the brain vasculature [15,294]. In AD pathogenesis, particular brain regions and cognitive faculties are affected in a reasonably well-defined pattern [20], suggesting that AD cannot be attributed to generalized cerebrovascular disease. The importance of altered lipid metabolism in the pathogenesis of neurodegenerative diseases other than AD also suggests that the significance of *APOE* variants lies in their effects on brain lipid metabolism, rather than vascular biology. This is further supported by genetic risk associated with *TREM2*, *PLD3*, *APOJ*, and *ABCA7* variants which, unlike *APOE* variants, have no clear association with cardiovascular disease. Nonetheless, this does not preclude cardiovascular disease, including effects of *APOE* genotype on cardiovascular health, as a key sensitizing influence for AD.

## The relationship between lysosomal storage diseases and dementias, including AD

The endosomal–lysosomal system is essential for the internalization, transport, recycling, and degradation of cellular macromolecules including proteins and lipids [295]. Enlarged early endosomes and autolysosomes are observed in neurons of AD brains, and this pathological feature is seen very early in the pathogenesis of the disease [296–298]. In first describing the disease in 1907, Alzheimer himself reported fatty deposits in glia that we now interpret as undegraded lipid storage material [299]. In recent years the biochemical and genetic evidence pointing to dysfunction of the endosomal–lysosomal network as a primary driver in the etiology of AD has grown considerably stronger [300–302].

Firstly, SNPs that modify disease risk have been identified in or near the *BIN1*, *SORL1*, *PICALM*, and *CD2AP* genes [48,61,64,65,303], which are all associated with endocytosis and vesicle sorting. In general, these genes are most highly expressed in microglia but they are present to varying degrees in all CNS cell types [94,101]. The *APOE4* variant produces an increase in the number and size of endosomes [296,301], while the β-C-terminal fragment of APP generated by β-secretase (β-CTF) causes endosomal enlargement and disrupts cellular protein traffic [304]. Gene variants in the *PLD3* gene that confer increased risk for AD may also affect endosomal trafficking and protein sorting [77].

Second, lipidomic analyses have provided some evidence supporting endosomal dysfunction in AD. Chan et al. reported accumulation of the late endosomal lipid marker bis(monoacylglycerol)phosphate (also known as lysobisphosphatidic acid) in AD entorhinal cortex [254], while Morel et al. reported selective accumulation of the endosomal sorting lipid phosphatidylinositol-3′-phosphate [165]. Accumulation of cholesterol esters in AD entorhinal cortex [254] may also be an indication of defective lipid catabolism in endosomes and lysosomes. Interestingly, inhibition or ablation of the enzyme that catalyses the transfer of fatty acyl chains to cholesterol, acyl-coenzyme A:cholesterol acyltransferase 1 (ACAT1), boosts autophagosome formation and lysosomal degradation of Aβ in microglia [305], as well as phospho-tau degradation in neurons [306].

Thirdly, LSDs characterized by perturbed lipid metabolism present with neuropathological features that are characteristic of AD and other dementias. NFTs containing hyperphosphorylated tau are observed in Niemann-Pick type C (NPC) disease and are essentially the same as those seen in AD brains [307,308]. NPC disease is caused by mutations in either the *NPC1* or *NPC2* genes, which encode proteins that act in concert to promote efflux of cholesterol, derived from lipoproteins, out of late endosomes and lysosomes [309]. Accumulation of the APP β-CTF and intraneuronal Aβ42 has been reported in some NPC cases [310]. Aβ deposits have been observed in NPC patients homozygous for the *APOE4* allele [308]. Intraneuronal Aβ has also been observed in the brains of Sandhoff disease (β-hexosaminidase B deficiency), GM1 gangliosidosis (β-galactosidase deficiency), and Tay–Sachs disease (β-hexosaminidase deficiency) patients [311], and Aβ peptide levels are higher in mouse models of Sandhoff disease, GM1 gangliosidosis, and NPC disease [311,312]. Thus, defective lipid catabolism in lysosomes is sufficient to produce the defective APP and tau proteostasis that is seen in AD. This has been verified experimentally with the demonstration that loading cells with complex sphingolipids or employing different LSD cell models promotes accumulation of APP-CTFs and enhances γ-secretase activity, leading to enhanced propensity to form Aβ [313]. Similarly, inhibition of lysosomal proteases produces cholesterol storage and amyloidogenic processing of APP [314].

## The example of GBA mutations in Parkinson's disease

The direct link between LSDs affecting lipid metabolism and neurodegenerative diseases of older age has been strengthened in recent years with the discovery that heterozygous mutations in the *GBA* gene, encoding the enzyme glucocerebrosidase (Fig. 23.5), are the most common genetic risk factor for PD and lower the average age at disease onset [11–13]. The connection was first made through the observation that patients with the non-neuronopathic form of Gaucher disease develop PD with ageing [315,316]. Homozygous loss of function mutations in *GBA* cause Gaucher disease. Like *APOE4* and other genetic risk factors for AD, heterozygous loss of function *GBA* mutations increase the risk for PD and DLB [14], but are distinct from higher penetrance inherited mutations that cause the rare familial forms of PD [317]. PD patients with heterozygous *GBA* mutations also have a more severe disease compared with noncarriers, mirroring the effect of *APOE4* on Aβ deposition in AD [318,319]. Heterozygous mutations in the *GRN* gene, encoding the secreted protein progranulin, are one of the most common genetic mutations causing another of the age-dependent neurodegenerative diseases, FTLD [320,321]. It was recently reported that homozygous *GRN* mutations cause the LSD neuronal ceroid lipofuscinosis [322], further reinforcing the link between LSDs and neurodegenerative diseases of ageing.

The case of *GBA* mutations in PD is a paradigmatic example of the relationship between lysosomal enzyme deficiencies and age-dependent neurodegeneration. Glucocerebrosidase catalyses the formation of ceramide from glucosylceramide, through removal of a glucose. This is an important enzymatic step in the catabolism of gangliosides in lysosomes. Glucocerebrosidase activity is reduced not only in carriers of heterozygous *GBA* mutations that cause PD, but also in sporadic PD patients without *GBA* mutations, indicating that reduced GBA activity is a general feature of PD [323–325]. Gaucher disease is characterized by marked accumulation of the substrates for GBA activity, glucosylceramide and glucosylsphingosine. Although there is some evidence for glucosylsphingosine accumulation in PD [325–327], other reports have indicated that there is no significant lipid storage in PD with heterozygous *GBA* mutations [315,328]. Thus, PD is robustly associated with GBA enzyme deficiency but not necessarily with age-dependent lipid accumulation.

Enhanced accumulation of α-synuclein deposits in PD and DLB patients with *GBA* mutations [329] suggests that this enzymatic deficit interferes sufficiently with lysosomal function to impede normal α-synuclein clearance. Several reports indicate that loss of GBA in mouse and cell culture models is sufficient to impede lysosomal protein degradation and drive α-synuclein accumulation [329,330]. In fact, glucosylceramide was reported to stabilize α-synuclein oligomers in vitro [329]. However, α-synuclein accumulation is not limited to GBA deficiency and Gaucher disease models. Accumulation of α-synuclein has been observed in brains of people with the LSDs Sandhoff disease, metachromatic leukodystrophy (arylsulfatase deficiency), and Tay–Sachs, as well as adrenoleukodystrophy, a disease characterized by a deficiency in peroxisomal degradation of very long chain fatty acids [331]. Sandhoff disease model mice also show accumulation of α-synuclein [311,332].

## How dysfunction of the endosomal and lysosomal systems leads to neurodegeneration

In age-dependent neurodegenerative conditions, there may be a protracted decline in endosomal–lysosomal function over many years that is exacerbated by gene variants adversely affecting lysosomal function. In this regard, GBA activity was reported to decrease with increasing age in the mouse brain [333], and in the substantia nigra and putamen of neurologically normal humans over the age of 50 [325]. There are a number of possible explanations for how neurodegeneration may ensue as a consequence of endosomal and lysosomal dysfunction. A simple blockade in endocytosis and recycling of macromolecules and organelles through autophagosomes and lysosomes could cause failure of neurotransmission and build-up of damaged mitochondria, leading to oxidative stress. However, Gegg et al. proposed that general lysosome function is not compromised despite reduced GBA activity in PD brains, since cathepsin D levels and activity of the enzyme β-hexosaminidase were unchanged [324]. Alternatively, cellular functions and viability may be disrupted by accumulation of undegraded lipids and proteins that eventually leak out of the lysosomes or cause lysosomal permeabilization (e.g., the lysolipid sphingosine will permeabilise the lysosomal membrane). Undegraded protein substrates include aggregation-prone species such as Aβ and hyperphosphorylated tau in the case of AD, and α-synuclein in PD and DLB, which are known to be neurotoxic. Undegraded lipid substrates such as glucosylsphingosine, which accumulates in GD, are also cytotoxic [326]. An important question in this regard is whether the high penetrance inherited mutations that cause dementias, such as APP and presenilin-1 mutations in AD, cause the disease through endosomal dysfunction or through the production of cytotoxic molecules such as Aβ. Mazzulli et al. showed that GBA deficiency selectively affects α-synuclein, not tau or huntingtin, and presented evidence indicating that neurotoxicity results from GBA-dependent α-synuclein aggregation [329].

Dysfunctional lipid catabolism could also explain the severe myelin depletion and white matter hyperintensities seen in AD, similar to inherited leukodystrophies. As mentioned above, myelin content declines with normal ageing, particularly in the frontal and temporal lobes that are heavily affected in AD [207]. The high lipid content of the brain and the need to turnover myelin as synapses and networks develop and remodel make it relatively unsurprising that the consequences of defects in lipid catabolic enzymes are most severe in the brain. A subtle defect in lysosomal lipid catabolism may therefore manifest as a significant loss of myelin integrity, as myelin turnover increases with normal ageing.

## Conclusions

The genetics of AD and other age-related neurodegenerative diseases, involvement of lipids in endocytosis and proteolytic processing of APP, and early deterioration of lipid-rich myelin all point to a central role for impaired lipid metabolism and function as a driver of AD pathogenesis. All major brain cell types appear to be involved in different aspects of the neurodegenerative process. ApoE and ApoJ are synthesized primarily by astrocytes, regulate lipid metabolism in neurons and oligodendrocytes, and deliver lipids to the brain's phagocytic cells, microglia, for catabolism. This pathway is also important for clearance of toxic Aβ. *APP* is most highly expressed at the RNA level in neurons and oligodendrocytes, whereas other major risk genes such as *TREM2* and *PLCG2* are highest in microglia [94,101]. Current evidence suggests that impaired catabolism of neuronal and oligodendrocyte lipids in microglia adversely affects neuronal health in the ageing brain, leading to AD. For reasons currently unknown, neurons of the entorhinal cortex and hippocampus are particularly affected in AD.

At the level of cellular lipid metabolism, there is convincing evidence for loss of myelin lipids and peroxisomal function, enhanced generation of ceramides and DAGs from catabolism of sphingomyelin and $PIP_2$, and loss of the protective signaling molecule S1P in AD. The evidence indicates that these are probably early events in AD pathogenesis, and the fact that altered lipid metabolism can precipitate aberrant accumulation of protein aggregates such as Aβ and NFTs points to altered lipid metabolism as a driver of the disease.

Lipid metabolic enzymes and receptors provide excellent targets for therapeutic or dietary intervention. Clinical trials aimed at increased intake of beneficial lipids and reducing cholesterol have not yet yielded improvements in cognitive outcomes, possibly because one must target specific aspects of lipid metabolism in the brain to improve outcomes for AD. Future studies should more intensively investigate the intersection between lipid metabolism, endosomal–lysosomal dysfunction, and genetic risk factors to provide novel insight into the pathogenesis of AD and other neurodegenerative diseases and bring the opportunity for early intervention a step closer.

# References


[1] O'Brien JS, Sampson EL. Lipid composition of the normal human brain: gray matter, white matter, and myelin. J Lipid Res 1965;6(4):537–44.

[2] Schmitt S, Castelvetri LC, Simons M. Metabolism and functions of lipids in myelin. Biochim Biophys Acta 2015;1851(8):999–1005.

[3] Snaidero N, Simons M. Myelination at a glance. J Cell Sci 2014;127(Pt. 14):2999–3004.

[4] Bradl M, Lassmann H. Oligodendrocytes: biology and pathology. Acta Neuropathol 2010;119(1):37–53.

[5] Platt FM, d'Azzo A, Davidson BL, Neufeld EF, Tifft CJ. Lysosomal storage diseases. Nat Rev Dis Primers 2018;4(1):27.

[6] Fuller M, Futerman AH. The brain lipidome in neurodegenerative lysosomal storage disorders. Biochem Biophys Res Commun 2018;504(3):623–38.

[7] Moser HW, Mahmood A, Raymond GV. X-linked adrenoleukodystrophy. Nat Clin Pract Neurol 2007;3(3):140–51.

[8] Pant DC, Dorboz I, Schluter A, Fourcade S, Launay N, Joya J, et al. Loss of the sphingolipid desaturase DEGS1 causes hypomyelinating leukodystrophy. J Clin Invest 2019;129(3):1240–56.

[9] Karsai G, Kraft F, Haag N, Korenke GC, Hanisch B, Othman A, et al. DEGS1-associated aberrant sphingolipid metabolism impairs nervous system function in humans. J Clin Invest 2019;129(3):1229–39.

[10] Corder EH, Saunders AM, Strittmatter WJ, Schmechel DE, Gaskell PC, Small GW, et al. Gene dose of apolipoprotein E type 4 allele and the risk of Alzheimer's disease in late onset families. Science 1993;261(5123):921–3.

[11] Clark LN, Ross BM, Wang Y, Mejia-Santana H, Harris J, Louis ED, et al. Mutations in the glucocerebrosidase gene are associated with early-onset Parkinson disease. Neurology 2007;69(12):1270–7.

[12] Neumann J, Bras J, Deas E, O'Sullivan SS, Parkkinen L, Lachmann RH, et al. Glucocerebrosidase mutations in clinical and pathologically proven Parkinson's disease. Brain 2009;132(Pt. 7):1783–94.

[13] Sidransky E, Nalls MA, Aasly JO, Aharon-Peretz J, Annesi G, Barbosa ER, et al. Multicenter analysis of glucocerebrosidase mutations in Parkinson's disease. N Engl J Med 2009;361(17):1651–61.

[14] Goker-Alpan O, Giasson BI, Eblan MJ, Nguyen J, Hurtig HI, Lee VM, et al. Glucocerebrosidase mutations are an important risk factor for Lewy body disorders. Neurology 2006;67(5):908–10.

[15] Alzheimer's Association Report. 2018 Alzheimer's disease facts and figures. Alzheimer's Dement 2018;14:367–429.

[16] Jellinger KA, Attems J. Challenges of multimorbidity of the aging brain: a critical update. J Neural Transm 2015;122(4):505–21.

[17] Holtzman DM, Morris JC, Goate AM. Alzheimer's disease: the challenge of the second century. Sci Transl Med 2011;3(77):77sr1.

[18] Pini L, Pievani M, Bocchetta M, Altomare D, Bosco P, Cavedo E, et al. Brain atrophy in Alzheimer's disease and aging. Ageing Res Rev 2016;30:25–48.

[19] Braak H, Braak E. Staging of Alzheimer's disease-related neurofibrillary changes. Neurobiol Aging 1995;16(3):271–8. [discussion 8–84].

[20] Hyman BT, Phelps CH, Beach TG, Bigio EH, Cairns NJ, Carrillo MC, et al. National Institute on Aging-Alzheimer's Association guidelines for the neuropathologic assessment of Alzheimer's disease. Alzheimers Dement. 2012;8(1):1–13.

[21] van der Kant R, Goldstein LS. Cellular functions of the amyloid precursor protein from development to dementia. Dev Cell 2015;32(4):502–15.

[22] Wang Y, Mandelkow E. Tau in physiology and pathology. Nat Rev Neurosci 2016;17(1):5–21.

[23] Ittner A, Chua SW, Bertz J, Volkerling A, van der Hoven J, Gladbach A, et al. Site-specific phosphorylation of tau inhibits amyloid-beta toxicity in Alzheimer's mice. Science 2016;354(6314):904–8.

[24] Villemagne VL, Dore V, Burnham SC, Masters CL, Rowe CC. Imaging tau and amyloid-beta proteinopathies in Alzheimer disease and other conditions. Nat Rev Neurol 2018;14(4):225–36.

[25] Scheff SW, Price DA, Schmitt FA, Mufson EJ. Hippocampal synaptic loss in early Alzheimer's disease and mild cognitive impairment. Neurobiol Aging 2006;27(10):1372–84.

[26] DeKosky ST, Scheff SW. Synapse loss in frontal cortex biopsies in Alzheimer's disease: correlation with cognitive severity. Ann Neurol 1990;27(5):457–64.

[27] Ferreira-Vieira TH, Guimaraes IM, Silva FR, Ribeiro FM. Alzheimer's disease: targeting the cholinergic system. Curr Neuropharmacol 2016;14(1):101–15.

[28] Heneka MT, Carson MJ, El Khoury J, Landreth GE, Brosseron F, Feinstein DL, et al. Neuroinflammation in Alzheimer's disease. Lancet Neurol 2015;14(4):388–405.
[29] Mosconi L, Pupi A, De Leon MJ. Brain glucose hypometabolism and oxidative stress in preclinical Alzheimer's disease. Ann N Y Acad Sci 2008;1147:180–95.
[30] Braak H, Braak E. Development of Alzheimer-related neurofibrillary changes in the neocortex inversely recapitulates cortical myelogenesis. Acta Neuropathol 1996;92(2):197–201.
[31] Nasrabady SE, Rizvi B, Goldman JE, Brickman AM. White matter changes in Alzheimer's disease: a focus on myelin and oligodendrocytes. Acta Neuropathol Commu 2018;6(1):22.
[32] Rajendran L, Honsho M, Zahn TR, Keller P, Geiger KD, Verkade P, et al. Alzheimer's disease beta-amyloid peptides are released in association with exosomes. Proc Natl Acad Sci USA 2006;103(30):11172–7.
[33] De Strooper B, Vassar R, Golde T. The secretases: enzymes with therapeutic potential in Alzheimer disease. Nat Rev Neurol 2010;6(2):99–107.
[34] Panza F, Lozupone M, Logroscino G, Imbimbo BP. A critical appraisal of amyloid-beta-targeting therapies for Alzheimer disease. Nat Rev Neurol 2019;15(2):73–88.
[35] Duyckaerts C, Braak H, Brion JP, Buee L, Del Tredici K, Goedert M, et al. PART is part of Alzheimer disease. Acta Neuropathol 2015;129(5):749–56.
[36] Berg L, McKeel DW Jr, Miller JP, Baty J, Morris JC. Neuropathological indexes of Alzheimer's disease in demented and nondemented persons aged 80 years and older. Arch Neurol 1993;50(4):349–58.
[37] McKee AC, Stein TD, Kiernan PT, Alvarez VE. The neuropathology of chronic traumatic encephalopathy. Brain Pathol 2015;25(3):350–64.
[38] Johnson VE, Stewart W, Smith DH. Widespread τ and amyloid-β pathology many years after a single traumatic brain injury in humans. Brain Pathol 2012;22(2):142–9.
[39] Revesz T, Holton JL, Lashley T, Plant G, Frangione B, Rostagno A, et al. Genetics and molecular pathogenesis of sporadic and hereditary cerebral amyloid angiopathies. Acta Neuropathol 2009;118(1):115–30.
[40] Graham WV, Bonito-Oliva A, Sakmar TP. Update on Alzheimer's disease therapy and prevention strategies. Ann Rev Med 2017;68:413–30.
[41] Jansen WJ, Ossenkoppele R, Knol DL, Tijms BM, Scheltens P, Verhey FR, et al. Prevalence of cerebral amyloid pathology in persons without dementia: a meta-analysis. JAMA 2015;313(19):1924–38.
[42] Couttas TA, Kain N, Daniels B, Lim XY, Shepherd C, Kril J, et al. Loss of the neuroprotective factor Sphingosine 1-phosphate early in Alzheimer's disease pathogenesis. Acta Neuropathol Commun 2014;2(1):9.
[43] Couttas TA, Kain N, Suchowerska AK, Quek LE, Turner N, Fath T, et al. Loss of ceramide synthase 2 activity, necessary for myelin biosynthesis, precedes tau pathology in the cortical pathogenesis of Alzheimer's disease. Neurobiol Aging 2016;43:89–100.
[44] Han X DMH, McKeel DW Jr, Kelley J, Morris JC. Substantial sulfatide deficiency and ceramide elevation in very early Alzheimer's disease: potential role in disease pathogenesis. J Neurochem 2002;82(4):809–18.
[45] Han X, Holtzman DM, McKeel DW Jr. Plasmalogen deficiency in early Alzheimer's disease subjects and in animal models: molecular characterization using electrospray ionization mass spectrometry. J Neurochem 2001;77(4):1168–80.
[46] He X, Huang Y, Li B, Gong CX, Schuchman EH. Deregulation of sphingolipid metabolism in Alzheimer's disease. Neurobiol Aging 2010;31(3):398–408.
[47] Bekris LM, Yu CE, Bird TD, Tsuang DW. Genetics of Alzheimer disease. J Geriatr Psychiatry Neurol 2010;23(4):213–27.
[48] Karch CM, Goate AM. Alzheimer's disease risk genes and mechanisms of disease pathogenesis. Biol Psychiatry 2015;77(1):43–51.
[49] Kelleher RJ, Shen J. Presenilin-1 mutations and Alzheimer's disease. Proc Natl Acad Sci 2017;114(4):629.
[50] Alzheimer's Association Report. 2019 Alzheimer's disease facts and figures. Alzheimer's Dement 2019;15:321–87.
[51] Kunkle BW, Grenier-Boley B, Sims R, Bis JC, Damotte V, Naj AC, et al. Genetic meta-analysis of diagnosed Alzheimer's disease identifies new risk loci and implicates Abeta, tau, immunity and lipid processing. Nat Genet 2019;51(3):414–30.
[52] Jonsson T, Atwal JK, Steinberg S, Snaedal J, Jonsson PV, Bjornsson S, et al. A mutation in APP protects against Alzheimer's disease and age-related cognitive decline. Nature 2012;488(7409):96–9.
[53] Liu CC, Kanekiyo T, Xu H, Bu G. Apolipoprotein E and Alzheimer disease: risk, mechanisms and therapy. Nat Rev Neurol 2013;9(2):106–18.

[54] Corder EH, Saunders AM, Risch NJ, Strittmatter WJ, Schmechel DE, Gaskell PC Jr, et al. Protective effect of apolipoprotein E type 2 allele for late onset Alzheimer disease. Nat Genet 1994;7(2):180–4.

[55] Dickson DW, Heckman MG, Murray ME, Soto AI, Walton RL, Diehl NN, et al. APOE epsilon4 is associated with severity of Lewy body pathology independent of Alzheimer pathology. Neurology 2018;91(12):e1182–95.

[56] Tsuang D, Leverenz JB, Lopez OL, Hamilton RL, Bennett DA, Schneider JA, et al. APOE epsilon4 increases risk for dementia in pure synucleinopathies. JAMA Neurol 2013;70(2):223–8.

[57] Berge G, Sando SB, Rongve A, Aarsland D, White LR. Apolipoprotein E epsilon2 genotype delays onset of dementia with Lewy bodies in a Norwegian cohort. J Neurol Neurosurg Psychiatry 2014;85(11):1227–31.

[58] Singleton AB, Wharton A, O'Brien KK, Walker MP, McKeith IG, Ballard CG, et al. Clinical and neuropathological correlates of apolipoprotein E genotype in dementia with Lewy bodies. Dement Geriatr Cogn Disord 2002;14(4):167–75.

[59] Grimm MOW, Michaelson DM, Hartmann T. Omega-3 fatty acids, lipids, and apoE lipidation in Alzheimer's disease: a rationale for multi-nutrient dementia prevention. J Lipid Res 2017;58(11):2083–101.

[60] Rebeck GW. The role of APOE on lipid homeostasis and inflammation in normal brains. J Lipid Res 2017;58(8):1493–9.

[61] Harold D, Abraham R, Hollingworth P, Sims R, Gerrish A, Hamshere ML, et al. Genome-wide association study identifies variants at CLU and PICALM associated with Alzheimer's disease. Nat Genet 2009;41(10):1088–93.

[62] Lambert JC, Heath S, Even G, Campion D, Sleegers K, Hiltunen M, et al. Genome-wide association study identifies variants at CLU and CR1 associated with Alzheimer's disease. Nat Genet 2009;41(10):1094–9.

[63] Foster EM, Dangla-Valls A, Lovestone S, Ribe EM, Buckley NJ. Clusterin in Alzheimer's disease: mechanisms, genetics, and lessons from other pathologies. Front Neurosci 2019;13:164.

[64] Lambert JC, Ibrahim-Verbaas CA, Harold D, Naj AC, Sims R, Bellenguez C, et al. Meta-analysis of 74,046 individuals identifies 11 new susceptibility loci for Alzheimer's disease. Nat Genet 2013;45(12):1452–8.

[65] Hollingworth P, Harold D, Sims R, Gerrish A, Lambert JC, Carrasquillo MM, et al. Common variants at ABCA7, MS4A6A/MS4A4E, EPHA1. CD33 and CD2AP are associated with Alzheimer's disease. Nature Genet 2011;43(5):429–35.

[66] Kim WS, Guillemin GJ, Glaros EN, Lim CK, Garner B. Quantitation of ATP-binding cassette subfamily-A transporter gene expression in primary human brain cells. Neuroreport 2006;17(9):891–6.

[67] Abe-Dohmae S, Ikeda Y, Matsuo M, Hayashi M, Okuhira K, Ueda K, et al. Human ABCA7 supports apolipoprotein-mediated release of cellular cholesterol and phospholipid to generate high density lipoprotein. J Biol Chem 2004;279(1):604–11.

[68] Cruchaga C, Karch CM, Jin SC, Benitez BA, Cai Y, Guerreiro R, et al. Rare coding variants in the phospholipase D3 gene confer risk for Alzheimer's disease. Nature 2014;505(7484):550–4.

[69] Tan MS, Zhu JX, Cao XP, Yu JT, Tan L. Rare variants in PLD3 increase risk for Alzheimer's disease in Han Chinese. J Alzheimers Dis 2018;64(1):55–9.

[70] van der Lee SJ, Holstege H, Wong TH, Jakobsdottir J, Bis JC, Chouraki V, et al. PLD3 variants in population studies. Nature 2015;520(7545):E2–3.

[71] Schulte EC, Kurz A, Alexopoulos P, Hampel H, Peters A, Gieger C, et al. Excess of rare coding variants in PLD3 in late- but not early-onset Alzheimer's disease. Hum Genome Var 2015;2:14028.

[72] Hooli BV, Lill CM, Mullin K, Qiao D, Lange C, Bertram L, et al. PLD3 gene variants and Alzheimer's disease. Nature 2015;520(7545):E7–8.

[73] Lambert JC, Grenier-Boley B, Bellenguez C, Pasquier F, Campion D, Dartigues JF, et al. PLD3 and sporadic Alzheimer's disease risk. Nature 2015;520(7545):E1.

[74] Heilmann S, Drichel D, Clarimon J, Fernandez V, Lacour A, Wagner H, et al. PLD3 in non-familial Alzheimer's disease. Nature 2015;520(7545):E3–5.

[75] Engelman CD, Darst BF, Bilgel M, Vasiljevic E, Koscik RL, Jedynak BM, et al. The effect of rare variants in TREM2 and PLD3 on longitudinal cognitive function in the Wisconsin Registry for Alzheimer's Prevention. Neurobiol Aging 2018;66. 177 e1-e5.

[76] Lin E, Tsai SJ, Kuo PH, Liu YL, Yang AC, Kao CF. Association and interaction effects of Alzheimer's disease-associated genes and lifestyle on cognitive aging in older adults in a Taiwanese population. Oncotarget 2017;8(15):24077–87.

[77] Mukadam AS, Breusegem SY, Seaman MNJ. Analysis of novel endosome-to-Golgi retrieval genes reveals a role for PLD3 in regulating endosomal protein sorting and amyloid precursor protein processing. Cell Mol Life Sci 2018;75(14):2613–25.

[78] Tan M, Li J, Ma F, Zhang X, Zhao Q, Cao X. PLD3 rare variants identified in late-onset Alzheimer's Disease affect amyloid-beta levels in cellular model. Front Neurosci 2019;13:116.

[79] Gavin AL, Huang D, Huber C, Martensson A, Tardif V, Skog PD, et al. PLD3 and PLD4 are single-stranded acid exonucleases that regulate endosomal nucleic-acid sensing. Nat Immunol 2018;19(9):942–53.

[80] Nibbeling EAR, Duarri A, Verschuuren-Bemelmans CC, Fokkens MR, Karjalainen JM, Smeets C, et al. Exome sequencing and network analysis identifies shared mechanisms underlying spinocerebellar ataxia. Brain 2017;140(11):2860–78.

[81] Ulland TK, Colonna M. TREM2 - a key player in microglial biology and Alzheimer disease. Nat Rev Neurol 2018;14(11):667–75.

[82] Jonsson T, Stefansson H, Steinberg S, Jonsdottir I, Jonsson PV, Snaedal J, et al. Variant of TREM2 associated with the risk of Alzheimer's disease. N Engl J Med 2013;368(2):107–16.

[83] Guerreiro R, Wojtas A, Bras J, Carrasquillo M, Rogaeva E, Majounie E, et al. TREM2 variants in Alzheimer's disease. N Engl J Med 2013;368(2):117–27.

[84] Sims R, van der Lee SJ, Naj AC, Bellenguez C, Badarinarayan N, Jakobsdottir J, et al. Rare coding variants in PLCG2, ABI3, and TREM2 implicate microglial-mediated innate immunity in Alzheimer's disease. Nat Genet 2017;49(9):1373–84.

[85] Borroni B, Ferrari F, Galimberti D, Nacmias B, Barone C, Bagnoli S, et al. Heterozygous TREM2 mutations in frontotemporal dementia. Neurobiol Aging 2014;35(4):934.e7–934.e10.

[86] Paloneva J, Manninen T, Christman G, Hovanes K, Mandelin J, Adolfsson R, et al. Mutations in two genes encoding different subunits of a receptor signaling complex result in an identical disease phenotype. Am J Hum Genet 2002;71(3):656–62.

[87] Poliani PL, Wang Y, Fontana E, Robinette ML, Yamanishi Y, Gilfillan S, et al. TREM2 sustains microglial expansion during aging and response to demyelination. J Clin Invest 2015;125(5):2161–70.

[88] Wang Y, Cella M, Mallinson K, Ulrich JD, Young KL, Robinette ML, et al. TREM2 lipid sensing sustains the microglial response in an Alzheimer's disease model. Cell 2015;160(6):1061–71.

[89] Yeh FL, Wang Y, Tom I, Gonzalez LC, Sheng M. TREM2 binds to apolipoproteins, including APOE and CLU/APOJ, and thereby facilitates uptake of amyloid-beta by microglia. Neuron 2016;91(2):328–40.

[90] Atagi Y, Liu CC, Painter MM, Chen XF, Verbeeck C, Zheng H, et al. Apolipoprotein E is a ligand for triggering receptor expressed on myeloid cells 2 (TREM2). J Biol Chem 2015;290(43):26043–50.

[91] Conway OJ, Carrasquillo MM, Wang X, Bredenberg JM, Reddy JS, Strickland SL, et al. ABI3 and PLCG2 missense variants as risk factors for neurodegenerative diseases in Caucasians and African Americans. Mol Neurodegener 2018;13(1):53.

[92] van der Lee SJ, Conway OJ, Jansen I, Carrasquillo MM, Kleineidam L, van den Akker E, et al. A nonsynonymous mutation in PLCG2 reduces the risk of Alzheimer's disease, dementia with Lewy bodies and frontotemporal dementia, and increases the likelihood of longevity. Acta Neuropathol 2019;138(2):237–50.

[93] Rhee SG. Regulation of phosphoinositide-specific phospholipase C. Annu Rev Biochem 2001;70:281–312.

[94] Zhang Y, Chen K, Sloan SA, Bennett ML, Scholze AR, O'Keeffe S, et al. An RNA-sequencing transcriptome and splicing database of glia, neurons, and vascular cells of the cerebral cortex. J Neurosci 2014;34(36):11929–47.

[95] Magno L, Lessard CB, Martins M, Lang V, Cruz P, Asi Y, et al. Alzheimer's disease phospholipase C-gamma-2 (PLCG2) protective variant is a functional hypermorph. Alzheimers Res Ther 2019;11(1):16.

[96] Pitas RE, Boyles JK, Lee SH, Hui D, Weisgraber KH. Lipoproteins and their receptors in the central nervous system. Characterization of the lipoproteins in cerebrospinal fluid and identification of apolipoprotein B,E(LDL) receptors in the brain. J Biol Chem 1987;262(29):14352–60.

[97] Rebeck GW, LaDu MJ, Estus S, Bu G, Weeber EJ. The generation and function of soluble apoE receptors in the CNS. Mol Neurodegener 2006;1:15.

[98] LaDu MJ, Gilligan SM, Lukens JR, Cabana VG, Reardon CA, Van Eldik LJ, et al. Nascent astrocyte particles differ from lipoproteins in CSF. J Neurochem 1998;70(5):2070–81.

[99] Elliott DA, Weickert CS, Garner B. Apolipoproteins in the brain: implications for neurological and psychiatric disorders. Clin Lipidol 2010;51(4):555–73.

[100] Koch S, Donarski N, Goetze K, Kreckel M, Stuerenburg HJ, Buhmann C, et al. Characterization of four lipoprotein classes in human cerebrospinal fluid. J Lipid Res 2001;42(7):1143–51.

[101] Zhang Y, Sloan SA, Clarke LE, Caneda C, Plaza CA, Blumenthal PD, et al. Purification and characterization of progenitor and mature human astrocytes reveals transcriptional and functional differences with mouse. Neuron 2016;89(1):37–53.

[102] Xu Q, Bernardo A, Walker D, Kanegawa T, Mahley RW, Huang Y. Profile and regulation of apolipoprotein E (ApoE) expression in the CNS in mice with targeting of green fluorescent protein gene to the ApoE locus. J Neurosci 2006;26(19):4985–94.

[103] Ignatius MJ, Gebicke-Harter PJ, Skene JH, Schilling JW, Weisgraber KH, Mahley RW, et al. Expression of apolipoprotein E during nerve degeneration and regeneration. Proc Natl Acad Sci USA 1986;83(4):1125–9.

[104] Messmer-Joudrier S, Sagot Y, Mattenberger L, James RW, Kato AC. Injury-induced synthesis and release of apolipoprotein E and clusterin from rat neural cells. Eur J Neurosci 1996;8(12):2652–61.

[105] White F, Nicoll JA, Horsburgh K. Alterations in ApoE and ApoJ in relation to degeneration and regeneration in a mouse model of entorhinal cortex lesion. Exp Neurol 2001;169(2):307–18.

[106] Fagan AM, Murphy BA, Patel SN, Kilbridge JF, Mobley WC, Bu G, et al. Evidence for normal aging of the septo-hippocampal cholinergic system in apoE (-/-) mice but impaired clearance of axonal degeneration products following injury. Exp Neurol 1998;151(2):314–25.

[107] Wahrle SE, Jiang H, Parsadanian M, Legleiter J, Han X, Fryer JD, et al. ABCA1 is required for normal central nervous system ApoE levels and for lipidation of astrocyte-secreted apoE. J Biol Chem 2004;279(39):40987–93.

[108] Kim WS, Rahmanto AS, Kamili A, Rye KA, Guillemin GJ, Gelissen IC, et al. Role of ABCG1 and ABCA1 in regulation of neuronal cholesterol efflux to apolipoprotein E discs and suppression of amyloid-beta peptide generation. J Biol Chem 2007;282(5):2851–61.

[109] Karten B, Campenot RB, Vance DE, Vance JE. Expression of ABCG1, but not ABCA1, correlates with cholesterol release by cerebellar astroglia. J Biol Chem 2006;281(7):4049–57.

[110] Minagawa H, Gong JS, Jung CG, Watanabe A, Lund-Katz S, Phillips MC, et al. Mechanism underlying apolipoprotein E (ApoE) isoform-dependent lipid efflux from neural cells in culture. J Neurosci Res 2009;87(11):2498–508.

[111] Krimbou L, Denis M, Haidar B, Carrier M, Marcil M, Genest J Jr. Molecular interactions between apoE and ABCA1: impact on apoE lipidation. J Lipid Res 2004;45(5):839–48.

[112] Hirsch-Reinshagen V, Zhou S, Burgess BL, Bernier L, McIsaac SA, Chan JY, et al. Deficiency of ABCA1 impairs apolipoprotein E metabolism in brain. J Biol Chem 2004;279(39):41197–207.

[113] Nordestgaard LT, Tybjaerg-Hansen A, Nordestgaard BG, Frikke-Schmidt R. Loss-of-function mutation in ABCA1 and risk of Alzheimer's disease and cerebrovascular disease. Alzheimers Dement 2015;11(12):1430–8.

[114] Michikawa M, Fan QW, Isobe I, Yanagisawa K. Apolipoprotein E exhibits isoform-specific promotion of lipid efflux from astrocytes and neurons in culture. J Neurochem 2000;74(3):1008–16.

[115] Hanson AJ, Bayer-Carter JL, Green PS, Montine TJ, Wilkinson CW, Baker LD, et al. Effect of apolipoprotein E genotype and diet on apolipoprotein E lipidation and amyloid peptides: randomized clinical trial. JAMA Neurol 2013;70(8):972–80.

[116] Heinsinger NM, Gachechiladze MA, Rebeck GW. Apolipoprotein E genotype affects size of ApoE complexes in cerebrospinal fluid. J Neuropathol Exp Neurol 2016;75(10):918–24.

[117] Hu J, Liu CC, Chen XF, Zhang YW, Xu H, Bu G. Opposing effects of viral mediated brain expression of apolipoprotein E2 (apoE2) and apoE4 on apoE lipidation and Abeta metabolism in apoE4-targeted replacement mice. Mol Neurodegener 2015;10:6.

[118] Cruchaga C, Kauwe JS, Nowotny P, Bales K, Pickering EH, Mayo K, et al. Cerebrospinal fluid APOE levels: an endophenotype for genetic studies for Alzheimer's disease. Hum Mol Genet 2012;21(20):4558–71.

v119] Yassine HN, Feng Q, Chiang J, Petrosspour LM, Fonteh AN, Chui HC, et al. ABCA1-mediated cholesterol efflux capacity to cerebrospinal fluid is reduced in patients with mild cognitive impairment and Alzheimer's disease. J Am Heart Assoc 2016;5(2):e002886.

[120] Castellano JM, Kim J, Stewart FR, Jiang H, DeMattos RB, Patterson BW, et al. Human apoE isoforms differentially regulate brain amyloid-beta peptide clearance. Sci Transl Med 2011;3(89):89ra57.

[121] Reiman EM, Chen K, Liu X, Bandy D, Yu M, Lee W, et al. Fibrillar amyloid-beta burden in cognitively normal people at 3 levels of genetic risk for Alzheimer's disease. Proc Natl Acad Sci USA 2009;106(16):6820–5.

[122] Shulman JM, Chen K, Keenan BT, Chibnik LB, Fleisher A, Thiyyagura P, et al. Genetic susceptibility for Alzheimer disease neuritic plaque pathology. JAMA Neurol 2013;70(9):1150–7.

[123] Dodart JC, Marr RA, Koistinaho M, Gregersen BM, Malkani S, Verma IM, et al. Gene delivery of human apolipoprotein E alters brain Abeta burden in a mouse model of Alzheimer's disease. Proc Natl Acad Sci USA 2005;102(4):1211–6.

[124] Bales KR, Verina T, Dodel RC, Du Y, Altstiel L, Bender M, et al. Lack of apolipoprotein E dramatically reduces amyloid beta-peptide deposition. Nat Genet 1997;17(3):263–4.

[125] DeMattos RB, Cirrito JR, Parsadanian M, May PC, O'Dell MA, Taylor JW, et al. ApoE and clusterin cooperatively suppress Abeta levels and deposition: evidence that ApoE regulates extracellular Abeta metabolism in vivo. Neuron 2004;41(2):193–202.

[126] Wahrle SE, Jiang H, Parsadanian M, Hartman RE, Bales KR, Paul SM, et al. Deletion of Abca1 increases Abeta deposition in the PDAPP transgenic mouse model of Alzheimer disease. J Biol Chem 2005;280(52):43236–42.

[127] Fitz NF, Cronican AA, Saleem M, Fauq AH, Chapman R, Lefterov I, et al. Abca1 deficiency affects Alzheimer's disease-like phenotype in human ApoE4 but not in ApoE3-targeted replacement mice. J Neurosci 2012;32(38):13125–36.

[128] Kim WS, Li H, Ruberu K, Chan S, Elliott DA, Low JK, et al. Deletion of Abca7 increases cerebral amyloid-beta accumulation in the J20 mouse model of Alzheimer's disease. J Neurosci 2013;33(10):4387–94.

[129] Sakae N, Liu CC, Shinohara M, Frisch-Daiello J, Ma L, Yamazaki Y, et al. ABCA7 deficiency accelerates amyloid-beta generation and Alzheimer's neuronal pathology. J Neurosci 2016;36(13):3848–59.

[130] Satoh K, Abe-Dohmae S, Yokoyama S, St George-Hyslop P, Fraser PE. ATP-binding cassette transporter A7 (ABCA7) loss of function alters Alzheimer amyloid processing. J Biol Chem 2015;290(40):24152–65.

[131] Tachibana M, Holm ML, Liu CC, Shinohara M, Aikawa T, Oue H, et al. APOE4-mediated amyloid-beta pathology depends on its neuronal receptor LRP1. J Clin Invest 2019;129(3):1272–7.

[132] Nathan BP, Bellosta S, Sanan DA, Weisgraber KH, Mahley RW, Pitas RE. Differential effects of apolipoproteins E3 and E4 on neuronal growth in vitro. Science 1994;264(5160):850–2.

[133] Teter B, Xu PT, Gilbert JR, Roses AD, Galasko D, Cole GM. Human apolipoprotein E isoform-specific differences in neuronal sprouting in organotypic hippocampal culture. J Neurochem 1999;73(6):2613–6.

[134] Mauch DH, Nagler K, Schumacher S, Goritz C, Muller EC, Otto A, et al. CNS synaptogenesis promoted by glia-derived cholesterol. Science 2001;294(5545):1354–7.

[135] Sen A, Alkon DL, Nelson TJ. Apolipoprotein E3 (ApoE3) but not ApoE4 protects against synaptic loss through increased expression of protein kinase C epsilon. J Biol Chem 2012;287(19):15947–58.

[136] Drouet B, Fifre A, Pincon-Raymond M, Vandekerckhove J, Rosseneu M, Gueant JL, et al. ApoE protects cortical neurones against neurotoxicity induced by the non-fibrillar C-terminal domain of the amyloid-beta peptide. J Neurochem 2001;76(1):117–27.

[137] Hayashi H, Campenot RB, Vance DE, Vance JE. Apolipoprotein E-containing lipoproteins protect neurons from apoptosis via a signaling pathway involving low-density lipoprotein receptor-related protein-1. J Neurosci 2007;27(8):1933–41.

[138] Liu L, MacKenzie KR, Putluri N, Maletic-Savatic M, Bellen HJ. The glia-neuron lactate shuttle and elevated ROS promote lipid synthesis in neurons and lipid droplet accumulation in glia via APOE/D. Cell Metab 2017;26(5):719–737.e6.

[139] Dik MG, Jonker C, Bouter LM, Geerlings MI, van Kamp GJ, Deeg DJ. APOE-epsilon4 is associated with memory decline in cognitively impaired elderly. Neurology 2000;54(7):1492–7.

[140] Farlow MR, He Y, Tekin S, Xu J, Lane R, Charles HC. Impact of APOE in mild cognitive impairment. Neurology 2004;63(10):1898–901.

[141] Jochemsen HM, Muller M, van der Graaf Y, Geerlings MI. APOE epsilon4 differentially influences change in memory performance depending on age. The SMART-MR study. Neurobiol Aging 2012;33(4):832.e15-22.

[142] Caselli RJ, Dueck AC, Osborne D, Sabbagh MN, Connor DJ, Ahern GL, et al. Longitudinal modeling of age-related memory decline and the APOE epsilon4 effect. N Engl J Med 2009;361(3):255–63.

[143] Caselli RJ, Dueck AC, Locke DE, Hoffman-Snyder CR, Woodruff BK, Rapcsak SZ, et al. Longitudinal modeling of frontal cognition in APOE epsilon4 homozygotes, heterozygotes, and noncarriers. Neurology 2011;76(16):1383–8.

[144] Mondadori CR, de Quervain DJ, Buchmann A, Mustovic H, Wollmer MA, Schmidt CF, et al. Better memory and neural efficiency in young apolipoprotein E epsilon4 carriers. Cereb Cortex 2007;17(8):1934–47.

[145] Reas ET, Laughlin GA, Bergstrom J, Kritz-Silverstein D, Barrett-Connor E, McEvoy LK. Effects of APOE on cognitive aging in community-dwelling older adults. Neuropsychology 2019;33(3):406–16.

[146] Yaffe K, Cauley J, Sands L, Browner W. Apolipoprotein E phenotype and cognitive decline in a prospective study of elderly community women. Arch Neurol 1997;54(9):1110–4.

[147] Salomon-Zimri S, Boehm-Cagan A, Liraz O, Michaelson DM. Hippocampus-related cognitive impairments in young apoE4 targeted replacement mice. Neurodegener Dis 2014;13(2–3):86–92.

[148] Andrews-Zwilling Y, Bien-Ly N, Xu Q, Li G, Bernardo A, Yoon SY, et al. Apolipoprotein E4 causes age- and Tau-dependent impairment of GABAergic interneurons, leading to learning and memory deficits in mice. J Neurosci 2010;30(41):13707–17.

[149] Boehm-Cagan A, Michaelson DM. Reversal of apoE4-driven brain pathology and behavioral deficits by bexarotene. J Neurosci 2014;34(21):7293–301.
[150] Boehm-Cagan A, Bar R, Liraz O, Bielicki JK, Johansson JO, Michaelson DM. ABCA1 agonist reverses the ApoE4-driven cognitive and brain pathologies. J Alzheimers Dis 2016;54(3):1219–33.
[151] Vanmierlo T, Rutten K, Dederen J, Bloks VW, van Vark-van der Zee LC, Kuipers F, et al. Liver X receptor activation restores memory in aged AD mice without reducing amyloid. Neurobiol Aging 2011;32(7):1262–72.
[152] Riddell DR, Zhou H, Comery TA, Kouranova E, Lo CF, Warwick HK, et al. The LXR agonist TO901317 selectively lowers hippocampal Abeta42 and improves memory in the Tg2576 mouse model of Alzheimer's disease. Mol Cell Neurosci 2007;34(4):621–8.
[153] Zhou W, Xu D, Peng X, Zhang Q, Jia J, Crutcher KA. Meta-analysis of APOE4 allele and outcome after traumatic brain injury. J Neurotrauma 2008;25(4):279–90.
[154] Nicoll JA, Roberts GW, Graham DI. Apolipoprotein E epsilon 4 allele is associated with deposition of amyloid beta-protein following head injury. Nat Med 1995;1(2):135–7.
[155] Plassman BL, Havlik RJ, Steffens DC, Helms MJ, Newman TN, Drosdick D, et al. Documented head injury in early adulthood and risk of Alzheimer's disease and other dementias. Neurology 2000;55(8):1158–66.
[156] Houlden H, Greenwood R. Apolipoprotein E4 and traumatic brain injury. J Neurol Neurosurg Psychiatry 2006;77(10):1106–7.
[157] Cao J, Gaamouch FE, Meabon JS, Meeker KD, Zhu L, Zhong MB, et al. ApoE4-associated phospholipid dysregulation contributes to development of Tau hyper-phosphorylation after traumatic brain injury. Sci Rep 2017;7(1):11372.
[158] Couttas TA, Kain N, Tran C, Chatterton Z, Kwok JB, Don AS. Age-dependent changes to sphingolipid balance in the human hippocampus are gender-specific and may sensitize to neurodegeneration. J Alzheimers Dis 2018;63(2):503–14.
[159] Bandaru VV, Troncoso J, Wheeler D, Pletnikova O, Wang J, Conant K, et al. ApoE4 disrupts sterol and sphingolipid metabolism in Alzheimer's but not normal brain. Neurobiol Aging 2009;30(4):591–9.
[160] Igarashi M, Ma K, Gao F, Kim HW, Rapoport SI, Rao JS. Disturbed choline plasmalogen and phospholipid fatty acid concentrations in Alzheimer's disease prefrontal cortex. J Alzheimers Dis 2011;24(3):507–17.
[161] Sharman MJ, Shui G, Fernandis AZ, Lim WL, Berger T, Hone E, et al. Profiling brain and plasma lipids in human APOE epsilon2, epsilon3, and epsilon4 knock-in mice using electrospray ionization mass spectrometry. J Alzheimers Dis 2010;20(1):105–11.
[162] Morishima-Kawashima M, Han X, Tanimura Y, Hamanaka H, Kobayashi M, Sakurai T, et al. Effects of human apolipoprotein E isoforms on the amyloid beta-protein concentration and lipid composition in brain low-density membrane domains. J Neurochem 2007;101(4):949–58.
[163] Zhu L, Zhong M, Elder GA, Sano M, Holtzman DM, Gandy S, et al. Phospholipid dysregulation contributes to ApoE4-associated cognitive deficits in Alzheimer's disease pathogenesis. Proc Natl Acad Sci USA 2015;112(38):11965–70.
[164] McIntire LB, Berman DE, Myaeng J, Staniszewski A, Arancio O, Di Paolo G, et al. Reduction of synaptojanin 1 ameliorates synaptic and behavioral impairments in a mouse model of Alzheimer's disease. J Neurosci 2012;32(44):15271–6.
[165] Morel E, Chamoun Z, Lasiecka ZM, Chan RB, Williamson RL, Vetanovetz C, et al. Phosphatidylinositol-3-phosphate regulates sorting and processing of amyloid precursor protein through the endosomal system. Nat Commun 2013;4:2250.
[166] Landman N, Jeong SY, Shin SY, Voronov SV, Serban G, Kang MS, et al. Presenilin mutations linked to familial Alzheimer's disease cause an imbalance in phosphatidylinositol 4,5-bisphosphate metabolism. Proc Natl Acad Sci USA 2006;103(51):19524–9.
[167] Berman DE, Dall'Armi C, Voronov SV, McIntire LB, Zhang H, Moore AZ, et al. Oligomeric amyloid-beta peptide disrupts phosphatidylinositol-4,5-bisphosphate metabolism. Nat Neurosci 2008;11(5):547–54.
[168] Wood PL, Barnette BL, Kaye JA, Quinn JF, Woltjer RL. Non-targeted lipidomics of CSF and frontal cortex grey and white matter in control, mild cognitive impairment, and Alzheimer's disease subjects. Acta Neuropsychiatr 2015;27(5):270–8.
[169] Wood PL, Medicherla S, Sheikh N, Terry B, Phillipps A, Kaye JA, et al. Targeted lipidomics of fontal cortex and plasma diacylglycerols (DAG) in mild cognitive impairment and Alzheimer's disease: validation of DAG accumulation early in the pathophysiology of Alzheimer's disease. J Alzheimers Dis 2015;48(2):537–46.
[170] Zhang D, Dhillon H, Prasad MR, Markesbery WR. Regional levels of brain phospholipase Cgamma in Alzheimer's disease. Brain Res 1998;811(1–2):161–5.

[171] Shimohama S, Sasaki Y, Fujimoto S, Kamiya S, Taniguchi T, Takenawa T, et al. Phospholipase C isozymes in the human brain and their changes in Alzheimer's disease. Neuroscience 1998;82(4):999–1007.
[172] Fields RD. White matter in learning, cognition and psychiatric disorders. Trends Neurosci 2008;31(7):361–70.
[173] Bjorkhem I, Meaney S. Brain cholesterol: long secret life behind a barrier. Arterioscler Thromb Vasc Biol 2004;24(5):806–15.
[174] Dietschy JM, Turley SD. Thematic review series: brain lipids. Cholesterol metabolism in the central nervous system during early development and in the mature animal. J Lipid Res 2004;45(8):1375–97.
[175] Coetzee T, Fujita N, Dupree J, Shi R, Blight A, Suzuki K, et al. Myelination in the absence of galactocerebroside and sulfatide: normal structure with abnormal function and regional instability. Cell 1996;86(2):209–19.
[176] Bosio A, Binczek E, Stoffel W. Functional breakdown of the lipid bilayer of the myelin membrane in central and peripheral nervous system by disrupted galactocerebroside synthesis. Proc Natl Acad Sci USA 1996;93(23):13280–5.
[177] Braverman NE, Moser AB. Functions of plasmalogen lipids in health and disease. Biochim Biophys Acta 2012;1822(9):1442–52.
[178] da Silva TF, Eira J, Lopes AT, Malheiro AR, Sousa V, Luoma A, et al. Peripheral nervous system plasmalogens regulate Schwann cell differentiation and myelination. J Clin Invest 2014;124(6):2560–70.
[179] Rodemer C, Thai TP, Brugger B, Kaercher T, Werner H, Nave KA, et al. Inactivation of ether lipid biosynthesis causes male infertility, defects in eye development and optic nerve hypoplasia in mice. Hum Mol Genet 2003;12(15):1881–95.
[180] Han XL, Gross RW. Plasmenylcholine and phosphatidylcholine membrane bilayers possess distinct conformational motifs. Biochemistry 1990;29(20):4992–6.
[181] Merrill AH Jr. Sphingolipid and glycosphingolipid metabolic pathways in the era of sphingolipidomics. Chem Rev 2011;111(10):6387–422.
[182] Wallin A, Gottfries CG, Karlsson I, Svennerholm L. Decreased myelin lipids in Alzheimer's disease and vascular dementia. Acta Neurol Scand 1989;80(4):319–23.
[183] Svennerholm L, Gottfries CG. Membrane lipids, selectively diminished in Alzheimer brains, suggest synapse loss as a primary event in early-onset form (type I) and demyelination in late-onset form (type II). J Neurochem 1994;62(3):1039–47.
[184] Cheng H, Wang M, Li J-L, Cairns NJ, Han X. Specific changes of sulfatide levels in individuals with pre-clinical Alzheimer's disease: an early event in disease pathogenesis. J Neurochem 2013;127(6):733–8.
[185] Hejazi L, Wong JW, Cheng D, Proschogo N, Ebrahimi D, Garner B, et al. Mass and relative elution time profiling: two-dimensional analysis of sphingolipids in Alzheimer's disease brains. Biochem J 2011;438(1):165–75.
[186] Ginsberg L, Rafique S, Xuereb JH, Rapoport SI, Gershfeld NL. Disease and anatomic specificity of ethanolamine plasmalogen deficiency in Alzheimer's disease brain. Brain Res 1995;698(1–2):223–6.
[187] Guan Z, Wang Y, Cairns NJ, Lantos PL, Dallner G, Sindelar PJ. Decrease and structural modifications of phosphatidylethanolamine plasmalogen in the brain with Alzheimer disease. J Neuropathol Exp Neurol 1999;58(7):740–7.
[188] Lizard G, Rouaud O, Demarquoy J, Cherkaoui-Malki M, Iuliano L. Potential roles of peroxisomes in Alzheimer's disease and in dementia of the Alzheimer's type. J Alzheimers Dis 2012;29(2):241–54.
[189] Goodenowe DB, Cook LL, Liu J, Lu Y, Jayasinghe DA, Ahiahonu PW, et al. Peripheral ethanolamine plasmalogen deficiency: a logical causative factor in Alzheimer's disease and dementia. J Lipid Res 2007;48(11):2485–98.
[190] Yamashita S, Kiko T, Fujiwara H, Hashimoto M, Nakagawa K, Kinoshita M, et al. Alterations in the levels of amyloid-beta, phospholipid hydroperoxide, and plasmalogen in the blood of patients with Alzheimer's disease: possible interactions between amyloid-beta and these lipids. J Alzheimers Dis 2016;50(2):527–37.
[191] Wood PL, Locke VA, Herling P, Passaro A, Vigna GB, Volpato S, et al. Targeted lipidomics distinguishes patient subgroups in mild cognitive impairment (MCI) and late onset Alzheimer's disease (LOAD). BBA Clin 2016;5:25–8.
[192] Fujino T, Yamada T, Asada T, Tsuboi Y, Wakana C, Mawatari S, et al. Efficacy and blood plasmalogen changes by oral administration of plasmalogen in patients with mild Alzheimer's disease and mild cognitive impairment: a multicenter, randomized, double-blind, placebo-controlled trial. EBioMedicine 2017;17:199–205.
[193] Cheng H, Jiang X, Han X. Alterations in lipid homeostasis of mouse dorsal root ganglia induced by apolipoprotein E deficiency: a shotgun lipidomics study. J Neurochem 2007;101(1):57–76.
[194] Han X, Cheng H, Fryer JD, Fagan AM, Holtzman DM. Novel role for apolipoprotein E in the central nervous system. Modulation of sulfatide content. J Biol Chem 2003;278(10):8043–51.

[195] Agosta F, Pievani M, Sala S, Geroldi C, Galluzzi S, Frisoni GB, et al. White matter damage in Alzheimer disease and its relationship to gray matter atrophy. Radiology 2011;258(3):853–63.
[196] Selnes P, Fjell AM, Gjerstad L, Bjornerud A, Wallin A, Due-Tonnessen P, et al. White matter imaging changes in subjective and mild cognitive impairment. Alzheimers Dement 2012;8(5 Suppl.):S112–21.
[197] Zhuang L, Sachdev PS, Trollor JN, Kochan NA, Reppermund S, Brodaty H, et al. Microstructural white matter changes in cognitively normal individuals at risk of amnestic MCI. Neurology 2012;79(8):748–54.
[198] Zhuang L, Sachdev PS, Trollor JN, Reppermund S, Kochan NA, Brodaty H, et al. Microstructural white matter changes, not hippocampal atrophy, detect early amnestic mild cognitive impairment. PLoS ONE 2013;8(3):e58887.
[199] Behrendt G, Baer K, Buffo A, Curtis MA, Faull RL, Rees MI, et al. Dynamic changes in myelin aberrations and oligodendrocyte generation in chronic amyloidosis in mice and men. Glia 2013;61(2):273–86.
[200] Desai MK, Sudol KL, Janelsins MC, Mastrangelo MA, Frazer ME, Bowers WJ. Triple-transgenic Alzheimer's disease mice exhibit region-specific abnormalities in brain myelination patterns prior to appearance of amyloid and tau pathology. Glia 2009;57(1):54–65.
[201] Baes M, Aubourg P. Peroxisomes, myelination, and axonal integrity in the CNS. Neuroscientist 2009;15(4):367–79.
[202] Kou J, Kovacs GG, Hoftberger R, Kulik W, Brodde A, Forss-Petter S, et al. Peroxisomal alterations in Alzheimer's disease. Acta Neuropathol 2011;122(3):271–83.
[203] Klosinski LP, Yao J, Yin F, Fonteh AN, Harrington MG, Christensen TA, et al. White matter lipids as a ketogenic fuel supply in aging female brain: implications for Alzheimer's disease. EBioMedicine 2015;2(12):1888–904.
[204] Ding F, Yao J, Rettberg JR, Chen S, Brinton RD. Early decline in glucose transport and metabolism precedes shift to ketogenic system in female aging and Alzheimer's mouse brain: implication for bioenergetic intervention. PLoS ONE 2013;8(11):e79977.
[205] Mosconi L, Berti V, Quinn C, McHugh P, Petrongolo G, Osorio RS, et al. Perimenopause and emergence of an Alzheimer's bioenergetic phenotype in brain and periphery. PLoS ONE 2017;12(10):e0185926.
[206] Ciavardelli D, Piras F, Consalvo A, Rossi C, Zucchelli M, Di Ilio C, et al. Medium-chain plasma acylcarnitines, ketone levels, cognition, and gray matter volumes in healthy elderly, mildly cognitively impaired, or Alzheimer's disease subjects. Neurobiol Aging 2016;43:1–12.
[207] Bartzokis G, Beckson M, Lu PH, Nuechterlein KH, Edwards N, Mintz J. Age-related changes in frontal and temporal lobe volumes in men: a magnetic resonance imaging study. Arch Gen Psychiatry 2001;58(5):461–5.
[208] Bartzokis G, Lu PH, Geschwind DH, Edwards N, Mintz J, Cummings JL. Apolipoprotein E genotype and age-related myelin breakdown in healthy individuals: implications for cognitive decline and dementia. Arch Gen Psychiatry 2006;63(1):63–72.
[209] Bartzokis G, Lu PH, Geschwind DH, Tingus K, Huang D, Mendez MF, et al. Apolipoprotein E affects both myelin breakdown and cognition: implications for age-related trajectories of decline into dementia. Biol Psychiatry 2007;62(12):1380–7.
[210] Cutler RG, Kelly J, Storie K, Pedersen WA, Tammara A, Hatanpaa K, et al. Involvement of oxidative stress-induced abnormalities in ceramide and cholesterol metabolism in brain aging and Alzheimer's disease. Proc Natl Acad Sci USA 2004;101(7):2070–5.
[211] Filippov V, Song MA, Zhang K, Vinters HV, Tung S, Kirsch WM, et al. Increased ceramide in brains with Alzheimer's and other neurodegenerative diseases. J Alzheimers Dis 2012;29(3):537–47.
[212] Satoi H, Tomimoto H, Ohtani R, Kitano T, Kondo T, Watanabe M, et al. Astroglial expression of ceramide in Alzheimer's disease brains: a role during neuronal apoptosis. Neuroscience 2005;130(3):657–66.
[213] Malaplate-Armand C, Florent-Bechard S, Youssef I, Koziel V, Sponne I, Kriem B, et al. Soluble oligomers of amyloid-beta peptide induce neuronal apoptosis by activating a cPLA2-dependent sphingomyelinase-ceramide pathway. Neurobiol Dis 2006;23(1):178–89.
[214] Choi JW, Chun J. Lysophospholipids and their receptors in the central nervous system. Biochim Biophys Acta 2013;1831(1):20–32.
[215] Mizugishi K, Yamashita T, Olivera A, Miller GF, Spiegel S, Proia RL. Essential role for sphingosine kinases in neural and vascular development. Mol Cell Biol 2005;25(24):11113–21.
[216] Ceccom J, Loukh N, Lauwers-Cances V, Touriol C, Nicaise Y, Gentil C, et al. Reduced sphingosine kinase-1 and enhanced sphingosine 1-phosphate lyase expression demonstrate deregulated sphingosine 1-phosphate signaling in Alzheimer's disease. Acta Neuropathol Commun 2014;2(1):12.

[217] Wang G, Dinkins M, He Q, Zhu G, Poirier C, Campbell A, et al. Astrocytes secrete exosomes enriched with proapoptotic ceramide and prostate apoptosis response 4 (PAR-4): potential mechanism of apoptosis induction in Alzheimer disease (AD). J Biol Chem 2012;287(25):21384–95.
[218] Cuvillier O, Pirianov G, Kleuser B, Vanek PG, Coso OA, Gutkind S, et al. Suppression of ceramide-mediated programmed cell death by sphingosine-1-phosphate. Nature 1996;381(6585):800–3.
[219] Kim S, Steelman AJ, Zhang Y, Kinney HC, Li J. Aberrant upregulation of astroglial ceramide potentiates oligodendrocyte injury. Brain Pathol 2012;22(1):41–57.
[220] Vidaurre OG, Haines JD, Katz Sand I, Adula KP, Huynh JL, McGraw CA, et al. Cerebrospinal fluid ceramides from patients with multiple sclerosis impair neuronal bioenergetics. Brain 2014;137(Pt. 8):2271–86.
[221] Vozella V, Basit A, Misto A, Piomelli D. Age-dependent changes in nervonic acid-containing sphingolipids in mouse hippocampus. Biochim Biophys Acta Mol Cell Biol Lipids 2017;1862(12):1502–11.
[222] Argraves KM, Argraves WS. HDL serves as a S1P signaling platform mediating a multitude of cardiovascular effects. J Lipid Res 2007;48(11):2325–33.
[223] Christoffersen C, Obinata H, Kumaraswamy SB, Galvani S, Ahnstrom J, Sevvana M, et al. Endothelium-protective sphingosine-1-phosphate provided by HDL-associated apolipoprotein M. Proc Natl Acad Sci USA 2011;108(23):9613–8.
[224] Liu M, Frej C, Langefeld CD, Divers J, Bowden DW, Carr JJ, et al. Plasma apoM and S1P levels are inversely associated with mortality in African Americans with type 2 diabetes mellitus. J Lipid Res 2019;60(8):1425–31.
[225] Sato K, Malchinkhuu E, Horiuchi Y, Mogi C, Tomura H, Tosaka M, et al. Critical role of ABCA1 transporter in sphingosine 1-phosphate release from astrocytes. J Neurochem 2007;103(6):2610–9.
[226] Sato K, Malchinkhuu E, Horiuchi Y, Mogi C, Tomura H, Tosaka M, et al. HDL-like lipoproteins in cerebrospinal fluid affect neural cell activity through lipoprotein-associated sphingosine 1-phosphate. Biochem Biophys Res Commun 2007;359(3):649–54.
[227] Kurz J, Parnham MJ, Geisslinger G, Schiffmann S. Ceramides as novel disease biomarkers. Trends Mol Med 2019;25(1):20–32.
[228] Summers SA. Could ceramides become the new cholesterol? Cell Metab 2018;27(2):276–80.
[229] Xing Y, Tang Y, Zhao L, Wang Q, Qin W, Zhang JL, et al. Plasma ceramides and neuropsychiatric symptoms of Alzheimer's disease. J Alzheimers Dis 2016;52(3):1029–35.
[230] Torretta E, Arosio B, Barbacini P, Casati M, Capitanio D, Mancuso R, et al. Particular CSF sphingolipid patterns identify iNPH and AD patients. Sci Rep 2018;8(1):13639.
[231] Han X, Rozen S, Boyle SH, Hellegers C, Cheng H, Burke JR, et al. Metabolomics in early Alzheimer's disease: identification of altered plasma sphingolipidome using shotgun lipidomics. PLoS ONE 2011;6(7):e21643.
[232] Mielke MM, Bandaru VV, Haughey NJ, Xia J, Fried LP, Yasar S, et al. Serum ceramides increase the risk of Alzheimer disease: the Women's Health and Aging Study II. Neurology 2012;79(7):633–41.
[233] Mielke MM, Bandaru VV, Haughey NJ, Rabins PV, Lyketsos CG, Carlson MC. Serum sphingomyelins and ceramides are early predictors of memory impairment. Neurobiol Aging 2010;31(1):17–24.
[234] Mielke MM, Haughey NJ, Han D, An Y, Bandaru VVR, Lyketsos CG, et al. The association between plasma ceramides and sphingomyelins and risk of Alzheimer's disease differs by sex and APOE in the Baltimore Longitudinal Study of Aging. J Alzheimers Dis 2017;60(3):819–28.
[235] Lingwood D, Simons K. Lipid rafts as a membrane-organizing principle. Science 2010;327(5961):46–50.
[236] Haughey NJ, Bandaru VV, Bae M, Mattson MP. Roles for dysfunctional sphingolipid metabolism in Alzheimer's disease neuropathogenesis. Biochim Biophys Acta 2010;1801(8):878–86.
[237] Harris B, Pereira I, Parkin E. Targeting ADAM10 to lipid rafts in neuroblastoma SH-SY5Y cells impairs amyloidogenic processing of the amyloid precursor protein. Brain Res 2009;1296:203–15.
[238] Hartl D, May P, Gu W, Mayhaus M, Pichler S, Spaniol C, et al. A rare loss-of-function variant of ADAM17 is associated with late-onset familial Alzheimer disease. Mol Psychiatry 2018;25:629–39.
[239] Ehehalt R, Keller P, Haass C, Thiele C, Simons K. Amyloidogenic processing of the Alzheimer beta-amyloid precursor protein depends on lipid rafts. J Cell Biol 2003;160(1):113–23.
[240] Xiong H, Callaghan D, Jones A, Walker DG, Lue L-F, Beach TG, et al. Cholesterol retention in Alzheimer's brain is responsible for high β- and γ-secretase activities and Aβ production. Neurobiol Disease 2008;29(3):422–37.
[241] Kakio A, Nishimoto SI, Yanagisawa K, Kozutsumi Y, Matsuzaki K. Cholesterol-dependent formation of GM1 ganglioside-bound amyloid beta-protein, an endogenous seed for Alzheimer amyloid. J Biol Chem. 2001;276(27):24985–90.

[242] Grimm MOW, Haupenthal VJ, Rothhaar TL, Zimmer VC, Grösgen S, Hundsdörfer B, et al. Effect of different phospholipids on α-secretase activity in the non-amyloidogenic pathway of Alzheimer's disease. Int J Mol Sci 2013;14(3):5879–98.
[243] Refolo LM, Malester B, LaFrancois J, Bryant-Thomas T, Wang R, Tint GS, et al. Hypercholesterolemia accelerates the Alzheimer's amyloid pathology in a transgenic mouse model. Neurobiol Dis 2000;7(4):321–31.
[244] Puglielli L, Ellis BC, Saunders AJ, Kovacs DM. Ceramide stabilizes beta-site amyloid precursor protein-cleaving enzyme 1 and promotes amyloid beta-peptide biogenesis. J Biol Chem 2003;278(22):19777–83.
[245] Schnaar RL. The biology of gangliosides. Adv Carbohydr Chem Biochem 2019;76:113–48.
[246] Svennerholm L. Gangliosides and synaptic transmission. Adv Exp Med Biol 1980;125:533–44.
[247] Matsuzaki K, Kato K, Yanagisawa K. Abeta polymerization through interaction with membrane gangliosides. Biochim Biophys Acta 2010;1801(8):868–77.
[248] Yanagisawa K, Odaka A, Suzuki N, Ihara Y. GM1 ganglioside-bound amyloid beta-protein (A beta): a possible form of preamyloid in Alzheimer's disease. Nat Med 1995;1(10):1062–6.
[249] Gellermann GP, Appel TR, Tannert A, Radestock A, Hortschansky P, Schroeckh V, et al. Raft lipids as common components of human extracellular amyloid fibrils. Proc Natl Acad Sci USA 2005;102(18):6297–302.
[250] Yamamoto N, Fukata Y, Fukata M, Yanagisawa K. GM1-ganglioside-induced Aβ assembly on synaptic membranes of cultured neurons. Biochim Biophys Acta 2007;1768(5):1128–37.
[251] Li H, Kim WS, Guillemin GJ, Hill AF, Evin G, Garner B. Modulation of amyloid precursor protein processing by synthetic ceramide analogues. Biochim Biophys Acta 2010;1801(8):887–95.
[252] Yanagisawa K, McLaurin J, Michikawa M, Chakrabartty A, Ihara Y. Amyloid β-protein (Aβ) associated with lipid molecules: immunoreactivity distinct from that of soluble Aβ. FEBS Lett 1997;420(1):43–6.
[253] Yanagisawa K, Ihara Y. GM1 ganglioside-bound amyloid β-protein in Alzheimer's disease brain. Neurobiol Aging 1998;19(1 Suppl. 1):S65–7.
[254] Chan RB, Oliveira TG, Cortes EP, Honig LS, Duff KE, Small SA, et al. Comparative lipidomic analysis of mouse and human brain with Alzheimer disease. J Biol Chem 2012;287(4):2678–88.
[255] Gizaw ST, Ohashi T, Tanaka M, Hinou H, Nishimura S-I. Glycoblotting method allows for rapid and efficient glycome profiling of human Alzheimer's disease brain, serum and cerebrospinal fluid towards potential biomarker discovery. Biochim Biophys Acta 2016;1860(8):1716–27.
[256] Virgolini MJ, Feliziani C, Cambiasso MJ, Lopez PH, Bollo M. Neurite atrophy and apoptosis mediated by PERK signaling after accumulation of GM2-ganglioside. Biochim Biophys Acta 2019;1866(2):225–39.
[257] Nakatsuji Y, Miller RH. Selective cell-cycle arrest and induction of apoptosis in proliferating neural cells by ganglioside GM3. Exp Neurol 2001;168(2):290–9.
[258] Tassoni D, Kaur G, Weisinger RS, Sinclair AJ. The role of eicosanoids in the brain. Asia Pac J Clin Nutr 2008;17(Suppl. 1):220–8.
[259] Bazan NG. Synaptic lipid signaling: significance of polyunsaturated fatty acids and platelet-activating factor. J Lipid Res 2003;44(12):2221–33.
[260] Serhan CN, Petasis NA. Resolvins and protectins in inflammation resolution. Chem Rev 2011;111(10):5922–43.
[261] Freedman SD, Katz MH, Parker EM, Laposata M, Urman MY, Alvarez JG. A membrane lipid imbalance plays a role in the phenotypic expression of cystic fibrosis in cftr(-/-) mice. Proc Natl Acad Sci USA 1999;96(24):13995–4000.
[262] Tully AM, Roche HM, Doyle R, Fallon C, Bruce I, Lawlor B, et al. Low serum cholesteryl ester-docosahexaenoic acid levels in Alzheimer's disease: a case-control study. Br J Nutr 2003;89(4):483–9.
[263] Golovko MY, Murphy EJ. An improved LC-MS/MS procedure for brain prostanoid analysis using brain fixation with head-focused microwave irradiation and liquid-liquid extraction. J Lipid Res 2008;49(4):893–902.
[264] Morris MC, Evans DA, Bienias JL, Tangney CC, Bennett DA, Wilson RS, et al. Consumption of fish and n-3 fatty acids and risk of incident Alzheimer disease. Arch Neurol 2003;60(7):940–6.
[265] Kyle DJ, Schaefer E, Patton G, Beiser A. Low serum docosahexaenoic acid is a significant risk factor for Alzheimer's dementia. Lipids 1999;34(Suppl.):S245.
[266] Barberger-Gateau P, Letenneur L, Deschamps V, Peres K, Dartigues JF, Renaud S. Fish, meat, and risk of dementia: cohort study. BMJ 2002;325(7370):932–3.
[267] Barberger-Gateau P, Raffaitin C, Letenneur L, Berr C, Tzourio C, Dartigues JF, et al. Dietary patterns and risk of dementia: the Three-City cohort study. Neurology 2007;69(20):1921–30.
[268] Huang TL, Zandi PP, Tucker KL, Fitzpatrick AL, Kuller LH, Fried LP, et al. Benefits of fatty fish on dementia risk are stronger for those without APOE epsilon4. Neurology 2005;65(9):1409–14.

[269] Schaefer EJ, Bongard V, Beiser AS, Lamon-Fava S, Robins SJ, Au R, et al. Plasma phosphatidylcholine docosahexaenoic acid content and risk of dementia and Alzheimer disease: the Framingham Heart Study. Arch Neurol 2006;63(11):1545–50.

[270] Laitinen MH, Ngandu T, Rovio S, Helkala EL, Uusitalo U, Viitanen M, et al. Fat intake at midlife and risk of dementia and Alzheimer's disease: a population-based study. Dement Geriatr Cogn Disord 2006;22(1):99–107.

[271] Harper CR, Jacobson TA. The fats of life: the role of omega-3 fatty acids in the prevention of coronary heart disease. Arch Intern Med 2001;161(18):2185–92.

[272] Breslow JL. n-3 fatty acids and cardiovascular disease. Am J Clin Nutr 2006;83(6 Suppl.):1477S–82S.

[273] Laurin D, Verreault R, Lindsay J, Dewailly E, Holub BJ. Omega-3 fatty acids and risk of cognitive impairment and dementia. J Alzheimers Dis 2003;5(4):315–22.

[274] Engelhart MJ, Geerlings MI, Ruitenberg A, Van Swieten JC, Hofman A, Witteman JC, et al. Diet and risk of dementia: Does fat matter?: The Rotterdam Study. Neurology 2002;59(12):1915–21.

[275] Begin ME, Plourde M, Pifferi F, Cunnane SC. What is the link between docosahexaenoic acid, cognitive impairment, and Alzheimer's disease in the elderly? In: Montmayeur JP, le Coutre J, editors. Fat detection: taste, texture, and post ingestive effects. Frontiers in Neuroscience. Boca Raton, FL: CRC Press/Taylor & Francis; 2010.

[276] Andrieu S, Guyonnet S, Coley N, Cantet C, Bonnefoy M, Bordes S, et al. Effect of long-term omega 3 polyunsaturated fatty acid supplementation with or without multidomain intervention on cognitive function in elderly adults with memory complaints (MAPT): a randomised, placebo-controlled trial. Lancet Neurol 2017;16(5):377–89.

[277] Quinn JF, Raman R, Thomas RG, Yurko-Mauro K, Nelson EB, Van Dyck C, et al. Docosahexaenoic acid supplementation and cognitive decline in Alzheimer disease: a randomized trial. JAMA 2010;304(17):1903–11.

[278] van de Rest O, Geleijnse JM, Kok FJ, van Staveren WA, Dullemeijer C, Olderikkert MG, et al. Effect of fish oil on cognitive performance in older subjects: a randomized, controlled trial. Neurology 2008;71(6):430–8.

[279] Soderberg M, Edlund C, Kristensson K, Dallner G. Fatty acid composition of brain phospholipids in aging and in Alzheimer's disease. Lipids 1991;26(6):421–5.

[280] Prasad MR, Lovell MA, Yatin M, Dhillon H, Markesbery WR. Regional membrane phospholipid alterations in Alzheimer's disease. Neurochem Res 1998;23(1):81–8.

[281] Corrigan FM, Horrobin DF, Skinner ER, Besson JA, Cooper MB. Abnormal content of n-6 and n-3 long-chain unsaturated fatty acids in the phosphoglycerides and cholesterol esters of parahippocampal cortex from Alzheimer's disease patients and its relationship to acetyl CoA content. Int J Biochem Cell Biol 1998;30(2):197–207.

[282] Skinner ER, Watt C, Besson JA, Best PV. Differences in the fatty acid composition of the grey and white matter of different regions of the brains of patients with Alzheimer's disease and control subjects. Brain 1993;116(Pt. 3):717–25.

[283] Lahoz C, Schaefer EJ, Cupples LA, Wilson PW, Levy D, Osgood D, et al. Apolipoprotein E genotype and cardiovascular disease in the Framingham Heart Study. Atherosclerosis 2001;154(3):529–37.

[284] Kivipelto M, Helkala E-L, Laakso MP, Hänninen T, Hallikainen M, Alhainen K, et al. Midlife vascular risk factors and Alzheimer's disease in later life: longitudinal, population based study. BMJ 2001;322(7300):1447–51.

[285] Hofman A, Ott A, Breteler MMB, Bots ML, Slooter AJC, van Harskamp F, et al. Atherosclerosis, apolipoprotein E, and prevalence of dementia and Alzheimer's disease in the Rotterdam Study. Lancet 1997;349(9046):151–4.

[286] Karjalainen JP, Mononen N, Hutri-Kahonen N, Lehtimaki M, Hilvo M, Kauhanen D, et al. New evidence from plasma ceramides links apoE polymorphism to greater risk of coronary artery disease in Finnish adults. J Lipid Res 2019;60(9):1622–9.

[287] Kivipelto M, Helkala E-L, Laakso MP, Hänninen T, Hallikainen M, Alhainen K, et al. Apolipoprotein E ε4 allele, elevated midlife total cholesterol level, and high midlife systolic blood pressure are independent risk factors for late-life Alzheimer disease. Ann Intern Med 2002;137(3):149–55.

[288] Wolozin B, Kellman W, Ruosseau P, Celesia GG, Siegel G. Decreased prevalence of Alzheimer disease associated with 3-hydroxy-3-methyglutaryl coenzyme A reductase inhibitors. Arch Neurol 2000;57(10):1439–43.

[289] Jick H, Zornberg GL, Jick SS, Seshadri S, Drachman DA. Statins and the risk of dementia. Lancet 2000;356(9242):1627–31.

[290] Tan ZS, Seshadri S, Beiser A, Wilson PW, Kiel DP, Tocco M, et al. Plasma total cholesterol level as a risk factor for Alzheimer disease: the Framingham Study. Arch Intern Med 2003;163(9):1053–7.

[291] Sano M, Bell KL, Galasko D, Galvin JE, Thomas RG, van Dyck CH, et al. A randomized, double-blind, placebo-controlled trial of simvastatin to treat Alzheimer disease. Neurology 2011;77(6):556–63.

[292] Trompet S, van Vliet P, de Craen AJM, Jolles J, Buckley BM, Murphy MB, et al. Pravastatin and cognitive function in the elderly. Results of the PROSPER study. J Neurol 2009;257(1):85.

[293] Group HPSC. MRC/BHF Heart Protection Study of cholesterol lowering with simvastatin in 20 536 high-risk individuals: a randomised placebocontrolled trial. Lancet 2002;360(9326):7–22.

[294] Jellinger KA. The enigma of vascular cognitive disorder and vascular dementia. Acta Neuropathol 2007;113(4):349–88.

[295] Colacurcio DJ, Pensalfini A, Jiang Y, Nixon RA. Dysfunction of autophagy and endosomal-lysosomal pathways: roles in pathogenesis of Down syndrome and Alzheimer's Disease. Free Radic Biol Med 2018;114:40–51.

[296] Cataldo AM, Peterhoff CM, Troncoso JC, Gomez-Isla T, Hyman BT, Nixon RA. Endocytic pathway abnormalities precede amyloid beta deposition in sporadic Alzheimer's disease and Down syndrome: differential effects of APOE genotype and presenilin mutations. Am J Pathol 2000;157(1):277–86.

[297] Nixon RA, Wegiel J, Kumar A, Yu WH, Peterhoff C, Cataldo A, et al. Extensive involvement of autophagy in Alzheimer disease: an immuno-electron microscopy study. J Neuropathol Exp Neurol 2005;64(2):113–22.

[298] Cataldo AM, Barnett JL, Pieroni C, Nixon RA. Increased neuronal endocytosis and protease delivery to early endosomes in sporadic Alzheimer's disease: neuropathologic evidence for a mechanism of increased beta-amyloidogenesis. J Neurosci 1997;17(16):6142–51.

[299] Strassnig M, Ganguli M. About a peculiar disease of the cerebral cortex: Alzheimer's original case revisited. Psychiatry (Edgmont) 2005;2(9):30–3.

[300] Gao S, Casey AE, Sargeant TJ, Makinen VP. Genetic variation within endolysosomal system is associated with late-onset Alzheimer's disease. Brain 2018;141(9):2711–20.

[301] Nuriel T, Peng KY, Ashok A, Dillman AA, Figueroa HY, Apuzzo J, et al. The endosomal-lysosomal pathway is dysregulated by APOE4 expression in vivo. Front Neurosci 2017;11:702.

[302] Nixon RA. Amyloid precursor protein and endosomal-lysosomal dysfunction in Alzheimer's disease: inseparable partners in a multifactorial disease. FASEB J 2017;31(7):2729–43.

[303] Seshadri S, Fitzpatrick AL, Ikram MA, DeStefano AL, Gudnason V, Boada M, et al. Genome-wide analysis of genetic loci associated with Alzheimer disease. JAMA 2010;303(18):1832–40.

[304] Xu W, Weissmiller AM, White JA2nd, Fang F, Wang X, Wu Y, et al. Amyloid precursor protein-mediated endocytic pathway disruption induces axonal dysfunction and neurodegeneration. J Clin Invest 2016;126(5):1815–33.

[305] Shibuya Y, Chang CC, Huang LH, Bryleva EY, Chang TY. Inhibiting ACAT1/SOAT1 in microglia stimulates autophagy-mediated lysosomal proteolysis and increases Abeta1-42 clearance. J Neurosci 2014;34(43):14484–501.

[306] Shibuya Y, Niu Z, Bryleva EY, Harris BT, Murphy SR, Kheirollah A, et al. Acyl-coenzyme A: cholesterol acyltransferase 1 blockage enhances autophagy in the neurons of triple transgenic Alzheimer's disease mouse and reduces human P301L-tau content at the presymptomatic stage. Neurobiol Aging 2015;36(7):2248–59.

[307] Auer IA, Schmidt ML, Lee VM, Curry B, Suzuki K, Shin RW, et al. Paired helical filament tau (PHFtau) in Niemann-Pick type C disease is similar to PHFtau in Alzheimer's disease. Acta Neuropathol 1995;90(6):547–51.

[308] Saito Y, Suzuki K, Nanba E, Yamamoto T, Ohno K, Murayama S. Niemann-Pick type C disease: accelerated neurofibrillary tangle formation and amyloid beta deposition associated with apolipoprotein E epsilon 4 homozygosity. Ann Neurol 2002;52(3):351–5.

[309] Infante RE, Wang ML, Radhakrishnan A, Kwon HJ, Brown MS, Goldstein JL. NPC2 facilitates bidirectional transfer of cholesterol between NPC1 and lipid bilayers, a step in cholesterol egress from lysosomes. Proc Natl Acad Sci USA 2008;105(40):15287–92.

[310] Jin LW, Shie FS, Maezawa I, Vincent I, Bird T. Intracellular accumulation of amyloidogenic fragments of amyloid-beta precursor protein in neurons with Niemann-Pick type C defects is associated with endosomal abnormalities. Am J Pathol 2004;164(3):975–85.

[311]Keilani S, Lun Y, Stevens AC, Williams HN, Sjoberg ER, Khanna R, et al. Lysosomal dysfunction in a mouse model of Sandhoff disease leads to accumulation of ganglioside-bound amyloid-beta peptide. J Neurosci 2012;32(15):5223–36.

[312] Boland B, Smith DA, Mooney D, Jung SS, Walsh DM, Platt FM. Macroautophagy is not directly involved in the metabolism of amyloid precursor protein. J Biol Chem 2010;285(48):37415–26.

[313] Tamboli IY, Hampel H, Tien NT, Tolksdorf K, Breiden B, Mathews PM, et al. Sphingolipid storage affects autophagic metabolism of the amyloid precursor protein and promotes Abeta generation. J Neurosci 2011;31(5):1837–49.

[314] Cermak S, Kosicek M, Mladenovic-Djordjevic A, Smiljanic K, Kanazir S, Hecimovic S. Loss of cathepsin B and L leads to lysosomal dysfunction. NPC-like cholesterol sequestration and accumulation of the key Alzheimer's proteins. PLoS ONE 2016;11(11):e0167428.
[315] Tayebi N, Walker J, Stubblefield B, Orvisky E, LaMarca ME, Wong K, et al. Gaucher disease with parkinsonian manifestations: does glucocerebrosidase deficiency contribute to a vulnerability to parkinsonism? Mol Genet Metab 2003;79(2):104–9.
[316] Neudorfer O, Giladi N, Elstein D, Abrahamov A, Turezkite T, Aghai E, et al. Occurrence of Parkinson's syndrome in type I Gaucher disease. QJM 1996;89(9):691–4.
[317] Klein C, Westenberger A. Genetics of Parkinson's disease. Cold Spring Harb Perspect Med 2012;2(1):a008888-a.
[318] Gan-Or Z, Amshalom I, Kilarski LL, Bar-Shira A, Gana-Weisz M, Mirelman A, et al. Differential effects of severe vs mild GBA mutations on Parkinson disease. Neurology 2015;84(9):880–7.
[319] Lesage S, Anheim M, Condroyer C, Pollak P, Durif F, Dupuits C, et al. Large-scale screening of the Gaucher's disease-related glucocerebrosidase gene in Europeans with Parkinson's disease. Human Mol Genet 2010;20(1):202–10.
[320] Baker M, Mackenzie IR, Pickering-Brown SM, Gass J, Rademakers R, Lindholm C, et al. Mutations in progranulin cause tau-negative frontotemporal dementia linked to chromosome 17. Nature 2006;442(7105):916–9.
[321] Cruts M, Gijselinck I, van der Zee J, Engelborghs S, Wils H, Pirici D, et al. Null mutations in progranulin cause ubiquitin-positive frontotemporal dementia linked to chromosome 17q21. Nature. 2006;442(7105):920–4.
[322] Smith KR, Damiano J, Franceschetti S, Carpenter S, Canafoglia L, Morbin M, et al. Strikingly different clinicopathological phenotypes determined by progranulin-mutation dosage. Am J Hum Genet 2012;90(6):1102–7.
[323] Alcalay RN, Levy OA, Waters CC, Fahn S, Ford B, Kuo SH, et al. Glucocerebrosidase activity in Parkinson's disease with and without GBA mutations. Brain 2015;138(Pt. 9):2648–58.
[324] Gegg ME, Burke D, Heales SJ, Cooper JM, Hardy J, Wood NW, et al. Glucocerebrosidase deficiency in substantia nigra of parkinson disease brains. Ann Neurol 2012;72(3):455–63.
[325] Rocha EM, Smith GA, Park E, Cao H, Brown E, Hallett P, et al. Progressive decline of glucocerebrosidase in aging and Parkinson's disease. Ann Clin Transl Neurol 2015;2(4):433–8.
[326] Taguchi YV, Liu J, Ruan J, Pacheco J, Zhang X, Abbasi J, et al. Glucosylsphingosine promotes α-synuclein pathology in mutant GBA-associated Parkinson's disease. J Neurosci 2017;37(40):9617–31.
[327] Rocha EM, Smith GA, Park E, Cao H, Brown E, Hallett P, et al. Progressive decline of glucocerebrosidase in aging and Parkinson's disease. Ann Clinical Transl Neurol 2015;2(4):433–8.
[328] Gegg ME, Sweet L, Wang BH, Shihabuddin LS, Sardi SP, Schapira AHV. No evidence for substrate accumulation in Parkinson brains with GBA mutations. Mov Disord 2015;30(8):1085–9.
[329] Mazzulli JR, Xu YH, Sun Y, Knight AL, McLean PJ, Caldwell GA, et al. Gaucher disease glucocerebrosidase and alpha-synuclein form a bidirectional pathogenic loop in synucleinopathies. Cell 2011;146(1):37–52.
[330] Xu YH, Sun Y, Ran H, Quinn B, Witte D, Grabowski GA. Accumulation and distribution of alpha-synuclein and ubiquitin in the CNS of Gaucher disease mouse models. Mol Genet Metab 2011;102(4):436–47.
[331] Suzuki K, Iseki E, Togo T, Yamaguchi A, Katsuse O, Katsuyama K, et al. Neuronal and glial accumulation of alpha- and beta-synucleins in human lipidoses. Acta Neuropathol 2007;114(5):481–9.
[332] Suzuki K, Iseki E, Katsuse O, Yamaguchi A, Katsuyama K, Aoki I, et al. Neuronal accumulation of alpha- and beta-synucleins in the brain of a GM2 gangliosidosis mouse model. Neuroreport 2003;14(4):551–4.
[333] Hallett PJ, Huebecker M, Brekk OR, Moloney EB, Rocha EM, Priestman DA, et al. Glycosphingolipid levels and glucocerebrosidase activity are altered in normal aging of the mouse brain. Neurobiol Aging 2018;67:189–200.

CHAPTER

# 24

# Role of Xenosterols in Health and Disease

*Babunageswararao Kanuri, Vincent Fong, Shailendra B. Patel*

Division of Endocrinology, Diabetes and Metabolism, University of Cincinnati, Cincinnati, OH, United States

OUTLINE



## Introduction

Sterols, also known as -steroid alcohols, are an essential class of organic molecules possessing higher biological relevance [1]. They occur naturally in all eukaryotes such as plants [2,3], animals [4], and fungi [5], as well as in some prokaryotic bacteria [6] to perform diverse biological functions with different organisms evolving their own preference for certain sterols over others. Cholesterol is evolutionarily preferred in all higher animals, including mammals, while ergosterol is the most abundant xenosterol in fungi, and the phytosterols such as sitosterol, campesterol, and stigmasterol are predominantly found in plants [4]. Other notable xenosterols include, avenasterol, cycloartenol and cycloartanol, gramisterol,

Lipid Signaling and Metabolism. http://dx.doi.org/10.1016/B978-0-12-819404-1.00024-5

fucosterol, and spinasterol in plants, and 22-dehydrocholesterol, isofucosterol, clionasterol, and 24-norcholesta-5,22-diene-3β-ol in mollusks [7,8]. Phytosterols are the natural constituents found in all plant-based foods [9,10], and possess the same cyclopentane phenanthrene ring, 3β-hydroxy group, and an 8-carbon side chain, as cholesterol [11]. However, they differ from cholesterol in the R tail, e.g., an additional methyl group at C-24 in campesterol, an extra ethyl group at C-24 in sitosterol, and an ethyl group at C-24 along with a double bond between C22 and C23 in stigmasterol [11]. The spectrum of plant sterols are extensive and many different plant sterol species have been reported, albeit at varying levels of abundance in different plant tissues and varying between different plant species. In general, the concentration of phytosterols in the human blood and tissue is extremely low (~1000 times lower than cholesterol) [12–15]. Factors such as low dietary absorbability as well as preferential excretion by sterolin pumps (ABCG5/ABCG8) result in a net average absorption of 0.4–5% for plant sterols compared with ~ 55% for cholesterol [11,16].

## Absorption of dietary sterols

Two distinct mechanisms, one in the intestinal enterocyte and other in the liver hepatocyte govern the absorption of the dietary cholesterol and xenosterols (Fig. 24.1) [11]. After enzymatic digestion to free cholesterol (FC) and fatty acids (FA) in the intestine lumen, dietary fats are transported into intestinal enterocytes as micelles where FC is re-esterified into cholesteryl esters (CE) with the help of Acyl coenzyme A: cholesterol acyltransferase-2, 4(ACAT-2), also

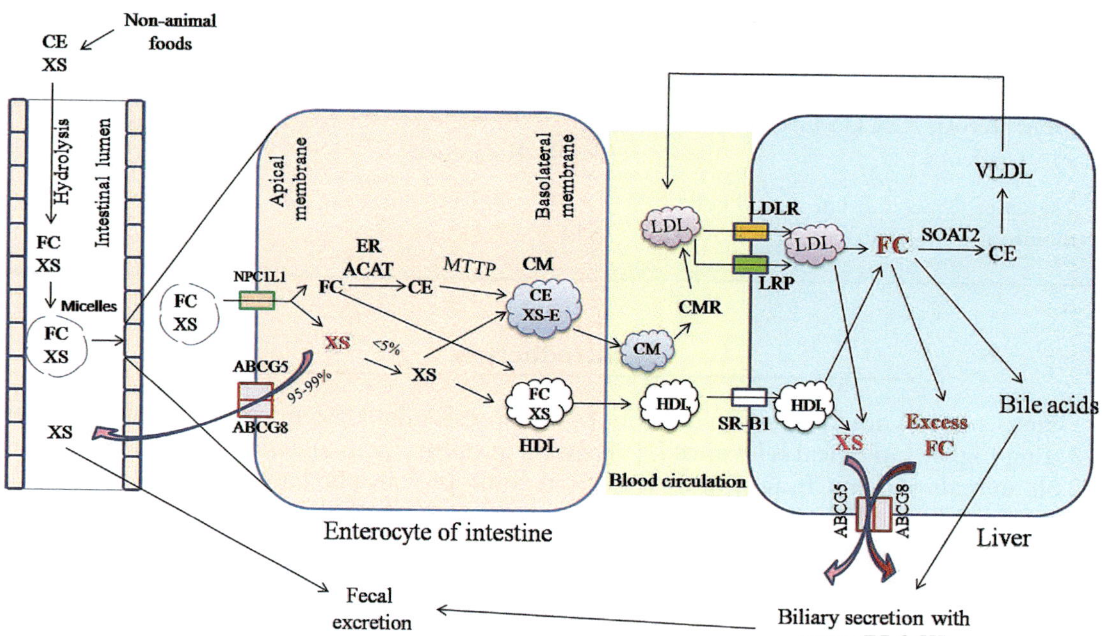


FIGURE 24.1 **A model depicting the fate of dietary xenosterols [11,19].**

called as sterol O-acyltransferase-2 (SOAT-2), and incorporated into apoB48 containing chylomicrons (CM) by microsomal triglyceride transfer protein (MTTP) [11,17]. These cholesterol-rich CMs are then secreted into the lacteals at the basolateral surface of the enterocytes, and eventually enter into the venous circulation [11,17]. Xenosterols (such as plant sterols) are also converted to free sterols in the intestinal lumen and like cholesterol, absorbed into enterocytes in micelles [11,17], but as they are poor substrates for sterol esterification by ACAT-2 in the endoplasmic reticulum [18], most of the absorbed xenosterols are preferentially secreted back to the intestinal lumen by ABCG5/G8 [19]. The remaining xenosterols that are not exported back into the intestinal lumen, have two possible fates; one, is to be incorporated into CM, and enter into circulation [20]; or to remain as free sterols and to be incorporated along with FC into apolipoprotein ApoA1 containing high density lipoproteins (HDL) [21–23]. While, transport of FC in HDL carriers through involvement of ABCA1 and ApoA1 proteins has been extensively studied both in vitro and in vivo [24,25], the possibility of xenosterols being transported through the same mechanism needs extensive investigations. Enhancement of basolateral efflux of β-sitosterol upon LXR/RXR activation in CaCo-2 cell line is first of its kind to support this idea of xenosterol secretion into nascent HDL carriers [23].

Xenosterols have been observed in all lipoproteins with highest concentrations found in low and HDL [26]. Both xenosterols and cholesterol have shown similar rates of esterification by lecithin:cholesterol acyltransferase (LCAT) in HDL, and the extent of their uptake from HDL is comparable with that shown in CM [18]. In the liver, cholesterol uptake into hepatocytes mainly occurs from HDL via scavenger receptor class B member 1 (SR-B1), and from chylomicron remnants (CMR) via low density lipoprotein receptors (LDLR) and LDLR-related protein 1 (LRP1) receptors [21]. Once in the hepatocyte, FC is esterified and incorporated into VLDL (very low density lipoprotein) for delivery to peripheral tissues [21]. While xenosterols in hepatocytes are preferentially excreted by ABCG5/G8 into bile for intestinal efflux [11], given the structural similarities to cholesterol, it is a distinct possibility that xenosterols might also be transported by the same or similar mechanisms as cholesterol. Although this has not been proven, a mechanism for loading xenosterol into VLDL particles would be very crucial in determining the extent of their delivery to peripheral tissues [21].

## Plant sterols as double-edged swords in various cellular processes

The beneficial role of xenosterols, such as plant sterols, in functional foods for effective management of cardiovascular (CVS) diseases has been well studied [9,26]; dietary plant sterols compete with dietary and biliary cholesterol for enterocyte absorption [26], which leads to a lower plasma LDL-C, and is therefore inferred to reduce CVS events [26]. Patients with autosomal recessive mutations in ABCG5/G8 have sitosterolemia, which is defined as the accumulation of high circulating levels of different xenosterols. It is also associated with platelet dysfunction, liver disease, xanthomas, and atherosclerosis [4,27]. Beneficial effects at low to moderate levels, and pathological significance at higher concentrations make these molecules “double-edged swords.” For years, plant sterols were extensively studied for their LDL-cholesterol lowering properties for prevention of CVS disorders as competitive inhibitors of intestinal cholesterol uptake, until recently when it was discovered that they are not inert molecules [9]. On the other hand, studies have also been conducted using sterol absorption inhibitors, like ezetimibe, for ameliorating the excess levels of circulatory xenosterols in cases of sitosterolemia [4,11]. An excellent review published by Plat et al. recently (2019) has discussed in depth the new paradigm understanding of phytosterols and how they may have important effects on diverse fields such as immunology, hepatology, pulmonology, and gastroenterology [9]. The further sections of this chapter will focus on the critical role of xenosterols, with main focus on phytosterols in maintaining good health, their impact on different disease states, and effects on (patho) physiological processes in the body and in cells.

## Phytosterols, ABCG5/G8 and sitosterolemia

The first incidence of elevated plant sterols was reported by Bhattacharyya and Connor in 1974 [27]. Two sisters showed signs and symptoms that were suggestive of familial hypercholesteremia, with arthralgia and tendon xanthomas but without elevated plasma cholesterol [28]. Sitosterol was most abundant plant sterol observed in them, consequently, the condition was coined as Sitosterolemia [28]. However, the name Sitosterolemia is misleading, given that ABCG5-ABCG8 heterodimer participate in efflux of around 20 different species of plant sterols, all of which were shown to be accumulated in patients with Sitosterolemia [27]. Our current understanding shows that these pumps help to keep out xenosterols from the body. Characterization of about 100 cases worldwide indicates that xenosterol accumulation can lead to dramatic phenotypes, including macrothrombocytopenia, platelet dysfunction, xanthomas, accelerated atherosclerosis, and occasionally arthritis, arthralgias, hemolysis, liver disease, etc.[29–38]. The severity of these phenotypes varies among patients based on the levels of individual plant sterol accumulation [19]. The significance of sterolin proteins in the pathophysiology of sitosterolemia is demonstrated by the number of different mutations across different loci in the ABCG5 or ABCG8 gene that result in the sitosterolemia phenotype [27,30]. Both ABCG5 and ABCG8 genes have 13 exons each and are located at the *STSL* (sitosterolemia locus) locus. Patients affected by this disease have been found to have mutations and polymorphisms in different exons [11,30] affecting amino acids close to the putative sterol binding sites [27,39]. An overview of the various mutations reported for sterolins ABCG5 and ABCG8 at the gene and protein levels were depicted in Fig. 24.2.

Different mice models targeting the *STSL* locus, including individual global knockouts for ABCG5 (Abcg5$^{-/-}$) [40], ABCG8 (Abcg8$^{-/-}$) [41], or double knockout for ABCG5 and ABCG8 (Abcg5$^{-/-}$,Abcg8$^{-/-}$) [42], as well as tissue-specific deletions or over-expressions [43–45], have been developed to investigate how these transporters impact different processes linked with plant sterols such as sterol absorption and secretion, sitosterolemia, and atherosclerosis [27]. Global knockout mice (Abcg5$^{-/-}$, Abcg8$^{-/-}$, and Abcg5$^{-/-}$,Abcg8$^{-/-}$) share most of the phenotypes observed in human Sitosterolemia [33,46–49], and additionally also demonstrate infertility, lipodystrophy, immune dysfunction, and cardiomyopathy [48–50]. The fact that Abcg5$^{-/-}$, Abcg8$^{-/-}$, or Abcg5$^{-/-}$, Abcg8$^{-/-}$ consistently develop sitosterolemia when fed on phytosterol enriched diet, demonstrated their inability to excrete xenosterols into biliary or intestinal tracts [40–42,46,49,51–53]. Furthermore, significantly lower levels of plant sterols in plasma and tissue in liver or intestine specific KO (knockout) mice (LiAbcg5$^{-/-}$,Abcg8$^{-/-}$or InAbcg5$^{-/-}$, Abcg8$^{-/-}$) compared to those with whole-body sterolin deficiency (Abcg5$^{-/-}$,A bcg8$^{-/-}$) indicate the role of ABCG5/G8 in both intestine and liver for maintaining sterol

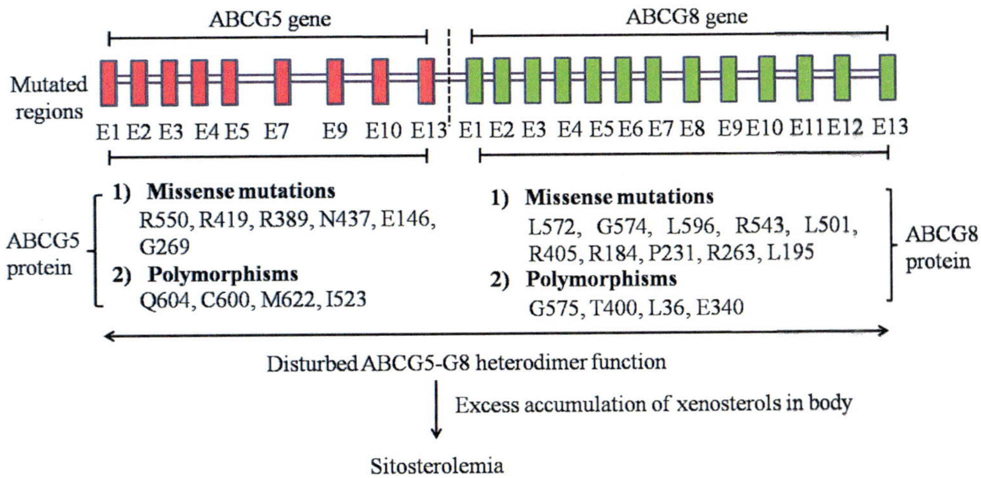


FIGURE 24.2 **Known mutations associated with ABCG5/G8 genes and proteins reported in different patients studies [11,30,39].** These missense mutations and nonsynonymous polymorphic variants have been mapped onto the ABCG5/G8 crystal structure [39].

homeostasis [43]. Indeed, in Abcg8$^{-/-}$ mice with only rescue of intestinal sterolin function, there was marked reduction of phytosterol levels, suggesting that intestinal ABCG5/G8 acts as first-pass gates in sterol efflux [49]. Moreover, mice over expressing both liver and intestinal ABCG5/G8 (LiInTgAbcg5/8) showed elevated fecal neutral sterol excretion (FNSE) and reduced fractional cholesterol absorption (FCA) but not in mice with ABCG5/G8 over expressed in only liver (LiTgAbcg5/8) [44,45,54,55]. Global KO for Abcg5$^{-/-}$ or Abcg8$^{-/-}$ or Abcg5$^{-/-}$, Abcg8$^{-/-}$ mice have shown reduced FNSE, while measurements of FCA varied according to experimental designs [40,42,52,56–60]. FCA and FNSE are key factors determining the net rate of cholesterol absorption, and thereby whole body sterol homeostasis [45,61]. A list of different studies involving mice with gain and loss of function of sterolin genes and their effects on understanding sitosterolemia and associated complications such as atherosclerosis are shown in Table 24.1.

As discussed earlier, important phenotypic characteristics observed in and specific to mouse models of sitosterolemia are infertility, cardiomyopathy, lipodystrophy, and immune dysfunction [48–50]. Cardiomyopathies in sterolin knockout mice (Abcg5$^{-/-}$ or Abcg8$^{-/-}$) are characterized by histiocytic infiltration, multifocal fibrosis, and tissue calcification along with phytosterol accumulation [46,48,50]. Tao et al. have shown that stigmasterol accumulation in heart tissue leads to cardiac toxicity promoting mortality [66]. The authors have proposed that CVS outcomes in clinical sitosterolemia could be primarily due to phytosterols-induced cardiac fibrosis rather than cholesterol-driven atherosclerosis [66]. Despite macrophage infiltration and cardiac fibrosis, phytosterolemic mice did not demonstrate signs of atherosclerotic plaque formation in the aortic root and descending aorta [66]. Additionally, phytosterol oxidation products haven't shown any acceleration of atherosclerosis in ApoE$^{-/-}$ mice despite being well absorbed [67], suggesting that plant sterols or their oxidized derivatives may not be part of the atherosclerotic complications. Nevertheless, premature atherosclerosis is one

**TABLE 24.1** An overview of different mice models studied for understanding various effects linked with sitosterolemia.

| S. No. | Mouse type | Effect | Reference |
|---|---|---|---|
| *Models for evaluation* | | | |
| 1 | $Abcg5^{-/-}$, $Abcg8^{-/-}$, or $Abcg5^{-/-}$,$Abcg8^{-/-}$ | Sitosterolemia | [40–42,46,49,51–53] |
| 2 | $InAbcg5^{-/-}$,$Abcg8^{-/-}$, or $LiAbcg5^{-/-}$,$Abcg8^{-/-}$ | Sitosterolemia | [43] |
| 3 | LiInTgAbcg5/8, or LiTgAbcg5/8 | Rescue some features observed in sitosterolemia | [44,45] |
| *Cholesterol absorption* | | | |
| 4 | LiInTgAbcg5/8 | Reduced FCA and Increased FNSE | [44,45,54] |
| 5 | LiTgAbcg5/8 | Similar FCA and FNSE | [55] |
| 6 | $Abcg5^{-/-}$, $Abcg8^{-/-}$, and $Abcg5^{-/-}$,$Abcg8^{-/-}$ | Similar FCA and Decreased FNSE | [40,42,52,56,57] |
| 7 | $Abcg5^{-/-}$,$Abcg8^{-/-}$ | Decreased FCA and FNSE | [58,59] |
| 8 | $Abcg8^{-/-}$ | Increased FCA and Decreased FNSE | [60] |
| 9 | $InAbcg5^{-/-}$,$Abcg8^{-/-}$ | No change in FCA and FNSE | [43] |
| *Bile cholesterol secretion* | | | |
| 10 | LiInTgAbcg5/8 | Increased | [45] |
| 11 | $Abcg5^{-/-}$, $Abcg8^{-/-}$ and $LiAbcg5^{-/-}$,$Abcg8^{-/-}$ | Decreased | [40–43] |
| *Atherosclerosis* | | | |
| 12 | LiInTgAbcg5/8; $Ldlr^{-/-}$ | Reduction | [55] |
| 13 | LiTgAbcg5/8; $Ldlr^{-/-}$and LiTgAbcg5/8; $ApoE^{-/-}$ | No change | [44] |
| 14 | LiTgAbcg5/8; $Ldlr^{-/-}$(addition of Ezetimibe) | Reduction | [62] |
| *Macrophage RCT* | | | |
| 16 | $Abcg5^{-/-}$, $Abcg8^{-/-}$ | No change | [63–65] |

*Note*: Relative changes depicted in the table refer to the comparisons done with respective controls used in studies referenced.

of the rare phenotypes observed in sitosterolemia patients [27], and the atherosclerosis progression in sitosterolemia has also been studied using atherosclerotic $Ldlr^{-/-}$or $ApoE^{-/-}$ mice models mice over expressing ABCG5/G8 [44,55]. Overexpression of ABCG5/G8 in the liver and intestine (LiInTgAbcg5/8; $Ldlr^{-/-}$), but not liver alone (LiTgAbcg5/8; $Ldlr^{-/-}$), reduced plasma cholesterol and aortic atherosclerosis in atherogenic $LDLR^{-/-}$ mice [44,55]. Furthermore, significant reduction in ApoB lipoproteins and atherosclerosis development in LiTgAbcg5/8; $Ldlr^{-/-}$ mice fed with a western diet containing ezetimibe showed that the combination of increased biliary cholesterol secretion with decreased intestinal cholesterol absorption increased the net sterol loss and reduced atherosclerosis [62]. Interestingly, $LDLR^{-/-}$ mice fed with diets high in plant sterols did not accumulate phytosterols, but $ApoE^{-/-}$ mice did, suggesting that remnant clearance may also be an important factor determining xenosterol accumulation [68]. Extensive studies involving the loss of function of sterolins need to be pursued to demonstrate whether the opposite holds true. Abcg8$^{-/-}$ mice on a sitosterol-enriched diet (HS, high sitosterol) demonstrated reduced body weight and abdominal fat loss, with a more severe phenotype in females [49]. Indeed, the females lost significant body weight after 4 weeks on HS feeding, and were prone to death after 8–9 weeks of this dietary regimen [49]. The mechanisms underlying the abdominal fat loss are not known and are an area of active basic research.

## Xenosterols accumulation and cell membrane dysfunction

Though molecular mechanisms behind the xenosterols induced disease progression have not been completely eludicated, one mechanism that has been well studied over years is their accumulation in cell membranes altering its function [47,49,69]. An important feature in the diagnosis of sitosterolemia is macrothrombocytopenia, which refers to abnormally large platelet size and a resultant tendency for bleeding [47]. Remarkably, it can be the only clinical phenotype that a patient may manifest [27]. Studies of Abcg5$^{-/-}$ and Abcg8$^{-/-}$ mice have shown platelet hyperreactivity induced by accumulation of plant sterols in platelet membranes and therefore platelet dysfunction through membrane αIIbβ3 complex internalization and filamin A degradation [47]. Studies of stroke prone spontaneously hypertensive (SHR-SP) rats, which has mutant ABCG5 and demonstrated a sitosterolemia phenotype, showed that accumulation of plant sterols may also increase fragility of red cells via preferential partitioning of phytosterols into the outer leaflet of the red blood cell membranes, and reduced RBC membrane deformability in animals fed diets high in phytosterols [70,71]. The accumulation of phytosterols in the membranes of ova and sperm has been hypothesized to be the cause of infertility in Abcg8$^{-/-}$ mice [49]. Abcg5$^{-/-}$ and Abcg8$^{-/-}$mice showed drastically increased plant sterol levels in mouse brain (7- to 16-fold) [72], and phytosterol mixtures have also shown to affect embryological development and levels of circulating hormones in studies on different fish species [73–78]. Moreover, unpublished RNA-Seq results from our lab have also demonstrated changes in expression of genes involved in biosynthesis of steroids in adrenals of Abcg8$^{-/-}$ mice fed with a high sterol diet. This opens up an intriguing possibility that phytosterol accumulation in cell membranes of each of these tissues may lead to their altered cellular functions (Fig. 24.3).

Independent of specific cell types, the accumulation of xenosterols in lipid membranes directly affects the properties of the membrane, and its resultant function [79,80]. Studies performed in vitro using model lipid membranes (whose composition may range among different lipid classes) demonstrated that addition of xenosterols can modify membrane condensation, ordering, and interactions [79,81,82]. The effect of individual plant sterols on the membrane structure and function has been partly explored over years using different techniques [80,83–85]. Sphingolipid membranes treated with campesterol, sitosterol, and stigmasterol revealed the differential involvement of free and conjugated phytosterols in the formation of ordered lipid domains [80]. The authors further suggested that as the diversity of plant lipids in phytosterol mixtures increases, so does the possibility of local combinations of lipid species and its effect on the membrane organization [80]. Though it was hypothesized that the alkylated side chains of plant sterols make them more efficient than cholesterol in ordering of phospholipids and thus reducing membrane fluidity, in fact cholesterol turned out to

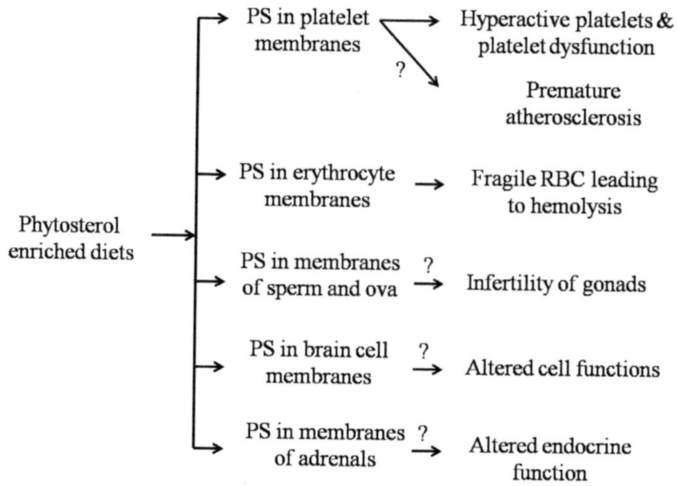


**FIGURE 24.3** **Proposed effects of plant sterol accumulation in membranes of different cells linked with sitosterolemia phenotype.**

possess higher order forming and membrane interacting abilities [81,86–89]. On the other hand studies comparing the major fungal xenosterol, ergosterol, and cholesterol, have shown contradictory results [83–85,90].

A more cognizant understanding of xenosterol effects on cell membrane properties may be gleaned from studies using an in vivo membrane system that can study the changes in such a complex system, and lipid rafts may be part of such a system [91,92]. Lipid rafts are dynamic microdomains of the plasma membrane mainly made of cholesterol and glycerosphingolipids, which acts as key players in membrane signal transduction mechanisms [93,94]. They are a key site in the plasma membrane for cholesterol deposition and perhaps a site where xenosterols may deposit and alter the membrane composition [91,92]. Indeed, depletion of plant sterols from plant plasma membrane disrupts lipid rafts through reduction of raft proteins and membrane sterols, indicating their affinity to participate in lateral structuring of plasma membranes [95]. Future proteomic studies on lipid rafts isolated from animals with xenosterols accumulation may expand our understanding of the effects of xenosterols on membrane composition and function. This may contribute to understanding the mechanisms responsible for clinical manifestations observed in sitosterolemia.

## Plant sterols and cardiovascular disorders

Besides pathological significance at higher concentrations, plant sterols also have physiological importance in CVS disease. Subtle changes in exogenous dietary cholesterol absorption and endogenous cholesterol synthesis can result from competition for these processes by phytosterols [96]. Plant sterols lower systemic LDL cholesterol levels through competitive inhibition of cholesterol uptake in the intestine [9,26]. As a result phytosterols have been investigated for their therapeutic potential in treating CVS diseases [9,16,26]. Around 30 studies have investigated their effects in various animal models of experimental atherosclerosis and most of them demonstrated protective effects, which include reduction in arterial lipid accumulation, inhibition of lesion initiation and progression [97–100]. In a study involving LDLR$^{-/-}$ mice, administration of high fat diet (HFD) with 2% w/w phytosterols for 16 weeks resulted in decreased plasma and arterial wall cholesterol with reduced atherosclerotic lesion and macrophage area [101]. Though plasma plant sterol (PS) levels were significantly higher in the 2% PS treated groups, PS content in the aortic wall remained similar between treated and non-treated groups, suggesting that the mechanism of PS induced protection against atherosclerosis is not associated with PS accumulation in the arterial wall [101]. However studies involving supplementation with plant sterol ester (PSE) demonstrated impaired endothelial function, aggravated ischemic brain injury, and increased atherosclerotic lesion in ApoE$^{-/-}$ mice fed with western type diet (WTD), and elevated plasma and aortic valve PS levels in humans [97]. In ApoE$^{-/-}$ mice plasma PS concentrations were strongly correlated with increased atherosclerotic lesion formation, raising concern about their use in CVS disease [97]. Meanwhile other studies having shown differing roles of different PS on atheroma development, with stigmasterol promoting cholesterol efflux and suppressing inflammatory cytokine secretion in response to lipid loading in macrophage foam cells, campesterol was largely inert, and sitosterol showing the inverse effects (increased pro-inflammatory cytokine secretion) [102]. One of the molecular mechanisms behind the protection from atherosclerosis is reduction of prostaglandin release from phytosterol accumulated macrophages was studied using cultured P388D1/MAB macrophages [103]. However, the mechanisms governing their protective effects on atherosclerosis development remain unclear and needs detailed investigation [26]. In the clinical context, the optimal dose of phytosterols to demonstrate the beneficial therapeutic effects remains to be determined. Till now, it is unclear whether or not is there a dose-dependent effect of plant sterols on decreasing LDL-C levels [104–106]. Additionally, there is debate as to whether the use of phytosterols may result in "pro-atherogenic effects" despite reduced circulatory cholesterol [9]. Given that these concerns have less scientific studies and that plant sterols can reduce LDL-C, the use of PS to reduce LDL-cholesterol for management of CVS complications has been recommended by both the American and British Heart Associations [107,108]. The value of phytosterol-enriched functional foods in reducing CVS events was also endorsed by European Commission, though there has not been a study to directly support this contention [109,110].

## Plant sterols and central nervous disorders disorders

Extensive research into different central nervous disorders (CNS) disorders has implicated disturbances of cholesterol metabolism in the pathogenesis of Alzheimer's disease (AD), multiple sclerosis (MS) and amyotrophic lateral sclerosis (ALS) [111,112]. Cholesterol metabolism in the brain is tightly regulated with all the available cholesterol synthesized in situ [112]. In contrast, dietary plant sterols have been consistently shown to cross the blood–brain barrier and accumulate in brain [72,113–115], albeit at very low levels. However, it is not clear whether accumulation of PS in brain has a specific temporal or spatial pattern or it has a biological impact [112]. Feeding C57BL/6NCrl mice with phytosterol-ester enriched diet for

6 weeks resulted in nearly doubled levels of phytosterols in the brain, while blocking their absorption for 6 months did not change their concentrations in the brain, likely due to their inability to efficiently exported out of the CNS [113]. Moreover, the concentration of sitosterol in cerebrospinal fluid (CSF), was found to be tenfold higher than 24(S)-hydroxycholesterol, the latter is efficiently exported out of the CNS [114,115]. Among individual phytosterols, those with a more complex hydrophobic side chain like sitosterol and stigmasterol, show less permeation across the endothelial barrier, while those with a lower side-chain complexity, like campesterol, can cross more easily [113,116,117]. Different earlier studies have indicated that sterols from plant origin may not enhance cognition in normo-cognitive settings, but they may have a potential for enhanced cognition during chemically induced and disease-related cognitive impairment [112]. Alzheimer's disease is a very common neurodegenerative disorder observed in elderly which is characterized by extracellular deposition of amyloid-β (Aβ) proteins, as senile plaques, and intracellular neurofibrillary tangles [112]. Effects of these phytosterols on amyloid precursor protein (APP) processing were studied using human neuroblastoma SH-SY5Y cells, and the results suggested that phytosterols, in general, are less amyloidogenic than cholesterol. Specifically, stigmasterol showed evidence of reduced Aβ generation through reduced β-secretase and γ-secretase activities [118]. However, due to limited brain absorption rates, the efficacy of stigmasterol to translate to a therapeutic drug is questionable and further research is needed [112].Another potential mechanistic link proposed that phytosterols may alter AD pathology by altering lipid rafts [119,120]. Earlier studies understanding the neuropathology of AD showed the amyloidogenic pathway is predominantly associated with lipid rafts, while non-amyloidogenic pathway is associated with non-lipid rafts [121,122]. Studies using sitosterol and stigmasterol to treat HT22 mouse hippocampal cells and SH-SY5Y human neuroblastoma cells demonstrated multiple interactions involving APP and lipid rafts [118,123]. Incorporation of sitosterol into the membrane of cultured hippocampal cells reduced cholesterol content in lipid rafts and prevented GOX (glucose oxidase)-induced oxidative stress and lipid peroxidation via estrogen receptor, PI3K, and Glycogen synthase kinase-3β through stimulation of antioxidant molecules, suggesting a possible beneficial role in AD [124]. In 12 and 18 months aged APP23 mice, a transgenic mouse model showing typical AD related pathological changes, significant accumulation of phytosterols, campesterol and sitosterol challenged their therapeutic importance in treating patients with AD [125]. Nevertheless, phytosterols were not found to be increased in the serum of patients with AD, in a recent study comparing men without AD ($n = 114$) to those with ($n = 18$) 'pure' AD [126]. There is some evidence that plant sterols may have beneficial effects in MS, however these studies only established associations and mechanistic links remain to be defined [112]. Whether phytosterols can affect other neurodegenerative disorders linked with altered cholesterol metabolism, such as Huntington disease, spinal cord injury, and Parkinson disease, remains to be studied.

## Conclusion

Xenosterols are the non-cholesterol sterols that form a major component of non-animal based foods. Over years, many aspects about their biology as well as pharmacology have been explored. We have moved beyond the old paradigm, which characterized them as inert molecules with minimal absorption into blood, to a new paradigm understanding that they have considerable absorption into blood, and may have biologically significant effects on health and disease. While Sitosterolemia demonstrated the pathological effects of excess xenosterol accumulation due to mutations in sterol transporters ABCG5 and ABCG8, there is growing data on the use of xenosterols such as plant sterols at lower concentrations for therapeutic interventions. While debate surrounding the use of plant sterols for treating patients with CVS and CNS disorders remains, data from many preclinical and/or clinical studies do suggest potential beneficial effects. Nevertheless, more rigorous and robust data with careful studies and involving well designed randomized clinical studies were required to prove their long-term efficacy and safety.

## References


[1] Kidambi S, Solca C, Patel S. Tracking the dietary cholesterol molecule. Future Lipidol 2006;1(3):357–71.

[2] Piironen V, et al. Plant sterols: biosynthesis, biological function and their importance to human nutrition. J Sci Food Agric 2000;80(7):939–66.

[3] Piironen V, et al. Plant sterols in vegetables, fruits and berries. J Sci Food Agric 2003;83(4):330–7.

[4] Patel SB. Sitosterolemia and Other Rare Sterol Disorders, in Dyslipidemias. Springer; 2015. 235-250.

[5] Mille-Lindblom C, von Wachenfeldt E, Tranvik LJ. Ergosterol as a measure of living fungal biomass: persistence in environmental samples after fungal death. J Microbiol Methods 2004;59(2):253–62.

[6] Wei JH, Yin X, Welander PV. Sterol Synthesis in Diverse Bacteria. Front Microbiol 2016;7:990.

[7] Moreau RA, Whitaker BD, Hicks KB. Phytosterols, phytostanols, and their conjugates in foods: structural diversity, quantitative analysis, and health-promoting uses. Prog Lipid Res 2002;41(6):457–500.

[8] Phillips K, et al. Sterol composition of shellfish species commonly consumed in the United States. Food Nutr Res 2012;56(1):18931.

[9] Plat, J., et al., Plant-based sterols and stanols in health & disease:"Consequences of human development in a plant-based environment?". Progress in lipid research, 2019.

[10] Trautwein EA, et al. LDL-cholesterol lowering of plant sterols and stanols-which factors influence their efficacy? Nutrients 2018;10(9).

[11] Patel S, Salen G, Sitosterolemia: xenophobia for the body. Evidence-Based Management of Lipid Disorders. Harley, England: TFM Publishing Ltd, 2010: p. 217–30.

[12] Weingartner O, Bohm M, Laufs U. Controversial role of plant sterol esters in the management of hypercholesterolaemia. Eur Heart J 2009;30(4):404–9.

[13] Weingartner O, et al. Plant sterol ester diet supplementation increases serum plant sterols and markers of cholesterol synthesis, but has no effect on total cholesterol levels. J Steroid Biochem Mol Biol 2017;169:219–25.

[14] Weingartner O, et al. The relationships of markers of cholesterol homeostasis with carotid intima-media thickness. PLoS One 2010;5(10):e13467.

[15] Fransen HP, et al. Customary use of plant sterol and plant stanol enriched margarine is associated with changes in serum plant sterol and stanol concentrations in humans. J Nutr 2007;137(5):1301–6.

[16] Kohler J, et al. Plant sterol enriched functional food and atherosclerosis. Br J Pharmacol 2017;174(11):1281–9.

[17] Patel SB. Plant sterols and stanols: their role in health and disease. J Clin Lipidol 2008;2(2):S11–9.

[18] Temel RE, et al. Compared with Acyl-CoA:cholesterol O-acyltransferase (ACAT) 1 and lecithin:cholesterol acyltransferase. ACAT2 displays the greatest capacity to differentiate cholesterol from sitosterol. J Biol Chem 2003;278(48):47594–601.

[19] Othman RA. S. B. Myrie, and P. J. Jones, Non-cholesterol sterols and cholesterol metabolism in sitosterolemia. Atherosclerosis 2013;231(2):291–9.

[20] Izar MC, et al. Phytosterols and phytosterolemia: gene–diet interactions. Genes Nutr 2011;6(1):17.

[21] van der Wulp MY, Verkade HJ, Groen AK. Regulation of cholesterol homeostasis. Mol Cell Endocrinol 2013;368(1–2):1–16.

[22] Jia L, Betters JL, Yu L. Niemann-pick C1-like 1 (NPC1L1) protein in intestinal and hepatic cholesterol transport. Annu Rev Physiol 2011;73:239–59.

[23] Field FJ, Born E, Mathur SN. LXR/RXR ligand activation enhances basolateral efflux of beta-sitosterol in CaCo-2 cells. J Lipid Res 2004;45(5):905–13.
[24] Kruit JK, et al. Emerging roles of the intestine in control of cholesterol metabolism. World J Gastroenterol 2006;12(40):6429–39.
[25] Brunham LR, et al. Intestinal ABCA1 directly contributes to HDL biogenesis in vivo. J Clin Invest 2006;116(4):1052–62.
[26] Gylling H, et al. Plant sterols and plant stanols in the management of dyslipidaemia and prevention of cardiovascular disease. Atherosclerosis 2014;232(2):346–60.
[27] Patel SB, Graf GA, Temel RE. ABCG5 and ABCG8: more than a defense against xenosterols. J Lipid Res 2018;59(7):1103–13.
[28] Bhattacharyya AK, Connor WE. Beta-sitosterolemia and xanthomatosis. A newly described lipid storage disease in two sisters. J Clin Invest 1974;53(4):1033–43.
[29] Rees DC, et al. Stomatocytic haemolysis and macrothrombocytopenia (Mediterranean stomatocytosis/macrothrombocytopenia) is the haematological presentation of phytosterolaemia. Br J Haematol 2005;130(2):297–309.
[30] Wang Z, et al. Specific macrothrombocytopenia/hemolytic anemia associated with sitosterolemia. Am J Hematol 2014;89(3):320–4.
[31] Ducrou W, Kimber RJ. Stomatocytes, haemolytic anaemia and abdominal pain in Mediterranean migrants. Some examples of a new syndrome? Med J Aust 1969;2(22):1087–91.
[32] Salen G, et al. Lethal atherosclerosis associated with abnormal plasma and tissue sterol composition in sitosterolemia with xanthomatosis. J Lipid Res 1985;26(9):1126–33.
[33] Kolovou G, et al. Coronary bypass grafts in a young girl with sitosterolemia. Eur Heart J 1996;17(6):965–6.
[34] Kawamura R, et al. Acute myocardial infarction in a 25-year-old woman with sitosterolemia. J Clin Lipidol 2018;12(1):246–9.
[35] Yagasaki H, et al. Carotid intima media thickness in a girl with sitosterolemia carrying a homozygous mutation in the ABCG5 gene. J Pediatr Endocrinol Metab 2017;30(9):1007–11.
[36] Mymin D, et al. Image in cardiovascular medicine. Aortic xanthomatosis with coronary ostial occlusion in a child homozygous for a nonsense mutation in ABCG8. Circulation 2003;107(5):791.
[37] Miettinen TA, et al. Liver transplantation in a patient with sitosterolemia and cirrhosis. Gastroenterology 2006;130(2):542–7.
[38] Bazerbachi F, et al. Cryptogenic Cirrhosis and Sitosterolemia: A Treatable Disease If Identified but Fatal If Missed. Ann Hepatol 2017;16(6):970–8.
[39] Lee JY, et al. Crystal structure of the human sterol transporter ABCG5/ABCG8. Nature 2016;533(7604):561–4.
[40] Plosch T, et al. Sitosterolemia in ABC-transporter G5-deficient mice is aggravated on activation of the liver-X receptor. Gastroenterology 2004;126(1):290–300.
[41] Klett EL, et al. A mouse model of sitosterolemia: absence of Abcg8/sterolin-2 results in failure to secrete biliary cholesterol. BMC Med 2004;2:5.
[42] Yu L, et al. Disruption of Abcg5 and Abcg8 in mice reveals their crucial role in biliary cholesterol secretion. Proc Natl Acad Sci U S A 2002;99(25):16237–42.
[43] Wang J, et al. Relative roles of ABCG5/ABCG8 in liver and intestine. J Lipid Res 2015;56(2):319–30.
[44] Wu JE, et al. Hepatic ABCG5 and ABCG8 overexpression increases hepatobiliary sterol transport but does not alter aortic atherosclerosis in transgenic mice. J Biol Chem 2004;279(22):22913–25.
[45] Yu L, et al. Overexpression of ABCG5 and ABCG8 promotes biliary cholesterol secretion and reduces fractional absorption of dietary cholesterol. J Clin Invest 2002;110(5):671–80.
[46] Chase TH, et al. The mouse mutation "thrombocytopenia and cardiomyopathy" (trac) disrupts Abcg5: a spontaneous single gene model for human hereditary phytosterolemia/sitosterolemia. Blood 2010;115(6):1267–76.
[47] Kanaji T, et al. Platelet hyperreactivity explains the bleeding abnormality and macrothrombocytopenia in a murine model of sitosterolemia. Blood 2013;122(15):2732–42.
[48] McDaniel AL, et al. Phytosterol feeding causes toxicity in ABCG5/G8 knockout mice. Am J Pathol 2013;182(4):1131–8.
[49] Solca C, Tint GS, Patel SB. Dietary xenosterols lead to infertility and loss of abdominal adipose tissue in sterolin-deficient mice. J Lipid Res 2013;54(2):397–409.
[50] Wilson DW, et al. Inflammatory dilated cardiomyopathy in Abcg5-deficient mice. Toxicol Pathol 2013;41(6):880–92.
[51] Kruit JK, et al. Plant sterols cause macrothrombocytopenia in a mouse model of sitosterolemia. J Biol Chem 2008;283(10):6281–7.

[52] Yu L, et al. Expression of ABCG5 and ABCG8 is required for regulation of biliary cholesterol secretion. J Biol Chem 2005;280(10):8742–7.

[53] Yu L, et al. Selective sterol accumulation in ABCG5/ABCG8-deficient mice. J Lipid Res 2004;45(2):301–7.

[54] Langheim S, et al. ABCG5 and ABCG8 require MDR2 for secretion of cholesterol into bile. J Lipid Res, 2005. 2005;46(8):1732–8.

[55] Wilund KR, et al. High-level expression of ABCG5 and ABCG8 attenuates diet-induced hypercholesterolemia and atherosclerosis in $Ldlr^{-/-}$ mice. J Lipid Res 2004;45(8):1429–36.

[56] van der Veen JN, et al. Activation of the liver X receptor stimulates trans-intestinal excretion of plasma cholesterol. J Biol Chem 2009;284(29):19211–9.

[57] Jakulj L, et al. Transintestinal Cholesterol Transport Is Active in Mice and Humans and Controls Ezetimibe-Induced Fecal Neutral Sterol Excretion. Cell Metab 2016;24(6):783–94.

[58] Nguyen TM, et al. Cholesterol esterification by ACAT2 is essential for efficient intestinal cholesterol absorption: evidence from thoracic lymph duct cannulation. J Lipid Res 2012;53(1):95–104.

[59] Zhang LS, et al. ABCG5/G8 deficiency in mice reduces dietary triacylglycerol and cholesterol transport into the lymph. Lipids 2015;50(4):371–9.

[60] Wang HH, et al. Quantifying anomalous intestinal sterol uptake, lymphatic transport, and biliary secretion in Abcg8(-/-) mice. Hepatology 2007;45(4):998–1006.

[61] Matthan NR, Lichtenstein AH. Approaches to measuring cholesterol absorption in humans. Atherosclerosis 2004;174(2):197–205.

[62] Basso F, et al. Hepatic ABCG5/G8 overexpression reduces apoB-lipoproteins and atherosclerosis when cholesterol absorption is inhibited. J Lipid Res 2007;48(1):114–26.

[63] Altemus JB, Patel SB, Sehayek E. Liver-specific induction of Abcg5 and Abcg8 stimulates reverse cholesterol transport in response to ezetimibe treatment. Metabolism 2014;63(10):1334–41.

[64] Calpe-Berdiel L, Escola-Gil JC, Blanco-Vaca F. New insights into the molecular actions of plant sterols and stanols in cholesterol metabolism. Atherosclerosis 2009;203(1):18–31.

[65] Dikkers A, et al. Hepatic ABCG5/G8 overexpression substantially increases biliary cholesterol secretion but does not impact in vivo macrophage-to-feces RCT. Atherosclerosis 2015;243(2):402–6.

[66] Tao C, et al. Stigmasterol accumulation causes cardiac injury and promotes mortality. Commun Biol 2019;2:20.

[67] Tomoyori H, et al. Phytosterol oxidation products are absorbed in the intestinal lymphatics in rats but do not accelerate atherosclerosis in apolipoprotein E-deficient mice. J Nutr 2004;134(7):1690–6.

[68] Nunes VS, et al. Phytosterol containing diet increases plasma and whole body concentration of phytosterols in apoE-KO but not in LDLR-KO mice. J Bioenerg Biomembr 2019;51(2):131–6.

[69] Eckstein J, Holzhutter HG, Berndt N. The importance of membrane microdomains for bile salt-dependent biliary lipid secretion. J Cell Sci 2018;131.(5).

[70] Ratnayake WM, et al. Vegetable oils high in phytosterols make erythrocytes less deformable and shorten the life span of stroke-prone spontaneously hypertensive rats. J Nutr 2000;130(5):1166–78.

[71] Nghiem-Rao TH, Patel SB. Investigating sitosterolemia to understand lipid physiology. Clin Lipidol 2013;8(3):649–58.

[72] Jansen PJ, et al. Dietary plant sterols accumulate in the brain. Biochim Biophys Acta 2006;1761(4):445–53.

[73] Gilman CI, et al. Effects of a phytosterol mixture on male fish plasma lipoprotein fractions and testis P450scc activity. Gen Comp Endocrinol 2003;130(2):172–84.

[74] Leusch FD, MacLatchy DL. In vivo implants of beta-sitosterol cause reductions of reactive cholesterol pools in mitochondria isolated from gonads of male goldfish (*Carassius auratus*). Gen Comp Endocrinol 2003;134(3):255–63.

[75] MacLatchy DL, Van Der Kraak GJ. The phytoestrogen beta-sitosterol alters the reproductive endocrine status of goldfish. Toxicol Appl Pharmacol 1995;134(2):305–12.

[76] Mattsson K, et al. Effects of wood-related sterols on the offspring of the viviparous blenny, *Zoarces viviparus* L. Ecotoxicol Environ Saf 2001;49(2):122–30.

[77] Nakari T, Erkomaa K. Effects of phytosterols on zebrafish reproduction in multigeneration test. Environ Pollut 2003;123(2):267–73.

[78] Sharpe RL, et al. Beta-sitosterol and 17beta-estradiol alter gonadal steroidogenic acute regulatory protein (StAR) expression in goldfish, *Carassius auratus*. Gen Comp Endocrinol 2007;151(1):34–41.

[79] Hac-Wydro K. The replacement of cholesterol by phytosterols and the increase of total sterol content in model erythrocyte membranes. Chem Phys Lipids 2010;163(7):689–97.

[80] Grosjean K, et al. Differential effect of plant lipids on membrane organization: specificities of phytosphingolipids and phytosterols. J Biol Chem 2015;290(9):5810–25.
[81] Hac-Wydro K, et al. The study on the interaction between phytosterols and phospholipids in model membranes. Chem Phys Lipids 2007;150(1):22–34.
[82] Gao WY, Quinn PJ, Yu ZW. The role of sterol rings and side chain on the structure and phase behaviour of sphingomyelin bilayers. Mol Membr Biol 2008;25(6–7):485–97.
[83] Xu X, et al. Effect of the structure of natural sterols and sphingolipids on the formation of ordered sphingolipid/sterol domains (rafts). Comparison of cholesterol to plant, fungal, and disease-associated sterols and comparison of sphingomyelin, cerebrosides, and ceramide. J Biol Chem 2001;276(36):33540–6.
[84] Oradd G, Shahedi V, Lindblom G. Effect of sterol structure on the bending rigidity of lipid membranes: A (2)H NMR transverse relaxation study. Biochim Biophys Acta 2009;1788(9):1762–71.
[85] Cournia Z, Ullmann GM, Smith JC. Differential effects of cholesterol, ergosterol and lanosterol on a dipalmitoyl phosphatidylcholine membrane: a molecular dynamics simulation study. J Phys Chem B 2007;111(7):1786–801.
[86] Hodzic A, et al. Differential modulation of membrane structure and fluctuations by plant sterols and cholesterol. Biophys J 2008;94(10):3935–44.
[87] Su Y, et al. Condensation effect of cholesterol, stigmasterol, and sitosterol on dipalmitoylphosphatidylcholine in molecular monolayers. Colloids Surf A: Physicochem Eng Aspects 2007;293(1–3):123–9.
[88] Haąc-Wydro K, Dynarowicz-Łaątka P. The impact of sterol structure on the interactions with sphingomyelin in mixed Langmuir monolayers. J Phys Chem B 2008;112(36):11324–32.
[89] Halling KK, Slotte JP. Membrane properties of plant sterols in phospholipid bilayers as determined by differential scanning calorimetry, resonance energy transfer and detergent-induced solubilization. Biochim Biophys Acta 2004;1664(2):161–71.
[90] Wu R, et al. Phase diagram of stigmasterol-dipalmitoylphosphatidylcholine mixtures dispersed in excess water. Biochim Biophys Acta 2006;1758(6):764–71.
[91] Lingwood D, Simons K. Lipid rafts as a membrane-organizing principle. Science 2010;327(5961):46–50.
[92] Binder WH, Barragan V, Menger FM. Domains and rafts in lipid membranes. Angew Chem Int Ed 2003;42(47):5802–27.
[93] Waheed AA, Freed EO. The role of lipids in retrovirus replication. Viruses 2010;2(5):1146–80.
[94] Simons K, Toomre D. Lipid rafts and signal transduction. Nat Rev Mol Cell Biol 2000;1(1):31–9.
[95] Roche Y, et al. Depletion of phytosterols from the plant plasma membrane provides evidence for disruption of lipid rafts. FASEB J 2008;22(11):3980–91.
[96] Cohen DE. Balancing cholesterol synthesis and absorption in the gastrointestinal tract. J Clin Lipidol 2008;2(2):S1–S13.
[97] Weingartner O, et al. Vascular effects of diet supplementation with plant sterols. J Am Coll Cardiol 2008;51(16):1553–61.
[98] Plat J, et al. Plant sterol or stanol esters retard lesion formation in LDL receptor-deficient mice independent of changes in serum plant sterols. J Lipid Res 2006;47(12):2762–71.
[99] Weingartner O, et al. Differential effects on inhibition of cholesterol absorption by plant stanol and plant sterol esters in apoE-/- mice. Cardiovasc Res 2011;90(3):484–92.
[100] Plat J, et al. Plant sterol or stanol esters retard lesion formation in LDL receptor-deficient mice independent of changes in serum plant sterols. J Lipid Res 2006;47(12):2762–71.
[101] Bombo RP, et al. Dietary phytosterol does not accumulate in the arterial wall and prevents atherosclerosis of LDLr-KO mice. Atherosclerosis 2013;231(2):442–7.
[102] Sabeva NS, et al. Phytosterols differentially influence ABC transporter expression, cholesterol efflux and inflammatory cytokine secretion in macrophage foam cells. J Nutr Biochem 2011;22(8):777–83.
[103] Awad AB, Toczek J, Fink CS. Phytosterols decrease prostaglandin release in cultured P388D1/MAB macrophages. Prostaglandins Leukot Essent Fatty Acids 2004;70(6):511–20.
[104] Ference BA, et al. Effect of long-term exposure to lower low-density lipoprotein cholesterol beginning early in life on the risk of coronary heart disease: a Mendelian randomization analysis. J Am Coll Cardiol 2012;60(25):2631–9.
[105] Musa-Veloso K, et al. A comparison of the LDL-cholesterol lowering efficacy of plant stanols and plant sterols over a continuous dose range: results of a meta-analysis of randomized, placebo-controlled trials. Prostaglandins, Leukotrienes and Essential Fatty Acids 2011;85(1):9–28.

[106] Demonty I, Ras RT, Trautwein EA. Comment on:" A comparison of the LDL-cholesterol lowering efficacy of plant stanols and plant sterols over a continuous dose range: results of a meta-analysis of randomized, placebo controlled trials" by Musa-Veloso K et al., Prostagl. Leucot. Essential Fatty Acids 2011, 85, 9-28. Prostaglandins, leukotrienes, and essential fatty acids 2011;85(1):7.
[107] Foundation, B.H., Reducing your blood cholesterol. 2014.
[108] Yang W, et al. The effectiveness and cost-effectiveness of plant sterol or stanol-enriched functional foods as a primary prevention strategy for people with cardiovascular disease risk in England: a modeling study. Eur J Health Econ 2018;19(7):909–22.
[109] EFSA Panel on Dietetic Products, N. and Allergies, Scientific opinion on the substantiation of a health claim related to 3 g/day plant sterols/stanols and lowering blood LDL-cholesterol and reduced risk of (coronary) heart disease pursuant to Article 19 of Regulation (EC) No 1924/2006. EFSA J, 2012. 10(5): p. 2693.
[110] Shortt C. Authorised EU health claims for phytosterols, in Foods, Nutrients and Food Ingredients with Authorised EU Health Claims, Vol. 2. Elsevier; 2015. 31-47.
[111] Anchisi L, et al. Cholesterol homeostasis: a key to prevent or slow down neurodegeneration. Front Physiol 2012;3:486.
[112] Vanmierlo T, et al. Plant sterols: Friend or foe in CNS disorders? Prog Lipid Res 2015;58:26–39.
[113] Vanmierlo T, et al. Dietary intake of plant sterols stably increases plant sterol levels in the murine brain. J Lipid Res 2012;53(4):726–35.
[114] Vanmierlo T, et al. The plant sterol brassicasterol as additional CSF biomarker in Alzheimer's disease. Acta Psychiatr Scand 2011;124(3):184–92.
[115] Popp J, et al. Cerebral and extracerebral cholesterol metabolism and CSF markers of Alzheimer's disease. Biochem Pharmacol 2013;86(1):37–42.
[116] Saeed AA, et al. Effects of a disrupted blood-brain barrier on cholesterol homeostasis in the brain. J Biol Chem 2014;289(34):23712–22.
[117] Smiljanic K, et al. Aging induces tissue-specific changes in cholesterol metabolism in rat brain and liver. Lipids 2013;48(11):1069–77.
[118] Burg VK, et al. Plant sterols the better cholesterol in Alzheimer's disease? A mechanistical study. J Neurosci 2013;33(41):16072–87.
[119] Korade Z, Kenworthy AK. Lipid rafts, cholesterol, and the brain. Neuropharmacology 2008;55(8):1265–73.
[120] Catani MV, et al. Essential dietary bioactive lipids in neuroinflammatory diseases. Antioxidants Redox Signal 2018;29(1):37–60.
[121] Beel AJ, et al. Direct binding of cholesterol to the amyloid precursor protein: An important interaction in lipid-Alzheimer's disease relationships? Biochim Biophys Acta 2010;1801(8):975–82.
[122] Ehehalt R, et al. Amyloidogenic processing of the Alzheimer beta-amyloid precursor protein depends on lipid rafts. J Cell Biol 2003;160(1):113–23.
[123] Wang J, Wu F, Shi C. Substitution of membrane cholesterol with β-sitosterol promotes nonamyloidogenic cleavage of endogenous amyloid precursor protein. Neuroscience 2013;247:227–33.
[124] Shi C, et al. Incorporation of beta-sitosterol into the membrane increases resistance to oxidative stress and lipid peroxidation via estrogen receptor-mediated PI3K/GSK3beta signaling. Biochim Biophys Acta 2013;1830(3):2538–44.
[125] Lutjohann D, et al. Profile of cholesterol-related sterols in aged amyloid precursor protein transgenic mouse brain. J Lipid Res 2002;43(7):1078–85.
[126] Sittiwet C, et al. Serum noncholesterol sterols in Alzheimer's disease: the Helsinki Businessmen Study. Transl Res 2018;202:120–8.

CHAPTER

# 25

# Adipose tissue development and metabolic regulation

*Hai P. Nguyen, Danielle Yi, Hei Sook Sul*

Department of Nutritional Sciences & Toxicology, Endocrinology Program, University of California, Berkeley, CA, United States

OUTLINE



## Function and importance of adipose tissues

In modern society, excess WAT leading to obesity has become an epidemic and is closely associated with metabolic diseases. However, WAT deficiency, such as lipodystrophy, also manifests as insulin resistance (IR) with ectopic lipid accumulation in other tissues, such as muscle and liver, underscoring the importance of maintaining the proper WAT mass [1,2]. The main function of white adipocytes is to store excess calories in the form of triglycerides, while secreting adipokines. The ability to store lipid effectively prevents lipotoxicity in other tissues including muscle, liver, and heart in which ectopic lipids deposits result in insulin resistance, nonalcoholic liver disease (NAFLD), and cardiovascular disease [1]. Additionally, WAT provides insulation in cold conditions. WAT also "wrap" around and protect critical internal organs such as the heart, adrenal glands, kidneys, and ovaries. Moreover, WAT provides

Lipid Signaling and Metabolism. http://dx.doi.org/10.1016/B978-0-12-819404-1.00025-7

cushioning in regions with high mechanical stress including palm, heel, and buttocks. Thus, white adipose tissue (WAT) is categorized anatomically into subcutaneous adipose depots and visceral adipose depots which are mainly for energy storage [2]. Subcutaneous WAT is found in interscapular and inguinal regions, while visceral WAT is found in perigonadal (epididymal), perirenal, epicardial, retroperitoneal, and mesenteric regions. These WAT depots are distinct and heterogeneous [3–6].

An additional class of adipose tissue referred as brown adipose tissue (BAT) found in many mammals has function distinct from WAT by burning energy for heat generation to maintain body temperature [7]. The major function of BAT is mediating adaptive thermogenesis or nonshivering thermogenesis [8]. BAT is enriched with a high number of mitochondria that possess a specialized inner mitochondrial $H^+$/fatty acid symporter, uncoupling protein 1 (UCP1) for thermogenesis, heat production. Since brown adipocytes are surrounded with many capillaries, heat generated by BAT can be immediately distributed by vascular system to maintain body temperature. In mice, expression of UCP1 is restricted to only BAT in unstimulated condition. However, upon cold exposure, $UCP1^+$ thermogenic adipocytes, termed "beige" or "brite" cells can arise in WAT depots, especially subcutaneous WAT, although whether this is from recruitment and/or transdifferentiation of white adipocytes is currently not well understood [9]. The presence of BAT or BAT-like tissues in human adults has now been established. BAT is detected mainly in the supraclavicular, paravertebral, and cervical regions in adult humans. These tissues have been shown to increase in adults upon cold exposure and are known to inversely correlate with adiposity. Although the underlying mechanism is not clear, BAT affects insulin sensitivity also [10]. BAT, not only has high capacity of glucose and fatty acid (FA) uptake, but also secretes adipokines, all potentially contributing to insulin sensitivity [11].

Bone marrow adipose tissue (BMAT) is an adipose depot with unique features [12]. Unlike other WATs, BMAT are more rigid due to structural composition of cortical bone. BMAT was demonstrated to support critical cell differentiation in bone marrow. There is evidence that BMAT is derived from distinct progenitors from both BAT and iWAT. BMAT is composed of two different populations including regulating BMAT (rMAT) that affect hematopoiesis and constitutive BMAT (cMAT) that is important for early vertebrate development. BMAT also participate in local and systematic metabolic processes.

These various adipose tissues critical not only for energy metabolism but also secrete adipokines that regulate various biological processes including appetite, metabolism, and insulin sensitivity. Origin, development and function of these adipose tissues have been studied extensively in recent years. A better understanding of development and function of adipose tissues may provide future therapeutic targets for obesity and related diseases, such as type 2 diabetes, hepatosteatosis, and cardiovascular diseases.

## Developmental origin of WAT

Adipose tissue is composed of adipocytes and a so-called stromal vascular fraction which includes preadipocytes, macrophages, endothelial cells and immune cells. Preadipocytes are thought to arise from mesenchymal cells which undergo a process called adipogenesis to commit and differentiate into mature adipocytes. Due to the heterogeneity of the stromal vascular fraction in adipose tissue, fluorescence-activated cell sorting (FACS) using stem cell markers to selectively enrich SVF for the progenitor/precursor populations by sorting for $Lin^-$ ($Cd31^-$, $Cd45^-$, $Ter119^-$) eliminated the majority of endothelial, hematopoietic cells and erythrocytes [13]. However, the selected progenitor/precursor population may be at different stages of differentiation, such as stem cells, committed preadipocytes, or those cells at early stage of differentiation. In order to select pure adipose precursors, Rodeheffer et al. isolated a $Lin^-$: $CD29^+$, $CD34^+$: $Sca1^+$: $CD24^+$ population of proliferating adipose precursors that gave rise to $Cd24^-$ cells in vivo during adipogenesis [13]. Additionally, $Cd24^-$ cells of the precursor population were shown to represent preadipocytes that express PPARγ and C/EBPα, key adipogenic transcription factors. Since $PGFFR\alpha^+$ labeled both $Cd24^+$ and $Cd24^-$ precursor populations in WAT, isolation of $Lin^-$: $PGFFR\alpha^+$ cells may represent a strategy to enrich for adipocyte precursors in adipose tissue. Using inducible Pref-1 promoter coupled with two fluorescent reporters—H2BGFP for transient labeling and Rosa26-flox-stop-flox-tdTomato for permanent labeling, Pref-1+ precursor cells was shown to first appear at E10.5 in mouse embryogenesis in the dorsal mesenteric region at the presumptive inguinal or dorsal subcutaneous depots. At E17.5, these precursor cells differentiated into lipid-containing adipocytes forming subcutaneous WAT. By E19.5, the number of lipid-filled cells more than doubled in this region, indicating hyperplasia as a mechanism for WAT expansion during embryogenesis. Interestingly, no Pref-1+ cells were detected in the visceral WAT development. These cells only appeared in visceral WAT postnatally. This is an evidence of that subcutaneous WAT starts its development perinatally, while visceral fat development takes place after birth. In humans, light microscope examination showed the WAT first appear at the 14th and 16th weeks of prenatal life. In the context of the adipogenic lineage, $Pref\text{-}1^+$ cells and showed their mesenchymal origin ($Sox9^+$, $Cd29^+$, $Sca1^+$, $Cd105^+$, and $Cd34^+$). The transiently labeled $Pref\text{-}1^+$ cells did not yet express the adipogenic transcription factor PPARγ or adipose commitment factor Zfp423, but were proliferative precursors based on expression of Cd24 and Ki-67 and incorporation of BrdU [14]. Some of the permanently labeled Pref-1 cells had lipid laden morphology and expressed Zfp423, PPARγ and C/EBPα indicating that Pref-1 cells indeed are adipose precursors. Additionally, using Pref-1-reverse tetracycline *trans*-activator (rtTA)/Tet-responsive element (TRE)-Cre, Sox9 ablation from WAT precursors in mice in an inducible manner diminishes the pool of proliferating $Pref\text{-}1^+$ cells by progressing these cells to become $PGFFR\alpha^+$ cells that do not proliferate but express early adipogenic genes. $Pref\text{-}1^+$: $Sox9^+$ cells are shown to appear prior to $PGFFR\alpha^+$ cells in the adipogenic pathway. To maintain early adipose precursors, Sox9 also activates Meis1, which in turn inhibits adipocyte differentiation [15]. There is clear evidence that adipose SVF contains a hierarchy of progenitor populations with different degree of progression from adipose commitment to differentiation. However, FACS using cell surface markers results in a biased sampling of known cell types. Recently, single-cell RNA sequencing allows to explore heterogeneity if cellular population in a completely unbiased and detailed manner. Using this method, it has recently been reported that myofibroblast arising from diet-induced obesity came from highly proliferative $Cd9^{high}$, $PDGFR\alpha^+$ cells with high Pref-1 expression. In contract, $Cd9^{low}$, $PDGFR\alpha^+$ population was enriched for PPARγ and C/EBPα, with low Pref-1 expression, having low proliferative capacity but high adipogenic potential [16]. Overall, the relationship between cell populations identified based on the expression of various markers needs further investigation. Identification and characterization of stage-specific markers may help to isolate and define various precursor populations and their relationship during WAT development.

Given the difference in the formation and function of subcutaneous WAT versus visceral fat, it is still unclear whether distinct progenitor populations give rise to subcutaneous WAT and visceral WAT. However, there are several studies supporting such hypothesis. It was reported that unlike subcutaneous WAT or BAT, there were six different visceral WAT depots occurred from cell expressing Wilms tumor 1 (Wt1) in late gestation [17]. It also suggested that WAT depots associated with visceral organs have a mesothelial layer to supply adipocyte progenitors. Subcutaneous adipocytes were derived from cells expressing Prx1, a homeobox transcription factor expressed in embryonic limb and bud [18]. Interestingly, lineage tracing using mT/mG with Myf5-Cre that was assumed to labeled only muscle and BAT, also labeled most cells of anterior subcutaneous WAT [19]. When neural crest cells were marked using Sox10-Cre and R26-YFP, Billon et al. found that only cephalic WAT around salivary glands was labeled, but not other WAT [20]. This study shows that craniofacial adipocytes arise from neural crest cells. Utilizing recent advantageous scRNA-seq, Min et al. showed that there were four subtypes of human adipocytes that derived from distinct progenitors. They were shown to have different gene-expression profiles associated with different adipocyte metabolic functions including lipogenesis specialized adipocytes, thermogenic/"brite" cells, and extracellular matrix and regulatory signaling sensitive adipocytes. Similarly, Gupta lab showed that there are two a distinct pro-inflammatory/pro-fibrogenic and thermogenic/brite progenitors within the visceral fat [21]. These studies, overall, highlight that a simplistic division of WAT into visceral and subcutaneous may need to be reconsidered. It is probable that different WAT depots may have different origins and even cells within the same adipose tissue may be heterogeneous in origin.

## Regulation of WAT development

In studying adipocyte differentiation in vitro, preadipocyte cell lines, such as 3T3-L1 and F442A or primary SVF cells from adipose tissue, are treated with a differentiation cocktail containing DEX, MIX, insulin and/or TZD. A cascade of transcription factors is involved in adipose differentiation. Notably, PPARγ and C/EBPα are the key drivers of adipogenesis, while C/EBPβ and C/EBPδ expressed early in differentiation induce C/EBPα and PPARγ [22,23]. Forced expression of PPARγ is sufficient to induce adipocyte differentiation in fibroblast. There has been no factor found to induce differentiation without PPARγ. The C/EBP family and Kruppel-like factors (KLFs) have been shown to induce PPARγ promoters. In contrast, GATA factors were shown to repress PPARγ expression [24]. Some C/EBP family members, including C/EBPα, C/EBPβ, and C/EBPδ are expressed highly in adipocytes [25]. Early induction of C/EBPβ and C/EBPδ results in increased C/EBPα expression. The KLFs family is composed of C2H2 zinc-finger transcription factors that regulate apoptosis, proliferation, and differentiation. KLF proteins were shown to involve in various steps during adipocyte differentiation. For instance, KLF15 can promote adipocyte differentiation and even induce glucose transporter 4 (GLUT4) while KLF5 is induced by C/EBPβ and C/EBPδ and in turn increase PPARγ2 [26,27]. In contrast, KLF2 and KLF7 both inhibit adipogenesis. KLF2 directly binds to PPARγ2 promoter and inhibits transcriptional activity in early development and KLF5 then binds to PPARγ promoter and replace KLF2 to activate PPARγ transcription [28]. There are other factors are involved in different stages of adipocyte differentiation including KRX20 (early growth response protein-2), LXRα and LXRβ (Liver X receptors), and SREBP1c [29].

Numerous soluble factors and growth factors regulate adipose differentiation also. For example, by binding to their cognate plasma membrane receptors, insulin/IGF1 enhances, whereas Wnt and TGFβ inhibit differentiation [27,30]. We originally cloned and identified preadipocyte factor-1 (Pref-1) as an adipose tissue-specific gene, expressed in preadipocytes, but not in mature adipocytes. Pref-1 is detected in several tissues during embryogenesis but extinguished postnatally and is restricted mainly to preadipocytes of adipose tissue, with an exception of certain neuroendocrine cells. Pref-1 is highly expressed before adipose conversion but is downregulated during adipogenesis, Pref-1 level inversely correlating with the degree of differentiation [13,31–34] and thus Pref-1 is used as a preadipocyte marker. By lineage tracing showed that Pref-1 marks very early adipose progenitors during embryogenesis and that they are required for adipose development and expansion in mice in adult stage. Pref-1 level increases greatly in murine models of lipodystrophies, while it is significantly lower in obesity models. Pref-1 is detected also in human preadipocytes and its level is higher in lipodystrophy-like syndrome, while lower in human obesity. In this regard, adipocytes have been reported to be derived from distinct cells [3,33–35]. In fact, recent studies characterizing SVF cells of WAT showed presence of early adipose progenitors with high-level Pref-1 expression, which progress into preadipocytes/adipose precursors having a low Pref-1 level. In addition, studies also found profibrogenic precursors that contain high Pref-1 level mainly in visceral WAT when mice were on high-fat diet (HFD), which, depending on environmental cues, inhibit preadipocyte differentiation or are converted into adipocytes.

In cultured 3T3-L1 cells, constitutively expressing Pref-1 prevents differentiation [35,36], and knockdown (KD) of Pref-1 enhances adipocyte differentiation. DEX, a component of adipogenic cocktail, downregulates Pref-1, allowing cells to undergo adipocyte differentiation [37]. Moreover, inhibitory function of Pref-1 in adipogenesis was unequivocally demonstrated in Pref-1-KO mice that showed increased adiposity [38]. Conversely, ectopic Pref-1 overexpression protected mice from diet-induced obesity, causing partial lipodystrophy [39,40]. In this regard, Pref-1 is synthesized as a transmembrane protein, having an N-terminal signal sequence and a single transmembrane-spanning domain [35]. The transmembrane form of Pref-1 is cleaved at the juxtamembrane region by TNFα converting enzyme (TACE/ADAM17) [41], releasing the ectodomain to become biologically active soluble Pref-1. Thus, treatment with soluble Pref-1 inhibits 3T3-L1 or primary preadipocyte differentiation, whereas the expression of the uncleavable membrane mutant is ineffective [41,42]. It has been shown that Pref-1 activates MEK/ERK in a time and dose-dependent manner [43]. By activating MEK/ERK, Pref-1 upregulates Sox9, which in turn suppresses C/EBPβ and C/EBPδ to inhibit adipocyte differentiation [44]. In understanding mechanism underlying Pref-1 function in adipogenesis, the most critical component missing at this time is the plasma membrane receptor.

## Transcriptional regulation of the thermogenic adipose program

Brown adipose tissue uniquely harbors UCP1, to dissipate chemical energy as heat. Upon cold exposure, sympathetic nervous system (SNS) releases norepinephrine via B3 adrenergic receptor-cAMP-PKA-p38 pathway to stimulate lipolysis to increase fatty acids to fuel mitochondria for oxidation. Early work describing regulation of the UCP1 promoter centered on the norepinephrine-β3AR-cAMP-cyclic AMP response element-binding protein (CREB)/p38

MAP kinase axis central to the response to cold. Thus, several target genes of this signaling pathway have been described such as PPARγ-coactivator 1α (PGC1α), CCAAT-enhancer binding protein β(C/EBPβ), deiodinase 2 (Dio2) as well as UCP1 itself [45].

In the work of identifying cell-autonomous transcriptional effectors of thermogenesis, the coregulatory PR domain containing 16 (PRDM16) is the first identified, BAT-enriched coregulator of the BAT gene program by RT-qPCR analysis of WAT and BAT [46]. Harms et al. showed that PRDM16 determines cell fate as PRDM16 overexpression in skeletal myotubules results in brown adipocyte differentiation [47]. Interestingly, PRDM16 is not regulated by cold and must rely on interacting factors to facilitate PRDM16-mediated BAT gene induction. Since the BAT program is induced upon cold exposure, DNA binding transcription factors that are enriched in BAT and also cold inducible may act by interacting with PRDM16. One such factor is Zfp516 which was found by screening known and putative transcription factors at a global level for activation of the UCP1 promoter. Zfp516 is BAT enriched, cold-induced, and Zfp516 binds and activates the proximal UCP1 promoter and other BAT enriched genes such as PGC1α. Furthermore, Zfp516 also interacts and recruits LSD1, a H3K9 demethylase, to UCP1 and other BAT enriched genes to promote thermogenesis. Furthermore, PRDM16 interacts with a wide variety of transcription factors and cofactors including C/EBPβ, C-terminal binding protein 1 and 2 (CTBP1, CTBP2), histone deacetylase 1 and 2 (HDAC1/2), mediator complex subunit 1 (MED1), PGC1α, PPARγ, and Zfp516 as well as epigenetic regulators-euchromatic histone lysine methyltransferase 1 (Ehmt1) and lysine-specific demethylase 1 (LSD1) [48,49]. Most PRDM16 interacting transcription factors bind at the −2.5 kb region of the UCP1 promoter, and Med1 interacts with PRDM16 to facilitate the ability of enhancer-bound transcription factors to recruit RNA pol II or writers of activation marks at BAT specific genes bringing distal enhancers to proximity to the transcription start site facilitating transcriptional activation.

PPARγ agonists have also been reported to increase mitochondrial biogenesis in white adipocytes and to induce brown adipose-selective markers including UCP1, PGC1α, and Cidea. A BAT specific transcription factor, early B cell factor 2 (EBF2) has been shown to bind at PPARγ sites of BAT-enriched genes to recruit PPARγ to promote PRDM16 transcription [50]. PPARγ coactivator-1α (PGC1α) is a transcriptional coactivator of PPARs and of many other transcription factors that are abundantly expressed in BAT and is further increased during cold exposure. Ectopic expression of PGC1α in mouse white adipocytes stimulates mitochondrial biogenesis and induces the expression of several genes involved in thermogenesis, such as UCP1 and Dio2 [51]. However, PGC1α-deficient adipose tissue shows reduced thermogenic defect. Another cold-inducible factor, interferon regulatory factor 4 (IRF4) has been reported to interact with PGC1α to upregulate thermogenesis, and Irf4-deficient mice have a thermogenic deficiency [52]. Recently, Zc3h10, a member of CCCH-type zinc finger proteins, has recently been a DNA-binding transcription factor that activates UCP1 and thermogenic gene program. Zc3h10 is brown fat enriched, cold inducible and overexpression of Zc3h10 increases thermogenic gene expressions in BAT and iWAT in vivo. Upon sympathetic signaling, Zc3h10 is phosphorylated by p38 MAPK to increase its binding to its target genes including Tfam and Nrf1, resulting in an increase in mitochondria number in BAT. Interestingly, while the majority of known transcription factors, such as ATF2, TR and PPARγ, all act through the −2.5 kb enhancer region of the UCP1, Zc3h10 binds to −4.6 kb UCP1 promoter (Fig. 25.1) [52a].

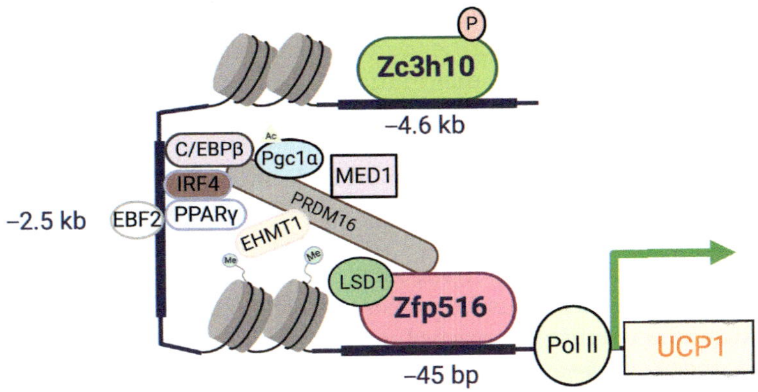


FIGURE 25.1 **Activation of thermogenic gene program.**

## Fat metabolism in WAT and BAT

Adipose tissue stores as triglycerides in a unilocular lipid droplet (LD), which encompasses most of the cell volume. While the LDs are unilocular in WAT, numerous smaller LDs are found in BAT. LD is composed of core neutral lipids mainly TAG and sterol esters, surrounded by a phospholipid monolayer. The surface of lipid droplets is decorated by a number of proteins which are involved in the regulation of lipid metabolism. During excessive energy or fed state, TAG is synthesized in the ER and immediately incorporated into LDs which then exit the ER membrane and reside in the cytosol. In contrast, during energy deprivation or fasted state, TAG is hydrolyzed on the LD surface and fatty acids (FAs) are then released into circulation for other tissues to use as energy source. In fasted stated, by B-adrenergic stimulation, lipolysis in WAT increases and causes lipid droplets to be fragmented. Lipolysis proceeds in an orderly and regulated manner, by converting TAG to DAG, MAG and then FAs. Each reaction is catalyzed by desnutrin/ATGL, hormone-sensitive lipase (HSL), and monoglyceride lipase (MGL), respectively [53]. The enzymatic steps carried out by these lipases are under tight hormonal regulation. During fasting, glucocorticoid is elevated and increase desnutrin/ATGL transcription. In addition, catecholamines bind to Gαs0-coupled B adrenergic receptors to activate adenylate cyclase, which increases cAMP levels activating protein kinase A (PKA). PKA then phosphorylates HSL, which translocates from the cytosol to its site of action on the lipid droplet. PKA also phosphorylates the lipid droplet associated protein, perilipin, and causing it to dissociate from the lipid droplet, exposing a greater surface for desnutrin/ATGL and MGL to access and function [54]. During fed state, lipolysis is inhibited by insulin binding to its receptor on adipocytes. Insulin signaling activates phosphodiesterase 3B by phosphorylation, which subsequently decreases cAMP and thus PKA activity. This causes reduced HSL/perilipin phosphorylation and eventually decreased lipolysis. Other cytokines, growth hormones, AMP-activated protein kinase (AMPK), and other molecules also affect lipolysis [55]. Thus, AMPK phosphorylates and activates desnutrin/ATGL to stimulate lipolysis. Beside endocrine effect of molecules such as insulin and catecholamines, lipolysis is regulated by autocrine/paracrine factors. TNF-α, secreted by adipocyte and macrophage, is showed to increase lipolysis. Prostaglandins have also been reported to affect lipolysis. Adipose-specific phospholipase A2 (AdPLA) has such an important role in adipocyte-derived PGE2 in the autocrine/paracrine regulation of lipolysis. AdPLA deficient mice exhibited resistance to high fat diet induced obesity and displayed an increase in lipolytic activity in white adipose tissue. In these mice, there was an increase in PEG2 production along with a lean phenotype even when on a high fat diet. This result indicates PGE2 is the predominant prostaglandin produced in WAT. In adipose tissue, PGE2 produced upon release of AA catalyzed by AdPLA acts through the Gi-alpha coupled EP3 to suppress lipolysis in a cAMP-dependent manner. AdPLA-PGE2-EP3 is composed of an autocrine-paracrine regulation of lipolysis in the adipose tissue [56].

Since, lipolysis is a critical metabolic process in both WAT and BAT, adipose overexpression of ATGL/desnutrin in mice increases lipolysis and FA oxidation, resulting in resistance to diet-induced obesity. These mice also show enhanced thermogenesis upon cold exposure having higher energy expenditure. Conversely, adipose-specific ablation of desnutrin/ATGL not only manifests in obesity but converts BAT to a WAT-like tissue. Thus, these mice also exhibited impaired thermogenesis with decreased expression of UCP1, with lower PPARα binding to its promoter, revealing the requirement of desnutrin/ATGL-catalyzed lipolysis in the maintenance of a BAT phenotype. In fact, due to decrease in lipolysis, intracellular fatty acid level was lower, resulting in lower ligand concentration of PPARα which is critical for BAT phenotype and is present at the highest level among PPAR family of transcription factors. Thus, PPARα binding and activation of UCP1 promoter was diminished upon desnutrin/ATGL KO BAT, and synthetic PPARα ligand, AICAR, could rescue the defective brown adipocyte morphology and thermogenesis from desnutrin/ATGL ablation [57].

## Fatty acid versus glucose metabolism for thermogenesis

Thermogenesis is a highly energetic process requiring a readily available fuel supply of glucose and FAs. In addition, BAT also contributes to whole body energy substrate homeostasis. After a meal. FAs and glucose are stored as triglycerides (TGs) in adipose tissue. Even though these nutrients can also be stored in other organs such as liver, muscle and heart, white adipose is the major storage of these fuels. Upon fasting or energy depletion, TGs are hydrolyzed to release FAs via lipolysis. FAs derived from lipolysis from the circulation are presumed to be used by active BAT [58–60]. Upon cold exposure, norepinephrine released from the sympathetic innervation in BAT stimulates β3-adrenergic receptor increasing lipolysis via ATGL and HSL. FAs thus produced are presumed to be used for oxidation and also for direct binding to UCP1 for activation.

Recently, however, it has been reported that FAs are taken up from circulation to be used for thermogenesis, especially in fasted condition and that lipolysis within BAT is not essential for thermogenesis [61,62]. Regardless, after cold exposure or β-adrenergic stimulation, glucose uptake increases in BAT, and BAT is the tissue with the highest glucose uptake [59,63]. In revealing the importance of glucose metabolism in thermogenesis, many proteins involved in glucose uptake and metabolism, including glucose transporter and glycolytic enzymes, are induced in BAT upon cold exposure and restricting glycolysis has been shown to impair thermogenesis [64–66]. In fact, it has recently been shown that glycolysis is essential for optogenetically induced thermogenesis [64]. The contribution of glucose in fueling thermogenesis, however, has not been well-established. The major portion of glucose is converted into pyruvate and lactate through glycolysis as evidenced by the secretion of both metabolites from BAT following norepinephrine stimulation. Glycolysis in BAT may run close to maximum capacity upon β-adrenergic stimulation. Glucose metabolized to pyruvate via glycolysis can also be transported into mitochondria to be oxidized via TCA cycle and citrate from TCA cycle can also be transported out to cytoplasm to be used for de novo lipogenesis. This can explain how lipogenesis and lipolysis are increased in parallel during thermogenesis [67,68]. Glucose utilization for BAT thermogenesis is particularly relevant in the fed state, when circulating glucose in high, with low NEFA released from WAT lipolysis due to insulin secretion [59].

In maintaining robust glycolysis, cytosolic NAD is critical for glyceraldehyde-3-phosphate dehydrogenase reaction in the glycolytic pathway. Inner mitochondrial membrane is impermeable to NADH or NAD. Thus, in most tissues, including liver and heart, malate-aspartate shuttle transfers NADH produced from glycolysis into mitochondria for ETC and return NAD to cytosol for glycolysis. Blocking malate-aspartate shuttle in many tissues causes inhibition of respiration [69,70]. However, in tissues, such as BAT, that require rapid ATP generation, glycerol-3-phosphate shuttle is used. First, cytosolic glycerol-3-phosphate dehydrogenase (cGPD) converts NADH to NAD while reducing dihydroxyacetone phosphate (DHAP) to glycerol-3-phosphate. Next, mitochondrial glycerol-3-phosphate dehydrogenase (mGPD) at the outer surface of the inner mitochondrial membrane facing cytosol, which is particularly high in BAT, oxidizes glycerol-3-phosphate back to DHAP, reducing FAD to FADH and then election from FADH is transferred to CoQ of mitochondrial ETC. However, there have been conflicting reports on the contribution of mGPD in thermogenesis [71–73]. The mGPD-KO mice did not show impaired thermogenesis, while a second independent mGPD-KO mouse model maintained on high-fat diet (HFD) exhibited a more rapid increase in body weight with lower energy expenditure, but this was due to higher circulating thyroid hormone. In this regard, lactate dehydrogenase (LDH) in the cytosol can regenerate NAD by converting pyruvate to lactate, which has been proposed to be important for glycolysis during thermogenesis [64]. However, lactate in BAT is used extensively for oxidation by conversion back to pyruvate to enter TCA cycle and lactate level in BAT does not increase, but decreases during thermogenesis [64].

NDHs (NADH dehydrogenase) that are mainly found in yeast, bacteria and plants are associated with the mitochondrial inner membrane and catalyze the same NADH oxidase reaction as complex I in ETC. There are two main classes of NDHs, NDI (internal NDH) facing the mitochondrial matrix and NDE (external) facing the intermembrane space [73a]. In yeast that lacks complex 1, NDI maintains mitochondrial NAD to ensure efficient TCA cycle, while NDE provides cytosolic NAD for glycolysis. NDHs also increase ETC activity and respiration by transferring electrons to CoQ. Thus, ablation of these enzymes leads to high NADH/NAD ratio causing an imbalance in cellular metabolism with defective mitochondrial [73b]. In fact, upon environmental cues, NDE has been shown to have much higher turnover numbers of NADH:UQ oxidase activity than mammalian complex I [73c]. In mammalian, Aifm2 (apoptosis inducing mitochondrion associated factor 2 (also called AMID, or Prg3) was discovered to function as an NDH. Aifm2 is a flavoprotein with a NADH/NAD oxidoreductase domain as a lipid-droplet associated protein that is highly and specifically expressed in BAT and is

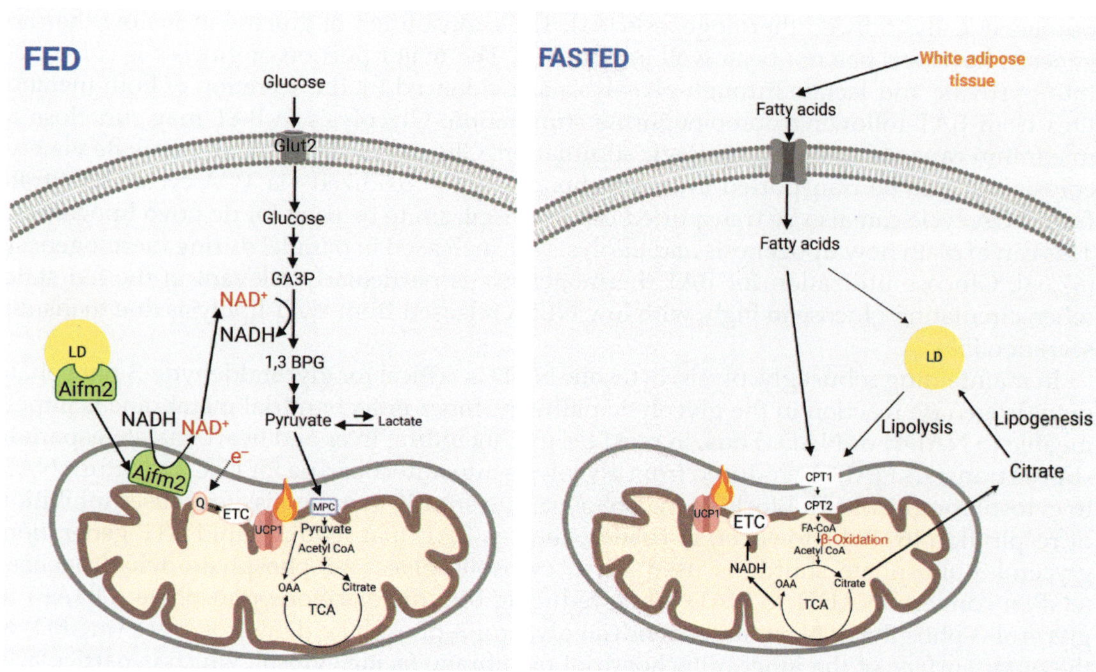


**FIGURE 25.2** **Metabolism during thermogenesis.**

induced upon cold exposure/β-adrenergic stimulation in BAT and iWAT. Upon stimulation, Aifm2 localizes to mitochondria for conversion of NADH to NAD to sustain robust glycolysis, while transferring electrons to mitochondrial ETC in fueling thermogenesis. Remarkably, Aifm2 not only has sequence similarity to yeast NDE1, but also shares the localization and the enzymatic activity. Yeast NDE1 can rescue impaired thermogenesis from Aifm2 deficiency, making Aifm2 to be a mammalian NDE specific to thermogenic tissues [73d]. Interestingly, search of GWAS database reveals multiple SNPs of Aifm2 to be associated with waist-hip ratio, body mass index and fasting glucose level-related insulin resistance, suggesting a potential role of Aifm2 in human obesity and type 2 diabetes (Fig. 25.2).

Several studies show glucose metabolism is critical for adipocyte differentiation. Wellen et al. showed that glucose, required for production of cytosolic acetyl-CoA is used for histone acetylation which results in changes in expression of differentiation programs [50]. Additionally, Shapiro et al. found that nicotinamide (NAM) is required for PPARγ expression in umbilical cord–derived mesenchymal stem cells differentiation to adipocytes [73e]. Similarly, Jackson et al. reported glucose and nicotinamide control adipogenesis in mouse 3T3-L1 adipocytes [74]. Aifm2 KO resulting in lower glucose oxidation and reduced level of cytosolic NAD could lead to decreased expression of PPARγ. Aifm2 KO BAT showed low level of thermogenic genes such as UCP1 and Dio2. Aifm2 KO results in higher adiposity and WAT-like brown adipocytes. Aifm2 may play an important role in the plasticity of adipocytes, conceivably by providing metabolites that regulate transcriptional activity of regulator of thermogenic programing in BAT.

## Conclusion

While WAT plays a critical role by serving as the major energy storage site and by secreting adipokines that control various biological processes, such as appetite, metabolism and insulin sensitivity, BAT has high capacity of glucose and fatty acid (FA) uptake for thermogenesis and all potentially contributing to insulin sensitivity. Understanding both depots development and the underlying to promote "browning" of WAT may provide targets for combatting and preventing obesity and associated diseases. Many studies identified several markers for adipose progenitors/precursors helping uncover the heterogeneity of adipose SVF. However, newly advanced sc-RNA seq would further allow to study of the possible mosaic development of WAT and BAT and to discover any new cell population that might contribute to adipose tissue homeostasis, function and development as well as potential mechanisms for "browing" of WAT. Furthermore, the relationship among transcription factors involved in transitioning from silent chromatin to poised/active chromatin in brown/beige adipocytes needs further investigation. With significant changes of chromatin structure during brown/beige development, the enzymes catalyzing these modifications remain to be elucidated.

## References

[1] Rosen ED, Spiegelman BM. What we talk about when we talk about fat. Cell 2014;156(1–2):20–44.
[2] Ghaben AL, Scherer PE. Adipogenesis and metabolic health. Nat Rev Mol Cell Biol 2019;20(4):242–58.
[3] Jeffery E, et al. Rapid depot-specific activation of adipocyte precursor cells at the onset of obesity. Nat Cell Biol 2015;17(4):376–85.
[4] Jeffery E, et al. The adipose tissue microenvironment regulates depot-specific adipogenesis in obesity. Cell Metab 2016;24(1):142–50.
[5] Kim SM, et al. Loss of white adipose hyperplastic potential is associated with enhanced susceptibility to insulin resistance. Cell Metab 2014;20(6):1049–58.
[6] Macotela Y, et al. Intrinsic differences in adipocyte precursor cells from different white fat depots. Diabetes 2012;61(7):1691–9.
[7] Harms M, Seale P. Brown and beige fat: development, function and therapeutic potential. Nat Med 2013;19(10):1252–63.
[8] Kajimura S, Spiegelman BM, Seale P. Brown and beige fat: physiological roles beyond heat generation. Cell Metab 2015;22(4):546–59.
[9] Inagaki T, Sakai J, Kajimura S. Transcriptional and epigenetic control of brown and beige adipose cell fate and function. Nat Rev Mol Cell Biol 2017;18(8):527.
[10] Stanford KI, et al. Brown adipose tissue regulates glucose homeostasis and insulin sensitivity. J Clin Invest 2013;123(1):215–23.
[11] Poher AL, et al. Brown adipose tissue activity as a target for the treatment of obesity/insulin resistance. Front Physiol 2015;6:4.
[12] Li Y, Meng Y, Yu X. The unique metabolic characteristics of bone marrow adipose tissue. Front Endocrinol (Lausanne) 2019;10:69.
[13] Rodeheffer MS, Birsoy K, Friedman JM. Identification of white adipocyte progenitor cells in vivo. Cell 2008;135(2):240–9.
[14] Gupta RK, et al. Transcriptional control of preadipocyte determination by Zfp423. Nature 2010;464(7288):619–23.
[15] Gulyaeva O, et al. Sox9-Meis1 inactivation is required for adipogenesis, advancing Pref-1(+) to PDGFRalpha(+) cells. Cell Rep 2018;25(4):1002–17. e4.
[16] Marcelin G, et al. A PDGFRalpha-mediated switch toward CD9(high) adipocyte progenitors controls obesity-induced adipose tissue fibrosis. Cell Metab 2017;25(3):673–85.
[17] Chau YY, et al. Visceral and subcutaneous fat have different origins and evidence supports a mesothelial source. Nat Cell Biol 2014;16(4):367–75.

[18] Sanchez-Gurmaches J, Hsiao WY, Guertin DA. Highly selective in vivo labeling of subcutaneous white adipocyte precursors with Prx1-Cre. Stem Cell Reports 2015;4(4):541–50.
[19] Sanchez-Gurmaches J, Guertin DA. Adipocytes arise from multiple lineages that are heterogeneously and dynamically distributed. Nat Commun 2014;5:4099.
[20] Billon N, et al. The generation of adipocytes by the neural crest. Development 2007;134(12):2283–92.
[21] Min SY, et al. Diverse repertoire of human adipocyte subtypes develops from transcriptionally distinct mesenchymal progenitor cells. Proc Natl Acad Sci U S A 2019;116(36):17970–9.
[22] Tontonoz P, Hu E, Spiegelman BM. Stimulation of adipogenesis in fibroblasts by PPAR gamma 2, a lipid-activated transcription factor. Cell 1994;79(7):1147–56.
[23] Wu Z, et al. Cross-regulation of C/EBP alpha and PPAR gamma controls the transcriptional pathway of adipogenesis and insulin sensitivity. Mol Cell 1999;3(2):151–8.
[24] Tong Q, et al. Function of GATA transcription factors in preadipocyte-adipocyte transition. Science 2000;290(5489):134–8.
[25] Kajimura S, et al. Initiation of myoblast to brown fat switch by a PRDM16-C/EBP-beta transcriptional complex. Nature 2009;460(7259):1154–8.
[26] Oishi Y, et al. Kruppel-like transcription factor KLF5 is a key regulator of adipocyte differentiation. Cell Metab 2005;1(1):27–39.
[27] Aggarwal A, et al. The circadian clock regulates adipogenesis by a Per3 crosstalk pathway to Klf15. Cell Rep 2017;21(9):2367–75.
[28] Wu Z, Wang S. Role of kruppel-like transcription factors in adipogenesis. Dev Biol 2013;373(2):235–43.
[29] Payne VA, et al. C/EBP transcription factors regulate SREBP1c gene expression during adipogenesis. Biochem J 2009;425(1):215–23.
[30] Ignotz RA, Massague J. Type beta transforming growth factor controls the adipogenic differentiation of 3T3 fibroblasts. Proc Natl Acad Sci U S A 1985;82(24):8530–4.
[31] Sarantopoulos CN, et al. Elucidating the preadipocyte and its role in adipocyte formation: a comprehensive review. Stem Cell Rev 2018;14(1):27–42.
[32] Tang W, et al. White fat progenitor cells reside in the adipose vasculature. Science 2008;322(5901):583–6.
[33] Tseng YH, et al. Prediction of preadipocyte differentiation by gene expression reveals role of insulin receptor substrates and necdin. Nat Cell Biol 2005;7(6):601–11.
[34] Fox KE, et al. Depletion of cAMP-response element-binding protein/ATF1 inhibits adipogenic conversion of 3T3-L1 cells ectopically expressing CCAAT/enhancer-binding protein (C/EBP) alpha, C/EBP beta, or PPAR gamma 2. J Biol Chem 2006;281(52):40341–53.
[35] Smas CM, Sul HS. Pref-1, a protein containing EGF-like repeats, inhibits adipocyte differentiation. Cell 1993;73(4):725–34.
[36] Smas CM, Chen L, Sul HS. Cleavage of membrane-associated pref-1 generates a soluble inhibitor of adipocyte differentiation. Mol Cell Biol 1997;17(2):977–88.
[37] Smas CM, et al. Transcriptional repression of pref-1 by glucocorticoids promotes 3T3-L1 adipocyte differentiation. J Biol Chem 1999;274(18):12632–41.
[38] Moon YS, et al. Mice lacking paternally expressed Pref-1/Dlk1 display growth retardation and accelerated adiposity. Mol Cell Biol 2002;22(15):5585–92.
[39] Villena JA, et al. Resistance to high-fat diet-induced obesity but exacerbated insulin resistance in mice overexpressing preadipocyte factor-1 (Pref-1): a new model of partial lipodystrophy. Diabetes 2008;57(12):3258–66.
[40] Lee K, et al. Inhibition of adipogenesis and development of glucose intolerance by soluble preadipocyte factor-1 (Pref-1). J Clin Invest 2003;111(4):453–61.
[41] Wang Y, Sul HS. Ectodomain shedding of preadipocyte factor 1 (Pref-1) by tumor necrosis factor alpha converting enzyme (TACE) and inhibition of adipocyte differentiation. Mol Cell Biol 2006;26(14):5421–35.
[42] Mei B, et al. Only the large soluble form of preadipocyte factor-1 (Pref-1), but not the small soluble and membrane forms, inhibits adipocyte differentiation: role of alternative splicing. Biochem J 2002;364(Pt 1):137–44.
[43] Kim KA, et al. Pref-1 (preadipocyte factor 1) activates the MEK/extracellular signal-regulated kinase pathway to inhibit adipocyte differentiation. Mol Cell Biol 2007;27(6):2294–308.
[44] Wang Y, Sul HS. Pref-1 regulates mesenchymal cell commitment and differentiation through Sox9. Cell Metab 2009;9(3):287–302.
[45] Cao W, et al. beta-Adrenergic activation of p38 MAP kinase in adipocytes: cAMP induction of the uncoupling protein 1 (UCP1) gene requires p38 MAP kinase. J Biol Chem 2001;276(29):27077–82.

[46] Iida S, et al. PRDM16 enhances nuclear receptor-dependent transcription of the brown fat-specific Ucp1 gene through interactions with Mediator subunit MED1. Genes Dev 2015;29(3):308–21.

[47] Harms MJ, et al. Prdm16 is required for the maintenance of brown adipocyte identity and function in adult mice. Cell Metab 2014;19(4):593–604.

[48] Ohno H, et al. EHMT1 controls brown adipose cell fate and thermogenesis through the PRDM16 complex. Nature 2013;504(7478):163–7.

[49] Sambeat A, et al. LSD1 interacts with Zfp516 to promote UCP1 transcription and brown fat program. Cell Rep 2016;15(11):2536–49.

[50] Wellen KE, et al. ATP-citrate lyase links cellular metabolism to histone acetylation. *Science* 2009;324(5930):1076–80.

[51] Kleiner S, et al. Development of insulin resistance in mice lacking PGC-1alpha in adipose tissues. Proc Natl Acad Sci U S A 2012;109(24):9635–40.

[52] Kong X, et al. IRF4 is a key thermogenic transcriptional partner of PGC-1alpha. Cell 2014;158(1):69–83.

[52a] Yi D, et al. Zc3h10 acts as a transcription factor and is phosphorylated to activate the thermogenic program. *Cell Rep* 2019;29(9):2621–33.

[53] Ahmadian M, Wang Y, Sul HS. Lipolysis in adipocytes. Int J Biochem Cell Biol 2010;42(5):555–9.

[54] Brasaemle DL. Thematic review series: adipocyte biology. The perilipin family of structural lipid droplet proteins: stabilization of lipid droplets and control of lipolysis. J Lipid Res 2007;48(12):2547–59.

[55] Duncan RE, et al. Regulation of lipolysis in adipocytes. Annu Rev Nutr 2007;27:79–101.

[56] Jaworski K, et al. AdPLA ablation increases lipolysis and prevents obesity induced by high-fat feeding or leptin deficiency. Nat Med 2009;15(2):159–68.

[57] Ahmadian M, et al. Desnutrin/ATGL is regulated by AMPK and is required for a brown adipose phenotype. Cell Metab 2011;13(6):739–48.

[58] Blondin DP, et al. Inhibition of intracellular triglyceride lipolysis suppresses cold-induced brown adipose tissue metabolism and increases shivering in humans. Cell Metab 2017;25(2):438–47.

[59] U Din M, et al. Postprandial oxidative metabolism of human brown fat indicates thermogenesis. Cell Metab 2018;28(2):207–16. e3.

[60] Lee J, Ellis JM, Wolfgang MJ. Adipose fatty acid oxidation is required for thermogenesis and potentiates oxidative stress-induced inflammation. Cell Rep 2015;10(2):266–79.

[61] Schreiber R, et al. Cold-induced thermogenesis depends on ATGL-mediated lipolysis in cardiac muscle, but not brown adipose tissue. Cell Metab 2017;26(5):753–63. e7.

[62] Shin H, et al. Lipolysis in brown adipocytes is not essential for cold-induced thermogenesis in mice. Cell Metab 2017;26(5):764–77. e5.

[63] Symonds ME, et al. Recent advances in our understanding of brown and beige adipose tissue: the good fat that keeps you healthy. F1000Res 2018;7.

[64] Jeong JH, Chang JS, Jo YH. Intracellular glycolysis in brown adipose tissue is essential for optogenetically induced nonshivering thermogenesis in mice. Sci Rep 2018;8(1):6672.

[65] Hao Q, et al. Transcriptome profiling of brown adipose tissue during cold exposure reveals extensive regulation of glucose metabolism. Am J Physiol Endocrinol Metab 2015;308(5):E380–92.

[66] Winther S, et al. Restricting glycolysis impairs brown adipocyte glucose and oxygen consumption. Am J Physiol Endocrinol Metab 2018;314(3):E214–23.

[67] Barquissau V, et al. White-to-brite conversion in human adipocytes promotes metabolic reprogramming towards fatty acid anabolic and catabolic pathways. Mol Metab 2016;5(5):352–65.

[68] Yu XX, et al. Cold elicits the simultaneous induction of fatty acid synthesis and beta-oxidation in murine brown adipose tissue: prediction from differential gene expression and confirmation in vivo. FASEB J 2002;16(2):155–68.

[69] Kauppinen RA, Sihra TS, Nicholls DG. Aminooxyacetic acid inhibits the malate-aspartate shuttle in isolated nerve terminals and prevents the mitochondria from utilizing glycolytic substrates. Biochim Biophys Acta 1987;930(2):173–8.

[70] Barron JT, Gu L, Parrillo JE. Malate-aspartate shuttle, cytoplasmic NADH redox potential, and energetics in vascular smooth muscle. J Mol Cell Cardiol 1998;30(8):1571–9.

[71] Brown LJ, et al. Normal thyroid thermogenesis but reduced viability and adiposity in mice lacking the mitochondrial glycerol phosphate dehydrogenase. J Biol Chem 2002;277(36):32892–8.

[72] DosSantos RA, et al. Evidence for a compensated thermogenic defect in transgenic mice lacking the mitochondrial glycerol-3-phosphate dehydrogenase gene. Endocrinology 2003;144(12):5469–79.

[73] Alfadda A, et al. Mice with deletion of the mitochondrial glycerol-3-phosphate dehydrogenase gene exhibit a thrifty phenotype: effect of gender. Am J Physiol Regul Integr Comp Physiol 2004;287(1):R147–56.

[73a] Yamashita T, et al. Ubiquinone binding site of yeast NADH dehydrogenase revealed by structures binding novel competitive- and mixed-type inhibitors. Sci Rep 2018;8(1):2427.

[73b] Melo AMP, et al. New insights into type II NAD(P)H:quinone oxidoreductases. Microbiol Mol Biol *Rev* 2004;68(4):603–16.

[73c] Elguindy MM, Nakamaru-Ogiso E. Apoptosis-inducing factor (AIF) and its family member protein, AMID, are rotenone-sensitive NADH:ubiquinone oxidoreductases (NDH-2). J Biol Chem 2015;290(34):20815–26.

[73d] Nguyen HP, et al. Aifm2, a NADH oxidase, supports robust glycolysis and is required for cold- and diet-induced thermogenesis. Mol cell 2020;77(3):600–17.

[73e] Shapiro ALB, et al. Nicotinamide promotes adipogenesis in umbilical cord-derived mesenchymal stem cells and is associated with neonatal adiposity: the Healthy Start BabyBUMP Project. PLoS ONE 2016;11(7):e0159575.

[74] Jackson RM, et al. Glucose availability controls adipogenesis in mouse 3T3-L1 adipocytes via up-regulation of nicotinamide metabolism. J Biol Chem 2017;292(45):18556–64.

# Index

Note: Page numbers followed by "f" indicate figures.

## T